CYTOCHROMES

LEMBERG, R
CYTOCHROMES

HCL QP99.3.L54

THE UNIVERSITY OF LIVERPOOL
HAROLD COHEN LIBRARY

CONDITIONS OF BORROWING

Members of Council, members and retired members of the University staff, and students registered with the University for higher degrees — 20 volumes for one month. All other readers entitled to borrow — 6 volumes for 14 days in term or for the vacation.

Books may be recalled after one week for the use of another reader.

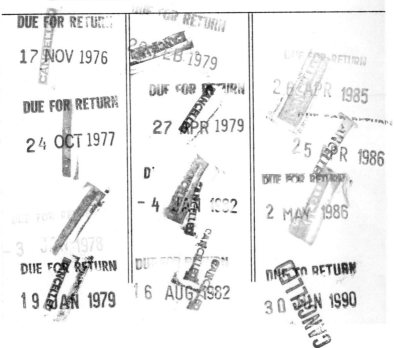

CYTOCHROMES

R. LEMBERG F.R.S. and J. BARRETT

Kolling Institute of Medical Research,
The Royal North Shore Hospital of Sydney,
Sydney, Australia

1973

ACADEMIC PRESS · LONDON AND NEW YORK

A subsidiary of Harcourt Brace Jovanovich, Publishers

ACADEMIC PRESS INC. (LONDON) LTD
24–28 Oval Road,
London NW1

U.S. Edition published by
ACADEMIC PRESS INC.
111 Fifth Avenue
New York, New York 10003

Copyright © 1973 by ACADEMIC PRESS INC. (LONDON) LTD

All Rights Reserved
No part of this book may be reproduced in any form by photostat, microfilm, or any other means, without written permission from the publishers

Library of Congress Catalog Card Number: 72-84450
ISBN: 0-12-443150-x

PRINTED IN GREAT BRITAIN BY
J. W. ARROWSMITH LTD, BRISTOL

Preface

In 1949, Lemberg and Legge predicted that "in a decade or so it may be necessary to give to each of the haematin enzymes as much space as to haemoglobin; later again it may become possible to treat all haemoproteins together from one particular physico-chemical aspect" ("Heme Compounds and Bile Pigments", p. 337). While the time estimate was wrong, the direction was correctly foretold. Even though our knowledge of other haemoproteins is still fragmentary if compared with that of haemoglobin, it is rapidly increasing and, at least with regard to cytochrome c, approaching that of haemoglobin. Today it would be beyond the power of a single author to treat the subject of haemoproteins alone successfully in a monograph. On the other hand, a new comparative biochemistry of the haemoproteins is rapidly coming in view in which the distinct characters and essential similarities of this "most varied and beautiful class of molecular machines known to the biochemist" (Epstein and Schechter, 1968) are more clearly revealed; and they cannot be even neglected in a monograph which deals only with one subclass, the cytochromes.

In the title of the historical review of David Keilin, edited by Joan Keilin, "The History of Cell Respiration and Cytochrome" (1966) there are three words which in themselves are of some historical significance, "cytochrome" and "cellular respiration". While the cytochromes—their number is continually increasing—are indeed of fundamental importance for cellular respiration, their importance is no longer restricted to this field of bioenergetics. Cytochromes are known to play a role too in the photochemical reactions of photosynthesis, in the primary photochemical reaction with chlorophyll and bacteriochlorophyll in green plants and algae, and in photosynthetic electron transfer. Their role in non-photosynthetic anaerobic and chemosynthetic bacterial reactions, e.g. in nitrate and sulphate reduction has also been recognized.

Bacterial cytochromes were studied as early as 1928 and particularly from 1933 onwards, but the number of publications on bacterial cytochromes has enormously increased in recent years. Not all of these publications are of equally high standard, and their critical discussion is an important task for the authors of this monograph. It has not only been shown that cytochromes are of greater general biological significance but also that their evolutionary significance extends back many hundreds of millions of years to times before the presence of molecular oxygen in the earth's atmosphere allowed respiratory processes to occur. They have also introduced a new dimension of variability into the knowledge of a single type of cytochrome, and have made evolutionary deductions possible which will become of increasing importance with the steadily growing success of purification, knowledge of amino acid composition and sequence, and crystallization to allow later x-ray analytical studies.

The main stress of this monograph will be directed towards a knowledge of the chemistry and function of individual cytochromes as enzymes. It appears rather pedantic to separate them from the enzymes just because they have two and not one substrate—in fact even hydrolytic enzymes have two substrates, the "substrate" and water. However, since the cytochromes typically react in "multienzyme systems" their interaction with themselves and other molecules (e.g. cytochrome reductases and dehydrogenases or with cytochrome peroxidases) cannot be neglected, nor can their interaction with molecular oxygen in the oxidases. Classical cytochromes which may act as electron (or hydrogen) carriers cannot at present be safely separated from cytochrome oxidases of haemoprotein natures, although their function can be differentiated from those of non-autoxidizable electron carrier cytochromes. Cytochromes and their role in electron transport also cannot entirely be separated from energy storage and oxidative phosphorylation, nor from ion transport, nor from the role they play in photosynthesis in their reaction with chlorophyll and bacteriochlorophyll and in photosynthetic electron transport; nor from the structure and mechanism of mitochondria and microsomes in which they are located, and the new field of "membranology". However, here the authors have exerted severe self-discipline or else these extensions into neighbouring fields would have exploded the frame of their monograph—and probably also their frame of knowledge. They must beg the indulgence of those who feel that too little or too much has been said in this connection and be content to allow the readers to seek further extension of their own studies, with help given to them by the references.

There is at present no comprehensive monograph on cytochromes available, although they have been treated in short reviews (Morton, 1958; Lemberg, 1958), in chapters of more extensive books and in comprehensive texts on enzymes. In "Methods of Enzymology" (S. P. Colowick and N. O. Kaplan, eds. in chief, vol. 10, edited by R. W. Estabrook and M. E. Pullman), there is valuable compilation of methods for the purification of many of the cytochromes. Smaller monographs dealing with individual classes of cytochromes are available; these will be quoted in the chapters of this book. A number of recent International Symposia (Canberra, 1959; Amherst, 1964; Philadelphia, 1966 and Osaka, 1967) are of unique value in that they supply interesting discussions between the contributors. These are often of greater value than the papers which are usually found in later scientific journals.

ACKNOWLEDGEMENTS

The authors wish to thank the Australian Research Grants Committee, the Hospital Commission of New South Wales and the Institute of Medical Research for support of their research and also for their contribution to library and secretarial facilities. A warm debt of gratitude is owed to the highly efficient work of the secretary, Mrs. Katherine Carson, and of the Librarian, Mrs. Mary Little. Thanks are also expressed to the late Dr. John Falk, Dr. N. K. Boardman and Dr. Cyril A. Appleby of the C.S.I.R.O. Division of Plant Industry, to Prof. M. Taylor of Sydney University and to our colleagues Dr. D. B. Morell, Dr. A. Nichol, Mr. W. H. Lockwood, Mrs. M. Cutler and Dr. Norma Newton.

This work would never have been written without the inspiration of the life work and discoveries of David Keilin.

March, 1973

R. LEMBERG
J. BARRETT

Contents

Preface		v
Chapter 1.	Introduction	1
Chapter 2.	Classification and momenclature of the cytochromes and their prosthetic groups	8
Chapter 3.	Cytochromes *a*	17
Chapter 4.	Cytochromes *b*	58
Chapter 5.	Cytochromes *c*	122
Chapter 6.	Bacterial cytochromes and cytochrome oxidases	217
Chapter 7.	The cytochromes in the respiratory chain	327
Chapter 8.	Some physiological aspects of cytochromes	385
Chapter 9.	Biosynthesis of cytochromes	446
Chapter 10.	The cytochromes in evolution	484
Appendix		499
Author index		511
Subject index		552

ABBREVIATIONS

"Haematin Enzymes" (J. E. Falk, R. Lemberg, R. K. Morton, eds.) A Symposium of the International Union of Biochemistry, organized by the Australian Academy of Science, Canberra, 1959. Pergamon Press, Oxford, 1961. 2 vols.
"Haematin Enzymes".

"Oxidases and Related Redox Systems". Proc. of a Symposium held in Amherst, Mass., 1964 (T. E. King, H. S. Mason, M. Morrison, eds.). John Wiley & Sons, New York, 1965, 2 vols.
"Amherst".

"Structure and Function of Cytochromes" (K. Okunuki, M. D. Kamen and I. Sekuzu, eds.) Proc. of the Symposium held at Osaka, 1967. Univ. of Tokyo Press and Univ. Park Press, Baltimore, 1968.
"Osaka".

"Methods in Enzymology", vol. 10 (R. W. Estabrook and M. E. Pullman, eds.), Academic Press, New York, 1967.
"Methods in Enzymol. 10".

"Comprehensive Biochemistry" (M. Florkin and E. H. Stotz, eds.). Elsevier, Amsterdam.
"Comprehensive Biochemistry".

"Hemes and Hemoproteins" (B. Chance, R. W. Estabrook and T. Yonetani, eds.). Academic Press, New York.
"Hemes and Hemoproteins".

To
HANNAH LEMBERG
and
JUDITH BARRETT

Chapter I

Introduction

A. Definition of cytochromes.
B. Historical Survey.
C. The role of cytochromes in bioenergetics.
   References.

A. DEFINITION OF CYTOCHROMES

In the 1961 Report of the Commission on Enzymes of the International Union of Biochemistry (cf. also Enzyme Nomenclature Recommendations, 1965) cytochromes are defined as "haemoproteins whose principal biological function is electron and/or hydrogen transport by virtue of a reversible valency change of their haem iron". This definition has biological as well as chemical features and is wide enough to compass non-autoxidizable and autoxidizable cytochromes, such as cytochrome oxidase of haemoprotein nature. It is, however, somewhat too wide in so far as peroxidases and catalase must be specifically excluded. They too are haemoproteins which are active in electron transfer. Restriction of the change of the valency of the iron to one between ferrous and ferric (cf. Nicholls, 1963) would exclude the hydroperoxidase enzymes and the word "principal" in front of function allows cytochrome oxidases to be included since compounds of these enzymes with a valency higher than ferric are, if existent, certainly unstable compounds. At this stage it may be permissible to pass by the claims of some authors who postulate that in fact no intermediates of iron valencies higher than three occur in the reactions of peroxidase and catalase and give other explanations to compounds of oxidation stages from II to VI (Peisach et al., 1968); these are not yet universally accepted. Similar considerations hold for the interpretation of oxyhaemoglobin and of myoglobin as ferric iron compounds (Weiss, 1964 and others). It may be, however, beneficial to remind us at this early stage that the borders between the various classes of haemoproteins are somewhat fluid. The same holds for the chemical definition of a cytochrome as a compound in which the ferrous-ferric valency change is between low spin ferrous and low spin ferric iron. While such a definition will satisfactorily exclude the non cytochrome haemoproteins, it may be too stringent in so far as it may exclude some cytochromes of oxidase nature which it is not desirable to separate from the other cytochromes at the present stage of our knowledge. This would e.g. separate cytochrome a_3 as oxidase from cytochrome a, which is certainly not yet desirable, or cytochromes "o" and "P-450" from cytochromes b, and would raise great difficulties with regard to a number of other cytochromes whose oxidase function is still far from clear, e.g. cytochromes cc'. In this region we can observe an almost continuous spectrum of differences between a "typical" cytochrome such as cytochromes c which do not react with carbon monoxide and are not autoxidizable, to cytochromes such as some of type b which are autoxidizable, but do not react with carbon monoxide, and to others which are readily autoxidizable and react with carbon monoxide, such as cytochrome a_3 and P-450. Again it may be useful to point out that strictly speaking, haemoglobin and myoglobin are slowly autoxidizable substances. Some of the cytochromes which will be discussed in this book are haemoproteins which do contain other constituents, e.g. Cu in cytochrome aa_3, flavin in cytochrome b_2 and some bacterial cytochromes c, and there may be others about whose non-haemoprotein constituents we have so far no knowledge. The writers do not agree with the claim of Dixon and Webb (1964) that cytochromes are not enzymes. Cytochrome P-450 is both a member of the b-type cytochromes and a hydroxylase. Cytochrome aa_3 though hardly an oxygenase (Hayaishi, 1962; Okunuki, 1962) is the enzyme cytochrome oxidase (EC 1.9.3.1), or O_2: cytochrome c oxidoreductase. Cytochrome c can equally well be considered as cytochrome aa_3: cytochrome c_1 oxidoreductase.

It will be discussed in later chapters that this gradation may be understood as due to a compromise between the direction of the iron orbitals of the haem, and possibly even the fine structure of the almost planar porphyrin rings on the one side, and the position of the ligands in the polypeptide chains of the protein. These may either be in a suitable sterical position for linkage without strain, or if less suitably placed, a compromise will have to be achieved by certain adaptations of the iron orbitals, the exact position of the iron atom in the porphyrin plane, and/or the polypeptide chain conformation, the so-called "rack"-mechanism of Lumry (1959), cf. also R. J. P. Williams (1956).

B. HISTORICAL SURVEY

The development of our knowledge on the cytochromes as respiratory catalysts of haemoprotein nature began with the spectroscopic observations of MacMunn (1886, 1887, 1889). In their attacks on these publications, Hoppe-Seyler (1890) and Levy (1889) overlooked that their claim of identity of myohaematins with myoglobin, partly justified by a premature attempt at isolation from muscle and the claim of absence of myoglobin from muscle could not possibly explain many of the observations of MacMunn; this is one of the occasions when a destructive criticism by a recognised authority delayed the development of a field for almost forty years and might have delayed it further but for Keilin's profound biological knowledge. Based on skilful observations with simple but well-suited instruments and on his own supreme power of observation on a great variety of cells and tissues David Keilin laid in the years 1925–1929 the foundations of our knowledge on the cytochromes. The early development of our knowledge on cellular respiration including his own work on cytochromes is excellently described in Keilin's posthumous book "The History of Cell Respiration and Cytochromes" edited by Joan Keilin (1966); cf. also Lemberg (1969, p. 50–51). Keilin spoke first of "cytochrome", but later recognized that the typical four-banded absorption spectrum was due to several different compounds which he called cytochromes a, b and c. He did not accept the name "histohaematins" because he recognized, with the then growing knowledge of haematin compounds at Cambridge, that these compounds were not simple haematin compounds, but haemoproteins of a class now known as ferro- and ferrihaemochromes (cf. Commission on Enzymes, 1961, 1965), the typical spectrum obtained on reduction being due to three types of ferrohaemochromes. He proved convincingly that the change of iron valency was connected with their respiratory function as electron carriers between molecular oxygen, activated by an enzyme, and substrates activated by specific dehydrogenases. The enzyme reacting with oxygen first described as Nadi- or indophenol oxidase and later as cytochrome c oxidase and assumed by Keilin to be a copper-enzyme, was later (Keilin, 1930; Keilin and Hartree, 1939) identified with Warburg's Atmungsferment and what is now called cytochrome aa_3. By ingenious use of the carbon monoxide inhibition of the respiration of yeast Warburg had demonstrated the haemoprotein nature of the oxidase by its photochemical action spectrum (Warburg and Negelein, 1929; Warburg, 1949). Thus Keilin's work finally closed the gap in the respiratory chain between the oxidase, the flavin enzymes, meanwhile discovered by Warburg et al., the pyridine nucleotides, also discovered by Warburg et al., and the substrate-specific dehydrogenases of Thunberg and Wieland. A chapter of biochemistry, then a great field of controversy and confusion, thus grew together to a cohesive story of the respiratory chain. While this book is by necessity restricted to deal primarily with the cytochromes, i.e. the haemoproteins which form the part of the respiratory chain near to the oxygen end, the cytochrome reductases of non-haemoprotein nature must also receive some attention (Chapter VII).

In the last twelve years or so purification of the cytochromes and their crystallization has made rapid progress. The crystallization of the cytochrome c from king penguin muscle (Bodo, 1955) was soon followed by the crystallization of a great number of cytochromes c, particularly by the Okunuki school (Hagihara et al., 1956a,b; 1958a,b,c; 1959a,b; Paléus and Theorell, 1957). Some cytochromes b have also been crystallized. The amino acid composition of several cytochromes has been established and the number of cytochromes, the complete amino acid sequence and primary structure of which is known steadily grows. Finally, the tertiary structure of mammalian cytochrome c has been established by x-ray diffraction (Dickerson et al., 1967); the binding of its haem iron between imidazole nitrogen and methionine-sulphur postulated by Harbury et al. (1965) is supported by the x-ray studies and has recently also been confirmed by NMR studies (McDonald, Phillips and Vinogradov, 1969).

While the development of our knowledge of the more complex structure of cytochrome aa_3 has not yet advanced that far, the structure of its prosthetic group has been established. Cytochrome aa_3 has been highly purified, the aa_3 concept of Keilin has been confirmed, and conformational differences found to exist between the "oxygenated" form of the enzyme and its stabilized ferric form have shown that alteration of the conformation of the cytochrome a_3 moiety by combination of its ferrous form with oxygen plays an important role in the function of the enzyme (Lemberg, 1969; Lemberg and Cutler, 1970).

An autoxidizable cytochrome b (P–450) has been shown to be an inductive enzyme of great importance for drug detoxication and steroid metabolism.

The study of bacterial cytochromes began as early as 1928 with a study of Yaoi and Tamiya (1928) on cytochrome "a_2", now ascribed to a different class of cytochromes (d) (cf. Barrett and Lemberg, 1954; Barrett, 1956), in *Escherichia coli* and *Shigella dysenteriae*. This has also been studied by Warburg and Negelein (1932) and Keilin (1933). Soon afterwards, a considerable number of bacteria were studied by Fujita and Kodama (1934), which contained among other cytochromes also b_1 and a_1, first found by Warburg and Negelein (1933) in acetic acid bacteria. The importance of these bacterial cytochromes not only lies in their greater variability, e.g. the different oxidases a_1, a_2 and o besides aa_3, and the unusual combinations of cytochromes found in them, but particularly in the fact that the role of cytochromes in realms of bioenergetics beyond aerobic metabolism has been opened up in these studies. Reviews of bacterial cytochromes have been written by L. Smith (1961), by Newton and Kamen (1961) and lately by Bartsch (1968).

C. THE ROLE OF CYTOCHROMES IN BIOENERGETICS

Even today many biochemists maintain that the major role of cytochromes is that of electron carriers to molecular oxygen in the respiratory chain. This is closely interwoven with the conversion of adenosine diphosphate (ADP) plus inorganic phosphate to adenosine triphosphate (ATP) in the process of "oxidative phosphorylation" firstly because the energy of this conversion ultimately stems from the energy of oxidation in the respiratory chain and in two of the three phosphorylation points of the chain cytochromes are involved; secondly the rate of the passage of electrons through the respiratory chain is greatly accelerated by the presence of ADP ("respiratory control") and the steady state oxidation of the cytochromes is greatly altered by its presence (Chance and Williams, 1956; Chance, 1961; Slater, 1958, 1966; Lehninger and Wadkins, 1962; Mahler and Cordes, 1968). Oxidative phosphorylation again is intimately connected with the process of proton migration (Mitchell and Moyle, 1967a,b,c,d) and with that of cation uptake in tissues and mitochondria (cf. Robertson, 1960, 1967). Again these processes are intimately connected with the locus, e.g. organelles (mitochondria), their structure and their membranes (Lehninger, 1962; Rothfield and Finklestein, 1968). It would be impossible to discuss all those relations in detail in a monograph on cytochromes, but they are too important to be entirely neglected.

The discovery of cytochromes in chloroplasts of green plants (cytochromes f and b_6) by R. Hill and co-workers (Hill and Scarisbrick, 1951; Davenport and Hill, 1952; Hill, 1954) led to the discovery of the role of cytochromes in photosynthetic reactions, not only in green plants and algae, but also in anaerobic photosynthetic bacteria. In these reactions they react with photochemically activated chlorophyll and bacteriochlorophyll and as electron carriers in the chain connecting various reactive centres. Chapters IVF and VC will be devoted to the role of the cytochromes in these reactions. The discovery of the role of cytochromes in anaerobic reactions is not restricted to photosynthetic reactions. They have been found essential in nitrate, nitrite and sulphate reduction (Sato and Egami, 1949; Egami *et al.*, 1961; Postgate, 1959, 1961). Thus cytochromes must be several hundreds of million years older than was previously believed and must have been essential catalysts at a time before photosynthetic processes gave the earth's atmosphere sufficient molecular oxygen to make cell respiration possible (see Chapter X).

REFERENCES

Barrett, J. and Lemberg, R. (1954). *Nature* **173**, 177.
Barrett, J. (1956). *Biochem. J.* **64**, 626–639.
Bartoch, R. (1968) *Ann. Rev. Microbiol.* **22**, 181–200.
Bartsch, R. and Kamen, M. D. (1958). *J. Biol. Chem.* **230**, 41–63.
Bodo, G. (1955). *Nature* **176**, 829–830.
Chance, B. (1961). "Haematin Enzymes", pp. 597–622.
Chance, B. and Williams, R. G. (1956). *Advances in Enzymol.* **17**, 65–134.
Commission on Enzymes of the International Union of Biochemistry (1961). Report, pp. 24–27, 47–48, 57–58, Pergamon Press, London.
Davenport, H. E., Hill, R. and Whatley, F. R. (1952). *Proc. Roy. Soc.* (London) B **139**, 327–345.
Dickerson, R. E., Kopka, M. L., Weinzierl, J., Varnum, J., Eisenberg, D. and Margoliash, E. (1967). *J. Biol. Chem.* **242**, 3015–18.
Dixon, M. and Webb, E. C. (1964). Second edition "Enzymes", Longman, Green, London.
Egami, F., Ishimoto, M. and Tamiguchi, S. (1961). "Haematin Enzymes" I, pp. 392–406.
Enzyme Nomenclature Recommendations (1965), pp. 18–24, Elsevier, Amsterdam.
Fujita, A. and Kodama, T. (1934). *Biochem. Z.* **273**, 186–197.
Hagihara, B., Horio, T., Yamashita, J., Nozaki, M. and Okunuki, K. (1956a). *Nature* (London) **178**, 629.
Hagihara, B., Morikawa, I., Sekuzu, I., Horio, T. and Okunuki, K. (1956b). *Nature* (London) **177**, 630.
Hagihara, B., Morikawa, I., Sekuzu, I. and Okunuki, K. (1958a). *J. Biochem.* (Tokyo) **45**, 551–563.
Hagihara, B., Tagawa, K., Morikawa, I., Shin, M. and Okunuki, K. (1958c). *J. Biochem.* (Tokyo) **45**, 725–735.
Hagihara, A., Tagawa, K., Sekuzu, I., Morikawa, I. and Okunuki, K. (1959a). *J. Biochem.* (Tokyo) **46**, 11–17.
Hagihara, B., Tagawa, K., Morikawa, I., Shin, M. and Okunuki, K. (1959b). *J. Biochem.* (Tokyo) **46**, 321–327.
Harbury, H. A., Cronin, J. R., Fanger, M. W., Hettinger, K., Murphy, A. J., Myer, Y. P. and Vinogradov, S. N. (1965). *Proc. Nat. Acad. Sci. U.S.* **54**, 1658–64.
Hayaishi, O. (1962). In "Oxygenases" (O. Hayaishi, ed.), pp. 1–29. Academic Press, New York and London.
Hill, R. (1954). *Nature* (London) **174**, 501–503.
Hill, R. and Scarisbrick, R. (1951). New Phytologist **50**, 98–111.
Hoppe-Seyler, F. (1890). *Z. physiol. Chem.* **14**, 106–108.
Keilin, D. (1930). *Proc. Roy. Soc.* (London B **106**, 418–444.
Keilin, D. (1933). *Nature* (London) **132**, 783.
Keilin, D. (1966). "The History of Cell Respiration and Cytochrome" (J. Keilin, ed.), Cambridge Univ. Press.
Keilin, D. and Hartree, E. F. (1939). *Proc. Roy. Soc.* (London) B **127**, 167–191.
Koshland, D. E. Jr. and Neet, K. E. (1968). *Ann. Rev. Biochem.* **37**, 359–410.
Lehninger, A. L. and Wadkins, C. L. (1962). *Ann. Rev. Biochem.* **31**, 47–78.
Lemberg, R. (1961). *Advances in Enzymology* **23**, 265–321.
Lemberg, R. (1969). *Physiol. Rev.* **49**, 50–121.
Lemberg, R. and Cutler, M. E. (1970). *Biochim. et Biophys. Acta* **197**, 1–10.
Levy, L. (1889). *Z. physiol. Chem.* **13**, 309–325.

Lumry, R. (1959). In "The Enzymes" (P. D. Boyer, H. Lardy, K. Myrbäck, eds.), vol. 1, pp. 177–231.
McDonald, C. C., Phillips, W. D. and Vinogradov, S. N. (1969). *Biochem. Biophys. Res. Commun.* **36**, 442–449.
MacMunn, C. A. (1886). *Philosophical Transactions Roy. Soc.* (London) **177**, 267–298.
MacMunn, C. A. (1887). *J. Physiol.* **8**, 51–65.
MacMunn, C. A. (1889). *Z. physiol. Chem.* **13**, 497–499.
Mahler, H. R. and Cordes, E. H. (1966). In "Biological Chemistry", pp. 606–619. Harper and Row, New York.
Mitchell, P. and Moyle, J. (1967a). *Nature* (London) **213**, 137–139.
Mitchell, P. and Moyle, J. (1967b). *Nature* (London) **214**, 1327–1328.
Mitchell, P. and Moyle, J. (1967c). *Biochem. J.* **104**, 588–600.
Mitchell, P. and Moyle, J. (1967d). *Biochem. J.* **104**, 1147–1162.
Newton, J. W. and Kamen, M. D. (1961). In "Bacteria" (I. C. Gunsalus and R. Y. Stanier, eds.). Vol. 2, pp. 397–424. Academic Press, New York.
Nicholls, P. (1963). In "The Enzymes" (P. D. Boyer, P. D. Lardy and K. Myrbäck eds.), vol. **8**, pp. 3–40. Academic Press, New York and London.
Okunuki, K. (1962). In "Oxygenases" (O. Hayaishi, ed.) chapter X, pp. 409–476.
Paléus, S. and Theorell, H. (1957). *Acta Chem. Scand.* **11**, 905 *et seq.*
Peisach, J., Blumberg, W. E., Wittenberg, B. A. and Wittenberg, J. B. (1968). *J. Biol. Chem.* **243**, 1871–1880.
Postgate, J. (1959). *Ann. Rev. Microbiol.* **13**, 505–520.
Postgate, J. (1961). "Haematin Enzymes", pp. 407–414.
Robertson, R. N. (1960). *Biol. Revs. Cambridge Philos. Soc.* **35**, 231–264.
Robertson, R. N. (1967). *Endeavour* **26**, 134–139.
Rothfield, L. and Finkelstein, A. (1968). *Ann. Rev. Biochem.* **37**, 463–496.
Sato, R. and Egami, F. (1949). *Bull. Chem. Soc. Japan* **22**, 137 *et seq.*
Slater, E. C. (1958). *Rev. Pure and Appl. Chem.* (Australia) **8**, 221–264.
Slater, E. C. (1966). "Comprehensive Biochemistry", Vol. **14**, pp. 327–396.
Smith, L. (1961). In "The Bacteria" (I. C. Gunsalus and R. Y. Stanier, eds.) vol. 2, pp. 365–396. Academic Press, New York.
Warburg, O. (1949). "Heavy Metal Prosthetic Groups and Enzyme Action". Clarendon Press, Oxford.
Warburg, O. and Negelein, E. (1929). *Biochem. Z.* **214**, 64–100.
Warburg, O. and Negelein, E. (1933). *Biochem. Z.* **262**, 237–238.
Weiss, J. (1964). *Nature* (London) **202**, 83–84; **203**, 183.
Williams, R. J. P. (1956). *Chem. Revs.* **56**, 299–328.
Yaoi, H. and Tamiya, H. (1928). *Proc. Imp. Acad. Tokyo* **4**, 436–439.

Chapter II

Classification and nomenclature of the cytochromes and their prosthetic groups

A. Classification and nomenclature.
B. The structure of the prosthetic groups.
   References.

A. CLASSIFICATION AND NOMENCLATURE

Four main groups of cytochromes can be distinguished by their prosthetic haem groups (Commission on Enzymes, 1961; Enzyme Nomenclature Recommendations, 1965).
1. Cytochromes *a*. Cytochromes in which the haem prosthetic group contains a formyl side chain, e.g. haem *a* (Fig. 1a).
2. Cytochromes *b*. Cytochromes with protohaem as prosthetic groups (Fig. 1b).
3. Cytochromes *c*. Cytochromes with covalent linkages between haem and protein (Fig. 1c). This group includes all cytochromes with prosthetic groups linked in this way. Today only thioether linkages are known and the prosthetic group can be termed "substituted mesohaems".
4. Cytochromes *d*. Cytochromes with a dihydroporphyrin (chlorin) iron as prosthetic group (Fig. 1d).

As a practical test it was suggested to use the position of the α-band of the pyridine ferrohaemochrome in alkali, 580–590 nm for cytochromes *a*, 556–558 nm for cytochromes *b*, 549–551 nm for cytochromes *c* and 600–620 nm for cytochromes *d*. This is due to the shift of the absorption maxima to longer wavelength by electrophilic side chains (formyl >

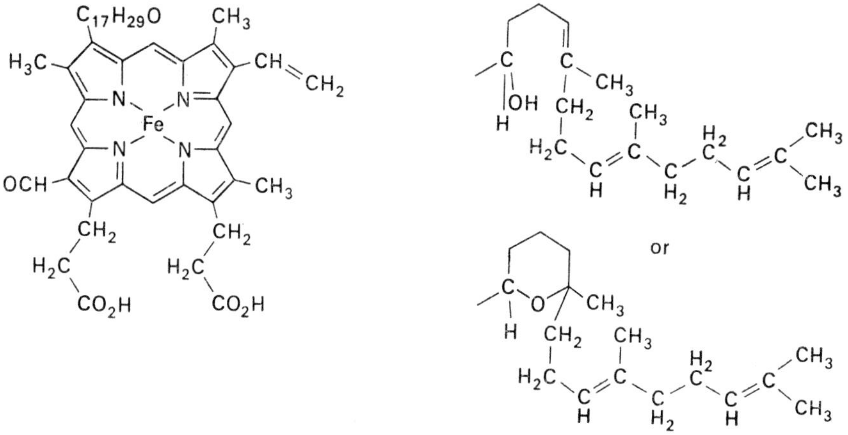

FIG. 1a. Haem *a* (Cytohaem *a*)

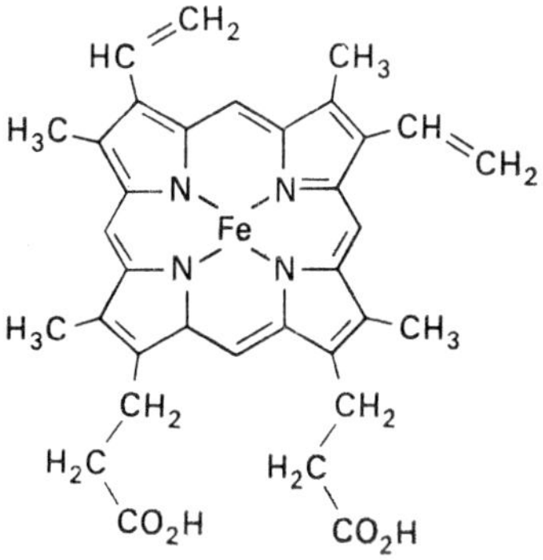

FIG. 1b. Protohaem (Cytohaem *b*).

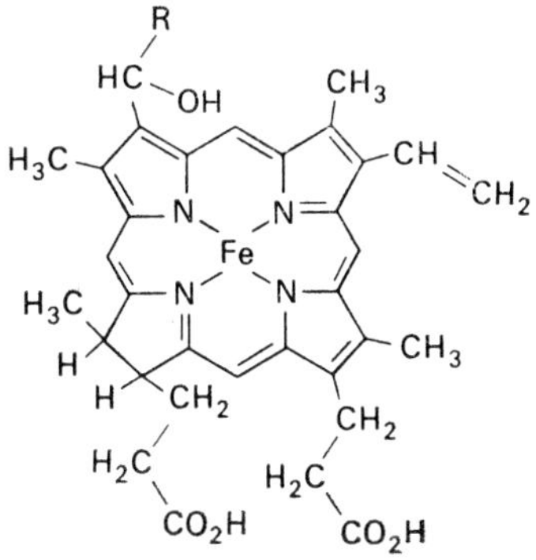

FIG. 1d. Cytohaem *d*.

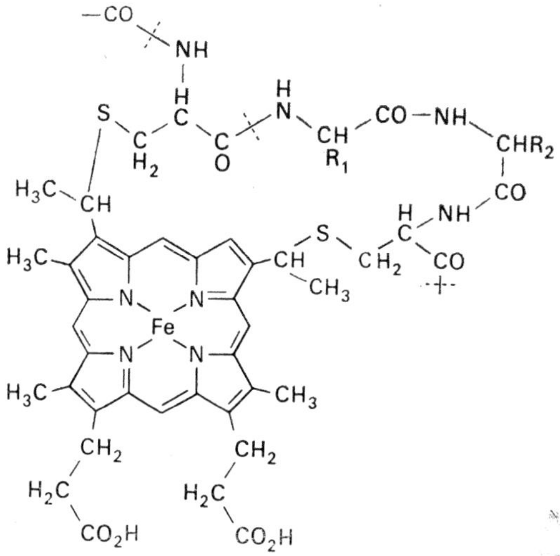

FIG. 1c. Haem *c* (Cytohaem *c*).

vinyl > substituted ethyl) or by addition of two hydrogens in β-position (porphyrin → chlorin). This is, indeed, a more reliable distinction than the position of the α-band of the cytochromes themselves, since those of cytochromes b and c overlap. It has been suggested that in order to avoid further multiplications of subscripts, some of which had been introduced without much justification, newly discovered cytochromes should be given interim names such as cytochrome b (562, *Escherichia coli*) or c (554, *Chlorobium*) in which the class name was followed in parentheses by the position of the α-band of the cytochrome in question and its source. At that time facilities for studying low temperature spectra were rarely available. However, not only does the position of this band depend on temperature, but in some instances a single band at normal temperature is revealed as due to several overlapping bands at the temperature of liquid nitrogen. The present nomenclature is therefore not entirely satisfactory and the real identity of some of the cytochromes of class b and particularly of class c is in doubt. Moreover, some other names have meanwhile been given to some cytochromes which are so widely used that they could only be altered by a decision of an authoritative committee. Unfortunately even the suggestions of the IUB Enzyme Commission have failed to provide this authority; and names such as cytochromes f (a cytochrome c) and a_2 (a cytochrome d) are still in use, and no new names have been suggested for cytochrome o and P-450. There seems little hope of remedying this position unless a collaboration of the editors of scientific journals enforces such rules.

Instead of the name "cytochromoid c" suggested in the earlier report, compounds such as the "RHP" (*Rhodospirillum* haemoprotein) of Vernon and Kamen (1954) are now named cytochromes cc' or c', the symbol c' denoting that a haem c prosthetic group is linked to the haem iron co-ordinately in a linkage which differs from that in haemochromes and the more typical cytochromes. These compounds have therefore a more haemoglobin-like spectrum and react with ligands such as carbon monoxide with which cytochrome c does not react. It has also been suggested (but as yet not accepted) that a cytochrome such as cytochrome a_3 is in fact an a' cytochrome, and cytochrome o a b' cytochrome.

However, there now appears to be an almost continuous gradation of compounds between cytochromes c' and c, and in cytochromes b a similar gradation appears to exist between cytochrome b, not autoxidizable and not binding ligands, and oxidases like cytochromes o and P-450. A strict separation of these compounds from cytochromes appears therefore inadvisable. It should also not be overlooked that there are haemochromes such as cytochrome a_1 which though having a typical haemochrome spectrum are able to react with ligands and probably to act as oxidases (Lemberg, 1969, p. 53).

On the whole the classification suggested in the Report has stood the test of time although some individual classifications may require reexamination. Thus it is possible that cytochrome h requires the redefinition of cytochromes b, and a new class may be necessary for the cytochromes of *Pseudomonas* oxidase (nitroreductase), at present included in class d.

Cytochromes a would include cytochrome aa_3, the cytochrome c oxidase of mitochondria of animals, plants, yeast and some bacteria, although this also contains copper, and cytochrome a_1 of certain bacteria. It does not include "cytochrome a_2" which belongs to class d (Barrett and Lemberg, 1954; Barrett, 1956). It would also include cryptohaem a (Parker, 1959; Lemberg *et al.*, 1961a; Barrett, 1961; Lemberg and Gilmour, 1968) if these can be shown to act as cytochromes. It is not impossible that the haemoprotein isolated by Morrison (1961) and Morrison, Reed and Hultquist (1970) from erythrocytes with a pyridine ferrohaemochrome band at 579 nm is such a compound. In contrast to Negelein (1932) no large amounts of cryptoporphyrin a could be isolated from pigeon breast muscle (Lemberg and Gilmour, 1968).

The structure of the prosthetic group, haem a of the Anglosaxon, or cytohaem of the German literature, has been recently reviewed by Lemberg (1969) and will be discussed in the next subchapter. The term "cytohaemin a" can be used as a compromise; the addition of a is certainly necessary to distinguish this from haem c or cytohaem c.

To the class of cytochromes b belong not only the cytochromes b and b_1–b_7 (cf. Enzyme Nomenclature Recommendations, 1965) but also a number of other cytochromes which according to the suggestions of the Enzyme Commission should receive preliminary names, as well as cytochrome P-450 (Omura and Sato, 1962), cytochrome o (Castor and Chance, 1959), helicorubin (Lemberg and Legge, 1949, p. 225). Whether or not cytochrome h (Keilin and Orlans, 1969) belongs to this group or to the cytochromes c depends upon whether or not the haematohaem prosthetic group is bound by covalent ester linkage; if not, it will be necessary to extend the cytochromes b to include haems with coordinately-bound prosthetic groups related to but different from protohaem. Cytochrome b_2 contains in addition to the protohaem group, flavin adenine mononucleotide as a second prosthetic group.

Cytochromes c have no clearly definable prosthetic group; their haem moiety is covalently linked by thioether linkages to cysteines built into the polypeptide chain of the protein. However, porphyrin c and haem c with two cysteines added by thioether linkage to two vinyl groups of protoporphyrin or protohaem may be considered as close to a prosthetic group (Hill and Keilin, 1930; Theorell, 1938; Zeile and Meyer, 1939). They include cytochromes c_1, c_2 and c_3, but in the 1965 report cytochromes c_4 and c_5 of *Azotobacter* have been omitted, because they do not appear sufficiently distinct from other bacterial cytochromes c in contrast to cytochromes c_2 and c_3. The justification for this now appears dubious. On the other hand, cytochrome f of chloroplasts had been included as cytochrome c_6.

In mammalian cytochrome c and probably in other cytochromes c the iron of the haem is held in co-ordinate linkage not as previously believed (Theorell and Akesson, 1941) between two protein nitrogens but between an imidazole-nitrogen and a methionine-sulphur (Harbury et al., 1965; Dickerson et al., 1967; McDonald et al., 1969). Cytochrome c itself is discussed in a separate chapter (Chapter V), the bacterial cytochromes c and c' in Chapter VI.

The prosthetic group of cytochromes d of *Pseudomonas* nitrate reductase (oxidase) (Horio et al., 1961; Yamanaka and Okunuki, 1963) and of *Micrococcus denitrificans* (Newton, 1967, 1969) differs from the prosthetic group of the membrane-bound cytochrome d (a_2) and may not be a dihydroporphyrin-haem.

B. THE STRUCTURE OF THE PROSTHETIC GROUPS

The structure of protohaem which is also the prosthetic group of haemoglobin, myoglobin, catalases and horse radish and cytochrome c peroxidases, is well known and was proved as early as 1926 by the elegant syntheses of H. Fischer (Fischer and Orth, 1937, p. 372; Lemberg and Legge, 1949, p. 56).

Strictly speaking, cytochrome c has no prosthetic group, since the haemin c of Hill and Keilin (1930) is built into the protein of the cytochromes c by peptide linkages as shown in Fig. 1c. Its structure was recognized by Theorell (1938, 1939) and by Zeile and Meyer (1939) as an adduct of two moles of cysteine to the two vinyl side chains of protohaem. By transforming it to mesoporphyrin IX dimethyl ester, Zeile and Reuter (1933; see also Davenport, 1952) showed the arrangement of the side chains to be identical with those of protohaem IX and the majority of other biological tetrapyrroles. Breakage of the thioether linkages can be achieved by hydrobromic acid to yield haematoporphyrin or more mildly by using silver or mercury salts (Paul, 1950, 1951; Barrett and Kamen, 1961). The haematohaemin thus obtained differs from synthetic haematohaemin by its optical activity (Paul, 1951). Robinson and Kamen have obtained porphyrin c by HF-catalysed synthesis from protoporphyrin and cysteine in liquid hydrofluoric acid. The ratio A_{406}/A_{553} of 16·5 was between that found by Neilands and Tuppy (1960) for their synthetic porphyrin c, 18·9, and the 15·0 found by Theorell (1938) for the natural split product of cytochrome c.

Protoporphyrinogen appears to add cysteine compounds more readily than protoporphyrin to its vinyl double bonds (Sano and Granick, 1961; Popper and Tuppy, 1963; Sano et al., 1964a,b; Sano, 1968).

Cytohaem a. Although Warburg and Negelein (1929, 1932) concluded early from the action spectrum of the Atmungsferment (cytochrome oxidase) that the prosthetic group contained a formyl side chain and somewhat resembled chlorocruorohaem, the structure of haem *a* (or cytohaem *a*) was a difficult problem which has been solved only in recent years. This development has been described by Lemberg (1969, pp. 60–63) in his review on cytochrome oxidase. A recent modification of the preparation of haemin *a* is described by Tuppy and Birkmeyer (1959). Only the essential steps of the growth of our knowledge of the structure are therefore repeated here. Warburg and Gewitz (1951) obtained a crystalline haemin *a* still containing traces of protohaemin and showed the presence of a larger side chain which gives the prosthetic group lipid character. Later Caughey and co-workers (Caughey and York, 1962; Caughey et al., 1966; York et al., 1967) separated the pyridine haemochromes of bovine heart muscle by chromatography on celite. Lemberg and co-workers (Lemberg, 1953; Lemberg and Stewart, 1955; Morell and Stewart, 1956; Morell et al., 1961b; Lemberg et al., 1961; Barrett, 1961) isolated porphyrin *a* free from other porphyrins using the inability of porphyrin *a* to be readily extracted by dilute HCl as a method of separation from protoporphyrin and cryptoporphyrin. Warburg and Gewitz (1951, 1953) by removing unsaturated, hydroxylated and α-carbonyl side chains in the resorcinol melt converted cytohaemin *a* into cytodeuterohaemin and cytodeuteroporphyrin, the correct structure of which was finally established by synthesis by Marks et al. (1960). Clezy and Barrett (1959, 1961) and Lemberg et al. (1961a) brought more stringent proof for the formyl nature of the carbonyl side chains and characterized the large side chain as an α-hydroxyalkyl side chain. After Piatelli (1960) had established the position of the formyl group in ring IV, the spectroscopic and chemical evidence of Lemberg et al. (1961a,b,c) proved the position of the large α-hydroxyalkyl group in 2 in ring I, that of the unsaturated group in 4 in ring II. Lynen and co-workers (Grassl et al., 1963a; Seyffert, Grassl and Lynen, 1966) established the nature of the C_1, side chain in position 2 as due to a reaction of farnesylpyrophosphate with a vinyl side chain of protohaem and brought evidence that the vinyl group in 4 was unsubstituted. Their formulation of the large alkyl side chain was $C_{15}H_{35}O$, while Caughey and co-workers (Caughey et al., 1966; York et al., 1967) and Lemberg (1965, 1966) found evidence that it contained four hydrogen atoms less, and the more recent evidence of Smythe and Caughey (1970) suggests that all three double-bonds of farnesyl are still present. In cytochrome oxidase the side chain has probably the structure indicated in Fig. 1a (see Lemberg, 1969) but Smythe and Caughey (1970) find no evidence for this ring and suggest esterification or labile esterification with an x group for which they do not, however, provide evidence. Optical activity of haematin *a* and haem *a* in phosphate buffer, but not in many other solvents, was found by King *et al.* (Yong and King, 1967; Yong *et al.*, 1968) and was confirmed by Nichol *et al.* (1968). This optical activity probably resides in the C_1 and C_5 atoms of the pyrane ring in Fig. 1d and is apparently liable to inversion and racemization. The ORD of the haematin and haem solutions in phosphate has recently been studied by Yong and King (1969). Under these conditions protohaematin and protohaem show no optical activity, while two different types of optical activity are revealed depending on the concentration of NaOH before buffering or on ionic strength. The authors postulate that the haems or haematins are linked by OH- and O-bridges between the haem irons as well as by H-bonding of the carboxyls. Asymmetric C-atoms in side chains, nonplanarity of ring, and di- or polymerization with coordination at positions 5 and 6 of iron are all assumed to be required for optical activity. A solution of haematin *a* in pyridine and a few other solvents able to form coordinate linkages with the haem iron, is automatically reduced to the ferrous form (Lemberg and Velins, 1965). The formyl groups of formyl-substituted porphyrins and haematins, including haem *a*, react with suitably placed ϵ-NH_2 groups of lysine of proteins in alkaline solution to form Schiff's bases (Lemberg and Newton, 1961).

Free porphyrin *a* has certainly a free α-hydroxyalkyl side chain, not closed to a pyrane or furane ring (Barrett, 1959; Clezy and Barrett, 1959, 1961). With acetic anhydride this is again closed into a ring (Nichol *et al.*, 1968; Lemberg, 1969, p. 62) or acetylated (Smythe and Caughey, 1970). By strong acid, porphyrin *a* ($a\alpha$) is reversibly transformed into porphyrin $a\beta$, which is far more readily extracted from ether into dilute HCl (3–5%) (Lemberg, 1953, 1955; Clezy and Barrett, 1961; Lemberg and Stewart, 1955; Morell and Stewart, 1956).

The structure of the prosthetic group of cytochrome *d* of *Aerobacter* and other organisms is probably that given in Fig. 1d (Barrett, 1956; Lemberg *et al.*, 1961a).

REFERENCES

Barrett, J. (1956). *Biochem. J.* **64**, 626–639.
Barrett, J. (1959). *Nature* (London) **183**, 1185–6.
Barrett, J. (1961). *Biochim. et Biophys. Acta* **54**, 580–582.
Barrett, J. and Kamen, M. D. (1961). *Biochim. et Biophys. Acta* **50**, 573–575.
Barrett, J. and Lemberg, R. (1954). *Nature* (London) **173**, 177.
Castor, L. N. and Chance, B. (1959). *J. Biol. Chem.* **234**, 1587–1592.
Caughey, W. S. and York, J. L. (1962). *J. Biol. Chem.* **237**, PC 2414–2416.
Caughey, W. S., York, J. L., McCoy, S. and Hollis, D. P. (1966). Hemes and Hemoproteins'', pp. 25–36.
Clezy, P. and Barrett, J. (1959). *Biochim. et Biophys. Acta* **33**, 584–586.

Clezy, P. and Barrett, J. (1961). *Biochem. J.* **78**, 798–806.
Commission on Enzymes of the International Union of Biochemistry, Report (1961). Chapter 5. The classification and nomenclature of cytochromes, pp. 24–27. Chapter 9, Summary of Recommendations, pp. 47–48. Appendix C. List of cytochromes, pp. 57–58.
Davenport, H. E. (1952). *Nature* (London) **169**, 75.
Dickerson, R. E., Kopka, M. L., Weinzierl, J., Varnum, J., Eisenberg, J. and Margoliash, E. (1967). *J. Biol. Chem.* **242**, 3015–3018.
Enzyme Nomenclature Recommendations (1965) pp. 18–24. Elsevier, Amsterdam.
Fischer, H. and Orth, H. (1937). "Die Chemie des Pyrrols". Pyrrolfarbstoffe II, Erste Hälfte. Akademische Verlagsgesellschaft, Leipzig.
Grassl, M., Augsburg, G., Coy, U. and Lynen, F. (1963a). *Biochem. Z.* **337**, 35–47.
Grassl, M., Coy, U., Seyffert, R. and Lynen, F. (1963b). *Biochem. Z.* **338**, 771–795.
Harbury, H. A., Cronin, J. R., Fanger, M. W., Hettinger, T. P., Murphy, A. J., Myer, Y. P. and Vinogradov, S. N. (1965). *Proc. Nat. Acad. Sci. U.S.* **54**, 1658–1664.
Hill, R. and Keilin, D. (1930). *Proc. Roy. Soc.* (London) B **107**, 286.
Horio, R., Sekuzu, I., Higashi, T. and Okunuki, K. (1961). "Haematin Enzymes", pp. 302–311.
Keilin, J. and Orlans, E. (1969). *Nature* (London) **233**, 304–306.
Lemberg, R. (1953). *Nature* (London) **172**, 619–620.
Lemberg, R. (1955). In "Biochemistry of Nitrogen", Suomalainen Tiedeakatemia (Acad. Sci. Fennica), pp. 165–173.
Lemberg, R. (1965). *Revs. Pure and Applied Chem.* (Australia) **15**, 125–136.
Lemberg, R. (1966). "Hemes and Hemoproteins", pp. 37–43.
Lemberg, R. (1969). *Physiol. Revs.* **49**, 48–121.
Lemberg, R., Clezy, P. and Barrett, J. (1961a). "Haematin Enzymes", vol. I, pp. 344–357.
Lemberg, R. and Gilmour, M. V. (1968). *Biochim. et Biophys. Acta* **156**, 407–408.
Lemberg, R. and Legge, J. W. (1949). "Hematin Compounds and Bile Pigments". Interscience, New York.
Lemberg, R., Morell, D. B., Newton, N. and O'Hagan, J. E. (1961b). *Proc. Roy. Soc.* (London) B **155**, 339–355.
Lemberg, R. and Newton, N. (1961). *Proc. Roy. Soc.* (London) B **155**, 364–373.
Lemberg, R., Newton, N. and O'Hagan, J. E. (1961c). *Proc. Roy. Soc.* (London) B **155**, 356–363.
Lemberg, R. and Stewart, M. (1955). *Austral. J. Exptl. Biol. Med. Sci.* **33**, 451–482.
Lemberg, R. and Velins, A. (1965). *Biochem. et Biophys. Acta* **104**, 487–495.
McDonald, C. C., Phillips, W. D. and Vinogradov, S. N. (1969). *Biochem. Biophys. Res. Commun.* **36**, 442–449.
Marks, G. S., Dougall, D. K., Bullock, E. and MacDonald, S. F. (1960). *J. Am. Chem. Soc.* **82**, 3183–3188.
Morell, D. B., Barrett, J. and Clezy, P. S. (1961a). *Biochem. J.* **78**, 793–797.
Morell, D. B., Barrett, J., Clezy, P. and Lemberg, R. (1961b). "Haematin Enzymes", vol. I, pp. 320–330.
Morell, D. B. and Stewart, M. (1956). *Austral. J. Exptl. Biol. Med. Sci.* **34**, 211–217.
Morrison, M., Reed, D. W. and Hultquist, D. E. (1970). *Biochim. et Biophys. Acta* **214**, 389–395.
Morrison, M. (1961). *Nature* (London) **189**, 765.
Morton, T. C. (1970). Ph.D. Thesis, Univ. of Melbourne, Australia.

Negelein, E. (1932). *Biochem. Z.* **248**, 243–245.
Neilands, J. B. and Tuppy, H. (1960). *Biochim. et Biophys. Acta* **38**, 351–353.
Newton, N. (1967). *Biochem. J.* **105**, 21–23C.
Newton, N. (1969). *Biochim. et Biophys. Acta* **185**, 316–331.
Nichol, A. W., Lemberg, R. and Harrap, B. S. (1968). *Biochim. et Biophys. Acta* **158**, 165–167.
Omura, T. and Sato, R. (1962). *J. Biol. Chem.* **237**, PC 1375–1376.
Parker, J. (1959). *Biochim. et Biophys. Acta* **35**, 496–509.
Paul, K. G. (1950). *Acta Chem. Scand.* **1**, 239–244.
Paul, K. G. (1951). *Acta Chem. Scand.* **5**, 389–405.
Piatelli, M. (1960). *Tetrahedron* **8**, 266–267.
Popper, T. L. and Tuppy, H. (1965). *Acta Chem. Scand.* **17**, Suppl. 1, 47–50.
Robinson, A. B. and Kamen, M. D. (1968). "Osaka", pp. 383–387.
Sano, S. (1968). "Osaka", p. 388.
Sano, S. and Granick, S. (1961). *J. Biol. Chem.* **236**, 1173–1180.
Sano, S., Ikeda, K. and Sakakibara, S. (1964a). *Biochem. Biophys. Res. Commun.* **15**, 284–289.
Sano, S., Nanzyo, N. and Rimington, C. (1964b). *Biochem. J.* **93**, 270–280.
Seyffert, R., Grassl, M. and Lynen, F. (1966). "Hemes and Hemoproteins", pp. 45–52.
Seyffert, R. and Lynen, F. (1968). "Osaka", p. 31.
Smythe, G. A. and Caughey, W. S. (1970). *J. Chem. Soc. D*, 809–811.
Theorell, H. (1938). *Biochem. Z.* **298**, 242–267.
Theorell, H. (1939). *Biochem. Z.* **301**, 201–209.
Theorell, H. and Åkesson, Å. (1961). *J. Am. Chem. Soc.* **63**, 1818–1820.
Tuppy, H. and Birkmayer, G. D. (1969). *Europ. J. Biochem.* **8**, 237–243.
Vernon, L. P. and Kamen, M. D. (1954). *J. Biol. Chem.* **211**, 643–662.
Warburg, O. and Gewitz, H.-S. (1951). *Z. Physiol. Chem.* **288**, 1–4.
Warburg, O. and Gewitz, H.-S. (1953). *Z. Physiol. Chem.* **292**, 174–177.
Warburg, O. and Negelein, E. (1929). *Biochem. Z.* **214**, 64–100.
Warburg, O. and Negelein, E. (1932). *Biochem. Z.* **244**, 9–32.
Yamanaka, T. and Okunuki, K. (1963). *Biochim. et Biophys. Acta* **67**, 379–416.
Yong, F. C. and King, T. E. (1967). *Biochem. Biophys. Res. Commun.* **27**, 59–62.
Yong, F. C., Bayley, P. M. and King, T. E. (1968). "Osaka", pp. 196–203.
Yong, F. C. and King, T. E. (1969). *J. Biol. Chem.* **244**, 509–514; 515–521.
York, J. L., McCoy, S., Taylor, D. N. and Caughey, W. S. (1967). *J. Biol. Chem.* **242**, 908–911.
Zeile, K. and Meyer, H. (1939). *Naturwiss.* **37**, 596–597.
Zeile, K. and Reuter, F. (1933). *Z. Physiol. Chem.* **221**, 101–116.

Chapter III

Cytochromes a

A. Introduction.
B. Cytochrome aa_3 (cytochrome o oxidase).
   1. Occurrence.
   2. Preparations.
   3. Activity assay.
   4. Chemical composition.
      a. Model compounds of haem a with nitrogenous ligands.
      b. The protein component.
      c. The copper.
   5. Reduced and oxidized cytochrome oxidase.
      a. Cytochrome aa_3.
      b. Ferrous oxidase.
      c. Ferric oxidase.
      d. Differences between ferrous and ferric oxidase.
      e. Oxidation reduction potentials.
      f. "Oxygenated" oxidase and evidence for conformation change.
      g. Kinetics of reduction and oxygenation.
   6. Compounds with CO, HCN and other inhibitors.
      a. CO-compound.
      b. NO-compound.
      c. Cyanide compounds.
      d. Azide compounds.
   7. The mechanism of cytochrome oxidase activity.

A. INTRODUCTION

As is shown in Table I, the most important cytochrome of this class is the cytochrome aa_3, or cytochrome c oxidase, EC 1.9.3.1, which forms the terminal oxidase of all higher organisms, animals and plants, and also of the yeasts, algae and some bacteria. In addition to two groups of haem

TABLE I
Cytochromes a

Name	Occurrence and function	M.W.	Solubility	Type of haem linkage	E'_0 mV	Isoel. point	Amino acid composition and sequence	Absorption spectra, nm (mM in parentheses)	Remarks	Chapter
aa_3	Mitochondria of animals	2 haems per 180,000	Bound to inner membrane. Detergent-solubilized and purified	See below	+290	4–5	Composition known but not sequence	Fe^{2+} α 604 (23)[1] 519 (low) γ 444 (120) Fe^{3+} 830 (1) (660)(sh) stable α 597 (11) γ 418 (88) "oxygenated" 830 (1) α 601 (13) γ 428 (90) ΔFe^{2+} minus Fe^{3+} 605 (12) γ 444 (85)	Cytochrome c oxidase EC 1.9.3.1. Also contains 2 Cu atoms per M.W. 180,000	III
a	Part of aa_3	—	—	Cytochrome c-like, no reaction with CO, HCN or O_2	+190	—	—	Fe^{2+} 605 (35) Fe^{3+} γ 444 (113) 605 (16) 444 (56) (not max.)	Calculated	III
a?	Isolated	1 haem per 180,000	—	Little reaction with CO	—	—	—	Fe^{2+} 602, 442 $\Delta COFe^{2+} - Fe^2$ +445, −417	Oxidase activity 10% of aa_3	III
a_3	Part of aa_3	—	—	Haemoglobin-like, reacts with CO and HCN and very rapidly with O_2	+395 (activated +300)	—	—	Fe^{2+} 605 (9) γ 444 (125) Fe^{3+} 605 (4·8) 444 (13·0) (not max.) CO α 590, γ 430	Calculated	III
a_1	Cytoplasmic membrane of bacteria. Terminal oxidase (?) and nitrate reductase	—	Membrane bound solubilized. Not purified	Haemochrome-like, reacts with CO, HCN, O_2	—	—	—	Fe^{2+} γ 589, γ 445	Not all bacterial "a_1" cytochromes may be identical with a_1 of *Acetobacter pasteurianum* of Warburg	III and IV

[1] Δ 605–630 nm, often quoted as 16·5 is in fact 19–20.

a, bound in a different way, probably to one and the same protein of MW of about 200,000, this enzyme contains two differently bound copper atoms and also is closely associated with lipids. The biological significance of this haemoprotein exceeds that of haemoglobin which, from this point of view, is only an auxiliary of the process of cell respiration carrying the oxygen into the tissues via the bloodstream. This process is essential for more bulky animals, in which diffusion of oxygen from the surface or from a tracheal system is insufficient, whereas the process of cellular respiration provides, through its intimate connection with oxidative phosphorylation (see Chapter VII), the energy essential for life processes of most aerobic beings.

Haemoglobin is, however, available in far larger amounts and has been investigated far more extensively. The difference of function of the oxygen-activating oxidases and of the oxygen-carrying haemoglobin is one of the most intriguing problems of biochemistry. The role of the haem groups in the function of cytochrome oxidase has been recognized for far longer than that of the copper. Even today the functional role of the copper is not yet entirely clear, but recent EPR evidence indicates a close interaction between the haems and the copper atoms. We face here a problem which is always intrinsically different—the relative role of different prosthetic groups in a "multiheaded" enzyme—a problem which we shall again encounter in discussing cytochrome b_2 (see Chapter IV). These two examples are not the only ones among the cytochromes. In cytochrome oxidase the problem is exacerbated by the presence of one and the same haem in two different types of binding, probably on one and the same protein molecule. Another example of this situation is found in the cytochromes cc' (see Chapter VI) and again other examples among the cytochromes may well be discovered. While no satisfactory separation of two independent cytochromes a and a_3 has been achieved, there is now convincing evidence for the presence of two haem a groups bound to protein in two different ways distinguishable by their reactions (or lack of reaction) with oxygen, carbon monoxide, cyanide, etc., in their reaction rates as well as by their absorption spectra (see Table I).

If two cytochromes a and a_3 could be separated, cytochrome a_3 alone would have to be considered as a terminal oxidase, since it alone is rapidly autoxidizable and reacts with the typical inhibitors of cellular respiration, while cytochrome a would be its first substrate carrying electrons from ferrocytochrome c (c^{2+}) to ferricytochrome a_3 (a_3^{3+}). The present evidence, however, does not justify the assumption of an independent existence of the two cytochromes. Although it has not been rigidly proved that the two haem a groups in a and a_3 are the same, so far no evidence to the contrary has been discovered, but there is good evidence for a different type of binding of haem a to the protein. Finally, recent studies have shown that it is necessary to assume two different conformational forms of oxidized cytochrome a_3, in what is known as "ferric" and as "oxygenated" oxidase.

Although the presence of some lipid appears to be essential for full activity of the oxidase, its role appears to be rather that of an unspecific matrix than of a specific constituent. We shall see below that a highly active solubilized purified oxidase can be obtained which contains only very small concentrations of an artificial, lipid-like detergent, but recent findings may indicate a specific role for a stoichiometric amount of the cardiolipin (see Chapter IIIB4b).

B. CYTOCHROME aa_3

1. Occurrence

In mammals, cytochrome aa_3 is found in the inner and cristal membranes of mitochondria (see Chapter VII). It has been discovered in rat thymus nuclei (Betel and Klouwen, 1967; Betel, 1967) but not in rat liver nuclei (Conover, 1967).

The occurrence of cytochrome aa_3 in rapidly contracting insect muscles was already known to MacMunn (1886) and Keilin (1925). It is evidently wide-spread in invertebrates, not only in insects (cf. the low temperature spectra of Goldwin and Farnsworth (1966) in *Drosophila*) but also in snails (Obuchowicz and Zerbe, 1962), in several marine invertebrates, including arthropods (Burrin and Beechey, 1963), cephalopods, molluscs and annelids (Ghiretti-Magaldi *et al.*, 1957, 1958; Ghiretti *et al.*, 1959; Pablo and Tappel, 1961) including unfertilized sea-urchin egg (Maggio and Ghiretti-Magaldi, 1958). In some shrimp muscles a concentration of up to 40 n moles of haem *a* per g. has been found approaching that in mammalian skeletal muscle. The abnormal CO-spectrum found by Pablo and Tappel in molluscs is certainly due to the presence of other CO-reacting pigments such as myoglobin (cf. Wittenberg *et al.*, 1965a,b) or perhaps cytochrome *o* (see Chapter VI).

In plants, cytochrome aa_3 is the major terminal oxidase and other enzymes such as the copper-containing oxidases play only a minor role (Lemberg and Legge, 1949, p. 379; Hartree, 1957; Smith and Chance, 1958; Chance and Hackett, 1959; Bonner, 1961; Mapson and Burton, 1962; Yonetani, 1963a; Chance and Bonner, 1965; Kasinsky *et al.*, 1966; Bonner, 1967). They have, however, been observed occasionally (Hackett, 1959; Bonner *et al.*, 1967; Storey and Bahr, 1969). The position of the adsorption bands of plant cytochrome aa_3 differs slightly from that in mammalian oxidase and the α-band may be double (Chance and Hackett, 1959; Chance and Bonner, 1965; Bendall and Bonner, 1966; Ducet *et al.*, 1970). This also holds for algae, e.g. *Euglena gracilis* (Perini *et al.*, 1964;

Raison and Smillie, 1969), where the latter have found the α-position of the band at 609 nm rather than the usual 605 nm. The denatured protein-haemochrome had the α-band at 597 nm (see below) and the position of the pyridine haemochrome at 586 nm was the normal one for pyridine haemochrome a, but it is remarkable that neither a band at 444 nm nor evidence for CO-combination could be detected and that a large part of the succinate oxidation was insensitive to cyanide. The cytochrome is thus possibly cytochrome a free from cytochrome a_3. A cyanide- and CO-inhibited respiration due to cytochrome aa_3 is found in zygospores of *Chlamydomonas* (Hommersand and Thimann, 1965).

The cytochrome aa_3 found in yeasts, in several *Saccharomyces* species as well as in *Candida krusei* (Yubisui et al., 1965; Okunuki, 1966; Duncan and Mackler, 1966; Palmer et al., 1967; Sekuzu et al., 1967; Okunuki et al., 1968; Yubisui, 1968) are similar to the mammalian oxidase, but some differences will be described later.

The bacterial cytochromes will be more fully discussed in Chapter VI. While much of our knowledge of bacterial cytochromes is still in the fact-gathering state, and the evidence is of unequal value and occasionally even contradictory, it is certain that a considerable number of bacteria contain a cytochrome aa_3 type of terminal oxidase. The evidence rests on the observation of the action spectra (Chance, 1953; Castor and Chance, 1955) as well as on the demonstration that a cytochrome of type aa_3 is bound to the cytoplasmic membrane and active as oxidase (Smith et al., 1966; Scholes and Smith, 1968; Broberg and Smith, 1967, 1968; Broberg et al., 1969). A great number of such organisms has been enumerated by Bartsch et al. (1969, pp. 192–194). The cytochrome aa_3 oxidase of *B. subtilis* has been isolated by Miki et al. (1967). It should be noted that the bacterial aa_3 oxidases do not always contain a and a_3 in the 1 : 1 ratio which is found in higher organisms and some bacteria (Appleby, 1969) and it has even been claimed that in some, e.g. *Micrococcus pyogenes*, var. *albus*, cytochrome a_3 is absent (Smith, 1955a). Absorption maxima of the α-band have been found to vary in position from 605 nm in mammalian cytochrome aa_3 to 598 nm in *Sarcina lutea* (Erickson and Parker, 1969). Cytochrome aa_3 has not only been found in strongly aerobic organisms such as *Sarcina lutea*, *Bacillus subtilis* and *Mycobacteria* but but under certain conditions in less respiring or even in facultative anaerobic bacteria such as *Staphylococci* (Smith, 1955b; Taber and Morrison, 1964) and *Streptomyces* (Niederpruem and Hackett, 1961). It has been found in fast-growing, but not slow-growing, *Corynebacterium diphtheriae* (Pappenheimer et al., 1962; Scholes and King, 1965), in which Rawlinson and Hale (1949) showed the presence of haem a as early as 1949. It has also been found in aerobically cultured *Rhizobia*, but not in bacteroids (Appleby, 1969) and in aerobically grown non-sulphur purple bacteria (Kikuchi and Motokawa, 1968) and *Nitrobacter agilis* (Straat and Nason, 1965). Cheah found a_1 (α-band 592 nm) together with cytochrome *o* in *H. halobium* and *H. salinarium*. It has been assumed for a long time that bacteria which contained cytochrome aa_3 did not contain cytochromes a_1, a_2 (=d) (see Chapter VI), or *o* (see Chapter IV). However, exceptions to this rule have now been found; thus *Xanthomones* (Arima and Oka, 1965) appears to contain aa_3 together with a_1, *Nitrobacter* aa_3 together with two cytochromes a_1 (Straat and Nason, 1965), *Pseudomonas riboflavine* and *Micrococcus denitrificans* aa_3 together with d type cytochromes (Arima and Oka, 1965; Smith et al., 1966). Cytochromes aa_3 and *o* can also occur together in *H. cutirubrum* (Cheah, 1969, 1970) but Lanyi (1968, 1969) found neither aa_3 nor *o* in this organism (see also Appleby, 1969; Revsin and Brodie, 1969; Ross et al., 1968). Some of these findings may be explained by intermediate phases between aerobic and anaerobic conditions (e.g. in *Rhodopseudomonas spheroides* cytochrome aa_3 is found under highly aerobic conditions, while later cytochrome *o* prevails or if light photosynthesis replaces dark respiration (Kikuchi et al., 1965; Kikuchi and Motokawa, 1968).

2. Preparations of cytochrome oxidase (cytochrome aa_3)

As material of mammalian cytochrome oxidation preparation usually beef heart muscle is used although that of other animals gives similar results. Either heart mitochondria (Crane et al., 1956; see also Blair, 1967) or a Keilin–Hartree (King, 1967) or related particulate preparation (Smith and Stotz, 1954) form the starting material. In fact, most of the methods are improved and simplified modifications of the pioneer methods of Straub (1939, 1941) and of Yakushiji and Okunuki (1940), cf. Okunuki (1962, pp. 414, 418). They are based on detergent extraction followed by fractional ammonium sulphate precipitation which separates cytochromes *c*, c_1 and *b* from cytochrome aa_3. Several detergents have been used (Wainio and Aronoff, 1955; Orii and Okunuki, 1965a; Detwiler et al., 1966; Kaniuga et al., 1966; Wharton and Tzagoloff, 1967); cholate appears to be less harmful than deoxycholate (Cooperstein, 1963c) and is preferred for the extraction. Non-ionic detergents, e.g. those of the Tween class are usually not as effective for extraction as the anionic detergents, but can be used to reactivate a cholate-inactivated preparation (Yonetani, 1959). However, Sun and Jacobs (1967) have developed a method of preparing solubilized oxidase with Triton X-100 without preceding cholate treatment. Membraneous oxidase has been obtained by removal of cholate from a cholate-solubilized preparation (McConnell et al., 1966), or directly by extraction of rat liver mitochondria with small concentrations of Triton X-114 (Jacobs et al., 1966a), and a solubilized oxidase is obtained from this with Triton X-100 (Jacobs et al., 1966b). Such green membrane preparations have also been studied by Seki et al. (1970), by Seki and Oda (1970) and by Sun et al. (1968). The membrane appears to contain two units of about 128,000 M.W. (46 × 56 Å) lying side by side horizontally to form the active dimer. Cationic detergents denature the oxidase irreversibly (see Chapter III B 3). An anionic detergent, sodium dodecyl sulphate also denatures, but in a very small concentration it has been found to cause an increase of activity of the purified preparation (Griffiths and Wharton, 1961; Orii and Okunuki, 1965). Greenwood (1963) has used snake venom to extract the lipoprotein. Recently, Person et al. (1969a,b) claim to have solubilized the oxidase by extraction with synthetic zeolites or at a pH 12. Their preparation certainly contains cytochromes b and c and its activity is probably low, but the finding itself is of interest in view of the known sensitivity of other oxidase preparations to a pH above 9·5 (Lemberg, 1964). If the observation can be confirmed and a reasonable activity of the enzyme can be preserved, it would indicate that in all cholate-treated preparations the oxidase has been modified to a certain extent but that this effect can be reversed by treatment with non-ionic detergents. This is supported by recent experiments of Stotz and co-workers (see B 4B).

In the experience of the authors, none of the other preparations matches in ease of preparation and particularly in the stability of the product the method of Yonetani (1967), particularly with some modifications introduced by Mansley et al. (1965) and Lemberg and Gilmour (1967). It does not precipitate or even become turbid on 1,000 fold dilution of the solution in Emasol-phosphate pH 7·4 with distilled water (Lemberg and Gilmour, 1967). Yonetani (1961) has obtained crystals which contained 90% of protein and only 10% of lipid + Emasol, but so far this crystallization does not appear to have been repeated. It has been shown by Lemberg et al. (1964) that there are no significant differences between the products obtained by this and by other methods (Griffiths and Wharton, 1961; Yonetani, 1960a; Smith, 1955a).

Although the Soret band of the stabilized ferric oxidase (B 5c) has been observed by us as well as by others at 418 nm not at 421–423 nm (Yonetani, 1967; Griffiths and Wharton, 1961; Lemberg et al., 1968) we have found that the maximum is shifted from 418 to 423 nm on prolonged standing of a solution at 0–4°. Differences in activity between various preparations appear to have been exaggerated (Orii and Okunuki, 1965b).

It is possible that the purified cytochrome aa_3 preparations contain a slightly altered enzyme. Firstly there is a slight shoulder in the 418–424 nm region (Gibson et al., 1965a; Van Gelder and Muijsers, 1966) even after complete reduction by dithionite which takes at least 30 minutes. In the best preparation the ratio A_{420nm}/A_{444nm} has then been found to be

0·39 (Gibson, Palmer and Wharton, 1965b) but has been found to vary between 0·40 and 0·50 in our preparations. Whether this is due to the presence of a "non-reducible" alteration product remains to be ascertained. Secondly, the percentage of total haem a in the oxidase preparation able to react with carbon monoxide has been found to vary from 36%—or even lower in preparations of Horie (1964a,b) and Morrison and Horie (1965)—to nearly 50% in the preparations of Mansley et al. (1966) and Nair and Mason (1967), while it is always 50% in the Keilin–Hartree preparation itself (Gibson, Greenwood et al., 1965; Vanneste, 1966). Preparations with a low A_{420}/A_{444} and a high percentage of CO-reactivity thus are least altered. Gibson, Palmer and Wharton (1965b) found only slight differences between the preparations of Yonetani and of Griffiths and Wharton in their rate of combination with oxygen and with carbon monoxide. They found a difference in their rate of reaction with ferrocytochrome c (but see Orii, Tsudzuki and Okunuki, 1965).

If the necessary corrections are applied for the measurement of haem a concentrations and the weight of fat-free protein, the good preparations all contain 8–11 nm moles of haem a/mg protein (Yonetani, 1961; Griffiths and Wharton, 1961; Wainio, 1964) corresponding to a molecular weight of about 100,000 which is in good agreement with the minimum molecular weight (see below). The presence of lipid can raise the apparent molecular weight to 130,000 if based on the ratio haem a/mg weight. The smallest fully active oxidase contains 2 haem a groups in a molecule twice the size (Griffiths and Wharton, 1961; Orii and Okunuki, 1965b).

The cytochrome aa_3 from yeasts and *Neurospora crassa* has also been purified (Sekuzu et al., 1964, 1967; Duncan and Mackler, 1966; Yubisui, 1968; Edwards and Woodward, 1970). The yeast enzyme contains non-haem iron and labile sulphide, in contrast to the heart enzymes and appears to have a slightly higher minimum molecular weight.

3. Activity assay

There are essentially two methods for the assay of activity, measuring the oxygen uptake using the whole respiratory chain from substrate to oxygen, or parts thereof (e.g. ascorbate-phenylene diamine oxidation) either manometrically or polarographically; or measuring the oxidation of ferrocytochrome c spectrophotometrically. These methods have been discussed by Yonetani (1962, 1963a) and by Lemberg (1969). The most frequently used method is the spectrophotometric method of Smith (1955c), cf. Yonetani (1967). It has been shown by Smith and Camerino (1963) and others (cf. Lemberg, 1969) that the apparent variations between the results are not as serious as was previously believed and that under certain conditions at least similar results can be achieved, so that the maximal turnover rate of cytochrome c approaches that of 450 sec^{-1}

which can be calculated from V_{max} for the purified oxidase preparations of Fowler et al. (1962). The maximal turnover rate of cytochrome c per mole of haem a is not far below the maximal interaction rate of cytochrome a_3 with cytochrome a, 750 sec^{-1} (Gibson et al., 1965a,b; Greenwood and Gibson, 1967; Gibson and Wharton, 1968).

The activity of purified solubilized cytochrome oxidase is not greatly inferior to that of the mitochondrial system under optimal conditions. The maximal second order-constant for purified oxidase preparations with oxygen, 10^7–10^8 M^{-1} sec^{-1}, found by Gibson et al. agrees well with those found by Chance and Schindler (1965) in intact rat liver mitochondria. Jacobs et al. (1966a,b) have found almost the same Q_{O_2} for their solubilized preparation with lipid added as for their membraneous oxidase preparation. Kimelberg and Nicholls (1969) found the turnover number for total haem a in the cytochrome c oxidation 90 (haem a)$^{-1}$ sec^{-1} for the purified oxidase preparation and 130 under optimal conditions in mitochondria of rat liver.

Data on the effects of pH of ionic strength and of cations and anions, of lipids and of the cytochrome c concentration have been summarized by Lemberg (1969).

It is well known that the activity increases with increased concentration of cytochrome c, but less than in proportion to it (Smith and Conrad, 1961, 1965). Lipids are essential for the activity but there does not appear to be any evidence for the need of a specific lipid in the oxidase. Thus the removal of lipids by cholates causes inactivation, but Tweens or Emasol, artificial lipids, can largely restore the activity (Lemberg et al., 1964) to about 30%, and Emasol is as effective in this as cardiolipin (Kimelberg and Nicholls, 1969).

Cytochrome aa_3 is destroyed by irradiation with blue light (Ninnemann et al., 1970; Epel and Butler, 1970) and preparations should not be exposed to strong light.

4. Chemical composition of cytochrome oxidase

(a) *Model compounds of haem a with nitrogeneous ligands.* The structure of the prosthetic group, haem a, has been described in Chapter II B. The study of the compounds of haem a and haematin a with nitrogenous ligands, while showing some similarities with the corresponding protohaem compounds, has also shown characteristic differences (Kiese and Kurz, 1958; Lemberg, 1961; Lemberg et al., 1961a,b,c; Morell et al., 1961). The pyridine ferrohaemochrome has its absorption band at 587 nm (ϵmM 30–33) (Lemberg and Stewart, 1955) in 20% pyridine–0·1N NaOH, while lower values e.g. 27·4 (Takemori and King, 1965) and partial alteration due to the interaction of the formyl group are found at lower pH (Lemberg, 1963; Vanderkooi and Stotz, 1965, 1966a); there is no distinct β-band and a high Soret band (γ) at 430 nm (ratio Aγ/Aα 4·6). Morrison and Horie (1965) found lower values (ϵ_{mM}^{587} 26) if the pyridine haemochrome was formed by the action of pyridine and alkali on cytochrome oxidase itself. Caughey and York (1962) and York et al (1967) reported spectroscopic differences between their pyridine haem a (directly extracted from heart muscle with pyridine–chloroform) and pyridine ferrohaemochrome a prepared from haemin a, but these have not been confirmed (Lemberg, 1963; Morrison, 1968; Vanderkooi and Stotz, 1966a). Lemberg and Velins (1965) found that strongly bound ligands such as pyridine and diethylamine cause spontaneous reduction of the haem iron; a solution of haemin a in pure pyridine thus contains the ferrohaemochrome. On dilution of the solution with water this becomes autoxidizable, but no evidence for a reversible oxygenation of pyridine ferrohaemochrome or of pyridine haem a could be observed by Lemberg (1963). Estimations of haem a in mixtures of pyridine haemochromes of haem a and protohaem are described by Lemberg et al. (1955) and by Williams (1964) while Lemberg and Parker (1955) have described methods to analyse mixtures of haem a, cryptohaem a and protohaem. Imidazole, 4(5)-methylimidazole, and histidine compounds of haem a have been studied by Lemberg (1961), Morell et al. (1961), Horie (1965) and Vanderkooi and Stotz (1965, 1966a,b). The methylimidazole haemochrome has its α and γ bands at 593–594 nm and 438–441 nm. The α-band is close to that of cytochrome a_1 and at much shorter wavelength than that of cytochrome aa_3 (605 nm), but at somewhat higher wavelength than of the pyridine ferrohaemochrome. Haematin a, like protohaematin, has a much higher affinity to imidazole than to pyridine. With imidazole itself Vanderkooi and Stotz found partly a similar position, but under slightly different conditions the position varied between 610 and 593 nm for the α-band and between 456 and 437 nm for the γ-band. These require further study (cf. Lemberg, 1969, p. 64).

The compounds of haem a and haematin a with various proteins, e.g. serum albumin, apomyoglobin, globin and apoperoxidase (horse radish) (Kiese and Kurz, 1958, 1959; Morell et al., 1961; Lemberg, 1961; Lemberg, Clezy and Barrett, 1961a; Lemberg, Morell et al., 1961b), show far less variations in both behaviour and spectra than the corresponding protohaem compounds. The globin compounds do not reversibly combine with oxygen, and the peroxidase activity of the apoperoxidase haematin a is weak. The ferrous α-bands were all in the 590–596 nm region, none as far towards the red as that of cytochrome aa_3 (605 nm), but close to the position of the α-band in alkali- or urea-modified cytochrome aa_3 (see below). The formyl group of haem a reacts with various carbonyl reagents (Lemberg, Clezy and Barrett, 1961a). Of particular interest is the formation of Schiff's bases (aldimines) with primary amines, e.g. the ϵ-NH$_2$ group of lysine, which occurs in alkali-denatured haemoproteins a (Lemberg, 1964; Lemberg and Newton, 1961) and with polylysine (Takemori and King, 1965; King et al., 1967). This explains the large shifts of absorption bands by the action of alkali that puzzled both Keilin and Warburg in their early studies. In native cytochrome oxidase at neutral pH, no suitable lysine group is apparently available in the close neighbourhood of the formyl group of the haem a. The formation of the aldimine bridge is reversed by acid, but can be made irreversible by borohydride reduction to an alkylamine bridge (King et al., 1967). Finally carbon monoxide shifts the α-band of all the haem a towards longer wavelength, except that of haem a itself in detergent solutions or in 50% alcohol (Morell et al., 1961; Lemberg, 1961; Lemberg, Newton and O'Hagan, 1961; Horie, 1965; Yong et al., 1968) and as we shall see, cytochrome a_3 behaves in the same way as haem a.

(b) *The protein component. Molecular weight.* It is necessary to distinguish between the minimum molecular weight (and this can be weight of the lipid-free or the lipid-containing protein per one mole of haem a), the minimum active molecular weight, and the weight in solution as measured by sedimentation. The first appears to be about 92,000 for fat-free protein (Matsubara et al., 1965, see below) although occasionally a smaller inactive subunit of as low as 67,000 has been observed (Orii and Okunuki, 1965; Criddle and Bock,* 1959). The lipid content can vary between 10% (Yonetani, 1961) and 30% (Griffiths and Wharton, 1961; Orii and Okunuki, 1967; Duncan and Mackler, 1966; Tzagoloff et al., 1965; Okunuki et al., 1968) so that the minimum molecular weight of protein plus lipid becomes 110–130,000. The minimal fully active weight is that of a dimer of 200–250,000 (Okunuki et al., 1968) while that in Emasol solution is probably a tetramer of M.W. 530,000 (Orii and Okunuki, 1965, 1967, but only 200,000 according to Love et al., 1970). In cholate the enzyme is polydisperse (Takemori et al., 1960b, 1961). It is of particular interest that the splitting of the tetramer (22 S) by guanidine hydrochloride or by small concentrations of sodium dodecylsulphate to the dimer (13–17 S) increased the activity (Okunuki et al., 1968) whereas higher concentrations of the dodecylsulphate decreasing the M.W. to 70,000 (6–8 S) destroy the activity and shift the α-band from 605 to 595 nm. For the complex IV of Green (membraneous cytochrome oxidase) dispersed in 0·1% taurodeoxycholate, Tzagoloff et al. (1965) found a molecular weight of 230,000 (12 S).

---

* The claim of F. F. Chuang and F. L. Crane (*Biochem. Biophys. Res. Commun.* 42, 1076–1081, 1971) of having isolated a cytochrome oxidase of M.W. 26,500 is obviously erroneous since the haem a and copper contents show a minimum M.W. of about 100,000. The error is probably due to the application of an unreliable method for estimating molecular weights (see J.-S. Tung and C. A. Knight, ibid. pp. 1117–1121).

Recently Love et al. (1970; see also Wilson and Greenwood, 1970) found that by raising the pH in non-ionic detergents to 9·5–11·0 the dimer oxidase is split into a monomer (or mixture of monomers!) of M.W. 100,000. This still retains some activity, although 80% of the haem a now reacts with CO. However, their identification of this modified oxidase with monomeric cytochrome a_3 must be rejected (see below).

The minimum molecular weight of the yeast enzyme appears to be somewhat higher (109–121,000) than that of the heart enzyme (92,000), possibly due to the presence of a ferredoxin-like compound in it (Yubisui, 1968).

Attempts at isolation of the haem-free apoenzyme have so far not been very successful. Gilmour and Volpe (1970) obtained a haem-free protein which still contained the copper. Reactivation by haematin a was, however, slight. Nevertheless such attempts should be continued in view of the demonstration by Tuppy and Birkmayer (1969) that an active oxidase can be obtained by haematin a addition to an apoprotein present in "petite" yeast mutants.

Amino acid composition. Three preparations have so far been studied, two from beef heart (Matsubara et al., 1965) and that from *Saccharomyces oviformis* M2 (Yubisui, 1968). The composition of the heart enzyme was Lys_{39}, His_{29-30}, Arg_{30-31}, Asp_{58-60}, Glu_{59-60}, Gly_{58-59}, Ala_{60-64}, Val_{49-52}, Ile_{43}, Leu_{86-88}, Ser_{53-55}, Thr_{52-53}, Pro_{46}, Met_{35}, Cys_7, Try_{30}, Phe_{46-47}, Tyr_{32-33}, $(NH_3)_{56}$, altogether 812–820 residues giving a M.W. of 92–93,000 for the lipid-free protein. While the basic residues slightly exceed the acidic residues (free aspartic and glutamic), the isoelectric point was found between pH 4–5 (Takemori et al., 1961), probably due to the fact that some of the basic residues are buried in the interior of the molecule. The leucine content is unusually high, the cysteine content is very low. The yeast enzyme contained less methionine, histidine, proline and amide, more aspartic and glutamic acids, alanine, valine, isoleucine and possibly tyrosine and cysteine than the mammalian enzyme. Its protein is thus more acidic.

The number of sulphydryl groups in native cytochrome oxidase is still in some doubt. Of the seven half-cystines found per mole haem a, all are found in the sodium-dodecylsulphate-modified enzyme. Orii et al. (1965) found at first only two of the sulphydryls of the native enzyme reactive with p-chloromercuribenzoate. Kirschbaum and Wainio (1966) found four reactive in the presence of EDTA, while Tsudzuki et al. (1967) found six, none of which appeared required for enzyme activity. The seventh SH-group was removed by p-chloromercuribenzoate only with difficulty, unless sodium dodecylsulphate was present. This group was assumed to be combined with copper as had already been postulated by Wainio (1961a,b). Cooperstein (1953a,b) had postulated the presence of an essential SS-bridge since SH compounds inactivated the enzyme, but this could not be confirmed by Orii and Okunuki (1967). Occasionally more than seven SH groups were found, but these are ascribed to impurities. Williams et al. (1968) found in three experiments 7·3, 4·4-(?) and 7·0 SH-groups reacting with p-chloromercuribenzoate in "ferric" oxidase, but on the average two more (9·1, 7·6 and 9·0) in "oxygenated" oxidase. The inhibition of aerobic electron transport by coeruloplasmin is ascribed to its reaction with an SH group of the oxidase (Brown and White, 1961).

Modification and denaturation. By exposure to high pH (above 9·5 in cholate) or to 8M urea cytochrome oxidase is converted into an inactive "modified" form (Lemberg, 1962; Lemberg and Pilger, 1964). This is accompanied by a shift of the α-band from 605 to 595 nm, with an isosbestic point at 602 nm, while the Soret band is shifted from 445 to 436 nm and increased. A similar change is caused by sodium dodecylsulphate (see above) accompanied by a decrease of the molecular weight, and by boiling (Person and Zipper, 1964). The removal of the copper from the oxidase is also accompanied by this modification (Wharton and Tzagoloff, 1964; Tsudzuki et al., 1967). In this modified form the whole of the haem *a* is reactive with carbon monoxide (Mansley et al., 1966) and the part which remains in solution after neutralization of an alkaline solution displays a CO-absorption band at 434 rather than 430 nm (Lemberg and Stanbury, unpublished).

This reaction is quite different from one which occurs with isosbestic points at 582·5 and 434·5 nm with cytochrome oxidase at a pH above 12 (Lemberg, 1961; Lemberg, 1964; Takemori and King, 1965) but with other haem *a*-proteins at the lower pH of 9·5. This reaction is due to the formation of a Schiff's base (aldimine) between the formyl group on the porphyrin and an ϵ-NH_2 of a lysine residue of the protein and has been discussed in B 4(a). Conformation changes by deoxycholate and ageing have been studied with CD-measurements by Myer and King (1968). The conversion of impure cytochrome oxidase into "mitochrome" is not fully understood, but may be due to a similar reaction. It has been discussed by Lemberg (1969, p. 71).

A different type of modification of the oxidase, occurring at pH 9·5–11·0 in non-ionic detergent solutions has recently been described by Wilson and Greenwood, 1970; Love et al., 1970; Chan et al., 1970 and Love and Auer, 1970. This appears to lead to little inactivation as contrasted with the same pH in cholate. From this and the fact that the CO-reaction was now 80–90% complete, Stotz and co-workers have concluded that the treatment converts the dimeric cytochrome aa_3 into monomeric cytochrome a_3, and that the properties ascribed to aa_3 are due to polymerization, seeing in this a support for the unitarian theory (more fully discussed by Lemberg, 1969) to which discussion the reader is referred. It can, however, be shown that this product of the alkali treatment is not a single compound, but a mixture of cytochrome a_3, some cytochrome a and an altered cytochrome a which has become reactive with CO (Lemberg and Geyer, unpublished). The altered CO-reactive cytochrome is comparable to modified cytochrome c although it differs from it by being monomeric, not dimeric. The kinetic differences between a_3 and CO-reactive a are still found, e.g. the Yonetani phenomenon, i.e. the far greater inhibition of the reduction of a_3^{3+} by cyanide than that of a^{3+}. The spectra, in particular the position of the α-band of the supposed a_3CO at 605 nm not at 590 nm (see the action spectrum and other evidence for this position of the a_3CO band) and its highly skew nature also disprove the identification of the product with cytochrome a_3.

Chromatography, electrophoresis and antibody formation. Only a few studies on the application of such methods are as yet available. Alumina $C_γ$ chromatography (Okunuki, Sekuzu et al., 1958) and Ca phosphate chromatography (Griffith and Wharton, 1961) considerably decrease the yield without achieving greater purity and have therefore been omitted in later preparations. Sephadex gel filtration has been used to remove dithionite and its oxidation products (Williams et al., 1968). Polyacrylamide gel electrophoresis applied under conditions under which the enzyme is denatured has given as many as eight different layers (Tzagoloff and MacLennan, 1966) whereas Stumm-Tegethoff (1967) under his special conditions found a single band of the Fowler et al. (1962) preparation which was the same as that of oxidase from eyes of a fruitfly. Kabel et al. (1968) and Mochan et al. (1970) have shown that oxidase is antigenic and that the antibody precipitate retains some oxidizing activity. There may be some conformational change in the antibody reaction, but it can only be slight since there is no spectral change and the haem a group reacts with the usual ligands. However, some changes in the reaction with azide have been found, both in the particulate and the soluble oxidase.

Study of the conformation of the oxidase by CD and ORD measurements. The earlier findings of King et al. (King and Schellman, 1966; Yong et al., 1968) and those of Urry et al. (1967) and Urry and Van Gelder (1968) gave rather contradictory results. Urry found the helical content did not exceed 30%, while King et al. found a large α-helical contribution, particularly at 193 nm. There is a large molar ellipticity in the region of the Soret band. This is not due to the optical activity of haem a itself which is inactive in Emasol, but depends on the iron valency. Urry concluded from the greater complexity of the CD spectra of the ferrous enzyme that the haem groups in the ferrous but not in the ferric enzyme were in close proximity. The later work of Myer and King (1968, 1969) however, indicates that this was an artifact due to the effect of deoxycholate on the ferrous enzyme (cf. also Willick, Schonbaum and Kay, 1969). Myer and King find evidence for two Gaussian bands in the CD spectrum of the ferrous enzyme corresponding to ferrocytochromes a and a_3, but four in that of the ferric enzymes and conclude that haem–haem interactions are absent but that the differences between ferrous and ferric enzyme are due to differences in the environment of the two haems, symmetrical in the ferrous, asymmetrical in the ferric enzyme; these in turn depend on conformational differences in the protein, and alteration of the intrinsic asymmetry of the chromophore. The CO-studies thus support the presence of two conformationally distinct and independent haem a moieties in the oxidase (see below). The different CD-spectra for the change of oxidation state of the component not reacting with CO or HCN is independent of the chemical nature of the central coordinating complex of the reacting form. The Cotton effect in the "aromatic" region appears to be much smaller than in cytochrome c, although the content of aromatic amino acids and particularly the 280 nm absorption of the oxidase is particularly high ($Fe^{3+}A\ 280/Fe^{2+}A_{445}$ abt. 2·5) (Lemberg et al., 1964).

Phospholipid. Previous findings, summed up by Lemberg, 1969, supported that phospholipids play only the role of an unspecific matrix for cytochrome oxidase activity. Recently, however, Awashti et al. (1970, 1971) have found that a small part of the cardiolipin is highly bound and not affected by phospholipid, and that this is essential for cytochrome oxidase activity. Chuang and Awashti (1970) propose that cytochrome oxidase forms a mosaic membrane with phospholipid bilayers interspersed between the protein globules.

(c) *The copper in cytochrome oxidase.* The presence of copper in cytochrome oxidase preparations had been known for many years. This has been reviewed by Lemberg (1969, p. 65). During the last few years it has been established that purified cytochrome oxidase free from extraneous copper always shows a ratio Cu/haem a of one or slightly above one, and that the minimal molecule of active native oxidase contains two copper atoms in addition to two haem groups (Wainio et al., 1959; Takemori, 1960; Takemori et al., 1960; Griffiths and Wharton, 1961; Morrison et al., 1963; Van Gelder and Slater, 1963; Van Gelder and Muijsers, 1964; Van Gelder, 1966; Beinert and Palmer, 1964, 1965; Beinert, 1966). Earlier contradictory findings on the valency of copper in the oxidase have been discussed by Beinert (1966) and Lemberg (1969). The copper is tightly bound to the protein and cannot be removed by cyanide chelators (bathocuproin, cyanide) without denaturation of the protein; claims that this was possible (Nair and Mason, 1962) have not been confirmed (Lemberg, Harris and Geyer, unpublished; Beinert et al., 1970; Mason and Ganapathy, 1970).

The valency of one of the two intrinsic copper atoms, Cu(α) or Cu(d)* is certainly cupric and can be distinguished from extrinsic copper by its

---

* "d" and "u" stand for ESR-detectable and undetectable Cu (Beinert et al., 1971).

EPR spectrum (g = 2·03) lacking hyperfine structure. This accounts only for about one half of the total copper. The valency of the second copper atom, Cu(β) or Cu(u)* is not yet certain. It may be either cuprous, or in the form suggested by Hemmerich (1966), an equilibrium form of Cu(I)-disulphide and Cu(II)-sulphydryl which may still undergo further reduction, or it may be cupric, its EPR-signal being suppressed by antiferromagnetic interactions of Cu(II) with a haematin a Fe(III) of cytochrome a_3 (Vanneste, 1968; Van Gelder and Beinert, 1969; Beinert and Orme-Johnson, 1969; Beinert et al., 1971). This may explain why the typical high spin ferric signal of a_3^{3+} is also absent and reappears only after initial reduction probably of Cu(β). The complete reduction of ferric oxidase by ferrocytochrome c requires about four reducing equivalents, two for those of two ferric haematins and two for the reduction of copper if both coppers are cupric; this occurs in pairs, first a^{3+} and one of the coppers, then more slowly of a_3^{3+} and the second copper (Lemberg and Cutler, 1969, 1970; Van Gelder and Beinert, 1969). The fast-reduction of copper accompanying that of a^{3+} is that of Cu(β), the EPR-non-detectable copper. Hence the EPR spectrum of Cu, due to Cu(α) disappears much more slowly on reduction by ferrocytochrome c or by dithionite (Beinert and Orme-Johnson, 1969). In the presence of cyanide only two reducing equivalents can be used since a_3^{3+}CN and Cu(α)—presumably also bound to cyanide—are not being reduced (Van Gelder and Beinert, 1969). From a kinetic point of view this appears to contradict the earlier assignations of Cu(α) to cytochrome a and of Cu(β) to cytochrome a_3. This had been based on spectrophotometric titrations of the ferric oxidase with NADH + phenazine methosulphate (Van Gelder and Slater, 1963; Van Gelder and Muijsers, 1964, 1966; Van Gelder, 1966) using the low absorption band at 830 nm as well as reducing equivalents not affecting the typical haem a absorption bands of the oxidase. However, for other non-kinetic reasons, discussed in the next section, Cu(β) still may be more closely connected with a_3 than with a.

The rather low 830 nm absorption band of ferric oxidase had been discovered by Griffiths and Wharton (1961) and confirmed by Lemberg et al. (1964) and Wharton and Tzagoloff (1964). As was pointed out by Lemberg (1969) who did not accept the nomenclature Cu(a_3) for Cu(α) and Cu(a) for Cu(β), it is still uncertain whether this band is due to both or only to one of the Cu-atoms, i.e. Cu(α), a conclusion also reached by Tzagoloff and MacLennan (1966). Cu(β) would then be neither recognizable by its EPR spectrum, nor by its spectroscopic features. The 830 nm band is destroyed by denaturation of the protein (Wharton and Tzagoloff, 1963; Gibson and Greenwood, 1965); the copper then shows hyperfine splitting of the EPR-spectrum and can be removed by dialysis against strong cyanide (Beinert and Palmer, 1964, 1965; Takemori, 1960a; Takemori et al., 1960), or, after reduction, by bathocuproine disulphonate (Beinert et al., 1962; Morrison et al., 1963; Yonetani, 1960). It has now become reducible by borohydride (Beinert and Palmer, 1964; Morrison et al., 1963), but is no longer reducible by ferrocytochrome c (Ehrenberg and Yonetani, 1961; Beinert and Palmer, 1964) although other copper chelates can oxidize ferrocytochrome c (Davison, 1968; Harris and Ritchie, 1969). From a thoroughly denatured enzyme MacLennan and Tzagoloff (1965) have isolated a copper-protein of M.W. 25,000 containing approximately all the copper, but little haem a. After modification of the enzyme at pH > 10 in non-ionic detergents, however, the 830 nm band is still present (Wilson and Greenwood, 1970) and cannot be removed by cyanide.

Although it is now clear that the intrinsic copper of the enzyme undergoes rapid valency changes on oxidation and reduction, the role of the copper in the electron transport still remains uncertain; copper may accelerate rather the electron transport between cytochromes a and a_3 than the reaction of the oxidase with molecular oxygen, although an important role of it in the latter can be postulated and will be discussed below. Gibson and Greenwood (1965) found that in the rapid reaction of reduced oxidase with molecular oxygen, the rate of oxidation of Cu, measured by the appearance of the 830 nm band, was slower than the oxidation of cytochrome a_3 measured by the decrease of the Soret band, but faster than the oxidation of cytochrome a measured by the decrease of the 605 nm band; it should be noted, however, that neither of these spectroscopic changes is a perfect measure of those of total Cu, of a_3, or of a. When only cytochrome a but not a_3 is reduced by partial reduction of ferric to ferrous oxidase (Lemberg and Gilmour, 1968) the 830-nm band disappeared completely, and on autoxidation it was restored before oxidation of cytochrome a occurred (Gilmour, 1967).

Using the 830-nm band as indicator Tzagoloff and MacLennan (1966) found an oxidation-reduction potential (Eó) of the reacting copper of +284 mV, and Wharton and Cusanovich (1969) of +274 V, with one electron reduction with a reduction mixture containing dithionite at 20°.

Yong and King (1970c) have shown that the valency state of the copper as observed by the 830 nm band and EPR, influences the $\Delta(a_3{}^{2+}CO - a_3{}^{3+}CN)$ spectrum. They conclude that this copper is sandwiched between cytochromes a and a_3 and that the conformation of a_3 is dependent on this copper.

5. Reduced and oxidized cytochrome oxidase

(a) *Cytochrome* aa_3. Reasons for the acceptance of the aa_3 theory and the rejection of the assumption of the "unitarian" theory of a single cytochrome A (Okunuki, 1966, Okunuki et al., 1968; Wainio, 1965, Wainio *et al.*, 1968) have been summarized by Lemberg (1969, pp. 85–88). Recent investigations of Van Gelder and Beinert (1969), Beinert and Orme-Johnson (1969), Tsudzuki and Okunuki (1969) and Myer and King (1968, 1969) have strengthened this evidence by EPR, CD and ORD data. Only a part of the haem a of cytochrome oxidase (a_3) reacts with CO and HCN, as shown by Keilin and Hartree (1939). By making use of the very slow velocity of reduction of a_3^{3+}CN by dithionite and other reductants, in contrast to the fast reduction of uncombined a^{3+}, difference spectra for ($a_3^{2+} - a_3^{3+}$) and ($a^{2+} - a^{3+}$) were obtained by Yonetani (1960a) and Lemberg *et al.* (1964). Even in the absence of cyanide a_3^{3+} of ferric oxidase is far more slowly reduced than a^{3+} (Lemberg and Mansley, 1965; Lemberg and Gilmour, 1967). By combining these difference spectra with the absorption spectrum of a_3^{2+}CO obtained from the photochemical action spectrum (Castor and Chance, 1955; Chance, 1961), Vanneste (1965, 1966a,b) and Vanneste and Vanneste (1965) have arrived at the absorption spectra for a^{2+}, a^{3+}, a_3^{2+} and a_3^{3+} which, with a minor modification, are given in Table II; this modification is the decrease of the value for the α-band of a^{2+} from ϵ mM 39·5 to ϵ mM 35. Vanneste's value was probably too high because a molar ratio a/a_3 of 1 was assumed while in fact in purified oxidase this ratio is usually above 1. The lower value is supported by the spectrophotometric results on the interaction of ferric cytochrome oxidase with ferrocytochrome c (Lemberg and Cutler, 1970). The other values of Vanneste probably require only minor corrections. The γ/α ratios, the isosbestic points for the difference spectra for a and a_3, and their absorption minima in the Soret band region agree with those of Lemberg and Mansley (1965) and Lemberg and Gilmour (1967) by direct observations at room temperature and are also similar to those of Gilmour, Wilson and Lemberg (1967) at low temperature. The contribution of a to the α-band is about 80%; the γ-band of the absolute spectrum is about 50% due to a, but its contribution to the difference spectrum is only 35% (Lemberg and Gilmour, 1967; Greenwood and Gibson, 1967). The possibility of some interaction between a and a_3 is not necessarily contradicted by these results, but they are spectroscopically inoperative as are those between the subunits of haemoglobin. In contrast to Horio and Ohkawa (1968) who were unable to explain the spectrophotometric data on the interaction of ferric cytochrome oxidase with ferrocytochrome c on the basis of two haems a and a_3 with the properties postulated by Vanneste, and who therefore assumed three different haems a-1, a-2 and a-3, Lemberg and Cutler (1970) found that their data were in good harmony with the aa_3 theory and Vanneste's values.

The failure of a to react with carbon monoxide and cyanide, the low γ/α ratio of the ferrous a, the relatively small difference between the a^{2+} and the a^{3+} Soret band position (Brill and Williams, 1961; Williams,

1961), the complete lack of a charge transfer band in 630–660 nm region, and the EPR spectra which characterize a^{3+} as a low spin complex, all indicate that cytochrome a is more closely related to cytochrome c than cytochrome a_3. Cytochrome a_3 reacts with carbon monoxide and cyanide, has a high γ/α ratio and a larger difference in the a_3^{2+} positions of the Soret band, possesses a charge transfer band in the 650–660 nm region, and a_3^{2+}, and at least in part a_3^{3+} are high spin complexes. The shift by CO to longer wavelength is among the haem a compounds only found with haem a itself in detergents, not with other haem a protein compounds. All these properties support the hypothesis that cytochrome a_3 is more closely related to haemoglobin or peroxidase-like haemoproteins, or to a "cytochromoid" (a') haemoprotein (cf. Lemberg, 1969, p. 83–84).

Attempts by Morrison and Horie to separate the two cytochromes have so far failed, although they have confirmed the aa_3 theory (Morrison and Horie, 1963, 1964; Horie and Morrison, 1964; King et al., 1965; Horie, 1967, 1968). It is possible to destroy a_3 by borohydride reduction and to remove largely the CO-reactive and autoxidizable part, leaving the haem a, however, in a molecule of hardly diminished size. The findings therefore support the theory of a "double-headed" (Lemberg, 1961) or "Siamese twin" (King et al., 1965) enzyme, with the haem a and a_3 group on one and the same protein, rather than the existence of two separable proteins, cytochromes a and a_3 (cf. Lemberg, 1969, p. 84–85).

(b) *Ferrous oxidase.* The absorption spectrum of ferrous oxidase ($a^{2+} + a_3^{2+}$) is given in Fig. 2 and millimolar, extinction coefficients for the maxima are found in Table I. Complete reduction by dithionite requires more than thirty minutes. Criteria of purity are given by Lemberg (1969, p. 72). For estimation of the haem a content use has often been made of $\Delta_{mM}^{605-630nm}$ 16·5, but the correct value is between 19 (Vanneste, 1965) and 21 (Lemberg and Gilmour, 1967). At 77°K the two absorption maxima are found at 601 and 442 nm (Gilmour et al., 1967). Absorption data for a^{2+} and a_3^{2+} are also given in Table I (cf. above). The spectrum of a_3^{2+}, but not of a^{2+}, shows a β-band (Vanneste, 1966). The γ/α ratio is 5·2 for $a^{2+}a_3^{2+}$, 3·3 for a^{2+} and 14 for a_3^{2+}. These figures are not so exceptional (Lemberg, 1961, 1966a, 1969) as had previously been believed, e.g. the γ/α ratio of ferrocytochrome c is 4·4 (Margoliash and Walasek, 1967) and that of the methylimidazole haemochrome a 3·3 (Morell et al., 1961) while that of haemoglobin is 10.

(c) *"Ferric" oxidase and its relation to ferrous oxidase.* The absorption spectrum of "ferric" oxidase, as obtained in the preparation of the oxidase is given in Fig. 2 and some absorption data are given in Table I. Low temperature spectra (77°K) are given by Gilmour, Wilson and Lemberg (1967). The rather flat α-band at 597–598 nm is less than half the strength of the ferrous α-band at 604 nm. We have found the Soret band of our preparations always at 418–419 nm but other authors have reported it in positions varying from 418 to 424 nm. So far it has not been established whether this is due to incomplete conversion of oxygenated oxidase (max. 427–428 nm see below) to the stable "ferric" oxidase, or to a secondary alteration of the ferric oxidase (Lemberg and Stanbury, 1967; Lemberg et al., 1968; discussion following the paper of Gibson and Wharton, 1968). Williams et al. (1968) have found evidence for the existence of another "ferric" oxidase with Soret band at 418 nm which is kinetically distinguishable from the stable form. Even the purest preparations of the oxidase contain a high absorption maximum at 280 nm

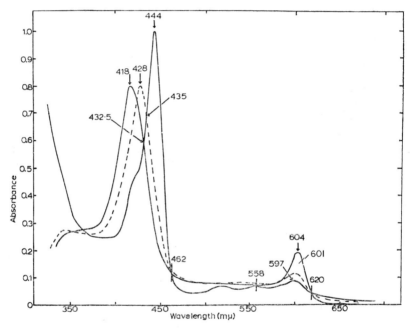

FIG. 2. Absorption spectra of ferrous (max. 604 and 444 nm), ferric (max. 597 and 418 nm), and oxygenated (broken line, max. 601 and 428 nm) cytochrome oxidase.

($A_{280}^{Fe3+}/A_{445}^{Fe2+}$ about 2·5). There is a weak band at 830 nm due to copper, and a shoulder of the α-band at 650–660 nm (Lemberg et al., 1964; Williams et al., 1968), which can be considered as a haematin charge transfer band (Williams, 1956). The data of Vanneste (1966a) indicate a stronger high spin character of a_3^{3+} than of a^{3+}. This is in agreement with magnetochemical data of Ehrenberg and Yonetani (1961) and Ehrenberg (1962) who found a^{3+} to be a low spin compound (1 paired electron, 2,400 cmu $\times 10^{-6}$ mole^{-1}), a_3^{2+} a fully high spin compound (5 paired electrons) while a_3^{3+} was not fully high spin (7,900 $\times 10^{-6}$ cmu mole^{-1}). It may thus be a mixture of high and low spin compounds. Thermal mixtures of high and low spin compounds have indeed been described for ferrihaemoprotein hydroxides by George et al. (1961), but Tsudzuki and Okunuki (1969) found no variation of paramagnetism between 77° and 273°K. The mixture may be thus one of two conformationally different high spin and low spin compounds. This would be in harmony with the relative weakness of the 650–640 nm band in "ferric" oxidase as compared with the strong 630–640 nm band found for other ferrihaemoproteins *a* (Lemberg, 1961; Morell et al., 1961), and also with EPR spectral data. EPR spectra can only be measured at very low temperature. The EPR spectrum of "ferric" oxidase has broad lines at g = 3, 2·0 and 1·5, typical for low spin ferric compounds. It lacks the g = 6 band typical for high spin ferric complexes.

(*d*) *Differences between ferrous and ferric oxidase.* Ferrous oxidase is more easily digested by proteinases than ferric oxidase, in contrast to the easier digestibility of ferric than of ferrous cytochrome *c* (Yamamoto and Okunuki, 1950). This is some indication of a conformational change (see also the next section).

A band at g = 6 which was not found in the ferric oxidase was, indeed, found on partial reduction of "ferric" oxidase with dithionite, dithiothreitol or ferrocytochrome *c* (Beinert et al., 1968; Van Gelder and Beinert, 1969; Beinert and Orme-Johnson, 1969; Tsudzuki and Okunuki, 1969). The explanation preferred for this is antiferromagnetic interaction between the Cu(II) and the a_3^{3+} haematin-Fe(III) which disappears if Cu(II) is reduced to Cu(I).

$$a_{\text{low spin}}^{3+}\text{Cu(II)}\alpha \qquad a_{\text{low spin}}^{2+}\text{Cu(II)}\alpha$$
$$\rightarrow$$
$$a_{3\ \text{high spin}}^{3+}\text{Cu(II)}\beta \qquad a_{3\ \text{high spin}}^{3+}\text{Cu(I)}\beta$$

If this interpretation is correct, there is a close relationship between a_3 and Cu(β) in contrast to the kinetic finding reported in IIIB4c. An alternative possibility, so far not excluded, is that "ferric" oxidase is, in fact, a ferryl complex which is initially reduced to the ferric. The Soret band position at 424 nm rather than 418 nm and the need for 2·5–3·0 reducing equivalents for its complete reduction, whereas Lemberg and Cutler (1970) found only 2·0 for "ferric" and 2·0–2·5 for oxygenated oxidase lends this assumption some support. Together with the g = 6 high spin form, another low spin form, probably of a_3^{3+} (g = 2·6, 2·2 and 1·88) appears on partial reduction (Beinert and Orme-Johnson, 1969; Tsudzuki and Okunuki, 1969).

Further evidence comes from CD and ORD data of King et al. (1968), Myer (1968, 1971), Myer and King (1968, 1969) and Yong and King (1970a,b,c) which supports the non-identity of *a* and a_3 haems and also the lack of direct haem–haem interaction between them. The CD maxima calculated by Myer (1968) for ferrocytochrome a (446 nm) and ferrocytochrome a_3 (444 nm) agree reasonably well with the optical maxima of Vanneste (1966a) at 444 nm and 442·5 nm respectively. Of the four CD extrema reported for "ferric" oxidase, 440, 428, 424 and 420 nm, the 428 nm maximum may be that of oxygenated oxidase (see below) or the a_{3x}^{3+} part of it, 424 nm that of a^{3+} and 420 nm that of a_3^{3+}, while that at 440 nm cannot be correlated. The spectra appear, however, to be dependent on both the haem and the copper chromophores and the Soret band of the ferric preparation of Myer was at 418–419 nm, not at 424 nm.

Some data of difference spectra ($Fe^{2+} - Fe^{3+}$) of aa_3 are found in Table I, while others as well as those for a and a_3 have been given by Vanneste (1966a). The γ/α ratios for the spectra of a and a_3 are even more different (2·9 and 24) than for the absolute spectra but again not entirely exceptional (Lemberg et al., 1964). Low temperature difference spectra at 77°K for a and a_3 have been studied by Gilmour, Wilson and Lemberg (1967). The spectrum for a shows an α-peak at 601 nm and a double Soret band with maxima at 447 and 442 nm and a minimum at 426 nm. a_3 has the α-maximum at 604 nm and a single Soret peak at 444 nm, with a minimum at 412 nm. No double Soret peak is seen with aa_3 since the a_3 peaks fill the trough between the two a Soret peaks. Similar observations had been previously made in plant mitochondria (Bendall and Bonner, 1966; Bonner and Plesnicar, 1967) and in sea-urchin sperm (Wilson and Epel, 1968). It is thus possible that a is still a mixture of two forms, but the low temperature splitting of the Soret band is not sufficient evidence for this; it may depend merely on two forms of a^{3+}.

(e) *Oxidation-reduction potentials.* The evidence as to the oxidation-reduction potentials of aa_3 of a and a_3 was until recently unsatisfactory. Measurements have frequently been carried out making use only of the α-band of the oxidase. This is mainly, but not entirely, due to cytochrome a and gives therefore no information on the potential of a_3. There is a moderately good agreement that the midpoint measured by the α-band lies between $+278$ and $+290$ mV 7·4, 10–25° (Ball, 1938; Wainio, 1955; Minnaert, 1965). It is disconcerting, however, that Minnaert has found the reaction with cytochrome c in the presence of ferroferricyanide to be $c^{2+} + e + 2a^{3+} \rightleftharpoons c^{3+} + 2a^{2+}$, whereas direct titrations of ferric oxidase with ferrocytochrome c (Van Gelder and Beinert, 1969; Lemberg and Cutler, 1969, 1970) show it to be $2c^{2+} + a^{3+} + Cu(II) \rightleftharpoons 2c^{3+} + a^{2+} + Cu(I)$. Horio and Ohkawa (1968) confirmed that, in the presence of ascorbate, $\log c^{2+}/c^{3+} = -\log K + 2 \log a^{2+}/a^3$ was linear in its middle range, but deviated from linearity at both extremes. They deduced by fitting of the curves the presence of three haems, $a - 1$, $a - 2$ and $a - 3$ with E'o $+ 360$ mV (n = 1·7), $+300$ mV (n = 1·7) and $+208$ mV (n = 1·5). The role of copper in the reaction is not taken into consideration, but obviously should not be neglected. From experiments in intact rat liver mitochondria, blocking reduction of cytochrome c and oxidation of cytochrome a by the use of inhibitors (rotenone, malonate and antimycin A for the first, cyanide for the second) and using as reductant TMPD (tetramethyl-p-phenylene diamine) and the Soret band, Caswell (1969) deduced an E'_0 for cytochrome a of $> +330$ mV. While it is not impossible that the redox potential in intact mitochondria differs from that of solubilized oxidase, the results cannot be accepted without reservations, since they rely on some assumptions which are open to doubt. It is assumed e.g. that cytochrome a_3 in the presence of cyanide does not contribute at all to the Soret band. While Wilson and Gilmour (1967) have shown that the $\Delta(a_3^{2+} - a_3^{3+})$ contribution is diminished by cyanide, it is in fact not abolished, but only halved (a_3 contributes about 66% of the Soret band, but the diminution by cyanide is only 33%). In plant mitochondria, Storey (1970a) found an E'_0 of $+300$ mV for cytochrome a, assuming an equilibrium between cytochromes a and c in the presence of cyanide.

Recently Wilson and Dutton (1970) and Dutton et al. (1970) have found that the redox potential of a_3 is $+395$ mV in the absence of ATP, but only $+300$ mV in its presence, while the potential of cytochrome a is as low as $+190$ mV. These results are in moderately good agreement with those of Horio and Ohkawa, particularly if one remembers that these estimations involve extrapolations from a complex S-shaped curve to tangential linear lines and cannot therefore lay claim to exactness. They confirm the claim of Wharton and Cusanovich (1969) that the copper potential in cytochrome oxidase lies between those of a_3 and a. Moreover, they can now be harmonized with the cytochrome aa_3 theory, if the "third" component of the oxidase is an energized form of cytochrome a_3. It is possible that this form will be found identical with "oxygenated" oxidase.

(*f*) *Oxygenated oxidase*. The problem of the oxidized oxidase is further complicated by the discovery of another form by Okunuki et al. (1958a). Its absorption spectrum is given in Fig. 2 and Table I. Its spectrum differs from that of ferric oxidase mainly by the position of the Soret band at 428 nm, of equal height as that of ferric oxidase at 418 nm (Lemberg and Gilmour, 1967). This band remains sharp and single at 77°K (Gilmour, Wilson and Lemberg, 1967; Tsudzuki and Okunuki, 1969). The 830 nm band is the same as that in "ferric" oxidase, but the 660 nm shoulder is missing (Lemberg and Gilmour, 1967; Williams et al., 1968). Kinetic differences in the reduction of "ferric" and "oxygenated" oxidase are discussed below. Although hydrogen peroxide can produce "oxygenated" oxidase (Orii and Okunuki, 1963b; Lemberg and Mansley, 1968) it arises by reaction of ferrous oxidase with molecular oxygen, even if the reduction is carried out by reductants which do not produce hydrogen peroxide by autoxidation (Lemberg and Stanbury, 1967; Gibson and Wharton, 1968; Davison and Wainio, 1968). However, an excess of ferrocytochrome c very rapidly shifts the Soret band from 428 to 418 nm (Okunuki, 1966, Lemberg et al., 1966; Lemberg and Gilmour, 1967; Gilmour et al., 1969). It is thus justified to use the term oxygenated oxidase as long as this is not presumed to mean a reversible ferrohaem-oxygen complex comparable to oxyhaemoglobin. This has been assumed by the schools of Okunuki (Okunuki, Sekuzu et al., 1958; Takemori et al., 1958; Sekuzu et al., 1960; Orii and Okunuki, 1963b; Orii et al., 1965) and of Wainio (Davison and Wainio, 1964, 1968; Wainio, 1965; Wainio et al., 1968). It has not been shown, however, that oxygenated oxidase contains dissociable oxygen nor has the claim that its oxygen can be directly replaced by carbon monoxide (Davison and Wainio, 1964; Wainio and Schulman, 1968) been confirmed (see Lemberg and Mansley, 1966; Lemberg and Stanbury, 1967; Lemberg and Gilmour, 1967). It was tempting to assume a closer relationship to another type of oxygenated haemoproteins, e.g. the compound III of horseradish peroxidase (Yamazaki et al., 1966, but see the Editors' Note on p. 326 of this reference), the "oxyperoxidase" of Wittenberg et al. (1967) or the oxygen compound of the tryptophan–tryptophan pyrrolase complex (Maeno and Feigelson, 1967; Ishimura et al., 1968). The chemical nature of these compounds has not yet been established and their formulation as $Fe^{3+}O_2$ compounds (Wittenberg et al., 1968) is of little help as long as the same structure is also assumed for oxyhaemoglobin and oxymyoglobin. More recently, findings of Williams et al. (1968) and of Lemberg and Cutler (1969, 1970) have made it unlikely that oxygenated oxidase still contains all the four electron-accepting positions of molecular oxygen. In anaerobic titrations of "ferric" and "oxygenated" oxidase, stabilized by Sephadex filtration, with ferrocytochrome c no clear cut difference between the two compounds was found with regard to the electrons required for complete reduction to ferrous oxidase. Unless other oxidizable groups in the protein are involved, oxygenated oxidase is either a ferric oxidase containing a conformationally different form of a_3, a_{3x} in $(a^{3+}a_{3x}^{3+})$ or it is $a^{3+}a_3^{4+}$, a ferryl form of a_3, or it is $a^{3+}a_{3x}^{4+}$ (Lemberg and Gilmour, 1967; Williams et al., 1968; Lemberg, 1969, pp. 90–91). It has been shown that it is the a_3 component which is different in "ferric" and "oxygenated" oxidase, not the a component as had been assumed by Nicholls (1968).

In contrast to the distinctly biphasic reduction of ferric oxidase by dithionite (see below) that of the oxygenated oxidase is not distinctly biphasic (Lemberg, 1966b; Lemberg and Gilmour, 1967; Lemberg et al., 1968; Lemberg and Cutler, 1970). The observation that the Soret band shift from 428 nm to 418 nm occurs before the re-establishment of biphasic reduction would appear to support the last of the three possibilities mentioned above, i.e. $a^{3+}a_{3x}^{4+}$ which by ferrocytochrome c is rapidly converted to $a^{3+}a_{3x}^{3+}$, then more slowly into $a^{3+}a_{3}^{3+}$ (see Fig. 10 of Williams et al., 1968); other evidence for this hypothesis has been adduced by Lemberg and Gilmour (1967) and Lemberg (1969, p. 90).

While the product of vigorous aeration or oxygenation of ferrous oxidase is oxygenated oxidase (Lemberg, 1966b) it has been shown by Lemberg and Stanbury (1967) and by Gilmour et al. (1969) that under conditions of low pO_2 the product is not only oxygenated oxidase. The maximum of the Soret band (Δ ferrous minus ferric) for "ferric" oxidase is at 412 nm, for "oxygenated" oxidase at 423 nm. The absorption curve 100 msec. after mixing a ferrous oxidase preparation, or dithionite-reduced cytochrome c-free mitochondria with small amounts of oxygen, indicated that the product was a mixture of ferric and oxygenated oxidase, to judge from the maximum (418 nm) and the shape of the band with a distinct shoulder at 426 nm. An alternative explanation may be that the product is predominantly $a^{3+}a_{3x}^{3+}$, the unstable conformationally different form of ferric oxidase the absorption spectrum of which remains to be studied. Finally, it may be assumed as do Beinert et al. (1971) that these "pseudo-oxygenated oxidases" represent intermediates of fully oxidized and partially reduced oxidase which are only slowly autoxidizable. According to Beinert et al. (1971) they differ from true "oxygenated" oxidase which is very rapidly formed from completely reduced ferrous oxidase. The experimental results of Lemberg et al. (1968) and Gilmour et al. (1969) agree with those of Gibson et al. (Greenwood and Gibson, 1967; Gibson and Wharton, 1968; Wharton and Gibson, 1968). Their interpretation based on the assumption of a dynamic as different from a static ferric oxidase spectrum was unacceptable to Lemberg (1968) as such, but the observed differences may be due to a conformationally and spectrally different form of a_3^{3+}, i.e. a_{3x}^{3+} which contains the same a_{3x} component as oxygenated oxidase, but in the ferric, not the ferryl form and/or as the "pseudo-oxygenated" form.

(g) *Kinetics of reduction and oxygenation.* The slow reduction of the a_3^{3+} component of "ferric" oxidase was discovered by Elliott (1961) and has been repeatedly confirmed. Lemberg and Gilmour (1967) found that the rate of reduction of a_3^{3+} is almost independent of dithionite concentration, whereas reduction of a^{3+} is proportional to it. Differences in reduction of the two components have also been found with ascorbate (Yonetani, 1960a), ascorbate-TMPD (Williams et al., 1968) and with cytochrome c (Gibson and Greenwood, 1965; Van Gelder and Beinert, 1969; Lemberg and Cutler, 1969, 1970). Under conditions under which only a^{3+}, but practically no a_3^{3+} is reduced (Minnaert, 1961; Nicholls, 1963; Lemberg and Mansley, 1966; Lemberg et al., 1968) no rapid autoxidation will occur, since a^{2+} is not autoxidizable and the exchange of valency between a_3^{3+} and a^{2+} is slow. This led the Okunuki school to postulate the need for cytochrome c for autoxidation of ferrous oxidase (Okunuki and Yakushiji, 1948; Okunuki, Sekuzu et al., 1958; Orii and Okunuki, 1965; Sekuzu et al., 1960), although Yonetani (1963b) already concluded that cytochrome c was not necessary for the autoxidation of a_3^{2+}. With oxygenated oxidase the dithionite reduction is practically monophasic. With very low dithionite concentrations this phase is preceded by a small lag period which is more marked with TMPD as reductant (Williams et al., 1968). The conversion of $a^{3+}a_{3x}^{4+}$ to $a^{3+}a_{3x}^{3+}$ accompanied by a shift of the Soret band to shorter wavelengths, precedes the conversion into $a^{3+}a_3^{3+}$, since the final reduction is still monophasic. This can also be observed with very low concentrations of dithionite as reductant. The autoxidation of an oxidase reduced by ascorbate-TMPD is thus complex which explains the results of Orii and Okunuki (1965a), cf. Lemberg (1969, p. 87). In the absence of cytochrome c, TMPD acting on ferric oxidase reduces primarily a^{3+} and the reduction of a_3^{3+} is much slower (Kimelberg and Nicholls, 1967, 1969). The much faster rate of a_3 reduction in the presence of cytochrome c is ascribed by the authors (1969) to a direct reaction of ferrocytochrome c with a_3^{3+} without passing through a. However, in the direct anaerobic reaction of ferrocytochrome c^{2+} with ferric oxidase (Lemberg and Cutler, 1970), the slow interaction with a_3^{3+} is also found. It is thus clear that the cause of the rapid interaction in the presence of oxygen must be due to the formation of a special active oxidation catalyst by the reaction of a_3^{2+} with oxygen which is not a_3^{3+}.

The autoxidation of ferrous cytochrome oxidase is so rapid even in the complete absence of cytochrome c that it can only be studied by the use of the rapid stop-flow split-beam technique of Chance and Gibson (Chance, 1957; Chance and Williams, 1955; Gibson and Milnes, 1964) or by the technique of Gibson and Greenwood (1963, 1964) in which a_3CO is dissociated to a_3^{2+} by light flash. The more rapid decrease of the Soret band at 445 nm than of the α-band at 605 nm shows that the oxidation of a_3^{2+} precedes that of a^{2+}.

Recent kinetic studies with computer technique by Pring (1969) and by Wohlrab (1969) on the kinetics of the autoxidation of ferrous oxidase show that the reaction is more complex than a single sequence of first order reactions previously assumed by Chance (1955) and that the rate of electron transport between a and a_3 depends on the presence or absence of molecular oxygen. This agrees with the findings discussed above, in which evidence is found for the importance of a conformational change in the cytochrome oxidase molecule.

6. Compounds with carbon monoxide, cyanide and other respiratory inhibitors

(a) *Carbon monoxide compound.* As only a_3 reacts with carbon monoxide and the α-band of cytochrome oxidase is mainly due to a, the α-band is only slightly diminished but a distinct shoulder at 590 nm due to a_3CO is found. The γ-band, however, has a far greater contribution of a_3 and CO shifts the maximum from 444 to 430 nm although a shoulder at 444 nm remains. The Δ spectrum ($FeCO - Fe^{2+}$) shows maxima at 590, 546–543 and 429 nm, and minima at 603–610, 565–562 and 445 nm. The peak to trough differences found by Lemberg et al. (1964) and by Vanneste (1966a,b) are in good agreement. The photochemical action spectrum of the oxidase, obtained by Warburg and Negelein (1929) and by Chance (1953) for beef heart muscle, and by Yonetani and Kidder (1963) for solubilized oxidase, is that of a_3CO. Owing to the less distinct band of ferric compounds, it closely resembles the ($a_3CO + a^{3+}$) spectrum obtained by exposing the product of reduction by dithionite + CO to brief action of ferricyanide which preferentially attacks the uncombined a (Horie and Morrison, 1963; Tzagoloff and Wharton, 1965); or the $\Delta(a_3CO - a_3^{3+}CN)$ spectrum of Nicholls (1963). Myer and King (1969) and Yong and King (1970c) have shown that the presence of $a_3^{2+}CO$ instead of $a_3^{3+}CN$ does not influence the $\Delta(a^{2+} - a^{3-})CD$ spectrum in the Soret band region whereas the presence of Cu as Cu(I) or Cu(II) has a distinct influence. On longer standing of the CO-compound of the oxidase unexplained changes occur which have previously been observed by Lemberg et al. (1964). Wainio and Greenlees (1960) found the Km for the combination of a_3^{2+} with CO to be 4×10^{-5} M. Beinert and Orme-Johnson (1968) found that CO, like cyanide or azide prevented the appearance of the g = 6 EPR band on partial reduction of their ferric oxidase.

In the Keilin–Hartree preparation the ratio of CO-unreactive a and CO-reactive a_3 is one, i.e. a_3 forms 50% of the haem a of the oxidase (Gibson and Greenwood, 1963; Gibson et al., 1965). With purified oxidase preparations sometimes slightly more a than a_3 has been found, but Mansley et al. (1966) found a ratio close to one.

Recent doubts on the identification of the 590 nm shoulder of the oxidase spectrum with a_3CO (Chan et al., 1970; Love and Auer, 1970) neglect the photochemical action spectrum of the oxidase and are based on an erroneous identification of an alkali-modified oxidase (see below) with cytochrome a_3. In this regard, it is also of interest that Caughey et al. (1970) found that the observed infrared band of the CO-compound of the oxidase was inconsistent with the assumption that CO served as a bridge ligand between cytochromes a and a_3 (see also Williams et al., 1968).

(b) *Nitric oxide compound.* The NO-compound of aa_3 resembles its CO-compound. Wainio (1961) reported the α-band at 601, 439 and 430 nm, Sekuzu et al. (1959) at 603 nm and 430 nm (single). The $\Delta(Fe^{2+}NO - Fe^{2+})$, probably $(a_3^{2+}NO - a_3^{2+})$ spectrum has maxima at 597, 545 and 426 nm. The high affinity of a_3^{2+} for NO is due to its fast rate of combination (Gibson and Greenwood, 1963; Gibson, Palmer et al., 1965).

(c) *Cyanide compounds.* Although much work has been done with the cyanide compounds, our knowledge of these reactions is still rather unsatisfactory. Cyanide does not appear to combine with a^{2+} or a^{3+} but combines with both a_3^{2+} and a_3^{3+}. Thus cyanide considerably slows down the reduction of a_3^{3+}, but not that of a^{3+}, the spectrum immediately after reduction being that of $a^{2+} + a_3^{3+}CN$ (Yonetani, 1960; Horie and Morrison, 1963; Lemberg et al., 1964; see III5(a)). Later $a_3^{3+}CN$ is, however, slowly reduced. The initial effect of cyanide on the ferric spectrum of aa_3 is small, but Camerino and King (1966) discovered a slow change of the Soret band on incubation with cyanide. This has been confirmed by Geyer and Lemberg (unpubl.) and consists of a shift to longer wavelengths, the final position being 427 nm. The spectrum of the ferrous oxidase compound $(a^{2+} + a_3^{2+}CN)$ greatly resembles that of the CO-compound in the α-region, with a distinct shoulder at 590 nm, but the Soret band is only slightly shifted to shorter wavelengths (441 nm) and decreased in intensity. While Lemberg et al. (1964) found this spectrum unaltered by a 50× decrease of cyanide concentration, it now appears that the shoulder at 590 nm and the shift of the Soret band to shorter wavelengths are less marked at very low cyanide concentrations. Gibson, Palmer et al. (1965) found that cyanide combines as rapidly with a_3^{2+} as does CO and that both ligands compete for the same site. According to Stannard and Horecker (1948) the enzymes combine with HCN, and not with CN^- ion as does methaemoglobin; buffering of the strongly alkaline cyanide solution is thus necessary. Observations on the Km of the cyanide inhibition of oxidase do not agree. Stannard and Horecker found 5×10^{-7} M, Albaum et al. 2×10^{-8}, Yonetani and Ray (1965) 4×10^{-6}; while Wainio and Greenlees (1960) found 5×10^{-6} as well as a higher affinity site on preincubation with ferrocytochrome c (3×10^{-8} M). The inhibition of ferrocytochrome c oxidation by cyanide is uncompetitive while ferricytochrome c inhibits it competitively. Both Wainio and Yonetani ascribe these results to a greater affinity of ferrous than of ferric oxidase to cyanide, and Yonetani distinguishes between cyanide-binding (a_3^{2+}) and cyanide-blocking sites (a_3^{3+}). In view of later findings this may have to be re-interpreted. Camerino and King observed, in contrast to previous assumptions, that the cyanide inhibition is reversible only if activity is measured by the standard manometric technique, not if the method is the spectrophotometric observation of oxidation of ferrocytochrome c. This may be explained in terms of the period of incubation in a reducing medium. They claim that only cyanide combined with haem a is dissociable by physical means such as dialysis, and that the irreversible shift of the Soret absorption maximum is due to a non-dissociable combination of cyanide with another site, probably copper. Such a conclusion is consistent with the earlier assumption by Lee and King (1962) that cytochrome oxidase binds cyanide irreversibly. Recent evidence (Beinert and Orme-Johnson, 1969) confirms copper as an alternative binding site; however, quantitative estimations (Geyer and Lemberg, unpublished) reveal that both cyanide complexes remain undissociated in rapid Sephadex filtration as well as by slow dialysis. On the other hand, the observation that complete reactivation takes place in the presence of ferrocytochrome c and ascorbate cannot be explained in the way suggested by Camerino and King. One would not expect removal of cyanide from copper under reducing conditions, since Cu(I) usually binds cyanide more firmly than Cu(II).

The inhibition due to cyanide may therefore be ascribed to its binding to the ferricytochrome a_3 iron with high affinity which, though spectroscopically hardly noticeable is demonstrated by the slow rate of reduction (see above). The secondary alteration accompanied by a shift of the Soret band to 427 nm is most probably due to a conformational change which resembles spectroscopically the formation of "oxygenated" oxidase.

Orii and Okunuki (1964) ascribe cyanide inhibition to the reaction of the formyl side chain of haem a with cyanide, with the formation of an oxynitrile; this has been rejected by Lemberg and Mansley (1966) for both spectroscopic and chemical reasons. Hydroxylamine (Utsumi and Oda, 1969) and hydrazine sulphate inhibit cytochrome oxidase, but the spectroscopic alteration by these reagents is also too small to be due to oxime or hydrazone formation of the formyl side chain.

Cyanide prevents not only the appearance of the $g = 6$ a_3^{3+} signal on partial reduction of "ferric" oxidase, but also the Cu(β) EPR signal; in the presence of cyanide only 1–1.5 reducing equivalents per haem a are required for the reduction of a^{3+} and (partly) of Cu(α) (Van Gelder and Beinert, 1969; Beinert and Orme-Johnson, 1969). Cyanide may cause a slight alteration of the low spin a_3^{3+} spectrum (Tsudzuki and Okunuki, 1969).

In plant mitochondria the oxidation of ferrous cytochrome oxidase is not inhibited by very low (< 0.3 mM) concentrations of cyanide, but is inhibited by higher concentrations (Storey, 1970a,b). This suggests that the affinity of ferric oxidase for cyanide is higher than that of ferrous oxidase. This agrees with our findings, but not with those of earlier investigators. Storey assumes two different binding sites, one in cytochrome a_3^{3+} and another in a_3^{2+} with dissociation constants of 2 μM and 30 μM respectively, but the different affinity may, in fact, be due to the different iron valency. Autoxidation of a_3^{2+}CN still occurs even at the higher cyanide concentration. The half-time for a_3^{2+} autoxidation is calculated to be 0·9 msec., that for a^{2+} oxidation 2 msec.

(d) *Azide compounds.* Azide can react with ferric haemoproteins in three different ways (Keilin and Nicholls, 1959). It can form high-spin complexes at low pH, reacting with HN_3 (see Stannard and Horecker, 1948) (I). It can form mixtures of high-spin and low spin complexes (Keilin and Hartree, 1951) (II). Finally azide can be oxidized by a haemoprotein-catalysed oxidation to NO and N_2O so that the product may be a ferrous NO complex as in the reaction of catalase, H_2O_2 and azide (Lemberg and Foulkes, 1948; Foulkes and Lemberg, 1949) (III). This may explain why the state of our knowledge of the reaction of cytochrome oxidase with azide is even less satisfactory than that of our knowledge of the reaction with cyanide.

It has been known for many years that the azide inhibition of respiration, at least under some circumstances, is far more complete than the Km derived from spectroscopic observations would warrant (Keilin, 1936; Stannard and Horecker, 1948; Keilin and Hartree, 1951; Minnaert, 1961; Yonetani and Ray, 1965; Kimelberg and Nicholls, 1967; Wilson and Chance, 1966). This has been ascribed by Palmieri and Klingenberg (1967) to the accumulation of azide in the mitochondria. It has long been assumed that azide inhibits the enzyme non-competitively by combining with a_3^{3+}, thus inhibiting its reduction. Azide slows down the reduction of ferric oxidase but not as effectively as cyanide (Horie and Morrison, 1963; Bendall and Bonner, 1966). The reaction of ferric oxidase is characterized by a shift of the Soret band from 418 to 428 nm, leaving a shoulder at 418 nm, and by an increase of the α-band; the $(a_3^{3+}N_3 - a_3^{3+})$ spectrum has a maximum at 437, a minimum at 419 nm (Muijsers *et al.*, 1966) attributed to the type II reaction, i.e. the formation of a mixture of high and low spin ferric compounds. This reaction was slow and the affinity low (6·3 mM). Muijsers *et al.* (1968) later observed a fast reaction of an affinity (0·1 mM) approaching the Ki of the oxidase 0·03–0·04 mM (Kimelberg and Nicholls, 1967). However, the shift of the Soret band $\Delta(a_3^{3+}N_3^- - a_3^{3+})$ maximum 415 nm, minimum 432 nm, was observed by Muijsers *et al.* with a preparation, the uncombined a_3 Soret band of which was at 424 nm ferric oxidase and we have found quite similar Δ spectra between a 424 nm ferric oxidase and a 418 nm ferric oxidase in the complete absence of azide; perhaps here conformational changes are induced by azide as by cyanide.

An interesting reaction has been noted if azide is added to an oxidase actively catalysing oxidation of substrate (Wilson and Chance, 1966, 1967; Wilson and Gilmour, 1967; Gilmour, Wilson and Lemberg, 1967). At

−190°C, in the frozen-in steady state an absorption maximum at 597 nm and a double Soret band at 447 and 440 nm have been ascribed to an a^{2+}-azide complex. This is of particular importance as Wilson and Chance (1966) have found strong azide inhibition only in state 3, i.e. in ADP stimulated respiration, and that uncouplers decrease azide sensitivity. They assume that the 597 nm compound is due to a special form of cytochrome a^{2+} only present if electron transport and oxidative phosphorylation are coupled (Chapter VII; but see Palmieri and Klingenberg, 1967). The inhibition would then be due to combination of a^{2+}, not of a_3^{3+} with azide and active turnover would be required to keep a^{2+} in a reactive form. This would require that purified cytochrome oxidase still maintains part of the energy-conservation mechanism, and that cytochrome a plays the essential role in it. A possible alternative explanation of the need for active turnover based on a type III (see above) reaction of azide has been discussed by Lemberg (1969). Kimelberg and Nicholls (1968) consider the interaction of azide with a high energy form of a^{2+} unlikely and ascribe the shift of the α-band to interaction between the a and a_3 haems. The reaction of azide with a_3^{3+} is supported by Van Gelder and Beinert (1969). They found that azide, like cyanide, prevents the appearance of the high spin ferric signal (g = 6) on partial reduction of their "ferric" oxidase, and forms a low spin $a_3^{3+}N_3$ complex (g = 2·9, 2·2, 1·67), while the low spin a^{3+} EPR bands at g = 3·0 remain unaltered. The observations of Nijs (1967) also support the reaction of azide with a_3^{3+}.

(e) *Fluoride, sulphide and other compounds.* Fluoride ion is well known to combine with haemoproteins to high spin complexes (Keilin and Hartree, 1951). Inhibition of cytochrome oxidase by fluoride was discovered by Borei (1954). With the oxidase fluoride increases absorption at 640 and 680 nm which can be ascribed to transformation of a mixed spin ferric oxidase into a more high-spin one. The Δ spectrum also shows a maximum at 400 nm, a minimum at 415 nm, and the spectroscopic dissociation constant (2·5 × 10^{-2} M) is close to Ki (1·6 × 10^{-2} M) (Muijsers et al., 1966). Tsudzuki and Okunuki (1969) found that fluoride slightly increased the high spin g = 6 signal. Some other effects of fluoride on the EPR signals were noted by Van Gelder and Beinert (1969).

Sulphide appears also to react with cytochrome a_3, preventing its reduction (Wilson and Gilmour, 1967; Gilmour et al., 1968), confirming the early assumption of Keilin and Hartree (1939) that the sulphide inhibition of the oxidase was due to this action.

Inhibition of the electron transport system by phenylhydrazine is only partially due to its action on cytochrome oxidase. The real inhibitor is its oxidation product phenyldymide (Asami, 1968; Itano, 1970) or a free radical formed from it. Inhibition by hydroxylamine is complex; it has been discussed by Yoshikawa and Orii (1970).

7. The mechanism of cytochrome oxidase activity

The most frequently assumed and "classical" theory is that developed by Keilin and supported by Chance and Gibson,

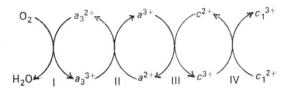

or, simplified $O_2 \to a_3 \to a \to c \to c_1$. Here we are only concerned with reactions I and II. The theory does not specify what happens in reaction I. Its weakest point is that reaction II observed in solubilized oxidase is too slow to play an essential role in the fast enzymic mechanism. It has been discussed above that this discrepancy can be resolved by assuming that it is not the stabilized a_3^{3+} found in "ferric" oxidase, but a more reactive conformation form a_{3x}^{3+} that is involved in the enzymic reaction. This is either present as such, or possibly in a ferryl (FeIV) form in oxygenated oxidase, and autoxidation of a_3^{2+} yields either of these, according to the pO_2 of the solution. While some, notably Yakushiji and Okunuki (1940) and Nicholls and Kimelberg (1968) have postulated a direct reaction of cytochrome c with a_3^{3+} by-passing cytochrome a, or a conformationally different form of a^{2+} (Nicholls, 1968), there is as yet no compelling reason for any of these assumptions, nor is there a sound basis for the theory of Okunuki (Orii et al., 1962; Takemori, 1965) that cytochrome c is needed for the autoxidation of the oxidase.

The above scheme has also to be modified to incorporate the valency changes of the oxidase copper. Whereas the role of the copper is still somewhat uncertain, rapid valency changes in it have been demonstrated beyond doubt. As an overall picture, the four electrons that must be added for the reduction of oxygen to water, come from a^{2+}, a_3^{2+} and two Cu^{1+} atoms of the oxidase. Though all these reactions occur at high speed, there is some evidence that the speed is not exactly the same. Some bound radical forms are therefore to be expected (see Beinert et al., 1971). So far neither free hydrogen peroxide nor free radicals have been demonstrated, although Fridovich and Handler (1961) adduce initiation of sulphite oxidation as evidence for a radical chain mechanism. Commoner and Hollocher (1960) found evidence for a free radical (g = 2·003) in heart muscle preparations oxidizing succinate, but they ascribed it to succinic dehydrogenase. The moderately stable superoxide ion $O_{\frac{1}{2}}$ (cf. Nilsson et al., 1969) has now frequently been observed in reactions of reduced flavoproteins, ferredoxins and other enzymes but not in the cytochrome oxidase reaction (Massey et al., 1969; Orme-Johnson and

Beinert, 1969). Lemberg (1969) has pointed out that the main function of the copper in the oxidase may be preventing, or at least minimizing monoelectron changes. The observation that erythrocuprein rapidly disproportions O_2^- (Orme-Johnson and Beinert, 1969; Ballou et al., 1969) provides evidence for such a role of copper. Antonini et al. (1970) assume that similar to laccase, cytochrome oxidase must be fully reduced to be rapidly autoxidizable. This is supported by findings of Beinert et al. (1970) but other experiments indicate that the reoxidation is rapid provided that a_3^{3+} has been reduced to a_3^{2+} independent of the complete reduction of the copper, and not only a^{3+} to a^{2+}; further studies on the rapid reoxidation of partially reduced oxidase are desirable.

REFERENCES

Albaum, H. G., Tepperman, J. and Bodansky, O. (1946). *J. Biol. Chem.* **163**, 641–647.
Antonini, E., Brunori, M., Greenwood, C. and Malmström, B. G. (1970). *Nature* **228**, 936–937.
Appleby, C. A. (1969). *Biochim. et Biophys. Acta* **172**, 88–105.
Arima, K. and Oka, T. (1965). *J. Biochem.* (Tokyo) **58**, 320–321.
Asami, K. (1968). *J. Biochem.* **63**, 425–433.
Awashti, Y. C., Chuang, T. F., Kleenan, T. W. and Crane, F. L. (1970). *Biochem. Biophys. Res. Commun.* **39**, 822–832; *Biochem. et Biophys. Acta* **226**, 42–52.
Ball, E. G. (1938). *Biochem. Z.* **295**, 262–264.
Ballou, D., Palmer, G. and Massey, Y. (1969). *Biochem. Biophys. Res. Commun.* **36**, 898–904.
Bartsch, R. G. (1968). *Ann. Rev. Microbiol.* **22**, 181–200.
Beinert, H. (1966). In "Copper in Biological Systems. The Biochemistry of Copper" (J. Peisach, P. Aisen and W. E. Blumberg, eds.) pp. 213–234. Academic Press, New York.
Beinert, H. and Orme-Johnson, W. H. (1969). *Ann. New York Acad. Sci.* **158**, 336–360.
Beinert, H. and Palmer, G. (1964). *J. Biol. Chem.* **239**, 1221–1227.
Beinert, H. and Palmer, G. (1965). "Amherst", vol. II, pp. 567–590.
Beinert, H., Griffiths, D. E., Wharton, D. C. and Sands, R. W. (1962). *J. Biol. Chem.* **237**, 2337–2346.
Beinert, H., Hartzell, C. R. and Orme-Johnson, W. R. (1971) in (B. Chance, C. P. Lee and T. Yonetani, eds.) "Structure and function of macromolecules and membranes". Academic Press, New York.
Beinert, H., Hartzell, C. R., Van Gelder, B. F., Ganapathy, K., Mason, H. S. and Wharton, D. C. (1970). *J. Biol. Chem.* **245**, 225–229.
Beinert, H., Van Gelder, B. F. and Hansen, R. E. (1968). "Osaka", pp. 141–150.
Bendall, D. S. and Bonner, W. D. (1966). "Hemes and Hemoproteins", pp. 485–502.
Betel, I. (1967). *Biochim. et Biophys. Acta* **143**, 62–69.
Betel, I. and Klouwen, H. M. (1967). *Biochim. et Biophys. Acta* **131**, 453–467.
Blair, P. V. (1967). "Methods in Enzymology", vol. **10**, pp. 78–81.
Bonner, W. D., Jr. (1961). "Haematin Enzymes", vol. II, pp. 479–497.
Bonner, W. D., Jr. and Plesnicar, M. (1967). *Nature* **214**, 616–617.

Bonner, W. D., Jr., Bendall, D. S. and Plesnicar, M. (1967). *Fed. Proc.* **26**, No. 2647; "Methods in Enzymol", vol. **10**, 126–137.
Borei, H. (1945). *Arkiv. Kemi, Mineral. Geol.* **20**A, No. 8.
Brill, A. S. and Williams, R. J. P. (1961). *Biochem. J.* **78**, 246 253.
Broberg, P. L. and Smith, L. (1967). *Biochim. et Biophys. Acta* **131**, 479–489.
Broberg, P. L. and Smith, L. (1968). "Osaka", pp. 182–187.
Broberg, P. L., Welsch, M. and Smith, L. (1969). *Biochim. et Biophys. Acta* **172**, 205–215.
Brown, F. C. and White, J. B., Jr. (1961). *J. Biol. Chem.* **236**, 911–913.
Burrin, D. H. and Beechey, R. B. (1963). *Biochem. J.* **87**, 48–53.
Camerino, P. W. and King, T. E. (1966). *J. Biol. Chem.* **241**, 970–979.
Castor, L. N. and Chance, B. (1955). *J. Biol. Chem.* **217**, 453–465.
Caswell, A. H. (1969). *J. Biol. Chem.* **243**, 5827–5836.
Caughey, W. S., Bayne, R. A. and McCoy, S. (1970). *J. Chem. Soc.* D. **15**, 950–951.
Caughey, W. S. and York, J. L. (1962). *J. Biol. Chem.* **237**, 2414–2416.
Chan, S. H. P., Love, B. and Stotz, E. (1970). *J. Biol. Chem.* **245**, 6668–6674.
Chance, B. (1953). *J. Biol. Chem.* **202**, 383–416.
Chance, B. (1955). *Discuss. Faraday Soc.* **20**, 205–216.
Chance, B. (1957). In "Methods in Enzymology" (S. P. Colowick and N. O. Kaplan, eds.), vol. **4**, pp. 273–329. Academic Press, New York.
Chance, B. (1961). "Haematin Enzymes", vol. **1**, p. 433.
Chance, B. and Bonner, W. (1965). *Plant Physiol.* **40**, 1198–1204.
Chance, B. and Hackett, D. P. (1959). *Plant Physiol.* **34**, 33–49.
Chance, B. and Schindler, F. (1965). "Amherst", vol. II, 921–929.
Chance, B. and Williams, G. R. (1955). *J. Biol. Chem.* **217**, 429–438.
Cheah, K. S. (1969). *Biochem. et Biophys. Acta* **180**, 320–333.
Cheah, K. S. (1970). *Biochim. et Biophys. Acta* **205**, 148–160.
Chuang, T. F. and Awashti, Y. C. (1970). *Fed. Proc.* **29**, Abstr. 1644.
Commoner, B. and Hollocher, T. C. (1960). *Proc. Natl. Acad. Sci. U.S.* **46**, 406–427.
Conover, T. E. (1967). 7th Internat. Congress Biochemistry, Tokyo, Abstract H-73.
Cooperstein, S. J. (1963a). *J. Biol. Chem.* **238**, 3750–3756.
Cooperstein, S. J. (1963b). *Biochim. et Biophys. Acta* **73**, 343–346.
Cooperstein, S. J. (1963c). *J. Biol. Chem.* **238**, 3606–3610.
Crane, F. L., Glenn, J. L. and Green, D. E. (1956). *Biochim. et Biophys. Acta* **22**, 475–487.
Criddle, R. S. and Bock, R. M. (1959). *Biochem. Biophys. Res. Commun.* **1**, 136–142.
Davison, A. J. (1968). *J. Biol. Chem.* **243**, 6064–6067.
Davison, A. J. and Wainio, W. W. (1964). *Fed. Proc.* **25**, Abstr. 1332.
Davison, A. J. and Wainio, W. W. (1968). *J. Biol. Chem.* **243**, 5023–5027.
Detwiler, T. C., Garrett, R. H. and Nason, A. (1966). *J. Biol. Chem.* **241**, 1621–1631.
Ducet, G., Diano, M. and Denis, M. (1970). *Compt. Rend.* **270**, 2289–2291.
Duncan, H. M. and Mackler, B. (1966). *J. Biol. Chem.* **241**, 1964–1967.
Dutton, P. L., Wilson, D. F. and Lee, C. P. (1970). *Biochemistry* **9**, 5077–5082.
Edward, D. L. and Woodward, D. O. (1970). *Fed. Proc.*, Abstr. 1642.
Ehrenberg, A. (1962). *Arkiv Kemi* **19**, No. 8, 119.
Ehrenberg, A. and Yonetani, T. (1961). *Acta Chem. Scand.* **15**, 1071–1080.
Elliott, W. B. (1961). *Fed. Proc.* **20**, 42.
Epel, B. L. and Butler, W. L. (1970). *Plant Physiol.* **45**, 728–734.
Erickson, S. K. and Parker, G. L. (1969). *Biochim. et Biophys. Acta* **180**, 56–62.

Foulkes, E. C. and Lemberg, R. (1949). *Enzymologia* **13**, 302–312.
Fowler, L. R., Richardson, S. W. and Hatefi, Y. (1962). *Biochim. et Biophys. Acta* **64**, 170–173.
Fridovich, I. and Handler, P. (1961). *J. Biol. Chem.* **236**, 1836–1840.
George, P., Beetlestone, J. and Griffiths, J. S. (1961). "Haematin Enzymes", vol. 1, pp. 105–139.
Ghiretti, F., Ghiretti-Magaldi, A. and Tosi, L. (1959). *J. Gen. Physiol.* **42**, 1185–1205.
Ghiretti-Magaldi, A., Guiditta, A. and Ghiretti, F. (1957). *Biochem. J.* **66**, 303–307.
Ghiretti-Magaldi, A., Guiditta, A. and Ghiretti, F. (1958). *J. Cell Comp. Physiol.* **52**, 389–492.
Gibson, Q. H. and Greenwood, C. (1963). *Biochem. J.* **86**, 541–554.
Gibson, Q. H. and Greenwood, C. (1964). *J. Biol. Chem.* **239**, 586–603.
Gibson, Q. H. and Greenwood, C. (1965). *J. Biol. Chem.* **240**, 2694–2698.
Gibson, Q. H. and Milnes, L. (1964). *Biochem. J.* **91**, 161–171.
Gibson, Q. H. and Wharton, D. C. (1968). "Osaka", pp. 5–19.
Gibson, Q. H., Greenwood, C., Wharton, D. C. and Palmer, G. (1965a). *J. Biol. Chem.* **240**, 888–894.
Gibson, Q. H., Palmer, G. and Wharton, D. C. (1965b). *J. Biol. Chem.* **240**, 915–920.
Gilmour, M. V. (1967). *Fed. Proc.* **26**, Abstr. 1107.
Gilmour, M. V. and Volpe, J. A. (1970). *Fed. Proc.* **29**, Abstr. 2752.
Gilmour, M. V., Lemberg, R. and Chance, B. (1969). *Biochim. et Biophys. Acta* **172**, 37–51.
Gilmour, M. V., Wilson, D. F. and Lemberg, R. (1967). *Biochim. et Biophys. Acta* **143**, 487–499.
Goldin, H. H. and Farnsworth, M. W. (1966). *J. Biol. Chem.* **241**, 3590–3594.
Greenwood, C. (1963). *Biochem. J.* **86**, 535–540.
Greenwood, C. and Gibson, Q. H. (1967). *J. Biol. Chem.* **242**, 1782–1787.
Griffiths, D. E. and Wharton, D. C. (1961). *J. Biol. Chem.* **236**, 1850–1856; 1857–1862.
Hackett, D. P. (1959). *Ann. Rev. Plant Physiol.* **10**, 113–146.
Harris, J. and Ritchie, K. (1969). *Ann. New York Acad. Sci.* **153**, 706–721.
Hartree, E. F. (1957). *Advances Enzymol.* **18**, 1–64.
Hemmerich, P. (1966). In "Copper in Biological Systems. The Biochemistry of Copper" (J. Peisach, P. Aisen and W. E. Blumberg, eds.), pp. 15–32, Academic Press, New York.
Hommersand, M. H. and Thimann, K. V. (1965). *Plant Physiol.* **40**, 1220–1227.
Horie, S. (1964a). *J. Biochem.* **56**, 57–66; 67–71.
Horie, S. (1964b). *J. Biochem.* **56**, 113–121.
Horie, S. (1965). *J. Biochem.* **57**, 147–154.
Horie, S. (1967). *7th Internat. Congr. Biochem. Tokyo. Abstr.* V H3.
Horie, S. (1968). "Osaka", 96–112.
Horie, S. and Morrison, M. (1964). *J. Biol. Chem.* **238**, 2859–2865.
Horio, T. and Ohkawa, J. (1968). *J. Biochem.* **64**, 383–404.
Ishimura, Y., Nozaki, N., Hayaishi, O., Tamura, M. and Yamazaki, I. (1968). "Osaka", pp. 188–193.
Itano, H. A. (1970). *Proc. Nat. Acad. Sci. U.S.* **67**, 485–492.
Jacobs, E. E., Andrews, E. C., Cunningham, W. and Crane, F. L. (1966a). *Biochem. Biophys. Res. Commun.* **25**, 87–95.
Jacobs, E. E., Kirkpatrick, F. A., Andrews, E. C., Cunningham, W. and Crane, F. L. (1966b). *Biochem. Biophys. Res. Commun.* **25**, 96–104.

Kabel, B. S., Lang, R. W. and Elliott, W. B. (1968). *Fed. Proc.* **27**, Abstr. 3456.
Kaniuga, Z., Gardas, A. and Jakubiak, M. (1966). *Biochim. et Biophys. Acta* **118**, 9–18.
Kasinsky, H. E., Shichi, H. and Hackett, D. P. (1966). *Plant Physiol.* **41**, 739–748.
Keilin, D. (1925). *Proc. Roy. Soc.* (London) B **98**, 312–339.
Keilin, D. (1936). *Proc. Roy. Soc.* (London) B **121**, 165–172.
Keilin, D. (1966). "The History of Cell Respiration and Cytochrome" (J. Keilin, ed.) Cambridge Univ. Press.
Keilin, D. and Hartree, E. F. (1939). *Proc. Roy. Soc.* (London) B **127**, 167–191.
Keilin, D. and Hartree, E. F. (1951). *Biochem. J.* **49**, 88–104.
Keilin, D. and Nicholls, P. (1959). *Biochim. et Biophys. Acta* **36**, 257–259.
Kiese, U. and Kurz, H. (1958). *Biochem. Z.* **330**, 177–192.
Kiese, U. and Kurz, H. (1959). *Biochem. Z.* **332**, 11–17.
Kikuchi, G. and Motokawa, Y. (1968). "Osaka", pp. 174–181.
Kikuchi, G., Saito, Y. and Motokawa, Y. (1965). *Biochim. et Biophys. Acta* **94**, 1–14.
Kimelberg, H. K. and Nicholls, P. (1967). *Fed. Proc.* **26**, Abstr. 1105.
Kimelberg, H. K. and Nicholls, P. (1969). *Arch. Biochem. Biophys.* **133**, 327–335.
King, T. E. (1967). "Methods in Enzymology", vol. **10**, pp. 202–208.
King, T. E. and Schellman, J. A. (1966). *Fed. Proc.* **25**, Abstr. 411.
King, T. E., Kuboyama, M. and Takemori, S. (1965). "Amherst", vol. **2**, 707–736.
King, T. E., Yong, F. C. and Bayley, P. M. (1968). "Osaka", pp. 204–209.
King, T. E., Yong, F. C. and Takemori, S. (1967). *J. Biol. Chem.* **242**, 819–829.
Kirschbaum, J. and Wainio, W. W. (1966). *Biochim. et Biophys. Acta* **118**, 643–644.
Kuboyama, M. and King, T. E. (1965). *Fed. Proc.* **24**, Abstr. 2302.
Lanyi, J. K. (1968). *Arch. Biochem. Biophys.* **128**, 716–724.
Lanyi, J. K. (1969). *J. Biol. Chem.* **244**, 2864–2869.
Lee, C. P. and King, T. E. (1962). *Biochim. et Biophys. Acta* **59**, 716–718.
Lemberg, R. (1961). *Advances in Enzymol.* **23**, 265–321.
Lemberg, R. (1962). *Nature* (London) **193**, 372–374.
Lemberg, R. (1963). *Biochem. Z.* **338**, 97–105.
Lemberg, R. (1964). *Proc. Roy. Soc.* (London) B **159**, 429–435.
Lemberg, R. (1965). *Revs. Pure and Applied Chem.* (Australia) **15**, 125–136.
Lemberg, R. (1966a). "Hemes and Hemoproteins", pp. 79–82.
Lemberg, R. (1966b). Ibid., pp. 477–484.
Lemberg, R. (1968). "Osaka", pp. 16–19.
Lemberg, R. (1969). *Physiol. Revs.* **49**, 48–121.
Lemberg, R. and Cutler, M. E. (1969). Proc. 12th Internat. Conf. on Co-ordination Chemistry (H. C. Freeman, ed.). Abstracts pp. 135–136. Science Press, Marrickville, N.S.W.
Lemberg, R. and Cutler, M. E. (1970). *Biochim. et Biophys. Acta* **197**, 1–10.
Lemberg, R. and Foulkes, E. C. (1948). *Nature* (London) **161**, 131.
Lemberg, R. and Gilmour, M. V. (1967). *Biochim. et Biophys. Acta* **143**, 500–517.
Lemberg, R. and Gilmour, M. V. (1968). *Biochem. et Biophys. Acta* **156**, 407–408.
Lemberg, R. and Legge, J. W. (1949). "Hematin Compounds and Bile Pigments", Interscience Publ., New York.
Lemberg, R. and Mansley, G. E. (1965). *Biochim. et Biophys. Acta* **96**, 187–194.
Lemberg, R. and Mansley, G. E. (1966). *Biochim. et Biophys. Acta* **118**, 19–35.
Lemberg, R. and Newton, N. (1961). *Proc. Roy. Soc.* (London) B **155**, 364–373.
Lemberg, R. and Parker, J. (1955). *Austral. J. Exptl. Biol. Med. Sci.* **33**, 483–490.
Lemberg, R. and Pilger, T. B. G. (1964). *Proc. Roy. Soc.* (London) B **159**, 436–448.
Lemberg, R. and Stanbury, J. T. (1967). *Biochim. et Biophys. Acta* **143**, 37–51.

Lemberg, R. and Stewart, M. (1955). *Austral. J. Exptl. Biol. Med. Sci.* **33**, 451–482.
Lemberg, R. and Velins, A. (1965). *Biochim. et Biophys. Acta* **104**, 487–495.
Lemberg, R., Bloomfield, B., Caiger, P. and Lockwood, W. H. (1955). *Austral. J. Exptl. Biol. Med. Sci.* **33**, 435–450.
Lemberg, R., Clezy, P. and Barrett, J. (1961a). "Haematin Enzymes", vol. **1**, pp. 344–357.
Lemberg, R., Gilmour, M. V. and Cutler, M. E. (1968). "Osaka", pp. 54–65.
Lemberg, R., Gilmour, M. V. and Stanbury, J. T. (1966). *Fed. Proc.* **25**, Abstr. 2582.
Lemberg, R., Morell, D. B., Newton, N. and O'Hagan, J. E. (1961b). *Proc. Roy. Soc.* (London) B **155**, 339–355.
Lemberg, R., Newton, N. and O'Hagan, J. E. (1961c). *Proc. Roy. Soc.* (London) B **155**, 356–373.
Lemberg, R., Pilger, T. B. G., Newton, N. and Clarke, L. (1964). *Proc. Roy. Soc.* (London) B **159**, 405–428.
Love, B. and Auer, H. E. (1970). *Biochem. Biophys. Res. Commun.* **41**, 1437–1442.
Love, B., Chan, S. H. P. and Stotz, E. (1970). *J. Biol. Chem.* **245**, 6664–6668.
McConnell, D. G., Tzagoloff, A., MacLennan, D. H. and Green, D. E. (1966). *J. Biol. Chem.* **241**, 2373–2382.
Maclennan, D. H. and Tzagoloff, A. (1965). *Biochim. et Biophys. Acta* **96**, 166–168.
MacMunn, C. A. (1886). *Phil. Trans. Roy. Soc. London* **177**, 267–298.
Maeno, H. and Feigelson, P. (1967). *J. Biol. Chem.* **242**, 596–601.
Maggio, R. and Ghiretti-Magaldi, A. (1958). *Exptl. Cell. Res.* **15**, 95–102.
Mansley, G. E., Stanbury, J. T. and Lemberg, R. (1966). *Biochim. et Biophys. Acta* **113**, 33–40.
Mapson, L. W. and Burton, W. G. (1962). *Biochem. J.* **82**, 19–25.
Margoliash, E. and Walasek, O. F. (1967). "Methods in Enzymology", vol. **10**, pp. 339–348.
Mason, H. S. and Ganapathy, K. (1970). *J. Biol. Chem.* **245**, 230–237.
Massey, B., Strickland, S., Mayhew, S. G., Howell, L. G., Engel, P. S., Matthews, R. G., Schuman, M. and Sullivan, P. A. (1969). *Biochem. Biophys. Res. Commun.* **36**, 891–897.
Matsubara, H., Orii, Y. and Okunuki, K. (1965). *Biochim. et Biophys. Acta* **97**, 61–67.
Miki, K., Sekuzu, I. and Okunuki, K. (1967). *Ann. Reports Biol. Works Faculty of Science*, Osaka Univ. **15**, 33–58.
Minnaert, K. (1961). *Biochim. et Biophys. Acta* **54**, 26–41.
Minnaert, K. (1965). *Biochim. et Biophys. Acta* **110**, 42–56.
Mochan, B. S., Lang, R. W. and Elliott, W. B. (1970). *Biochim. et Biophys. Acta* **216**, 96–105; 106–121.
Morell, D. B., Barrett, J., Clezy, P. and Lemberg, R. (1961). "Haematin Enzymes", vol. 1, pp. 320–330.
Morrison, M. (1968). "Osaka", pp. 89–95.
Morrison, M. and Horie, S. (1965a). *J. Biol. Chem.* **240**, 1359–1364.
Morrison, M. and Horie, S. (1963). *Biochem. Biophys. Res. Commun.* **10**, 160–164
Morrison, M. and Horie, S. (1964). *J. Biol. Chem.* **239**, 1432–1437.
Morrison, M. and Horie, S. (1965). *Anal. Biochem.* **12**, 77–82.
Morrison, M., Horie, S. and Mason, H. S. (1963). *J. Biol. Chem.* **238**, 2220–2224.
Muijsers, A. O., Slater, E. C. and Van Buuren, K. J. H. (1968), "Osaka", pp. 129–137.
Muijsers, A. O., Van Gelder, B. F. and Slater, E. C. (1966). "Hemes and Hemoproteins", pp. 467–475.

Myer, Y. P. (1968). *Fed. Proc.* **27**, Abstr. 3141.
Myer, Y. P. (1971). *J. Biol. Chem.* **246**, 1241-1248.
Myer, Y. P. and King, T. E. (1968). *Biochem. Biophys. Res. Commun.* **33**, 43-48.
Myer, Y. P. and King, T. E. (1969). *Biochem. Biophys. Res. Commun.* **34**, 170-175.
Nair, P. M. and Mason, H. S. (1967). *J. Biol. Chem.* **242**, 1406-1415.
Newton, N. (1967). *Biochem. J.* **105**, 21C.
Newton, J. W. and Kamen, M. D. (1961). In "The Bacteria" (I. C. Gunsalas and R. Y. Stanier, eds.), vol. 2, pp. 397-423. Academic Press, New York.
Nicholls, P. (1963). *Biochim. et Biophys. Acta* **73**, 667-670.
Nicholls, P. (1968). "Osaka", pp. 76-86.
Nicholls, P. and Kimelberg, H. K. (1968). *Biochim. et Biophys. Acta* **162**, 11-21.
Niederpruem, D. J. and Hackett, D. P. (1961). *J. Bacteriol.* **81**, 557-563.
Nijs, P. (1967). *Biochim. et Biophys. Acta* **143**, 454-461.
Nilsson, R., Pick, F. M., Bray, R. C. and Fielden, M. (1969). *Acta Chem. Scand.* **23**, 2554-2556.
Ninnemann, H., Butler, W. L. and Epel, B. L. (1970). *Biochim. et Biophys. Acta* **205**, 499-506; 507-512.
Obuchowitz, L. and Zerbe, T. (1962). *Bull. Soc. Amis Sci. Lettres Poznan*, Ser. D3, 39-46.
Okunuki, K. (1962). In "Oxygenases" (O. Hayaishi, ed.) chapter 10, pp. 409-467. Academic Press, New York.
Okunuki, K. (1966). In "Comprehensive Biochemistry" (M. Florkin and E. Stotz, eds.), vol. **14**, chapter 5, pp. 232-308. Elsevier, Amsterdam.
Okunuki, K. and Yakushiji, E. (1948). *Proc. Japan Imper. Acad.* **24**, 12-14.
Okunuki, K., Hagihara, B., Sekuzu, I. and Horio, T. (1958a). *Proc. Internat. Symposium Enzyme Chem.*, Tokyo, Kyoto, 1957 (K. Ichihara, ed.), pp. 264-272. Academic Press, New York.
Okunuki, K., Sekuzu, I., Orii, Y., Tsudzuki, T. and Matsumura, Y. (1968). "Osaka", pp. 35-50.
Okunuki, K., Sekuzu, I., Yonetani, T. and Takemori, S. (1958b). *J. Biochem.* **45**, 847-854.
Orii, Y. and Okunuki, K. (1963a). *J. Biochem.* **53**, 489-499.
Orii, Y. and Okunuki, K. (1963b). *J. Biochem.* **54**, 207-213.
Orii, Y. and Okunuki, K. (1964). *J. Biochem.* **55**, 37-48.
Orii, Y. and Okunuki, K. (1965a). *J. Biochem.* **57**, 45-54.
Orii, Y. and Okunuki, K. (1965b). *J. Biochem.* **58**, 561-568.
Orii, Y. and Okunuki, K. (1967). *J. Biochem.* **61**, 388-403.
Orii, Y., Sekuzu, I. and Okunuki, K. (1962). *J. Biochem.* **51**, 204-215.
Orii, Y., Tsudzuki, T. and Okunuki, K. (1965). *J. Biochem.* **58**, 373-384.
Orme-Johnson, W. H. and Beinert, H. (1969). *Biochem. Biophys. Res. Commun.* **36**, 905-911.
Pablo, I. S. and Tappel, A. L. (1961). *J. Cellular Comp. Physiol.* **58**, 185-194.
Palmer, G., Mackler, B. and Duncan, H. M. (1967). *Biochim. et Biophys. Acta* **143**, 636-638.
Palmieri, F. and Klingenberg, M. (1967). *European J. Biochem.* **1**, 439-446.
Pappenheimer, A. M., Jr., Howland, J. L. and Miller, P. A. (1962). *Biochim. et Biophys. Acta* **64**, 229-242.
Perini, F., Schiff, J. A. and Kamen, M. (1964). *Biochim. et Biophys. Acta* **88**, 74-90.
Person, P. and Zipper, H. (1964). *Biochim. et Biophys. Acta* **92**, 605-607.
Person, P., Felton, J. H., O'Connell, D. J., Zipper, H. and Philpott, D. E. (1969a). *Arch. Biochem. Biophys.* **131**, 470-477.

Person, P., Zipper, H. and Felton, J. H. (1969b). *Arch. Biochem. Biophys.* **131**, 457–469.
Pring, M. (1969). *Fed. Proc.* **28**, Abstr. 3546.
Raison, J. K. and Smillie, R. M. (1969). *Biochim. et Biophys. Acta* **180**, 500–508.
Rawlinson, W. A. and Hale, J. H. (1949). *Biochem. J.* **45**, 247–255.
Revsin, B. and Brodie, A. F. (1969). *J. Biol. Chem.* **244**, 3101–3104.
Ross, A. J., Schoenhoff, M. I. and Aleem, M. I. (1968). *Biochem. Biophys. Res. Commun.* **32**, 301–306.
Scholes, P. B. and King, H. K. (1965). *Biochem. J.* **97**, 754–765.
Scholes, P. B. and Smith, L. (1968). *Biochim. et Biophys. Acta* **153**, 350–362; 363–375.
Seki, S., Hayaishi, H. and Oda, T. (1970). *Arch. Biochem. Biophys.* **138**, 110–131.
Sekuzu, I., Mizushima, H. and Okunuki, K. (1964). *Biochim. et Biophys. Acta* **85**, 516–519.
Sekuzu, I., Mizushima, H., Hirota, S., Yubisui, T., Matsumura, X. and Okunuki, K. (1967). *J. Biochem.* **62**, 710–718.
Sekuzu, I., Takemori, S., Orii, Y. and Okunuki, K. (1960). *Biochim. et Biophys. Acta* **37**, 64–71.
Sekuzu, I., Takemori, S., Yonetani, T., and Okunuki, K. (1959). *J. Biochem.* **46**, 43–49.
Smith, L. (1954). *Bacteriol. Rev.* **18**, 106–130.
Smith, L. (1955a). *J. Biol. Chem.* **215**, 833–846.
Smith, L. (1955b). *J. Biol. Chem.* **215**, 847–857.
Smith, L. (1955c). In "Methods of Enzymology" (S. P. Colowick and N. O. Kaplan, eds.), vol. **2**, 732–740. Academic Press, New York.
Smith, L. (1961). In "The Bacteria" (I. C. Gunsalus and R. Y. Stanier, eds.), vol. **2**, pp. 365–396, Academic Press, New York.
Smith, L. and Camerino, P. W. (1963). *Biochemistry* **2**, 1428–1432.
Smith, L. and Chance, B. (1958). *Ann. Rev. Plant Physiol.* **9**, 449–482.
Smith, L. and Conrad, H. (1956). *Arch. Biochem. Biophys.* **63**, 403–413.
Smith, L. and Conrad, H. (1961). "Haematin Enzymes", pp. 260–274.
Smith, L. and Stotz, E. (1954). *J. Biol. Chem.* **209**, 819–828.
Smith, L., Newton, N. and Scholes, P. (1966). "Hemes and Hemoproteins", pp. 395–403.
Stannard, J. N. and Horecker, B. L. (1948). *J. Biol. Chem.* **172**, 599–608.
Storey, B. T. (1970a). *Plant Physiology* **45**, 447–454.
Storey, B. T. (1970b). *Plant Physiol.* **45**, 455–460.
Storey, B. T. and Bahr, J. T. (1969). *Plant Physiol.* **44**, 126–134.
Straat, P. A. and Nason, A. (1965). *J. Biol. Chem.* **240**, 1412–1426.
Straub, F. B. (1939). *Biochem. J.* **33**, 787–793.
Straub, F. B. (1941). *Z. Physiol. Chem.* **268**, 217–234.
Stumm-Tegethoff, B. (1967). *J. Chromatogr.* **30**, 284–286.
Sun, F. F. and Jacobs, E. E. (1967). *Biochim. et Biophys. Acta* **143**, 639–641.
Sun, F. F., Pozbindowski, K. S., Crane, F. L. and Jacobs, S. S. (1968). *Biochim. et Biophys. Acta* **153**, 804–818.
Taber, H. W. and Morrison, M. (1964). *Arch. Biochem. Biophys.* **105**, 367–379.
Takemori, S. (1960). *J. Biochem.* **47**, 382–390.
Takemori, S. (1965). "Amherst", vol. **2**, p. 561.
Takemori, S. and King, T. E. (1956). *J. Biol. Chem.* **240**, 504–513.
Takemori, S., Sekuzu, I. and Okunuki, K. (1960a). *Biochim. et Biophys. Acta* **38**, 158–160.
Takemori, S., Sekuzu, I. and Okunuki, K. (1960b). *J. Biochem.* **48**, 569–578.

Takemori, S., Sekuzu, I. and Okunuki, K. (1961). *J. Biochem.* **51**, 464–472.
Takemori, S., Sekuzu, I., Yonetani, T. and Okunuki, K. (1958). *Nature* **182**, 1306–1307.
Tsudzuki, T. and Okunuki, K. (1969) *J. Biochem.* **66**, 281–283.
Tsudzuki, T., Orii, Y. and Okunuki, K. (1967). *J. Biochem.* **62**, 37–45.
Tuppy, H. and Birkmeyer, G. D. (1969). *Europ. J. Biochem.* **8**, 237–243.
Tzagoloff, A. and MacLennan, D. H. (1966). In "The Biochemistry of Copper" (J. Peisach, P. Aisen and W. E. Blumberg, eds.), pp. 253–265. Academic Press, New York and London.
Tzagoloff, A. and Wharton, D. C. (1965). *J. Biol. Chem.* **240**, 2628–2633.
Tzagoloff, A., Yang, P. C., Wharton, D. C. and Rieske, J. S. (1965). *Biochim. et Biophys. Acta* **96**, 1–8.
Urry, D. W. and Van Gelder, B. F. (1968). "Osaka", pp. 210–214.
Urry, D. W., Wainio, W. W. and Grebner, D. (1967). *Biochem. Biophys. Res. Commun.* **27**, 625–631.
Utsumi, K. and Oda, T. (1969). *Arch. Biochem. Biophys.* **131**, 67–73.
Vanderkooi, G. and Stotz, E. (1965). *J. Biol. Chem.* **240**, 3418–3424.
Vanderkooi, G. and Stotz, E. (1966a). *J. Biol. Chem.* **241**, 2260–2267.
Vanderkooi, G. and Stotz, E. (1966b). *J. Biol. Chem.* **241**, 3316–3323.
Van Gelder, B. F. (1966). *Biochim. et Biophys. Acta* **118**, 36–46.
Van Gelder, B. F. and Beinert, H. (1969). *Biochim. et Biophys. Acta* **189**, 1–24.
Van Gelder, B. F. and Muijsers, A. O. (1964). *Biochim. et Biophys. Acta* **81**, 504–507.
Van Gelder, B. F. and Muijsers, A. O. (1966). *Biochim. et Biophys. Acta* **118**, 47–57.
Van Gelder, B. F. and Slater, E. C. (1963). *Biochim. et Biophys. Acta* **73**, 603–605.
Van Gelder, B. F., Orme-Johnson, W. H., Hansen, R. E. and Beinert, H. (1967). *Proc. Nat. Acad. Sci. U.S.* **58**, 1073–1079.
Vanneste, W. H. (1965). *Biochem. Biophys. Res. Commun.* **18**, 563–568.
Vanneste, W. H. (1966a). *Biochem.* **5**, 838–848.
Vanneste, W. H. (1966b). *Biochim. et Biophys. Acta* **113**, 175–186.
Vanneste, W. H. (1968). *Arch. Internat. Physiol. et Biochim.* **76**, 394–396.
Vanneste, W. H. and Vanneste, M. T. (1965). *Biochem. Biophys. Res. Commun.* **19**, 182–186.
Wainio, W. W. (1955). *J. Biol. Chem.* **216**, 593–594.
Wainio, W. W. (1961a). Moscow Internat. Congress of Biochem., Abstracts, p. 472.
Wainio, W. W. (1961b). "Haematin Enzymes", vol. **1**, pp. 281–300.
Wainio, W. W. (1965). "Amherst", vol. 2, pp. 622–633.
Wainio, W. W. and Aronoff, S. (1955). *Arch. Biochem. Biophys.* **57**, 115–123.
Wainio, W. W. and Greenlees, J. (1960). *Arch. Biochem. Biophys.* **90**, 18–21.
Wainio, W. W. and Schulman, J. (1968). *Fed. Proc.* **27**, Abstr. 3173.
Wainio, W. W., Grebner, D. and O'Farrell, A. (1968). "Osaka", pp. 66–75.
Wainio, W. W., Van der Wende, C. and Skimp, N. F. (1959). *J. Biol. Chem.* **234**, 2433–2436.
Warburg, O. and Negelein, E. (1929). *Biochem. Z.* **214**, 26–63; 64–100.
Wharton, D. C. and Cusanovich, M. A. (1969). *Biochem. Biophys. Res. Commun.* **37**, 111–113.
Wharton, D. C. and Gibson, Q. H. (1968). *J. Biol. Chem.* **243**, 702–706.
Wharton, D. C. and Tzagoloff, A. (1963). *Biochem. Biophys. Res. Commun.* **13**, 121–125.
Wharton, D. C. and Tzagoloff, A. (1964). *J. Biol. Chem.* **239**, 2036–2041.

Wharton, D. C. and Tzagoloff, A. (1967). "Methods in Enzymology", vol. 10, pp. 245–250.
Williams, G. R., Lemberg, R. and Cutler, M. E. (1968). *Canad. J. Biochem.* **46**, 1371–1379.
Williams, J. N., Jr. (1964). *Arch. Biochem. Biophys.* **107**, 537–543.
Williams, R. J. P. (1961). "Haematin Enzymes", vol. **1**, pp. 41–52.
Williams, R. J. P. (1956). *Chem. Revs.* **56**, 299–328.
Willick, G. E., Schonbaum, G. R. and Kay, C. M. (1969). *Biochemistry* **8**, 3729–3734.
Wilson, D. F. and Chance, B. (1966). *Biochem. Biophys. Res. Commun.* **23**, 751–756.
Wilson, D. F. and Chance, B. (1967). *Biochim. et Biophys. Acta* **131**, 421–430.
Wilson, D. F. and Dutton, P. L. (1970). *Arch. Biochem. Biophys.* **136**, 583–585.
Wilson, D. F. and Epel, D. (1958). *Arch. Biochem. Biophys.* **126**, 83–90.
Wilson, D. F. and Gilmour, M. V. (1967). *Biochim. et Biophys. Acta* **143**, 52–61.
Wilson, M. and Greenwood, C. (1970). *Biochem. J.* **116**, 17 P.
Wittenberg, J. B., Brown, P. K. and Wittenberg, B. A. (1965a). *Biochim. et Biophys. Acta* **109**, 518–529.
Wittenberg, J. B., Noble, R. W., Wittenberg, B. A., Antonini, E., Brunori, M. and Wyman, J. (1967). *J. Biol. Chem.* **242**, 626–634.
Wittenberg, B. A., Wittenberg, J. B., Stolzberg, S. and Valenstein, E. (1965b). *Biochim. et Biophys. Acta* **109**, 530–535.
Wittenberg, B. A., Kampa, L., Wittenberg, J. B., Blumberg, W. E. and Peisach, J. (1968). *J. Biol. Chem.* **243**, 1863–1871.
Wohlrab, H. (1969). *Biochem. Biophys. Res. Commun.* **35**, 560–564.
Yakushiji, E. and Okunuki, K. (1940). *Proc. Imper. Acad. Tokyo*, **16**, 299–302.
Yakushiji, E. and Okunuki, K. (1941). *Proc. Imper. Acad. Tokyo* **17**, 38–40.
Yamamoto, T. and Okunuki, K. (1970). *J. Biochem.* **67**, 505–506.
Yamazaki, I., Yokota, K. and Tamura, M. (1966). "Hemes and Hemoproteins", pp. 319–325.
Yonetani, T. (1959). *J. Biochem.* **46**, 917–924.
Yonetani, T. (1960a). *J. Biol. Chem.* **235**, 845–852; 3138–3143.
Yonetani, T. (1960b). *Biochem. Biophys. Res. Commun.* **3**, 549–553.
Yonetani, T. (1961). *J. Biol. Chem.* **236**, 1680–1688.
Yonetani, T. (1962). *J. Biol. Chem.* **237**, 550–559.
Yonetani, T. (1963a). In "The Enzymes", 2nd ed. (P. D. Boyer, H. Lardy and K. Myrbäck, eds.), vol. **8**, pp. 41–79. Academic Press, New York and London.
Yonetani, T. (1963b). *Proc. 5th Internat. Congress Biochem.* **5**, 396–415.
Yonetani, T. (1967). In "Methods in Enzymology", vol. **10**, pp. 332–335.
Yonetani, T. and Kidder, G. W. (1963). *J. Biol. Chem.* **238**, 386–388.
Yonetani, T. and Ray, G. S. (1965). *J. Biol. Chem.* **240**, 3392–3398.
Yong, F. C., Bayley, M. P. and King, T. E. (1968). "Osaka", p. 202.
Yong, F. C. and King, T. E. (1970a). *Biochem. Biophys. Res. Commun.* **38**, 940–946.
Yong, F. C. and King, T. E. (1970b). *Biochem. Biophys. Res. Commun.* **40**, 1445–1451.
Yong, F. C. and King, T. E. (1970c). *Fed. Proc.* **29**, Abstr. 871.
York, J. L., McCoy, S., Taylor, D. N. and Caughey, W. S. (1967). *J. Biol. Chem.* **242**, 908–911.
Yoshikawa, S. and Orii, Y. (1970). *J. Biochem.* **68**, 145–156.
Yubisui, T. (1968). *Ann. Report of Biol. Works, Fac. of Sci.*, Osaka Univ. **16**, 35–60.
Yubisui, T., Mizushima, H. and Okunuki, K. (1965). *Biochim. et Biophys. Acta* **118**, 442–444.

Chapter IV

Cytochromes b

A. Introduction.
B. Cytochrome b. References.
C. Cytochrome b_5. References.
D. Cytochrome P-450.
    1. Discovery and Occurrence.
    2. Purification.
    3. Spectroscopic properties.
    4. Spectral alterations caused by combustion with substrates of hydroxylation reactions.
    5. Role in sterol and drug metabolism.
    6. Induction and breakdown.
    7. Reaction mechanism and chemical structure.
    8. Relation of P-450 to haemoglobin catabolism.
    References.
E. Cytochromes "h" and helicorubin of invertebrates.
    1. Introduction.
    2. Isolation and purification.
    References.
F. Cytochromes in higher plants and algae.
    1. Introduction.
    2. Respiratory cytochromes b.
        a. in mitochondria.
        b. in microsomes.
    3. Cytochromes b in the electron transport of photosynthesis.
    References.
G. Cytochrome c_2 of Yeast ($L(+)$-lactate cytochrome c oxidoreductase).
    1. Introduction.
    2. Purification.
    3. Composition.
    4. Absorption spectra and other physical properties.
    5. Reaction mechanism, acceptor and substrate specificity and kinetics.
    6. Biological function.
    References.

A. INTRODUCTION

Cytochromes of type b are widely distributed in nature. They are found in animals, plants, yeasts and bacteria. They are protohaem-proteins, and with pyridine they give pyridine-protohaemochrome with its α-band at 556 nm. The only exception is "cytochrome h" or "enterochrome 556", which is not a protohaem, but probably a haematohaem-protein. It is dealt with in this chapter since the lack of covalent thioether linkages makes it belong to cytochromes b rather than to the cytochromes c. Table II gives a short summary of the occurrence, the main absorption bands, some of their essential properties and their biological function, which is not yet known in all instances with certainty. Cytochrome b is found widespread in animal and plant mitochondria, in yeast and bacteria; it is an insoluble enzyme and it is now known that it comprises more than a single entity. In some specialized mitochondria, e.g. those of adrenals and ovaries, cytochrome P-450 is found and in some plant mitochondria b_7. In animal microsomes, cytochromes b_5 and P-450 occur, in invertebrates special types of b and b_5 as well as cytochrome h, helicorubin and enterochromes; plant tissues contain cytochromes b and b_7, in mitochondria, cytochromes b_3 in microsomes, and b_6 and b_{559} in chloroplasts. Cytochrome b_2 has only been found in yeasts. Bacteria contain b_1 and b_{562} in $E.\ coli$, as well as cytochrome o and P-450. The function of these compounds differs greatly. Cytochromes b, b_1, b_2, b_3 and b_5 are electron carriers, while b_7 and P-450 and o are probably terminal oxidases; b_6, b_{559} and b_3 are electron carriers in photosynthetic processes, while P-450 is a hydroxylating enzyme of special importance for steroid metabolism and drug detoxication. It should be noted that the biological role of some of these cytochromes is not known with certainty, and that some of them may turn out to be mixtures of several compounds.

B. CYTOCHROME b

Cytochrome b, with absorption bands at 564, 530 and 432 nm in the ferrous state, was described by Keilin (1925) as a component of the respiratory chain in many aerobic organisms and characterized as a protohaemoprotein of haemochrome spectroscopic type (cf. Chapter II). It is not, or only very slowly, autoxidizable, and does not react with carbon monoxide or cyanide. It is present in mitochondria in a membrane-bound form, but can be solubilized by extraction with cholate or deoxycholate and separated from cytochrome aa_3 by fractional precipitation with ammonium sulphate. However, these extracts still contain mixtures of cytochrome b with c_1. Particles of bc_1 have been isolated as part of the respiratory chain by several workers (Goldberger and Green, 1963;

TABLE II
Cytochromes b

Name	Occurrence	Solubility	Absorption spectrum of Fe^{2+} cpd; in parentheses mM and mM ($Fe^{2+} - Fe^{3+}$)			Purification	E'_0 pH 7 mV	Autoxidizability	CO reaction	Function	Chapter and Remarks
b	Widespread in mitochondria of animals, plants, yeasts and some bacteria	Membrane-bound; solubilized but modified; in complex III with cyt. c_1	563 (21)	531	429 (114)	Not separable from c_1 without modification. Modified b: one haem per 18,000–30,000	+70 Modified −340	Not or slowly; Modified fast	No Modified +	Electron carrier in respiratory chain between dehydrogenases or flavoproteins to cyt. c_1. Involved in oxid. phosphorylation	IV B. Consists of two forms, b and b_T (or b_1) plus one (or two?) energized forms. Oxidation inhibited by antimycin A or quinoline-N-oxides, reduction by rotenone or amytal
	In blowfly larvae mitochondria	Soluble	563 555	530 528	428·5 424	M.W. 23,000 Cryst. M.W. 13,700	+0·00	Slowly	No		IV B. Differs from adult fly cytochrome
b_1	In bacteria, e.g. E. coli, Streptomyces	Membrane-bound in cytoplasmic membrane	559	529	427	Modified in purification M.W. 13,000?	+250	Fast	No	Electron carrier in respiratory chain and nitrate reductase	VI. May comprise several compounds
b_{562}	E. coli	Soluble	562 (24·1) (Δ14·4)	532	427	M.W. about 12,000		Not	No		VI.
b_2	In Saccharomyces	Soluble	557 (39)	528	423 (230)	Crystalline tetramer M.W. 235,000	0·00	Not	No	L-lactic dehydrogenase EC 1.1.2.3, electron carrier between flavoproteins and cyts. c or c_1	IV G. "Double headed enzyme" with protohaem and FMN as prosthetic groups. Frequently contains DNA
b_3	In plant microsomes	Membrane-bound but solubilizable	559	529	425	Partially purified M.W. 40,000	+40	Fast		Electron carrier, or oxidase in cyanide-insensitive respiration	IV F.

Cytochromes b

	Now recognized as abnormal type c cytochrome							
c^* b_4								
b_5	In animal microsomes; in midgut of silkworm	Soluble, rather thermostable	556 (26)	423 (171)	M.W. 11,000–12,000 crystalline 1 haem; amino acid sequence known	Fast	No	Electron carrier in microsomal respiration between b_5-reductase (EC 1.6.2.21) and cyt. c or b-420 oxidase	IV C. Most intact cytochromes b_5 contain, about 93 amino acid residues, but the tryptic core of 83 residues has the same catalytic and spectral properties
b_6	In chloroplasts of green plants		563				No	Electron carriers in photosynthesis	IV F.
b_{559}			559		+360				
b_7	In *Arum* spadix mitochondria	Soluble	560		−30	Fast	No	Oxidase of CN-insensitive respiration or electron carrier to unknown oxidase	IV F.
h enterochrome-556	In hepato-pancreas or gut of molluscs	Soluble	556	527	422 M.W. 18,000				IV E. Prosthetic group haem with saturated side chains, not protohaem.
Helicorubin; enterochrome-566	In gut of molluscs	Soluble Soluble	562 566	530	427	+200			IV E. Derived from cytochrome h?
o	In some bacteria	Membrane-bound	CO cpd. 568				Fast	+	Terminal oxidase	VI. Frequently with cyt a_1 or cyt d
P-450	In animal microsomes and in mitochondria of adrenals and ovaries; in bacteria	Membrane-bound, but solubilizable; soluble in *Pseudomonas putida* and *Rhizobium*	Characteristic CO-compound band 450 (114)	Soret	Only a bacterial P-450 of *Pseudomonas putida* has been crystallized	About −410 (apparent)	Fast	+	Hydroxylation of steroids and hormone regulation; hydroxylation and detoxification of drugs	IV D and VI (Bacterial). Easily transformed to inactive P-420. Forms two types of compounds with substrates

Rieske et al., 1967; King, 1967; Oda, 1968). The separation of the two cytochromes has not been achieved with certainty without modification of cytochrome b. The preparation by Feldman and Wainio (1960) using a snake venom phospholipase and a pancreatic lipase followed by ammonium sulphate fractionation probably comes nearest to it. It gives a high M.W. cytochrome b (abt. 175,000) of an oxidation-reduction potential close to that found for the cytochrome b of heart muscle preparations by Slater and Colpa-Boonstra (1961), Straub and Colpa-Boonstra (1962) and Urban and Klingenberg (1969) by succinate-fumarate titration, E'_0 pH7 + 77 mV. A somewhat lower potential (0·00 mV or slightly lower) had been found in earlier studies by Ball (1938), Hill (1954) and Bendall and Hill (1956) (see, however, below for "energized" cytochrome b). The absorption spectrum deviated little from that of the cytochrome in the mitochondria, with a γ/α ratio of 5·5. The absorption however was rather low and the preparation may have contained extraneous colourless protein of the enzymes used in the preparation.

Other preparations obtained by the use of proteinases, sodium dodecyl sulphate, or cationic detergent (Bomstein et al., 1960; Goldberger et al., 1961, 1962; Ohniski, 1966a; Shichi and Kuroda, 1967; Rieske and Tisdale, 1967b; Yamashita and Racker, 1969) were modified cytochromes b, although some of them (Sekuzu and Okunuki, 1956; Ohnishi, 1966a; Rieske and Tisdale, 1967a) were obtained crystalline. They rapidly reacted with atmospheric oxygen and with carbon monoxide, their absorption bands were shifted to somewhat shorter wavelengths, and their oxidation-reduction potential was lowered, in some instances to as low as -34 mV, although this could be raised by the addition of structural protein (Goldberger et al., 1962). Although these preparations contain a cytochrome b with altered tertiary structure they are nevertheless of great interest. Not only is the primary structure probably unaltered, but Yamashita and Racker (1969) have used such an altered cytochrome b in the reconstitution of an active bc_1 complex (see Chapter VII, E 1).

A number of observations led some earlier workers to doubt the role of cytochrome b in electron transport and others to postulate more than one form of cytochrome b (see Wainio et al., 1968). Whereas the former has now been disproved, the latter has been established. Doeg et al. (1960) and Ziegler and Doeg (1962) considered cytochrome b merely a structural constituent of succinic dehydrogenase which did not participate in electron transport. Chance and Williams (1955a,b) and Chance (1958, 1961) found it active as electron carrier only in phosphorylating, but not in non-phosphorylating mitochondria. Shore and Wainio (1965) found only one part of cytochrome b (b') readily reducible by NADH and succinate, while the other was only slowly reduced by NADH and only in the presence of antimycin by succinate. Kirschbaum and Wainio (1966) found non-haem-iron protein essential for only one third of the reduction of cytochrome c_1 by cytochrome b (see also Baginsky and Hatefi, 1968). Thus Wainio et al. (1968) concluded that multiple forms of cytochrome b were possible, though not yet proved.

Meanwhile some of the contradictions have been solved. One factor of confusion was that it was not known then that sonically-produced submitochondrial particles are turned inside-out (Lee and Ernster, 1966; Malviya et al., 1968; Tyler, 1970). The second was the realization that under certain conditions cytochrome b can be displaced from the bc_1 particle to form a closer combination with succinic dehydrogenase than with cytochrome c_1 (Ernster et al., 1969; cf. also Nicholls, 1969). Thus functionally modified forms of cytochrome b arise as a result of structural damage or by antimycin treatment. In these the

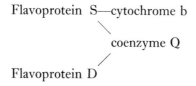

part of the respiratory chain becomes separated from the $c_1 \to c \to aa_3$ part, or the flavoprotein-coenzyme Q part from the $b \to c_1 \to c \to aa_3$ part. In the first instance, cytochrome b joins the succinic dehydrogenase and has a structural effect on it (Bruni and Racker, 1968).

Antimycin A, which has been frequently used to block the reactions between cytochromes b and c_1 (see Chapter VII, F4) affects, in very low concentration, the properties of cytochrome b but not those of cytochrome c_1 (Chance, 1952, 1958; Slater and Colpa-Boonstra, 1961; Pumphrey, 1962; Deul and Thorn, 1962; Shore and Wainio, 1965; Rieske et al., 1967; Bryla and Kaniuga, 1968; Rieske, 1969; Ernster et al., 1969; Bryla et al., 1969a,b; Kováč et al., 1970). $\Delta\epsilon$ mM ($Fe^{2+} + Fe^{3+}$) at 563–577 nm for the total cytochrome b is reported to be 25·6 (Berden and Slater, 1970). The α-band is shifted to 566 nm by antimycin and the completeness of reduction by succinate or NADH is decreased. Cholate displaces the α-band in a direction opposite to that by antimycin (Pumphrey, 1962); thus the shift may depend upon a change of the hydrophobic environment of the protohaem. The recent results of the schools of Chance (Chance et al., 1970; Chance and Wilson, 1970; Wilson and Dutton, 1970; Dutton et al., 1970) and of Slater (Berden and Slater, 1970; Slater 1970b; Slater, Berden and Bertina, 1970; Slater et al., 1970a,b; Wegdam et al., 1970) show that of the two molecules of cytochrome b present in the bc_1 complex (Chapter VII, E1, E2) only one reacts with antimycin. This is called cytochrome b_T by Chance, cytochrome b_i by Slater. This is shown by the stoichiometry of the antimycin spectral shifts as well as by the quenching of the antimycin fluorescence (Berden and Slater, 1970).

The existence of an energized form of cytochrome b (see Chapter VII) was first claimed by Chance and Schoener (1966a,b) and by Chance et al. (1966). They observed in pigeon heart mitochondria a "cytochrome b_{555}" which they ascribed to an energy-rich form of ferrocytochrome b. The 555 nm band was observed under conditions in which terminal respiration was inhibited by sulphide and succinate reduction by malonate or oxaloacetate. Under these conditions the bands of ferrous cytochrome c_1 and cytochrome b (563 nm) are decreased, but that of 555 nm is increased, while uncouplers decreased the 555 nm band. Cytochrome b_{555} had a life-time of about 16 min. at 25° (Azzi and Chance, 1969). It should be noted that this band is in fact the $\Delta(Fe^{2+} - Fe^{3+})$ band at *low* temperature (77°K), the band at room temperature being close to 560 nm. Since difference spectra depend on those of the ferric as well as of the ferrous forms, it is not inconceivable that the spectroscopic difference is due to two conformationally different forms of the oxidized cytochrome b, similar to those found for cytochrome a_3 (see Chapter III). The biphasic cytochrome b reduction in phosphorylating rat liver mitochondria, in contrast to the fast monophasic reduction after uncoupling (Hommes, 1964; Lee et al., 1969) is perhaps another parallel to a_3. Tyler et al. (1966) distinguish between a non-energy-linked form of cytochrome b reducible only by succinate, and an energy-linked form reducible also by TMPD-ascorbate in reversed electron transport.

It is, however, difficult to harmonize these findings with the recent findings of both Slater and Chance even if one assumes that the b_{555} of Chance and Schoener was in fact, energized cytochrome b. According to Chance et al. only cytochrome b_T, not cytochrome b is present in an energized form, and since ATP shifts the absorption band towards the red, this should not be found at 555 nm. The energized form b_T has a redox potential of $+245$ mV, in contrast to the non-energized b_T the potential of which is at $+35$ mV, while that of b lies at about $0\cdot0$ mV, or possibly at -55 mV (Wilson and Dutton, 1970; Dutton et al., 1970; Chance, et al., 1970). According to the Slater school both b and b_i can be energized, the former in the b^{2+}, the latter in the b^{3+} form. This energization can be achieved without ATP (the action of which can be blocked by oligomycin) by active respiration. Slater et al. assume an intramolecular electron exchange between b and b_i:

$$b^{2+} \sim X \cdot b_i{}^{3+} \sim X + ADP + P_i = b^{3+}b_i{}^{2+} + 2X + ATP$$

It is not necessary to assume an "energy-rich" binding of the $b \sim X$ type, or even a change of ligands; the energization may be merely the result of conformational changes in the cytochromes (see Chapter VII, H). At present, there is no evidence for an alteration of the redox potential of the cytochrome b moiety on energization.

Cytochrome b forms with an α absorption maximum at shorter wavelengths than that normal for cytochrome b have been observed by others. Sato and Hagihara (1968) found two forms of cytochrome b in liver mitochondria, of which the one with the maximum 559 nm (at room temperature) was prevalent in cancer cells. This did not react with carbon monoxide, differing in this from the b_{558} (b_5?) of Shinagawa et al. (1966) from rabbit neutrophil leucocytes, and the b_{559} from bovine heart microsomes (Shichi and Kuroda, 1967).

Like antimycin A, uncoupling agents in relative high concentration increase the oxidation of cytochrome b of yeast (Kováč et al., 1970) even anaerobically.

Cytochrome b in invertebrates. Invertebrates frequently contain the normal respiratory chain with cytochrome b (Kawai, 1959). Okunuki and co-workers have isolated two cytochromes b from fly (*Musca domestica*) larvae, which appear to differ from the cytochrome b of the adult animal (Estabrook and Sacktor, 1958a,b; Chance and Sacktor, 1958). Both are soluble enzymes. The one, with an absorption spectrum close to that of mammalian cytochrome b (Fe^{2+} 563, 530 and 428·5 nm) (Okada and Okunuki, 1970b) has been obtained crystalline, of M.W. 23,000 and, contrary to the modified mammalian crystalline cytochrome b does not combine with carbon monoxide and is only slightly autoxidizable (Ohnishi, 1966b). It was reduced by yeast L-lactate dehydrogenase but not by succinate or NADH. It resembled the soluble cytochrome b from etiolated mung bean seedlings (Shichi and Hackett, 1962) and that from the fungus *Sclerotina libertiana* (Yamanaka et al., 1960) as well as the cytochrome b_{562} from *E. coli* (Hager and Itagaki, 1967). The other (Okada and Okunuki, 1970a) with absorption bands at 555, 528 and 424 nm, with a double α-band at 556 and 552 nm at 77°K is also little autoxidizable but appears to react with CO or cyanide to judge from the CD spectra. Its M.W. was 13,700 and the redox potential E'_0 (pH 7, 12°C) + 6 mV. It is probably an invertebrate cytochrome b_5. The slow moving muscles of the worm *Ascaris lumbricoidis* or the tapeworm *Monieza expansa* which contain some cytochrome aa_3, are far richer in cytochrome b and c, including possibly some cytochrome o (Chance and Parsons, 1963; Cheah, 1968). Cytochromes of type b have been frequently observed spectroscopically in marine invertebrates, e.g. in the *Octopus* (Ghiretti-Magaldi et al., 1957) and in the worm *Spirographis spallanzani* (Ghiretti et al., 1959a,b). (Cytochrome h, enterochromes and helicorubin are described in Chapter IV, E.)

REFERENCES

Azzi, A. and Chance, B. (1969). *Biochim. et Biophys. Acta* **189**, 141–151.
Baginsky, M. L. and Hatefi, Y. (1968). *Biochem. Biophys. Res. Commun.* **32**, 945–950.
Ball, E. G. (1938). *Biochem. Z.* **295**, 262–264.
Bendall, D. S. and Hill, R. (1956). *New Phytol.* **55**, 206–212.
Berden, J. A. and Slater, E. C. (1970). *Biochim. et Biophys. Acta* **216**, 237–249.
Bomstein, R., Goldberger, R. and Tisdale, H. (1960). *Biochem. Biophys. Res. Commun.* **2**, 234–238.
Bruni, A. and Racker, E. (1968). *J. Biol. Chem.* **234**, 962–967.
Bryla, J. and Kaniuga, Z. (1968). *Biochim. et Biophys. Acta* **153**, 910–913.
Bryla, J., Kaniuga, Z. and Slater, E. C. (1969a). *Biochim. et Biophys. Acta* **189**, 317–326.
Bryla, J., Kaniuga, Z. and Slater, E. C. (1969b). *Biochim. et Biophys. Acta* **189**, 327–336.
Chance, B. (1952). *Nature* (London) **169**, 215.
Chance, B. (1958). *J. Biol. Chem.* **233**, 1223–1229.
Chance, B. (1961). "Haematin Enzymes" vol. **2**, p. 593.
Chance, B. and Parsons, D. F. (1963). *Science* **142**, 1176–1179.
Chance, B. and Sacktor, B. (1958). *Arch. Biochem. Biophys.* **76**, 509–531.
Chance, B. and Schoener, B. (1966a). *J. Biol. Chem.* **241**, 4567–4573.
Chance, B. and Schoener, B. (1966b). *J. Biol. Chem.* **241**, 4577–4587.
Chance, B. and Williams, G. R. (1955a). *J. Biol. Chem.* **217**, 409–427.
Chance, B. and Williams, G. R. (1955b). *J. Biol. Chem.* **217**, 429–438.
Chance, B., Erecinska, M., Wilson, D., Dutton, C. and De Vault, D. (1970). 8th Internat. Congress Biochemistry, Switzerland, Abstract 4, p. 239–240.
Chance, B., Lee, C. P. and Schoener, B. (1966). *J. Biol. Chem.* **241**, 4574–4576.
Chance, B., Wilson, D. F., Dutton, P. L. and Erecinska, M. (1970). *Proc. Nat. Acad. Sci. U.S.* **66**, 1175–1182.
Cheah, K. S. (1968). *Biochim. et Biophys. Acta* **153**, 718–720.
Deul, D. H. and Thorn, M. B. (1962). *Biochim. et Biophys. Acta* **59**, 426–436.
Doeg, K. A., Krüger, S. and Ziegler, D. M. (1960). *Biochim. et Biophys. Acta* **41**, 491–497.
Dutton, P. L., Wilson, D. F. and Lee, C. P. (1970). *Biochemistry* **9**, 5077–5082.
Ernster, L., Lee, I. -Y., Norling, B. and Persson, B. (1969). *European J. Biochem.* **9**, 299–310.
Estabrook, R. W. and Sacktor, B. (1958a). *Arch. Biochem. Biophys.* **76**, 532–545.
Estabrook, R. W. and Sacktor, B. (1958b). *J. Biol. Chem.* **233**, 1014–1019.
Feldman, D. and Wainio, W. W. (1960). *J. Biol. Chem.* **235**, 3635–3639.
Ghiretti, F., Ghiretti-Magaldi, A. and Tosi, L. (1959a). *Bull. Soc. Ital. Biol. Sper.* **35**, 2216–2218.
Ghiretti, F., Ghiretti-Magaldi, A. and Tosi, L. (1959b). *J. Gen. Physiol.* **42**, 1185–1205.
Ghiretti-Magaldi, A., Guiditta, A. and Ghiretti, F. (1957). *Biochem. J.* **66**, 303–307.
Goldberger, R. and Green, D. E. (1963). In "The Enzymes", 2nd ed. (P. D. Boyer, H. Lardy, and K. Myrbäck, eds.), Vol. **8**, 81–95. Academic Press, New York and London.
Goldberger, R., Pumphrey, A. N. and Smith, A. L. (1962). *Biochim. et Biophys. Acta* **58**, 307–313.
Goldberger, R., Smith, A. L., Tisdale, H. and Bomstein, R. (1961). *J. Biol. Chem.* **236**, 2788–2793.

Hager, L. P. and Itagaki, E. (1967). "Methods in Enzymol.", Vol. 10, pp. 373–378.
Hill, R. (1954). *Nature* (London) 174, 501–503.
Hommes, F. H. (1964). *Arch. Biochem. Biophys.* 107, 78–81.
Kawai, K. (1959). *Biol. Bull.* 117, 125–132.
Kawai, K. (1961). *J. Biochem.* 49, 427–435.
Keilin, D. (1925). *Proc. Roy. Soc.* (London) B 98, 312–339.
King, T. E. (1963). *J. Biol. Chem.* 238, 4037–4051.
King, T. E. (1967). "Methods in Enzymol.", Vol. 10, pp. 216–225.
Kirschbaum, J. and Wainio, W. W. (1966). *Biochim. et Biophys. Acta* 113, 27–32.
Kovac, L., Smigan, P., Hrusovska, E. and Hess, B. (1970). *Arch. Biochem. Biophys.* 139, 370–379.
Lee, C. P. and Ernster, L. (1966). In "Regulation of Metabolic Processes in Mitochondria" (J. M. Tager, S. Papa, E. Quagliariello and E. C. Slater, eds.), pp. 218–234. Elsevier, Amsterdam.
Lee, C. P., Ernster, L. and Chance, B. (1969). *European J. Biochem.* 8, 153–163.
Lee, C. P., Estabrook, R. W. and Chance, B. (1965). *Biochim. et Biophys. Acta* 99, 32–45.
Malviya, A. N., Parsa, B., Yodaiken, R. E. and Elliott, W. B. (1968). *Biochim. et Biophys. Acta* 162, 195–209.
Nicholls, P. (1969). *Fed. Proc.* 28, Abstr. 1198.
Oda, T. (1968). "Osaka", pp. 500–515.
Ohnishi, K. (1966a). *J. Biochem.* 59, 1–8, 17–23.
Ohnishi, K. (1966b). *J. Biochem.* 59, 9–16.
Okada, Y. and Okunuki, K. (1969). *J. Biochem* 65, 581–596.
Okada, Y. and Okunuki, K. (1970a). *J. Biochem.* 67, 487–496.
Okada, Y. and Okunuki, K. (1970b). *J. Biochem.* 67, 603–605.
Pumphrey, H. M. (1962). *J. Biol. Chem.* 237, 2384–2390.
Rieske, J. S. (1969). *Fed. Proc.* 28, Abstr. 1197.
Rieske, J. S. and Tisdale, H. D. (1967a). "Methods in Enzymol." Vol. 10, pp. 349–353.
Rieske, J. S. and Tisdale, H. D. (1967b). "Methods in Enzymol." Vol. 10, pp. 353–356.
Rieske, J. S., Baum, H., Stoner, C. D. and Lipton, H. J. (1967). *J. Biol. Chem.* 242, 4854–4866.
Sato, N. and Hagihara, B. (1968). *J. Biochem.* 64, 723–726.
Sekuzu, I. and Okunuki, K. (1956). *J. Biochem.* 43, 107–109.
Shichi, H. and Hackett, D. P. (1962). *J. Biol. Chem.* 237, 2955–2958.
Shichi, H. and Kuroda, Y. (1967). *Arch. Biochem. Biophys.* 118, 682–692.
Shinagawa, Y., Tanaka, C., Teraoka, A. and Shinagawa, Y. (1966). *J. Biochem.* 59, 622–624.
Shore, J. P. and Wainio, W. W. (1965). *J. Biol. Chem.* 240, 3165–3169.
Slater, E. C. (1970). In (J. M. Tager, S. Papa, E. Quagliariello and E. C. Slater, eds.), "Electron Transport and Energy Conservation", pp. 363–369. Adriatica Ed., Bari.
Slater, E. C., Berden, J. A. and Bertina, R. M. (1970b). 8th Internat. Congress Biochemistry, Switzerland, Sympos. 3, pp. 157–159.
Slater, E. C. and Colpa-Boonstra, J. P. (1961). "Haematin Enzymes", Vol. 2, 575–592.
Slater, E. C., Lee, C. P., Berden, J. A. and Wegdam, H. J. (1970b). *Biochim. et Biophys. Acta* 223, 354–364.
Straub, J. P. and Colpa-Boonstra, J. P. (1962). *Biochim. et Biophys. Acta* 60, 650–652.

Tyler, D. D. (1970). *Biochem. J.* **116**, 30P–31P.
Tyler, D. D., Estabrook, R. W. and Sanadi, D. R. (1966). *Arch. Biochem. Biophys.* **114**, 239–251.
Urban, P. F. and Klingenberg, M. (1969). *European J. Biochem.* **9**, 519–528.
Wainio, W. W., Kirschbaum, J. and Shore, J. D. (1968). "Osaka", pp. 713–722.
Wegdam, H., Berden, J. A. and Slater, E. C. (1970). *Biochim. et Biophys. Acta* **223**, 365–373.
Wilson, D. F. and Dutton, P. L. (1970). *Biochem. Biophys. Res. Commun.* **39**, 59–64.
Yamanaka, T., Horio, T. and Okunuki, K. (1960). *Biochim. et Biophys. Acta* **40**, 349–351.
Yamashita, S. and Racker, E. (1969). *J. Biol. Chem.* **244**, 1220–1227.
Ziegler, D. M. and Doeg, K. A. (1962). *Arch. Biochem. Biophys.* **97**, 41–50.

C. CYTOCHROME b_5

The name cytochrome b_5 was given by Chance and Williams (1954) to a cytochrome from rat liver microsomes at a time when it was believed that cytochrome "b_4" was a b-type. Its properties closely resemble those of the cytochrome of *Cecropia* (silkworm) larvae (Chance and Pappenheimer, 1954) which probably differs from the liver enzyme only by species-specificity. The same cytochrome had previously been called b_1 by Keilin and Hartree (1940), a name now applied only to a similar bacterial cytochrome, b' by Yoshikawa (1951), "m" by C. F. Strittmatter and Ball (1952) or "x" by Sanborn and Williams (1950). It has been observed in the mammary gland by Bailie and Morton (1955). The earlier work has been fully summarized by C. F. Strittmatter (1961) and by P. Strittmatter (1963). It is the enzyme of the endoplasmic reticulum, found in highest concentration in the microsomes of the liver, but also in many other organs or tissues (Mangum et al., 1970), probably also in erythrocytes (Hultquist et al., 1969). The similar cytochrome from rabbit neutrophil leucocytes has, however, been found to react with carbon monoxide (Shinagawa et al., 1966). Finer differences exist between the cytochromes b_5 of various species (Krisch and Staudinger, 1959; Inouye et al., 1966; Tsugita et al., 1968, 1970; Ozols and Strittmatter, 1968b; Nobrega et al., 1969; Nobrega and Ozols, 1970), and even between the cytochrome b_5 of the adrenal medulla or adrenal cortex (Ichikawa and Yamano, 1965; Cooper et al., 1965). Cytochrome b_5 has also been observed in the outer membranes (Davis and Kreil, 1968) of liver mitochondria (Raw et al., 1959, 1960; Raw and Mahler, 1959) from which source it was first crystallized (Raw and Colli, 1959).

Its biological role is still uncertain, although it is connected by a special reductase (Strittmatter, 1964; Loverde and Strittmatter, 1968; see also Chapter VII) with oxidation of NADH and NADPH; the latter, however, is oxidized more efficiently by a special NADPH–P-450 reductase (see Chapter VII, Section VC E3). It is also reduced by an ascorbate-cytochrome

b_5-oxidoreductase; and can be reoxidized by semidehydroascorbate (Everling *et al.*, 1969). Cytochrome b_5 has been considered a storage protohaemprotein (Strittmatter, 1961), an electron carrier in microsomal respiration, or a terminal oxidase in it, responsible e.g. for slow or cyanide-insensitive oxygen uptake (Krisch and Staudinger, 1959). It is slightly autoxidizable or may carry electrons to P-450 or to another unknown terminal oxidase. While it is readily oxidized by ferricytochrome c, and also by methaemoglobin in erythrocytes (Petragnani *et al.*, 1959; Passon and Hultquist, 1970) there is no evidence for the presence of these compounds in microsomes. Schmeling *et al.* (1969) and Mayer *et al.* (1968) suggested its involvement in kynurenine-3-hydroxylase in rat liver mitochondria.

In contrast to the lack of definite knowledge of the biological function of cytochrome b_5, its chemical structure (or at least that of its active core) is better known than that of any other cytochrome except cytochrome c.

Preparation. The cytochrome is bound to lipoprotein particles, but is solubilized by pancreatic lipase, though not by cabbage leaf phospholipase D (Lumper *et al.*, 1969). It is also solubilized by proteinases, e.g. trypsin, which attacks the enzyme itself less than the apoenzyme, chymotrypsin, or by the *Bacillus subtilis* proteinase (Ito and Sato, 1969). It is then purified by fractional ammonium sulphate precipitation and chromatography on DEAE cellulose and Sephadex gel (Strittmatter and Ozols, 1966; Strittmatter, 1967; Poltaratsky-Bois and Chaix, 1966). During the exposure to the proteolytic enzymes, however, the molecular weight is somewhat decreased. Estabrook (1968) finds the molecular weight of the enzymes extracted with deoxycholate and Triton-X 100 to be 120,000, treated with urea 25,000 (Ito and Sato, 1968), while the M.W. of the isolated crystalline cytochromes b_5 from calf, rat or pig liver has been 11,500 and from rabbit liver 12,800 (Strittmatter, 1963; Ehrenberg and Bois-Poltoratsky, 1968; Kajihara and Hagihara, 1968). However, trypsin removes only nine amino acid residues from the C-terminal and a dipeptide from the N-terminal end of the enzyme giving the resistant core with 82 amino acid residues as the simplest unit which still has unaltered spectral and catalytic properties (Strittmatter and Huntley, 1970). Trypsin solubilization is also used in the large scale preparations of b_5 from bovine liver microsomes by Omura and Takesue (1970). Both calf-, rat- and rabbit-enzymes have yielded two crystallizable fractions (Ozols and Strittmatter, 1966; Archakov *et al.*, 1969; Kajihara and Hagihara, 1968). Though of smaller molecular weight, the activity of the crystalline fractions is unaltered.

Properties. Cytochrome b_5 is slightly autoxidizable, but does not combine with carbon monoxide, cyanide, or azide. It is rather stable to temperature, acid and alkali (to about pH 9·5). The absorption spectrum and extinction coefficients given in table II are those found by Strittmatter and Velick (1956) and Garfinkel (1958). At low temperature (liquid nitrogen) the α-band is double (Estabrook, 1961). The molecule is slightly ellipsoid ($f/f_0 = 1.19$), similar to that of cytochrome c. Both ferrous and ferric forms are low spin, but the ferric forms contain a small amount of high spin form (Watari et al., 1967; Bois-Poltoratsky and Ehrenberg, 1967). E_0' pH 7 is +20 mV (Velick and Strittmatter, 1956). Both rabbit and calf liver cytochrome b_5 crystallize in group $P2_12_12_1$, but rabbit b_5 has 2 moles per unit cell, calf b_5 only one (Kretsinger et al. 1970). A lower value (−12 to −14 mV) has been found for the enzyme in beef liver microsomes (C. F. Strittmatter and Ball, 1952; Kawai et al., 1963), but it may be doubted whether oxidation-reduction values for cytochromes bound in membranes have any definite meaning (see also Weber et al., 1971). Its half-life in rat liver microsomes is 45 hrs (Greim, 1970).

Amino acid sequence and linkage of haem-iron. A complete amino acid sequence for the calf-liver enzyme trypsin-resistant core of 93 residues

TABLE III
Amino acid sequence for calf liver cytochrome b_5
(according to Ozols and Strittmatter, 1967, 1968, 1969)
(Haem-binding histidines underlined)

	1	2	3	4	5	6	7	8	9	10	11
NH$_2$	(Ser.	Lys.)	Ala.	Val.	Lys.	Tyr.	Tyr.	Thr.	Leu.	Glu.	Gln.
	12	13	14	15	16	17	18	19	20	21	22
	Glu.	Ile.	Lys.	His.	Asn.	Asn.	Ser.	Lys.	Ser.	Thr.	Trp.
	23	24	25	26	27	28	29	30	31	32	33
	Leu.	Ile.	Leu.	His.	Tyr.	Lys.	Val.	Tyr.	Asp.	Leu.	Thr.
	34	35	36	37	38	39	40	41	42	43	44
	Lys.	Phe.	Leu.	Glu.	Gln.	His.	Pro.	Gly.	Gly.	Glu.	Glu.
	45	46	47	48	49	50	51	52	53	54	55
	Val.	Leu.	Arg.	Glu.	Gln.	Ala.	Gly.	Gly.	Asp.	Ala.	Thr.
	56	57	58	59	60	61	62	63	64	65	66
	Glu.	Asp.	Phe.	Glu.	Asp.	Val.	Gly.	His.	Ser.	Thr.	Asp.
	67	68	69	70	71	72	73	74	75	76	77
	Ala.	Arg.	Glu.	Leu.	Ser.	Lys.	Thr.	Phe.	Ile.	Ile.	Gly.
	78	79	80	81	82	83	84	85	86	87	88
	Glu.	Leu.	His.	Pro.	Asp.	Asp.	Arg.	(Ser.	Lys.	Ile.	Thr.
	89	90	91	92	93						
	Lys.	Pro.	Ser.	Glu.	Ser.	CO$_2$H)					

In parentheses residues not present in tryptic core.

has been established by Ozols and Strittmatter (1967, 1968, 1969). Tryptic and chymotryptic splitting of apocytochrome b_5 yields 12 peptides which are further analysed by carboxypeptidases and Edman degradations. The apoenzyme was first prepared by Strittmatter, 1960. The amino acid sequence of the calf-liver enzyme is given in Table III. The apoenzyme readily combines with various iron-porphyrins (Ozols and Strittmatter, 1964) to enzymically active haemoproteins, but only weakly with iron-free porphyrins. Protohaem protects the enzyme against trypsin. The amino acid sequence in the rabbit liver enzyme is also known (Kajihara and Hagihara, 1968; Tsugita et al., 1970).

While initially one histidine-nitrogen and one α- or ε-lysine amino group had been assumed to bind the haem-iron, Ozols and Strittmatter (1964, 1968a,b) and Nobrega et al. (1969) found the linkage through two histidine-imidazoles. Ozols and Strittmatter point out the similarity of the histidine residues 63 and 80 and their surroundings in the calf-liver enzyme to the histidines 63 and 92 in the larger β-chain of haemoglobin; this is confirmed by the work of Strittmatter and Huntley (1970). Neither lysyl groups (Ozols and Strittmatter, 1966; Loverde and Strittmatter, 1968; Nagishi and Omura, 1970) nor tryptophane or tyrosine, nor methionine are involved in the binding of the haem; in fact apart from the human enzyme the other liver enzymes have been found to contain no methionine (Nobrega et al., 1969).

The similarity of the haem binding in cytochrome b_5 and haemoglobin makes it important to find out what finer differences cause the different functional behaviour of the haem group. This will be finally only settled by x-ray diffraction studies of cytochrome b_5 crystals. Castro and Davis (1969) have speculated that the position of the haem in cytochromes b may be in a crevice not allowing carbon monoxide to penetrate to the iron, but electrons to be allowed passage via the π-system of the porphyrin sticking out of the crevice. This is indeed the situation found for cytochrome c but it should not be forgotten that, in contrast to cytochromes b, cytochrome c is not autoxidizable by molecular oxygen so that other factors must be of importance. Although the 3-68 fragment combines with protohaem, the resulting compound is spectroscopically different, unable to combine with the cytochrome b_5 reductase (Chapter VII, E3) and catalytically inactive.

Tsugita et al. (1970) have studied the amino acid sequence in man, the bovine (cow and calf), the rabbit and chicken. They found 63 of the 82 residues of the tryptic core invariable including the five histidines. Calf and cow b_5 differ (see also Ozols, 1970). Rabbit b_5 crystals have a unit cell of $111 \cdot 1 \times 36 \cdot 3 \times 51 \cdot 4$ Å of the group $P2_12_12_1$. Intact rabbit b_5 has 11 more residues (Ozols, 1970), but there is good agreements of his and the Tsugita et al. findings. Ozols observed heterogeneity of residues 10 and 95 in rabbit b_5. Nobrega and Ozols (1970) also reported the amino acid sequence in the cytochromes b_5 of monkey and pig.

Making use of the susceptibility of apocytochrome b_5 to trypsin and the greater stability of cytochrome b_5 itself (Ozols and Strittmatter, 1968a),

Hara and Minakami (1970) demonstrated the presence of apocytochrome b_5 in liver microsomes (see also Chapter IX).

REFERENCES

Archakov, A. I., Devichensky, V. U. and Severina, V. A. (1969). *Biokhimiya* **34**, 782–790.
Bailie, M. and Morton, R. K. (1955). *Nature* (London) **176**, 111–113.
Bois-Poltoratsky, R. and Ehrenberg, A. (1967). *Europ. J. Biochem.* **2**, 361–365.
Castro, C. E. and Davis, H. F. (1969). *J. Amer. Chem. Soc.* **91**, 5405–5407.
Chance, B. and Pappenheimer, A. M. (1954). *J. Biol. Chem.* **209**, 931–943.
Chance, B. and Williams, G. R. (1954). *J. Biol. Chem.* **209**, 945–951.
Cooper, D. Y., Narasimhulu, J., Rosenthal, O. and Estabrook, R. W. (1965). "Amherst", Vol. **2**, pp. 838–854.
Davis, K. A. and Kreil, G. (1968). *Biochim. et Biophys. Acta* **162**, 627–630.
Ehrenberg, A. and Bois-Poltoratsky, R. (1968). "Osaka", pp. 594–600.
Estabrook, R. W. (1961). "Haematin Enzymes", Vol. **2**, pp. 436–457.
Estabrook, R. W. (1968). *Science* **160**, 1368–1376.
Everling, F. B., Weiss, W. and Staudinger, Hj. (1969). *Z. Physiol. Chem.* **350**, 1485–1492.
Garfinkel, D. (1958). *Arch. Biochem. Biophys.* **77**, 493–509.
Greim, H. (1970). *Naunyn-Schmiedeberg Arch. Pharmakol.* **266**, 261–275.
Hara, T. and Minakami, S. (1970). *J. Biochem.* **67**, 541–543.
Hultquist, D. E., Reed, D. W., Passon, P. G. and Andrews, W. E. (1971). *Biochim. et Biophys. Acta* **229**, 33–41.
Ichikawa, Y. and Yamano, T. (1965). *Biochem. Biophys. Res. Commun.* **20**, 263–268.
Inouye, A., Shinagawa, Y. and Shinagawa, Ya. (1966). *J. Neurochem.* **13**, 385–390.
Ito, A. and Sato, R. (1968). *J. Biol. Chem.* **243**, 4822–4923.
Ito, A. and Sato, R. (1969). *J. Cell Biol.* **40**, 179–189.
Kajihara, T. and Hagihara, B. (1968a). "Osaka", pp. 581–593.
Kajihara, T. and Hagihara, B. (1968b). *J. Biochem.* **63**, 453–461.
Kawai, Y., Yoneyama, Y. and Yoshikawa, H. (1963). *Biochim. et Biophys. Acta* **67**, 522–524.
Keilin, D. and Hartree, E. F. (1940). *Proc. Roy. Soc. London B* **129**, 277–306.
Kretsinger, R. H., Hagihara, B. and Tsugita, A. (1970). *Biochim. et Biophys. Acta* **200**, 421–422.
Krisch, K. and Staudinger, Hj. (1959). *Biochem. Z.* **331**, 195–209.
Loverde, A. and Srittmatter, P. (1968). *J. Biol. Chem.* **243**, 5777–5787.
Lumper, L., Zubrzycki, Z. and Staudinger, Hj. (1969). *Z. physiol. Chem.* **350**, 163–172.
Mangum, V. M., Klingler, M. D. and North, J. A. (1970). *Biochem. Biophys. Res. Commun.* **40**, 1520–1525.
Mayer, G., Ullrich, V. and Staudinger, Hj. (1968). *Z. physiol. Chem.* **349**, 459–464.
Nagishi, M. and Omura, T. (1970). *J. Biochem.* **67**, 745–747.
Nobrega, F. G., Aranjo, P. S., Pasetto, M. and Raw, I. (1969). *Biochem. J.* **115**, 849–856.
Nobrega, F. G. and Ozols, J. (1970). *Fed. Proc.* **29**, Abstr. 3994.
Omura, T. and Takesue, S. (1970). *J. Biochem.* **67**, 249–257.
Ozols, J. (1970). *J. Biol. Chem.* **245**, 4863–4874.
Ozols, J. and Strittmatter, P. (1964). *J. Biol. Chem.* **239**, 1018–1023.
Ozols, J. and Strittmatter, P. (1966). *J. Biol. Chem.* **241**, 4793–4797.
Ozols, J. and Strittmatter, P. (1967). *Proc. Nat. Acad. Sci. U.S.* **58**, 264–267.

Ozols, J. and Strittmatter, P. (1968a). *J. Biol. Chem.* **243**, 3367–3381.
Ozols, J. and Strittmatter, P. (1968b). "Osaka", pp. 576–580.
Ozols, J. and Strittmatter, P. (1969). *J. Biol. Chem.* **244**, 6617–6619.
Passon, P. G. and Hultquist, D. E. (1970). *Fed. Proc.* **29**, Abstr. 2751.
Petragnani, N., Nogueira, O. C. and Raw, I. (1959). *Nature* **184**, 1651.
Poltoratsky-Bois, R. and Chaix, P. (1966). *Bull. Soc. Chim. Biol.* **48**, 449–452.
Raw, I. and Colli, W. (1959). *Nature* **184**, 1798–1799.
Raw, I. and Mahler, H. R. (1959). *J. Biol. Chem.* **234**, 1867–1873.
Raw, I., Molinari, R., do Amaral, D. F. and Mahler, H. R. (1959). *J. Biol. Chem.* **233**, 225–229.
Raw, I., Petragnani, N. and Nogueira, O. C. (1960). *J. Biol. Chem.* **235**, 1517–1520.
Sanbon, R. C. and Williams, C. M. (1950). *J. Gen. Physiol.* **33**, 579–588.
Schmeling, U., Mayer, G., Diehl, H., Ullrich, V. and Staudinger, Hj. (1969). *Z. physiol. Chem.* **350**, 349–350.
Shinagawa, Y., Tanaka, C., Teroaka, A. and Shinagawa, Y. (1966). *J. Biochem.* **59**, 622–624.
Strittmatter, C. F. (1961). "Haematin Enzymes", Vol. **2**, pp. 461–470.
Strittmatter, C. F. and Ball, A. G. (1952). *Proc. Nat. Acad. Sci. U.S.* **38**, 19–25.
Strittmatter, P. (1960). *J. Biol. Chem.* **235**, 2492–2497.
Strittmatter, P. (1963). In "The Enzymes" (P. D. Boyer, H. Lardy, K. Myrbäck, eds.), Vol. **8**, pp. 113–145. Academic Press, New York and London.
Strittmatter, P. (1964). *J. Biol. Chem.* **239**, 3043–3050.
Strittmatter, P. (1967). "Methods, in Enzymol.", Vol. **10**, pp. 553–556.
Strittmatter, P. and Huntley, T. E. (1970). 8th Internat. Congress Biochemistry, Switzerland. Sympos. **1**, 21–22.
Strittmatter, P. and Ozols, J. (1966). *J. Biol. Chem.* **241**, 4782–4792.
Strittmatter, P. and Velick, S. F. (1956). *J. Biol. Chem.* **221**, 253–275.
Tsugita, A., Kobayashi, M., Kajihara, T. and Hagihara, B. (1968). *J. Biochem.* **64**, 727–730.
Tsugita, A., Kobayashi, M., Tani, S., Kyo, S., Rashid, M. A., Yoshida, Y., Kagihara, T. and Hagihara, B. (1970). *Proc. Nat. Acad. Sci. U.S.* **67**, 442–447.
Weber, H., Weiss, W. and Staudinger, Hj. (1971). *Z. physiol. Chem.* **352**, 109–110.
Yoshikawa, H. (1959). *J. Chem.* (Tokyo) **38**, 1.

D. CYTOCHROME P-450

1. Discovery and occurrence

The presence of a pigment reacting with carbon monoxide and accompanying cytochrome b_5 in microsomes was discovered by G. R. Williams in 1955 (see Cooper *et al.*, 1965a) and Garfinkel (1958) in the Johnson Foundation and by Klingenberg (1958). Sato *et al.* (1965) therefore suggested the name "reticulochrome", but P-450 is not restricted to microsomal membranes. It is characterized by an unusual Soret band of the carbon monoxide compound at 450 nm, hence the name P-450. It was further described by Sato and Omura (Sato and Omura, 1961; Omura and Sato, 1962, 1964) who established the protohaem nature of the prosthetic group, after Klingenberg had already shown that it gave with pyridine a haemochrome spectroscopically indistinguishable from protohaemochrome. The presence of P-450 in microsomes explains why cytochrome b_5 forms only the smaller part of microsomal protohaem, which corresponds to their sum, only little haemoglobin being present (Sato et al., 1965). Cytochrome P-450 was also found in the mitochondria of rat adrenals by Harding et al. (1964, 1966), and in mitochondria from the bovine adrenal cortex by Kinoshita et al. (1966), Cammer and Estabrook (1967), Whysner and Harding (1968) and Oldham et al. (1968). Although these mitochondria contain cytochrome aa_3, the concentration of P-450 is considerably stronger. P-450 like cytochrome oxidase is found in the inner mitochondrial membrane of the adrenal cortex (Satre et al., 1969). Yohro and Horie (1967) found it also in the mitochondria of ovaries. In the adrenal mitochondria, the locus of P-450 is in the inner membrane (Sottocasa and Sandri, 1970; Satre et al., 1969; Yago et al., 1970). This P-450 differs from that of the adrenal microsomes (Sweat et al., 1970).

P-450 also occurs in *Saccharomyces cerevisiae* (Lindenmeyer and Smith, 1964; Smith, 1965; Ishidate et al., 1969). A soluble P-450 has been found in *Nitrosomonas europaea* (Rees and Nason, 1965), in *Pseudomonas putida* (Katagiri et al., 1968) and in the bacteroids of *Rhizobium japonicum* (Appleby, 1968, 1969). The latter has been partially purified.

The biological function of "mixed function oxidases", or hydroxylases or monooxygenases (Hayano, 1962), hydroxylating various sterols, drugs, aromatic hydrocarbons, and aliphatic hydrocarbons and carboxylic acids (Mason, 1957; Hayano, 1962; Mason et al., 1965a; Staudinger et al., 1965) was known before the role of cytochrome P-450 in these reactions was demonstrated. In these reactions, in which NADPH is the reductant (Cammer and Estabrook, 1967; Simpson and Estabrook, 1968), P-450 functions as terminal oxidase. Its function is, however, not so much general energy provision, as specific biochemical reactions, as in the metabolism of sterols and steroid hormones (Estabrook et al., 1963) and drug detoxication (Cooper et al., 1965b). Even in microsomes, cytochrome b_5 is of greater importance for energy provision than P-450 (Cooper et al., 1965a) and cytochrome b_5 does not directly interact with P-450.

2. Purification of cytochrome P-450

Cytochrome P-450 is readily separated from cytochrome b_5 (Sato et al., 1965; Imai and Sato, 1968a) but is easily modified to another compound, P-420, which spectrally resembles other cytochromes b, but is autoxidizable and readily reacts with CO. Cytochrome b_5 can be solubilized with pancreatic lipase, while cytochrome P-450 remains attached to particles. Initially this was partially converted to P-420, and this alteration is irreversible (Sato) or at least only partly reversible (Ichikawa et al., 1969). However, more recently P-450 almost completely free from cytochrome b_5 and P-420 has been obtained (Nishibayashi et al., 1968; Ichikawa and Yamano,

1970). Maclennan et al. (1967) achieved this by tertiary amylalcohol treatment, followed by cholate and ammonium sulphate extraction of beef liver microsomes. The induction of P-450 synthesis by phenobarbital (Remmer et al., 1967) has been used by Orrenius and Ernster (1964) and Kinoshita and Horie (1967). Miyake et al. (1968a) and Jefcoate et al. (1969) used a non-ionic detergent for extraction. Finally, sonicated adrenal mitochondrial particles (Cooper et al., 1967; Omura et al., 1966; Horie et al., 1966; and Cooper, cf. Nishibayashi et al., 1968) contain little cytochrome b_5 and are a good starting material. The P-450 from *Pseudomonas putida* has been obtained crystalline (Yu and Gunsalus, 1970).

3. Spectroscopic properties

The earlier studies on the spectra of microsomal pigments with particular reference to P-450 have been summarized by Cooper et al. (1965a) and Sato et al. (1965). The absorption spectra of derivatives of purified P-450 have been studied by Horie et al. (1966), Miyake et al. (1968a,b; 1969), Ullrich et al. (1968), Nishibayashi and Sato (1967, 1968) and Nishibayashi et al. (1968), and those of the isocyanide compounds by Nishibayashi et al. (1966) and Imai and Sato (1968a). In contrast to the two forms of the isocyanide complexes, cyanide gives only the normal spectrum of b-type cytochromes (Miyake et al., 1969). The data given in Table IV

TABLE IV

Absorption maxima in nm (ϵ mM in parentheses) of cytochrome P-450

Compound	
Fe^{2+}	555(11·7) 420(85)
$\Delta(Fe^{2+} - Fe^{3+})$	$+553 - 557 + 433 - 413$
Fe^{3+}	650(2·4) 570(7·3) 535(8·5) 415(108)
$Fe^{2+}CO$	555(11·4) 449 (114)
	$\Delta 449 - 480$ nm (91)
$Fe^{2+}EtNC$	455 and 430[1]
$Fe^{3+}NO$	575 543 436
$\Delta(Fe^{2+}NO - Fe^{2+})$	$+585 - 422$

[1] Depending on conditions, e.g., pH, see below.

are mostly those given by Nishibayashi et al. (1968) while some data on the $\Delta Fe^{2+} - Fe^{3+}$ and the $Fe^{3+}NO$ and the $\Delta(Fe^{2+}NO - Fe^{2+})$ spectra are those given by Ullrich et al. (1968).

A new form with absorption bands at 440 nm has been observed by Estabrook et al. (1970) in the steady state of hydroxylation of type I substrates (see below) together with a free radical signal at g = 2·00 (O^{2-}?).

The absorption spectra of *Rhizobium* P-450 are similar, but not the same, e.g. the spectrum of Fe^{2+} is 541, 408 (116).

The absorption maximum of the CO-compound is also found as the maximum of action spectrum of microsomal 17-hydroxy-progesterone oxidation (Estabrook *et al.*, 1963) and of hydroxylating drug demethylation (Estabrook, 1965; Cooper *et al.*, 1965b; 1967; Wilson and Harding, 1970). The partition coefficient K_{O_2}/K_{CO} which is 0·02 for haemoglobin and 10 for cytochrome a_3 is about 1 for P-450 (Estabrook, 1965). However, Wilson and Harding (1970) report a distribution coefficient of 17 for total P-450 (spectroscopically) and 5 for the 11 β-hydroxylase system (cf. also Wada *et al.*, 1969a and Ichikawa *et al.*, 1967). These differences and also a shoulder at 420 nm disappear when extracts of acetone powders of bovine adrenal cortical mitochondria are studied, indicating a multiplicity of P-450 species in adrenal mitochondria, with the 11 β-hydroxylase being the least CO-sensitive one. The light-sensitivity of the P-450 CO compound is only about 1/100 of that of the cytochrome aa_3 oxidase. For the *Rhizobium* enzyme, Appleby (1968) found an affinity for CO of 0·3 μM. The K_m of oxidation by oxygen was found 0·1 μM by Estabrook (1965) and 2 μM by Ullrich *et al.* (1968) for the microsomes enzyme. One mole of P-450 is oxidized by 1 mole of O_2. The enzyme is inhibited by CO, but not by cyanide or azide. Ethyl isocyanide combines to give two interconvertible forms, the ratio of which is e.g. altered by pH, a low pH favouring the form with a maximum at 455 nm, a high pH that with a maximum at 430 nm (Imai and Sato, 1968b). The ethyl isocyanide complex of the high pH is spectroscopically similar to that of P-420, but assumed to be different from it (see below). The spectrum of the nitrosobenzene compounds has also been described (Ichikawa and Yamano, 1970).

P-420. By a variety of reagents, e.g. deoxycholate, P-450 is gradually changed into P-420 which has been purified by ammonium sulphate fractionation, and filtration through calcium phosphate and Sephadex gels (Sato *et al.*, 1965, Ichikawa *et al.*, 1968). This is nearly homogeneous in the ultracentrifuge, with a molecular weight of about 150,000 (monomer at pH 9·5). The absorption maxima of its ferrous form at 559, 530 and 527 nm (ε mM 149) are those of other cytochromes *b*, but it is still able to react with CO giving a higher absorption (ϵ_{mM}^{421} 213) and is readily autoxidizable. Its oxidation-reduction potential was found to be −20 mV at pH 7. The preparation still contained 1–2 atoms of non-haem iron per mole of protohaem. The "S (stroma)"-haemoprotein of erythrocytes is probably a P-420, formed by the breakdown of P-450 from immature red cells (Hultquist *et al.*, 1971).

The microsomal Fe_x. By EPR studies Hashimoto *et al.* (1962) and Mason *et al.* (1965b) discovered a low spin ferric haemoprotein in rabbit liver microsomes which they called Fe_x. Its major EPR bands are at $g = 2\cdot41$, $2\cdot25$ and $1\cdot94$ (cf. also Oldham et al., 1968; Schleyer et al., 1970; Whysner et al., 1970). Autoxidation of its ferrous form gives a radical signal $g = 2\cdot003$. It has now been established that Fe_x is the low spin form of cytochrome P-450 (Wada et al., 1964; Miyake et al., 1968a,b; Peisach and Blumberg, 1970).

Jefcoate and Gaylor (1969a) observed an EPR band at $g = 6\cdot6$ which may be the high spin form which had not been found by other workers even when the high spin type was predominant (Katagiri et al., 1968), although a weak $g = 6$ band had been described by Hildebrandt et al. (1968). Partial alterations into a high-spin form by combination with substrate has also been found by Whysner et al. (1970) for the liver P-450 and by Tsai et al. (1970) for the bacterial camphor hydroxylase.

4. Spectral alterations caused by combination of cytochrome P-450 with substrates of hydroxylation reactions

Cytochrome P-450 catalyses hydroxylation reactions of steroids, drugs and other compounds (see Chapter IV, E5). It has been found that P-450 combines with these compounds causing frequently spectral effects. These are of two main types. In type I there is a minimum at about 420 nm and a maximum at 385–395 nm; while in type II there is a maximum at about 430 nm and a minimum at 390–395 nm. $CO + Na_2S_2O_4$ prevents the binding and release bound substrate (Orrenius and Ernster, sterone (McIntosh and Salhanick, 1969; Sweat et al., 1969a,b), progesterone, cholesterol (Young et al., 1970a) and several drugs, hexobarbital (Sies and Brauser, 1970), phenobarbital, and aminopyrine (Remmer et al., 1966; Schenkman et al., 1967; but see Flynn et al., 1970), although under certain conditions phenobarbital can cause a type II change (Imai and Sato, 1967; Jefcoate et al., 1969); unsaturated fatty acids (Di Augustine and Fouts, 1969); ethyl-morphine (Gigon et al., 1968; Thorgeirsson and Davies, 1971); phenacetine (Imai and Sato, 1967); benzphetamine (Lu et al., 1969b) and 3,4-benzpyrene (Schenkman, 1970b), the latter being accompanied in the induction with a type II change. It is remarkable that CCl_4 should cause a type I spectral change (McLean, 1969) reversed by aniline (type II). Type II substrates are aniline (Imai and Sato, 1967), acetanilide (Leibman et al., 1969), methylaniline, phenylhydrazine, 3-methylcholanthrene (Hildebrandt et al., 1968), aliphatic alcohols (Jefcoate et al., 1969); of sterols estradiol, 17-hydroxy pregnenolone and 20α-hydroxycholesterol (Whysner and Harding, 1968). "Metopirone" (Williamson and O'Donnell, 1969) or "metyrapone" (Wilson et al., 1969; Colby and Brownie, 1970; Sies and Brauser, 1970) and related drugs produce no spectral change but shift the EPR from $g = 2\cdot42$ to $g = 2\cdot47$

(Wilson et al., 1969), stimulate type II hydroxylation of acetanilide and inhibit hydroxylation of type I substrates (McIntosh and Salhanick, 1969; Hildebrandt et al., 1969; Wilson et al., 1969; Colby and Brownie, 1970). It displaces deoxycorticosterone (Williamson and O'Donnell, 1969; Sweat et al., 1969b). In this it resembles type II substrates like aniline (Leibman et al., 1969) which is a competitive inhibitor of the phenobarbital-caused spectral change. These authors found no inhibition between type II and other type II substrates, but competitive inhibition between type I and other type I substrates. These results contradict, however, those of Mitani and Horie (1969). Both types of substrates protect P-450 against destruction and react with the ferric forms of P-450 (Schenkman et al., 1969). Type I substrates accelerate (Kupfer and Orrenius, 1970), type II substrates decelerate the reduction to the ferrous form; type I substrates do not affect re-oxidation while type II substrates may accelerate it (Gigon et al., 1968; Gillette and Sasame, 1970). Schenkman (1970b) has recently adduced evidence that most type II changes are accompanied by type I changes and become enlarged and symmetrical if corrected for the type I changes. Ligation of the bile duct impairs binding of substrates of type I and their metabolism, but not binding or metabolism of type II substrates (Hutterer et al., 1970).

It has been debated whether the existence of these two types of spectral alteration by the various substrates can be explained on the basis of a single cytochrome P-450 with two, a low spin and a high spin form of the enzyme, the proportion of which can be altered by the substrate, as was shown for the ethyl isocyanide complex, or whether it is necessary to postulate two different cytochromes P-450. While the majority of workers support the first hypothesis (Remmer et al., 1966; Hildebrandt et al., 1968; Jefcoate and Gaylor, 1969a; Leibman et al., 1969; Mitani and Horie, 1969; Williamson and O'Donnell, 1969; Kupfer and Orrenius, 1970), Staudinger et al. (1965) and Sladek and Mannering (1966) favour the second. There does not appear to be agreement on whether type I or type II spectral alterations favour the high spin form; according to Jefcoate and Gaylor (1969a) and Mitani and Horie (1969) this is type II (methyl-cholanthrene or pregnenolone resp.), but Whysner et al. (1969) report opposite findings, while Hildebrandt et al. (1968) find aniline, a type II substrate, to favour the formation of the low spin form in methyl-cholanthrene-induced P-450.

Staudinger et al. (1965) find two different forms of P-450, one with high O_2 affinity (K_m 2×10^{-6}), more high spin, insensitive to cyanide and less inhibited by CO, inhibited by EDTA, but not by copper reagents; and a second form of lower O_2 affinity (K_m $1 \cdot 5 \times 10^{-4}$ M), predominantly low spin, inhibited by Cu reagents, but not by EDTA, more sensitive to CO and cyanide. Levin and Kuntzman (1969) adduce evidence for a biphasic decrease of radioactivity of P-450, both in control and after methylcholanthrene induction in rats, the more stable form being the one induced by methylcholanthrene. The Soret band of the carbon monoxide is also shifted by methylcholanthrene or benzanthracene from 450 to 446–448 nm (Nebert and Gelboin, 1968; Alvares et al., 1967; Kuntzman et al., 1968; Schenkman et al., 1969; Kuntzman, 1969; Nebert, 1969; Glaumann et al., 1969). A difference in spin form can hardly be the correct explanation of this difference, which may either be due to an allosteric binding of the substrate (Wilson et al., 1969) or to the induction of the synthesis of a different protein. P-446–448 has been termed P_1-450 by Bidleman and Mannering (1970). The fact that its maximum is not shifted back to 450 nm by removal of benzpyrene (Nebert, 1970) and that the maximum is not shifted in the absence of protein synthesis supports the second alternative. Nebert also found that "P_1-450" is less readily converted to P-420 than P-450.

There is now a great deal of evidence for multiple forms of P-450 so that this name means rather a class of cytochromes or cytochromoids rather than a single cytochrome (Nishibayashi et al., 1966; Imai and Sato, 1966; Sladek and Mannering, 1966; Alvares et al., 1967; Coney, 1967; Schenkman et al., 1969; Glaumann et al., 1969; Kato and Takanaka, 1969). Recent work has produced further evidence for multiplicity.* Wada et al. (1969b) found steroid 7-d-hydroxylase activity not to parallel the whole P-450 content. Phenobarbital probably induces two different types of P-450 (Schenkman, 1970b). There can now be little doubt that there are at least two types of steroid hydroxylases in adrenocortical mitochondria, 11β-hydroxylase and cholesterol-side chains splitting enzyme or enzymes (Jefcoate et al., 1970; Peisach and Blumberg, 1970; Schenkman, 1970a; Wilson and Harding, 1970; Young et al., 1970a). They differ in stability to acetone treatment of the mitochondria, affinity to CO, O_2/CO distribution coefficients and light sensitivity of the CO-compound, and in rate of reduction. P-420 from normal and methylcholanthrene induced rats differ (Shoeman et al., 1970), ϵ mM of the P-420 from normal rats being 110, from methylcholanthrene-treated 134. Acetanilide hydroxylases in liver microsomes from normal and methyl-cholanthrene-treated rats differ in the retention of deuterium in the hydroxylation reaction (Daly et al., 1969). In the rat kidney cortex microsomes fatty acids induce a P-454 which does not hydroxylate testosterone. Laurate causes a type I spectral change in both kidney cortex and liver microsomes, but testosterone produces a type II spectral change only in the kidney microsomes (Jacobson et al., 1970). Thus the enzymes in kidney and liver microsomes differ. The P-450

---

* The P_1-450 induced by polycyclic aromatic hydrocarbons differs from P-450 in substrate specificity, spectrum with ethyl isocyanide and drug binding properties (Parli and Mannering, 1970; Shoeman et al. (1969).

obtained from microsomes of normal and DDT-resistant flies differ in activity with different substrates and in the ethyl isocyanide spectra (Matthews and Casida, 1970, cf. also Perry and Buchner, 1970). The bacterial P-450$_{Cam}$ undoubtedly is a distinct cytochrome.

5. The role of cytochrome P-450 in sterol and drug metabolism

The purpose of this subchapter cannot be more than a brief survey of this subject and a guide to the vast and quite recent literature. Not only biochemists, but also endocrinologists interested in hormone metabolism and pharmacologists interested in drug metabolism have contributed to this literature. Perhaps it is worth while mentioning that studies such as these have also provided a greater insight in the general problems of relation between structure and function of drugs than was possible before the specific enzyme systems were known.

The role of "mixed function oxygenases" or "hydroxylases" in sterol metabolism was reviewed by Hayano (1962) before the role of P-450 was clearly defined. P-450 plays a role in both cholesterol synthesis and decomposition to sterol hormones and finally to bile acids. While there is good evidence that it is required for the transformation of squalene to cholesterol (Boyd, 1969; Wada et al., 1969a) the specific locus of its action is still unknown. Several hydroxylases are active in cholesterol breakdown. A 7α-hydroxylase converts cholesterol to cholest-5-ene-3β, 7γ-diol (Danielsson and Einarsson, 1966; Bryson and Sweat, 1968; Scholan and Boyd, 1968; Wada et al., 1968b; Boyd, 1969). A 6β-hydroxylase of taurochenodeoxycholate is also increased by phenobarbital induction (Voigt et al., 1970). A cholesterol-20α-hydroxylase (Simpson and Boyd, 1967; Meigs and Ryan, 1968; Boyd, 1969) occurs in the human placenta and in hog adrenal mitochondria (Ichii et al., 1967). 20α-hydroxycholesterol combines with P-450 (Whysner and Harding, 1968) and is further hydroxylated at C_{22} by a 22-ε-hydroxylase (Whysner et al., 1969) leading to the pregnenolones (see also Van Lier and Smith, 1970; Wilson and Harding, 1970; Young and Hall, 1969). The cholesterol side-chain-cleavage enzymes of adrenal mitochondria (EC 1.14.1.8) depends on a P-450 which is more sensitive to acetone treatment than the 11-β-hydroxylase (Young and Hall, 1968; Young et al., 1970b). In the adrenals, 11-β-hydroxylase (EC 1.14.1.6) converts deoxycorticosterone to cortisone (Omura et al., 1965; Cooper et al., 1965c; Mitani and Horie, 1969; Sih, 1969); 11-deoxycortisol is also oxidized (Oldham et al., 1968). Deoxycorticosterone and 11-deoxycortisol combine with P-450 (Whysner and Harding, 1968); cortisol is also oxidized (Burstein, 1967). Adrenal mitochondria have two cytochrome chains, the one with cytochrome aa_3, the other with cytochrome P-450 as terminal oxidase, which compete with one another (Cammer and Estabrook, 1967). Thus deoxycorticosterone inhibits ATP formation in adrenal mitochondria (Cooper and Thomas, 1970). Estabrook *et al.* (1963) identified the C_{21}-hydroxylase of Ryan and Engel (1957) (see also Young *et al.* 1970a) with P-450 (but cf. Matthysen and Mandel, 1970). This hydroxylates progesterone to deoxycorticosterone, 17-hydroxyprogesterone to cortexolone. Sweat *et al.* (1969a) found that adrenodoxin, an NHI-protein, inhibits the 20-22 hydroxylation, but not the 11-β-hydroxylation (see also Jefcoate *et al.*, 1970 and Sauer and Mulrow, 1969). P-450 is also involved in the C_{18}-hydroxylation of corticosterone and its conversion to aldosterone (Greengard *et al.*, 1967; Ruhmann-Wennhold *et al.*, 1970). There is, as yet, no evidence for a C_{17}-hydroxylase which could convert C_{21}-steroids to androgens and oestrogens. P-450 is involved in the oxidation of testosterone in 6β, 7α and 16 (Conney *et al.*, 1968; see also Orrenius *et al.*, 1969), but not in the aromatization of androst-4-ene-3,17-dione (Narasimhulu *et al.*, 1965; Meigs and Ryan, 1968), nor in the demethylation of 4-methyl sterols (Gaylor and Mason, 1968). Androgens inhibit 11-β-hydroxylation and formation of corticosteroids (Brownie *et al.*, 1970). Testosterone binds P-450 and lowers its level (Rushmann-Wennhold *et al.*, 1970).

A great variety of aliphatic and aromatic compounds including many drugs are substrates of hydroxylations catalysed by the P-450 system (Gillette, 1966; Conney, 1967). Among these are aliphatic hydrocarbons (Cardine and Jurtschuk, 1968; Frommer *et al.*, 1970), e.g. n-octane hydroxylation; or that of n-bromopentane (Grasse *et al.*, 1970); cyclohexane (Diehl *et al.*, 1969); polycyclic aromatic hydrocarbons, e.g. 3-methylcholanthrene (Conney and Gilman, 1963; Orrenius, 1965; Sladek and Mannering, 1966; Jefcoate and Gaylor, 1969a) or benzanthracene and benzpyrene (Baron and Tephley, 1969; Nebert, 1969, 1970; safrole (Parke and Rabman, 1970)); aliphatic fatty acids undergoing ω-hydroxylation (Wada *et al.*, 1967, 1968a,b,c; Lu and Coon, 1968; Lu *et al.*, 1969a,b; Orrenius, 1969); barbiturates (Kato *et al.*, 1962; Remmer and Merker, 1963; Orrenius and Ernster, 1969; Degwitz *et al.*, 1969; Garner and McLean, 1969); aniline and acetanilide (Estabrook, 1965; Staudinger *et al.*, 1965; Mason *et al.*, 1965a,b); aminophenols (Staudinger *et al.*, 1965); naphthoquinones (Sato *et al.*, 1965); DDT (Remmer *et al.*, 1967; Bunyan *et al.*, 1970) and coumarin (Kratz and Staudinger, 1968). Bacterial P-450 of *Pseudomonas putida* acts as methylene hydroxylase of camphor (Gunsalus *et al.*, 1965; Katagiri *et al.*, 1968). Drugs like metopirone (2,3-bis(3-pyridyl)-2-methylpropan-3-one) competitively inhibit 11β-steroid hydroxylation (Williamson and O'Donnell, 1969; Hildebrandt *et al.*, 1969; Wilson *et al.*, 1969). Both ^{14}C-labelled phenobarbital and aniline have been shown to be bound to liver microsomes. Another important field of drug detoxication is that of oxidative demethylation of N-, S- and O-methyl compounds. This has recently been briefly reviewed by Davies (1969). Substrates are codeine and monomethylaminopyrine (Estabrook et al., 1963; Orrenius et al., 1964; Estabrook, 1965; Sladek and Mannering, 1966) and benzphetamine (Lu et al., 1969b) or N-methylhydrazines (Wittkop and Brough, 1969). Oxidative demethylation by P-450 inhibits lipid peroxidation (Orrenius and Ernster, 1964). In the developing avian liver the increase of oxidative N-demethylation by phenobarbital paralleled that of P-450 (Strittmatter and Umberger, 1969). It is of interest that carbon tetrachloride, a well-known liver poison, inhibits oxidative demethylation (Orrenius and Ernster, 1964; Ernster and Orrenius, 1965; McLean, 1969; Smuckler et al., 1970). The synthesis of ergot alkaloids in *Claviceps purpurea* is stimulated by P-450 (Ambike et al., 1970). The P-450 action on some aromatic hydrocarbons such as benz[a]anthracene may be a protective mechanism against their carcinogenic action (Nebert and Gelboin, 1968).

6. Induction and breakdown of cytochrome P-450

Cytochrome P-450 is an inducible enzyme, its formation in animal tissues being induced by phenobarbital (Remmer et al., 1967; Garner and McLean, 1969; Kuntzmann et al., 1968) and by 3-methylcholanthrene and other polycyclic hydrocarbons (Conney and Gilman, 1963; Sladek and Mannering, 1966; Long, 1969; Jefcoate and Gaylor, 1969a; Nebert, 1969). The increase is a true increase rather than decreased breakdown (Garner and McLennan, 1969; Greim et al., 1970; Orrenius, 1969; Orrenius et al., 1969). It requires protein synthesis (Orrenius et al., 1965; Marshall and McLean, 1968, 1969) and leads to the formation of specific apoproteins of P-450 depending on the inducing substrate (Sladek and Mannering, 1966; Long, 1969). This is of importance in so far as the dependence of induction on protein synthesis might have been due to more indirect causes. Thus induction of δ-ALA-synthetase is required for the increase of haem synthesis (Baron and Tepley, 1969a,b; 1970) and the label of labelled δ-ALA is found in P-450 (Levin, Kuntzman et al., 1970; cf. also Held and von Oldershausen, 1966). Haematin prevents as repressor, the induction of microsomal protein synthesis by drugs (Marver et al., 1968). In yeast, P-450 decreases with aerobiosis and mitochondrial biogenesis and cytochrome formation (Ishidate et al., 1969). Allylisopropyl-acetamide (De Matteis, 1970) and a dihydrocollidine derivative (Wada et al., 1968; Watersfield et al., 1969) can decrease or increase P-450 synthesis; both are porphyrinogenic agents, probably dependent on inhibition of haem synthesis in the liver and feedback on δ-ALA-synthetase.

An inducible enzyme must, on the other hand, also undergo rapid breakdown in the absence of an inducer. This has been shown for P-450 by the findings of Schmid et al. (Schmid et al., 1966; Robinson et al.,

1966; Robinson, 1969), that the major source of the formation of "early labelled bile pigment" is cytochrome P-450 breakdown in the liver. The breakdown of P-450 has been found to be biphasic, the fast phase having a half-time of 7–8 hours, the slow one of 47 hours, independent of whether induction had been caused by phenobarbital or methylcholanthrene (Levin and Kuntzman, 1969), the slow phase, however, being more abundant after methylcholanthrene induction. An apocytochrome P-450 is present in the microsomes (Levin et al., 1970a).

The chemical structure of the enzyme and its relation to its function is still poorly elucidated. The "abnormal" CO-reaction giving a Soret band at 450 nm rather than at 420 nm, is perhaps not quite so anomalous as believed. Jefcoate and Gaylor (1969b) have confirmed the early findings of Holden and Lemberg (1939) that under certain conditions (low pyridine concentration) the Soret band of a CO-pyridine protohaemochrome is found at 440 nm rather than at 418 nm, and have ascribed this and similar band positions with butylamine and imidazole to an association of the haem to a dimer or polymer which is long known for solutions of protohaem and protohaematin itself (Lemberg and Legge, 1949). From model experiments of protohaem-ethyl isocyanide compounds, Imai and Sato (1968b) have also concluded that one of its two bands, at 428 and 455 nm, the latter increased by higher concentration of protohaem and low pH, is due to aggregation. Although the pH effect is the reverse with P-450-ethyl isocyanide compounds they assumed a reversible conversion of a conformational form of P-450 in which two haem groups were able to interact with one another to another conformational form in which each haem could interact with the ligand separately. This form was assumed to be similar to but different from P-420, in which latter the hydrophobic linkages maintaining the special state of P-450 had been more extensively and irreparably destroyed. It is, however, unlikely that cytochrome P-450 contains two haems in one molecule. Ichikawa et al. (1969) drew similar conclusions from the conversion of P-450 to P-420. Protohaem is, indeed, readily removed from P-450 by haem-combining proteins such as apomyoglobin and haemopexin, at least from bacterial P-450 (Muller-Eberhardt et al., 1969). The explanation of the P-450 to P-420 conversion as due to a loss of protein linkages, possibly RSH groups from the β-positions of a pyrrole ring and, thus converting from a d-type to a b-type cytochrome (Williams, 1968) appears less attractive. Nor is there good evidence for a haem-CO-Cu type of linkage in CO-P-450. Mason et al. (1965a,b) observed that copper chelators accelerate the conversion of P-450 to P-420 and therefore assumed that Cu plays a role in the binding of CO. Appleby (1968) also postulated a role of copper in holding one mole of carbon monoxide between two haem groups. However, MacLennan et al. (1967) found that traces of copper in their enzyme preparation could be removed by cyanide and

EDTA without affecting the CO-reaction, while Staudinger et al. (1965) found that only one of two P-450 oxidases could be inhibited by Cu-reagents. Mason et al. (1965b) found strong inhibition of acetanilide hydroxylation by p-chloromercuribenzoate or mersalyl, and could reverse this inhibition by reduced glutathione. Iodine converts P-450 more rapidly and completely to P-420 than mercurials (Ullrich et al., 1968). Jefcoate and Gaylor (1969a) postulated a linkage of the protohaem between a histidine imidazole and an SH-compound, possibly methionine, assuming that the Fe-S linkage was split on reduction and provided an open cleft for haem-haem interactions to occur and for oxygen to enter. However, a methionine-sulphur linkage is also present in cytochrome c, it is not split by reduction, and cytochrome c is not autoxidizable. P-450, like cytochrome o (see below) should be classified as cytochrome b' rather than cytochrome b (see Chapter II). Roeder and Bayer (1969) have obtained haematin-mercaptide complexes the EPR values of which are similar to those of oxidized P-450 (see also Schenkman and Sato, 1968, and Ullrich, 1969).

The redox-potential of P-450 is extraordinarily low, about -410 mV (Schenkman, 1970a; Schleyer et al., 1970; Waterman and Mason, 1970; Cooper et al., 1970), not only in the adrenal 11β-hydroxylase which contains adrenodoxin, an NHI-protein, but also in the liver microsomal enzyme which does not. Schleyer et al. (1970) have resolved the enzyme system into P-450, a specific flavoprotein and NHI iron-sulphur protein which are all essential for the activity. Even in the presence of excess NADPH this P-450 is only 20% reduced. Phospholipid is essential for the activity of the microsomal liver P-450 (Strobel et al., 1970). The low potential is difficult to harmonize with its function as a terminal oxidase in the hydroxylation system.

Hill et al. (1969) assume that the curious Δ spectra ($Fe^{2+} - Fe^{3+}$) of P-450, according to Schleyer et al. 1970, Fe^{3+} maximum 416 nm, Fe^{2+} 419 nm are due to an unusual spin state and the weakening or breaking of Fe-S bonds on reduction. ESR spectra show that P-450 is the most rhombically distorted of all haem compounds (Peisach and Blumberg, 1970; cf. also Tsai et al., 1970). Substrates of 11β-hydroxylation induce EPR changes which show a decrease of the predominantly low spin form ($g = 2\cdot42$, $2\cdot24$, $1\cdot91$) and a rise of one signal ($g = 7\cdot9$), not the other ($g = 6$) of the high spin form (Schleyer et al., 1970; Whysner et al., 1970).

There is not much more than speculations on the binding site of the haem iron, although Hlavica (1970) agrees with Jefcoate and Gaylor (1969a) and Hill et al. (1969) that haem is liganded either to histidine or to a cysteine SH-group. Extrinsic Cotton effects have been found only in the ferric or the FeCO form, but not in the ferrous form, or in any form of P-420 (Yong et al., 1970). This supports the hypothesis that the functional activity of P-450 depends on a conformational change (see Chapter 7, VIII).

7. Reaction mechanism and chemical structure of cytochrome P-450

As a "mixed function" oxidase P-450 obeys the general formula of Mason (1957):

$$H^+ + AH_2 + NADPH + O_2 \rightarrow AHOH + NAPD^+ + H_2O.$$

Sih (1969) has written the results of Omura et al. (1965a,b), Kimura and Suzuki (1967) and Nakamura et al. (1966) on the hydroxylation reaction in the form

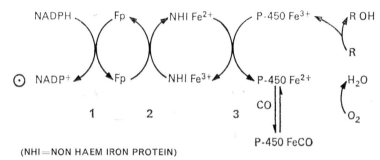

(NHI = NON HAEM IRON PROTEIN)

The NADPH-cyt P-450 reductase (Orrenius and Ernster, 1964) differs from the NADPH-cyt c-reductase of Kamin et al. (1965) (see also Lu et al., 1969b). He has, however, postulated that NADPH enters the reaction twice, once as reductant of ferric P-450 and then again transforming the initial oxygen adduct of ferrous P-450 into an active hydroperoxocomplex which then oxidized the bound substrate molecule after transformation into a ferryl form:

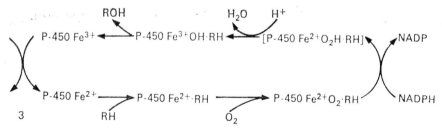

The postulated mechanism is hypothetical and none of the assumed intermediary compounds have been demonstrated. Cooper et al. (1970) have recently disputed that NADPH is a specific hydrogen donor for the 11-β-hydroxylation and have found that it can be replaced by dithionite, whereas a specific flavoprotein and non-haem-iron proteins are both essential.

It is well supported that cytochrome P-450 acts as a terminal oxidase, thus resembling cytochrome a_3 rather than "normal" cytochromes, and that both it and a soluble reducing system, which can be separated from it (Nakamura and Tamaoka, 1970) are essential for the hydroxylation system. The adrenal reducing system is specific and cannot be replaced by the reducing system of liver microsomes. Davies et al. (1969) have found the species differences in the demethylation to parallel differences in NADPH-cytochrome P-450 reductase more closely than those in NADPH-cytochrome c reductase or P-450 content.

An active oxygen complex of P-450 has been postulated by several investigators (cf. Davies et al., 1969). Estabrook et al. (1971) have recently found a spectroscopically different form of P-450 with a broad absorption maxima at 590 and 440 nm which is formed in liver microsomes on oxygenation of the reduced form in the presence of NADPH and substrate. It remains to be seen whether this is an oxygenated form, or like "oxygenated" cytochrome oxidase a conformationally different form of the enzyme. They have also observed the transformation of this form into a peroxidic-ferric form by one electron reduction by cytochrome $b_5{}^{2+}$; again its formulation is dubious. This compound differs from oxygenated bacterial P-450$_{\text{Cam}}$ (Ishimura et al., 1971).

P-450 and cytochrome o (see below), should be considered as cytochromes b' rather than cytochromes b, as cytochrome a_3 is now considered a cytochrome a' and some compounds related to cytochrome c as cytochromes c'. Later a separation of terminal oxidases, including all the a', b' and c' compounds, from typical cytochromes may become possible. Their mechanism may thus have some features in common with that of cytochrome a_3, e.g. the existence of states of a valency higher than ferric and of conformational changes in the oxygenation reaction.

8. Relation of P-450 to haemoglobin catabolism

In a recent paper, Tenhunen et al. (1969) have claimed that P-450 plays an essential role in the catabolism of haematins and haemoglobin to bile pigment, particularly in the spleen. It is difficult to harmonize this with the previous findings from Schmid's laboratory quoted above that the P-450 of the liver is rapidly catabolized and is the source of the "early-labelled" bile pigment, and that haematin is a repressor of microsomal cytochrome. This has been confirmed by Coburn (1970). It would be unusual and unlikely for a less stable haemoprotein to catalyse the breakdown of more stable ones such as haemoglobin or methaemalbumins. Schmid and co-workers (see Tenhunen et al., 1969) were led to their hypothesis by their findings that NADPH is an essential hydrogen donor in the biliverdin formation in the spleen, and that carbon monoxide, more than cyanide or azide inhibits the reaction. The first product of the reaction, assumed to be due to an α-methylene oxygenase, was shown to be biliverdin, reduced by a specific NADPH-dehydrogenase to bilirubin.

However, in a recent study of biliverdin formation in cell cultures of chicken blood macrophages, Nichol (1970) found that this reaction was not inhibited by carbon monoxide, nor by compounds such as phenobarbital acetanilide, type I and II inhibitors of cytochrome P-450. The fact that carbon monoxide is a stoichiometric product of haem breakdown is another, perhaps less important difficulty for a hypothesis assuming a role of P-450 in haemoglobin breakdown. Levitt et al. (1968) found that early bilirubin production is little influenced by increased hepatic porphyrin syntheses or by P-450 changes.

What has been discussed in the preceding subchapter allows us to harmonize the new observations with the now classical model of Lemberg and Legge (1949, Chapters 10 and 11) and Lemberg (1956), i.e. the coupled oxidation of haemoglobin or other haem compounds and hydrogen donor systems with haem itself acting as a catalyst of its own destruction, and without assuming the special role of cytochrome P-450 or another specific enzyme. The following features of bile pigment formation are no longer in doubt:

1. The reaction involves the initial hydroxylation of a methine bridge (Lemberg and Legge, 1949, p. 91, 92; Lemberg, 1956; Clezy and Nichol, 1965, Clezy et al., 1966; Jackson et al., 1965, 1966, 1968a,b; Bonnet et al., 1969); see also Morell and Nichol (1969) for the explanations of the difference between pseudohaem and oxyhaem). The reaction involves predominantly the IXα methine bridge between the vinyl-substituted pyrroles.

2. This is followed by the removal of this methine-carbon as carbon monoxide (Sjöstrand, 1951, 1952).

3. The primary product is biliverdin (Lemberg and Barcroft, 1932; Lemberg and Legge, 1949, Chapters 10 and 11; Wise, 1964; Wise and Drabkin, 1964, 1965; Tenhunen et al., 1968, 1969).

4. The reduction of biliverdin to bilirubin is by the action of dehydrogenase systems (Lemberg and Wyndham, 1963), of which NADPH-dehydrogenase (Tenhunen et al., 1970) and NADH-dehydrogenase (Wise and Drabkin, 1965) are probably the most important ones.

What is still uncertain is the nature of the haem compound predominant as primary source, the relative importance of enzymic or non-enzymic hydrogen donors, the cause of the stereospecificity of the reaction and the main locus of bile pigment formation in the organisms. These areas in doubt do not decrease the value of the model system of "coupled oxidation". It is the purpose of a model to reveal the essential chemical principle of a biological reaction without necessarily providing all the details; stereochemical specificity, e.g. may be superadded (see Bonnett and McDonagh, 1970). Moreover it appears unlikely that the details of the mechanism should be the same in bile pigment formation inside the red cell (Lemberg et al., 1941; Lemberg and Legge, 1942; Lemberg, 1949)*; for formation from circulating haemoglobin-haptoglobin and methaemalbumin; for formation in the dog's placenta (Lemberg and Barcroft, 1932), the haemophageous organ of Wise and Drabkin (1965); for its formation in the liver, not only from haemoglobin but also from catalase (Lemberg et al., 1939; Lemberg and Legge, 1943); for its formation in the spleen (Tenhunen et al., 1968 in microsomes; Gray and Nakajima, 1969 in mitochondria); and for its formation in macrophages (Nichol, 1970), and that differences in detail should be smaller than differences from the reaction in haemoglobin or methaemalbumin solution (Lemberg et al., 1938, 1939). These can all be considered as variations of one and the same theme.

The hypothesis of "coupled oxidation" has been attacked by Gray et al. (Gray et al., 1958; Petryka et al., 1962; Nakajima and Gray, 1967; by Tenhunen et al., 1968, 1970; and by Colleran and O'Carra (1969) (but see O'Carra and Colleran (1969)) on the basis that coupled oxidation of pyridine haemochrome with hydrazine and with ascorbate does not exclusively lead to the natural IXα-isomer of biliverdin. This has been confirmed by Morell and Nichol (1969) and Bonnett and McDonagh (1970). Recently it has been shown, however, by Nichol and Morell (1969) for the coupled oxidation of haemoglobin with ascorbate, and by O'Carra and Colleran (1969) for that of myoglobin, that the principal product of these reactions is, indeed, biliverdin IXα. The same probably holds for coupled oxidations of methaemalbumin; the stereospecificity of its reaction has, however, so far only been investigated in particulate systems (Wise and Drabkin, 1965; Snyder and Schmid, 1965; Tenhunen et al., 1969). It therefore appears likely that only the coupled oxidation of pyridine haemochrome has less stereospecificity than that of haemoglobin, and that the stereospecificity is caused by the protein moiety of the haemoproteins rather than by an enzymic mechanism. It must not be forgotten that the haemoproteins themselves resemble enzymes in their specificity, and that the stereodirective influence of their protein moiety would be expected, if only for reasons of propinquity, to be greater than that of any external enzyme.

The confusion caused by this lack of chemical understanding is revealed by the subsequent history of the studies on haem catabolism. Nakajima et al. (1963) and Nakajima and Gray (1967) claimed the presence of a specific "haem α-methenyl oxygenase" in guinea pig liver that catalysed the conversion of protohaemochrome into a formylbiliverdin yielding only biliverdin IXα on hydrolysis. The enzyme was supposed to be heat-labile,

---

* It appears e.g. likely that the small amount of biliverdin the initial formation of which cannot be prevented by the inhibition of macrophage activity is preformed biliverdin in the red cell.

active in 20% pyridine, and to attack also haemoglobin-haptoglobin and myoglobin, but not oxyhaemoglobin (Nakajima, 1964). The claims of Nakajima have now been refuted by several workers. Levin (1967) and Murphy et al. (1967) found no enzymic formation, but only non-enzymic formation of verdohaemochrome with a heat-stable, dialysable factor (cf. also Nakajima and Gray, 1969) present in liver and other extracts.

The evidence of Tenhunen et al. (1968, 1969, 1970) for a haem-α-methylene oxygenase in spleen microsomes depending on NADPH is more convincing, but open to a different interpretation. The reactions 1, 2, 3 of P-450 reductase (see the text figures in subchapter 7) may well be as postulated by Tenhunen et al., but the oxidative mechanism after reaction 4 may be autocatalytic in haem catabolism, replacing P-450 in their scheme by the haem-protein itself. We are then back to the mechanism of coupled oxidation. It is possible that as in the formula of Sih (1969) this involves two points of attack for NADPH, once for the primary reduction of the haem, and then again in the oxygen-activating mechanism; and one, or the other of these may be replaced by another hydrogen donor. In the reaction mechanism of the P-450 flavins have been postulated; in the haem reductase active in the macrophage bile pigment formation, flavoproteins are also involved as the mepacrin inhibition shows. It will be of great interest for further insights into both the mechanism of P-450 enzymic activity and that of haem catabolism to compare the two reactions further, and in particular the reactions occurring in the oxygen activation.

REFERENCES

Alvares, A. P., Schilling, G., Levin, W. and Kuntzman, R. (1967). *Biochem. Biophys. Res. Commun.* **29**, 521–526.
Ambike, S. H., Baker, R. M. and Zalind, N. D. (1970). *Phytochem.* **9**, 1953–1962.
Appleby, C. A. (1968). "Osaka", pp. 666–679.
Appleby, C. A. (1969). *Biochim. et Biophys. Acta* **172**, 71–87.
Baron, J. and Tephley, T. R. (1969). *Biochem. Biophys. Res. Commun.* **36**, 526–532; *Mol. Pharmacol.* **5**, 10–20.
Baron, J. and Tephley, T. R. (1970). *Arch. Biochem. Biophys.* **139**, 410–420.
Bidleman, K. and Mannering, G. J. (1970). *Mol. Pharm.* **6**, 697–701.
Bonnett, R. and McDonagh, A. F. (1970). *J. Chem. Soc.* D **327**, 238–239.
Bonnett, R., Dimsdale, M. J. and Stephenson, G. F. (1969). *J. Chem. Soc.* C, 564–570.
Boyd, G. S. (1969). *Biochem. J.* **115**, 24–25.
Brownie, A. C., Simpson, E. R., Skelton, F. R., Elliott, W. B. and Estabrook, R. W. (1970). *Arch. Biochem. Biophys.* **141**, 18–25.
Bryson, M. L. and Sweat, M. L. (1968). *J. Biol. Chem.* **243**, 2799–2804.
Bunyan, P. J., Taylor, A. and Towsend, M. G. (1970). *Biochem. J.* **118**, 51–52P.
Burstein, S. (1967). *Biochem. Biophys. Res. Commun.* **26**, 697–703.
Cammer, W. and Estabrook, R. W. (1967). *Arch. Biochem. Biophys.* **122**, 721–747.
Cardini, G. and Jurtschuk, P. (1968). *J. Biol. Chem.* **243**, 6070–6072.

Clezy, P. S. and Nichol, A. W. (1965). *Austral. J. Chem.* **11**, 1835–1845.
Clezy, P. S., Looney, F. D., Nichol, A. W. and Smythe, G. A. (1966). *Austral. J. Chem.* **19**, 1481–1486.
Coburn, R. F. (1970). *New England J. Med.* **283**, 512–515.
Colby, H. D. and Brownie, A. C. (1970). *Arch. Biochem. Biophys.* **138**, 632–639.
Colleran, E. and O'Carra, P. O. (1969). *Biochem. J.* **115**, 13P.
Conney, A. H. (1967). *Pharmacol. Rev.* **19**, 317–366.
Conney, A. H. and Gilman, A. G. (1963). *J. Biol. Chem.* **238**, 3682–3685.
Conney, A. H., Levin, W., Ikeda, M., Kuntzman, R., Cooper, D. Y. and Rosenthal, O. (1968). *J. Biol. Chem.* **243**, 3912–3915.
Cooper, D. Y., Levin, S., Narasimhulu, S., Rosenthal, O. and Estabrook, R. W. (1965a). *Science* **147**, 402.
Cooper, D. Y., Levin, S. S. and Rosenthal, O. (1970). *Fed. Proc.* **29**, Abstr. 3530.
Cooper, D. Y., Narasimhulu, S., Rosenthal, O. and Estabrook, R. W. (1965b). "Amherst", Vol. **2**, pp. 838–854.
Cooper, D. Y., Narasimhulu, S., Slade, A., Raich, W., Toroff, O. and Rosenthal, O. (1965c). *Life Sciences* **4**, 2109–2114.
Cooper, D. Y., Novack, B., Toroff, O., Slade, A., Saunders, E., Narasimhulu, S. and Rosenthal, O. (1967). *Fed. Proc.* **26**, Abstr. 478.
Cooper, D. Y., Schleyer, M. D. and Rosenthal, O. (1970b). *Ann. New York Acad. Sci.* **174**, 205–217.
Cooper, J. M. and Thomas, P. (1969). *Biochem. J.* **117**, 24P.
Daly, J., Jarina, F., Farnsworth, J. and Guroff, G. (1969). *Arch. Biochem. Biophys.* **131**, 238–244.
Danielsson, H. and Einarsson, K. (1966). *J. Biol. Chem.* **241**, 1449–1454.
Davies, D. S. (1969). *Biochem. J.* **115**, 23P–24P.
Davies, D. S., Gigon, P. L. and Gillette, J. R. (1969). *Life Sciences* **8**, Part II, 85–91.
Degkwitz, E., Ullrich, V., Staudinger, Hj. and Rummel, W. (1969). *Z. physiol. Chem.* **350**, 547–553.
De Matteis, F. (1970). *FEBS Letters* **6**, 343–345.
Di Augustine, R. P. and Fouts, J. R. (1969). *Biochem. J.* **115**, 547–554.
Diehl, H., Capilna, S. and Ullrich, V. (1969). *FEBS Letters* **4**, 99–102.
Ernster, L. and Orrenius, S. (1965). *Fed. Proc.* **24**, Part 5, 1190–1199.
Estabrook, R. W. (1965). "Amherst", Vol. **2**, pp. 832–833, 855–858.
Estabrook, R. W., Cooper, D. Y. and Rosenthal, O. (1963). *Biochem. Z.* **338**, 741–755.
Estabrook, R. W., Franklin, M. R. and Hildebrandt, A. G. (1970b). *Ann. New York Acad. Sci.* **174**, 218–232.
Estabrook, R. W., Hildebrandt, A. G., Baron, J., Netter, K. J. and Leibman, K. (1971). *Biochem. Biophys. Res. Commun.* **42**, 132–140.
Estabrook, R. W., Netter, K., Pawar, S., Hildebrandt, A. and Leibman, K. (1970a). *Fed. Proc.* **29**, Abstr. 3567.
Estabrook, R. W., Schenkman, J. B., Cammer, W., Remmer, H., Cooper, D. Y., Narasimhulu, S. and Rosenthal, O. (1966). "Biological and Chemical Aspects of Oxygenases" (K. Bloch and O. Hayaishi, eds.) p. 153f, Maruzen, Tokyo.
Flynn, E., Lynch, M. and Zannoni, V. G. (1970). *Fed. Proc.* **29**, Abstr. 3142.
Frommer, U., Ullrich, V. and Staudinger, Hj. (1970). *Z. physiol. Chem.* **351**, 913–918.
Garfinkel, D. (1958). *Arch. Biochem. Biophys.* **77**, 493–509.
Garner, R. C. and McLean, A. E. M. (1969). *Biochem. J.* **114**, 7P.
Gaylor, J. L. and Mason, H. S. (1968). *J. Biol. Chem.* **243**, 4966–4972.

Gigon, P. L., Gram, T. E. and Gillette, J. R. (1968). *Biochem. Biophys. Res. Commun.* **31**, 558–562.
Gillette, J. R. (1966). *Advances in Pharmacol.* **4**, 219–261.
Gillette, J. R. and Sasame, H. A. (1970). *Fed. Proc.* **29**, Abstr. 2788.
Glaumann, H., Kuylenstierna, B. and Dallner, G. (1969). *Life Sciences* **8**, 1309–1315.
Grasse, F. R., James, S. P. and Waring, R. H. (1970). *Biochem. J.* **119**, 51P–52P.
Gray, C. H., Nicholson, D. C. and Nicolaus, R. A. (1958). *Nature* (London) **181**, 183–185.
Gray, C. H. and Nakajima, O. (1969). *Biochem. J.* **111**, 23P.
Greengard, P., Psychoyos, S., Tallan, H. H., Cooper, D. Y., Rosenthal, O. and Estabrook, R. W. (1967). *Arch. Biochem. Biophys.* **121**, 298–303.
Greim, H., Schenkman, J. B., Klotzbücher, M. and Remmer, H. (1970). *Biochim. et Biophys. Acta* **201**, 20–25.
Gunsalus, T. C., Conrad, H. E. and Trudgill, P. W. (1965). "Amherst", Vol. **1**, 418–442.
Harding, B. W., Wong, S. H. and Nelson, D. H. (1964). *Biochim. et Biophys. Acta* **92**, 415–417.
Harding, B. W., Nelson, D. H. and Bell, J. B. (1966). *J. Biol. Chem.* **241**, 2212–2229.
Hashimoto, Y., Yamano, T. and Mason, H. S. (1962). *J. Biol. Chem.* **237**, PC 3842–3844.
Hayaishi, O. and Nozaki, M. (1969). *Science* **164**, 389–396.
Hayano, M. (1962). In "Oxygenases" (O. Hayaishi, ed.) 181–240.
Held, H. and von Oldershausen, H. F. (1969). *Klin. Wochsohr.* **47**, 1234–1235.
Hildebrandt, A., Leibman, K. C. and Estabrook, R. W. (1969). *Biochem. Biophys. Res. Commun.* **37**, 477–485.
Hildebrandt, A., Remmer, H. and Estabrook, R. W. (1968). *Biochem. Biophys. Res. Commun.* **30**, 607–612.
Hill, H. A. O., Röder, A. and Williams, R. J. P. (1969). *Biochem. J.* **115**, 59P–60P; *Naturwiss.* (1970) **57**, 69.
Hlavica, P. (1970). *Biochem. Biophys. Res. Commun.* **40**, 212–217.
Holden, H. F. and Lemberg, R. (1939). *Austral. J. Exptl. Biol. Med. Sci.* **17**, 133–143.
Horie, S., Kinoshita, T. and Shimazono, N. (1966). *J. Biochem.* **60**, 660–673.
Hultquist, D. E., Reed, D. W., Passon, P. G. and Andrews, J. E. (1971). *Biochim. et Biophys. Acta* **229**, 33–41.
Hutterer, F., Denk, H., Bacchin, P. G., Schenkman, J. B., Schaffner, J. and Popper, H. (1970). *Life Sciences* **9**, 877–887.
Ichii, S., Kobayashi, Sh. and Omata, S. (1967). *J. Biochem.* **62**, 740–745.
Ichikawa, Y., Hagihara, B. and Yamano, T. (1967). *Arch. Biochem. Biophys.* **120**, 204–213.
Ichikawa, Y., Uemura, T. and Yamano, T. (1968). "Osaka", pp. 634–644.
Ichikawa, Y., Yamano, T. and Fujishima, H. (1969). *Biochim. et Biophys. Acta* **171**, 32–46.
Ichikawa, Y. and Yamano, T. (1970). *Biochim. et Biophys. Acta* **200**, 220–240.
Imai, Y. and Sato, R. (1966). *Biochem. Biophys. Res. Commun.* **25**, 80–86.
Imai, Y. and Sato, R. (1967). *J. Biochem.* **62**, 239–249.
Imai, Y. and Sato, R. (1968a). *J. Biochem.* **63**, 270–273, 370–389; **64**, 147–159.
Imai, Y. and Sato, R. (1968b). "Osaka", pp. 626–633.
Ishidate, K., Kawaguchi, K. and Tagawa, K. (1969). *J. Biochem.* **65**, 385–392.
Ishimura, Y., Ullrich, V. and Peterson, J. A. (1971). *Biochem. Biophys. Res. Commun.* **42**, 140–146.

Jackson, A. H., Kenner, G. W., McGillivray, G. and Sach, G. S. (1965). *J. Amer. Chem. Soc.* **87**, 676–677.
Jackson, A. H., Kenner, G. W. and Smith, K. M. (1966). *Amer. J. Chem. Soc.* **88**, 4539–4541.
Jackson, A. H., Kenner, G. W., McGillivray, G. and Smith, K. M. (1968a). *J. Chem. Soc.* **C**, 294–302.
Jackson, A. H., Kenner, G. W. and Smith, K. M. (1968b). *J. Chem. Soc.* **C**, 302–310.
Jacobsson, S., Thor, S. and Orrenius, S. (1970). *Biochem. Biophys. Res. Commun.* **39**, 1073–1080.
Jefcoate, C. R. E. and Gaylor, J. L. (1969a). *Biochemistry* **8**, 3464–3472.
Jefcoate, C. R. E. and Gaylor, J. L. (1969b). *J. Amer. Chem. Soc.* **91**, 4610–4611.
Jefcoate, C. R. E., Gaylor, J. L. and Calabrese, R. L. (1969). *Biochemistry* **8**, 3455–3463.
Jefcoate, C. R., Hume, R. and Boyd, G. S. (1970). *FEBS Letters* **9**, 41–44.
Kamin, H., Masters, B. S. S., Gibson, Q. H. and Williams, C. H. (1965). *Fed. Proc.* **24**, Part 5, pp. 1164–1171.
Katagiri, M., Ganguli, B. N. and Gunsalus, I. C. (1968). *J. Biol. Chem.* **243**, 2543–2546.
Kato, R. and Takanaka, A. (1969). *Jap. J. Pharmacol.* **19**, 171–173.
Kato, R., Chiesara, E. and Vassanelli, P. (1962). *Biochim. Pharmacol.* **11**, 913–922.
Kimura, T. and Suzuki, K. (1966). *J. Biol. Chem.* **242**, 485–491.
Kinoshita, T. and Horie, S. (1967). *J. Biochem.* **61**, 26–34.
Kinoshita, T., Horie, S., Shimazono, N. and Yohno, T. (1966). *J. Biochem.* **60**, 391–404.
Klingenberg, M. (1958). *Arch. Biochem. Biophys.* **75**, 376–386.
Kratz, F. and Staudinger, Hj. (1968). *Z. physiol. Chem.* **349**, 455–458.
Kuntzman, R. (1969). *Ann. Rev. Pharmacol.* **9**, 21–36.
Kuntzman, R., Levin, W., Jacobson, M. and Conney, A. H. (1968). *Life Sciences* **7**, 215–224.
Kupfer, D. and Orrenius, S. (1970). *Europ. J. Biochem.* **14**, 317–322; *Mol. Pharmacol.* **6**, 221–230.
Leibman, K. C., Hildebrandt, A. G. and Estabrook, R. W. (1969). *Biochem. Biophys. Res. Commun.* **36**, 789–794.
Lemberg, R. (1949). *Nature* (London) **163**, 97.
Lemberg, R. (1956). *Revs. Pure Appl. Chem.* **6**, 1–23.
Lemberg, R. and Barcroft, J. (1932). *Proc. Roy. Soc.* (London) B **110**, 362–373.
Lemberg, R. and Legge, J. W. (1942). *Austral. J. Exptl. Biol. Med. Sci.* **20**, 65–68.
Lemberg, R. and Legge, J. W. (1943). *Biochem. J.* **37**, 117–127.
Lemberg, R. and Legge, J. W. (1949). "Hematin Compounds and Bile Pigments", Interscience, New York. p. 499.
Lemberg, R. and Wyndham, R. A. (1936). *Biochem. J.* **30**, 1147–1170.
Lemberg, R., Legge, J. W. and Lockwood, W. H. (1938). *Nature* (London) **142**, 148.
Lemberg, R., Legge, J. W. and Lockwood, W. H. (1939). *Biochem. J.* **33**, 754–758.
Lemberg, R., Lockwood, W. H. and Legge, J. W. (1941). *Biochem. J.* **35**, 363–379.
Levin, E. Y. (1967). *Biochim. et Biophys. Acta* **136**, 155–158.
Levin, W. and Kuntzman, R. (1969). *J. Biol. Chem.* **244**, 3671–3676.
Levin, W., Alvares, A. P. and Kuntzman, R. (1970a). *Arch. Biochem. Biophys.* **139**, 230–235.
Levin, W., Kuntzman, R., Alvares, A. and Conney, A. H. (1970b). *Fed. Proc.* **29**, Abstr. 2783.

Levitt, M., Schachter, B. A., Zipurski, A. and Israels, L. G. (1968). *J. Clin. Invest.* **47**, 1281–1299.
Lindenmayer, A. and Smith, L. (1964). *Biochim. et Biophys. Acta* **93**, 445–461.
Long, R. T. (1969). *Biochem. J.* **115**, 26P.
Lu, A. Y. H. and Coon, M. J. (1968). *J. Biol. Chem.* **243**, 1331–1332.
Lu, A. Y. H., Junk, K. W. and Coon, M. J. (1969a). *J. Biol. Chem.* **244**, 3714–3721.
Lu, A. Y. H., Strobel, H. W. and Coon, M. J. (1969b). *Biochem. Biophys. Res. Commun.* **36**, 545–551; (1970) *Mol. Pharmacol.* **6**, 213–220.
McIntosh, E. N. and Salhanick, H. A. (1969). *Biochem. Biophys. Res. Commun.* **36**, 552–558.
MacLennan, D. H., Tzagoloff, A. and McConnell, D. G. (1967). *Biochim. et Biophys. Acta* **131**, 59–80.
Marshall, W. J. and McLean, A. E. M. (1968). *Biochem. J.* **107**, 15P.
Marshall, W. J. and McLean, A. E. M. (1969). *Biochem. Pharmacol.* **18**, 153–157.
Marver, H. S., Schmid, R. and Schützel, H. (1968). *Biochem. Biophys. Res. Commun.* **33**, 969–974.
Mason, H. S. (1957). *Advances in Enzymol.* **19**, 79–233.
Mason, H. S., North, J. C. and Vanneste, M. (1965a). *Fed. Proc.* **24**, Part 5, pp. 1172–1180.
Mason, H. S., Yamano, T., North, J. E., Hashimoto, Y., and Sakagishi P. (1965b). "Amherst", Vol. **2**, pp. 879–901.
Matthews, H. B. and Casida, J. E. (1970). *Life Sciences* **9**, 981–1001.
Matthijsen, C. and Mandel, J. E. (1970). *Steroids* **15**, 541–547.
Meigs, R. A. and Ryan, K. J. (1968). *Biochim. et Biophys. Acta* **165**, 476–482.
Mitani, F. and Horie, S. (1969). *J. Biochem.* **65**, 269–280; **66**, 139–149.
Miyake, Y., Gaylor, J. L. and Mason, H. S. (1968a). "Osaka", pp. 656–657.
Miyake, Y., Gaylor, J. L. and Mason, H. S. (1968b). *J. Biol. Chem.* **243**, 5788–5797.
Miyake, Y., Mori, K. and Yamano, T. (1969). *Arch. Biochem. Biophys.* **133**, 318–323.
Morell, D. B. and Nichol, A. W. (1969). *J. Chem. Soc.* C, 517–519.
Muller-Eberhard, Y., Liem, H. H., Yu, A. C. and Gunsalus, I. C. (1969). *Biochem. Biophys. Res. Commun.* **35**, 229–235.
Murphy, R. F., O'hEocha, C. and O'Carra, P. (1967). *Biochem. J.* **104**, 6C.
Nakajima, O. (1964). *Proc. Imper. Acad. Japan* **40**, 576–581.
Nakajima, O. and Gray, C. H. (1967). *Biochem. J.* **104**, 20–22.
Nakajima, O. and Gray, C. H. (1969). *Biochem. J.* **111**, 23P.
Nakajima, H., Takemura, T., Nakajima, O. and Yamaoka, K. (1963). *J. Biol. Chem.* **238**, 3784–3796.
Nakamura, T. and Tamaoki, B.-I. (1970). *Biochim. et Biophys. Acta* **218**, 532–538.
Nakamura, Y., Otsuka, H., Tamaoki, V.-I. (1966). *Biochim. et Biophys. Acta* **122**, 34–42.
Narasimhulu, S., Cooper, D. Y. and Rosenthal, O. (1965). *Life Sciences* **4**, 2101–2108
Nebert, D. W. (1969). *Biochem. Biophys. Res. Commun.* **36**, 885–890.
Nebert, D. W. (1970). *J. Biol. Chem.* **245**, 519–527.
Nebert, D. W. and Gelboin, H. V. (1968). *J. Biol. Chem.* **243**, 6242–6249.
Nichol, A. W. (1970). *Biochim. et Biophys. Acta* **222**, 28–40.
Nichol, A. W. and Morell, D. B. (1969). *Biochim. et Biophys. Acta* **184**, 173–180.
Nishibayashi, H. and Sato, R. (1967). *J. Biochem.* **61**, 491–496.
Nishibayashi, H. and Sato, R. (1968). *J. Biochem.* **63**, 766–779.
Nishibayashi, H., Omura, T. and Sato, R. (1966). *Biochim. et Biophys. Acta* **118**, 651–654.

Nishibayashi, H., Omura, T., Sato, R. and Estabrook, R. W. (1968). "Osaka", pp. 658–665.
O'Carra, P. and Colleran, E. (1969). *Biochem. J.* **115**, 13–14P.
Oldham, S. B., Wilson, D. L., Landgraf, W. L. and Harding, B. W. (1968). *Arch. Biochem. Biophys.* **123**, 484–495.
Omura, T. and Sato, R. (1962). *J. Biol. Chem.* **237**, PC1375.
Omura, T. and Sato, R. (1964). *J. Biol. Chem.* **239**, 2379–2385.
Omura, T., Sato, R., Cooper, D. Y., Rosenthal, R. W. and Estabrook, R. W. (1965a). *Fed. Proc.* **24**, Part 5, pp. 1181–1189.
Omura, T., Sanders, E., Cooper, D. Y., Rosenthal, O. and Estabrook, R. W. (1965b). In "Non-haem iron proteins" (A. San Pietro, ed.). Kettering Sympos., Dayton, Ohio, 401–412.
Omura, T., Sanders, E., Estabrook, R. W., Cooper, D. Y. and Rosenthal, O. (1966). *Arch. Biochem. Biophys.* **117**, 660–673.
Orrenius, S. (1965). *J. Cell Biol.* **26**, 725–733.
Orrenius, S. (1969). *Biochem. J.* **115**, 25–26P.
Orrenius, S. and Ernster, L. (1964). *Biochem. Biophys. Res. Commun.* **16**, 60–65.
Orrenius, S. and Ernster, L. (1967). *Life Sciences* **6**, 1473–1482.
Orrenius, S., Dallner, G. and Ernster, L. (1964). *Biochem. Biophys. Res. Commun.* **14**, 329–334.
Orrenius, S., Das, M. and Gnosspelius, Y. (1969). In "Microsomes and Drug Oxidations" (J. R. Gillette, A. H. Conney, G. J. Cosmides, R. W. Estabrook, R. J. Fouts and G. J. Mannering, eds.) p. 251, Academic Press, New York and London.
Orrenius, S., Ericsson, J. L. E. and Ernster, L. (1965). *J. Cell Biol.* **25**, 627–639.
Orrenius, S., Kupfer, D. and Ernster, L. (1970). *FEBS Letters* **6**, 249–252.
Parke, D. V. and Rahman, H. (1970). *Biochem. J.* **119**, 53P.
Parli, C. J. and Mannering, G. J. (1970). *Mol. Pharmacol.* **6**, 178–183.
Pawar, S. S. and Estabrook, R. W. (1969). *Fed. Proc.* **28**, No. 3528.
Peisach, J. and Blumberg, W. H. (1970). *Proc. Nat. Acad. Sci. U.S.* **67**, 172–179.
Perry, A. S. and Buchner, A. J. (1970). *Life Sciences* **9**, 335–350.
Petryka, Z., Nicholson, D. C. and Gray, C. H. (1962). *Nature* (London) **194**, 1047–1048.
Rees, M. and Nason, A. (1965). *Biochem. Biophys. Res. Commun.* **21**, 248–256.
Remmer, H. and Merker, H. J. (1963). *Science* **142**, 1657–1658.
Remmer, H. and Merker, H. J. (1965). *Ann. New York. Acad. Sci.* **123**, 79–97.
Remmer, H., Greim, H., Schenkman, J. B. and Estabrook, R. W. (1967). "Methods in Enzymol.", Vol. **10**, pp. 703–708.
Remmer, H., Schenkman, J., Estabrook, R. W., Sasame, H., Gillette, J., Narasimhulu, S., Cooper, D. Y. and Rosenthal, O. (1966). *Mol. Pharmacol.* **2**, 187–190.
Robinson, S. H. (1969). *J. Lab. Clin. Med.* **73**, 668–676.
Robinson, S. H., Tsong, M., Brown, B. W. and Schmid, R. (1966). *J. Clin. Invest.* **45**, 1569–1586.
Roeder, A. and Bayer, E. (1969). *Europ. J. Biochem.* **11**, 89–92.
Ruhmann-Wennhold, A., Johnson, L. V. R. and Nelson, D. M. (1970). *Biochim. et Biophys. Acta* **223**, 206–209.
Ryan, K. J. and Engel, L. L. (1957). *J. Biol. Chem.* **225**, 103–114.
Sato, R. and Omura, T. (1961). *Moscow Internat. Biochem. Congr.* 1961.
Sato, R., Omura, T. and Nishibayashi, H. (1965). "Amherst", Vol. **2**, pp. 861–875.
Satre, M., Vignais, P. V. and Idelman, C. S. (1969). *FEBS Letters* **5**, 135–140.
Sauer, L. A. and Mulrow, P. J. (1969). *Arch. Biochem. Biophys.* **134**, 486–496.

Schenkman, J. B. (1970a). *Science* **168**, 612–613.
Schenkman, J. B. (1970b). *Biochemistry* **9**, 2081–2089.
Schenkman, J. B. and Sato, R. (1968). *Mol. Pharmacol.* **4**, 613–620.
Schenkman, J. B., Frey, I., Remmer, H. and Estabrook, R. W. (1967). *Mol. Pharmacol.* **3**, 113–123, 516–525.
Schenkman, J. B., Greim, H., Zange, M. and Remmer, H. (1969). *Biochim. et Biophys. Acta* **171**, 23–31.
Schleyer, H., Cooper, D. Y. and Rosenthal, O. (1970). *Fed. Proc.* **29**, Abstr. 3871.
Schmid, R., Marver, H. S. and Hammaker, L. (1966). *Biochem. Biophys. Res. Commun.* **24**, 319–328.
Scholan, N. A. and Boyd, G. S. (1968). *Z. physiol. Chem.* **349**, 1628–1630.
Shoeman, D. W., Chaplin, M. D. and Mannering, G. J. (1969). *Mol. Pharmacol.* **5**, 412–419.
Shoeman, D. W., Vane, F. and Mannering, G. J. (1970). *Fed. Proc.* **29**, Abstr. 2785.
Sies, H. and Brauser, B. (1970). *Europ. J. Biochem.* **15**, 531–549.
Sih, C. J. (1969). *Science* **163**, 1297–1300.
Simpson, E. R. and Boyd, G. S. (1967). *Europ. J. Biochem.* **2**, 275–285.
Simpson, E. R. and Estabrook, R. W. (1968). *Arch. Biochem. Biophys.* **126**, 977–978.
Sjoestrand, T. (1951). *Nature* **168**, 1118.
Sjoestrand, T. (1952). *Acta Physiol. Scand*, **26**, 328–344.
Sladek, N. E. and Mannering, G. J. (1966). *Biochem. Biophys. Res. Commun.* **24**, 668–674.
Sladek, N. E. and Mannering, G. J. (1969). *Mol. Pharmacol.* **5**, 174–199.
Smith, L. (1965). "Amherst", Vol. **2**, p. 901.
Smuckler, E. A., Orrenius, S. and Hultin, H. O. (1972). *Biochem. J.* (In Press).
Snyder, A. L. and Schmid, R. (1965). *J. Lab. Clin. Med.* **65**, 817–824.
Sottocasa, G. and Sandri, G. (1970). *Biochem. J.* **116**, 16–17P.
Staudinger, Hj. and Ullrich, V. (1964). *Z. f. Naturfschg.* **19B**, 409–413.
Staudinger, Hj., Kerekjarto, B., Ullrich, V. and Zubrzycki, Z. (1965). "Amherst", Vol. **2**, pp. 815–832.
Strittmatter, C. F. and Umberger, F. T. (1969). *Biochim. et Biophys. Acta* **180**, 18–21.
Strobel, H. W., Lee, A. Y. H., Heidema, J. and Coon, M. J. (1970). *J. Biol. Chem.* **245**, 4851–4854.
Sweat, M. L., Dutcher, J. S., Young, R. B. and Bryson, M. J. (1969a). *Biochemistry* **8**, 4956–4963.
Sweat, M. L., Young, R. B. and Bryson, M. J. (1969b). *Arch. Biochem. Biophys.* **130**, 66–69.
Sweat, M. L., Young, R. B. and Bryson, M. J. (1970). *Biochim. et Biophys. Acta* **223**, 105–114.
Tenhunen, R., Marver, H. S. and Schmid, R. (1968). *Proc. Natl. Acad. Sci. U.S.* **61**, 748–755.
Tenhunen, R., Marver, H. S. and Schmid, R. (1969). *J. Biol. Chem.* **244**, 6388–6394.
Tenhunen, R., Marver, H. S. and Schmid, R. (1970). *J. Lab. Clin. Med.* **75**, 410–421.
Tenhunen, R., Ross, M. E., Marver, H. S. and Schmid, R. (1970). *Biochemistry* **9**, 9298–9303.
Thorgeirsson, S. S. and Davies, D. S. (1971). *Biochem. J.* **122**. 30P.
Tsai, R., Yu, C. A., Gunsalus, I. C., Peisach, J., Blumberg, W., Orme-Johnson, W. H. and Beinert, H. (1970). *Proc. Natl. Acad. Sci. U.S.* **66**, 1157–1163.
Ullrich, V. (1969). *Z. f. Naturfschg.* **24B**, 699–704.

Ullrich, V., Cohen, B., Cooper, D. Y. and Estabrook, R. W. (1968). "Osaka", pp. 649–655.
Van Lier, J. E. and Smith, L. L. (1970). *Biochim. et Biophys. Acta* **210**, 153–163.
Voigt, W., Fernandez, F. P. and Hsia, S. L. (1970). *Proc. Soc. Exp. Biol. Med.* **113**, 158–161.
Wada, F., Higashi, T., Tada, K. and Sakamoto, Y. (1964). *Biochim. et Biophys. Acta* **88**, 655–657.
Wada, F., Hirata, K. and Sakamoto, Y. (1967). *Biochim. et Biophys. Acta* **143**, 273–275.
Wada, F., Hirata, K., Nakao, K. and Sakamoto, Y. (1968a). *J. Biochem.* **64**, 109–113.
Wada, F., Hirata, K., Nakao, K. and Sakamoto, Y. (1968b). *J. Biochem.* **64**, 415–417.
Wada, F., Hirata, K., Nakao, K. and Sakamoto, Y. (1969a). *J. Biochem.* **66**, 699–703.
Wada, F., Hirata, K. and Sakamoto, Y. (1969b). *J. Biochem.* **65**, 171–175.
Wada, F., Shibata, M., Goto, M. and Sakamoto, Y. (1968c). *Biochim. et Biophys. Acta* **162**, 518–524.
Wada, O., Yano, Y., Urata, G. and Nakao, K. (1968). *Biochem. Pharmacol.* **17**, 595–603.
Waterman, M. K. and Mason, H. S. (1970). *Biochem. Biophys. Res. Commun.* **39**, 450–454.
Watersfield, M. D., Del Favero, A. and Gray, C. H. (1969). *Biochim. et Biophys. Acta* **184**, 470–473.
Whysner, J. A. and Harding, B. W. (1968). *Biochem. Biophys. Res. Commun.* **32**, 921–927.
Whysner, J. A., Ramseyer, J., Kazini, G. M. and Harding, B. (1969). *Biochem. Biophys. Res. Commun.* **36**, 795–801.
Whysner, J., Ramseyer, J. and Harding, B. W. (1970). *Fed. Proc.* **29**, Abstr. 3693.
Williams, R. J. P. (1968). "Osaka", pp. 645–648.
Williamson, D. G. and O'Donnell, V. J. (1969). *Biochemistry* **8**, 1306–1311.
Wilson, L. D. and Harding, B. W. (1970). *Biochemistry* **9**, 1615–1620, 1621–1625.
Wilson, L. D., Oldham, S. B. and Harding, B. W. (1969). *Biochemistry* **8**, 2975–2980.
Wise, C. D. (1964). Ph.D. Dissertation, Univ. of Philadelphia.
Wise, C. D. and Drabkin, D. L. (1964). *Fed. Proc.* **23**, Abstr. 736.
Wise, C. D. and Drabkin, D. L. (1965). *Fed. Proc.* **24**, Abstr. 519.
Wittkop, J. A. and Prough, R. A. (1969). *Arch. Biochem. Biophys.* **134**, 308–315.
Yago, N. and Ichii, S. (1969). *J. Biochem.* **65**, 215–224.
Yago, N., Kobayashi, S., Sekiyama, S., Kurokawa, H., Iwai, Y., Suzuki, I. and Ichii, S. (1970). *J. Biochem.* **68**, 775–783.
Yohro, T. and Horie, S. (1967). *J. Biochem.* **61**, 515–517.
Yong, F. C., King, T. E., Oldham, S., Waterman, M. R. and Mason, H. S. (1970). *Arch. Biochem. Biophys.* **138**, 96–100.
Young, D. G. and Hall, P. F. (1968). *Biochem. Biophys. Res. Commun.* **31**, 925–931.
Young, D. G. and Hall, P. F. (1969). *Biochemistry* **8**, 2987–2996.
Young, D. G., Bryson, M. J. and Sweat, M. L. (1970a). *Biochim. et Biophys. Acta* **215**, 119–124.
Young, D. G., Holroyd, J. D. and Hall, P. F. (1970b). *Biochem. Biophys. Res. Commun.* **38**, 184–190.
Yu, Chang-an and Gunsalus, I. C. (1970). *Biochem. Biophys. Res. Commun.* **40**, 1431–1436.

E. CYTOCHROMES "h" AND "HELICORUBIN" OF INVERTEBRATES

1. Introduction

Cytochromes quite similar to or identical with those of vertebrates have been found in many tissues of invertebrates (see Chapters III, IVb, IVc and V). There is, however, one class of cytochromes typical for invertebrates, represented by "helicorubin", "cytochrome h" or "enterochrome-556" which are found in the hepatopancreas of gastropods and the "helicorubin" or "enterochrome-566" which are found in the gut fluid of gastropods, marine as well as terrestrial, cephalopods, echinoderms, crustaceans and polychaetes (Keilin, 1957, 1960; Kawai, 1959, 1960, 1961; Ghiretti–Magaldi and Ghiretti, 1958; Tosi and Ghiretti, 1959; Tosi et al., 1961; Reddy and Swami, 1968).

"Helicorubin" was discovered by Sorby (1876) in the crop and gut fluid of the European garden snail and given its name by Krukenberg (1882). The early work of Joan Keilin on it and on cytochrome h (Keilin, 1956) has been described by Lemberg and Legge (1949) and Keilin (1956, 1957). There now appears no valid reason to exclude helicorubin from the cytochromes, and to call it a haemochrome. Most of the "typical" cytochromes are haemochromes, and helicorubin is chemically a typical cytochrome b. Its removable prosthetic group is protohaem, it does not react with carbon monoxide and has a haemochrome spectrum. While there appears to be no biological function for it in the gut fluid, a very similar haemoprotein, enterochrome-566, has been found to function in the electron transport of the hepatopancreas of Japanese snails (Kawai, 1960, 1961). Whether or not cytochrome h and enterochrome-556 should also be included in the ranks of cytochromes b, or form a separate class, depends on whether the structure of the prosthetic group, which is not protohaem, but a haem with saturated side chains, is considered the more important criterion, or its type of linkage, which is dissociable and not by covalent linkage, as it is in cytochromes c. The need for a nomenclature reconsideration has already been mentioned by J. Keilin (1968).

For the purposes of this book it appeared desirable to discuss the two cytochromes h and helicorubin together, although the earlier hypothesis that helicorubin is the solubilized cytochrome h can probably no longer be maintained (Kawai, 1961; J. Keilin, 1968; J. Keilin and Orlans, 1969). "Helicorubin" differs from other cytochromes b so far discussed, including cytochrome b_5 of the *Cecropia* midgut.

2. Isolation and purification of helicorubin and cytochrome h

Both cytochromes have been isolated from gut fluid and hepatopancreas of *Helix pomatio*, respectively, in almost pure form, but have not yet been crystallized (Keilin, 1956, 1968; Keilin and Orlans, 1969). Steps of the isolation were lead acetate precipitation, keiselguhr filtration, ammonium sulphate precipitation, heat inactivation of accompanying cellulose, and paper electrophoresis, later replaced by Sephadex and DEAE cellulose chromatography. Both cytochromes, which can also be separated by the last-named procedure, are rather heat-stable and stable in the pH range of 4–13, do not react with carbon monoxide or as ferric compounds with cyanide, fluoride or NO, and have a low isoelectric point, 4·3. They are slowly autoxidizable at a pH 6·0, but faster at a lower pH. The M.W. of helicorubin is 11,500, that of cytochrome h 12,000. The amino acid composition is similar, though different in some amino residues. It is remarkable that in both methionine is absent and lysine low, though slightly higher in helicorubin, and that proline is higher than in other haemoproteins. The E_0' (pH 7) of both helicorubin, cytochrome h and enterochromes has been found at +20 mV (Keilin, 1957).

Helicorubin. The absorption spectra of helicorubin and cytochrome h are given in Table V and compared with those of cytochrome c (Keilin

TABLE V

Absorption spectra of helicorubin, cytochrome h and enterochromes

	Maxima in nm, in parentheses ϵ mM	$A_{276}/A_\gamma Fe^{2+}$
Helix pomatio "Helicorubin"[1]	Fe^{2+} 563(28·9) 531(14·6) 428(176) Fe^{3+} 416(111) pyridine ferrohaemochrome 558, 421	0·267
Euhydra "Enterochrome-566"[2]	Fe^{2+} 566 530 427 Fe^{3+} 416 pyridine ferrohaemochrome 557, 418	
Helix pomatio "Cytochrome h"[1]	Fe^{2+} 556(32·8) 526(14·4) 422·5(190) Fe^{3+} 409(106) pyridine ferrohaemochrome 551	0·253
Euhydra "Enterochrome-556"[2]	Fe^{2+} 556 525–526 424 Fe^{3+} 412 pyridine ferrohaemochrome 553	
Aplysia "Cytochrome h^1"[3]	Fe^{2+} 556 527 423·5 Fe^{3+} 411	
Cytochrome c	Fe^{2+} 550(27·7) 520 (15·9) 416(129) Fe^{3+} 410(106)	0·219

[1] Keilin and Orlans, 1969.
[2] Kawai, 1960, 1961.
[3] Tosi and Ghiretti, 1961.

and Orlans, 1969). This table also includes data on the "enterochromes" and on the "cytochrome h^1" of *Aplysia* and *Cryptomphalus*, assumed to differ in minor aspects from cytochrome h. In contrast to helicorubin, "enterochrome-566" is also found particle-bound in the hepatopancreas with little difference from the solubilized form in the gut. At low temperature the α-band of *Helix* helicorubin is split into two bands at 562 and 556 nm; it is similarly split at room temperature at a pH below 4–5. While the α-bands of all the cytochromes in Table V appear to be of nearly the same height, the Soret bands of cytochromes h and helicorubin are higher than those of cytochrome c increasing the γ/α ratio from 4·6 to about 6. Helicorubin has only one absorption band at 276 nm in the distant ultraviolet.

Cytochrome h. The properties of the compounds called cytochrome h from *Helix* of "enterochrome-556" or cytochrome h^1 are so similar (see Table V) that it appears likely that all are species-specific cytochromes of the gastropod hepatopancreas cytochrome h. Enterochrome-556 has been found in both microsomes and mitochondria of the hepatopancreas (Kawai, 1960). While, however, cytochrome h from *Helix* is not excreted into the gut, that of other organisms, e.g. the enterochromes-556 of Japanese land snails is with no or only minor alterations. Its prosthetic group, to judge from the absorption band of the pyridine haemochrome is probably haematohaem, with an admixture of either monovinyl-monohydroxyethyl deuterohaem or even protohaem. In contrast to helicorubin, cytochrome h has a more complex low wavelength band with maxima at 282, 276 and 260 nm (Keilin, 1956) and a somewhat different δ-band. It is probably of evolutionary significance that cytochrome h and helicorubin are able to react with exogenous cytochrome c, but without it do not react with cytochrome oxidase (Keilin, 1957), while they react with the bacterial cytochrome c oxidase of *Acetobacter oxidans* (Tissières, 1956). Kawai (1960, 1961; see also Tosi and Ghiretti, 1959, 1961) found, however, enterochrome-556 able to react with the oxidase of hepatopancreas and to replace cytochrome c. The reduction of enterochrome-556 by succinate is antimycin-sensitive, and there is a high affinity to NADH (Kawai, 1961c). The cytochrome c_1 reported by Ghiretti-Magaldi and Ghiretti (1958) in *Sepia officinalis* may have been, in fact, cytochrome h.

REFERENCES

Ghiretti-Magaldi, A. and Ghiretti, F. (1958). *Experientia* **14**, 170–172.
Kawai, K. (1959). *Biol. Bull.* **117**, 125–132.
Kawai, K. (1960). *Biochim. et Biophys. Acta* **43**, 349–351; **44**, 202–204.
Kawai, K. (1961). *J. Biochem.* **52**, 241–247; 248–253.
Keilin, J. (1956). *Biochem. J.* **64**, 663–676.
Keilin, J. (1957). *Nature* **180**, 427–429.

Keilin, J. (1960). *Acta Biochim. Polonica* **7**, 367–375.
Keilin, J. (1968). "Osaka". pp. 691–700.
Keilin, J. and Orlans, P. (1969). *Nature* (London). **223**, 304–306.
Krukenberg, C. F. W. (1882). "Vergleichende Physiologische Studien", Series 2, part 2, p. 63. Heidelberg.
Lemberg, R. and Legge, J. W. (1949). "Haematin Compounds and Bile Pigments", p. 225. Interscience, New York.
Reddy, R. and Swami, K. S. (1968). *Arch. Internat. Physiol. Biochem.* **76**, 26–46.
Sorby, H. C. (1876). *Quart. J. Micro. Sci.* **16**, 76.
Tissières, A. (1956). *Biochem. J.* **64**, 582–589.
Tosi, L. and Ghiretti, F. (1959). *Experientia* **15**, 18–19.
Tosi, L. and Ghiretti, F. (1961). *Arch. Biochem. Biophys.* **93**, 399–406.

F. CYTOCHROMES b IN HIGHER PLANTS AND ALGAE

1. Introduction

Cytochromes of colourless parts of plants were studied spectroscopically as early as those of animals by MacMunn and Keilin (1925). Their closer study and attempts at isolation, however, began only 26 years later by Hill and Scarisbrick (1951). They isolated a soluble cytochrome b from variegated elder and other leaves which they called cytochrome b_3, together with a type c cytochrome (see Chapter VC) which they called cytochrome f. Later Hill discovered a cytochrome of type b which he called cytochrome b_6 in chloroplasts (Davenport, 1952; Hill, 1954). Even today, none of the plant cytochromes b has been purified, and only three, cytochrome b_3, cytochrome b_7 from mitochondria of the *Arum spadix*, and cytochrome b-555 from microsomes of Mung bean seedlings, have been partially purified. The earlier investigations have been reviewed by Hartree (1957).

There is still a great deal of confusion in this field, as the review of Chance *et al.* (1968) shows. One and the same cytochrome has received several names, e.g. b_3 and b-559 for a chloroplast cytochrome (Lundegårdh, 1961, 1962a; Boardman and Anderson, 1967), while the same name, b_3 has been applied to a soluble cytochrome, probably derived from microsomes and to a chloroplast one. Other plant cytochromes b now appear to be only species-specific variants of the animal cytochromes met with in subchapters C and D. These are cytochromes b (Bhagvat and Hill, 1951; James and Leech, 1960, 1964), cytochrome b_5 (as b-555) of plant microsomes and P-450 (Moore, 1967). It is therefore doubtful whether the plant cytochromes b are really a much more complex group than the animal ones. There is, of course, the fact that plants contain not only respiratory cytochromes but also in their chloroplasts or chromatophores cytochromes b essential for photosynthetic photochemical reactions which differ from the respiratory cytochromes. The complete separation of chloroplasts from mitochondria was achieved by Leech (1963) by centrifugation through a glycerol-sucrose gradient. De Kouchkovsky (1959) has also obtained chloroplasts free from cytochrome oxidase.

The writers prefer, when in doubt, to err rather on the conservative side and to distinguish as few cytochrome species as possible. Not only will they thereby make the task of the reader less difficult, but omissions can later be readily remedied when distinctions can be convincingly demonstrated, whereas errors of unwarranted duplication lead to confusion and are less easily eradicated, requiring convincing proof of identity. Stress on spectroscopic observations was necessitated by the inadequate methods of purification, and has indeed been of great value in this field. Even the greater resolving power of low temperature spectroscopy evolved by Estabrook (1961) and by Bonner (1961) cannot, however, overcome lack of chemical data, and can even introduce "pseudo-species". Under certain conditions bands measured at low temperature are doubled, and it is not always easy to decide whether this is due to the presence of a mixture of two cytochromes, or a physical phenomenon (see e.g. Chance and Hackett, 1959; Lance and Bonner, 1968). Even in one and the same laboratory bands differing by about 2 nm have been reported for the same cytochrome, both at low and at room temperature (Lance and Bonner, 1968; Storey, 1969). Thus differences of this order between different laboratories are hardly a sound basis for identification of cytochromes isolated in different laboratories. Correlation of low and room temperature spectra is another difficulty; as a rule of thumb it may be accepted that the low temperature maxima at 77°K are usually 2–3 nm shifted to shorter wavelengths, but there may be exceptions. The same difficulties will be encountered in the field of bacterial cytochromes (Chapter VI). The decision of the Cytochrome Nomenclature Subcommittee of the Nomenclature Commission on Enzymes to recommend *preliminary* naming of new enzymes on the basis of the position of their α-band maxima at room temperature had the purpose of preventing undue extension of a series of names b, $b_1 \ldots b_x$, but this purpose would be frustrated if this series is merely replaced by another one, b-565, b-564 ... b-555.

There are other reasons to be cautious in the field of plant cytochromes. Methods of separation of plant mitochondria and microsomes are more recent (Bonner, 1967; Wiskich, 1967) and perhaps less satisfactory than those for animal mitochondria and microsomes. While it is possible to choose colourless parts of plants for the study of respiratory cytochromes b, it is not always clear whether a colourless mutant of an alga, or an artificially decolourized alga is free of chloroplast cytochromes; or that colourless chloroplasts are not contaminated with mitochondrial or microsomal cytochromes. Finally, plant cells contain protohaemoproteins other than cytochromes b. The cytochrome "dh" found by Lundegårdh

(1954) in wheat roots was, e.g. later recognized as a peroxidase (Martin and Morton, 1956; Lundegårdh, 1958; Tagawa and Shin, 1959). A cytochrome b-557, different from bacterial cytochrome b_1 (Chapter VI) and the cytochrome b of the fungus *Sclerotina libertiana* (Yamanaka *et al.*, 1960) has been found in *Neurospora crassa* by Garette and Nason (1967) and has been claimed to act as assimilatory nitrate reductase. This contains dissociable flavin and may, in fact be cytochrome b_2.

2. Respiratory cytochromes b

(a) *In mitochondria*. Plant mitochondria probably contain three different cytochromes b (Bonner and Plesnicar, 1967; Bonner and Slater, 1970). They contain more cytochromes b than c (James and Lundegårdh, 1959).

Cytochrome b. The α-band of the ferrous compound has been found at 563–566 nm, at low temperature at 561–562 nm. At steady state it is more reduced than other cytochromes of type b and its reduction is still further increased by energization. Antimycin combines preferentially with the energized cytochrome. The band position has been frequently reported at slightly lower wavelength than that of the animal enzyme (James and Leech, 1960; Wiskich *et al.*, 1960). Pollen is rich in a cytochrome b with an α-band at 561 nm (Okunuki, 1939); its position may be either due to admixture of other cytochromes b, or to a slight structural difference, hardly justifying excluding it from cytochrome b. Dawson *et al.* (1968) found that a part of cytochrome b in plant mitochondria could not be readily reduced in the presence of malate, and concluded that this form was not in the respiratory chain, and may have another function. A similar situation for mammalian cytochrome b has been discussed in Chapter 5, II.

In algae, cytochromes with a band position 563–561 nm have been reported frequently. Some of these are, however, forms of photosynthetic cytochrome b_6 rather than of b. There are exceptions. Streptomycin-bleached *Euglena* contains a cytochrome b-561 (Raison and Smillie, 1969) which appears to be a respiratory cytochrome b distinct from the b-561 of *Euglena* chloroplasts (Perini *et al.*, 1964). A water-soluble b-562 has been found by Katoh and San Pietro (1965) in green algae; according to Bendall and Hill (1968), this has an oxidation reduction-potential of 130 mV. It may be possible to distinguish cytochromes b from cytochromes b_6 on the basis of the greater lability of b to cold acetone. Finally, cytochromes of type b have been demonstrated in colourless *Cyanophyta* (Webster and Hackett, 1966) in which cytochrome o was the oxidase.

Cytochrome b-559 (mitochondrion) $= b_7$? Cytochrome b_7 was isolated from the spadix of *Arum maculatum* (Bendall and Hill, 1956; Bendall, 1958, 1968) partially purified. Its oxidation-reduction potential was found at -30 mV. It is more rapidly autoxidizable than other cytochromes b, but like most of them, does not react with CO or cyanide. It is not yet clear whether the b_7 is formed in the maturing spadix, or whether the concentration of a preformed b-559 demonstrated in the mitochondria of many plants (see Lance and Bonner, 1968; Storey, 1969) is only increased. The identification of b-559 or b-557 ($\Delta Fe^{2+} - Fe^{3+}$ at $-190°C$) of Mung bean mitochondria with b_7 appears more likely than its identification with b_3 (Shichi et al., 1963b), although Storey (1970a) reports differences between b-557 and b_7 in the behaviour to antimycin. The half-time of its oxidation is 5·8 msec, independent of the presence or absence of 0·3 mM cyanide.

Cytochrome b-555. Its α-band lies at 555–556 nm at room temperature, at 552–553 nm at low temperature (Lance and Bonner, 1968; Storey, 1969). This is a cytochrome b, not one of type c (f) as assumed by Lundegårdh (1958) and, preliminarily, by Wiskich et al. (1960), since antimycin A (which inhibits oxidation of cytochromes b by cytochromes c) causes appearance of the band by reduction. Only a single band at 428 nm was observed for the three mitochondrial b components distinguishable by their α-bands, but other cytochromes absorbing in the Soret band region are present. A similar cytochrome has been reported by Colmano and Wolken (1963) in dark-grown *Euglena*.

(b) *In microsomes.* Three cytochromes spectroscopically quite similar to those in the mitochondria have been found in microsomes, and it has not yet been finally proved that they differ from them. Shichi and Hackett (1962c) found a cytochrome b-like pigment with an α-band 561 nm (at low temp. 558·5 nm), $E_0' = 0$ mV, in small amounts in Mung bean seedlings (microsomes?). The "soluble" cytochrome b_3 (α-band 559 nm), E_0' +40 mV of Hill and Scarisbrick (1951) (see also Lundegårdh, 1962), in wheat roots is probably identical with the low temp. 557 nm microsomal cytochrome of Lance and Bonner (1968). The microsomal cytochrome was already called cytochrome b_3 by Martin and Morton (1955) who found the α-band at room temperature at 559·5 nm and reported its presence in microsomes (1957). This cytochrome can be reduced by ascorbate. Shichi et al. (1963a) reported that a solubilized preparation of it had a M.W. of 28,000. Shichi (1969) found it spectroscopically similar to, if not identical with, P-420 of heart microsomes.

The third microsomal cytochrome has an α-band at 555 nm (Morton, 1961; Shichi and Hackett, 1962a,b; Kasinsky et al., 1966; Shichi et al., 1966; Lance and Bonner, 1968). It is again not a cytochrome f, but a cytochrome b, and has a molecular weight of 13,500 and an E_0' of 0·00 V. According to Moore (1967) this is cytochrome b_5. At low temperature the α-band is split to 559 + 552 nm as found for cytochrome b_5 by Estabrook (1961). This was also found for b-559 of plant microsomes by Chance and

Hackett (1959) and Shichi and Hackett (1962a,b), but not by Lance and Bonner (1968) who found only the band at 552–3 nm. Cytochrome P-450 has also been reported in plant microsomes (Moore, 1967).

Since plant respiration is largely insensitive to cyanide as well as to antimycin A, particularly the respiration of the *Arum spadix*, and since cytochrome b_7 is readily autoxidizable, it was thought that cytochrome b_7 may be the terminal oxidase. In spite of much discussion (James and Beevers, 1950; James, 1957, p. 308; Lundegårdh, 1955, 1958; Martin and Morton, 1955; Yocum and Hackett, 1957; Simon, 1957; Bendall and Hill, 1958; Bendall, 1958; Hackett, 1957, 1959; Hackett and Haas, 1958; Hackett *et al.*, 1960; Chance and Hackett, 1959; Bonner, 1961; Chance and Bonner, 1965; Bonner *et al.*, 1967), the problem is still unsolved (Bonner and Bendall, 1968; Storey and Bahr, 1968; Storey, 1970a; Erecinska and Storey, 1970).

3. Cytochromes b in the electron transport in photosynthesis of green plants and algae

There are at least two, possibly three cytochromes of type b involved in this electron transport together with cytochrome f, and found in chloroplasts. The longest known of these is cytochrome b_6 discovered by Davenport (1952) and Hill (1954), which has been further studied by Hill and Bendall (1960), James and Leech (1964), Boardman and Anderson (1967) and Bendall (1968). Its oxidation-reduction potential was found at 0 to -60 mV. It is autoxidizable. Probably the same cytochrome, with band positions varying between 563 and 561 nm has been observed in several species of green algae (Kok, 1957; Chance and Sager, 1957; Chance, Schleyer and Legallais, 1963; Smillie and Levine, 1963; Müller *et al.*, 1963; Katoh and San Pietro, 1965; Hind and Olson, 1966; Bendall and Hill, 1968; Sugimura *et al.*, 1968; Hiyama, 1969; Powls *et al.*, 1969), in *Euglena* (Olson and Smillie, 1963; Perini *et al.*, 1964; Ikegami *et al.*, 1968) and in red algae (Nishimura, 1968).

Cytochrome b-559 (chloroplast). This cytochrome of chloroplasts was first described by Lundegårdh (1964a,b) under the name "cytochrome b_3" but since Boardman and Anderson (1967) found b-559 tightly bound in chloroplast lamellae, while cytochrome b_3 is a microsomal or soluble cytochrome it is unlikely that the two are identical. In contrast to cytochrome b_6, b-559 is reduced by ascorbate. Boardman and Anderson (1964, 1967) found that differential centrifugation of digitonin-treated chloroplast particles allowed the separation of a lighter fraction, sedimented at 144,000 × g which contains only cytochromes f and b_6, from a heavier fraction, sedimented at 10,000 × g which contained cytochrome b-559, more cytochrome b, less P-700 type of chlorophyll a and more chlorophyll b than chloroplasts of the light fraction. Probably the same cytochrome (b-560) was found to accompany b-563 in spinach chloroplasts (Cramer and Butler, 1967), and the two behaved differently on illumination with light of a wavelength above and below 700 nm (see below). Similar cytochromes with absorption bands at 558–559 nm were found in green algae (Hind and Olson, 1966; Hind et al., 1967; Givan and Levine, 1969; Hiyama, 1969). The redox potential at pH 7–7·2 was found by Boardman et al. (1970) +350 mV, while Bendall (1968) and Ben Hayyim and Avron (1970a) report +370 mV and Knaff and Arnon (1969b) in chloroplasts and Ikegawa et al. (1968) in *Euglena* +300–320 mV. Much lower potentials (from +80 to −65 mV) have been found by Fan and Cramer (1970) and by Hind and Nakatani (1970) but as they are further decreased by acetone treatment of the chloroplast they are probably due to a modification.

Cytochrome b-559 has also been found in the colourless alga *Prototheca zopfii* and has been shown to be sensitive to irradiation with blue light (Epel and Butler, 1970).

The position of the cytochromes b in the electron transport chain of photosynthesis. It would be beyond the frame of this review to discuss the whole theory of electron transport in photosynthesis. The reader must be referred to recent reviews, e.g. those of Vernon and Avron (1965), Arnon (1967), Avron (1967), Boardman (1968), Bendall and Hill (1968), Döring et al. (1969), Walker (1970) and Walker and Crofts (1970). The short summary in this paragraph is only intended to give the reader sufficient background to enable him to obtain a picture of how the chloroplast cytochromes function in this electron transport, although this picture can only be considered as preliminary and subject to later modifications. Some of this will have to be discussed again in the chapter on cytochrome f (Vc) and on bacterial cytochromes (VI). By the pioneer studies of Emerson, Blinks, Duysens, Witt et al., Kok and others, it has been shown that photosynthesis in higher plants and algae, even in primitive blue-green algae (Biggins, 1967; Rurainski et al., 1970) depends on the co-action of at least two photochemical systems, the one (system I, hvI) specifically stimulated by light of longer wavelength than 700 nm, absorbed by a special form of chlorophyll a (P-700) and the second (system II, hvII) more strongly stimulated particularly by red light of a wavelength about 650 nm absorbed by the remainder of the chlorophyll. According to a frequently adopted scheme (Boardman, 1968) hvI enables electron transfer against the chemical gradient between cytochrome f (potential +0·36 V) or plastocyanin, a Cu-protein (potential +0·37 V) for the reduction of $NADP^+$ (−0·34 V) via a hypothetical primary reductant (Z) of photooxidized $P-700^+$; whereas light energy hvII absorbed by chlorophyll a enables electron transfer against the chemical gradient from the water to oxygen reaction via a primary reductant Q (or E, or C-550) having an absorption band at 500 nm in reduced form, but probably not a cytochrome, to cytochrome b-559 (but see below). Photosystem II is absent in bacteria and also in agranal bundle sheath of chloroplasts (Woo et al., 1970) and in the heterocysts of blue-green algae (Thomas, 1970). This hypothesis and the different roles which cytochromes b_6, b-559 and f were assumed to play in the electron transfer in photosynthesis have been outlined by Boardman (1968), see also Rumberg Witt et al. (1963, 1965, 1967). Boardman and co-workers first succeeded in a partial separation of the two reaction centres by fractional centrifugation in digitonin suspensions; this has been confirmed by others, e.g. Knaff and Arnon (cf. Briantais, 1969). The scheme of Boardman has been adopted by others (Ben Hayyim and Avron, 1970; for the green algae *Chlamydomonas* by Levine, 1966; Levine and Gorman, 1966; Levine et al., 1966; Hind, 1968; Givan and Levine, 1969; Garnier and Maroc, 1970; Hiyama et al., 1970); whereas Hiyama et al. (1969) had found only cytochromes b_6 and f. Weikard (1968) found b-563 (b_6) in the cyclic electron transport stimulated by photosystem I.

Photosystems II and I were assumed to be connected by an electron transfer chain including cytochrome b-559, near photosystem II, to cytochrome f and plastocyanin, near photosystem I. Originally, cytochrome b_6 had been assumed to play the role ascribed in this scheme to b-559 (Bendall and Hill, 1960), but there is now good evidence that cytochrome b_6 is involved in cyclic electron transport stimulated by photosystem I, carrying electrons back from ferredoxin to cytochrome f (cf. Hind and Olson, 1969; Hiyama et al., 1970). It is rapidly, but transiently reduced on illumination with the shorter λ light of photosystem II (Bendall, 1968). Hiyama et al. (1970) postulate two cyclic pathways only one involving cytochrome b_6 in a pale green mutant of *Chlamydomonas*.

Boardman also confirmed the original findings of Knaff and Arnon (1969a,b) that cytochrome b-559 was rapidly photooxidized at very low temperatures by light of photosystem II. In fact, the low temperature photooxidation of cytochrome f observed by Chance et al. (1967) was probably photooxidation of b-559, not of f. The photooxidation of b-559 depends on photosystem II, is inhibited by DCMU, an inhibitor of photosystem II, and is absent in agranal bundle sheath chloroplasts which lack it (Woo et al., 1970).

However, recently doubts have arisen about the position of cytochrome b-559. The 2-photosystem theory had postulated that the cyclic phosphorylation system I stimulated by long λ light, formed part of the non-cyclic photosystem II, stimulated by shorter λ light. However, Arnon and co-workers (Arnon et al., 1965, 1967; Arnon, 1967, 1970; McSwain and Arnon, 1968; Knaff and Arnon, 1969a,b, 1970; Arnon et al., 1970) suggested that there are three photosystems. Systems IIa (stimulated by short λ light) and IIb (stimulated by long λ light) were connected in series by one electron chain, but cyclic phosphorylation I (also stimulated by long λ light) was different from IIb and parallel to it. It had also a different electron chain. According to their scheme C-550 (better called P-550 or E), b-559 and plastocyanin formed the electron transfer zig-zag chain between photosystems IIa and IIb, whereas the electron chain of system I included cytochromes b_6 and f but not plastocyanin (cf. however,

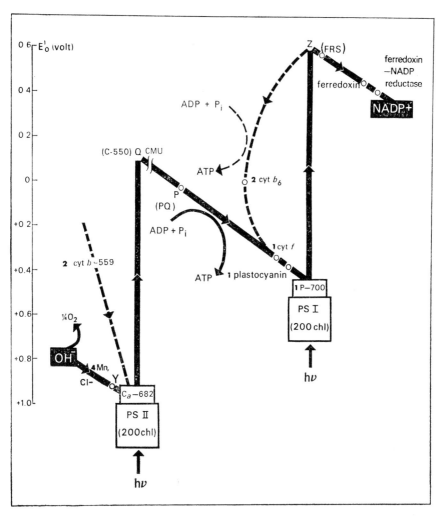

FIG. 3. Scheme of photosynthesis and the role of cytochromes in it according to Boardman et al., 1971.

Hauska et al., 1970 who found plastocyanin in System I). Not only are there difficulties in harmonizing this scheme with the quantum requirements for photosynthesis, but it fails to connect the system I and II, and the antagonistic effects of long and short λ irradiation on oxidation and reduction of cytochrome f are not easy to understand (Rumberg, 1965; Nishimura, 1967; Levine et al., 1966; Butler, 1967; Forti et al., 1963; Kok et al., 1964; Krendeleva et al., 1969). Hauska et al. (1970) found formation of 1 mole ATP between cytochromes b-559 and f, and another associated with cyclic electron flow in system I.

Boardman et al. (1971) have recently also thrown some doubt on their original scheme and the role of cytochrome b-559 (Fig. 3). They observed that after illumination at room temperature and rapid freezing, no photo-oxidation of b-559 could be observed. It is thus possible that b-559 interacts with E only after an alteration of the chloroplast on freezing has occurred. Their present scheme maintains the zig-zag electron transfer chain between photosystems II and I from E (C-550, which they confirm) to cytochrome f and plastocyanine (see Chapter Vc), but cytochrome b-559 is placed on a side path, which possibly also connects photosystems II and I.

There is thus no compelling reason to abandon the familiar zig-zag scheme of photosynthesis, but the position of the individual cytochromes in the electron transfer chains cannot as yet be considered established. Data on the relative amounts of cytochromes $b_6 + b$-559 or cytochrome f

TABLE VI

Molar ratios of chlorophyll to cytochromes
(according to Boardman, 1968)

Fraction	$\dfrac{Chl^a}{Cyt\ b^b}$	$\dfrac{Cyt\ b^b}{Cyt\ f}$	$\dfrac{Chl^a}{Cyt\ f}$	$\dfrac{Chl}{P\text{-}700}$
Chloroplasts	118	3·6	430	440
10,000 g^c	120	6·1	730	690
144,000^{c,d}	390	2·3	900	205

[a] Chl = Chl a + Chl b.
[b] Cyt b = Cyt b_6 + Cyt b-559.
[c] In digitonin separation.
[d] Cyt b mainly b_6.

to chlorophyll in chloroplasts and subchloroplast fragments are found in a paper of Boardman (1971) (see Table VI). The sites of photophosphorylation have been discussed by Izawa et al. (1967), Jagendorf and Uribe (1967) and Avron and Chance (1967).

REFERENCES

Arnon, D. I. (1967). *Physiol. Revs.* **47**, 317–358.
Arnon, D. I. (1970). *8th Internat. Congress Biochem., Switzerland*, **1970**. *Sympos.* **4**, pp. 137–139.
Arnon, D. I., Chain, R. K., McSwain, B. D., Tsujimoto, H. Y. and Knaff, D. B. (1970). *Proc. Nat. Acad. Sci. U.S.* **67**, 1404–1409.

Arnon, D. I., Tsujimoto, H. Y., Swain, B. D. (1965). *Nature* (London) **207**, 1367–1372.
Arnon, D. I., Tsujimoto, H. Y. and McSwain, B. D. (1967). *Nature* (London) **214**, 562–567.
Avron, M. (1967). In (D. R. Sanadi, ed.) "Current Topics in Bioenergetics", Academic Press, New York.
Avron, M. and Chance, B. (1967). *Brookhaven Sympos. Biol.* **19**, 149–160.
Baker, J. E. and Borchett, P. (1965). *Plant Physiol.* **40**, IV.
Bendall, D. S. (1958). *Biochem. J.* **70**, 381–390.
Bendall, D. S. (1968). *Biochem. J.* **109**, 46P.
Bendall, D. S. and Hill, R. (1956). *New Phytologist* **55**, 206–212.
Bendall, D. S. and Hill, R. (1968). *Ann. Rev. Plant Physiol.* **19**, 167–186.
Ben Hayyim, G. and Avron, M. (1970a). *Europ. J. Biochem.* **14**, 205–213.
Ben Hayyim, G. and Avron, M. (1970b). *Europ. J. Biochem.* **15**, 155–160.
Bhagvat, K. and Hill, R. (1951). *New Phytologist* **50**, 112–120.
Biggins, J. (1967). *Plant Physiol.* **42**, 1447–1456.
Boardman, N. K. (1968). *Advances in Enzymol.* **30**, 1–79.
Boardman, N. K. and Anderson J. M. (1964). *Nature* (London) **203**, 166–167.
Boardman, N. K. and Anderson, J. M. (1967). *Biochim. et Biophys. Acta* **143**, 187–203.
Boardman, N. K., Anderson, J. M. and Hiller, R. G. (1971). *Biochim et. Biophys. Acta* **234**, 126–136.
Bonner, W. D. (1961). "Haematin Enzymes" Vol. **2**, pp. 479–497.
Bonner, W. D. (1967). "Methods, in Enzymol.", Vol. **10**, 126–137.
Bonner, W. D. and Bendall, D. S. (1968). *Biochem. J.* **109**, 47P.
Bonner, W. D. and Plesnicar, M. (1967). *Nature* (London) **214**, 616–617.
Bonner, W. D. and Slater, E. C. (1970). *Biochim. et Biophys. Acta* **223**, 349–353.
Bonner, W. D., Bendall, D. S. and Plesnicar, M. (1967). *Fed. Proc.* **26**, No. 2647.
Briantais, J. M. (1969). *Physiol. Veg.* **7**, 135–180.
Butler, W. L. (1967). *Brookhaven Sympos. Biol.* **19**, p. 184.
Chance, B. and Bonner, D. W. (1956). *Plant Physiol.* **40**, 1198–1204.
Chance, B., De Vault, D., Hildreth, W. W., Parsons, W. W. and Nishimura, M. (1967). *Brookhaven Sympos. Biol.* **19**, 115–125.
Chance, B. and Hackett, D. P. (1959). *Plant Physiol.* **34**, 33–49.
Chance, B. and Sager, R. (1957). *Plant Physiol.* **32**, 548–561.
Chance, B., Bonner, W. D. and Storey, B. T. (1968). *Ann. Rev. Plant Physiol.* **19**, 295–320.
Chance, B., Schleyer, H. and Legallais, V. (1963). "Microalgae and Photosynthetic Bacteria", pp. 337–346.
Colmano, G. and Wolken, J. J. (1963). *Nature* (London) **198**, 783–784.
Cramer, W. A. and Butler, W. L. (1967). *Biochim. et Biophys. Acta* **143**, 332–339.
Davenport, H. E. (1952). *Nature* (London) **170**, 1112–1114.
Dawson, A. P., Cox, G. F. and Selwyn, M. J. (1968). *Biochem. Biophys. Res. Commun.* **32**, 579–587.
De Kouchkovsky, Y. (1959). *Compt. Rend.* **248**, 3597–3599.
Doering, G., Bailey, J. L., Kreutz, W., Weikard, J., Witt, H. T., Vater, J., Renger, G., Stiehl, H. H., Seifert, K., Reinwald, E., Siggel, U. and Rumberg, B. (1968). *Naturwiss.* **55**, 217–224.
Epel, B. L. and Butler, W. L. (1970). *Plant Physiol.* **45**, 728–734.
Erecinska, M. and Storey, B. T. (1970). *Plant Physiol.* **46**, 618–624.
Estabrook, R. W. (1961). "Haematin Enzymes", Vol. **2**, pp. 436–457.

Fan, H. N. and Cramer, W. A. (1970). *Biochim. et Biophys. Acta* **216**, 200–207.
Forti, G., Bertole, M. L. and Parisi, B. (1963). *Biochem. Biophys. Res. Commun.* **10**, 384–389.
Garette, R. H. and Nason, A. (1967). *Proc. Nat. Acad. Sci.* **58**, 1603–1610.
Garnier, J. and Maroc, J. (1970). *Biochim. et Biophys. Acta* **205**, 205–219.
Givan, A. L. and Levine, R. P. (1969). *Biochim. et Biophys. Acta* **189**, 404–410.
Hackett, D. P. (1957). *J. Exptl. Biol.* **8**, 157–171.
Hackett, D. P. (1959). *Ann. Rev. Plant. Physiol.* **10**, 113–146.
Hackett, D. P. and Haas, D. W. (1958). *Plant Physiol.* **33**, 27–32.
Hackett, D. P., Haas, D. W., Griffiths, S. K. and Niederpruem, D. J. (1960). *Plant Physiol.* **35**, 8–19.
Hartree, E. F. (1957). *Advances in Enzymol.* **18**, 1–64.
Hauska, G. A., McCarty, R. E. and Racker, E. (1970). *Biochim. et Biophys. Acta* **197**, 206–218.
Hill, R. (1954). *Nature* (London) **174**, 501–503.
Hill, R. and Bendall, F. (1960). *Nature* **186**, 136–137.
Hill, R. and Scarisbrick, R. (1951). *New Phytologist* **50**, 98–111.
Hind, G. (1968). *Photochem. and Photobiol.* **7**, 369–375.
Hind, G. and Nakatani, H. Y. (1970). *Biochim. et Biophys. Acta* **216**, 223–225.
Hind, G. and Olson, J. M. (1966). *Brookhaven Sympos. Biol.* **19**, 188–193.
Hind, G. and Olson, J. M. (1968). *Ann. Rev. Plant Physiol.* **19**, 249–282.
Hind, G., Nakatani, H. Y. and Izawa, S. (1967). *Fed. Proc.* **26**, Abstr. 2649.
Hiyama, T. (1969). *Fed. Proc.* **28**, Abstr. 1500.
Hiyama, T., Nishimura, M. and Chance, B. (1969). *Plant Physiol.* **44**, 527–534.
Hiyama, T., Nishimura, M. and Chance, B. (1970). *Plant Physiol.* **46**, 163–168.
Ikegami, I., Katoh, S. and Takamiya, A. (1968). *Biochim. et Biophys. Acta* **162**, 604–606.
Ikuma, H. and Bonner, W. D. (1967). *Plant Physiol.* **42**, 67–75.
Izawa, S., Connolly, T. N., Winget, G. D. and Good, N. E. (1966). *Brookhaven Sympos. Biol.* **19**, 169–184.
Jagendorf, A. T. and Uribe, E. (1967). *Brookhaven Sympos.* **19**, 215–241.
James, W. O. (1957). *Advances in Enzymol.* **18**, 281–318.
James, W. O. and Beevers, H. (1950). *New Phytologist* **49**, 353–374.
James, W. O. and Leech, R. M. (1960). *Endeavour* **19**, 108–114.
James, W. O. and Leech, R. M. (1964). *Proc. Roy. Soc. London B* **160**, 13–24.
James, W. O. and Lundegårdh, H. (1959). *Proc. Roy. Soc. London B* **150**, 7–12.
Jolchine, G., Reiss-Husson, F. and Kamen, M. D. (1969). *Proc. Nat. Acad. Sci. U.S.* **64**, 650–653.
Kasinsky, H. E., Shichi, H. and Hackett, D. P. (1966). *Plant Physiol.* **41**, 739–748.
Katoh, S. and San Peitro, A. (1965). *Biochem, Biophys. Res. Commun.* **20**, 406–410.
Keilin, D. (1925). *Proc. Roy. Soc. London* **98**, 312–339.
Knaff, D. B. and Arnon, D. I. (1969a). *Proc. Nat. Acad. Sci. U.S.* **63**, 956–962, 963–969.
Knaff, D. B. and Arnon, D. I. (1969b). *Proc. Nat. Acad. Sci. U.S.* **64**, 715–722.
Knaff, D. B. and Arnon, D. I. (1970). *Biochim. et Biophys. Acta* **223**, 201–203.
Kok, B. (1957). *Nature* (London) **179**, 583.
Kok, B. (1966). *Brookhaven Sympos. Biol.* **19**, 446–458.
Kok, B., Rurainski, H. J. and Harmon, E. A. (1964). *Plant Physiol.* **39**, 513–520.
Krendeleva, T. E., Korshunova, V. S. and Rubin, A. B. (1969). *Biofizika* **14**, 427–434.
Lance, C. and Bonner, W. D. (1968). *Plant Physiol.* **43**, 756–766.
Leech, P. M. (1963). *Biochim. et Biophys. Acta* **71**, 253–265.

Levine, R. P. (1969). *Ann. Rev. Plant. Physiol.* **20**, 523–540.
Levine, R. P. and Gorman, D. S. (1966). *Plant Physiol.* **41**, 1293–1300.
Levine, R. P., Gorman, D. S., Avron, M. and Butler, W. L. (1966). *Brookhaven Sympos.* **19**, 143–147.
Lundegårdh, H. (1954). *Physiol. Plantarum* **7**, 375–382.
Lundegårdh, H. (1955). *Physiol. Plantarum* **8**, 95–105.
Lundegårdh, H. (1958). *Nature* (London) **181**, 28–30.
Lundegårdh, H. (1961). *Nature* (London) **192**, 243–248.
Lundegårdh, H. (1962a). *Biochim. et Biophys. Acta* **57**, 352–368.
Lundegårdh, H. (1962b). *Physiol. Plantarum* **15**, 390–408.
Lundegårdh, H. (1964a). *Biochim. et Biophys. Acta* **88**, 37–56.
Lundegårdh, H. (1964b). *Proc. Nat. Acad. Sci. U.S.* **52**, 1587–1590.
McSwain, B. D. and Arnon, D. I. (1968). *Proc. Nat. Acad. Sci. U.S.* **61**, 959–966.
Martin, E. M. and Morton, R. K. (1955). *Nature* (London) **176**, 113–114.
Martin, E. M. and Morton, R. K. (1957). *Biochem. J.* **65**, 404–413.
Moore, C. W. D. (1967). Doctor Diss., Cambridge.
Morton, R. K. (1961). "Haematin Enzymes", Vol. **2**, pp. 498–499.
Müller, A., Rumberg, B. and Witt, H. T. (1963). *Proc. Roy. Soc. B* **157**, 313–322.
Nishimura, M. (1967). *Brookhaven Sympos. Biol.* **19**, 132–142.
Nishimura, M. (1968). *Biochim. et Biophys. Acta* **153**, 838–847.
Okunuki, K. (1939). *Acta Phytochim.* **11**, 27, 65.
Olson, J. M. (1955). *Nat. Acad. Sci.—Nat. Res. Council*, p. 174–178.
Olson, J. M. and Smillie, R. M. (1963). *Nat. Acad. Sci.—Nat. Res. Council*, **1145**, 42–56.
Perini, F., Kamen, M. D. and Schiff, J. A. (1964). *Biochim. et Biophys. Acta* **88**, 74–98.
Powls, R., Wong, J. and Bishop, N. I. (1969). *Biochim. et Biophys. Acta* **180**, 490–499.
Raison, J. K. and Smillie, R. M. (1969). *Biochim. et Biophys. Acta* **180**, 500–508.
Rumberg, B. (1965). *Biochim. et Biophys. Acta* **102**, 354–360.
Rurainski, H. J., Randles, J. and Hoch, G. E. (1970). *Biochim. et Biophys. Acta* **205**, 254–262.
Shichi, H. (1969). *Biochim. et Biophys. Acta* **188**, 339–341.
Shichi, H. and Hackett, D. P. (1962a). *Nature* (London) **193**, 776–777.
Shichi, H. and Hackett, D. P. (1962b). *J. Biol. Chem.* **237**, 2955–2958.
Shichi, H. and Hackett, D. P. (1962c). *Fed. Proc.* **21**, p. 46.
Shichi, H. and Hackett, D. P. (1966). *J. Biochem.* **59**, 84–88.
Shichi, H., Hackett, D. P. and Funatsu, G. (1963a). *J. Biol. Chem.* **238**, 1156–1161.
Shichi, H., Kasinsky, N. E. and Hackett, D. P. (1963b). *J. Biol. Chem.* **238**, 1162–1166.
Simon, E. W. (1957). *J. Exptl. Bot.* **8**, 20–35.
Smillie, R. M. and Levine, R. P. (1963). *J. Biol. Chem.* **238**, 4058–4062.
Storey, B. T. (1969). *Plant Physiol.* **44**, 413–421.
Storey, B. T. (1970). *Plant Physiol.* **45**, 447–454.
Storey, B. T. and Bahr, J. T. (1969). *Plant Physiol.* **44**, 115–125; 126–134.
Sugimura, Y., Toda, F., Murata, T. and Yakushiji, E. (1968). "Osaka", pp. 452–458.
Tagawa, K. and Shin, M. (1959). *J. Biochem.* **46**, 865–881.
Thomas, J. (1970). *Nature* (London) **228**, 181–183.
Vernon, L. P. and Avron, M. (1965). *Ann. Rev. Biochem.* **34**, 269–296.
Walker, D. A. (1970). *Nature* (London) **226**, 1204–1208.
Walker, D. A. and Crofts, A. R. (1970). *Ann. Rev. Biochem.* **39**, 389–428.

Webster, D. A. and Hackett, D. P. (1966). *Plant Physiol.* **41**, 599–605.
Weikard, J. (1968). *Z. f. Naturfschg.* **23**, 235–238.
Wiskich, J. T. (1967). "Methods in Enzymol.", vol. **10**, pp. 122–126.
Wiskich, J. T., Morton, R. K. and Robertson, R. N. (1960). *Austral. J. Biol. Sci.* **13**, 109–122.
Witt, H. T., Döring, G., Rumberg, B., Schmidt-Mende, P., Siggel, U. and Stiehl, H. A. (1967). *Brookhaven Sympos. Biol.* **19**, 161–167.
Witt, H. T., Müller, A. and Rumberg, B. (1963). *Nature* (London) **197**, 987–991.
Witt, H. T., Rumberg, B., Schmidt-Mende, P., Siggel, U., Skerra, B., Vater, J. and Weikard, J. (1965). *Angew. Chem.* **77**, 799–819.
Woo, K. C., Anderson, J. M., Boardman, N. K., Downton, W. J. S., Osmond, C. B. and Thorne, S. W. (1970). *Proc. Nat. Acad. Sci. U.S.* **67**, 18–25.
Yamanaka, T., Horio, T. and Okunuki, K. (1960). *Biochim. et Biophys. Acta* **40**, 349–351.
Yocum, C. S. and Hackett, D. P. (1957). *Plant Physiol.* **32**, 186–191.

G. CYTOCHROME b_2 OF YEAST L(+)LACTATE: CYTOCHROME c OXIDOREDUCTASE

1. Introduction

Oxidation of lactate by yeast is long known (Fürth and Lieben, 1922). Cytochrome c-linked dehydrogenases were first observed by Bernheim (1928) and by Green and Brosteaux (1936). Cytochrome b_2 was first characterized and purified by Dixon and co-workers (Dixon and Zerfas, 1939; Bach *et al.*, 1942, 1946) who also established that the haem was protohaem. This enzyme (EC 1.1.2.3, L(+)lactate:cytochrome c oxidoreductase) differs by its haem content from both animal lactate dehydrogenases (EC 1.1.1.27, 1.1.1.28) and yeast D(−)lactate:cytochrome c cytochrome, but is a flavoprotein (Slonimski and Tysarowski, 1958; Labeyrie *et al.*, 1958; Nygaard, 1959b; Singer *et al.*, 1960; Gregolin *et al.*, 1961) the L(+)lactate enzyme is a double-headed enzyme containing both protohaem and flavin-mononucleotide (FMN) as prosthetic groups (Appleby and Morton, 1954; Boeri *et al.*, 1955a,b,c; Nygaard, 1959a; Morton *et al.*, 1961; Boeri and Rippa, 1961a; Dickens *et al.*, 1961). In view of Morton's early death, the review by one of his earlier co-workers (Armstrong, 1965) is particularly valuable. In Chapter III cytochrome oxidase (aa_3) had been described as a double-headed enzyme containing two differently bound haem a groups. Bacterial cytochrome c containing, like cytochrome b_2, haem and flavin as prosthetic groups will be described in Chapter VI.

A certain confusion has arisen by Okunuki and co-workers (Okunuki, 1957; Yamanaka *et al.*, 1958; Yamashita and Okunuki, 1962) applying the name "cytochrome b_2" to a flavin-free and enzymically inactive split product of the enzyme. This crystalline haemoprotein has a molecular weight of about 20,000 and appears to be similar to but different from the "cytochrome b_2-core" of M.W. 11,000 obtained by tryptic digestion by Labeyrie et al. (1966, 1967).

2. Purification and crystallization

Cytochrome b_2 is a labile enzyme which easily loses its FMN on autoxidation and denaturation. There is no flavin fluorescence in the intact enzyme and its appearance is a sign of detachment of the flavin prosthetic group. The substrate, lactate, affords a protection of the enzyme and must be present in its purification (Morton et al., 1961). The presence of FMN is necessary for the substrate protection (Capeillère-Blandin, 1969). This is, however, not due as was then believed to the reduction of the enzyme by its substrate, but to a conformational change, since oxalate, a competitive inhibitor, affords a similar protection (Iwatsubo and Capeillère, 1967). Methods for purification have been worked out by Appleby and Morton (1959a), Morton et al. (1961), Boeri and Rippa (1961a,b), Rippa (1961) and Nygaard (1958, 1959b). Some of the observations of Nygaard (1960, 1961) must, however, be ascribed to an admixture of the D(−)lactate to the L(+)lactate enzyme.

Cytochrome b_2 has been crystallized in two forms. The first, square-shaped plates, obtained by slow precipitation at low ionic strength, "type I", still contains some deoxyribonucleic acid (DNA) (Appleby and Morton, 1954, 1959a; Rippa, 1961; Morton and Shepley, 1963b). This DNA was at first believed to be a characteristic constituent of cytochrome b_2 since it appeared to differ from total yeast DNA by composition and low molecular weight. It has later been shown, however, that it is not necessary for crystallization nor is it necessary for activity. By dialysis against strong ammonium sulphate solution, or by chromatography on DEAE-cellulose DNA-free crystals of active enzyme, "type II", are obtained in the form of hexagonal plates (Armstrong and Morton, 1961; Morton and Shepley, 1961, 1963a; Symons, 1965). Type I crystals can be reconstituted from type II crystals with DNA (Burgoyne and Symons, 1966). There is no evidence for a combination of cytochrome b_2 with DNA in solution. The DNA in type I crystals does not appear to be a random degradation product of total yeast DNA (Symons and Ellery, 1967) but is probably heterogeneous (Jackson et al., 1965) and, contrary to the earlier assumptions (Appleby and Morton, 1960; Montague and Morton, 1960; Mahler and Pereira, 1962) of it being a DNA of small molecular weight (10,000) is a DNA of high molecular weight ($> 100,000$) (Armstrong et al., 1963b; Burgoyne and Symons, 1966).

A slight modification of the enzyme during crystallization is indicated by the slight increase of K_m for L(+)lactate from 0·4–0·6 to 1·4–1·7 mM (Labeyrie and Slonimski, 1964; Nicholls et al., 1966).

3. Composition of cytochrome b_2

The molar ratio FMN to haem remains 1 : 1 in 7,000× purification. The enzyme contains 2 moles of protohaem and 2 moles of FMN in a molecular weight of 161,000 (DNA-free) or 172,000 in type I crystals (Appleby and Morton, 1959b; Baudras and Labeyrie, 1968). However, recently Jacq and Lederer (1970) have found the minimum M.W. per haem to be 53,000 not 80,000 and to form tetramers, not dimers. The amino acid composition has been determined by Appleby et al. (1960) and Morton et al. (1961). Remarkable is the complete absence of methionine which is a typical haem-binding constituent in cytochromes c. $S°_{20w}$ was found 8·36 (Armstrong et al., 1960) or 8·48 (Morton et al., 1961) somewhat dependent on concentration. The molecule is nearly spherical (f/fo = 1·18).

Horio et al. (1961) assumed it to be a complex of a haemoprotein and a flavoprotein, but it loses free FMN more readily than a flavoprotein (Armstrong et al., 1963a,b; Morton and Shepley, 1963a; Iwarsubo and Capeillère, 1967). It contains 0·077% of haem-iron, and protohaem is removed by acid acetone (Morton et al., 1961); the excess of non-haem iron found by Boeri and Tosi (1956) must have been present as an impurity. FMN is bound by its phosphate group and probably by hydrogen linkages of the isoalloxazine ring of the flavin to carboxyl and arginine-guanidine residues of the protein, with the lactate bound stereospecifically with its carboxyl group to the guanidine and its H and OH to the isoalloxazine ring (Dikstein, 1959). Morton et al. (1961) and Armstrong et al. (1963a) postulated a sulphhydryl linkage between flavin and protein. The reduced enzyme contains six, the oxidized five SH groups (Armstrong et al., 1963b; Pajot et al., 1967; Tsong and Sturtevant, 1969b). An SH-group was found to become accessible after FMN dissociation and the protein is aggregated unless this SH group is blocked by p-chloromercuriphenyl-sulphonate. Although this mercurial (Morton et al., 1961) removes the flavin (Pajot et al., 1967), the effect is not produced by all SH-reagents and is assumed to be directly caused by a conformational change. Although enzyme activity can be almost completely restored by FMN + mercaptoethanol (Sturtevant and Tsong, 1968a), the SH groups are not directly involved in the binding of FMN or the catalytic mechanism. FMN can also be removed by modification with 2·3–3 M urea (Armstrong et al., 1963; Boeri and Rippa, 1961b; Baudras and Labeyrie, 1968; see also Palamarczyk and Pacheka, 1968; and Rippa and Traniello, 1963), but in higher concentration it causes irreversible inactivation.

Under certain circumstances, it appears, however, possible to split cytochrome b_2 into a haemoprotein and a flavoprotein. This is indicated by sedimentation studies of Baudras and Labeyrie (1968). In 2 M guanidine, e.g. subunits of M.W. 17,000 (1·70 S) and 34,000 (2·30 S) were observed, smaller than the haem-flavin monomer of M.W. 53,000–80,000.

An interesting finding is the conversion of cytochrome b_2 into a c-type cytochrome on denaturation under specific conditions (Morton, 1961; Morton and Shepley, 1965). At least one SS-group must be in close neighbourhood of one vinyl group of the protohaem so as to condense with it forming a thioether bond and shifting the α-band to 550 nm. This SS group must be unmasked by heat, urea or p-chloromercuriphenylsulphonate previous to acetone precipitation. The prosthetic group is now firmly bound and can only be removed by Ag_2SO_4-acetic acid yielding haematohaem. The M.W. is 74–83,000 (but see below) and the E_0' is increased to $+280$ mV.

4. Absorption spectra and other physical properties

In crystals of type I, Appleby and Morton (1959a) and Morton et al. (1961) found the following absorption maxima (λ in nm and ϵ mM in parentheses): $\alpha 556·6$ (38·8) $\beta 528$ (18·6) $\gamma 423$ (232) $\delta 330$ (52·2) 265 (199). The α and γ bands are unusually high for cytochromes b and Pajot and Groudinsky (1970) found ϵ_{mM}^{γ} to be in fact, 183, not 232, identical with that for the "b_2" core (Labeyrie et al., 1966). The flavin contribution itself to these bands and also the DNA contribution is small (Appleby and Morton, 1954) but interaction between haem and flavin possibly increases the absorbancies. In contrast to previous observations of Chance et al. (1956) and Lindenmayer and Estabrook (1958), the α-band is not split into two at the temperature of liquid air but is only shifted to 555·2 nm. The oxidized enzyme has a ferrihaemochrome spectrum (γ-band 413 nm).

DNA-free cytochrome b_2 crystallizes in the trigonal system (unit cell $a = 168·2$ Å, $c = 112·0$ Å), whereas DNA-containing b_2 crystallizes in the tetragonal system (unit cell $a = 99·1$ Å, $c = 87·01$ Å), both having a M.W. of 235,000 (tetramer) (Monteilhet and Risler, 1970) or $92 \times 82 \times 26$ Å in square plates (Burgoyne et al., 1967).

The EPR spectrum of both cytochrome b_2 itself and "b_2-core" greatly resemble those of cytochrome b_5 ($g = 2·99$, 2·28, 1·49) and confirm the low spin type, with a small amount of high spin compound ($g = 6·2$) in the ferric compound (Watari et al., 1967). FMN has no influence on the EPR haem spectrum.

As has been stated above, the intact enzyme does not fluoresce and fluorescence after alteration is due to free FMN.

Recent results in ORD and CD spectra (Iwatsubo and di Franco, 1968; Iwatsubo et al., 1968; Iwatsubo and Risler, 1969; Sturtevant and Tsong, 1968a, 1969; Tsong and Sturtevant, 1969a,b) agree in many points. The results in the short U.V. λ-range (190 nm) indicate an α-helix content of 20–25%; the CD in this region does not depend on iron valency or FMN.

The b_3-core, however, shows random coiling and a high proline content. In the Soret region there is a considerable effect of the haem iron valency on the CD amplitude, indicative of a conformational change; and there is also a similar amplitude change by the substrate. The most marked effect, however, is that caused by FMN. Only in its presence (DNA being absent) is the Soret CD effect negative, while the effect is positive in FMN-free enzyme whether produced by urea, by high pH (9·0) or by trypsin. Only after removal of FMN does cytochrome b_2 resemble other haemoproteins. The French and Yale workers have, however, drawn very different conclusions from their experiments. The Yale authors conclude that there is direct interaction between FMN and haem, essential for both activity and spectrum, and that both must be in close contact so that direct electron transfer from FMN to haem occurs (see also Morton and Sturtevant, 1964). The French authors, however, conclude from experiments on the quenching of fluorescence of tryptophan and tyrosine, that haem and flavin are rather distant from one another, probably even in different subunits of cytochrome b_2, so that there is only slow electron transfer between them (Iwatsubo et al., 1968; see also Chance, 1961).

Data on the oxidation-reduction potential of cytochrome b_2 are contradictory. Boeri and Cutolo (1957) found a rather high potential ($+220$ mV) while Hasegawa and Ogura (1961, 1962a) report $+120$ mV at pH 6–8, with $\Delta -60$ mV/pH above pH 8. Finally Labeyrie et al. (1966) report $+0$ mV for b_2, -28 mV for b_2-core.

5. Reaction mechanism, acceptor and substrate specificity and kinetics

In the overall reaction with L(+) lactate as electron donor, ferricytochrome c reacts with almost the same rate as ferricyanide or dyes as acceptors (Morton et al., 1961). The activation energies found by Appleby and Morton (1959) for ferricytochrome c, ferricyanide and methylene blue, were 7·2, 7·6 and 8·7 kcals respectively. The rate of the reaction with molecular oxygen is far slower, only about 7% as effective (Morton et al., 1961; Boeri and Rippa, 1961a). In agreement with these findings Mahler et al. (1964) have shown that the activity of cytochrome b_2 in "petite yeast" which lacks cytochrome oxidase, was practically the same as in normal, wild-type yeast. The turnover number in moles ferricyanide per min. per mole enzyme is 16–17,000 (Boeri and Rippa, 1961a; Rippa, 1961). The overall reaction can be written: 3 lactate + 2 FMN + haem $Fe^{3+} \rightarrow 3$ pyruvate + 2 $FMNH_2$ + haem Fe^{2+} (Morton and Shepley, 1963a). The electron chain appears to be: lactate $\rightarrow$ FMN $\rightarrow$ haem $\rightarrow$ cyt. c, but ferricyanide can apparently react directly with FMN or with haem according to the conditions (Appleby and Morton, 1954; Boeri and Tosi, 1956; Morton, 1961; Ogura and Hasegawa, 1962; Hasegawa, 1962b;

Hinkson and Mahler, 1963; Ogura and Nakamura, 1966). The reaction with ferricytochrome c follows first order kinetics, that with ferricyanide depending on conditions first- or zero-order. Ferricyanide inhibits the reduction of ferricytochrome c by lactate (Morton and Sturtevant, 1964) but it can be shown by the stopped-flow method that reduction of the haem of b_2 is not essential for ferricyanide reduction. Km for cytochrome c was found to be 47 μM by Morton et al. (1961), but only 1·85 μM by Boeri et al. (1955). In contrast to succinate and NADH oxidation, that of L(+) lactate is caused by external cytochrome c (Mahler et al., 1964). Amytal inhibits the reaction with ferricytochrome c, but not with ferricyanide (Morton and Shepley, 1963a), while antimycin A does not inhibit the reaction with ferricytochrome c (Appleby and Morton, 1954).

The Km for L(+) lactate has been found to be 300–600 μM (Boeri et al., 1955b; Morton et al., 1961; Hinkson and Mahler, 1963; Labeyrie and Slonimski, 1964; Nicholls et al., 1966) before crystallization of the enzyme.

Various kinetics have been discussed by Morton et al. (1961), Hasegawa (1962b) and Hinkson and Mahler (1963). A ternary complex with substrate and acceptor appears to be formed, but with the great variety of possibilities in a double-headed enzyme, details remain problematical (Hiromi and Sturtevant, 1968). Suzuki and Ogura (1970) found the enzyme with the semi-quinone form of FMN to be an active intermediate by the stopped flow method. It was not found in the steady state on ferricyanide oxidation, but rapidly produced on autoxidation of the substrate-reduced enzyme.

Cytochrome b_2 does not react with D(−) lactate. Boeri et al. (1955b) found no inhibition of oxidation of L(+) lactate by D(−) lactate, but competitive inhibition was found by Dikstein (1959) and Hinkson and Mahler (1963). In contrast to this marked stereospecificity, there is a rather wide specificity between a variety of L(+)α-hydroxy acids (Yamashita, 1960; Yamashita et al., 1958). An increase of the size of R in R.CHOH.CH$_2$H decreases activity (Morton et al., 1961; Hasegawa and Ogura, 1961; Mahler and Huennekens, 1953; Dikstein, 1959) and carboxyl substituents appear to decrease the activity strongly. The function of L-glycerate as substrate and the inhibitory effects of D- and L-malate are still in doubt (Boeri and Rippa, 1961a, p. 563).

6. Biological function of cytochrome b_2

Cytochrome b_2 is not involved in oxidative phosphorylation, and its biological function remains uncertain, particularly as long as the formation of L(+) lactate or other stereochemically similar α-hydroxy acids in the yeast has not been clearly demonstrated. However, the occurrence of the enzyme in "petite" and other respiratory defective yeasts (Gregolin and Ghiretti-Magaldi, 1961; Sherman and Slonimski, 1964; Mahler et al.,

1964; Lorenc and Palomarczyk, 1968) increases the possibility that it has an enzyme function. At least some of the enzyme of yeast is found in mitochondria (Vitols and Linnane, 1961). The results of Singer et al. (1961) and Gregolin et al. (1961) and Mahler et al. (1964) do not substantiate the earlier assumption of Slonimski and Tysarovska, 1958; Labeyrie et al., 1959; Kattermann and Slonimski, 1960; and Nygaard, 1960b that the $D(-)$ enzyme is a precursor of cytochrome b_2.

It is of interest that a 5–6× more active $L(+)$ dehydrogenase, similar to, but not identical with, cytochrome b_2 of yeast, is found in the yeast *Hansenula anomala* (Iwatsubo and Risler, 1969; Sturtevant and Tsong, 1969). Morita and Mifuchi (1970) found a respiration-deficient mutant induced by a carcinogenic agent (4-nitroquinoline) to contain cytochrome b_2. The haemoprotein obtained by Yoshida and Kumaoka (1969) from anaerobically grown yeast and called "C_{556}" was probably cytochrome b_2. It had previously been observed by many workers, sometimes labelled b_1, sometimes b_5 (see e.g. Ephrussi and Slonimski, 1950; Lindemayer and Estabrook, 1958; Labbe and Chaix, 1964; Ishidata et al., 1969).

REFERENCES

Appleby, C. A. and Morton, R. K. (1954). *Nature* (London) **173**, 749–750.
Appleby, C. A. and Morton, R. K. (1959a). *Biochem. J.* **71**, 492–499.
Appleby, C. A. and Morton, R. K. (1959b). *Biochem. J.* **73**, 539–550.
Appleby, C. A. and Morton, R. K. (1960). *Biochem. J.* **75**, 258–269.
Appleby, C. A., Morton, R. K. and Simmonds, D. H. (1960). *Biochem. J.* **75**, 72–76.
Armstrong, J. McD. (1965). *Annals. New York Acad. Sci.* **119**, 877–887.
Armstrong, J. McD. and Morton, R. K. (1961). *Austral. J. Sci.* **24**, 137.
Armstrong, J. McD., Coates, J. H. and Morton, R. K. (1960). *Nature* **186**, 1033–1034.
Armstrong, J. McD., Coates, J. H. and Morton, R. K. (1961). "Haematin Enzymes", Vol. **2**, pp. 569–572.
Armstrong, J. McD., Coates, J. H. and Morton, R. K. (1963a). *Biochem. J.* **86**, 136–145.
Armstrong, J. McD., Coates, J. H. and Morton, R. K. (1963b). *Biochem. J.* **88**, 266–276.
Bach, S. J., Dixon, M. and Zerfas, L. G. (1942). *Nature* **149**, 21.
Bach, S. J., Dixon, M. and Zerfas, L. G. (1946). *Biochem. J.* **40**, 229–239.
Baudras, A. and Labeyrie, F. (1968). "Osaka", pp. 601–612.
Bernheim, F. (1928). *Biochem. J.* **22**, 1178–1192.
Boeri, E. and Cutolo, E. (1957). *Boll. Soc. Ital. biol. sper.* **33**, 1711–1714.
Boeri, E. and Rippa, M. (1961a). "Haematin Enzymes", Vol. **2**, pp. 524–533, 563–564, 568–569.
Boeri, E. and Rippa, M. (1961b). *Arch. Biochem. Biophys.* **94**, 336–341.
Boeri, E. and Tosi, L. (1956). *Arch. Biochem. Biophys.* **60**, 463–475.
Boeri, E., Cutolo, E. and Tosi, L. (1955a). *Boll. Soc. Ital. biol. sper.* **31**, 1392–1393.
Boeri, E., Cutolo, E., Luzzati, M. and Tosi, L. (1955b). *Arch. Biochem. Biophys.* **56**, 487–499.

Boeri, E., Cutolo, E., Luzzati, M. and Tosi, L. (1955c). *Congr. Intern. Biochim.*, *Resumes commun.*, *3rd Congr.* Brussels, p. 56.
Burgoyne, L. A. and Symons, R. A. (1966). *Biochim. et Biophys. Acta* **129**, 502–510.
Burgoyne, L. A., Dyer, P. Y. and Symons, R. H. (1967). *J. Ultrastruct. Res.* **20**, 20–32.
Capeillère-Blandin, C. (1969). *FEBS Letters* **4**, 311–315.
Chance, B. (1961). "Haematin Enzymes", Vol. **2**, pp. 564–566.
Chance, B., Klingenberg, M. and Boeri, E. (1956). *Fed. Proc.* **15**, p. 231.
Dickens, F. (1961). "Haematin Enzymes", Vol. **2**, pp. 558–560.
Dikstein, S. (1959). *Biochim. et Biophys. Acta* **36**, 397–401.
Dixon, M. and Zerfas, L. G. (1939). *Nature* (London) **143**, 557.
Ephrussi, B. and Slonimski, P. P. (1950). *Biochim. Acta* **6**, 256–267.
Fürth, O. and Lieben, F. (1922). *Biochem. Z.* **128**, 144–168.
Green, D. E. and Brosteaux, J. (1936). *Biochem. J.* **30**, 1389–1408.
Gregolin, C. and Ghiretti-Magaldi, A. (1961). *Biochim. et Biophys. Acta* **54**, 62–66.
Gregolin, C., Singer, T. P., Kearney, E. B. and Boeri, E. (1961). *Annals New York Acad. Sci.* **94**, 780–797.
Hasegawa, H. (1962a). *J. Biochem.* **52**, 5–15.
Hasegawa, H. (1962b). *J. Biochem.* **52**, 207–213.
Hasegawa, H. and Ogura, Y. (1961). "Haematin Enzymes", Vol. **2**, pp. 534–543.
Hinkson, J. W. and Mahler, H. R. (1963). *Biochemistry* **2**, 209–216.
Hiromi, K. and Sturtevant, J. M. (1968). In (K. Yagi, ed.) "Flavins and Flavoproteins". Biol. Conf. 1967 CA **71**(6), 45791.
Horio, T., Yamashita, J., Yamanaka, T., Nozaki, M. and Okunuki, K. (1961). "Haematin Enzymes', Vol. **2**, pp. 552–558, 560–562.
Ishidate, K., Kawaguchi, K., Tagawa, B. and Hagihara, B. (1969). *J. Biochem.* **65**, 375–383.
Iwatsubo, M. and Capeillère, Ch, (1967). *Biochim. et Biophys. Acta* **146**, 349–366.
Iwatsubo, M. and di Franco, A. (1968). "Osaka", pp. 613–625.
Iwatsubo, M., Baudras, A., di Franco, A., Capeillère, Ch. and Labeyrie, F. (1968). In "Flavins and Flavoproteins" (K. Yagi, ed.) p. 41, Univ. Park Press, Baltimore.
Iwatsubo, M. and Risler, J. L. (1969). *Europ. J. Biochem.* **9**, 280–285.
Jackson, J. F., Kornberg, R. D., Berg, P., Rajehandary, U. L., Stuart, A., Khorana, H. G. and Kornberg, A. (1965). *Biochim. et Biophys. Acta* **108**, 243–248.
Jacq, C. and Lederer, F. (1970). *Europ. J. Biochem.* **12**, 154–157.
Kattermann, R. and Slonimski, P. P. (1960). *Compt. rend.* **250**, 220–221.
Labbe, R. F. and Chaix, P. (1964). *Comptes. rend.* **258**, 1645–1647.
Labeyrie, F. and Slonimski, P. P. (1964). *Bull. soc. Chim. biol.* **46**, 1793–1828.
Labeyrie, F., Slonimski, P. P. and Naslin, L. (1959). *Biochim. et Biophys. Acta* **34**, 262–265.
Labeyrie, F., di Franco, A., Iwatsubo, M. and Baudras, A. (1967). *Biochemistry* **6**, 1791–1797.
Labeyrie, F., Groudinski, O., Jacquet-Armand, Y. and Naslin, L. (1966). *Biochim. et Biophys. Acta* **128**, 492–503.
Lindenmayer, A. and Estabrook, R. W. (1958). *Arch. Biochem. Biophys.* **78**, 66–82.
Lorenc, R. and Palamarczyk, G. (1968). *Bull. Acad. Polon. des Sciences, Serie Sci. Biol.* **6**, 83–87.
Mahler, H. R. and Huennekens, F. M. (1953). *Biochim. et Biophys. Acta* **11**, 575–583.
Mahler, H. R. and Pereira, H. da S. (1962). *J. Mol. Biol.* **5**, 325–347.

Mahler, H. R., Mackler, B., Grandchamp, S. and Slonimski, P. P. (1964). *Biochemistry* **3**, 668–671.
Montague, M. D. and Morton, R. K. (1960). *Nature* **187**, 916–917.
Monteilhet, C. and Risler, J. L. (1970). *Europ. J. Biochem.* **12**, 165–169.
Morita, R. and Mifuchi, I. (1970). *Biochem. Biophys. Res. Commun.* **38**, 191–196.
Morton, R. K. (1961). *Nature* **192**, 727–731.
Morton, R. K. and Shepley, K. (1961). *Nature* **192**, 639–641.
Morton, R. K. and Shepley, K. (1963a). *Biochem. Z.* **338**, 122–139.
Morton, R. K. and Shepley, K. (1963b). *Biochem. J.* **89**, 257–262.
Morton, R. K. and Shepley, K. (1965). *Biochim. et Biophys. Acta* **96**, 349–356.
Morton, R. K. and Sturtevant, J. M. (1964). *J. Biol. Chem.* **239**, 1614–1624.
Morton, R. K., Armstrong, J. McD. and Appleby, C. A. (1961). "Haematin Enzymes", Vol. **2**, pp. 501–523, 562–563, 566–568.
Nicholls, R. G., Atkinson, M. R., Burgoyne, L. A. and Symons, R. N. (1966). *Biochim. et Biophys. Acta* **122**, 14–21.
Nygaard, A. P. (1958). *Biochim. et Biophys. Acta* **30**, 450.
Nygaard, A. P. (1959a). *Biochim. et Biophys. Acta* **33**, 517–521.
Nygaard, A. P. (1959b). *Biochim. et Biophys. Acta* **35**, 212–216.
Nygaard, A. P. (1960a). *Biochim. et Biophys. Acta* **40**, 85–92.
Nygaard, A. P. (1960b). *Arch. Biochem. Biophys.* **86**, 317.
Nygaard, A. P. (1961a). "Haematin Enzymes", Vol. **2**, pp. 544–551.
Nygaard, A. P. (1961b). *Annals. New York Acad. Sci.* **94**, 774–779.
Nygaard, A. P. (1961c). *J. Biol. Chem.* **236**, 1585–1588.
Nygaard, A. P. (1961d). *J. Biol. Chem.* **236**, 2779–2782.
Ogura, Y. and Hasegawa, H. (1962). *J. Biochem.* **52**, 275–278.
Ogura, Y. and Nakamura, T. (1966). *J. Biochem.* **60**, 77–85.
Okunuki, K. (1957). *Nature* (London). **179**, 959.
Pajot, P. and Groudinsky, O. (1970). *Europ. J. Biochem.* **12**, 158–167.
Pajot, P., Pell, K. S. and Sturtevant, J. M. (1967). *J. Biol. Chem.* **242**, 2555–2562.
Palamarczyk, H. and Pachecke, J. (1968). *Bull. Acad. Polon. Sci., Sci. Biol.* **16**, 89–95.
Rippa, M. (1961). *Arch. Biochem. Biophys.* **94**, 333–335.
Rippa, M. and Traniello, M. (1963). *Giorn. Biochim.* **12**, 211–221.
Sherman, F. and Slonimski, P. P. (1964). *Biochim. et Biophys. Acta* **90**, 1–15.
Singer, T. P., Kearney, E. B., Gregolin, C., Boeri, E. and Rippa, M. (1960). *Biochem. Biophys. Res. Commun.* **3**, 428–434.
Singer, T. P., Kearney, E. B., Gregolin, C., Boeri, E. and Rippa, M. (1961). *Biochim. et Biophys. Acta* **54**, 52–61.
Slonimski, P. P. and Tysarowski, W. (1958). *Compt. rend.* **246**, 1111–1114.
Sturtevant, J. M. and Tsong, T. Y. (1968a). *Fed. Proc.* **27**, Abstr. 1734.
Sturtevant, J. M. and Tsong, T. Y. (1968b). *J. Biol. Chem.* **243**, 2359–2366.
Sturtevant, J. M. and Tsong, T. Y. (1969). *J. Biol. Chem.* **244**, 4942–4950.
Suzuki, H. and Ogura, Y. (1970). *J. Biochem.* **67**, 277–289, 291–295.
Symons, R. H. (1965). *Biochim. et Biophys. Acta* **103**, 298–310.
Symons, R. H. and Ellery, R. W. (1967). *Biochim. et Biophys. Acta* **145**, 368–377.
Tsong, T. Y. and Sturtevant, J. M. (1969a). *J. Amer. Chem. Soc.* **91**, 2382–2384.
Tsong, T. Y. and Sturtevant, J. M. (1969b). *J. Biol. Chem.* **244**, 2397–2402.
Vitols, E. and Linnane, A. W. (1961). *J. Biophys. Biochem. Cytol.* **9**, 701–710.
Watari, H., Groudinsky, O. and Labeyrie, F. (1967). *Biochim. et Biophys. Acta* **151**, 592–594.
Yamanaka, T., Horio, T. and Okunuki, K. (1958). *J. Biochem.* **45**, 291–298.
Yamashita, J. (1960). *J. Biochem.* **48**, 525–538.

Yamashita, J. and Okunuki, K. (1962). *J. Biochem.* **52**, 117–124.
Yamashita, J., Higashi, T., Yamanaka, T., Nozaki, M., Mizushima, M., Matsubara, H., Horio, T. and Okunuki, K. (1957). *Nature* (London) **179**, 959.
Yamashita, J., Horio, T. and Okunuki, K. (1958). *J. Biochem.* **45**, 707–715.
Yoshida, Y. and Kumaoka, H. (1969). *Biochim. et Biophys. Acta* **189**, 461–463.

Chapter V

Cytochromes c

A. Introduction.
B. Cytochromes c of eukaryotic organisms from man to yeast.
 1. Preparation, purification and crystallization.
    References.
 2. Amino acid composition and sequence.
    a. Haem peptides.
    b. Amino acid composition.
    c. Linear amino acid sequence.
    d. Attempts at a complete synthesis.
    References.
 3. Tertiary structure.
    a. Haem linkage.
    b. Conformational change with change of iron valency.
    c. X-ray crystallography.
    References.
 4. Physicochemical properties.
    a. pH stability.
    b. Optical properties.
    c. Magnetochemical properties.
    d. Oxidation-reduction potentials.
 5. Compounds with iron ligands.
    References.
 6. Modifications and reactions.
    a. Modification and denaturation.
    b. Modification of amino acid side chains.
    c. Other reactions of cytochrome c.
    d. Immunological reactions.
    e. Lipid cytochrome c.
    References.
 7. Cytochromes c in various tissues and cells.
    a. In animal organs and cell organelles.
    b. In fungi.
    c. Mitochondrial cytochrome c in plants.

C. Cytochrome f and similar cytochromes of algal chromatophores.
   1. Cytochrome f.
   2. Algal cytochromes of type f.
   3. Models of photosynthesis.
   References.
D. Cytochrome c_1.
   1. Introduction.
   2. Preparation of cytochrome c_1.
   3. Properties.
   4. Occurrence and biological role.
   References.

A. INTRODUCTION

Cytochrome c was named and described in the classical work of D. Keilin (1925, 1926) which established its wide occurrence in cells from mammals to invertebrates and yeast. Its relative stability led to its early isolation in nearly pure state by Keilin and Theorell and their co-workers in the years 1935–1945, and finally to the first crystallization by Bodo in 1955 (see sub-chapter B1). Lemberg and Legge (1949) drew attention to the fact that "cytochrome c", like "haemoglobin", was a collective name covering a variety of species-specific cytochromes c.

The biological role of cytochromes in cellular respiration was established by Keilin (see Keilin, 1966), but today we know that it is not restricted to processes of cellular respiration. Cytochromes c play also an important role in photosynthetic processes, first established by R. Hill in 1951–1953 (see sub-chapter C) and in anaerobic dark processes of bacteria such as nitrate reduction and sulphate reduction (see Chapter VI). A short summary of the occurrence and of some properties of the cytochromes of type c is given in Table VII; the nature of its haem group and its covalent linkage to the protein has already been discussed in Chapter II.

Our knowledge of the individual cytochromes varies from a few spectroscopic observations for some to a complete knowledge of the primary structure and a great deal, though still incomplete, knowledge of the tertiary structure of many eukaryotic cytochromes c from man to yeast. The large Chapter V, B deals with them. The far less well known cytochromes c_1, so far mainly studied in the cellular respiration of mammals, but perhaps also present in lower organisms, are discussed in sub-chapter V, D, the cytochromes of type c engaged in photosynthetic processes, cytochrome f (or c_6) of chloroplasts and similar cytochromes of algal chromatophores in V, C, while Chapter VI will deal with the bacterial cytochromes. The chapter headings above and the large Chapter V, B are indicative of the imbalance of our knowledge of cytochromes c. Again while the comparative biochemistry and evolutionary significance of the

TABLE VII

Cytochromes c

Cytochrome	Occurrence	Main absorption bands (c^{2+}); γ in nm; in parentheses mM	Solubility and protein	E'_0 pH 7 mV.	Function	Remarks
c	Widespread, typical mitochondria of eukaryotic organisms from yeast to man	550 521 415 (30) (135)	Soluble, rather stable to heat and acids, MW 12–13,000; isoelectric point above 10. Many crystallized both as c^{2+} and c^{3+}. Complete amino acid sequence known for many	+255	Electron carrier in respiratory chain between cytochromes c_1 and aa_3. Not autoxidizable, does not react with CO	V B
c_1	In mitochondria of animals, possibly also present in plants and some bacteria	553 524 418	Membrane-bound, but solubilizable, thermolabile M.W. 37,000 per one Fe	+226	Electron carrier in respiratory chain between cytochromes b and c, not autoxidizable, does not react with CO	V D
c_2	In photosynthetic non-sulphur purple bacteria	551 521 417	Soluble, stable	+310 to +340	Electron carrier in photosynthesis. Not autoxidizable, does not react with CO. Reacts only slowly with cytochrome aa_3	VI
c_3	In *Desulphovibrio desulphuricans*	553 525 419	Soluble, thermostable M.W. 13,000 with 2 haems	−205	Electron carrier in sulphate reduction, autoxidizable	VI

TABLE VII (continued)

				E*	
c_4	In *Azotobacter vinelandii*	551 522 416	Soluble, stable. M.W. 12,000	+300	Electron carrier in the respiratory chain, but does not react with cytochrome aa_3. Not autoxidizable	VI
c_5	In *Azotobacter vinelandii*	555 526 420	As for c_4	+320	As for c_4	VI
$c_6 = f$	Chloroplasts of higher plants	554 525 423	Bound in chlorpolast thylakoids, solubilizable, M.W. 110,000 with 2 haems	+350 to 365	Electron carrier in plant photosynthesis, does not react with cytochrome aa_3	V C
f-like	Algal chromatophores in green, brown, red and blue algae	552 520 415 to 556 521 419 frequently 553	More readily extractable than cytochrome f. M.W. 16,500 to 23,000	+340 to +390	Electron carriers in algal photosynthesis, do not react with cytochrome aa_3, but with cytochrome cd of *Pseudomonas*	V C
c_x	Some blue algae	549	Autoxidizable, reacts with CO. M.W. 34,000?	−260	? oxidase	V C
c_x	Bacterial cytochromes of *E. coli*, *Pseudomonas*; photosynthetic non-sulphur bacteria, *Mycobacteriaceae*	Variable 551–554	Variable. Some (*Chromatium*, *Chlorobium*) contain flavin	from −200 to +350	Variable, respiratory photosynthetic, nitrate and sulphate reductases	VI
c' and cc'	Bacteria ("RHP")	Ferrous spectrum resembling that of ferromyoglobin	Autoxidizable, reacts with CO; soluble, M.W. 25,000?		? oxidase	VI

eukaryotic cytochromes c is well developed, only the outlines of an overall evolutionary biochemistry of cytochromes c become apparent (Chapter X).

Today, the structure of the eukaryotic (i.e. in organisms the cell of which contain nuclei) and a few other cytochromes c is one of the best known protein structures rivalling that of the haemoglobins, but extending over a far wider biological field. The establishment of the amino acid sequence, the primary structure, has become almost a routine procedure, though a laborious one, which can be carried out with less than 100 mg of a pure protein. This owes more than is always mentioned to the pioneers (Bergmann, Fruton and others) who established the way of action of proteolytic enzymes, in particular trypsin and chymotrypsin, and the chromatographic separation of peptides developed by Martin and Synge. This led in the hands of Sanger (insulin), Moor and Stein (ribonuclease) and Braunitzer, Konigsberg and Hill, Schroeder, and Ingram (haemoglobin) to the establishment of the primary structure of proteins, which for a cytochrome c was first established by Margoliash *et al*. A very close analogue of this cytochrome c, with one amino acid of 104 replaced by another, has now been synthesized (see B, 2d). The haem iron is now known, at least in the eukaryotic cytochromes c, to be linked to one nitrogen of histidine-imidazole and one sulphur of methionine (see B, 3). This is supported by studies of the physicochemical, optical and magnetochemical properties and in particular NMR studies of cytochromes c (see B, 4c) and of compounds of their peptides (see B, 2a), by studies on modification by reactions of the amino acid residues (see B, 6b) and by x-ray crystallographic studies (see B, 3c). Thus the knowledge of the tertiary structure of some cytochromes c may soon be as firmly established as that of myoglobin (Kendrew).

Some reviews of the subject have been written by Theorell (1943, 1951), Lemberg and Legge (1949), Paul (1951, 1960), Morton (1958), Lemberg (1958), Keilin (1966), Margoliash and Schejter (1966) and Paléus (1967).

B. THE CYTOCHROME C OF EUKARYOTIC ORGANISMS, FROM MAN TO FUNGI

1. Preparation, purification and crystallization

The remarkable stability of cytochrome c to drying, heat and acids (cf. e.g. Paul, 1948a,b), together with its great solubility in concentrated ammonium sulphate solution which allows the removal of many more easily precipitated protein impurities, was used in the pioneer experiments of Keilin (1930, 1933), Keilin and Hartree (1937) and Theorell (1935, 1936) for the preparation of cytochromes c from ox heart and yeast. Plant cytochromes were first purified by Goddard (1944). From our present knowledge of the moelcular weight of cytochromes c (see B2) (horse c 12,350, beef c 12,200, yeast c about 13,000) and the iron content of these preparations (0·34%) it can be concluded that they were about 64% pure. Keilin et al. used trichloroacetic acid, Theorell, Paléus and others 0·1 N sulphuric acid for extraction; Zeile and Reuter (1933) using sulphuric acid extraction and kaolin absorption obtained a yeast cytochrome of similar purity. Some of the impurity in these preparations was later identified as sialic (acetyl-neuraminic) acid by Henderson and Paléus (1963) and by Henderson and Ada (1965). Purer preparations of 0·43% iron, about 95% pure, were then obtained by Theorell and Åkeson (1939, 1941) and by Keilin and Hartree (1945) from beef heart (see also Tsou, 1951), by Paléus (1954) from beef, chicken and salmon hearts, and by Li and Tsou (1956) from yeast. This purification was achieved by electrophoresis by Theorell and Åkeson, and by ammonium sulphate fractionation at pH 10, near the isoelectric point by Keilin and Hartree. Electrophoresis and paper-electrophoresis were also used by Tint and Reiss (1950), Hagihara et al. (1956), Yamanaka et al. (1959), Paléus (1952, 1960) and Paléus and Porath (1963), but nowadays the preferred method is column chromatography through cationic exchange resins (Paléus and Neilands, 1950, see below) or filtration through Sephadex G-25 (Flatmark, 1964). Henderson and Rawlinson (1956) (cf. also Paléus and Paul, 1963) drew attention to the fact that iron content can be used as a standard of purity only if the presence of extraneous iron can be excluded. Dialysis against dilute ammonia instead of distilled water, e.g. fails to remove such an iron-rich impurity. The difference in molecular weight should also not be forgotten, a content of 0·43% iron indicates, e.g. 100% purity of yeast cytochrome c but only 95% of beef heart c. The iron content of electrophoretically purified beef heart c (0·456%) found by Tint and Reiss (1950) is in excellent agreement with the theoretical value of 0·457%, as in the 0·43% value for yeast c by Li and Tsou (1956); but some iron values of Paléus and Neilands (1950), Leaf et al. (1958) and Paléus (1962) are probably slightly too high.

More recently, Margoliash and Walasek (1967) distinguished between "purity" and homogeneity, the former denoting absence of colourless or iron-free proteins or other admixed compounds, but giving no indication of the presence of altered cytochrome c, as long as the alteration did not noticeably affect the molecular size. Two types of such alterations are now known: polymerization (Margoliash, 1962; Margoliash and Lustgarten, 1961) and the hydrolysis of asparagine and glutamine residues to aspartic acid and glutamic acid residues (Margoliash, 1962, Flatmark, 1964, 1966; Margoliash and Schejter, 1966). The stability of cytochrome c to heat, drying and acids is now known to be less complete than had previously been assumed and the intactness of the absorption spectrum is no certain sign of the absence of alterations. Even the absolute height of the α-band at 550 nm (ϵmM 27–30) (Massey, 1959; Van Gelder and Slater, 1962; Flatmark, 1964) appears to vary in cytochromes of different species. The use of the refinement by measurements of the ratio $E^{c^{2+}}_{550}/E^{c^{3+}}_{280}$ (Cameron, 1965; Margoliash and Walasek, 1967) is a better measurement for the absence of extraneous protein, but cannot be used indiscriminately, as it depends on the presence of aromatic acid residues, in particular tryptophan. The ratio appears to be always 1·25–1·28 in pure mammalian cytochrome c which contains only one trytophan residue (Hagihara et al., 1958g; Matsubara and Yasunobu, 1961; Sutherland, 1962; Flatmark, 1964), whereas it is 1·00–1·07 in those containing two tryptophans, e.g. the fish cytochromes (Kreill, 1963; Gürtler and Horstman, 1970). For the removal of alteration products of cytochrome c, their removal by chromatographic processes or electrophoresis is required.

The earlier methods of extraction with acids or treatment with organic solvents (still used by Morrison et al., 1960) are now known to cause such alterations (Margoliash 1954; Nozaki et al., 1957; Nozaki, 1960; Hagihara et al., 1958a,h; Yamanaka et al., 1959). They have been replaced by extraction by salt solutions, sulphite, ammonium sulphate, potassium chloride with (Hagihara et al., 1958a) or without preceding disorganization of the tissue with acetic acid (Margoliash, 1952, 1955; Boardman and Partridge, 1953; Hagihara et al., 1956b; Paléus and Theorell, 1957; Margoliash and Walasek, 1967; Lenaz and MacLennan, 1967). Margoliash and Walasek (1967) use aluminium sulphate solutions at pH 4·5. Even then it is not always possible to avoid alterations, particularly if autolytic processes, e.g. autolysis with ethyl acetate, have to be used as in the extraction of plant or yeast cytochromes (Hagihara et al., 1956a; Sels and Zanen, 1965).

After extraneous proteins have been removed by their readier precipitation by ammonium sulphate, the dialysed solutions are then subjected to chromatographic processes. With cytochromes of high isoelectric point near or above 10, this is achieved by cationic ion exchange resins such as Amberlite IRC50, CG50, XE-64. or Duolite CS-101, although zeolites have also been used (Paléus and Neilands, 1950; Margoliash, 1952; Boardman and Partridge, 1955; Tsou and Li, 1956; Hagihara et al., 1956c; Nozaki et al., 1957; Paléus and Theorell, 1957; Paléus, 1960; Soru and Rudescu, 1969). For more acidic cytochromes, such as the cytochrome f-like algal cytochromes and bacterial cytochromes c, diethylamino-ethyl-cellulose (Katoh, 1960), carboxymethyl-cellulose (Sutherland, 1962; Dixon and Thompson, 1968), Duolite CS-101 (Horio et al., 1960) are preferable. According to Yamanaka et al., (1959) low reducibility by ascorbate characterizes intact cytochrome c, but there appears to be no straightforward relationship (Flatmark, 1964; Skov and Williams, 1968; see also B 5b). Cytochrome c^{2+} is readily obtained from c^{3+} by reduction with ascorbate or dithionite followed by filtration through a Sephadex G-25 gel column (Horton, 1968). Cytochrome c^{2+} is usually less easily soluble than c^{3+}, and reduction is thus used for the crystallization of cytochromes which, as c^{3+}, are soluble in saturated ammonium sulphate solution.

Many cytochromes c have been crystallized and their number is increasing. Both c^{2+} and c^{3+} of beef heart (Kuby et al., 1956; Hagihara et al., 1956b; Paléus and Theorell, 1957; Hagihara et al., 1958a; Paléus, 1960) and of beef kidney (Hagihara et al., 1958e,f,g; 1959a) crystallize in plates or rosettes of plates. These are probably identical since Margoliash and Stewart (1965) have shown the identity of cytochrome c from heart, liver, kidney, brain and skeletal muscle of the pig, and pig cytochrome c does not differ from beef cytochrome c in its primary structure (see B2c).Cytochrome c of horse (heart and skeletal muscle) c^{2+} and c^{3+} was crystallized by Hagihara et al., (1958a), Blow et al., (1964) and Dickerson et al., (1967, 1968) in prisms, those of human heart in needles (Matsubara and Yasunobu, 1961; Paléus, 1962). Whale heart c has been crystallized by Minakami et al. (1958). Of bird cytochromes c, that of the king penguin was the first c to be crystallized by Bodo (1955) in thin plates. Cytochromes c^{2+} and c^{3+} of pigeon breast muscle crystallize in needles and plates (Hagihara et al., 1958b,d) and chicken heart c has been obtained in crystals by Paléus (1964). Of fish cytochromes c, those of tunny and bonito crystallize in thick plates, both c^{2+} and c^{3+} (Blow et al., 1964; Hagihara et al., 1957, 1958c; Takano et al., 1968), that of the primitive hagfish (*Myxine glutinosa*) in hexagonal plates (Paléus et al., 1969). From invertebrates, cytochrome c of the flight muscles of the moth, *Samia cynthia*, has been crystallized in large prisms by Chan and Margoliash (1966). Wheat germ c^{2+} and c^{3+} crystallize in prisms (Hagihara, 1958c,d, 1959b), rice embryo c in dodecahedra (Morita and Ida, 1968). Corn pollen c has been obtained crystalline by Podgieter (see Margoliash and Schejter, 1966) and soy bean c in good purity by Fridman et al. (1968). Yeast cytochromes of *Saccharomyces cerevisiae* c^{2+} and c^{3+}, have been obtained in needles or rods by Hagihara et al., (1956a), that of the yeast *Candida krusei* by Shirasaka et al. (1968). Some of the cytochrome f-like algal cytochromes have been crystallized, e.g. the c-553 of the red alga *Porphyra tenera* in needles and that of the brown alga *Petalonia fascia* in prisms (Yakushiji et al., 1959; Katoh, 1960; Sugimura et al., 1968), while *Euglena gracilis* cytochrome c crystallizes in large square plates (Mitsui and Tsushima, 1968). Finally, some bacterial cytochromes, e.g. cytochrome c_2, have been crystallized (see Chapter VI).

Crystallization is, however, no absolutely reliable evidence for purity and homogeneity. Crystalline whale heart cytochrome c still contained some myoglobin. The crystalline beef heart cytochrome c consists of a mixture of intact c and three other cytochromes c, some of the asparagine or glutamine residues of which have been hydrolysed (Flatmark, 1964). Blow et al. (1964) had found horse heart c^{3+} to crystallize in $P4_12_12$ space groups in needles.

According to Dickerson et al. (1967, 1968) pure horse c^{3+} crystallizes only in one form, of tetragonal $P4_1$ space group with only one molecule per unit cell of a = b = 58·45 Å and c = 42·34 Å. The crystals are square prisms elongated along the c axis, with the molecules packed around the screw axis in a spiral with a repeat at each fourth molecule. Only 45% of the crystal by weight is protein, the remainder inter-molecular liquid of crystallization. Hori and Morimoto (1970) have studied by EPR spectroscopy the orientation of the haem plate in single crystals of the bonito cytochrome c^{3+} which crystallizes in the space group $P2_12_12_1$.

REFERENCES

Blow, D. M., Bodo, G., Rossmann, M. G. and Taylor, C. P. S. (1964). *J. Mol. Biol.* **8**, 606–609.
Boardman, N. K. and Partridge, S. M. (1953). *Nature* (London) **171**, 208.
Boardman, N. K. and Partridge, S. M. (1955). *Biochem. J.* **59**, 543–552.
Bodo, G. (1955). *Nature* (London) **176**, 829–830.
Cameron, B. F. (1965). *Analyt. Biochem.* **11**, 164–169.
Chan, S. K. and Margoliash, E. (1966). *J. Biol. Chem.* **241**, 335–348.
Dickerson, R. E., Kopka, M. L., Borders, C. L., Varnum, J., Weinzierl, J. E. and Margoliash, E. (1967). *J. Mol. Biol.* **29**, 77–95.
Dickerson, R. E., Kopka, M. L., Weinzierl, J. E., Varnum, J. C., Eisenberg, D. and Margoliash, E. (1968). "Osaka", pp. 225–251.
Dixon, H. B. F. and Thompson, C. M. (1968) *Biochem. J.* **107**, 427–431.
Flatmark, T. (1964). *Acta. Chem. Scand.* **18**, 1517–1527.
Flatmark, T. (1966). *Acta. Chem. Scand.* **20**, 1487–1496.
Fridman, C., Lis, H., Sharon, N. and Katchalski, E. (1968). *Arch. Biochem. Biophys.* **126**, 299–304.
Goddard, D. R. (1944). *Amer. J. Bot.* **31**, 270–276.
Gürtler, L. and Horstmann, H. J. (1970). *Europ. J. Biochem.* **12**, 48–57.
Hagihara, B., Horio, T., Yamashita, J., Nozaki, M. and Okunuki, K. (1956a). *Nature* (London) **178**, 629.
Hagihara, B., Morikawa, I., Sekuzu, I., Horio, T. and Okunuki, K. (1956b). *Nature* (London) **178**, 630.
Hagihara, B., Horio, T., Nozaki, M., Sekuzu, I., Yamashita, J. and Okunuki, K. (1956c). *Nature* (London) **178**, 631.
Hagihara, B., Tagawa, K., Nozaki, M., Morikawa, I., Yamashita, J. and Okunuki, K. (1957). *Nature* (London) **179**, 249.
Hagihara, B., Morikawa, I., Sekuzu, I. and Okunuki, K. (1958a). *J. Biochem.* **45**, 551–563.
Hagihara, B., Yoneda, M., Tagawa, K., Morikawa, I. and Okunuki, K. (1958b). *J. Biochem.* **45**, 565–574.
Hagihara, B., Tagawa, K., Morikawa, I., Shin, M. and Okunuki, K. (1958c). *J. Biochem.* **45**, 725–735.
Hagihara, B., Sekuzu, I., Tagawa, K., Yoneda, M. and Okunuki, K. (1958d). *Nature* (London) **181**, 1588.
Hagihara, B., Sekuzu, I., Tagawa, M., Yoneda, M. and Okunuki, K. (1958e). *Nature* (London) **181**, 1589.
Hagihara, B., Tagawa, K., Sekuzu, I., Morikawa, I. and Okunuki, K. (1958f). *Nature* (London) **181**, 1590.

Hagihara, B., Morikawa, K., Tagawa, K. and Okunuki, K. (1958g). *Biochem. Preps.* **6**, 1–7.
Hagihara, B., Tagawa, K., Morikawa, I., Shin, M. and Okunuki, K. (1958h). *Proc. Japan Acad.* **34**, 169–171.
Hagihara, B., Tagawa, K., Sekuzu, I., Morikawa, I. and Okunuki, K. (1959a). *J. Biochem.* **46**, 11–17.
Hagihara, B., Tagawa, K., Morikawa, I., Shin, M. and Okunuki, K. (1959b). *J. Biochem,.* **46**, 321–327.
Henderson, R. W. and Ada, G. L. (1965). *Biochim. et Biophys. Acta* **77**, 513–515.
Henderson, R. W. and Paléus, S. (1963). *Acta Chem. Scand.* **17**, 110–116.
Henderson, R. W. and Rawlinson, W. A. (1956). *Biochem. J.* **62**, 21–29.
Hori, H. and Morimoto, H. (1970). *Biochim. et Biophys. Acta* **200**, 581–583.
Horio, T., Higashi, T., Sasagawa, M., Kusai, M., Nakai, M. and Okunuki, K. (1960). *Biochem. J.* **77**, 194–201.
Horton, A. A. (1968). *Analyt. Biochem.* **23**, 334–335.
Katoh, S. (1960). *Nature* (London) **186**, 138–139.
Keilin, D. (1925). *Proc. Roy. Soc. London B* **98**, 312–339.
Keilin, D. (1926). *Proc. Roy. Soc. London B* **100**, 129–151.
Keilin, D. (1930). *Proc. Roy. Soc. London B* **106**, 418–444.
Keilin, D. (1933). *Ergebn. d. Enzymfschg.* **2**, 239–271.
Keilin, D. (1966). "The History of Cell Respiration and Cytochrome" (J. Keilin, ed.) Cambridge Univ. Press.
Keilin, D. and Hartree, E. F. (1937). *Proc. Roy. Soc. London B* **122**, 298–308.
Keilin, D. and Hartree, E. F. (1945). *Biochem. J.* **39**, 289–292.
Kreil, G. (1963). *Z. physiol. Chem.* **334**, 154–166.
Kuby, S. A., Paléus, S., Paul, K. G. and Theorell, H. (1956). *Acta Chem. Scand.* **10**, 148.
Leaf, G., Gillies, N. E. and Pirrie, R. (1958) *Biochem. J.* **69**, 605–611.
Lemberg, R. and Legge, J. W. (1949). In "Hematin Compounds and Bile Pigments", pp. 347–358. Interscience, New York.
Lemberg, R. (1958). *Atti della Accad., Anat. Chirurg. di Perugia* **49**, 304–366.
Lenaz, G. and MacLennan, D. W. (1967). "Methods in Enzymol.", pp. 499–504.
Li, W. -C. and Tsou, S. -L. (1956). *Scientia Sinica* **5**, 663–674.
Margoliash, E. (1952). *Nature* (London) **170**, 1014.
Margoliash, E. (1954). *Biochem. J.* **56**, 529–534, 535–543.
Margoliash, E. (1955). *Biochem. Preps.* **5**, 33–37.
Margoliash, E. (1962). *Brookhaven Sympos. Biol.* **15**, 266–280.
Margoliash, E. (1972). Keilin Memorial Lecture, in print.
Margoliash, E. and Lustgarten, J. (1961). *Annals New York Acad. Sci.* **94**, 731–740.
Margoliash, E. and Schejter, A. (1966). *Adv. Protein Chem.* **21**, 113–286.
Margoliash, E. and Stewart, J. W. (1965). *Canad. J. Biochem.* **43**, 1187–1206.
Margoliash, E. and Walasek, O. F. (1967). "Methods of Enzymol.", Vol. **10**, pp. 339–348.
Massey, V. (1959). *Biochim. et Biophys. Acta* **34**, 255–256.
Matsubara, H. and Yasunobu, K. T. (1961). *J. Biol. Chem.* **236**, 1701–1705.
Minakami, S., Titani, K., Ishihura, H. and Takahashi, K. (1958). *J. Biochem.* **45**, 547–549.
Mitsui, A. and Tsushima, K. (1968). "Osaka", pp. 459–466.
Morita, Y. and Ida, S. (1968). *Agricult. and Biol. Chem.* (Japan) **32**, 441–447.
Morrison, M., Hollocher, T., Murray, R., Marinetti, G. and Stotz, E. (1960). *Biochim. et Biophys. Acta* **41**, 334–338.
Morton, R. K. (1958). *Revs. Pure and Applied Chem.* (Australia) **8**, 161–220.

Nozaki, M. (1960). *J. Biochem.* **47**, 592–599.
Nozaki, M., Yamanaka, T., Horio, T. and Okunuki, K. (1957). *J. Biochem.* **44**, 453–464.
Paléus, S. (1952). *Acta Chem. Scand.* **6**, 969–970.
Paléus, S. (1954). *Acta Chem. Scand.* **8**, 971–984.
Paléus, S. (1960). *Acta Chem. Scand.* **14**, 1743–1748.
Paléus, S. (1962). *Arch. Biochem. Biophys.* **96**, 60–62.
Paléus, S. (1964). *Acta. Chem. Scand* **18**, 1324–1325.
Paléus, S. (1967). *Bull. Soc. Chem. Biol.* **49**, 917–933.
Paléus, S. and Neilands, J. B. (1950). *Acta Chem. Scand.* **4**, 1024–1030.
Paléus, S. and Paul, K. G. (1963). In "The Enzymes", 2nd ed. (P. D. Boyer, H. Lardy and K. Myrbäck, eds.) vol. **8**, 97–112. Academic Press, New York and London.
Paléus, S. and Porath, J. (1963). *Acta Chem. Scand.* **17**, 57–61.
Paléus, S. and Theorell, H. (1957). *Acta Chem. Scand.* **11**, 905.
Paléus, S., Tota, B. and Liljeqvist, G. (1969). *Compar. Biochem. Physiol.* **31**, 813–817.
Paul, K. G. (1948a). *Acta Chem. Scand.* **2**, 430–439.
Paul, K. G. (1948b). *Acta Chem. Scand.* **2**, 557–560.
Paul, K. G. (1951). In "The Enzymes" (J. B. Sumner and K. Myrbäck, eds.), vol. 2, pp. 369–396. Academic Press, New York and London.
Paul, K. G. (1960). In 'The Enzymes" 2nd ed. (P. D. Boyer, H., Lardy and K. Myrbäck, eds.) vol. 3, pp. 216–238. Academic Press, New York and London.
Sels, A. A. and Zanen, J. (1965). *Arch. Internat. de Physiol. et de Biochim.* **73**, 532–543.
Shirasaka, M., Nakayama, N., Endo, A., Haneishi, R. and Okazaki, H. (1968). *J. Biochem.* **63**, 417–424.
Skov, K. and Williams, G. R. (1968). "Osaka", pp. 349–352.
Soru, E. and Rudescu, K. (1969) *J. Chromat.* **41**, 236–241.
Sugimura, Y., Toda, F., Murata, E. and Yakushiji, E. (1968). "Osaka", pp. 452–458.
Sutherland, I. W. (1962). *Canad. J. Biochem. Physiol.* **40**, 547–548.
Takano, T., Sugijara, A., Ando, O., Ashida, T., Kakudo, N., Horio, T., Sasada, Y. and Okunuki, K. (1968). *J. Biochem.* **63**, 808–810.
Theorell, H. (1935). *Biochem. Z.* **279**, 463–464.
Theorell, H. (1936). *Biochem. Z.* **285**, 207–218.
Theorell, H. (1943). *Ergebn. d. Enzymfschg.* **9**, 239–256.
Theorell, H. (1951). In "The Enzymes" (J. B. Sumner and K. Myrbäck, eds.), vol. 2, pp. 397–427.
Theorell, H. and Åkeson, Å. (1939). *Science* **90**, 67.
Theorell, H. and Åkeson, Å. (1941). *J. Amer. Chem. Soc.* **63**, 1804–1812.
Tint, H. and Reiss, W. (1950). *J. Biol. Chem.* **182**, 385–396, 397–403.
Tsou, C. L. (1951). *Biochem. J.* **49**, 362–367.
Tsou, C. L. and Li, W.-C. (1956). *Scientia Sinica* **5**, 253–262.
Van Gelder, B. F. and Slater, E. C. (1962). *Biochim. et Biophys. Acta* **58**, 593–595.
Yakushiji, E., Sugimura, Y., Sekuzu, I., Morikawa, I. and Okunuki, K. (1968). *Nature* **185**, 105–106.
Yamanaka, T., Mizushima, T., Nozaki, M., Horio, T. and Okunuki, K. (1959). *J. Biochem.* **46**, 121–132.
Zeile, K. and Reuter, F. (1933). *Z. physiol. Chem.* **221**, 101–116.

2. Amino acid composition and sequence of cytochromes c

(a) *The haem peptides.* Studies on the digestion of cytochrome c with proteolytic enzymes, pepsin, trypsin, chymotrypsin and papain were first carried out by Tsou (1949, 1951a,b). Attempts were then directed towards the isolation of peptides containing the covalently linked haem known to be bound to two cysteine residues by thioether linkages (see Chapter II). After Tuppy and Bodo (1954a) had obtained a tetrapeptide Lys-Cys-Ala-Gln and a tri-peptide Gln-Cys-His by sulphuric acid hydrolysis, these authors isolated a haem nonapeptide Cys-Ala-Gln-Cys-His-Thr-Val-Glu-Lys (amino acid residues 14-22, see Table VIIIa) by tryptic hydrolysis of horse cytochrome c followed by hydrochloric acid hydrolysis at 37° and a decapeptide (residues 14-23, with Lys in position 13) from horse, cow and pig cytochrome c (Tuppy and Bodo, 1954c). By pepsin, Tuppy and Paléus (1955) and Ehrenberg and Theorell (1955a) obtained an undecapeptide (residues 11-21), see also Leaf and Gillies (1955), Harbury and Loach (1959), Margoliash et al. (1959), Margoliash (1961). Paléus (1954) recognized that the haem linkage in the peptides differed from that in cytochrome c itself. The peptides lacked cytochrome c activity in the succinoxidase system, but were very active oxidases of cytochrome c and as peroxidases (Baba et al. 1969; Tu et al., 1968). They were autoxidizable and combined with carbon monoxide. Different from cytochrome c, they were high spin iron complexes (Paléus et al., 1955) in the ferric state, but reacted with histidine to give low spin compounds (Ehrenberg and Theorell, (1955b)). In the aggregated form, however, low spin iron complexes were found by EPR at low temperature (Der Vartanian, 1970). It is now known that there are always two amino acid residues (15 and 16) between the two haem-combined cysteines; they are usually Ala and Gln, but not invariably so, e.g. Ala is replaced by Ser in chicken cytochrome c. His in position 18 is invariable and combined by its imidazole to the haem-iron. The oxidation-reduction potential of the haem-peptides was low; Minakami et al. (1957) found -300 mV, while Harbury and Loach (1959) reported -195 mV for the undecapeptide. The linkage of haem-iron to amino acid residues will be discussed in Section B3. It is now known that the haem iron is bound to the thioether-sulphur of methionine rather than to a second histidine-imidazole as had long been assumed. Neither a second His nor Met are present in the haem-peptides and the only nitrogen which may combine with haem iron of the undecapeptide on the side opposite to that of His-18 imidazole is the ϵ-NH_2-group of Lys-14 (Margoliash et al., 1959; Harbury and Loach, 1960a). By digestion with trypsin the undecapeptide was converted to the octapeptide Cys-Ala-Gln-Cys-His-Thr-Val-Glu (residues 14-21) which no longer contains lysine. The only nitrogenous group able to combine with haem-iron apart from the His-18 imidazole was here the free α-NH_2 group of Cys-14 and treatment by nitrous acid indeed removes this iron-ligand. Only the ϵ-NH_2 group of lysine in the undecapeptide can cause aggregation to a dimer in which two haems with their imidazoles outside are combined by the ϵ-NH_2 group in a head to tail arrangement (Harbury and Loach, 1960a,b; Urry, 1966).

By chymotryptic digestion of horse cytochrome c, Margoliash and Smith (1961, 1962) obtained a large haemopeptide of 26 residues, and observed that there was no free N-terminal NH_2 group. It was shown by Kreil and Tuppy (1961) and by Margoliash et al. (1961) that the N-terminal group is acetylglycine. A hexadecapeptide comprising the amino acid residues 11–26 (Val-His) was isolated from horse cytochrome c (Margoliash, 1962a,b) as well as from turtle cytochrome c (Chan and Margoliash, 1966a,b). Paléus and Tuppy (1959) listed 6 tryptic haemopeptides of 11–13 residues isolated from the cytochromes c of horse, ox, pig, chicken, salmon, silkworm, yeast and the purple non-sulphur bacterium *Rhodospirillum rubrum*. A short haemopeptide of only 6 residues, obviously residues 14–19 (Cys-Glu-Leu-Cys-His-Thr) has been obtained by Titani and Narita
⎣————haem————⎦
(1964) from yeast cytochrome c by digestion with trypsin and pronase. Attempts at synthesis of the haem octapeptide have begun (Theodoropoulos and Souchleris, 1966). From cytochrome c_1 Wada et al. (1968) obtained by chymotrypsin a peptide containing 20 residues ($Gly_2Ala_1Leu_1Ile_1Cys_2$ $Met_1Pro_1Ser_3Thr_1Asp_2Glu_2Lys_1His_1$ without Arg, Phe or Tyr).

These haem peptides are of importance not only for the establishment of part of the amino acid sequence of several cytochromes c close to the linked haem, but also for several other reasons. Firstly, they provide valuable models for the study of the combination of the haem-iron with amino acid residue ligands in cytochrome c. Secondly, since they do not contain any aromatic acid residues (phenylalanine, tyrosine or tryptophan) they reveal that part of the absorption ascribed to them at 280 nm is, in fact, the property of haem itself, as was postulated early by Drabkin (1961). Even more important was the recognition that a part of the absorption at 240–250 nm so far ascribed to the α-helix was also due to haem (Harbury et al. 1966; Urry, 1967a,b) and that the percentage of α-helix in cytochrome c had been overestimated. This gradually changed our earlier conceptions on the role of the α-helix for the structure of cytochrome c (Theorell and Agner, 1943; Theorell, 1956; Ehrenberg and Theorell, 1955b; Margoliash and Hill, 1961), so that it is now recognized that the α-helix plays only a minor role in the conformation of cytochrome c (see below). In the undecapeptide the α-helix contribution is certainly less than 10% (Urry, 1967a).

Both the α-NH_2 group of cysteine in the octapeptide and the ϵ-NH_2 group of lysine in the undecapeptide, are replaceable by α-benzoylhistidine

(Tuppy and Bodo, 1954; Paléus, 1961) or by N-acetylmethionine. It will be seen in Section B3 that acetylmethionine gives compounds spectroscopically more closely resembling cytochrome c than imidazole.

In NMR studies on cytochrome c and its haemopeptides Kowalsky (1962, 1964, 1965, 1966, 1969) found that only one type of the two anomalous resonances observable in cytochrome c^{3+} was present in the ferric haem peptides. The anomalously high field absorption of cytochrome c ($+$ 33·9τ) was absent, but the anomalously low field resonances ($-$ 21·6, $-$ 24·3τ) were still present though somewhat shifted. The author explains the anomalous resonances as due to delocalization of the unpaired electron in the low spin complex over the carbon skeleton transmitted through σ-bonds to adjacent hydrogens of the carbon skeleton with consequent hyperfine contact interaction between the electron and certain protons. The protons of the iron-bound -S-CH$_3$ group of methionine are probably responsible for the anomalous shift. With the N-acetylmethionine compound of the haem octapeptide, Harbury et al. (1965, 1966) found a clearcut double resonance, $\tau = 7·83$, due to acetyl, and $\tau = 7·90$ (due to -S-CH$_3$) (see also Section B5).

These findings have been further supported by investigations on circular dichroism (CD) and optical rotatory dispersion (ORD) of the haem peptides and their derivatives (Harbury, 1966; Harbury et al., 1966; Ulmer, 1966; Urry, 1966; 1967a; Urry and Pettegrew, 1967; Myer, 1968). Comparing the increase of the Soret band of the chymotryptic hexadecapeptide (Margoliash, 1962b) by exposure to 20% ethylene glycol in water to that of cytochrome c^{3+}, Stellwagen (1967) concluded that only a portion of the haem in cytochrome c is exposed to the solvent. This agrees with x-ray crystallographic findings (see below). The concentration of the glycol was not high enough to alter the conformation of cytochrome c. The effect of cyanide on its conformation was indicated by its effect on the Soret band after ethylene glycol treatment.

The peptides 50–56 and 78–84 of horse heart cytochrome c have been synthesized by Wolman et al. (1970a,b).

(b) *Amino acid composition.* In Table VIII the amino acid compositions of 9 mammalian, 7 vertebrate non-mammalian, 1 invertebrate, 2 higher plant, 3 yeast and 1 bacterial cytochrome are summarized. Cytochrome c usually contain eighteen amino acids, apart from lysine-betaine (trimethyl-lysine) found only in plant and fungal cytochromes (Smith, 1970) in horse c serine and in *Rhodospirillum c* arginine is missing. The composition is on the whole remarkably constant. Thus all contain only one tryptophanyl (except fish cytochromes c which sometimes contain two), two haem-bound cysteines (except yeast cytochrome c and that of the Annelid *Tubifex* which contains an additional free cysteine) or frog cytochrome c which contains four. Lysine is always the predominant basic amino acid. Only 2–3 His are

TABLE VIII
Amino acid composition of 36 eukaryotic cytochromes

	Gly	Ala	Val	Leu	Ile	Pro	Cys	Met	Ser	Thr	Phe	Tyr	Trp	Asp	Glu	Lys*	His	Arg
11 mammals[1] (104 residues)	12-15	6-7	3-4	6	6-8	3-4	2	2-3	0-2	7-10	3-4	4-5	1	7-10	9-12	18-19	3	2
8 non-mammalian vertebrates[2]	13-14	5-11	3-5	6	6-8	3-5	2-4	1-2	2-4	7-10	3-4	4-5	1	7-10	9-12	14-18	2-3	2-3
3 fishes[3]	12-13	7	4-6	6-7	3-4	3-4	2	1-2	4-5	6-8	3-4	5	1-2	9-10	9-11	16-17	2-3	2-3
5 invertebrates[4] (107 residues)	10-14	10-12	4-5	5-8	4-5	5-6	2-3	1-3	1-4	6-9	4-6	3-5	1-2	9-10	8-10	13-15	1-3	3-4
5 higher plants[5] (111-112 residues)	10-12	7-14	3-5	7-8	2-4	4-8	2	1-3	4-7	7-8	3-4	5-6	1	8-11	8-12	12-16	2-4	2-4
4 yeasts[6] (107-109 residues)	12-15	7-12	1-5	6-8	2-5	3-7	2-3	1-3	3-6	7-9	3-6	4-5	1	8-13	8-12	12-16	2-4	3-4

[1] Man, chimpanzee, rhesus monkey, horse, ox, pig, sheep, rabbit, dog.
[2] Chicken, turkey, turtle, rattlesnake, penguin, duck, pigeon, frog.
[3] Fishes: tunny, carp and shark (*Squalus*).
[4] *Samia cynthia* moth, *Drosophila*, Tobacco horn worm, Screw worm fly, *Tubifex*.
[5] Wheat, Mung beans, *Helianthus*, Sesame, Castor.
[6] *Saccharomyces*, *Debaromyces*, *Candida*, *Neurospora*.
* 2 Lys replaced by trimethyl-lysine in higher plants and one in yeasts.

present (except in some fungal cytochromes c which have four and the *Tubifex* cytochrome c which has only the one in position 18 (Hennig *et al.* 1969). The number of Gly, Val, Leu, Met, Thr, Phe, Tyr, Glu, His, Arg and Lys are somewhat less variable than those of Ala, Ile, Pro, Ser, and Asp. There are more glycine residues in animals than in plants and fewer alanine residues in vertebrate animals than in invertebrates, in plants, yeast and bacteria. Of leucine there are usually 6, more rarely 7, residues in animals, but more in plants, yeast and bacteria, while there is more isoleucine in mammals, birds and reptiles than in fish and invertebrates. There is less proline in the cytochromes c of vertebrate animals than in those of plants and yeasts. There is less serine relative to threonine in the cytochromes c of animals than those in plants, yeast and bacteria. Lysine is more abundant in vertebrate animals than in invertebrates, plants and yeasts. Final proof for these relationships will have, however, to wait for a larger number of samples. The excess of basic over acidic amino acid residues is still greater than indicated by the table since some of the acidic residues are present as asparagine and glutamine. This is in harmony with the high isoelectric point of the eukaryotic cytochromes c. The data given for the amine acid composition in earlier work do not always tally with those which can now be derived from the amino acid sequence (cf. e.g. Gillies and Pirrie, 1958; Leaf *et al.*, 1958; Matsubara *et al.*, 1963; Takahashi *et al.*, 1959a,b).

(c) *The primary structure (amino acid sequence) of cytochromes c.* Table IX summarizes the amino acid sequence in 24 cytochromes of vertebrates, 12 from ammmals, 5 from birds, 3 from reptiles and amphibia, and 4 from fishes. The total number of residues is 104, except in two fishes, in which the number is 102. N-acetylglycine is always the N-terminal residue. The abbreviations for amino acid residues used in Tables IX and X are those recommended by the IUPAC-IUB Nomenclature Commission (1968).

In *mammals* 84 positions have been found invariant, and in 9 others only "conservative" replacements have been found, i.e. uncharged hydrophobic versus uncharged hydrophobic, basic versus basic, and acidic versus acidic. Only in 11 positions "radical" replacements have been found, and in 5 of these (28, 29, 33, 88 and 103) only a single deviant mammalian cytochrome c has been observed. "Polyvariant" positions are residues 12 (only for primates), 44 (only for rabbit and whale), 50 (only for primates), 60, 66 (only for rabbit and whale), 89 and 92. It should be noted, however, that there is no clearcut division between "conservative" and radical replacements. It is not certain whether alanine, serine, threonine and tyrosine can be considered as large hydrophobic residues. The role of glutamine and asparagine residues is uncertain. Even lysine cannot always be regarded as basic, if only eight of the ϵ-amino groups are directed towards the outside, while the remainder is probably hydrogen-bonded inside the molecule.

TABLE IX (a)
Amino acid sequence of cytochromes c from Vertebrates

Source		1	2	3	4	5	6	7	8	9	10	11	12	13	14	15	
1, 2 Man, chimpanzee	MAMMALS	Acet-Gly	Asp	Val	Glu	Lys	Gly	Lys	Lys	Ile	Phe	Ile	Met	Lys	Cys	Ser	Matsubara and Smith, 1962, 1963; Matsubara et al., 1962.
3 Rhesus monkey		,,	,,	,,	,,	,,	,,	,,	,,	,,	,,	,,	,,	,,	,,	,,	Rothfus and Smith, 1965.
4, 5, 6 Cattle, sheep, pig		,,	,,	,,	,,	,,	,,	,,	,,	,,	,,	Val	Gln	,,	,,	Ala	Yasunobu et al., 1963; Margoliash et al., 1963; Stewart and Margoliash, 1965; Nakashima et al., 1966.
7 Grey whale		,,	,,	,,	,,	,,	,,	,,	,,	,,	,,	,,	,,	,,	,,	,,	Goldstone and Smith, 1966.
8 Kangaroo		,,	,,	,,	,,	,,	,,	,,	,,	,,	,,	,,	,,	,,	,,	,,	Nolan and Margoliash, 1966.
9 Horse		,,	,,	,,	,,	,,	,,	,,	,,	,,	,,	,,	,,	,,	,,	,,	Margoliash et al., 1961, 1962, 1963; Margoliash, 1962, a, b.
10 Donkey		,,	,,	,,	,,	,,	,,	,,	,,	,,	,,	,,	,,	,,	,,	,,	Margoliash and Fitch, 1968.
11 Dog		,,	,,	,,	,,	,,	,,	,,	,,	,,	,,	,,	,,	,,	,,	,,	McDowall and Smith, 1965.
12 Rabbit		,,	,,	,,	,,	,,	,,	,,	,,	,,	,,	,,	,,	,,	,,	,,	Needleman and Margoliash, 1966; Margoliash, 1963.
13 Penguin	BIRDS	,,	,,	Ile	,,	,,	,,	,,	,,	,,	,,	,,	,,	,,	,,	Ala	Fitch and Margoliash, 1968.
14, 15 Chicken, turkey		,,	,,	,,	,,	,,	,,	,,	,,	,,	,,	,,	,,	,,	,,	Ser	Chan et al., 1963; Chan and Margoliash, 1966, a, b.
16 Duck		,,	,,	Val	,,	,,	,,	,,	,,	,,	,,	,,	,,	,,	,,	Ala	Fitch and Margoliash, 1968.
17 Pigeon		,,	,,	Ile	,,	,,	,,	,,	,,	,,	,,	,,	,,	,,	,,	,,	Fitch and Margoliash, 1968.
18 Turtle	REPTILES	,,	,,	Val	,,	,,	,,	,,	,,	,,	,,	,,	,,	,,	,,	,,	Chan et al., 1963, 1966
19 Snake		,,	,,	,,	,,	,,	,,	,,	,,	,,	,,	Ile	Thr	,,	,,	Ser	Rattlesnake: Bahl and Smith, 1960.
20 Frog	AMPHIBIANS	,,	,,	,,	Ala	,,	,,	,,	,,	,,	,,	Ala	Glu	,,	,,	,,	Chan et al., 1967.
21 Tuna	FISHES	,,	,,	,,	,,	,,	,,	,,	,,	Thr	,,	Val	Gln	,,	,,	Ala	Kreil, 1963, 1965; Margoliash, 1963.
22 Carp		,,	,,	,,	Glu Asp Glu	,,	,,	,,	,,	Val	,,	,,	,,	,,	,,	,,	Gürtler and Horstmann, 1970.
23 Shark		,,	,,	,,	,,	,,	,,	,,	,,	,,	,,	,,	,,	,,	,,	,,	Goldstone and Smith, 1967.
24 Bonito		,,	,,	Ile	Ala	,,	,,	,,	,,	Thr	,,	,,	,,	,,	,,	,,	Nakayama et al. (quoted by

Cytochromes c

	Source	16	17	18	19	20	21	22	23	24	25	26	27	28	29	30	
MAMMALS	1, 2 Man, chimpanzee	Gln	Cys	His	Thr	Val	Glu	Lys	Gly	Gly	Lys	His	Lys	Thr	Gly	Pro	Matsubara and Smith, 1962, 1963; Matsubara et al., 1963.
	3 Rhesus monkey	,,	,,	,,	,,	,,	,,	,,	,,	,,	,,	,,	,,	,,	,,	,,	Rothfus and Smith, 1965.
	4, 5, 6 Cattle, sheep, pig	,,	,,	,,	,,	,,	,,	,,	,,	,,	,,	,,	,,	,,	,,	,,	Yasunobu et al., 1963 Margoliash et al., 1963; Stewart and Margoliash, 1965; Nakashima et al., 1966.
	7 Grey whale	,,	,,	,,	,,	,,	,,	,,	,,	,,	,,	,,	,,	,,	,,	,,	Goldstone and Smith, 1966.
	8 Kangaroo	,,	,,	,,	,,	,,	,,	,,	,,	,,	,,	,,	,,	,,	,,	,,	Nolan and Margoliash, 1966.
	9 Horse	,,	,,	,,	,,	,,	,,	,,	,,	,,	,,	,,	,,	,,	,,	,,	Margoliash et al., 1961, 1962, 1963; Margoliash, 1962, a, b.
	10 Donkey	,,	,,	,,	,,	,,	,,	,,	,,	,,	,,	,,	,,	,,	,,	,,	Margoliash and Fitch, 1968.
	11 Dog	,,	,,	,,	,,	,,	,,	,,	,,	,,	,,	,,	,,	,,	,,	,,	McDowall and Smith, 1965.
	12 Rabbit	,,	,,	,,	,,	,,	,,	,,	,,	,,	,,	,,	,,	,,	,,	,,	Needleman and Margoliash, 1966; Margoliash, 1963.
BIRDS	13 Penguin	,,	,,	,,	,,	,,	,,	,,	,,	,,	,,	,,	,,	,,	,,	,,	Fitch and Margoliash, 1968.
	14, 15 Chicken, turkey	,,	,,	,,	,,	,,	,,	,,	,,	,,	,,	,,	,,	,,	,,	,,	Chan et al., 1963, Chan and Margoliash, 1966, a, b.
	16 Duck	,,	,,	,,	,,	,,	,,	,,	,,	,,	,,	,,	,,	,,	,,	,,	Fitch and Margoliash, 1968.
	17 Pigeon	,,	,,	,,	,,	,,	,,	,,	,,	,,	,,	,,	,,	,,	,,	,,	Fitch and Margoliash, 1968.
REPTILES	18 Turtle	,,	,,	,,	,,	,,	,,	,,	,,	,,	,,	,,	,,	,,	,,	,,	Chan et al., 1963, 1966.
	19 Snake	,,	,,	,,	,,	,,	,,	,,	,,	,,	,,	,,	,,	,,	,,	,,	Rattlesnake: Bahl and Smith, 1960.
AMPHIBIANS	20 Frog	,,	,,	,,	,,	,,	,,	,,	,,	,,	,	,,	,,	Val	,,	,,	Chan et al., 1967.
FISHES	21 Tuna	,,	,,	,,	,,	,,	Glx	Asn	,,	,,	,,	,,	,,	Thr	,,	,,	Kreil, 1963, 1965; Margoliash, 1963.
	22 Carp	,,	,,	,,	,,	,,	,,	Asx	,,	,,	Asp	,,	,,	Val	,,	,,	Gürtler and Horstmann, 1970.
	23 Shark	,,	,,	,,	,,	,,	Glu	Asn	,,	,,	Lys	,,	,,	Thr	,,	,,	Goldstone and Smith, 1967.
	24 Bonito	,,	,,	,,	,,	,,	,,	,	,,	,,	,,	,	,,	Val	,,	,,	Nakayama et al. (quoted by Dickerson et al., 1970.)

TABLE IX (Continued) (c)

Source		31	32	33	34	35	36	37	38	39	40	41	42	43	44	45	
	1, 2 Man, chimpanzee	Asn	Leu	His	Gly	Leu	Phe	Gly	Arg	Lys	Thr	Gly	Gln	Ala	Pro	Gly	Matsubara and Smith, 1962, 1963; Matsubara et al., 1963.
MAMMALS	3 Rhesus monkey	,,	,,	,,	,,	,,	,,	,,	,,	,,	,,	,,	,,	,,	,,	,,	Rothfus and Smith, 1965.
	4, 5, 6 Cattle, sheep, pig	,,	,,	,,	,,	,,	,,	,,	,,	,,	,,	,,	,,	,,	,,	,,	Yasunobu et al., 1963, Margoliash et al., 1963; Stewart and Margoliash, 1965; Nakasuma et al., 1966.
	7 Grey whale	,,	,,	,,	,,	,,	,,	,,	,,	,,	,,	,,	,,	,,	Val	,,	Goldstone and Smith, 1966.
	8 Kangaroo	,,	,,	Asn	,,	Ile	,,	,,	,,	,,	,,	,,	,,	,,	Pro	,,	Nolan and Margoliash, 1966.
	9 Horse	,,	,,	His	,,	Leu	,,	,,	,,	,,	,,	,,	,,	,,	,,	,,	Margoliash et al., 1961, 1962, 1963; Margoliash, 1962, a, b.
	10 Donkey	,,	,,	,,	,,	,,	,,	,,	,,	,,	,,	,,	,,	,,	,,	,,	Margoliash and Fitch, 1968.
	11 Dog	,,	,,	,,	,,	,,	,,	,,	,,	,,	,,	,,	,,	,,	,,	,,	McDowall and Smith, 1965.
	12 Rabbit	,,	,,	,,	,,	,,	,,	,,	,,	,,	,,	,,	,,	,,	Val	,,	Needleman and Margoliash, 1966; Margoliash, 1963.
BIRDS	13 Penguin	,,	,,	,,	,,	,,	,,	,,	,,	,,	,,	,,	,,	,,	Glu	,,	Fitch and Margoliash, 1968.
	14, 15 Chicken, turkey	,,	,,	,,	,,	,,	,,	,,	,,	,,	,,	,,	,,	,,	,,	,,	Chan et al., 1963, Chan and Margoliash, 1966, a, b.
	16 Duck	,,	,,	,,	,,	,,	,,	,,	,,	,,	,,	,,	,,	,,	,,	,,	Fitch and Margoliash, 1968.
	17 Pigeon	,,	,,	,,	,,	,,	,,	,,	,,	,,	,,	,,	,,	,,	,,	,,	Fitch and Margoliash, 1968.
REPTILES	18 Turtle	,,	,,	Asn	,,	,,	Ile	,,	,,	,,	,,	,,	,,	,,	,,	,,	Chan et al., 1963, 1966.
	19 Snake	,,	,,	His	,,	,,	Phe	,,	,,	,,	,,	,,	,,	,,	Val	,,	Rattlesnake: Bahl and Smith, 1960.
AMPHIBIANS	20 Frog	,,	,,	Tyr	,,	,,	Ile	,,	,,	,,	,,	,,	,,	,,	Ala	,,	Chan et al., 1967.
FISHES	21 Tuna	,,	,,	Trp	,,	,,	,,	,,	,,	,,	,,	,,	,,	,,	Glu	,,	Kreil, 1963, 1965; Margoliash, 1963.
	22 Carp	,,	,,	,,	,,	,,	,,	,,	,,	,,	,,	,,	,,	,,	Pro	,,	Gürtler and Horstmann, 1970.
	23 Shark	,,	,,	Ser	,,	,,	,,	,,	,,	,,	,,	,,	,,	,,	Gln	,,	Goldstone and Smith, 1967.
	24 Bonito	,,	,,	Trp	,,	,,	,,	,,	,,	,,	,,	,,	,,	,,	Glu	,,	Nakayama et al. (quoted by Dickerson et al., 1970.)

Cytochromes c

TABLE IX (Continued) (d)

Source	46	47	48	49	50	51	52	53	54	55	56	57	58	59	60	References
MAMMALS																
1, 2 Man, chimpanzee	Tyr	Ser	Tyr	Thr	Ala	Ala	Asn	Lys	Asn	Lys	Gly	Ile	Ile	Trp	Gly	Matsubara and Smith, 1962, 1963; Matsubara et al., 1963.
3 Rhesus monkey	"	"	"	"	"	"	"	"	"	"	"	"	Thr	"	"	Rothfus and Smith, 1965.
4, 5, 6 Cattle, sheep, pig	Phe	"	"	"	Asp	"	"	"	"	"	"	"	"	"	"	Yasunobu et al., 1963, Margoliash et al., 1963; Stewart and Margoliash, 1965; Nakashima et al., 1966.
7 Grey whale	"	"	"	"	"	"	"	"	"	"	"	"	"	"	"	Goldstone and Smith, 1966.
8 Kangaroo	"	Thr	"	"	"	"	"	"	"	"	"	"	Ile	"	"	Nolan and Margoliash, 1966.
9 Horse	"	"	"	"	"	"	"	"	"	"	"	"	Thr	"	Lys	Margoliash et al., 1961, 1962, 1963; Margoliash, 1962, a, b.
10 Donkey	"	Ser	"	"	"	"	"	"	"	"	"	"	"	"	"	Margoliash and Fitch, 1968.
11 Dog	"	"	"	"	"	"	"	"	"	"	"	"	"	"	Gly	McDowall and Smith, 1965.
12 Rabbit	"	"	"	"	"	"	"	"	"	"	"	"	"	"	"	Needleman and Margoliash, 1966; Margoliash, 1963.
BIRDS																
13 Penguin	"	"	Lys	"	"	"	"	"	"	"	"	"	"	"	"	Fitch and Margoliash, 1968.
14, 15 Chicken, turkey	"	"	"	"	"	"	"	"	"	"	"	"	"	"	"	Chan et al., 1963, Chan and Margoliash, 1966, a, b.
16 Duck	"	"	"	"	"	"	"	"	"	"	"	"	"	"	"	Fitch and Margoliash, 1968.
17 Pigeon	"	"	"	"	"	"	"	"	"	"	"	"	"	"	"	Fitch and Margoliash, 1968.
REPTILES																
18 Turtle	"	"	Tyr	"	Glu	"	"	"	"	"	"	"	"	"	"	Chan et al., 1963, 1966.
19 Snake	Tyr	"	"	"	Ala	"	"	"	"	"	"	"	Ile	"	"	Rattlesnake: Bahl and Smith, 1960.
AMPHIBIANS																
20 Frog	Phe	"	"	"	Asp	"	"	"	"	"	"	"	Thr	"	"	Chan et al., 1967.
FISHES																
21 Tuna	Tyr	"	"	"	"	"	"	"	Ser	"	"	"	Val	"	Asn	Kreil, 1963, 1965; Margoliash, 1963.
22 Carp	Phe	"	"	"	"	"	"	"	"	"	"	"	"	"	Asx	Gürtler and Horstmann, 1970.
23 Shark	"	"	"	"	"	"	"	"	"	"	"	"	Thr	"	Gln	Goldstone and Smith, 1967.
24 Bonito	Tyr	"	"	"	"	"	"	"	"	"	"	"	Val	"	Asn	Nakayama et al., (quoted by Dickerson et al., 1970.

TABLE IX (Continued) (e)

	Source	61	62	63	64	65	66	67	68	69	70	71	72	73	74	75	
MAMMALS	1, 2 Man, chimpanzee	Glu	Asn	Thr	Leu	Met	Glu	Tyr	Leu	Glu	Asn	Pro	Lys	Lys	Thr	Ile	Matsubara and Smith, 1962, 1963; Matsubara et al., 1963.
	3 Rhesus monkey	,,	,,	,,	,,	,,	,,	,,	,,	,,	,,	,,	,,	,,	,,	,,	Rothfus and Smith, 1965.
	4, 5, 6 Cattle, sheep, pig	,,	Glu	,,	,,	,,	,,	,,	,,	,,	,,	,,	,,	,,	,,	,,	Yasunobu et al., 1963, Margoliash et al., 1963; Stewart and Margoliash, 1965; Nakashima et al., 1966.
	7 Grey whale	,,	,,	,,	,,	,,	,,	,,	,,	,,	,,	,,	,,	,,	,,	,,	Goldstone and Smith, 1966.
	8 Kangaroo	,,	Asp	,,	,,	,,	,,	,,	,,	,,	,,	,,	,,	,,	,,	,,	Nolan and Margoliash, 1966.
	9 Horse	,,	Glu	,,	,,	,,	,,	,,	,,	,,	,,	,,	,,	,,	,,	,,	Margoliash et al., 1961, 1962, 1963; Margoliash, 1962, a, b.
	10 Donkey	,,	,,	,,	,,	,,	,,	,,	,,	,,	,,	,,	,,	,,	,,	,,	Margoliash and Fitch, 1968.
	11 Dog	,,	,,	,,	,,	,,	Gly	,,	,,	,,	,,	,,	,,	,,	,,	,,	McDowall and Smith, 1965.
	12 Rabbit	,,	Asp	,,	,,	,,	,,	,,	,,	,,	,,	,,	,,	,,	,,	,,	Needleman and Margoliash, 1966; Margoliash, 1963.
BIRDS	13 Penguin	,,	Glu	,,	,,	,,	Glu	,,	,,	,,	,,	,,	,,	,,	,,	,,	Fitch and Margoliash, 1968.
	14, 15 Chicken, turkey	,,	,,	,,	,,	,,	,,	,,	,,	,,	,,	,,	,,	,,	,,	,,	Chan et al., 1963, Chan and Margoliash, 1966, a, b.
	16 Duck	,,	,,	,,	,,	,,	,,	,,	,,	,,	,,	,,	,,	,,	,,	,,	Fitch and Margoliash, 1968.
	17 Pigeon	,,	,,	,,	,,	,,	,,	,,	,,	,,	,,	,,	,,	,,	,,	,,	Fitch and Margoliash, 1968.
REP-TILES	18 Turtle	,,	,,	,,	,,	,,	,,	,,	,,	,,	,,	,,	,,	,,	,,	Tyr	Chan et al., 1963, 1966.
	19 Snake	Asp	,,	,,	,,	,,	,,	,,	,,	,,	,,	,,	,,	,,	,,	,,	Rattlesnake: Bahl and Smith, 1960.
AMPHI-BIANS	20 Frog	,,	,,	,,	,,	,,	,,	,,	,,	,,	,,	,,	,,	,,	,,	,,	Chan et al., 1967.
FISHES	21 Tuna	Asn	Asp	,,	,,	,,	,,	,,	,,	,,	,,	,,	,,	,,	,,	,,	Kreil, 1963, 1965; Margoliash, 1963.
	22 Carp	Glx	Glx	,,	,,	,,	,,	,,	,,	,,	,,	,,	,,	,,	,,	,,	Gürtler and Horstmann, 1970.
	23 Shark	Gln	Glu	,,	,,	Arg	Ile	,,	,,	,,	,,	,,	,,	,,	,,	,,	Goldstone and Smith, 1967.
	24 Bonito	Glu	Asn	,,	,,	Met	Glu	,,	,,	,,	,,	,,	,,	,,	,,	,,	Nakayama et al. (quoted by Dickerson et al., 1971.)

Cytochromes c

Source	76	77	78	79	80	81	82	83	84	85	86	87	88	89	90	
1, 2 Man, chimpanzee	Pro	Gly	Thr	Lys	Met	Ile	Phe	Val	Gly	Ile	Lys	Lys	Lys	Glu	Glu	Matsubara and Smith, 1962, 1963; Matsubara et al., 1963.
3 Rhesus monkey	,,	,,	,,	,,	,,	,,	,,	,,	,,	,,	,,	,,	,,	,,	,,	Rothfus and Smith, 1965.
4, 5, 6 Cattle, sheep, pig	,,	,,	,,	,,	,,	,,	,,	Ala	,,	,,	,,	,,	,,	Gly	,,	Yasunobu et al., 1963, Margoliash et al., 1963; Stewart and Margoliash, 1965; Nakashima et al., 1966.
7 Grey whale	,,	,,	,,	,,	,,	,,	,,	,,	,,	,,	,,	,,	,,	,,	,,	Goldstone and Smith, 1966.
8 Kangaroo	,,	,,	,,	,,	,,	,,	,,	,,	,,	,,	,,	,,	,,	,,	,,	Nolan and Margoliash, 1966.
9 Horse	,,	,,	,,	,,	,,	,,	,,	,,	,,	,,	,,	,,	,,	Thr	,,	Margoliash et al., 1961, 1962, 1963; Margoliash, 1962, a, b.
10 Donkey	,,	,,	,,	,,	,,	,,	,,	,,	,,	,,	,,	,,	,,	,,	,,	Margoliash and Fitch, 1968.
11 Dog	,,	,,	,,	,,	,,	,,	,,	,,	,,	,,	,,	,,	Thr	Gly	,,	McDowall and Smith, 1965.
12 Rabbit	,,	,,	,,	,,	,,	,,	,,	,,	,,	,,	,,	,,	Lys	Asp	,,	Needleman and Margoliash, 1966; Margoliash, 1963.
13 Penguin	,,	,,	,,	,,	,,	,,	,,	,,	,,	,,	,,	,,	,,	Ser	,,	Fitch and Margoliash, 1968.
14, 15 Chicken, turkey	,,	,,	,,	,,	,,	,,	,,	,,	,,	,,	,,	,,	,,	,,	,,	Chan et al., 1963, Chan and Margoliash, 1966, a, b.
16 Duck	,,	,,	,,	,,	,,	,,	,,	,,	,,	,,	,,	,,	,,	,,	,,	Fitch and Margoliash, 1968.
17 Pigeon	,,	,,	,,	,,	,,	,,	,,	,,	,,	,,	,,	,,	,,	Ala	,,	Fitch and Margoliash, 1968.
18 Turtle	,,	,,	,,	,,	,,	,,	,,	,,	,,	,,	,,	,,	,,	,,	,,	Chan et al., 1963, 1966.
19 Snake	,,	,,	,,	,,	,,	,,	,,	,,	Thr	,,	Leu	Ser	,,	Lys	,,	Rattlesnake: Bahl and Smith, 1960.
20 Frog	,,	,,	,,	,,	,,	,,	,,	,,	Ala	,,	Ile	Lys	,,	Glu	,,	Chan et al., 1967.
21 Tuna	,,	,,	,,	,,	,,	,,	,,	,,	,,	,,	,,	,,	,,	Gly	,,	Kreil, 1963, 1965; Margoliash, 1963.
22 Carp	,,	,,	,,	,,	,,	,,	,,	,,	,,	,,	,,	,,	,,	,,	,,	Gürtler and Horstmann, 1970.
23 Shark	,,	,,	,,	,,	,,	,,	,,	,,	,,	,,	Leu	,,	,,	Ser	,,	Goldstone and Smith, 1967.
24 Bonito	,,	,,	,,	,,	,,	,,	,,	,,	,,	,,	Ile	,,	,,	Gly	,,	Nakayama et al. (quoted by Dickerson et al., 1970.

MAMMALS — BIRDS — REPTILES — AMPHIBIANS — FISHES

144

TABLE IX (Continued) (g)

Source	91	92	93	94	95	96	97	98	99	100	101	102	103	104	
MAMMALS															
1, 2 Man, chimpanzee	Arg	Ala	Asp	Leu	Ile	Ala	Tyr	Leu	Lys	Lys	Ala	Thr	Asn	Glu	Matsubara and Smith, 1962, 1963; Matsubara et al., 1963.
3 Rhesus monkey	,,	,,	,,	,,	,,	,,	,,	,,	,,	,,	,,	,,	,,	,,	Rothfus and Smith, 1965.
4, 5, 6 Cattle, sheep, pig	,,	Glu	,,	,,	,,	,,	,,	,,	,,	,,	,,	,,	,,	,,	Yasunobu et al., 1963, Margoliash et al., 1963; Stewart and Margoliash, 1965; Nakashima et al., 1966.
7 Grey whale	,,	,,	,,	,,	,,	,,	,,	,,	,,	,,	,,	,,	,,	,,	Goldstone and Smith, 1966.
8 Kangaroo	,,	Ala	,,	,,	,,	,,	,,	,,	,,	,,	,,	,,	,,	,,	Nolan and Margoliash, 1966.
9 Horse	,,	Glu	,,	,,	,,	,,	,,	,,	,,	,,	,,	,,	,	,,	Margoliash et al., 1961, 1962, 1963; Margoliash, 1962, a, b.
10 Donkey	,,	,,	,,	,,	,,	,,	,,	,,	,,	,,	,,	,,	,,	,,	Margoliash and Fitch, 1968.
11 Dog	,,	Ala	,,	,,	,,	,,	,,	,,	,,	,,	,,	,,	Lys	,,	McDowall and Smith, 1965.
12 Rabbit	,,	,,	,,	,,	,,	,,	,,	,,	,,	,,	,,	,,	Asn	,,	Needleman and Margoliash, 1966; Margoliash, 1963.
BIRDS															
13 Penguin	,,	,,	,,	,,	,,	,,	,,	,,	,,	Asp	,,	,,	Ser	Lys	Fitch and Margoliash, 1968.
14, 15 Chicken, turkey	,,	Val	,,	,,	,,	,,	,,	,,	,,	,,	,,	,,	,,	,,	Chan et al., 1963, Chan and Margoliash, 1966, a, b.
16 Duck	,,	Ala	,,	,,	,,	,,	,,	,,	,,	,,	,,	,,	Ala	,,	Fitch and Margoliash, 1968.
17 Pigeon	,,	,,	,,	,,	,,	,,	,,	,,	,,	Glu	,,	,,	,,	,,	Fitch and Margoliash, 1968.
REPTILES															
18 Turtle	,,	,,	,,	,,	,,	,,	,,	,,	,,	Asp	,,	,,	Ser	,,	Chan et al., 1963, 1966.
19 Snake	,,	Thr	,,	,,	,,	,,	,,	,,	,,	Glu	Lys	,,	Ala	Ala	Rattlesnake: Bahl and Smith, 1960.
AMPHIBIANS															
20 Frog	,,	Gly	,,	,,	,,	,	,,	,,	,,	Asp	Ala	?	Cys	Lys	Chan et al., 1967.
FISHES															
21 Tuna	,,	Gln	,,	,,	Val	,,	,,	,,	,,	Ser	Ala	Thr	Ser	—	Kreil, 1963, 1965; Margoliash, 1963.
22 Carp	,,	Ala	,,	,,	Ile	,,	,,	,,	,,	Ser	,,	Thr	Ser	—	Gürtler and Horstmann, 1970.
23 Shark	,,	Gln	,,	,,	,,	,,	,,	,,	,,	Lys	Thr	Ala	Ala	Ser	Goldstone and Smith, 1967.
24 Bonito	,,	,,	,,	,,	Val	,,	,,	,,	,,	Ser	Ala	Thr	Ser	?	Nakayama et al. (quoted by Dickerson et al., 1970.)

TABLE X (a)

Amino acid sequence of cytochromes c from Invertebrates, Plants, Fungi and a few bacteria

Group	Source	-8	-7	-6	-5	-4	-3	-2	-1	1	2	3	4	5	6	7	References
INVERTEBRATES	Moth *Samia cynthia*				(H₂N)Gly		Val	Pro	Ala	Gly	Asn	Ala	Glu	Asn	Gly	Lys	Chan and Margoliash, 1966a.
	Tobacco horn worm moth				,,	,,	,,	,,	,,	,,	,,	,,	Asp	,,	,,	,,	Dickerson et al., 1970; Chan, 1970.
	Fruit fly *Drosophila*				,,	,,	,,	,,	,,	,,	Asp	Val	Glu	Lys	,,	,,	Dickerson et al., 1970.
	Screw worm fly				,,	,,	,,	,,	Pro	,,	,,	,,	,,	,,	,,	,,	Dickerson et al., 1970.
PLANTS	Wheat	Acet-Ala	Ser	Phe	Ser Ala	Glu	Ala	,,	Pro	,,	Asn	Pro Ser	Asp Ala	Ala Lys	,,	Ala Met	Stevens et al., 1967; De Lange et al., 1960, †Boulter et al., 1970.
	Bean (*Phaseolus aureus*)		,,	,,	Asx	,,	,,	,,	,,	,,	Asp Asn	Ser	Lys Asp	Ser	,,	Glu Ala	De Lange et al., 1969, 1970; Thompson et al., 1970, a, b; Boulter et al., 1970.
	Helianthus annuis		,,	,,	Ala	,,	,,	,,	Ala	,,	Asp	Pro	Thr	Thr	,,	Ala	Ramshaw et al., 1970; Boulter et al., 1970.
	Sesame		,,	,,	Asx	,,	,,	,,	Pro	,,	,,	Val	Lys	Ser	,,	Glu	Boulter et al., 1970.
	Castor		,,	,,	Asx	,,	,,	,,	Pro	,,	,,	,,	,,	Ala	,,	,,	Boulter et al., 1970.
FUNGI	*Saccharomyces*				(H₂N)Thr	,,	Phe	Lys	Ala	,,	Ser	Ala	,,	Lys	,,	,,	Narita et al., 1963 a, b; Sato et al., 1966; De Lange et al., 1970.
	Neurospora crassa				(H₂N)Gly	,,	,,	Ser	,,	,,	Asp	Ser	,,	,,	,,	,,	Heller and Smith, 1965; De Lange et al., 1969.
	Candida krusei			(H₂N)Pro	Ala	Pro	,,	Glu	Gln	,,	Ser	Ala	,,	,,	,,	,,	Narita and Sugeno, 1968; Narita et al., 1968; De Lange et al., 1970.
	Debaromyces kloeckeri		,,	,,	,,	,,	Tyr	,,	Lys	,,	,,	Glu	,,	.	,,	,,	Narita and Sugeno, 1968; Narita et al., 1968; De Lange et al., 1970.
BACTERIA	*Rhodospirillum rubrum* c₂									(H₂N)Glu	Glu	Gly	Asp	Ala	Ala	Ala	Dus and Sletten, 1968.
	Pseudomonas fluorescens c-551									(H₂N)Glu	,,	,,	,,	Pro	Gln	Val	Ambler, 1962, 1963.
	Desulphovibrio vulgaris c₃		−16 Ala Pro Lys Ala Pro Ala Asp Gly Leu Lys Met Glu Ala Thr Lys Gln							Pro	Val	Val	Phe	Asn	His	Ser	Ambler, 1968; Ambler et al., 1969.
	Desulphovibrio gigas c₃		−19 Val Asp Val Pro Ala Asp Gly Ala Lys Ile Asp Phe Ile Ala Gly Gly Glu Lys Asn							Leu	,,	,,	,,	,,	,,	,,	Ambler, 1968; Ambler et al., 1969.

TABLE X (Continued) (b)

	Source	8	9	10	11	12	13	14	15	16	17	18	19	20	21	22	23	24	25	
										Haem										
INVERTE-BRATES	Moth *Samia cynthia*	Lys	Ile	Phe	Val	Gln	Arg	Cys	Ala	Gln	Cys	His	Thr	Val	Glu	Ala	Gly	Gly	Lys	Chan and Margoliash, 1966a.
	Tobacco horn worm moth	,,	,,	,,	,,	,,	,,	,,	,,	,,	,,	,,	,,	,,	,,	,,	,,	,,	,,	Dickerson et al., 1970; Chan, 1970.
	Fruit fly *Drosophila*	,,	Leu	,,	,,	,,	,,	,,	,,	,,	,,	,,	,,	,,	,,	,,	,,	,,	,,	Dickerson et al., 1970.
	Screw worm fly	,,	Ile	,,	,,	,,	,,	,,	,,	,,	,,	,,	,,	,,	,,	,,	,,	,,	,,	Dickerson et al., 1970.
PLANTS	Wheat	,,	,,	,,	Lys	Thr	Lys	,,	,,	,,	,,	,,	,,	,,	Asp	Ala / Lys	,,	Ala	Gly	Stevens et al., 1967; De Lange et al., 1960, †Boulter et al., 1970.
	Bean (*Phaseolus aureus*)	,,	,,	,,	,,	,,	,,	,,	,,	,,	,,	,,	,,	,,	,,	Lys	,,	,,	,,	De Lange et al., 1969, 1970; Thompson et al., 1970, a, b; Boulter et al., 1970.
	Helianthus annuis	,,	,,	,,	,,	,,	,,	,,	,,	,,	,,	,,	,,	,,	Glu	,,	,,	,,	,,	Ramshaw et al., 1970; Boulter et al., 1970.
	Sesame	,,	,,	,,	,,	,,	,,	,,	,,	,,	,,	,,	,,	,,	Asp	,,	,,	,,	,,	Boulter et al., 1970.
	Castor	,,	,,	,,	,,	,,	,,	,,	,,	,,	,,	,,	,,	,,	Glu	,,	,,	,,	,,	Boulter et al., 1970.
FUNGI	*Saccharomyces*	Thr	Leu	,,	,,	,,	Arg	,,	Glu	Leu	,,	,,	Gly	Glu	Glu	Gly	,,	Gly	Pro	Narita et al., 1963, a, b; Sato et al., 1966; De Lange et al., 1970.
	Neurospora crassa	Asn	,,	,,	,,	,,	,,	,,	Ala	Gln	,,	,,	,,	Glu	Gly	Gly	Asn	Leu	Thr	Heller and Smith, 196 ; De Lange et al., 1969.
	Candida krusei	Thr	,,	,,	,,	,,	,,	,,	,,	,,	,,	,,	Thr	Ile	Glu	Ala	Gly	Gly	Pro	Narita and Sugeno, 1968; Narita et al., 1968; De Lange et al., 1970.
	Debaromyces kloecheri	Asn	,,	,,	,,	,,	,,	,,	Glu	Leu	,,	,,	,,	Val	,,	,,	,,	,,	,,	Narita and Sugeno, 1968; Narita et al., 1968; De Lange et al., 1970.
BACTERIA	*Rhodospirillum rubrum* c_2	Gly	Gln	Lys	Val	Ser	Lys	,,	Leu	Ala	,,	,,	,,	Phe	Asp	Gln	Gly	Gly	Ala	Dus and Sletten, 1968.
	Pseudomonas fluorescens c—551	Leu	Phe	,,	Asn	Lys	Gly	,,	Val	,,	,,	,,	Ala	Ile	,,	Thr	Lys	Met	Val	Ambler, 1962, 1963.
	Desulphovibrio vulgaris c_3	Thr	His	,,	Ser	Val	Lys	,,	Gly	Asp	,,	,,	His	Pro	Val	Asn	Gly	Lys	Gln	Ambler, 1968; Ambler et al., 1969.
	Desulphovibrio gigas c_3	,,	,,	,,	Asp	Val	Lys	,,	Gly	Asp	,,	,,	His	Glx	Pro	Gly	Asx	Lys	Gln	Ambler, 1968; Ambler et al., 1969.
										Haem										

TABLE X (Continued) (c)

	Source	26	27	28	29	30	31	32	33	34	35	36	37	38	39	40	41	42	43	
INVERTE-BRATES	Moth *Samia cynthia*	His	Lys	Val	Gly	Pro	Asn	Leu	His	Gly	Phe	Tyr	Gly	Arg	Lys	Thr	Gly	Gln	Ala	Chan and Margoliash, 1966a.
	Tobacco horn worm moth	,,	,,	,,	,,	,,	,,	,,	,,	,,	,,	Phe	,,	,,	,,	,,	,,	,,	,,	Dickerson et al., 1970; Chan, 1970.
	Fruit fly *Drosophila*	,,	,,	,,	,,	,,	,,	,,	,,	,,	Leu	Ile	,,	,,	,,	,,	,,	,,	,,	Dickerson et al., 1970.
	Screw worm fly	,,	,,	,,	,,	,,	,,	,,	,,	,,	,,	Phe	,,	,,	,,	,,	,,	,,	,,	Dickerson et al., 1970.
PLANTS	Wheat	,,	,,	Gln	,,	,,	,,	,,	,,	,,	,,	,,	,,	,,	Gln	Ser	,,	Thr	Thr	Stevens et al., 1967; De Lange et al., 1969, †Boulter et al., 1970.
	Bean (*Phaseolus aureus*)	,,	,,	,,	,,	,,	,,	,,	Asn	,,	,,	,,	,,	,,	,,	,,	,,	,,	,,	De Lange et al., 1969, 1970; Thompson et al., 1970, a, b; Boulter et al., 1970.
	Helianthus annuus	,,	,,	,,	,,	,,	,,	,,	,,	,,	,,	,,	,,	,,	,,	,,	,,	,,	,,	Ramshaw et al., 1970; Boulter et al., 1970.
	Sesame	,,	,,	,,	,,	,,	,,	,,	,,	,,	,,	,,	,,	,,	,,	,,	,,	,,	,,	Boulter et al., 1970.
	Castor	,,	,,	,,	,,	,,	,,	,,	,,	,,	,,	,,	,,	,,	,,	,,	,,	,,	,,	Boulter et al., 1970.
FUNGI	*Saccharomyces*	Gln	,,	Val	,,	,,	Ala	,,	His	,,	Ile	,,	,,	,,	His	,,	,,	Gln	Ala	Narita et al., 1963, a, b; Sato et al., 1966; De Lange et al., 1970.
	Neurospora crassa	,,	,,	Ile	,,	,,	,,	,,	,,	,,	Leu	,,	,,	,,	Lys	Thr	,,	Ser	Val	Heller and Smith, 1965; De Lange et al., 1969.
	Candida krusei	His	,,	Val	,,	,,	Asn	,,	,,	,,	Ile	,,	Ser	,,	His	Ser	,,	Gln	Ala	Narita and Sugeno, 1968; Narita et al., 1968; De Lange et al., 1970.
	Debaromyces kloecheri	,,	,,	,,	,,	,,	,,	,,	,,	,,	Val	—	—	,,	Thr	,,	,,	,,	,,	Narita and Sugeno, 1968; Narita et al., 1968; De Lange et al., 1970.
BACTERIA	*Rhodospirillum rubrum* c_2	Asn	Lys	Val	Gly	Pro	,,	,,	Phe	,,	,,	Phe	Glx	Asn	,,	Ala	Ala	His	Lys	Dus and Sletten, 1968.
	Pseudomonas fluorescens c—551	Gly	Pro	Ala	Tyr	Lys	Asp	Val	Ala	Ala	Lys	Phe	Ala	Gly	Gln	Ala	Gly	Ala	Glu	Ambler, 1962, 1963.
	Desulphovibrio vulgaris c_3	Asp	Tyr	Arg	Lys	Cys	Gly	Thr	Ala	Gly	Lys	His	Asp	Ser	Met	Asp	Lys	Lys	Asp	Ambler et al., 1968; Ambler et al., 1969.
	Desulphovibrio gigas c_3	Tyr	Ala	Gly	Cys	Thr	Thr	Asp	Gly	Cys	His	Asn	Ile	Leu	Asp	Lys	Ala	Asp	Lys	Ambler, 1968; Ambler et al., 1969.

TABLE X (Continued) (d)

	Source	44	45	46	47	48	49	50	51	52	53	54	55	56	57	58	59	
INVERTE-BRATES	Moth *Samia Cynthia*	Pro	Gly	Phe	Ser	Tyr	Ser	Asn	Ala	Asn	Lys	Ala	Lys	Gly	Ile	Thr	Trp	Chan and Margoliash, 1966a.
	Tobacco horn worm moth	,,	,,	,,	,,	,,	,,	,,	,,	,,	,,	,,	,,	,,	,,	,,	,,	Dickerson et al., 1970; Chan, 1970.
	Fruit fly *Drosophila*	Ala	,,	,,	Ala	,,	Thr	,,	,,	,,	,,	,,	,,	,,	,,	,,	,,	Dickerson et al., 1970.
	Screw worm fly	,,	,,	,,	,,	,,	,,	,,	,,	,,	,,	,,	,,	,,	,,	,,	,,	Dickerson et al., 1970.
PLANTS	Wheat	,,	,,	Tyr	Ser	,,	Ser	Ala	,,	,,	Lys	Asn	,,	Ala	Val	Glu	,,	Stevens et al., 1967; Le Lange et al., 1960; †Boulter et al., 1970.
	Bean (*Phaseolus aureus*)	,,	,,	,,	,,	,,	,,	Thr	,,	,,	,,	,,	Met	,,	,,	Ile	,,	De Lange et al., 1969, 1970; Thompson et al., 1970, a, b; Boulter et al., 1970.
	Helianthus annuis	,,	,,	,,	,,	,,	,,	Ala	,,	,,	,,	,,	,,	,,	,,	,,	,,	Ramshaw et al., 1970; Boulter et al., 1970.
	Sesame	Pro	,,	,,	,,	,,	,,	,,	,,	,,	,,	,,	,,	,,	,,	,,	,,	Boulter et al., 1970.
	Castor	Ala	,,	,,	,,	,,	,,	,,	,,	,,	,,	,,	,,	,,	,,	Gln	,,	Boulter et al., 1970.
FUNGI	*Saccharomyces*	Glu	,,	,,	,,	,,	Thr	Asp	,,	,,	Ile	Lys	Lys	Asn	,,	Leu	,,	Narita et al., 1963, a, b; Sato et al., 1966; De Lange et al., 1970.
	Neurospora crassa	Asp	,,	,,	Ala	,,	,,	,,	,,	,,	Lys	Gln	,,	Gly	Ile	Thr	,,	Heller and Smith, 1965 De Lange et al., 1969.
	Candida krusei	Glu	,,	,,	Ser	,,	,,	,,	,,	,,	,,	Arg	Ala	,,	Val	Glu	,,	Narita and Sugeno, 1968; Narita et al., 1968; De Lange et al., 1970.
	Debaromyces kloeckeri	,,	,,	Phe	,,	,,	Ser	Glu	Ser	Tyr	,,	Lys	—	,,	,,	,,	,,	Narita and Sugeno, 1968; Narita et al., 1968; De Lange et al., 1970.
BACTERIA	*Rhodospirillum rubrum* c_2	Asp	Asn	Tyr	Ala	,,	Arg	Ile	Lys	Asn	Thr	Ser	Met	Lys	(Ala	Lys	,,	Dus and Sletten, 1968.
	Pseudomonas fluorescens c—551	Ala	Glu	Leu	Ala	Glu	Tyr	Tyr	His	Val	Gly	Ser	Glu	Gly	Val	Trp	Gly	Ambler, 1962, 1963.
	Desulphovibrio vulgaris c_3	Lys	Ser	Ala	Lys	Gly	Arg	Ile	Val	Val	Met	His	Asp	Lys	Asn	Thr	Lys	Ambler, 1968; Ambler et al., 1969.
	Desulphovibrio gigas c_3	Ser	Val	Asn	Ser	Trp	Tyr	Lys	Val	Val	His	His	Asp	Ala	Lys	Gly	Ala	Ambler, 1968; Ambler et al., 1969.

TABLE X (Continued) (e)

Source	60	61	62	63	64	65	66	67	68	69	70	71	72	73	74	
INVERTE-BRATES																
Moth *Samis cynthia*	Gly	Asp	Asp	Thr	Leu	Phe	Glu	Tyr	Leu	Glu	Asn	Pro	Lys	Lys	Tyr	Chan and Margoliash, 1966a.
Tobacco horn worm moth	Gln	,,	,,	,,	,,	,,	,,	,,	,,	,,	,,	,,	,,	,,	,,	Dickerson *et al.*, 1970; Chan, 1970.
Fruit fly *Drosophila*	,,	,,	,,	,,	,,	,,	,,	,,	,,	,,	,,	,,	,,	,,	,,	Dickerson *et al.*, 1970.
Screw worm fly	,,	,,	,,	,,	,,	,,	,,	,,	,,	,,	,,	,,	,,	,,	,,	Dickerson *et al.*, 1970.
PLANTS																
Wheat	Glu	Glu	Asn	,,	,,	Tyr	Asp	,,	,,	Leu	,,	,,	Trim Lys	,,	,,	Stevens *et al.*, 1967; De Lange *et al.*, 1960; †Boulter *et al.*, 1970.
Bean (*Phaseolus aureus*)	,,	,,	Lys	,,	,,	,,	,,	,,	,,	,,	,,	,,	,,	,,	,,	De Lange *et al.*, 1969, 1970; Thompson *et al.*, 1970, a, b; Boulter *et al.*, 1970.
Helianthus annuis	Glu	,,	Asn	,,	,,	,,	,,	,,	,,	,,	,,	,,	,,	,,	,,	Ramshaw *et al.*, 1970; Boulter *et al.*, 1970.
Sesame	Gly	,,	,,	,,	,,	,,	,,	,,	,,	,,	,,	,,	,,	,,	,,	Boulter *et al.*, 1970.
Castor	,,	,,	,,	,,	,,	,,	,,	,,	,,	,,	,,	,,	,,	,,	,,	Boulter *et al.*, 1970.
FUNGI																
Saccharomyces	Asp	,,	Asn	Asn	Met	Ser	Glu	,,	,,	Thr	,,	,,	Trim Lys	,,	,,	Narita *et al.*, 1963, a, b; Sato *et al.*, 1966; De Lange *et al.*, 1970.
Neurospora crassa	,,	,,	,,	Thr	Leu	Phe	,,	,,	,,	Glu	Asp	,,	Trim Lys	,,	,,	Heller and Smith, 1965; De Lange *et al.*, 1969.
Candida krusei	Ala	,,	Pro	,,	Met	Ser	Asp	,,	,,	,,	,,	,,	Trim Lys	,,	,,	Narita and Sugeno, 1968; Narita *et al.*, 1968; De Lange *et al.*, 1970.
Debaromyces kloecheri	Thr	,,	—	Asp	Leu	,,	,,	,,	,,	,,	,,	,,	Trim Lys	,,	,,	Narita and Sugeno, 1968; Narita *et al.*, 1968; De Lange *et al.*, 1970.
BACTERIA																
Rhodospirillum rubrum c_2	Gly	Ala)	Thr	Glu	Ala	Asn	Leu	(Ser	Lys	Lys)	Ser	Gly	Asp	Pro	Lys	Dus and Sletten, 1968.
Pseudomonas fluorescens c − 551	Pro	Ile	Pro	Met	Pro	Pro	Asn	Ala	Val	Ser	Asp	Asp	Glu	Ala	Gln	Ambler, 1962, 1963.
Desulphovibrio vulgaris c_3	Phe	Lys	Ser	Cys	Val	Gly	Cys	His	Val	Glu	Val	Ala	Gly	Ala	Asp	Ambler, 1968; Ambler *et al.*, 1969.
Desulphovibrio gigas c_3	Lys	Pro	Thr	Cys	Ile	Ser	Cys	His	Lys	Asp	Lys	Ala	Gly	Asp	Asp	Ambler, 1968; Ambler *et al.*, 1969.

TABLE X (Continued) (f)

	Source	75	76	77	78	79	80	81	82	83	84	85	86	87	83	89	
INVERTE-BRATES	Moth *Samia cynthia*	Ile	Pro	Gly	Thr	Lys	Met	Val	Phe	Ala	Gly	Leu	Lys	Lys	Ala	Asn	Chan and Margoliash, 1966a.
	Tobacco horn worm moth	,,	,,	,,	,,	,,	,,	,,	,,	,,	,,	,,	,,	,,	,,	,,	Dickerson *et al.*, 1970; Chan, 1970.
	Fruit fly *Drosophila*	,,	,,	,,	,,	,,	,,	Ile	,,	,,	,,	,,	,,	,,	Pro	,,	Dickerson *et al.*, 1970.
	Screw worm fly	,,	,,	,,	,,	,,	,,	,,	,,	,,	,,	,,	,,	,,	,,	,,	Dickerson *et al.*, 1970.
PLANTS	Wheat	,,	,,	,,	,,	,,	,,	Val	,,	Pro	,,	,,	Trim Lys	,,	,,	Gln	Stevens *et al.*, 1967; De Lange *et al.*, 1960; †Boulter *et al.*, 1970.
	Bean (*Phaseolus aureus*)	,,	,,	,,	,,	,,	,,	,,	,,	,,	,,	,,	,,	,,	,,	,,	De Lange *et al.*, 1969, 1970; Thompson *et al.*, 1970, a, b; Boulter *et al.*, 1970.
	Helianthus annuis	,,	,,	,,	,,	,,	,,	,,	,,	,,	,,	,,	,,	,,	,,	,,	Ramshaw *et al.*, 1970; Boulter *et al.*, 1970.
	Sesame	,,	,,	,,	,,	,,	,,	,,	,,	,,	,,	,,	,,	,,	,,	,,	Boulter *et al.*, 1970.
	Castor	,,	,,	,,	,,	,,	,,	,,	,,	,,	,,	,,	,,	,,	,,	,,	Boulter *et al.*, 1970.
FUNGI	*Saccharomyces*	,,	,,	,,	,,	,,	,,	Ala	,,	Gly	,,	,,	Lys	,,	Glu	Lys	Narita *et al.*, 1963, a, b; Sato *et al.*, 1966; De Lange *et al.*, 1970.
	Neurospora crassa	,,	,,	,,	,,	,,	,,	,,	,,	,,	,,	,,	,,	,,	Asp	,,	Heller and Smith, 1965; De Lange *et al.*, 1969.
	Candida krusei	,,	,,	,,	,,	,,	,,	,,	,,	,,	,,	,,	,,	,,	Ala	,,	Narita and Sugeno, 1968; Narita *et al.*, 1968; De Lange *et al.*, 1970.
	Debaromyces kloecheri	,,	,,	,,	,,	,,	,,	,,	,,	,,	,,	,,	,,	,,	,,	,,	Narita and Sugeno, 1968; Narita *et al.*, 1968; De Lange *et al.*, 1970.
BACTERIA	*Rhodospirillum rubrum* c_2	Gly	Thr	Leu	(Ile	Glu	Ala	Asp	Ala	Gly	Ser)	Tyr	Ala	Lys	Val	Lys	Dus and Sletten, 1968.
	Pseudomonas fluorescens c—551	Thr	Leu	Ala	Lys	Trp	Val	Leu	Ser	Gln	Lys	,,	,,	,,	,,	,,	Ambler, 1962, 1963.
	Desulphovibrio vulgaris c_3	Ala	Ala	Lys	Lys	Lys	Asp	Leu	Thr	Gly	Cys	Lys	Lys	Ser	Cys	Cys	Ambler, 1968, Ambler *et al.*; 1969.
	Desulphovibrio gigas c_3	Lys	Glu	Leu	Lys	Lys	Lys	Leu	Thr	Gly	Cys	Lys	Gly	Ser	Ala	Cys	Ambler, 1968; Ambler *et al.*, 1969.

Cytochrome c

TABLE X (Continued) (g)

Source	90	91	92	93	94	95	96	97	98	99	100	101	102	103	104	
INVERTEBRATES																
Moth *Samia cynthia*	Glu	Arg	Ala	Asp	Leu	Ile	Ala	Tyr	Leu	Lys	Glu	Ser	Thr	Lys	—	Chan and Margoliash, 1966a.
Tobacco horn worm moth	,,	,,	,,	,,	,,	,,	,,	,,	,,	,,	Gln	Ala	,,	,,	—	Dickerson et al., 1970; Chan, 1970.
Fruit fly *Drosophila*	,,	,,	Gly	,,	,,	,,	,,	,,	,,	,,	Ser	,,	,,	,,	—	Dickerson et al., 1970.
Screw worm fly	,,	,,	,,	,,	,,	,,	,,	,,	,,	,,	,,	,,	,,	,,	—	Dickerson et al., 1970.
PLANTS																
Wheat	Asp	,,	Ala	,,	,,	,,	,,	,,	,,	,,	Val Lys	,,	Thr	Ser	Ser	Stevens et al., 1967; De Lange et al., 1960; †Boulter et al., 1970.
Bean (*Phaseolus aureus*)	,,	,,	,,	,,	,,	,,	,,	,,	,,	,,	Glu	Ser	,,	Ala	—	De Lange et al., 1969, 1970; Thompson et al., 1970, a, b; Boulter et al., 1970.
Helianthus annuis	Glu	,,	,,	,,	,,	,,	,,	,,	,,	,,	Thr	,,	,,	,,	—	Ramshaw et al., 1970; Boulter et al., 1970.
Sesame	,,	,,	,,	,,	,,	,,	,,	,,	,,	,,	Glu	Ala	,,	,,	—	Boulter et al., 1970.
Castor	Asp	,,	,,	,,	,,	,,	,,	,,	,,	,,	Gln	,,	,,	,,	—	Boulter et al., 1970.
FUNGI																
Saccharomyces	,,	,,	Asn	,,	,,	,,	Thr	,,	,,	,,	Lys	Ala	Cys	Glu		Narita et al., 1963, a, b; Sato et al., 1966; De Lange et al., 1970.
Neurospora crassa	,,	,,	,,	,,	Ile	,,	,,	Phe	Met	,,	Glu	,,	Thr	Ala		Heller and Smith, 1965; De Lange et al., 1969.
Candida krusei	,,	,,	,,	,,	,,	Val	,,	Tyr	,	Leu	,,	,,	Ser	Lys		Narito and Sugeno, 1968; Narita et al., 1968; De Lange et al., 1970.
Debaromyces kloecheri	,,	,,	,,	,,	,,	Ile	,,	,,	Leu	Val	Lys	,,	Thr	,,		Narita and Sugeno, 1968; Narita et al., 1968; De Lange et al., 1970.
BACTERIA																
Rhodospirillum rubrum c_2	Asn	Pro	Lys	Ala	Phe	Val	Leu	Glu	Lys	Met	Thr	Phe	Lys	Leu	*	Dus and Sletten, 1968.
Pseudomonas fluorescens c – 551		Glu														Ambler, 1962, 1963.
Desulphovibrio vulgaris c_3	His															Ambler, 1968; Ambler et al., 1969.
Desulphovibrio gigas c_3	His	Pro	Ser													Ambler, 1968; Ambler et al., 1969.

† The findings of Boulter et al. differ from those of the earlier workers.
* Thr-Lys-Asp-Asp-Glu-Ile-Glu-Val-Ile-Asn-Ala-Tyr-Leu-Lys-Thr-Lys.

The importance of the hydrophobic residues for the screening off of the haem will be discussed in connection with x-ray crystallographic results in B3c. Attention has been drawn by Margoliash (1966) see also Margoliash et al. (1961), to the fact that the hydrophobic and basic residues frequently occur in clusters, e.g. hydrophobic residues in 9–11, 35–36, 57–59, 74–75, 81–82, 94–98, basic residues in 26–27, 38–39, 72–73, 86–87. The two lysines 72 and 73 have their ϵ-NH_2 groups directed towards the outside (Margoliash, 1962a,b) and are probably essential for the binding of cytochrome oxidase (Ando et al., 1966).

Some mammalian species have the same cytochrome c (man and chimpanzee; cattle, sheep and pig). Some closely related species (man and monkey; horse and donkey) differ by only one residue; others by two (cattle and grey whale; cattle and dog), three (cattle and horse), up to 12 residues (man and horse). It is interesting to note the close structural similarity of cattle and whale cytochrome c. For primates Ile in 11, Met in 12, Ser in 15, Tyr in 46, Ala in 50 and Val in 83 appear to be distinctive; for equines Lys in 60 and Thr in 89; for the kangaroo Asn in 33 and Ile in 35 and 58; for the dog Thr in 88 and Lys in 103; for the rabbit Val in 44 and for both rabbit and dog Gly in 66.

In *vertebrates* 66 positions are invariant, Gly in eleven positions, 1, 6, 23, 24, 34, 37, 41, 45, 56, 77 and 84; Ala in 43, 51, 96; Val in 20; Leu in 32, 64, 68, 98; Ile in 57, 75, 81; Pro in 30, 71, 76; Cys (haem-binding) in 14, 17; Met (haem-iron binding) in 80; Thr in 19, 40, 49, 63, 78; Phe in 10, 82; Tyr in 67, 97; Trp in 59; Asp in 2, 93; Asn in 31, 52, 70; Glu in 21, 69, 90; Gln in 16, 42; Lys in 13 positions, 5, 7, 8, 13, 27, 39, 53, 55, 72, 73, 79, 87, 99; His in 18 (haem-iron binding) and 26; Arg in 38 and 91. The large number of invariant glycines, together with the prolines interrupt the formation of α-helical stretches, for the stabilization of which, at least, five to six residues are required; the shortest α-helical sequence in myoglobin has seven residues. In addition to the 66 invariant positions, there are 17 with conservative replacements only. Of the remaining 20–21 positions in which "radical" substitutions occur, in six only a single deviant has been found among the 19 species, leaving 14 "polyvariant" positions (12, 22, 23, 44, 48, 50, 60, 66, 74, 89, 92, 100, 101, 103). Three of these belong to the C-terminal group which is not essential for the activity of cytochrome c; Titani et al. (1959) have shown that the last seven C-terminal residues can be removed by carboxypeptidase without loss of cytochrome c activity in the succinoxidase system. Margoliash et al. (1963) pointed out that there are only a few changes in the positions 16–43 and 63–82. Although now a greater variability in these regions (see above) has been found, the regions 16–27, 30–32, 34–43, 51–59, 63–65 and 67–82 show, indeed, only few variations.

There are 12 differences between rabbit and chicken (or turtle), 10 between turtle and frog, but 20 between rattlesnake and frog. Between frog and fish cytochromes there are about 20 differences, and fish cytochromes differ from one another in 10–19 residues. In chicken, Ile in 3 appears to be characteristic. Frog c is unique in having two additional cysteine residues at 20 and 23 connected by an intramolecular -S-S-bond (Margoliash et al., 1968). In fish cytochromes, 22 is always Asn and 54 Ser. In tuna and carp, a second Trp replaces the His in 33.

In the cytochromes c from invertebrates, plants and fungi (Table X, B), the acetyl-glycine position 1 in vertebrates is replaced by up to eight residues preceding glycine which in the table are numbered from − 8 to − 1 (Smith and Margoliash, 1964; Smith, 1968). This alignment shows the full degree of similarity which is in marked contrast to the far greater variability in myoglobin (Fitch and Margoliash, 1967b). Deletions of residues rarely occur in cytochromes except in the C-terminal region. For all cytochromes c from yeast to man, the "eukaryotic" cytochromes c, 34 residues remain invariant, Gly 1, 6, 29, 34, 41, 45, 77, 84; Ala 51; Leu 32, 68; Ile 75; Pro 30, 71, 76; Cys 14, 17 (haem-binding); Met 80 (haem-iron binding); Thr 78; Phe 10, 82; Tyr 46, 67, 74, 78; Trp 59; Asn 52; Lys 27, 73, 79, 87; His 18 (haem-iron binding) and Arg 38, 91. Residues 70–85 show only a few, mostly "conservative" replacements. The total number of "conservative" replacements amounts to about 25. This leaves less than half of the number of residues with one (5) or more (41) radical replacements. These can obviously be replaced without harm to activity (Margoliash and Schejter, 1966). The cytochrome c of the *Samia* moth differs from that of tuna in 32, from that of wheat in 52 positions, while wheat and bean differ in 14. Plant cytochromes c differ from one another in only 21 positions, and only 9 of these show radical alterations. Finally, the cytochrome c of wheat differs from those of fungal cytochromes in 47 residues. If rabbit is used as an example of mammalian cytochrome c, other mammalian cytochromes differ by 4–11 residues, chicken c by 13, frog c by 18, tuna c by 21, moth c in 27, wheat c in 41 (from man only 35) and yeast c in 45 residues.

The insect cytochromes c appear to begin at − 4 to − 6 with a free N-terminal group and thus have 108–110 residues. Lys in 13 is replaced by Arg, Met in 65 by Phe. The silkworm c has also been studied by Tuppy (1958). In the plant cytochromes, the N-terminal is acetylalanine in − 8 and the number of residues 111–112. Lys in 7 is replaced by Ala or Met, and in 72 and 82 by trimethyllysine, its betaine. The latter maintains, however, the positive charge of lysine. Fungal cytochromes c have also trimethyllysine in 72 but not in 82 (De Lange et al., 1970). The specific trimethylating system is restricted to *Ascomycetes* and higher plants. The amino acid composition of the soybean cytochrome c has been established by Fridman et al. (1968). Fungal cytochromes c have a free NH_2-terminal group (Thr, Gly or Pro) in positions -4 to -6, and a total number of 107–113 residues, Lys in 4, Ala in 7. As in insects, but not in plants, Lys in 13 is replaced by Arg; 83 is always Gly, 92 Asn, 93 Asp and 96 Thr. Only *Saccharomyces* isocytochrome-1 (Narita *et al.*, 1963; see also Narita and Titani, 1969; Titani and Narita, 1969; Titani *et al.*, 1964) has an additional single Cys in 103 and thus can be dimerized by $CuSO_4$ and other oxidants (Narita *et al.*, 1964; Little and O'Brien, 1966, 1967). Iso-cytochrome -2 of yeast differs from iso-1 (Sherman *et al.*, 1965, 1966, 1968a,b; Slonimski *et al.*, 1965; Stewart *et al.*, 1966) in 13 residues, e.g. 16 Glu for Leu, 20 Ile for Val, 22 Gly for Lys, 26 Asn for His. These yeast iso-cytochromes are thus quite different from variants of cytochromes in beef or yeast (Flatmark, 1966; Flatmark and Sletten, 1968) which differ only in the hydrolysis of acid amide residues. Mutants of iso-cytochrome-1 have also been studied. Fungal cytochromes differ one from the other by 24–29 residues, but the difference between yeast and *Neurospora c* is surprisingly large (40 residues).

The great evolutionary significance of these studies, which have also been reviewed recently by Nolan and Margoliash (1968), will be discussed in Chapter X. The close similarity of all eukaryotic cytochrome *c* makes a common ancestor virtually certain, from which the various cytochromes have been formed by homologous evolution, not by convergent analogous evolution (Margoliash *et al.*, 1963; Smith and Margoliash, 1964; Fitch and Margoliash, 1967, 1968). These authors have calculated the branching times between different families and genera in millions of years from the number of replacements, or more correctly from the number of mutations in the coding DNA. This can be based on a wider evidence than that obtained from cytochrome *c* alone, but cytochrome *c* is of special importance as it covers a much larger span of evolutionary period than, say, haemoglobin or myoglobin.

The amino acid sequence of bacterial cytochromes *c* is far less completely known. Three examples are given in Table X. There is certainly a far greater variation between them and eukaryotic cytochromes *c*, as well as between different bacterial cytochromes *c*, with 82, 106 and 118 residues, the N-terminal being Gln at $+2$ in *Rhodospirillum* c_2, Glu at $+3$ in *Pseudomonas* *c*-551, and Ala at -15 in *Desulphovibrio* c_3. All have, however, the characteristic Cys $--$ Cys-His arrangement which links the side chains and one side of the haem-iron. This occurs once or even twice, as in cytochrome c_3. This His is the only one found in *Pseudomonas* *c*-551.

(*d*) *Attempts at a chemical synthesis of a cytochrome c.* Sano and Granick (1961; cf. also Sano and Rimington, 1963) found the formation of protoporphyrin and protohaem by the action of a mitochondrial coproporphyrinogen oxidase on coproporphyrinogen. Sano *et al.* (1964a) also found that protoporphyrinogen, but not protoporphyrin reacts with cysteine or other

SH-compounds to form porphyrin c type compounds in which the vinyl groups of protoporphyrinogen add RSH. This study did not only show the way in which cytochromes c may be synthesized *in vivo*, but it also opened a way to their total chemical synthesis in which the haem prosthetic group was finally introduced into the polypeptide (Sano and Tanaka, 1964; Sano et al., 1964) synthesizing a cytochrome a haemopeptide. Making use of the peptide synthesis solid phase method of Merrifield (1965), Sano and Kurihara (1969), (cf. also Sano et al., 1968) finally succeeded in synthesizing a cytochrome c which differs only in one amino acid residue (Phe instead of Trp in position 59) from that of horse heart c. This was unavoidable because the Trp-containing peptide proved too unstable to be used in further addition of amino acid residues. This synthetic cytochrome c had typical low spin spectra but had only low activity as oxidase of succinic dehydrogenase. Evidently tryptophan, which is always present in eukaryotic cytochromes c (Smith and Margoliash, 1964) is essential for the cytochrome activity, at least for its reaction with the reductase.

REFERENCES

Ambler, R. P. (1962). *Biochem. J.* **82**, 30P.
Ambler, R. P. (1963). *Biochem. J.* **89**, 341–349, 349–378.
Ambler, R. P. (1968). *Biochem. J.* **109**, 47P.
Ambler, R. P., Bruschi, M. and Le Gall, J. (1969). *FEBS Letters* **5**, 115–117.
Ando, K., Matsubara, H. and Okunuki, K. (1966). *Biochim. et Biophys. Acta* **118**, 256–267.
Bahl, O. P. and Smith, E. L. (1965). *J. Biol. Chem.* **240**, 3585–3593.
Baba, Y., Mizushima, H. and Watanabe, H. (1969). *Chem. Abstr.* **69**(3), 16221.
Boulter, D., Thompson, E. W., Ramshaw, J. A. M. and Richardson, M. (1970). *Nature* (London) **228**, 552–554.
Chan, S. K. (1970). *Biochim. et Biophys. Acta* **221**, 497–501.
Chan, S. K. and Margoliash, E. (1966a). *J. Biol. Chem.* **241**, 335–348.
Chan, S. K. and Margoliash, E. (1966b). *J. Biol. Chem.* **241**, 507–515.
Chan, S. K., Needleman, S. B., Stewart, J. W., Walasek, O. F. and Margoliash, E. (1963). *Fed. Proc.* **22**, Abstr. 2972.
Chan, S. K., Tullross, I. and Margoliash, E. (1966). *Biochemistry* **5**, 2586–2597.
Chan, S. K., Walasek, O. F., Barlow, G. H. and Margoliash, E. (1967). *Fed. Proc.* **26**, Abstr. 2603.
De Lange, R. J., Glazer, N. and Smith, E. L. (1969). *J. Biol. Chem.* **244**, 1385–1388.
De Lange, R. J., Glazer, A. N. and Smith, E. L. (1970). *J. Biol. Chem.* **245**, 3325–3327.
Der Vartanian, D. V. (1970). *Biochem. Biophys. Res. Commun.* **41**, 932–937.
Dickerson, R. E., Takang, T., Eisenberg, D., Kallai, D. B., Samson, L., Cooper, A. and Margoliash, E. (1971). *J. Biol. Chem.* **246**, 1511–1535.
Drabkin, D. (1961). "Haematin Enzymes", Vol. I, pp. 171–172.
Dus, K. and Sletten, K. (1968) "Osaka" pp. 293–303.
Ehrenberg, A. and Theorell, H. (1955a). *Acta Chem. Scand.* **9**, 1119–1125.
Ehrenberg, A. and Theorell, H. (1955b). *Acta Chem. Scand.* **9** 1193–1205.
Fitch, W. and Margoliash, E. (1967a). *Science* **155**, 279–284.

Fitch, W. and Margoliash, E. (1967b). *Biochem. Gen.* I, 65–71.
Fitch, W. and Margoliash, E. (1968). *Brookhaven Sympos. Biol.* **21**, 217–232.
Flatmark, T. (1966). "Hemes and Hemoproteins". pp. 411–413.
Flatmark, T. and Sletten, K. (1968). *J. Biol. Chem.* **243**, 1623–1629.
Fridman, C., Lis, H., Sharon, N. and Katchalski, E. (1968). *Arch. Biochem. Biophys.* **126**, 299–304.
Gillies, N. E., and Pirrie, R. (1958). *Biochem. J.* **69**, 605–611.
Goldstone, A. and Smith, E. L. (1966). *J. Biol. Chem.* **241**, 4480–4486.
Goldstone, A. and Smith, E. L. (1967). *J. Biol. Chem.* **242**, 4702–4710.
Gürtler, L. and Horstmann, H. J. (1970). *European J. Biochem.* **12**, 48–57.
Harbury, H. A. (1966). "Hemes and Hemoproteins" pp. 391–393.
Harbury, H. A., Cronin, J. R., Fanger, J. R., Hettinger, M. W., Murphy, T. P., Myer, A. J. and Vinogradov, S. N. (1965). *Proc. Nat. Acad. Sci.* **54**, 1658–1664.
Harbury, H. A. and Loach, P. A. (1959). *Proc. Nat. Acad. Sci.* **45**, 1344–1359.
Harbury, H. A. and Loach, P. A. (1960a). *J. Biol. Chem.* **235**, 3640–3645.
Harbury, H. A. and Loach, P. A. (1960b). *J. Biol. Chem.* **235**, 3646–3653.
Harbury, H. A., Myer, Y. P., Murphy, A. J. and Vinogradov, S. N. (1966). "Hemes and Hemoproteins", pp. 415–426.
Heller, J. and Smith, E. L. (1965). *Proc. Nat. Acad. Sci.* **54**, 1621–1625.
Hennig, A., Horstman, H. -J. and Gürtler, L. (1969). *Z. physiol. Chem.* **350**, 1103–1110.
IUPAC-IUB Commission on Biochemical Nomenclature (1968). *Biochim. et Biophys. Acta* **168**, 6–10.
Kowalsky, A. (1962). *J. Biol. Chem.* **237**, 1807–1819.
Kowalsky, A. (1964). *Fed. Proc.* **23**, Abstr. 730.
Kowalsky, A. (1965). *Biochemistry* **4**, 2382–2388.
Kowalsky, A. (1966). "Hemes and Hemoproteins", pp. 529–535.
Kowalsky, A. (1969). *Fed. Proc.* **28**, Abstr. 1919.
Kreil, G. (1963). *Z. physiol. Chem.* **334**, 154–166.
Kreil, G. (1965). *Z. physiol. Chem.* **340**, 86–87.
Kreil, G. and Tuppy, H. (1961). *Nature* (London) **192**, 1123–1125.
Leaf, G. and Gillies, N. E. (1955). *Biochem. J.* **61**, 7P.
Leaf, G., Gillies, N. E. and Pirrie, R. (1958). *Biochem. J.* **69**, 605–611.
Little, C. and O'Brien, P. J. (1967). *Biochem. J.* **101**, 11–12P.
Little, C. and O'Brien, P. J. (1966). *Arch. Biochem. Biophys.* **122**, 406–410.
McDowall, M. A. and Smith, E. L. (1965). *J. Biol. Chem.* **240**, 4635–4637.
Margoliash, E. (1961). "Haematin Enzymes", Vol. **II**, pp. 383–384.
Margoliash, E. (1962a). *J. Biol. Chem.* **237**, 2161–2174.
Margoliash, E. (1962b). *Brookhaven Sympos. Biol.* **15**, 266–281.
Margoliash, E. (1963). *Proc. Nat. Acad. Sci.* **50**, 372–379.
Margoliash, E. (1966). "Hemes and Hemoproteins", pp. 371–391.
Margoliash, E. and Fitch, W. M. (1968). *Annals. New York Acad. Sci.* **151**, 359–381.
Margoliash, E. and Hill, R. (1961). "Haematin Enzymes" Vol. **II**, pp. 383–384.
Margoliash, E. and Schejter, A. (1966). *Adv. in Protein Chem.* **21**, 113–286.
Margoliash, E. and Smith, E. L. (1961). *Nature* (London) **192**, 1121–1123.
Margoliash, E. and Smith, E. L. (1962). *J. Biol. Chem.* **237**, 2151–2160.
Margoliash, E., Fitch, W. M. and Dickerson, R. E. (1968). *Brookhaven Sympos. Biol.* **21**, 259–305.
Margoliash, E., Frowirth, N. and Wiener, E. (1959). *Biochem. J.* **71**, 559–570.
Margoliash, E., Kimmel, J. R., Hill, R. L. and Schmidt, W. R. (1962). *J. Biol. Chem.* **237**, 2148–2150.

Margoliash, E., Needleman, S. B. and Stewart, J. W. (1963). *Acta Chem. Scand.* **17**, S250–256.
Margoliash, E., Smith, E. L., Kreil, G. and Tuppy, H. (1961). *Nature* (London) **192**, 1125–1127.
Matsubara, H. and Smith, E. L. (1962). *J. Biol. Chem.* **237**, PC 3575–3576.
Matsubara, H. and Smith, E. L. (1963). *J. Biol. Chem.* **238**, 2732–2753.
Matsubara, H., Chu, T. L. C. and Yasunobu, K. T. (1963). *Arch. Biochem. Biophys.* **101**, 209–214.
Minakami, S., Titani, K. and Ishikura, H. (1957). *J. Biochem.* **44**, 535–536.
Myer, Y. P. (1968). *Biochemistry* **7**, 765–776.
Nakashima, T., Higa, H., Matsubara, H., Benson, A. and Yasunobu, T. (1966). *J. Biol. Chem.* **241**, 1166–1177.
Nakayama, T., Titani, K. and Narita, K. (unpubl.) see Dickerson *et al.* 1970.
Narita, K. and Sugeno, K. (1968). "Osaka", pp. 304–307.
Narita, K. and Titani, K. (1969). *J. Biochem.* **65**, 259–267.
Narita, K., Sugeno, K., Mizoguchi, T. and Hamaguchi, K. (1968). "Osaka" pp. 362–369.
Narita, K., Titani, K., Yaoi, Y., Murakami, H., Kimura, M. and Vanecek, J. (1963a). *Biochim. et Biophys. Acta* **73**, 670–673.
Narita, K., Titani, K., Yaoi, Y., and Murakami, H. (1963b). *Biochim. et Biophys. Acta* **77**, 688–690.
Needleman, S. B. and Margoliash, E. (1966). *J. Biol. Chem.* **241**, 853–863.
Nolan, C. and Margoliash, E. (1966). *J. Biol. Chem.* **241**, 1049–1059,
Nolan, C. and Margoliash, E. (1968). *Ann. Rev. Biochem.* **37**. 727–790.
Paléus, S. (1954). *Acta Chem. Scand.* **8**, 971–984.
Paléus, S. and Tuppy, H. (1959). *Acta Chem. Scand.* **13**, 641–646.
Paléus, S., Ehrenberg, A. and Tuppy, H. (1955). *Acta Chem. Scand.* **9**, 365–374.
Ramshaw, J. A., Thompson, E. W. and Boulter, D. (1970). *Biochem. J.* **119**, 535–539.
Rothfus, J. A. and Smith, E. L. (1965). *J. Biol. Chem.* **240**, 4277–4283.
Sano, S. and Granick, S. (1961). *J. Biol. Chem.* **236**, 1173–1180.
Sano, S. and Tanaka, K. (1964). *J. Biol. Chem.* **239**, PC 3109–3110.
Sato, C., Titani, K. and Narita, K. (1966). *J. Biochem.* **60**, 682–690.
Sherman, F., Taber, H. and Campbell, W. (1965). *J. Mol. Biol.* **13**, 21–35.
Sherman, F., Stewart, J. W., Margoliash, E., Parker, J. and Campbell, W. (1966). *Proc. Nat. Acad. Sci.* **55**, 1498–1505.
Sherman, F., Stewart, J. W., Parker, J., Putterman, G. P. and Margoliash, E. (1968a). "Osaka", pp. 257–268.
Sherman, F., Stewart, J. W., Parker, J. H., Inhaber, E., Shipman, N. A., Putterman, G. J., Gardisky, R. L. and Margoliash, E. (1968b). *J. Biol. Chem.* **243**, 5446–5456.
Slonimski, P. P., Acher, R., Rere, G., Sels, A. and Somlo, M. (1965). Mecanisms de Regulation des Activites Cellulaires chez les Microorganisms. (Centre Nat. de la Recherche Scientifique, eds.) p. 435. Paris.
Smith, E. L. (1968). "Osaka", pp. 282–288.
Smith, E. L. (1970). *8th Internat. Congress Biochem., Switzerland, Sympos.* **1**, 15–16.
Smith, E. L. and Margoliash, E. (1964). *Fed. Proc.* **23**(6), 1243–1247.
Stellwagen, E. (1967). *J. Biol. Chem.* **242**, 602–606.
Stevens, F. C., Glazer, A. N. and Smith, E. L. (1967). *J. Biol. Chem.* **242**, 2764–2779.
Stewart, J. M. and Margoliash, E. (1965). *Canad. J. Biochem.* **43**, 1187–1206.

Stewart, J. M., Margoliash, E. and Sherman, F. (1966). *Fed. Proc.* **25**, Abstr. 258.
Takahashi, K., Titani, K. and Minakami, S. (1959a). *J. Biochem.* **46**, 1323–1330.
Takahashi, K., Titani, K. and Minakami, S. (1959b). *J. Biochem.* **46**, 1437–1440.
Theodoropoulos, D. and Souchleris, I. (1966). *J. Organ. Chem.* **31**, 4009–4013.
Theorell, H. (1956). *Science* **124**, 467.
Theorell, H. and Agner, K. (1943). *Ergebnisse d. Enzymfschg.* **9**, 231–296.
Thompson, E. W., Laycock, M. V., Ramshaw, J. A. and Boulter, D. (1970). *Biochem. J.* **117**, 183–192.
Thompson, E. W., Richardson, M. and Boulter, D. (1970b). *Biochem. J.* **121**, 439–446.
Titani, K. and Narita, K. (1964). *J. Biochem.* **56**, 241–256.
Titani, K. and Narita, K. (1969). *J. Biochem.* **65**, 247–257.
Titani, K., Ishikura, H. and Minakami, S. (1959). *J. Biochem.* **46**, 151–154.
Titani, K., Kimura, M., Vanecek, J., Murakami, H. and Narita, K. (1964). *J. Biochem.* **56**, 230–240.
Titani, K., Narita, K. and Okunuki, K. (1962). *J. Biochem.* **51**, 350–358.
Tsou, C. L. (1949). *Nature* (London) **164**, 1134–1135.
Tsou, C. L. (1951a). *Biochem. J.* **49**, 362–367.
Tsou, C. L. (1951b). *Biochem. J.* **49**, 367–374.
Tu, A. T. T., Reinosa, J. A. and Hsiao, Y. Y. (1968). *Experientia* **24**, 219–221.
Tuppy, H. (1958). *Z. Naturfsch.* **12b**, 784–788.
Tuppy, H. and Bodo, G. (1954a). *Monatshefte d. Chem.* **85**, 807–821.
Tuppy, H. and Bodo, G. (1954b). *Monatshefte d. Chem.* **85**, 1024–1045.
Tuppy, H. and Bodo, G. (1954c). *Monatshefte d. Chem.* **85**, 1182–1186.
Tuppy, H. and Paléus, S. (1955). *Acta Chem. Scand.* **9**, 353–364.
Ulmer, D. D. (1966). *Proc. Nat. Acad. Sci.* **55**, 894–899.
Urry, D. W. (1966). "Hemes and Hemoproteins", pp. 435–445.
Urry, D. W. (1967a). *J. Biol. Chem.* **242**, 4441–4448.
Urry, D. W. (1967b). *J. Am. Chem. Soc.* **89**, 4190–4196.
Urry, D. W. and Pettegrew, J. W. (1967). *J. Am. Chem. Soc.* **89**, 5276–5283.
Wada, K., Matsubara, H. and Okunuki, K. (1968). "Osaka", pp. 309–310.
Wolman, Y., Schwarzberg, M. and Frankel, M. (1970a). *Israel J. Chem.* **8**, 53–59.
Wolman, Y., Ladkany, D. and Frankel, M. (1970b). *Israel J. Chem.* **8**, 59.
Yasunobu, K. T., Nakashima, T., Higa, H., Matsubara, A. and Benson, A. (1963). *Biochim. et Biophys. Acta* **78**, 791–794.

3. Tertiary structure

(a) *Coordinate linkage between haem-iron and ligands in the protein.* The spectral and magnetochemical similarities between cytochrome *c* and meso- or haemato-haemochromes supported the hypothesis that the haem-iron was linked on either side to a nitrogenous ligand. These were assumed to be imidazole-nitrogens of two histidine residues by Theorell and his co-workers (Theorell and Åkeson, 1941a,b; Paléus, 1954; Paul, 1951a,b; Boeri *et al.*, 1953; Ehrenberg and Theorell, 1955). This has been accepted for a long time. Horinishi *et al.* (1965) assumed that in horse heart cytochrome *c* the two histidines were His-18 and His-26 (see also Margoliash, 1955). However, while it is now generally accepted that His-18 is one of two ligands, it had become increasingly difficult to accept that the ligand on the other side of the haem plate is also histidine-imidazole, even before it was discovered by Harbury et al. (1965) that thioether-sulphur could combine with haem to give haemochromes. His-18 is invariant in all cytochromes and its imidazole-N is in a sterical position to combine with the haem iron without stress. However, no second histidine residue was found in salmon cytochrome c (Paléus, 1954; cf. also Margoliash et al. 1959). Stellwagen (1966) showed that the spectroscopic similarity of the benzoylhistidine complex of the tryptic haemopeptide of cytochrome c (Tuppy and Bodo, 1954) could not be considered as good evidence for a second imidazole linkage, although only His was found to be reactive with bromoacetate at neutral pH. Cyanide which must replace one of the two protein-iron ligands in cytochrome c, should then expose a second His to carboxymethylation, whereas in fact it exposed a second methionine. While Margoliash (1966) still assumed that His-imidazole was replaced by cyanide, there is now evidence that the methione-sulphur iron linkage is replaced by it. The carboxymethylation of His involves His 33, and if this is absent as in tuna cytochrome c no His carboxymethylation is found (Harbury, 1966). Only one His, readily destroyed by photooxidation with methylene blue (Nakatani, 1960) is essential for activity and absorption spectrum of cytochrome c. Finally, the second His while present in yeast isocytochrome c-1 was absent in isocytochrome c-2, although the ORD changes in both could be correlated with the disappearance of the 695 nm absorption band, now known to be due to the sulphur-iron linkage (Mirsky and George, 1967b).

For similar reasons, Margoliash et al. (1959) and Margoliash (1964) assumed that the haem-iron in c was bound between an imidazole -N and the ϵ-NH_2 group of lysine. Polylysine had been shown to form haemochromes by Blauer and Ehrenberg (1963). However, all lysine ϵ-NH_2 groups can be guanidinated or trifluoroacetylated without loss of the typical absorption spectrum or activity in the succinoxidase system (Fanger and Harbury, 1964; Hettinger and Harbury, 1965; Vinogradov and Harbury, 1967; Fanger et al., 1967; see also Margoliash and Schejter, 1966). It appears unlikely that the indolenitrogen of tryptophan can be a haem-iron ligand as suggested by Sano et al. (1968), since it combines only with Fe^{3+}, not Fe^{2+}. Although one Trp has been found in all eukaryocytic cytochromes c, it is absent in some bacterial ones (see Table X).

Methionine-80. The methionine residue of horse c in position 65 can be carboxymethylated with bromoacetate at neutral pH together with His 33, without loss of activity or typical spectrum. If, however, Met-80 is also carboxymethylated, e.g. after addition of cyanide, activity or spectrum cannot be reconstituted by removal of the cyanide (Ando et al., 1965, 1966; Tsai et al., 1965). Met-65 is more exposed at the surface than is Met-80. Harbury et al. (1965) showed that N-acetylmethionine, its methyl ester, diethylsulphide and other thioethers formed haemochromes with haem and the haem octapeptide of cytochrome c. They suggested that cytochromes c were mixed imidazole-N-Fe-S-Met haemochromes. The first S-compounds of haems had been observed by Falk (1961) with mercapto-ethanol. The ORD spectrum in the 550 nm region is, indeed, indicative of a highly asymmetrical binding of haem in the 5 and 6 positions (Eichhorn and Cairns, 1958); this was later confirmed by Mössbauer spectral studies of McDermott et al. (1967).

The first experiments of Harbury et al. (1965) and Hettinger et al. (1966) were carried out with *Pseudomonas* cytochrome c-551 which has only one His-18 and two Met (24 and 63, in our way of counting, cf. Table IX). Met-63 is the iron-binding residue of the bacterial cytochrome, corresponding to Met-80 in the eukaryotic cytochromes c (Hettinger et al., 1966; Fanger et al., 1967; Fanger and Harbury, 1965). The dicarboxymethylated CN-free compound was of high-spin type with a distinct 620 nm band (cf. Stellwagen, 1966). It can also be obtained by carboxymethylation of the dimer of cytochrome c. It does not activate the respiration of cytochrome c depleted rat liver mitochondria, but forms a complex with oxygen (Schejter and Aviram, 1970). Harbury et al. suggested that, in contrast to the assumption of Margoliash (1966), cyanide does not replace the imidazole, but the weaker methionine-sulphur linkage. This has also been assumed by Schejter and George (1965) and has been confirmed by Tsai and Williams (1965a,b) with iodoacetate, by Schejter and Frowirth (1966), Stellwagen (1966), Shechter and Saludjian (1967), Wüthrich (1969), Aviram and Schejter (1969), and with yeast isocytochrome c-1 by Schejter and Aviram (1969). In contrast to the carboxymethylation of methionine, that of His-33 is independent of the presence or absence of cyanide, and even low concentrations of free imidazole replace the Met 80-Fe linkage. Its splitting is accompanied by spectroscopic changes, in particular the disappearance of the 695 nm absorption band of c^{3+} (Mirsky and George, 1967b) and in the Soret band region (Harbury, 1966; Shechter and Saludjian, 1967). Thus the N-acetylmethionine complex of c-undecapeptide but not its imidazole complex show the 695 nm band, a Soret band at 410 nm of ϵmM 106, and a high absorption (ϵmM 285) at 360 nm (see also O'Brien, 1969). The absorption spectrum of polymethionine cytochrome c undecapeptide resembles that of cytochrome c more than the polyglycyl or polyvalyl peptide (Wainio et al., 1970). Such findings are more important evidence than those of Warme and Hager (1970a,b) with the model compounds derived from mesohaem-sulphuric anhydrides. In these the spectra of bis-His compounds resembled those of c^{2+} more closely than of the His-Met compound; the protein conformation holding Met-80 in position may produce a closer contact of its sulphur with the iron than in these models. To this conformation-enforced contact is also probably due the high E_0' of eukaryotic cytochromes c and the low-spin iron of their ferric form. The slight deviation of the α-band position of haem c-methionine haemochrome at 552·5 nm from that at 550 nm for cytochrome c, found by Nanzyo and Sano (1968) may be explained similarly. The cyanide effect is similar in horse and yeast cytochrome c (isocytochrome c-1) but the methionine-iron linkage is weaker by 1 k cal in the yeast cytochrome (Aviram and Schejter, 1969; Schejter et al., 1970).

The role of this methionine-S-iron linkage is also supported by the photooxidation experiments of Kowalsky (1965), and Jori et al., (1970), and particularly by the NMR studies of Wüthrich (1966) and McDonald et al. (1969). The latter experiments, carried out with the cytochromes of horse, ox, hog, chicken, turkey, duck, tuna and yeast, show proton resonances at $-3·15$, $-3·55$, $-2·45$ and $-1·75$ p.p.m. The first is ascribed to the freely rotating SCH_3 group of methionine, the second and third to the two protons of the γ-CH_2, and the last to one of the β- or α-carbons of methionine

$$S\diagdown\begin{matrix}\overset{\gamma}{CH_2} - \overset{\beta}{CH_2} - \overset{\alpha}{CH.}\\ CH_3\end{matrix}$$ These are close enough to the haem group ring current to shift the values to such high negative values, which the authors do not expect to be due to α- or -methine protons of iron-attached imidazole of His-18.

The oxidizing effect of N-bromosuccinimide is ascribed by Stellwagen (1968), by Stellwagen and Van Rooyan (1968) and by O'Brien (1969) to the oxidation of the methionine thioether group to a sulphoxide group rather than to the oxidation of tryptophan indole to oxindole. The spectral effects with disappearance of the 695 nm band resemble those of methionine carboxymethylation, but modifications of tryptophan 59 by formulation or of tyrosine 67 by mutation also effect this absorption band (Schejter et al., 1970).

(b) *Change of iron valency and conformation.* Evidence for a difference of the protein conformation in c^{2+} and c^{3+} was obtained first by Zeile and Reuter (1933). Jonxis (1939) found differences in their spreading on a water-air surface. Differences also exist in their crystal structure (see B2), in their thermostability and their ORD and CD, NMR and Mössbauer spectra (Kowalsky, 1962, 1965) (see B4). Okunuki et al. found c^{3+} more easily digested by proteinases than c^{2+} (Nozaki et al., 1957, 1958; Mizushima et al., 1958). Ulmer and Kägi (1968) found a constant difference of about 10% between c^{3+} and c^{2+} in the percentage of peptide hydrogens which undergo a rapid exchange with deuterium. It is, however, still uncertain whether these differences are due to the binding of Fe^{2+} and Fe^{3+} in c^{2+} and c^{3+} to different ligands of the protein, or to other conformation changes. Lumry et al. (1962) and Lumry and Sullivan (1963) found that the paramagnetism of both, c^{2+} and c^{3+} increased on drying. They assumed that a change of protein conformation caused by the removal of water placed one of the liganded protein groups (they then assumed one imidazole) in a position less suitable to combine with the haem iron (the so-called rack-mechanism of Lumry and Eyring (1954). A compromise solution between the ligand position favoured by the iron orbitals, the iron position relative to the porphyrin plane, and the ligand position demanded by the protein conformation was suggested as a possible cause of the two free electrons (S = $\frac{3}{2}$) found in the dried c^{3+}, due to lowering of the energy of the d_{z^2} orbital. Whether this explanation is acceptable or not, a conformational difference in c^{2+} and c^{3+} may occur without difference in the iron ligands of the protein due to stress in the methionine sulphur-iron linkage caused by the valency change and partly met by adjustment of the protein part, of rather poor folded stability which carries Met-80.

Harbury et al. (1965) found that different from imidazole, methionine is bound less strongly to the Fe^{3+} form of the haem octapeptide than to its Fe^{2+} form. With monoiodoacetate c^{3+} reacts faster than c^{2+} (Matsubara et al., 1965) in the carboxymethylation of His-33. Harbury et al. (1966) have drawn attention to the great differences in the ORD spectrum in the Soret region of eukaryotic cytochromes c^{2+} and c^{3+} which were absent in *Pseudomonas* c-551. Mirsky and George (1967a) concluded from ORD spectra of di-carboxymethylated c that a ligand change was involved in the reaction with c^{2+}, but not with c^{3+}. The di-carboxymethylation product of c^{2+}NO (after removal of NO) differed from that obtained from c^{3+}CN (after removal of cyanide), suggesting that different groups may have been available for carboxymethylation (Schejter and Frowirth, 1966). Mössbauer studies suggested different types of binding of the haem-iron in c^{2+} and c^{3+} (McDermott et al., 1967). In 1966, Margoliash still considered it possible that ϵ-NH_2 was involved in the iron-binding in c^{3+}. However, Stellwagen (1964) as well as Wüthrich (1969) found that cyanide replaces the Met-S-Fe linkage and that this linkage is present in c^{3+} as well as in c^{2+}. Of the 5 tyrosyls of tuna cytochrome c, 3 (probably residues 46, 67, 74) are iodinated under conditions which do not cause a gross change of protein structure and the marked changes of their pK on reduction to c^{2+} suggests that these are involved in the conformational change between c^{3+} and c^{2+}; tryptophane 59 also is exposed in c^{3+} (Stellwagen and McGowan, 1970).

(c) *X-ray crystallographic studies on cytochrome c.* Although crystalline cytochrome c has been obtained as early as 1955 methods of obtaining well-developed crystals suitable for x-ray diffraction methods on account of size and uniformity have been obtained only recently (see above). Horse heart cytochrome c^{3+} has now been studied successfully with 4 Å resolution by Dickerson et al. (1967a,b; 1968) and Margoliash et al. (1968), and recently a resolution at the 2·7 Å has been reported (Dickerson, 1970a,b).

Less complete x-ray studies have been carried out with bonito cytochrome c (*Katsuwonus pelamis*, Lesson) (Dickerson *et al.*, 1970a,b; Takano *et al.*, 1970) and of *Rhodospirillum rubrum* cytochrome c_2, c^{3+} and c^{2+}, by Kraut *et al.* (1968). The bonito cytochrome c seems to have a crystal structure close to that of horse cytochrome c.

The importance of these studies is firstly the confirmation of the evidence available from other studies, but also their extension to a number of other findings and very interesting and important working hypotheses. That the haem of cytochrome c is included in a crevice of the protein molecule had already been postulated by Ehrenberg and Theorell (1955) and had been strongly supported by the studies of George *et al.* (George and Lyster, 1958; George and Tsou, 1952; George *et al.* 1967) and by Stellwagen (1967). The linkage of the haem in this crevice between His-18 and Met-80 discussed above, has now been confirmed.

The crystals were obtained from 90–95% saturated ammonium sulphate solution containing 0·5–1·0 M NaCl or in 4·5–5 M phosphate buffer, allowed to grow slowly over three months. They grew in the form of tetragonal prisms of space group P_{4_1} (or possibly P_{4_3}) showing the faces (100), (010) and (111). The molecules are packed around the screw axis in a spiral with four molecules of M.W. 12,400 per unit cell, 1 molecule per asymmetrical unit, a = b = 58·45 Å with a repeat at 42·34 Å (cf. Kraut *et al.*, 1962). Only 45% of the crystal by weight is protein, the remainder being intermolecular liquid of crystallization. The molecules are prolate spheroids of 25 × 25 × 37 Å, without their exterior side chains, approximating "hydrophobic drop" spheroids. Heavy metal labelling without alteration of crystal structure was achieved by K_2PtCl_4 and by mersalyl-Hg. The centre of the spheroidal molecule is lacking in density indicating a region of loosely packed hydrophobic side chains closely surrounding the haem (Phe 10, Pro 30, Leu 32, Leu 35, Phe 46, Tyr 48, Trp 59, Leu 64, Tyr 67, Leu 68, Pro 71, Ile 85, Leu 94, Ile 95, Leu 98). This region is in turn surrounded by a dense shell of the coils of the main polypeptide chain, the total length of which is about 365 Å (104 residues). The haem which is recognized as the organizing principle sits sideways in a crevice between the right side of the molecule (residues 1–46) and the left (residues 47–91) (Fig. 4), while the residues 92–104 return over the top of the molecule to the right, outside the residues 1–46 and probably bridged by the invariant Lys-13 residue. There is little evidence for α-helical parts, with the exception of the C-terminal residues, which are shielded from the haem. The haem group is clearly visible (Fig. 5, see also several figs. in Dickerson *et al.*, 1971) showing the attachment of Cys-14 to thioether bridge to position 2 in pyrrole ring I and Cys-17 to position 4 in pyrrole ring II, proving the direction of the chain from Lys-13—Cys-14—Ala-15—Gln-16—Cys-17—His-18—Thr-19, with the imidazole ring of His-18 perpendicular to the

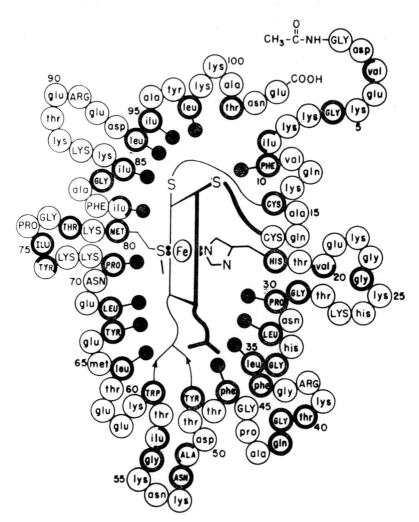

FIG. 4. Haem packing diagram of the cytochrome c molecule (Dickerson et al., 1971, Fig. 7). Heavy circles indicate side chains that are buried in the interior of the molecule, and attached black dots mark residues, the side chains of which are packed against the haem. Light circles indicate side chains on the outside of the molecule and dark half circles show groups that are half-buried at the surface. Arrows from tryptophan 59 and tyrosine 48 to the buried propionic acid groups represent hydrogen bonds. Residues in capitals are totally invariant among the proteins of 29 species.

haem plane. The haem plane forms an angle of about 20° with the Z axis of the crystal (Eaton and Hochstrasser, 1967). The haem is inserted sideways in a channel of approximately 21 Å length in such a manner that the side between rings II and II (methine bridge β) is exposed to the surface,

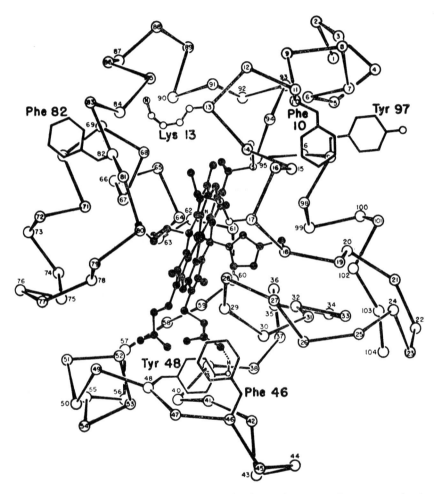

FIG. 5. Stereoscopic α-carbon diagram of the horse heart cytochrome c molecule (Dickerson et al., 1970b, Fig. 2).

with the propionic acid group in position 6, pyrrole ring III, probably free to the surface (cf. Lemberg, 1967), but the other propionic acid group (7 of ring IV) inside, made less hydrophobic by hydrogen bonding to Tyr 48 and Trp 59 (Fig. 6). In agreement with the studies reported in subchapter B3, the sulphur of methionine 80 is supported as the haem-iron binding group on the left hand side, where a largely invariant sequence of residues 67–82 includes this as well as the two lysine residues 72 and 73, the positively charged ε-amino groups of which are directed towards the outside and are probably concerned with the linkage to cytochrome aa_3 (cf. Okunuki et al., 1965), cytochrome bc_1 and perhaps also to cytochrome c

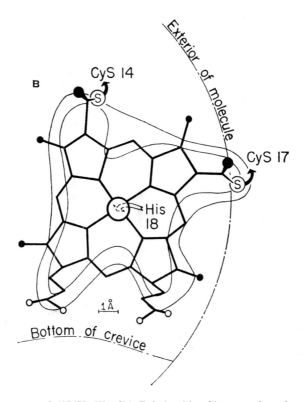

FIG. 6. Dickerson et al. (1967b, Fig. 2b). Relationship of haem to the polypeptide chain, to the bottom of its crevice and to the surface of the molecule.

reductases. The many invariant glycines are probably necessitated by lack of space for side chains in a closely packed molecule. There are two "pseudo-channels" into the interior, probably also packed by hydrophobic side chains, one at the top of the molecule, near His-18, another at the lower left rear, which are possibly points of attachment to the structure of the N-terminal sequence as a cross-linked hairpin bend suggested by Dickerson et al. (1968), Fig. 7. A different structure is suggested by the authors in a more recent publication (Fig. 8). The stability of the loop is

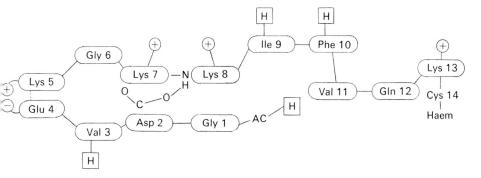

FIG. 7. Structure of cytochrome *c* according to Dickerson *et al.*, 1968.

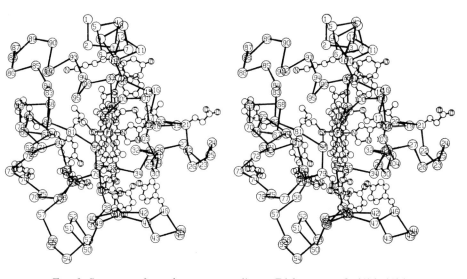

FIG. 8. Structure of cytochrome *c* according to Dickerson *et al.*, 1970, 1971.

strengthened by the hydrophobic interactions of Val-11 with the acetyl of Gly-1, and other hydrophobic interactions at Val-3, Ile-9 and Phe-10. The haem is attached to the polypeptide chain as it comes off the ribosomes and is the organizing principle making α-helical structure unnecessary (Margoliash *et al.*, 1968). It is, however, not yet certain whether a short α-helix span of about two turns could explain the ORD and CD data reported below. The charged groups are predominantly grouped in two positive batches, and one negative immediately beside where the hydrophobic channel reaches the surface.

REFERENCES

Ando, K., Matsubara, H. and Okunuki, K. (1965a). *Proc. Japan Acad.* **41**, 79–82.
Ando, K., Matsubara, H. and Okunuki, K. (1966). *Biochim. et Biophys. Acta* **118**, 240–255
Ando, K., Orii, Y., Takemori, S. and Okunuki, K. (1965b). *Biochim. et Biophys. Acta* **111**, 540–552.
Aviram, I. and Schejter, A. (1969). *J. Biol. Chem.* **244**, 3773–3778.
Blauer, G. and Ehrenberg, A. (1963). *Acta Chem. Scand.* **17**, 8–12.
Boeri, E., Ehrenberg, A., Paul, K. G. and Theorell, H. (1953). *Biochim. et Biophys. Acta* **12**, 273–282.
Dickerson, R. E., Eisenberg, D., Takano, T., Battfay, O. and Samson, L. (1970). *Symposium I. 8th Internat. Congress of Biochem.*, Switzerland, p. 13.
Dickerson, R. E., Kopka, M. L., Borders, C. L., Varnum, J., Weinzierl, J. E. and Margoliash, E. (1967a). *J. Mol. Biol.* **29**, 77–95.
Dickerson, R. E., Kopka, M. L., Weinzierl, J., Varnum, J., Eisenberg, D. and Margoliash, E. (1967b). *J. Biol. Chem.* **242**, 3015–3017.
Dickerson, R. E., Kopka, M. L., Weinzierl, J. E., Varnum, J. C., Eisenberg, D. and Margoliash, E. (1968). "Osaka", pp. 225–251.
Dickerson, R. E., Takano, T., Wisenberg, D., Kallai, O. B., Samson, L. and Margoliash, E. (1971). *J. Biol. Chem.* **246**, 1511–1535.
Drott, H. R., Lee, C. P. and Yonetani, T. (1970). *J. Biol. Chem.* **248**, 5875–5879.
Eaton, W. A. and Hochstrasser, R. M. (1967). *J. Chem. Phys.* **46**, 2533–2539.
Ehrenberg, A. and Theorell, H. (1955). *Acta. Chem. Scand.* **9**, 1193–1205.
Eichhorn, G. L. and Cairns, J. F. (1958). *Nature* (London) **181**, 994.
Falk, J. E. (1961). "Haematin Enzymes", Vol. **I**, pp. 74–76.
Fanger, M. W. and Harbury, H. A. (1965). *Biochemistry* **4**, 2541–2555.
Fanger, M. W., Hettinger, T. P. and Harbury, H. A. (1967). *Biochemistry* **6**, 713–720.
George, P. and Lyster, R. L. J. (1958). *Proc. Nat. Acad. Sci.* **44**, 1013–1029.
George, P. and Tsou, C. L. (1952). *Biochem. J.* **50**, 440–441.
George, P., Clauser, S. C. and Schejter, A. (1967). *J. Biol. Chem.* **242**, 1690–1695.
Harbury, H. A. (1966). "Hemes and Hemoproteins", pp. 391–393.
Harbury, H. A., Cronin, J. R., Fanger, M. W., Hettinger, T. P., Murphy, A. J., Myer, Y. P. and Vinogradov, S. (1965). *Proc. Nat. Acad. Sci.* **54**, 1658–1664.
Hettinger, T. P. and Harbury, H. A. (1965). *Biochemistry* **4**, 2585–2590.
Hettinger, T. P., Fanger, M. W., Vinogradov, S. N. and Harbury, H. A. (1966). *Fed. Proc.* **25**, Abstr. 2588.
Horinishi, H., Kurihara, K., and Shibata, K. (1965). *Arch. Biochem. Biophys.* **111**, 520–528.
Jonxis, J. H. P. (1939). *Biochem. J.* **33**, 1743–1751.
Jori, G., Gennari, G., Galliazzo, G. and Scoffone, E. (1970). *FEBS Letters* **6**, 267–269.
Kowalsky, A. (1962). *J. Biol. Chem.* **237**, 1807–1819.
Kowalsky, A. (1965). *Biochemistry* **4**, 2382–2388.

Kraut, J., Sieker, L. C., High, D. F. and Freer, S. T. (1962). *Proc. Nat. Acad. Sci.*, **48**, 1417–1423.
Kraut, J., Singh, S. and Alden, R. A. (1968). "Osaka" pp. 252–256.
Lemberg, R. (1967). *Enzymol.* **32**, 18–36.
Lumry, R. and Eyring, H. (1954). *J. Phys. Chem.* **58**, 110–120.
Lumry, R. and Sullivan, J. (1963). *Fed. Proc.* **22**, Abstr. 2544.
Lumry, R., Solbakken, A., Sullivan, J. and Reyerson, L. H. (1962). *J. Amer. Chem. Soc.* **84**, 142–149.
McDermott, P., May, L. and Orlando, J. (1967). *Biophys. J.* **7**, 615–620.
McDonald, C. C., Phillips, W. D. and Vinogradov, S. N. (1969). *Biochem. Biophys. Res. Commun.* **36**, 442–449.
Margoliash, E. (1955). *Nature* (London) **175**, 293.
Margoliash, E. (1961). "Haematin Enzymes", Vol. **II**, pp. 383–384.
Margoliash, E. (1962). *Brookhaven Sympos. Biol.* **15**, 266–281.
Margoliash, E. (1964). *Canad. J. Biochem.* **42**, 745–753.
Margoliash, E. (1966). "Hemes and Hemoproteins", pp. 371–391.
Margoliash, E. and Schejter, A. (1966). *Advances in Protein Chem.* **21**, 113–286.
Margoliash, E., Fitch, W. M. and Dickerson, R. E. (1968). *Brookhaven Sympos. Biol.* **21**, 259–305.
Margoliash, E., Frohwirth, N. and Wiener, E. (1959). *Biochem. J.* **71**, 559–570.
Matsubara, H., Ando, K. and Okunuki, K. (1965). *Proc. Jap. Acad.* **41**, 408–413.
Mirsky, R. and George, P. (1967a). *Biochemistry* **6**, 1872–1875.
Mirsky, R. and George, P. (1967b). *Biochemistry* **6**, 3671–3675.
Mizushima, H., Nozaki, M., Horio, T. and Okunuki, K. (1958). *J. Biochem.* **45**, 845–846.
Nakatani, M. (1960). *J. Biochem.* **48**, 633–638.
Nanzyo, N. and Sano, S. (1968). *J. Biol. Chem.* **243**, 3431–3440.
Nozaki, M., Yamanaka, T., Horio, T. and Okunuki, K. (1957). *J. Biochem.* **44**, 453–464.
Nozaki, M., Mizushima, H., Horio, T. and Okunuki, K. (1958). *J. Biochem.* **45**, 815–823.
O'Brien, P. J. O. (1969). *Biochem. J.* **113**, 13P.
Okunuki, K., Wada, K., Matsubara, H. and Takemori, S. (1965). "Amherst", vol. II, pp. 549–560.
Paléus, S. (1954). *Acta Chem. Scand.* **8**, 971–984.
Paul, K. G. (1951a). *Acta. Chem. Scand.* **5**, 379–388.
Paul, K. G. (1951b). In "The Enzymes. Chemistry and Mechanism of Action" (J. B. Summer and K. Myrbäck, eds.), Vol. **II**, pp. 357–398. Academic Press, New York and London.
Sano, S., Kurihara, M., Nishimura, O. and Yajima, H. (1968) "Osaka", pp. 370–379.
Schejter, A. and Aviram, I. (1969). *Biochemistry* **8**, 149–153.
Schejter, A. and Aviram, I. (1970). *J. Biol. Chem.*, **245**, 1552–1557.
Schejter, A. and Frowirth, N. (1966). *Israel J. Chem.* **4**, 80.
Schejter, A. and George, P. (1965). *Nature* (London) **206**, 1150.
Schejter, A., Aviram, I., Margalit, R. and Sokolovsky, M. (1970). *8th Internat. Congress Biochem. Switzerland, Sympos.* **1**, 18–19.
Shechter, E. and Saludjian, P. (1967). *Biopolymers* **5**, 788–790.
Stellwagen, E. (1964). *Biochemistry* **3**, 919–923.
Stellwagen, E. (1966). *Biochem. Biophys. Res. Commun.* **23**, 29–33.
Stellwagen, E. (1967). *J. Biol. Chem.* **242**, 602–606.
Stellwagen, E. (1968). *Biochemistry* **7**, 2496–2501.

Stellwagen, E. and McGowan, E. B. (1970). *Biochemistry* **9**, 3765–3767.
Stellwagen, E. and Van Rooyan, S. (1967). *J. Biol. Chem.* **242**, 4801–4805.
Takano, T., Ueki, T., Sugihara, A., Tokuma, Y., Tsukihara, T., Ashida, T. and Kakuda, M. (1970). *8th Internat. Congres Biochem. Switzerland, Sympos.* **1**, 19–21.
Theorell, H. and Åkeson, Å. (1941a). *J. Am. Chem. Soc.* **63**, 1804–1811.
Theorell, H. and Åkeson, Å. (1941b). *J. Am. Chem. Soc.* **63**, 1818–1820.
Tsai, H. J. and Williams, G. P. (1965a). *Canad. J. Biochem.* **43**, 1409–1415.
Tsai, H. J., Tsai, H. and Williams, G. R. (1965b). *Canad. J. Biochem.* **43**, 1995–1996.
Tuppy, H. and Bodo, G. (1954). *Monatshefte d. Chem.* **85**, 1024–1045.
Ulmer, D. and Kägi, J. H. R. (1968). *Biochemistry* **7**, 2710–2717.
Vinogradov, S. N. and Harbury, H. A. (1967). *Biochemistry* **6**, 709–712.
Wainio, W. W., Krunsz, L. M. and Hillman, K. (1970). *Biochem. Biophys. Res. Commun.* **39**, 1134–1139.
Warme, P. K. and Hager, L. P. (1970a). *Biochemistry* **9**, 1599–1605.
Warme, P. K. and Hager, L. P. (1970b). *Biochemistry* **9**, 1606–1614.
Wüthrich, K. (1969). *Proc. Nat. Acad. Sci.* **63**, 1071–1078.
Zeile, K. and Reuter, F. (1933). *Z. physiol. Chem.* **221**, 101–116.

4. Physicochemical properties of cytochrome c

(a) *Effects of pH on mammalian cytochrome c^{3+}*. Horse heart cytochrome c has an isoelectric point of 10·5–10·8 (Tint and Reiss, 1950). Flatmark (1966) found the isoelectric point of bovine c to vary slightly with the number of asparagine residues hydrolysed to aspartic acid, from 10·78 for $c^{3\pm}$-1 and 10·80 for $c^{2\pm}$-1 to 10·58 for $c^{3\pm}$-2 and 10·60 for $c^{2\pm}$-2, and 10·36 for $c^{3\pm}$-3 and 10·22 for $c^{2\pm}$-3, but of these only c-1 can be considered negative. Fungal cytochromes c have their isoelectric point also above pH 9, but bacterial cytochromes with this point lying well below pH 7 are known (see Chapter VI). The electrophoretic behaviour of c^{3+} varies somewhat with nature and concentration of the buffer anions (Barlow and Margoliash, 1966).

This "neutral" form of cytochrome c^{3+} is stable between pH 2·5 and 9·35, but reversible alteration is possible in an even wider pH-range, although some irreversible alterations occur at the pH extremes near − 0·3 and 13·8 of Theorell and Åkeson (1941) (see above in B2; Keilin and Hartree, 1937; Paul, 1948a,b; Boeri *et al.*, 1953; Butt and Keilin,1962). The various forms of c^{3+} interpreted on the basis of the di-imidazole theory, were given by Lemberg and Legge (1949), together with spectrophotometric and magnetochemical data, but they now require re-interpretation after the more readily dissociable ligand has been found to be the methionine-sulphur, not a second histidine-imidazole.

Czerlinski *et al.* (1967) and Czerlinski (1969) have shown that at pH 7, contrary to the findings of Theorell and Åkeson, no proton is released on oxidation of c^{2+}. The role of Cl^- in these reactions (Boeri *et al.*, 1953) is not yet clear, although Fung and Vinogradov (1968) have found that Cl^- stabilizes the two low pH forms. There is some evidence for Cl^- present in cytochrome c from the x-ray data of Dickerson et al., pH-difference spectra of c^{3+} (pH 9·4–8·4) show peaks at 532, 458 and 361 nm with a trough at 394 nm (Greenwood, 1968).

(b) *Optical properties.* The typical double-banded haemochrome-type spectrum of c^{2+} in the visible part of the spectrum has long been accepted as a criterion of unmodified cytochrome c. It has now been recognized that it is in fact a rather poor criterion, less sensitive than others (cf. e.g. Yamanaka et al., 1962). The most reliable values for the height of the α-band of c^{2+} (ϵ_{mM}^{550}) are probably those of Massey (1959) and by Van Gelder and Slater (1962) who found values close to 29, although Flatmark (1964) reports 30·5 and other workers values between 27 and 28 (Paléus and Neilands, 1950; Tint and Reiss, 1950; Henderson and Rawlinson, 1956; Margoliash and Frowirth, 1956; Schejter et al., 1963; Vinogradov and Harbury, 1966). The value $\epsilon_{550}^{mM} \Delta(c^{2+} - c^{3+})$ is about 21 (cf. Lemberg and Cutler, 1970). The spectra are given for the visible part by Henderson and Rawlinson (1956a,b) and include the ultraviolet absorptions given by Paul and Theorell (1954); Margoliash and Frowirth (1959) and Butt and Keilin (1962). They show in the c^{2+} spectrum the α-band at 550 nm, the β-band at 520·5 (ϵmM 15·9), the γ-band at 416 nm (ϵmM 129), a δ-band at 315·5 nm (ϵmM 33·6) and a band at about 270 nm (ϵmM 31·8). The ratio E_{550}/E_{590} is more sensitive to alteration than the α/β ratio and near to or above 50 (Henderson and Rawlinson, 1956; Margoliash and Frowirth, 1959). The γ/α ratio is about 4·66. The c^{3+} spectrum has, apart from the weak bands in the red part of the spectrum (see below) a maximum at about 529 nm (ϵmM 11·2), the γ-band at 410 nm (ϵmM 106), a δ-band at 360 nm (ϵmM 28·5), (cf. also Theorell, 1936 and Shechter and Saludjian, 1967) and isosbestic points with the c^{2+} spectrum at 551, 542, 526 and 404–405 nm. $E_{550}^{c^{2+}}/E_{220}^{c^{3+}}$ has been found by most workers between 1·20 and 1·30 (Paléus and Neilands, 1950; Henderson and Rawlinson, 1956; Paléus, 1960, 1962; Morrison et al., 1960; Matsubara and Yasunobu, 1961; Flatmark, 1964; Vinogradov and Harbury, 1966), a high value being a criterion of absence of colourless protein impurities.

Low temperature spectra were first studied by Keilin and Hartree (1949) and extensively by Estabrook (1956, 1957, 1961, 1966), cf. also Morrison et al. (1960). For approximate estimation of ϵmM see Elliott and Doebbler (1966). It is of interest that the fine structure of the α and β bands at low temperature in frozen glycerol-water mixtures is a sign of unmodified cytochrome and not of a mixture. Any modification diminishes the fine structure. The α-band at low temperature is split into three (549, 546, 538 nm), the β-band into six (stronger bands at 525, 519, 208 nm, weaker at 529, 515 and 512 nm). The Soret band is only shifted by 1 nm to shorter λ and develops a shoulder at 432 nm. The sharpness of the bands allows better distinction of cytochromes of some species (cf. Estabrook, 1966), e.g. of c from bacterial c_4 (Estabrook, 1961). Caution must be used, however, when assuming a distinct entity merely on the basis of low temperature spectra (cf. Chapter IV, F1).

The stability of the absorption spectrum of c^{2+} to high temperatures (100°) has been studied by Yamanaka et al. (1959) and Butt and Keilin (1962). The haemochrome-type spectrum remains, but the bands in the visible become less sharp and the α-band moves from 550 to 552 nm. This alteration is reversible. On heating c^{3+}, the ferrihaemochrome spectrum becomes more diffuse, and the alteration is largely reversible.

In the red, unaltered c^{3+} shows a band at 695 nm with a shoulder at 650 nm. This was discovered by Theorell and Åkeson (1939). The band appears to be a highly sensitive criterion of intactness of cytochrome c. Horecker and Kornberg (1946) and Horecker and Stannard (1948) found that the ligands CN and N_3 abolish the band. It is also abolished by dimerization, heat, denaturation, high pH and imidazole (Schejter et al., 1963; Schejter and George, 1964; Margoliash and Schejter, 1966) and is not restored when the dimer is monomerized by guanidine. The band is not strong (ϵmM 0·81) but of relatively high ellipticity. c^{2+} shows only weak bands in the infrared region (Eaton and Charney, 1969), again of relatively high ellipticity. As has been shown above, the band is due to the linkage of the haem iron with the thioether sulphur of methionine (Met-80). Greenwood and Palmer (1965) and Greenwood (1968) found a correlation of the presence of the c^{3+} band at 695 nm with the reducibility of it by ascorbate or tetrachlorohydroquinone, but this does not appear to hold under all conditions (Skov and Williams, 1968). On di-carboxymethylation of c^{3+} at pH 3, e.g., the band disappears, but ascorbate reducibility remains. The band is present in the crystals subjected to the x-ray diffraction, and there is evidence that it is present in intact mitochondria, and does not only arise on extraction (Chance et al., 1968). To judge from the disappearance of the band, yeast cytochrome c^{3+} appears to be somewhat more sensitive to low pH and increased temperature than horse c^{3+} (Aviram and Schejter, 1969). At 38° more than half of the band of yeast c is lost, but only 15% of that of horse c. Thus the closed crevice structure is weaker in the yeast cytochromes. The changes involved in the removal of the Met-80 iron linkage also appear to affect the interactions of the haem with the tryptophan and tyrosine residues in the region of the peptide chains from Trp-59 to Met-80. Eaton and Hochstrasser (1966) found the 695 nm band polarized perpendicularly to the haem plane, and probably due to an electron from a non-porphyrin-ligand, such as the $-\underset{\cdot\cdot}{S}-$ of methionine entering an iron orbital (cf. also Williams, 1966). According to Stellwagen (1968), the di-carboxymethylation of Met-80 at pH 3 precedes the disappearance of the 695 nm band, but parallels the appearance of a high-spin band at 620 nm. Protonation of tyrosine residues also appear to play a role in the maintenance of the 695 nm band (Skov et al., 1969; Schejter et al., 1970). The band persists in a complex of cytochrome c with the phospholipid phosvitin (Taborsky, 1970).

The ellipticity of the bands of c^{2+} and c^{3+} has been extensively studied by ORD (optical rotatory dispersion), MORD (magnetic optical rotatory dispersion) and CD (circular dichroism) studies. The measuring of ORD curves was initiated by the work of Eichhorn and Cairns (1958) and continued by many workers (Ulmer, 1965, 1966a,b; Urry, 1965 and Urry and Doty, 1965; Myer and Harbury, 1965; Myer et al., 1966; Cronin and Harbury, 1965; Harbury et al., 1966; Mirsky and George, 1966, 1967a,b; Greenwood, 1968).

It is occasionally possible to identify the absorption band to which the positive and negative extrema of the ORD curve are due, by its agreement with the point where the curve cuts the zero-line, but in cytochrome c when multiple overlapping Cotton effects influence each other and the existence of limbs of opposite signs on each site of the transition further complicates the situation, a specific analysis is often impossible. This makes some of the conclusions drawn from ORD uncertain, and this is probably also the cause of divergencies between the result of some workers. In order to enhance the small ORD effects of the visible bands, Shashoua (1964, 1965) Shashoua and Estabrook (1966) and Morrison (Morrison, 1965; Morrison and Duffield, 1966) used MORD, i.e. ORD measurements in a strong magnetic field. At ph 10 MORD α-band of yeast cytochrome c is higher than that of other cytochromes (Shashoua and Estabrook, 1966). The development of suitable apparatus, particularly the Cary-60 Spectropolarimeter has, however, made the measurement of circular dichroism, i.e. the ellipticity of the absorption bands directly possible, and the results from CD measurements have been more clear-cut (Myer, 1970). It should be realized that these studies are only a beginning and that the conclusions are often only tentative.

For the purpose of this discussion the spectrum is divided into six regions:

(1) the region from 250–180 nm. It has been shown that ellipticities in this region are due to optical activity of the protein with no contribution from the asymmetry of the haem-protein linkage (intrinsic). In all the other regions we have effects due to the interaction of these two factors (Urry, 1965; Ulmer, 1966a). Whereas earlier results indicated a contribution of the haem to the typical ellipticities in this region too (Urry and Doty, 1965; Myer and Harbury, 1966; Harbury et al., 1966), in the later studies of Flatmark and Robinson (1968), Zand and Vinogradov (1968) and Myer (1968a,c) no effect on change of the haem iron valency was found (but see

such effects on the CD spectrum of bacterial cytochrome c_4, Van Gelder et al., 1968). The negative CD bands at 222 and 210 nm, and the positive at 190 nm are ascribed to the α-helix, while shoulders at 218 and 195 nm are ascribed to a smaller contribution of the pleated β-structure of the protein. These studies indicate an α-helical contribution of about 30% which is in contradiction to the much smaller helical contribution which can be assumed from the x-ray data. Some caution as to quantitative calculations of α-helical content from the optical activities of these wavelengths seems to be warranted and recent studies of Myer (1970) indicate a low α-helical content.

(2) The so-called "aromatic" region includes the region from about 300 to 250 nm. This includes contribution from the aromatic amino acid residues, tyrosine and tryptophane which absorb in this region as well as contributions from haem asymmetry. Vibrational fine structure of the bands of aromatic residues contributes to the complexity. The main ORD extrema are negative and found at 298–287 nm and about 280 nm (Urry and Doty, 1965; Myer and Harbury, 1965; Ulmer, 1966a; Cronin and Harbury, 1965). These have been variously ascribed to tryptophan (Ulmer, 1966a; Myer, 1968b), based on the comparison of tuna cytochrome c (with two Trp) to mammalian c (with one), or based on the effect of N-bromosuccinimide; or they have been ascribed to tyrosine (Cronin and Harbury, 1965) on the basis of acetylation effects. The CD spectra show multiple positive bands extending from 261 to 295 nm for c^{2+}, while for c^{3+} the 289 and 282 nm bands are negative, those at 251 and 263 nm positive. Urry (1968) found the 263 and 278 nm bands also in ferric haem c undecapeptide, and the 253 nm bands in its ferrous form, and concludes that these bands are due to haem, or to its co-ordinated imidazole; Myer (1968b) however, found only the 262 nm band in the haem peptide CD spectra; he ascribes the 251 nm band in c^{3+} tentatively to tyrosine (but see Hamaguchi et al., 1968). Haem bands contribute more to the c^{2+} than to the c^{3+} spectrum and may be n-π* transitions involving non-bonding pyrrole nitrogen electrons and iron orbitals (Urry, 1968). Flatmark and Robinson (1968) found that only the 263 nm band is abolished when the iron of cytochrome c is removed by hydrogen fluoride and ascribe it therefore to the iron-imidazole linkage. This, however, makes it difficult to understand why the band should be absent in the c^{2+} form of Pseudomonas c-551. Urea abolishes the 288 and 282 nm bands and replaces the double band of c^{3+} of 263 and 251 nm by an unsymmetrical higher band in between (Myer, 1968b). At pH 10 the band at 262 nm is lost, that at 251 nm increased. From pH 3–11·6 there appear to be only changes in haem-protein interaction, but no unfolding of the peptide chain. The various cytochromes c display similar CD spectra in this region with the exception of tuna cytochrome c (Myer, 1970).

(3) The region from about 370–300 nm appears to be dominated by the haem δ-band (see above). A negative ORD extremum is found for c^{2+} at 366–370 nm (Ulmer, 1965, 1966) with a positive extremum at 311 nm, while for c^{3+} the former has been reported at 357, the latter at 321 nm. The CD bands are similar for c^{2+} and c^{3+}, with two negative bands at about 375 and 335 nm (Urry, 1968; Flatmark and Robinson, 1968). Urea, but not heat, eliminates the former (Myer, 1968b). In cytochrome c_4 (*Azotobacter vinelandii*) Van Gelder et al. (1968) found, however, a positive CD band at 320 nm.

(4) The most interesting region, obviously influenced by the Soret band of the haem, is that near 400 nm. The ORD spectra of c^{3+} show a positive extremum at 410 nm with a greater ellipticity than the two negative extrema at 458 nm and 390 nm which surround it (Ulmer, 1965; Mirsky, 1960; Harbury et al., 1966), in contrast to c^{2+} which shows two negative extrema. Guanidination or trifluoroacetylation of lysine has only small effects. The ORD spectra of the c haemopeptides have been studied by Urry, 1966; the more complex spectra of the Fe^{3+} undecapeptide ($+$ 430, $-$ 415, $+$ 405, $-$ 380) and of the Fe^{2+} undecapeptide ($+$ 430, $-$ 425, $+$ 420, $-$ 414) are simplified by the addition of imidazole or lysine (Fe^{3+} $+$ 416, $-$ 380; Fe^{2+} $+$ 423, $-$ 400). Most noticeable is the effect of the iron valency on the CD spectra of cytochrome c (Zand and Vinogradov, 1967; Flatmark, 1967; Flatmark and Robinson, 1968) which causes the c^{2+} and c^{3+} CD bands to be almost mirror images. Bovine c^{3+} has a strong negative band at 418 nm (molar ellipticity θ about 4×10^4 degrees $cm^2 d$ $mole^{-1}$) and a positive band at 405 nm ($92 \cdot 7 \times 10^{-4}$) whereas c^{2+} has a positive band at 424 nm ($94 \cdot 7$) with a weak positive band at 410 nm and a negative band at about 390 nm which becomes a shoulder of the 376 nm band. It is equally remarkable that the CD spectrum of cytochrome c_2 (*Rhodospirillum rubrum*) does not show this inversion. Here both c^{2+} band (410 nm. $98 \cdot 5 \times 10^{-4}$) and c^{3+} (417 nm, $99 \cdot 2 \times 10^{-4}$) are positive. The CD spectra of the cytochrome c of *Candida krusei* (Hamaguchi et al., 1968) show the same inversion with iron valency, a negative band at 418 nm for c^{3+}, and a positive maximum at 432 nm for c^{2+}. However, the cytochrome c_4 (*A. vinelandii*) has the positive CD band for ferrous and ferric like that of c_2. Myer (1968b,c) concludes that the change of valency of the haem-iron must either be connected with a change of the protein ligand, or the haem-iron must be displaced from one side of the haem plane to the other (cf. also Williams, 1966, who found such an inversion also with corrin derivatives).

Urea denaturation of mammalian c^{3+} increases the main Soret band ellipticity but abolishes the negative bands at 440 and 417 nm. Dimerization of cytochrome c only decreases the ORD positive extremum of c^{3+} and shifts the whole ORD spectrum towards more negative Cotton effects

(Mirsky, 1966), and partial conversion of the asparagine of bovine cytochrome c to aspartic acid (Flatmark, 1967) has the same effect.

Changes in the ORD spectrum of the c^{3+} form of horse c, as well as yeast isocytochromes-1 and -2 with pH changes from 7 to 9·5, can be correlated with the disappearance of the 695 nm band (Mirsky and George, 1967b). ORD curves of mono-carboxymethylated horse c differ little from those of normal c^{2+} or c^{3+}, but di-carboxymethylation while not affecting the ORD of c^{3+} at pH 7, does alter the ORD curve of c^{2+} (Mirsky and George, 1967a).

(5) Bands in the visible from 600–500 nm include the α and β ferrohaemochrome bands. The ORD spectra of c^{2+} and c^{3+} show some weak Cotton effects in the green part of the spectrum (Eichhorn and Cairns, 1958; Ulmer, 1965; Urry, 1965; Mirsky and George, 1967a,b). The effects are greatly increased by the use of MORD (Morrison, 1965; Shashoua, 1964, 1966; Morrison and Duffield, 1966). The negative extremum of c^{2+} at 549 nm becomes, by the application of this method, almost as high as that at the Soret peak and reaches 0·23 of the absorption band. In addition, there are 5 weak positive and 5 weakly negative extrema in this region of the spectrum from 560 to 507 nm. c^{3+} shows only weak negative extrema at 559 and 530 nm. The CD spectra show for c^{2+} positive peaks at 553, 527 and 511 nm, for c^{3+} a structured positive band at 558 . . . 533 . . . 495 nm. The CD spectrum of *R. rubrum* ferrocytochrome c_2 resembles that of the horse cytochrome c, but the ferric c_2 CD spectrum, while similar to the ferrous c_2 spectrum, differs from that of horse c^{3+}.

(6) The abolition of the absorption band of mammalian c^{3+} at 695 nm by modification of the protein is accompanied by CD changes in other parts of the spectrum (cf. above). On carboxymethylation of Met-80 or at high pH the positive ORD Cotton effect at 610-700 nm disappears (Mirsky and George, 1966, 1967a,b); the Cotton effect of c^{2+} in this region is negative and is not changed by high pH up to 10. The z-polarized 695 nm band (Schejter and George, 1964; Shechter and Saludjian, 1967; Eaton and Hochstrasser, 1966, 1967) is ascribed to promotion of a porphyrin πa_{2u} orbital electron into an empty d_{z^2} iron orbital. Eaton and Charney (1969) have studied the near-infrared absorption and CD bands of c^{2+} and ascribe them to magnetic dipole allowed d → d transitions of crystal field and molecular orbital theories. The anisotropy factor for the weak absorption bands in the 10,000–17,000 cm^{-1} region is comparatively high. It is suggested that the ligand field of porphyrin in c^{2+} is weaker than in myoglobin, possibly due to the iron being further out of the porphyrin plane in c^{2+}. Day *et al.* (1967a,b) have investigated polarized single crystal spectra. The transitions are all polarized in the haem plane, the spectrum is not changed on crystallization, but the intensity of the bands is somewhat decreased. This is ascribed to some low spin to high spin change.

In c^{3+} there is a weak band at 8,000 cm^{-1} and another at 10,700 cm^{-1} indicating a few percent of high spin complex.

(c) *Magnetochemistry*. Theorell and Åkeson (1941) showed that c^{2+} was diamagnetic at neutral pH and c^{3+} had one unpaired electron at neutral pH. The latter became, however, paramagnetic with five unpaired electrons at very low pH (cf. Boeri et al., 1953; Paléus et al., 1955). Similar changes can be produced by dehydration (Lumry et al., 1962) in both c^{2+} and c^{3+}. The ESR spectra of c^{3+} at 77°K show the low-spin type spectrum $g_1 = 3 \cdot 0$, $g_2 = 2 \cdot 26$, $g_3 = 2 \cdot 0$ (Rein et al., 1968). This remains low spin at higher pH ($g_1 = 2 \cdot 73$, $g_2 = 2 \cdot 14$, $g_3 = 1 \cdot 7$). At pH 0·7 the spectrum is mainly of high-spin type (g ~ 6·3) with another g near 2. The crevice structure favours the low spin type. Mössbauer spectra have been studied by McDermott et al. (1967) and by Cooke and Debrunner (1968), those of certain bacterial cytochromes by Moss et al. (1968). These confirm the low-spin state of the haem iron. The high quadrupole splitting suggests that the iron atom is in an asymmetric ligand field and reduction increases this asymmetry. Freeze-drying changed the quadrupole splitting of c^{2+} little, but decreased that of c^{3+}, indicating a change in conformation; the haem crevice structure is more stable in c^{2+} than in c^{3+} (see also Lang et al., 1968, for cytochrome c of *Torula utilis*). Williams (1966) has pointed out that here the Mössbauer spectrum provides less information than optical spectroscopy. Drott (1969) has spin-labelled cytochrome c using N-(1-oxyl-2,2,5,5-tetramethyl-3 pyrrolidinyl) bromoacetamide. Its ESR spectrum is rather insensitive to iron valency and the spin label is only weakly immobilized indicating alkylation on the surface of the molecule. The NMR studies of Kowalsky (1962, 1965) have been discussed above. The relationship between high and low spin states and absorption spectra, while on the whole certain, has still some areas of uncertainty, in particular with regard to the bands in the red part of the spectrum often described as charge transfer bands (Williams, 1966). While such bands are often characteristic for high spin complexes, they can also be found in low spin complexes such as c^{3+} or oxymyoglobin.

(d) *Oxidation-reduction potentials*. The oxidation-reduction potential at pH 7 (E_0') of purified ox and horse cytochrome c at 25° is + 255 mV (Henderson and Rawlinson, 1956a, 1961) in remarkably good agreement with the earlier values of Coolidge (1932) on yeast c and of Rodkey and Ball (1950) on beef c (cf. also Minakami, 1955, and Tosi, 1957). The pH dependence of potential between pH 7·0 and 9·4 has been studied by Brandt et al. (1966). pH has no effect on electron transport itself, but high pH causes a slow modification of c^{3+}, thus decreasing the concentration of c^{3+} which effectively participates in the electron process. This may explain the earlier findings of a pK at 6·68 in the potential (Paul, 1947; but see George et al., 1968).

5. Compounds of cytochrome c with iron ligands

The formation of a compound of cytochrome c^{3+} with cyanide was discovered by Potter (1941) and confirmed by Horecker and Kornberg (1946). The spectroscopic alterations include a shift of the 530–535 nm maximum to longer wavelengths by 5 nm and a lowering of the extinction; more distinct is the abolition of the 695 nm band. The c^{2+}-CN compound is very unstable and can only be observed as a passing phase in the reduction of c^{3+}-CN (Schejter and George, 1962; Nicholls and Mochan, 1968); its α-band lies at 555 nm. It cannot be obtained from c^{2+} directly with cyanide. The rate of formation of c^{3+}CN is slow, but the affinity is quite high. George and Tsou (1952) found that two reactions occur; c^{3+} + HCN → c^{3+}CN + H$^+$ and c^{3+} + CN$^-$ → c^{3+}CN, k_{HCN} being 5·43, k_{CN} 15·4 M^{-1} sec^{-1} at 25° with activation energies of 18·4 and 17·0 kcal/mole resp. The equilibrium constants are $K_1 = 4·8 \times 10^{-4}$ and $K_2 = 1·22 \times 10^6 M^{-1}$ with heat content changes of 9·2 kcal and 1·1 kcal, respectively, and entropy changes of + 15·7 and + 31·3 e.u. This is for exogenous cytochrome c, while endogenous does not react with cyanide (Tsou, 1951c). The dissociation of c^{2+}CN is about 400 × faster than that of c^{3+}CN, mostly due to entropy differences which indicate a difference of protein conformation of c^{3+} and c^{2+}.

George et al. (1967) concluded that the bond broken by cyanide was an iron-imidazole bond, but the evidence of Stellwagen (1966, 1968) indicates instead that cyanide replaced the methionine bond and the x-ray analyses of Dickerson et al. (see above) and NMR studies have confirmed this (see above). The crevice structure in yeast isocytochrome c-1 is by about 1 kcal less firm than that in horse c. $\Delta F^=$ for isocytochrome-1 was found − 7·6 kcal, $\Delta H^=$ − 12·9 kcal, $\Delta S^=$ − 17·8 e.u., i.e. the enthalpy is more favourable, the entropy less favourable than for horse c (Aviram and Schejter, 1969; Schejter and Aviram, 1969).

The azide complex of cytochrome c^{3+} was discovered by Horecker and Stannard (1948). The maximum at 530 nm is shifted to 540 nm, and a small new band appears at 570 nm; again the 695 nm is abolished. In contrast to the combination with CN$^-$ that with N$_3^-$ is fast, but leads to a more dissociable complex (K = 0·15). Azide does not combine appreciably with c^{2+} and probably replaces the methionine residue in its combination with c^{3+} (Gupta and Redfield, 1970). This is supported by NMR studies.

Only modified c^{2+} reacts with carbon monoxide, and the same appears to be true for NO, although a c^{2+}NO-compound is obtainable in an indirect way. Keilin and Hartree (1937; cf. also Butt and Keilin, 1962) observed two compounds, cpd. I formed by the action of NO at pH 6·8 on c^{3+} with absorption maxima at 562 (ϵ_{mM} 12·8), 529 (ϵ_{mM} 12·5) and 417 nm (ϵ_{mM} 139). This was found to be diamagnetic by Ehrenberg and

Szczepkowski (1960) and interpreted as $c^{2+}NO^+$ (ferrocytochrome c-nitrosyl). Butt and Keilin raise the objection against this interpretation that compound I can be obtained even in the presence of high concentrations of ferricyanide, and is reconverted to free c^{3+} by removing excess of NO. It is, however, difficult to understand the diamagnetic character of the compound on any other basis. Compound II is obtained from c^{2+} with NO only at pH 13, but not in neutral solution. It shows two less distinct bands at 567 (ϵmM 9·8), 538·5 (ϵmM 10·2), and the Soret band 411 nm (ϵmM 27). The compound once formed can, however, be maintained at neutral pH and is interpreted as $c^{2+}NO$ complex. The compound has been studied by EPR spectroscopy by Kon (1969) and Kon and Kataoka (1969) who could not confirm earlier results of Gordy and Rexroad (1961, quoted from the paper of Kon). Compound II is paramagnetic and no change is caused during the neutralization. The results show an extra hyperfine structure due to the ^{14}N nucleus in the histidine-imidazole on the opposite axial position to the NO, and thus show that NO has replaced the methionine-S, not the histidine-imidazole-N. On the basis of study of other haem-NO and haemoglobin-NO compounds, Kon and Kataoka (1969) conclude that the unpaired electron in compound II occupies the out of antibonding plane σ orbinal consisting of a Fe d_{z^2} orbital rather than being completely localized in a orbital of the NO group.

Fluoride reacts with c^{3+} only a low pH (2–3) to give a high spin complex (Butt and Keilin, 1962).

REFERENCES

Aviram, I. and Schejter, A. (1969). *J. Biol. Chem.* **244**, 3773–3778.
Barlow, G. H. and Margoliash, E. (1966). *J. Biol. Chem.* **241**, 1473–1477.
Boeri, E., Ehrenberg, A., Paul, K. G. and Theorell, H. (1953). *Biochim. et Biophys. Acta* **12**, 273–282.
Brandt, K. G., Parks, P. C., Czerlinski, G. H. and Hess, G. P. (1966). *J. Biol. Chem.* **241**, 4180–4185.
Butt, W. D. and Keilin, D. (1962). *Proc. Roy. Soc. London* B **156**, 429–458.
Chance, B. and Williams, G. R. (1955). *J. Biol. Chem.* **217**, 429–438.
Chance, B., Lee, C. P., Mela, L. and Wilson, D. F. (1968). "Osaka", pp. 353–356.
Cooke, R. and Debrunner, P. (1968). *J. Chem. Phys.* **48**, 4532–4537.
Coolidge, T. B. (1932). *J. Biol. Chem.* **98**, 755–764.
Cronin, J. R. and Harbury, H. H. (1965). *Biochem. Biophys. Res. Commun.* **20**, 503–508.
Czerlinski, G. (1969). *Fed. Proc.* **28**, Abstr. 3284.
Czerlinski, G., Darr, K. and Hess, G. P. (1967). *Fed. Proc.* **26**, Abstr. 2330.
Day, P., Smith, D. W. and Williams, R. J. P. (1967a). *Biochemistry* **6**, 1563–1566.
Day, P., Smith, D. W. and Williams, R. J. P. (1967b). *Biochemistry* **6**, 3747–3750.
Drott, H. R. (1969). *Fed. Proc.* **28**, Abstr. 1920.
Eaton, W. A. and Charney, E. (1969). *J. Chem. Phys.* **51**, 4502–4505.
Eaton, W. A. and Hochstrasser, R. M. (1966). "Hemes and Hemoproteins", pp. 581–583.

Eaton, W. A. and Hochstrasser, R. M. (1967). *J. Chem. Phys.* **46**, 2533–2539.
Ehrenberg, A. and Szczepkowski, T. W. (1960). *Acta Chem. Scand.* **14**, 1684–1692.
Eichhorn, G. L. and Cairns, J. F. (1958). *Nature* (London) **181**, 994.
Elliott, W. B. and Doebbler, G. F. (1966). *Anal. Biochem.* **15**, 463–469.
Estabrook, R. W. (1956). *Biochim. et Biophys. Acta* **19**, 184.
Estabrook, R. W. (1957a) *J. Biol. Chem.* **223**, 781–794.
Estabrook, R. W. (1957b). *J. Biol. Chem.* **227**, 1093–1108.
Estabrook, R. W. (1961). "Haematin enzymes", Vol. II, pp. 436–457.
Estabrook, R. W. (1966). "Hemes and Hemoproteins", pp. 405–409.
Flatmark, T. (1964). *Acta Chem. Scand.* **18**, 1517–1527.
Flatmark, T. (1966). *Acta Chem. Scand.* **20**, 1476–1486.
Flatmark, T. (1967). *J. Biol. Chem.* **242**, 2454–2549.
Flatmark, T. and Robinson, A. B. (1968). "Osaka", pp. 318–327.
Fung, D. and Vinogradov, S. (1968). *Biochem. Biophys. Res. Commun.* **31**, 596–602.
George, P. and Hanania, G. I. H. (1955). *Nature* (London) **175**, 1034.
George, P. and Schejter, A. (1964). *J. Biol. Chem.* **239**, 1504–1508.
George, P. and Tsou, C. L. (1952). *Biochem. J.* **50**, 440–441.
George, P., Eaton, W. A. and Trachtman, M. (1968). *Fed. Proc.* **27**, Abstr. 1736.
George, P., Glauser, S. C. and Schejter, A. (1967). *J. Biol. Chem.* **242**, 1690–1695.
Greenwood, C. (1968). "Osaka", pp. 340–348.
Greenwood, C. and Palmer, G. (1965). *J. Biol. Chem.* **240**, 3660–3663.
Gupta, R. K. and Redfield, A. G. (1970). *Biochem. Biophys. Res. Commun.* **41**, 273–281.
Hamaguchi, K., Ikeda, K., Sakai, H., Sugeno, K. and Narita, K. (1967). *J. Biochem.* **62**, 99–104.
Hamaguchi, K., Ikeda, K. and Narita, K. (1968). "Osaka", pp. 328–334.
Harbury, H. A., Myer, Y. P., Murphy, A. J. and Vinogradov, S. N. (1966). "Hemes and Hemoproteins", pp. 415–426, 444–445.
Henderson, R. W. and Rawlinson, W. A. (1956a). *Biochem. J.* **62**, 21–29.
Henderson, R. W. and Rawlinson, W. A. (1956b). *Nature* (London) **177**, 1180–1181.
Henderson, R. W. and Rawlinson, W. A. (1961). "Haematin Enzymes", Vol. I, pp. 370–382.
Holzwarth, G. and Doty, P. (1965). *J. Am. Chem. Soc.* **87**, 218–228.
Horecker, B. L. and Kornberg, A. (1946). *J. Biol. Chem.* **165**, 11–20.
Horecker, B. L. and Stannard, J. N. (1948). *J. Biol. Chem.* **172**, 589–597.
Keilin, D. and Hartree, E. F. (1937). *Proc. Roy. Soc. London B* **122**, 298–308.
Keilin, D. and Hartree, E. F. (1949). *Nature* (London) **164**, 254–259.
Kon, H. (1969). *Biochem. Biophys. Res. Commun.* **35**, 423–427.
Kon, H. and Kataoka, N. (1969). *Biochemistry* **8**, 4757–4762.
Kowalsky, A. (1962). *J. Biol. Chem.* **237**, 1807–1819.
Kowalsky, A. (1965). *Biochemistry* **4**, 2382–2388.
Lang, G., Herbert, D. and Yonetani, T. (1968). *J. Chem. Phys.* **49**, 944–950.
Lemberg, R. and Cutler, M. E. (1970). *Biochim. et Biophys. Acta* **197**, 1–10.
Lemberg, R. and Legge, J. W. (1949). "Hematin Compounds and Bile Pigments", Table III, pp. 354. Interscience Publishers, New York.
Lumry, R. and Sullivan, J. (1963). *Fed. Proc.* **22**, Abstr. 2544.
Lumry, R., Solbakken, A., Sullivan, J. and Reyerson, L. H. (1962). *J. Am. Chem. Soc.* **84**, 142–149.
McDermott, P., May, L. and Orlando, J. (1967). *Biophys. J.* **7**, 615–620.
Margalit, R. and Schejter, A. (1970). *FEBS Letters* **6**, 278–280.
Margoliash, E. (1954). *Biochem. J.* **56**, 529–534.

Margoliash, E. and Frowirth, N. (1959). *Biochem. J.* **71**, 570–572.
Margoliash, E. and Schejter, A. (1966). *Adv. in Protein Chem.* **21**, 113–286.
Massey, V. (1959). *Biochim. et Biophys. Acta* **34**, 255–256.
Matsubara, H. and Yasunobu, K. T. (1967). *J. Biol. Chem.* **236**, 1701–1705.
Minakami, S. (1955). *J. Biochem.* **42**, 749–756.
Minakami, S., Titani, K. and Ishikura, H. (1957). *J. Biochem.* **44**, 535–536.
Mirsky, R. (1966). "Hemes and Hemoproteins", pp. 444–445.
Mirsky, R. and George, P. (1966). *Proc. Nat. Acad. Sci.* **56**, 222–229.
Mirsky, R. and George, P. (1967a). *Biochemistry* **6**, 1872–1875.
Mirsky, R. and George, P. (1967b). *Biochemistry* **6**, 3671–3675.
Morrison, M. (1965). "Amherst", pp. 414–415.
Morrison, M. and Duffield, J. (1966). "Hemes and Hemoproteins", pp. 431–434.
Morrison, M., Hollocher, T., Murray, R., Marinetti, G. and Stotz, E. (1960). *Biochim. et Biophys. Acta* **41**, 334–338.
Moss, T. H., Bearden, A. J., Bartsch, R. G. and Cusanovich, M. A. (1968). *Biochemistry* **7**, 1883–1890.
Myer, Y. P. (1968a). *Biochim. et Biophys. Acta* **154**, 84–90.
Myer, Y. P. (1968b). *J. Biol. Chem.* **243**, 2115–2122.
Myer, Y. P. (1968c). *Biochemistry* **7**, 765–776.
Myer, Y. P. (1970). *Biochim. et Biophys. Acta* **214**, 94–106.
Myer, Y. P. and Harbury, H. A. (1965). *Proc. Nat. Acad. Sci.* **54**, 1391–1398.
Myer, Y. P. and Harbury, H. A. (1966). *J. Biol. Chem.* **241**, 4299–4303.
Myer, Y. P., Murphy, A. J. and Harbury, H. A. (1966). *J. Biol. Chem.* **241**, 5370–5374.
Nicholls, P. (1964). *Arch. Biochem. Biophys.* **106**, 25–48.
Nicholls, P. and Mochan, E. (1968). *Biochim. et Biophys. Acta* **131**, 397–400.
Paléus, S. (1960). *Acta Chem. Scand.* **14**, 1743–1748.
Paléus, S. (1962). *Arch. Biochem. Biophys.* **96**, 60–62.
Paléus, S. and Neilands, J. B. (1950). *Acta. Chem. Scand.* **4**, 1024–1030.
Paléus, S., Ehrenberg, A. and Tuppy, H. (1955). *Acta Chem. Scand.* **9**, 365–374.
Paul, K. G. (1947). *Arch. Biochem.* **12**, 441–450.
Paul, K. G. (1948a). *Acta Chem. Scand.* **2**, 430–439.
Paul, K. G. (1948b). *Acta Chem. Scand.* **2**, 557–560.
Paul, K. G. (1951). *Acta Chem. Scand.* **5**, 389–405.
Paul, K. G. (1960). In "The Enzymes", 2nd ed. (P. D. Boyer, H. Lardy and K. Myrbäck, eds.) Vol. 3, pp. 277–328. Academic Press, New York and London.
Paul, K. G. and Theorell, H. (1954). *Acta Chem. Scand.* **8**, 1087–1088.
Potter, V. R. (1941). *J. Biol. Chem.* **137**, 13–20.
Rein, H., Ristau, O. and Jung, F. (1968). *Experientia* **24**, 797–798.
Rodkey, F. L. and Ball, E. G. (1947). *Fed. Proc.* **6**, 286.
Rodkey, F. L. and Ball, E. G. (1950). *J. Biol. Chem.* **182**, 17–28.
Schejter, A. and George, P. (1962). *Fed. Proc.* **21**, 46.
Schejter, A. and George, P. (1964). *Biochemistry* **3**, 1045–1049.
Schejter, A. and George, P. (1965). *Nature* (London) **206**, 1150–1151.
Schejter, A. and Margalit, R. (1970). FEBS Letters **10**, 179–181.
Schejter, A., Glauser, S. C., George, P. and Margoliash, E. (1963). *Biochim. et Biophys. Acta* **73**, 614–643.
Shashoua, V. E. (1964). *Nature* (London) **203**, 972–973.
Shashoua, V. E. (1965). *Arch. Biochem. Biophys.* **111**, 550–558.
Shashoua, V. E. and Estabrook, R. W. (1966). "Hemes and Hemoproteins", pp. 427–430.

Shechter, E. and Saludjian, P. (1967). *Biopolymers* **5**, 788–790.
Skov, K. and Williams, G. R. (1968). "Osaka", pp. 349–352.
Skov, K., Hofmann, T. and Williams, G. R. (1969). *Canad. J. Biochem.* **47**, 750–752.
Stellwagen, E. (1966). *Biochem. Biophys. Res. Commun.* **23**, 29–33.
Stellwagen, E. (1968a). *Biochemistry* **7**, 2496–2501.
Stellwagen, E. (1968b). *Biochemistry* **7**, 2893–2898.
Taborsky, G. (1970). *Biochemistry* **9**, 3768–3773.
Theorell, H. (1936). *Biochem. Z.* **285**, 207–218.
Theorell, H. and Åkeson, Å. (1939). *Science* **90**, 67.
Theorell, H. and Åkeson, Å. (1941). *J. Am. Chem. Soc.* **63**, 1804–1811.
Tint, H. and Reiss, W. (1950). *J. Biol. Chem.* **182**, 385–396, 397–403.
Tosi, L. (1957). *Pubbl. staz. zool., Napoli*, **29**, 425–433.
Tsou, C. L. (1951c). *Biochem. J.* **49**, XLVII.
Ulmer, D. D. (1965). *Biochemistry* **4**, 902–907.
Ulmer, D. D. (1966a). *Biochemistry* **5**, 1886–1892.
Ulmer, D. D. (1966b). *Proc. Nat. Acad. Sci.* **55**, 894–899.
Ungar, G., Aschheim, E., Psochoyos, S. and Romano, D. V. (1957). *J. Gen. Physiol.* **40**, 635–652.
Urry, D. W. (1965). *Proc. Nat. Acad. Sci.* **54**, 640–648.
Urry, D. W. (1966). "Hemes and Hemoproteins", pp. 435–445.
Urry, D. W. (1968). "Osaka", pp. 311–317.
Urry, D. W. and Doty, P. (1965). *J. Am. Chem. Soc.* **87**, 2756–2758.
Van Gelder, B. F. and Slater, E. C. (1962). *Biochim. et Biophys. Acta* **58**, 593–595.
Van Gelder, B. F., Urry, D. W. and Beinert, H. (1968). "Osaka", pp. 335–339.
Vinogradov, S. V. and Harbury, H. A. (1966). *Biochim. et Biophys. Acta* **115**, 494–495.
Williams, R. J. P. (1966). "Hemes and Hemoproteins", pp. 557–576, 585–587.
Yamanaka, T., Horio, T. and Okunuki, K. (1962). *J. Biochem.* **51**, 426–430.
Yamanaka, T., Mizushima, H., Nozaki, M., Horio, T. and Okunuki, K. (1959). *J. Biochem.* **46**, 121–132.
Zand, R. and Vinogradov, S. (1967). *Biochem. Biophys. Res. Commun.* **26**, 121–127.
Zand, R. and Vinogradov, S. (1968). *Arch. Biochem. Biophys.* **125**, 94–97.

6. Modification and reactions of cytochrome c

(*a*) *Modification and denaturation.* Modification of cytochrome c by acids, polymerization and proteolysis has already been mentioned frequently in the preceding sections of this chapter (B,4a). Abolition of the 695 nm absorption bands of c^{3+} (see B3a and 4b) and reactivity with CO (Slater, 1949a; Ben–Gershom, 1961; Butt and Keilin, 1962) are sensitive indicators. One type of modification is combined with dimerization (Nozaki, 1960; Margoliash and Lustgarten, 1962; Schejter *et al.*, 1963). The dimer has been obtained crystalline but is inactive in the succinoxidase system. Dimerization is caused by acids and with mammalian, rather than yeast cytochrome c^{3+}, and by surface activity of gas-liquid and liquid-liquid interfaces several polymers are formed (Havez *et al.* 1966). The polymers inhibit sulphite cytochrome c reductase (Howell and Fridovich, 1968). The peroxide-produced polymers (O'Brien, 1967) cannot be due to the formation of disulphide bridges, and they appear to differ from acid-produced polymers by the failure of protein denaturants (4 M urea, guanidine hydrochloride, or 30% alcohol) to revert them partially into the monomer.

The tertiary structure of cytochrome c described in Chapter V, B3c makes it very resistant to irreversible denaturation and uncoiling (cf. also V, B4a). Myer (1968) found no uncoiling of peptide between pH 3 and 11·6, or in 3 M urea, but only alteration of the haem-protein interaction. Not even in 6–9 M urea was there any evidence for it. Partial unfolding, however, occurred at 53°, and became predominant at 82°. Heat alteration of c^{2+} was largely reversible even at 100° (Butt and Keilin, 1962). The shift of the absorption bands to longer wavelengths caused by heat, was also caused by guanidine (Yamanaka et al., 1962). Kowalsky (1962) found that urea affected only the structure of c^{3+}, not of c^{2+}, but even c^{3+} remained predominantly in the low-spin form. The great viscosity increase by 4–8 M urea (Stellwagen et al., 1964, 1967, 1968), however, indicates some unfolding. The Soret band is shifted to 403 nm and increased. These alterations still remain reversible. Organic solvents can even partially reverse polymerization, but Tsushima and Miyama (1956) found modification (perturbation) by benzoate and salicylate, at first to increase reducibility by ascorbate and then later binding of CO. Kaminsky and Davison (1969) and Kaminsky et al. (1971) describe the effects of various alcohols on Soret band and autoxidizibility. Small concentrations of (2 mM) of sodium dodecylsulphate increase the rate of aerobic oxidation of ascorbate in the presence of TMPD (Hill, 1970); this alteration is reversible. By studying hydrogen-deuterium exchange of c^{2+} and c^{3+}, Kaegi and Ulmer (1968) found evidence for a conformational change at low pH (1·5) (see also Bull and Breese, 1966), but not at high pH. However, from enthalpy changes accompanying the oxidation of c^{2+} in the pH range 6–11, Watt and Sturtevant (1969) concluded that a conformational change occurred at pH 9·3 and calculated its enthalpy and entropy. At pH 9·5 proton release with ferryicyanide was altered prior to the 695 nm absorption, particularly in chicken cytochrome c (Czerlinski, 1967, 1969); they found a biphasic proton release between pH 8 and 9·5. Brumer et al. (1966) found that treatment at pH 11 and 65° altered cytochrome c; the product inhibited the oxidation of native c^{2+} in the succinoxidase, but not in the NADH oxidase reaction; it had increased ascorbate oxidase activity.

(b) *Modification of amino acid side chains*. Many of such modifications have already been discussed, particularly in Chapter V B3, since they gave important clues for the tertiary structure and the binding of haem iron, particularly trifluoroacetylation, trinitrophenylation, guanidation and carboxymethylation.

The hydrolysis of asparagine to aspartic acid has no effect on activity in the succinoxidase reaction except an increase of the K_M' values (Flatmark,

1966c, 1967) and does not cause a reaction with CO, but increases autoxidizability and decreases reducibility by ascorbate or by NADH-reductase (EC 1.6.2.1). The large temperature effect on the rate of deamination and some changes in light absorption and ORD or CD curves indicate some conformational change.

The *acetylation* of cytochrome c has been studied by Minakami et al. (1958a,b); Takemori et al. (1962a,b); Conrad-Davies et al. (1964); Cronin and Harbury (1965); Ulmer (1966) and Wada and Okunuki (1968, 1969). Acetylated cytochrome c is spectroscopically unchanged and does not react with CO, but it is more autoxidizable and less active in the succinoxidase system. Acetylation of 6 ϵNH_2-groups of lysine destroys the activity. The evidence against lysine being involved in the binding of haem iron has been discussed in Section B5. Succinylation is even more active than acetylation in decreasing cytochrome c activity, but guanidation of these groups has no effect, their electrostatic positive charge being the essential requirement (Takahashi et al., 1958; Takemori et al., 1962b). With more acetic anhydride all 18 lysines and also 4 tyrosyl hydroxyls are acetylated. This results in profound changes of spectrum, CO-reactivity and autoxidizability. This fully acetylated (22-acetyl) cytochrome c^{3+} is at room temperature in the high spin form, though still in low spin form at the temperature of liquid nitrogen. Partial hydrolysis of the fully guanidated and twice acetylated c^{3+} to the monoacetylated form by added free imidazole restores the lost electron transport activity (Cronin and Harbury, 1965).

All four *tyrosine* groups have abnormal pK's above pH 11. They are thus buried and their environment is not altered by changes in iron valency. On loosening the structure with urea, they ionize normally. Flatmark (1964a), however, found two tyrosines with almost normal pK (10·8) and only two of pK 12·2. Urea did not affect the tyrosine absorption band at 246 nm. Rupley (1964) found one normal pK at 10·0, three abnormal ones at pH 11·0, 12·35 and 13·0 by studying the 243 nm band.

Iodination of the tyrosine residues was studied by Ishikura et al. (1959) and McGowan and Stellwagen (1970). With up to 1 atom of iodine per mole the spectrum is not altered and no reaction with CO is produced. More iodine, however, causes such alterations. At pH 9·5 two of the four tyrosines are di-iodinated. McGowan and Stellwagen found Tyr-97 and Tyr-48 unreactive, but Tyr-67 and Tyr-74 reactive with iodine (cf. also Narita et al., 1968, for iodination of *Candida* cytochrome c). Since in c^{2+} none of the tyrosines is exposed (Margoliash and Schejter, 1966) the conversion of c^{2+} into c^{3+} appears to result in conformational changes at the tyrosines 67 and 74. By tetranitromethane the two tyrosines nitrated are, however, Tyr-48 and Tyr-67 (Skov et al., 1969), and the authors believe that the acetylation by N-acetylimidazole (Ulmer, 1965) is more likely to affect these tyrosines than 48 and 74, assumed by Ulmer, since acetylated c^{3+} cannot be nitrated (see also Sokolovsky et al., 1970). The nitrate cytochrome c is inactive, and at low pH (3–4·5) but not at neutral pH still shows the 695 nm band (see also Schejter et al., 1970b). There appear to be two forms of the cytochrome c, both nitrated in tyrosine 67 but with different positions of the nitro group, differing in their reducibility by ascorbate (Schejter et al., 1970a).

Trifluoroacetylation, guanidation, carboxymethylation and trinitrophenylation have been discussed in Chapter V B3. Trifluoroacetylation affects all the ϵ-NH_2-groups of lysine and inactivates it but leaves the spectrum unaltered. Guanidation (see above), unless used at high concentration (Yamanaka et al., 1962) does not alter cytochrome c spectroscopically, and leaves it unable to combine with CO and active, though it somewhat decreases the stability. In contrast to the slow electron transfer between native c^{2+} and c^{3+}, that between guanidated c^{2+} and c^{3+} is rapid (Fridovich, 1966).

Whereas carboxymethylation of Met-65 and His-33 (in beef cytochrome c) by bromo- or iodoacetate at neutral pH causes little alteration, that at pH 3 or at neutral pH in the presence of cyanide, which affects the iron-bound Met-80, transforms cytochrome c into a myoglobin-like compound (Schejter and George, 1965; Atanasov et al., 1969). This is inactive in the succinoxidase system and reacts with CO, but has still as c^{3+} a low-spin type spectrum at a pH of 4·7. According to Schejter (1966) it forms an unstable oxygenated complex. This compound is of interest as a model for the bacterial cytochromes cc' and c' (see Chapter VI).

Trinitrophenylsulphonate affects only ϵ-NH_2 groups of lysine (Takemori, 1962b; Ando et al., 1965b). It decreases the reaction of cytochrome c with the oxidase, but does not prevent it entirely (Okunuki et al., 1965; Wada and Okunuki, 1968, 1969). All ϵ-NH_2 groups of lysine in cytochrome c can be made to react with salicylaldehyde to form aldimines (Williams and Jacobs, 1966, 1968) at room temperature and pH 9·6. The aldimines are readily split at pH below 3 or above 11. The product of the reaction contains polymers of cytochrome c. It is insoluble and does not react with NADH-cytochrome c reductase; removal of salicylaldehyde restores this reaction only partially.

N-bromosuccinimide has been introduced as a reagent for tryptophan by Yonetani (1968a,b) and Stellwagen (1968). Mirsky and George (1966) believe that Trp-59 plays a vital structural role in cytochrome c and is vicinal to the haem group, at least in c^{3+}. This is confirmed by Schejter et al. (1970b) and Aviram and Schejter (1971) who used formylation with formic acid-HCl. Dimerization of c^{3+} may cause a decreased interaction of haem with Trp-59. Tryptophan appears also to be oxidized by H_2O_2 at pH 7 (O'Brien, 1966b). The Soret band is shifted to shorter wavelengths and somewhat diminished in intensity. The ferric compound becomes high-spin at a pH less than 7 and mixed spin from pH 3–6. This is confirmed by EPR spectra (Yonetani and Schleyer, 1968). The ferrous compound reacts with CO, and has a Soret band shifted 2 nm towards shorter wavelengths and decreased α and β bands. Methionine in cytochrome c is, however, also readily oxidized to its sulphoxide (O'Brien, 1966a, 1969; Stellwagen and Van Rooyan, 1967) which is unable to form ferrihaemochromes, and this is assumed by them a more likely explanation of the alteration. N-bromosuccinimide caused dimerization of c at pH 4, but not at pH 8 (O'Brien, 1967). Cyanogen bromide splits the molecule at Met-65 (Corradin and Harbury, 1970).

(c) *Other reactions of cytochrome c.* The autoxidation of c^{2+} is greatly increased by modifications of the protein. Boeri and Tosi (1954) found high ionic strength increasing its rate. Davison (1970) reported that decreasing the pH from 4·5 to 2·5, while increasing the rate of autoxidation, gradually increased the activation energy. He ascribes the acceleration at low pH to the more favourable entropy of the reaction. Cytochrome oxidase also acts mainly by favourable entropy. In contrast, the peroxidative oxidation of c^{2+} by cytochrome peroxidase at pH 7·0 showed a large decrease of the activation energy by the enzyme and changes in ionic strength had no effect.

Shashoua (1965) observed complicated kinetics of the oxidation of c^{2+} by H_2O_2. According to Mochan and Degn (1969) c^{3+} catalyses the oxidation of c^{2+} by H_2O_2 which thus has autocatalytic character. They find, however, no spectroscopic evidence for the existence of a complex between cytochrome c and H_2O_2. Degradation of c^{3+} by H_2O_2 is more rapid than that of c^{2+}. The amino acid residues attacked by H_2O_2 are histidine < methionine < cysteine < tyrosine and tryptophan (O'Brien, 1966b). The methionine is oxidized to the sulphoxide and the sulphone. The oxidation of ferro-cytochrome c by cytochrome c peroxidase of yeast (Yonetani, 1970) has been studied in detail by Mochan and Nicholls (1971) and Nicholls and Mochan (1971). Unlike cytochrome oxidase, the peroxidase is located in the intracristal space, not in the inner membrane. The kinetic behaviour of the peroxidase, however, resembles that of the oxidase, although there are some differences. Complexes are formed by the peroxidase with cytochrome c which are able to form a ternary complex with H_2O_2. Polycations competitively inhibit the complex formation. The interaction is between protein and protein, not between haem and protein. While gross conformational changes in the complex formation are not indicated by the lack of alteration of fractional ratios (f/fo), finer conformational changes such as those made likely for the oxidase-cytochrome c interaction (see Chapter VII, H) appear likely. Cytochrome c itself has peroxidative activity towards pyrogallol (Flatmark, 1962, 1963, 1966a,b) and is progressively destroyed in the reaction (1964b). Lipoperoxidases also damage cytochrome c (Dinescu, 1969).

The oxidation of c^{2+} by ferricyanide has been studied by Sutin and Christman (1961), George et al. (1968) and by Chance (1970). Insufficient attention may, however, have been paid to possible effects of the ferrocyanide formed in the reaction which is a strong combinant with basic proteins (Kossel and Kutscher, 1900; Kitahara et al., 1968; Geyer and Lemberg, 1971). In reactions of dihydroxyphenylalanine with polyphenoloxidase, c^{3+}, if present, is reduced to c^{2+} (Kertesz, 1968). Xanthine oxidase reduces c^{3+} to c^{2+} in a reaction which is inhibited by myoglobin if this is added or present as an impurity (Fridovich, 1962; Quinn and Pearson, 1964; Muraoka et al., 1967). The reduction is ascribed by McCord and Fridovich (1968, 1970), by Fridovich (1970) and by Nakamura and Yamazaki (1969) to the formation of the superoxide anion ($O\dot{\overline{2}}$) in the reaction of xanthine oxidase with its substrate. This can be trapped as lactoperoxidase compound III or by catechols (Miller, 1970). There is possibly a relation of this to the fact that reduction of c^{3+} by hydroquinones (Williams, 1963; Wosilait, 1961; Morrison et al., 1968) is accelerated by quinones. The reduction of c^{3+} to c^{2+} by ionizing radiations has, however, been ascribed by Masuda et al. (1968) to the OH radical.

Only intact cytochrome c catalyses oxidation of cysteine by cytochrome oxidase (Boeri et al., 1953). GSSG catalyses the reduction of c^{3+} by GSH (Froede and Hunter, 1970; Painter and Hunter, 1970) in a reaction independent of the presence of metal and greatly inhibited by higher ionic strength. There is, however, also a metal-catalysed reaction. Cytochrome c^{3+} is able to oxidize a great variety of substances but there is no evidence that these reactions are of biological significance, as e.g. the oxidation of tocopherols (Detwiler and Nason, 1966), of cortisol (Monder, 1968) of catecholamines (Laparra and Blanquet, 1966) and of some carbohydrates, e.g. glucosamine (Mora et al., 1965). The iron of c^{3+} is rapidly reduced by Cr^{2+} which is firmly bound to the protein (Kowalsky, 1967, 1969). While this reduction is of no significance in the biological electron transport itself, it is studied to serve as a model throwing possibly some light on it. According to Taube (1955) Cr may be used to study active sites in biological systems. At neutral pH ferrous ion salts reduce cytochrome c^{3+} while at low pH EDTA is required for the reduction (Kurihara and Sano, 1970).

By xanthydrol 3 xanthyl groups are introduced into cytochrome c (Westcott and Dickmann, 1954). Xanthyl-c^{2+} is readily oxidized by the succinic oxidase system, but xanthyl-c^{3+} is not reduced by it. The data suggest that the points of attachment of c to cytochrome oxidase differ from the points of attachment of it to the reducing part of the respiratory chain, probably the bc_1 complex. The respiratory chain in which cytochrome c plays such an essential role will be discussed in Chapter VII.

The iron can be removed from cytochrome c by hydrofluoric acid without essential alteration of the apoprotein or its linkage to the porphyrin (Robinson and Kamen, 1968).

(d) *Immunological reactions.* For a long period it was assumed that cytochromes c are not antigenic and allergic reactions were ascribed to impurities (Storck *et al.*, 1964; Jonsson and Paléus, 1966). Indeed, commercial preparations of cytochrome c contain such impurities, removable by Sephadex filtration (Chidlow *et al.*, 1967). However, the antigenic effect of cytochromes c has now been established, e.g. by repeated injections of native cytochromes c for prolonged periods in rabbits (Reichlin *et al.*, 1966). Antibody formation in rabbits immunized with several heterologous cytochromes c have been studied also (Watanabe *et al.*, 1967; Nisinoff *et al.*, 1967a,b; Margoliash *et al.*, 1968; Nanni and Ferro, 1968; and Noble *et al.*, 1969). The rabbit is usually the experimental animal, but mouse antibodies against dimeric yeast cytochrome c have been observed by Watanabe *et al.* Dimers which are less readily excreted than native c have stronger antigenicity (Margoliash *et al.*, 1968; Reichlin *et al.*, 1970), whether produced by acids or ethanol (Margoliash and Lustgarten, 1962), by polymerizing oxidation of yeast c (Okada *et al.*, 1964) or by glutaraldehyde (Reichlin *et al.*, 1968); however, human c is effective also as a monomer. Another method for producing antibodies is the conjugation of the cytochromes c by covalent linkage to acetylated γG-globulins (Nisinoff *et al.*, 1967a,b; Margoliash *et al.*, 1970). Such an antibody reacted even with cytochrome c of the same rabbit. Human cytochrome c is a better antigen and forms precipitating antibodies, while tuna and turkey c yield non-precipitating antibodies. The former is probably due to the presence of several effective antigenic sites. Cross-reactions are somewhat variable (Nisinoff *et al.*, 1967b). Cross-reactions of 26 different cytochromes with antibodies prepared against tuna, chicken, turkey, kangaroo, horse and human c have been studied by Margoliash *et al.* (1968, 1970). No cross reactions were found between yeast and bovine or yeast and horse c (Okada *et al.*, 1964; Watanabe *et al.*, 1967), although yeast and tuna c and tuna and beef c cross-reacted. Nanni and Ferro (1968) found no cross-reaction between horse and beef c by immunoelectrophoretic methods. However, the γG-globulins formed in rabbit from tuna or horse c cross-reacted with numerous other cytochromes (Reichlin *et al.*, 1966; Nisinoff *et al.*, 1967a; Noble *et al.*, 1969). About 60–70% of closely related cytochromes c of man and *Macaca* monkey, or of horse and donkey cross-reacted though differing in a single residue (Nisinoff *et al.*, 1967b, 1970; Margoliash *et al.*, 1968; Noble *et al.*, 1969). Such cross-reactions were established either by competition experiments using iodine-labelled homologous c with a large amount of unlabelled competing c; or by the complement fixation reaction (Margoliash *et al.*, 1968). Some antisera differentiate between c^{2+} and c^{3+} though not in chicken c. No antigenic differences exist between c from different organs of the same species (Stewart and Margoliash, 1965). Attempts to identify the specific antigenic site of cytochromes c have been made by Margoliash and co-workers (Nisinoff et al., 1967b, 1970; Margoliash et al., 1968). The cytochromes contain a limited number of antigenic determinants which involve large hydrophobic residues such as tryptophane, tyrosine, phenylalanine, leucine and isoleucine. One such determinant is Ile-58 in human c; Ile-58 in human and kangaroo c are antigenically identical. Antihuman c sera adsorbed to monkey cytochrome c (which lacks Ile in this position) are able to react with human and kangaroo c which contains it but not with other cytochromes c. One of the four antibodies to human c greatly decreases the rate of reaction of it with cytochrome oxidase (Smith et al., 1970) in stoichiometric ratio, while a different antibody probably affects the interaction of cytochrome c^{3+} with the succinate-cytochrome c reductase.

(e) *Lipid cytochrome c.* Interest in the combination of cytochrome c with phospholipids was increased by the observations of Widmer and Crane (1958) that a complex of cytochrome c and phospholipid in the heptane extract of mitochondria which he called "lipid cytochrome c" restored the NADH- and succinate oxidase activities of "electron transport particles" of beef mitochondria. It had been previously found by Tsou (1952) (cf. also Slater, 1949b) that soluble cytochrome c re-added to saline-extracted mitochondria did not fully restore the oxidase activity lost in the extraction. He distinguished exogenous from the endogenous c present in the mitochondria and Keilin-Hartree preparation (cf. Lemberg, 1969 and Chapter VII, D1).

Complexes formed from cytochrome c with phospholipids have been studied extensively by Fleischer et al. (1961, 1962), Green and Fleischer (1963a,b), Das and co-workers (Das and Crane, 1964; Das et al., 1962a,b, 1964a,b, 1965), Reich and Wainio (1961a,b) and Hart et al. (1969). Reich and Wainio obtained a precipitate of M.W. 85,000 containing 15% cytochrome c and 85% ethanolamine phosphatide. One has to distinguish precipitates or liquid crystals (Kimelberg and Lee, 1969) in aqueous suspensions or in water-organic solvent interfaces from isooctane or heptane-dissolved proteolipids. The latter require for solubility the presence of unsaturated fatty acids (Das et al., 1962a) and unsaturated phospholipids predominate in mitochondria. Getz et al. (1968) studying the phospholipids in organs and organelles found mitochondria to contain more ethanolamine phosphatides and cardiolipin which also predominates in cytochrome oxidase (Fleischer, Klouwen and Brierley, 1961) than other phospholipids. The combination of cytochrome c with phospholipids is mainly, but not entirely, based on electrostatic forces. Ethanolamine phosphatides are mainly bound by their phosphoric acid groups to ϵ-NH_2

groups of lysine in cytochrome c; they cannot be bound to acetylated or succinylated, in contrast to guanidated cytochrome c. Lecithin, however, with its strongly basic choline, is weakly bound to the acidic cytochromes c and binding does not depend on the presence of free amino groups in cytochrome c. Inositol phosphatide- and lecithin-complexes are inactive, and the latter even inhibitory to the oxidase reaction. There is no clear-cut stoichiometric binding, and Van der Waals forces and dipole attraction between phosphatide molecules themselves play also a role. The predominantly electrostatic character of the binding is supported by the prevention or abolition of it by salt (Jacobs and Sanadi, 1960; Green and Fleischer, 1961b, 1963; Ulmer et al., 1965; MacLennan et al., 1966; Morrison et al., 1966; Gonzales-Cadavid and Campbell, 1967; Gulik-Krzywicki et al., 1967, 1969) reminiscent of the extraction of cytochrome c from mitochondria by saline.

It was initially suggested that "lipid cytochrome c" plays an important role in the interaction between cytochrome c and the oxidase, but it is now clear (Sun and Crane, 1969) that phospholipid is only important as a matrix of the interaction, that its action is not specific and that lipid c + lipid-poor oxidase is less active a system than free cytochrome c + lipid-containing oxidase. It remains to be shown that "lipid c" has unique properties in the respiratory chain. Aggregates of lipid c formed at, or below pH 5·5 are inactive.

The differences in extractability of cytochrome c from mitochondria and from submitochondria particles is now known to be due to the fact (Jacobs and Sanadi, 1960; Kimelberg and Lee, 1969) that cytochrome oxidase is bound on the outer surface of the inner mitochondrial membrane and that in sonicated submitochondrial particles cytochrome c is entrapped in phospholipid vesicles which make it also hardly accessible to large anions such as ascorbate or dithionite. "Lipid cytochrome c" is thus more closely connected with these anomalies than with the normal cytochrome c-oxidase interaction.

The behaviour of cytochrome c on ethanolamine and cardiolipin monolayers has been studied by Quinn and Dawson (1969a,b, 1970) using ^{14}C-labelled cytochrome c. There is no evidence of stoichiometric interaction, nor of extensive uncoiling of cytochrome c. Undenatured c molecules lie under expanded lipid monolayers rather than being physically dissolved in the film. Such unfolding as is necessary must remain reversible. The initial effect of 1 M NaCl supports electrostatic reaction but later 1 M NaCl causes only partial desorption. A dramatic decrease of adsorption in the monolayer is found at pH 8·5 when the positive charge of the lysine ϵ-NH$_2$ groups of cytochrome c is removed. Gulik-Krzywicki et al. (1969) observed CD spectra, which they ascribe to differences in the orientation of haem groups rather than to conformational transitions. Azzi et al. (1969) found structural changes associated with valency change of c as determined by 8-aniline-1-naphthalene sulphonic acid as fluorescence probe to occur only in the presence of phospholipid, in particular cardiolipin; there is, however, ample evidence for conformational differences between c^{2+} and c^{3+} in the absence of lipid. Galactolipid (Chang and Lundin, 1965) stimulates the effect of cytochrome c on the photoreduction by fresh intact chloroplasts. As DCMU completely inhibits the effect of galactolipid this appears to participate in photoreaction II. Hardesty and Mitchell (1963a,b) observed accumulation of free fatty acid in the poky mutant of *Neurospora crassa*. The fatty acids make cytochrome c less easily reducible, more autoxidizable and more susceptible to destructive peroxidations.

REFERENCES

Ando, K., Orii, Y., Takemori, S. and Okunuki, K. (1965). *Biochim. et Biophys. Acta* **111**, 540–552.
Atanasov, B., Volkenstein, M. V. and Sharonov, Y. A. (1969). *Molekulyarnaya Biol.* **3**, 518–526.
Aviram, I. and Schejter, A. (1970). *Biochim. et Biophys. Acta* **229**, 113–118.
Azzi, A., Fleischer, S. and Chance, B. (1969). *Biochem. Biophys. Res. Commun.* **36**, 322–327.
Ben-Gershom, E. (1961). *Biochem. J.* **78**, 218–223.
Boeri, E. and Tosi, L. (1954). *Arch. Biochem.* **52**, 83–92.
Boeri, E., Baltscheffsky, H., Bonnichsen, R. and Paul, K. G. (1953). *Acta Chem. Scand.* **7**, 831–844.
Brumer, P., Levine, W. G., Peisach, J. and Strecker, H. (1966). *Arch. Biochem. Biophys.* **117**, 232–238.
Bull, H. B. and Breese, K. (1966). *Biochem. Biophys. Res. Commun.* **24**, 74–78.
Butt, W. D. and Keilin, D. (1962). *Proc. Roy. Soc. London* B **156**, 429–458.
Chance, B. (1970). *Fed. Proc.* **29**, Abstr. 2762.
Chang, S. B. and Lundin, K. (1965). *Biochem. Biophys. Res. Commun.* **21**, 424–431.
Chidlow, J. W., Stephen, J. and Smith, H. (1967). *Biochem. J.* **104**, 4–5P.
Conrad-Davies, H. C., Smith, L. and Wasserman, A. R. (1964). *Biochim. et Biophys. Acta* **85**, 238–246.
Corradin, G. and Harbury, H. A. (1970). *Biochim. et Biophys. Acta* **221**, 489–496.
Cronin, J. R. and Harbury, H. A. (1965). *Biochem. Biophys. Res. Commun.* **20**, 503–508.
Czerlinski, G. (1967). *J. Theor. Biol.* **17**, 343–382.
Czerlinski, G. (1969). *Fed. Proc.* **28**, Abstr. 3284.
Das, M. L. and Crane, F. L. (1964). *Biochemistry* **3**, 696–700.
Das, M. L., Crane, F. L. and Machinist, J. M. (1964a). *Biochem. Biophys. Res. Commun.* **17**, 593–596.
Das, M. L., Haak, D. and Crane, F. L. (1965). *Biochemistry* **4**, 859–865.
Das, M. L., Hiratsuka, H., Machinist, H. M. and Crane, F. L. (1962a). *Biochim. et Biophys. Acta* **60**, 433–434.
Das, M. L., Machinist, J. M. and Crane, F. L. (1962b). *Fed. Proc.* **21**, 154.
Das, M. L., Myers, D. E. and Crane, F. L. (1964b). *Biochim. et Biophys. Acta* **84**, 618–620.
Davison, A. J. (1970). *Fed. Proc.* **29**, Abstr. 3541.

Detwiler, T. D. and Nason, A. (1966). *Biochemistry* **5**, 3936–3946.
Dinescu, G. (1969). *Rev. Roum. Biochim.* **6**, 111–116. CA 72(1) 124.
Flatmark, T. (1962). *Nature* **196**, 894–895.
Flatmark, T. (1963). *Nature* **200**, 112 113.
Flatmark, T. (1964a). *Acta Chem. Scand.* **18**, 1796–1798.
Flatmark, T. (1964b). *Acta Chem. Scand.* **18**, 2269–2279.
Flatmark, T. (1966a). *Acta Chem. Scand.* **19**, 2059–2064.
Flatmark, T. (1966b). *Acta Chem. Scand.* **20**, 1470–1475.
Flatmark, T. (1966c). *Acta Chem Scand.* **20**, 1476–1486.
Flatmark, T. (1967). *J. Biol. Chem.* **242**, 2454–2459.
Fleischer, S., Brierley, G., Klouwen, H. and Slautterback, D. B. (1962). *J. Biol. Chem.* **237**, 3264–3272.
Fleischer, S., Klouwen, H. and Brierley, G. (1961). *J. Biol. Chem.* **236**, 2936–2941.
Fridovich, I. (1962). *J. Biol. Chem.* **237**, 584–586.
Fridovich, I. (1966). *Biochim. et Biophys. Acta* **118**, 419–421.
Fridovich, I. (1970). *J. Biol. Chem.* **245**, 4053–4057.
Froede, H. C. and Hunter, F. E. Jr. (1970). *Biochem. Biophys. Res. Commun.* **38**, 954–961.
George, P., Eaton, W. A. and Trachtman, M. (1968). *Fed. Proc.* **27**, Abstr. 1736.
Getz, G. S., Bartley, W., Lurie, D. and Notton, B. M. (1968). *Biochim. et Biophys. Acta* **152**, 325–339.
Geyer, D. and Lemberg, R. (1971). *Biochim. et Biophys. Acta* **229**, 284–285.
Gonzales-Cadavid, N. F. and Campbell, P. N. (1967). *Biochem. J.* **105**, 427–442.
Green, D. E. and Fleischer, S. (1963). *Biochim. et Biophys. Acta* **70**, 554–582.
Green, D. E. and Fleischer, S. (1961). In "Horizons in Biochemistry" (Kaska, M. and Pullman, B. eds.) Academic Press, p. 381.
Gulik-Krzywicki, T., Rivas, E. and Luzzati, V. (1967). *J. Mol. Biol.* **27**, 303–322.
Gulik-Krzywicki, T., Shechter, E., Luzzati, V. and Faure, M. (1969). *Nature* **223**, 1116–1121.
Hardesty, B. A. and Mitchell, H. K. (1963a). *Arch. Biochem. Biophys.* **100**, 1–8.
Hardesty, B. A. and Mitchell, H. K. (1963b). *Arch. Biochem, Biophys.* **100**, 330–334.
Hart, C. J., Leslie, R. B., Davis, A. F. and Lawrence, G. A. (1969). *Biochim. et Biophys. Acta* **193**, 308–318.
Havez, K., Hayem-Levy, A., Mizou, J. and Biserte, J. (1966). *Bull. soc. chim. biol.* **38**, 117–131.
Hill, J. M. (1970). *Biochem. J.* **118**, 677–678.
Howell, L. G. and Fridovich, I. (1968). *J. Biol. Chem.* **243**, 5941–5947.
Ishikura, H., Takahashi, K., Titani, K. and Minakami, S. (1959). *J. Biochem.* **46**, 719–724.
Jacobs, E. E. and Sanadi, D. R. (1960). *J. Biol. Chem.* **235**, 531–534.
Jonsson, J. and Paléus, S. (1966). *Biochemistry* **7**, 2718–2724.
Kaegi, J. H. and Ulmer, D. D. (1968). *Biochemistry* **7**, 2718.
Kaminsky, L. S., Wright, R. L. and Davison, A. J. (1971). *Biochemistry* **10**, 458–462.
Kaminsky, L. S. and Davison, A. J. (1969). FEBS Letters **3**, 338–340.
Kertesz, D. (1968). *Biochim. et Biophys. Acta* **167**, 250–256.
Kimelberg, H. K. and Lee, C. P. (1969). *Biochem. Biophys. Res. Commun.* **34**, 784–790.
Kitahara, N., Endo, A., Mizushima, A. and Okazaki, H. (1968). *J. Jap. Biochem. Soc.* **39**, 161–166. CA 67(7) 61143.

Kossel, A. and Kutscher, F. (1900). *Z. physiol. Chem.* **31**, 165–214.
Kowalsky, A. (1962). *Ann. Chem. Soc. Meetg. Atlantic City,* **1962**, Abstr. 1424.
Kowalsky, A. (1967). *Fed. Proc.* **26**, Abstr. 2329.
Kowalsky, A. (1969). *J. Biol. Chem.* **244**, 6619–6625.
Kurihara, M. and Sano, S. (1970). *J. Biol. Chem.* **245**, 4804–4806.
Laparra, J. and Blanquet, P. (1965, 1966). *Compt. rend.* **261**, 4887–4900; **262**, 2298–2300.
Lemberg, R. (1969). *Physiol. Revs.* **49**, pp. 59, 94.
McCord, J. M. and Fridovich, I. (1968). *J. Biol. Chem.* **243**, 5753–5760.
McCord, J. M. and Fridovich, I. (1970). *J. Biol. Chem.* **245**, 1374–1377.
McGowan, E. B. and Stellwagen, E. (1970). *Biochemistry* **9**, 3047–3052.
MacLennan, D. H., Lenaz, G. and Szarkowska, L. (1966). *J. Biol. Chem.* **241**, 5251–5279.
Margalit, R. and Schejter, A. (1970). *FEBS Letters* **6**, 278–280.
Margoliash, E. and Lustgarten, J. (1962). *J. Biol. Chem.* **237**, 3397–3405.
Margoliash, E. and Schejter, A. (1966). *Adv. Protein Chem.* **21**, 113–286.
Margoliash, E., Reichlin, M. and Nisonoff, A. (1968). "Osaka", pp. 269–280.
Margoliash, E., Nisinoff, A. and Reichlin, M. (1970). *J. Biol. Chem.* **245**, 931–939.
Masuda, T., Yamanaka, T., Matsuda, K., Murayama, K. and Kondo, M. (1968). "Osaka", pp. 429–436.
Miller, R. W. (1970). *Canad. J. Biochem.* **48**, 935–939.
Minakami, S., Titani, K. and Ishikura, H. (1958a). *J. Biochem.* **45**, 341–348.
Minakami, S., Titani, K., Ishikura, H. and Takahashi, K. (1958b). *Proc. Internat. Sympos. Enzymol. Chem.* Tokyo and Kyoto, 1957, pp. 211–215.
Mirsky, R. and George, P. (1966). *Proc. Nat. Acad. Sci.* **56**, 222–229.
Mochan, E. (1970). *Biochim. et Biophys. Acta* **216**, 80–95.
Mochan, E. and Degn, H. (1969). *Biochim. et Biophys. Acta* **189**, 354–359.
Mochan, E. and Nicholls, P. (1971). *Biochem. J.* **121**, 69–82.
Monder, E. (1968). *Biochim. et Biophys. Acta* **164**, 369–380.
Mora, P. T., Creshoff, E. and Person, P. (1965). *Science* **149**, 642–645.
Morrison, M., Bright, J. and Rouser, G. (1966). *Arch. Biochem. Biophys.* **114**, 50–55.
Morrison, M., Steele, W.-F. and Chowdhury, D. M. (1968). "Osaka", pp. 380–382.
Muraoka, S., Enomoto, H., Sugiyama, M. and Yamasaki, H. (1967). *Biochim. et Biophys. Acta* **143**, 408–415.
Myer, Y. P. (1968). *Biochemistry* **7**, 765–776.
Nakamura, S. and Yamazaki, I. (1969). *Biochim. et Biophys. Acta* **189**, 29–37.
Nanni, G. and Ferro, M. (1968). *Ital. J. Biochem.* **17**, 274–277.
Narita, K., Sugeno, K., Mizoguchi, T. and Hamaguchi, K. (1968). "Osaka", pp. 362–369.
Nicholls, P. and Mochan, E. (1971). *Biochem. J.* **121**, 55–67.
Nisinoff, A., Margoliash, E. and Reichlin, M. (1967a). *Science* **155**, 1273–1275.
Nisinoff, A., Reichlin, M. and Margoliash, E. (1967b). *Fed. Proc.* **26**, Abstr. 2493.
Nisinoff, A., Reichlin, M. and Margoliash, E. (1970). *J. Biol. Chem.* **245**, 940–946.
Noble, R. W., Reichlin, M. and Gibson, Q. H. (1969). *J. Biol. Chem.* **244**, 2403–2411.
Nozaki, M. (1960). *J. Biochem.* **47**, 592–599.
O'Brien, P. J. (1966a). *Biochem. J.* **101**, 11P.
O'Brien, P. J. (1966b). *Biochem. J.* **101**, 12P.
O'Brien, P. J. (1967). *Biochem. J.* **102**, 28P.
O'Brien, P. J. (1969). *Biochem. J.* **113**, 13P.
Okada, Y., Watanabe, S. and Yamamura, Y. (1964). *J. Biochem.* **55**, 342–343.

Okunuki, K., Wada, K., Matsubara, H. and Takemori, S. (1965). "Amherst", pp. 549–564.
Painter, A. A. and Hunter, E. E. Jr. (1970). *Fed. Proc.* **29**, Abstr. 3805.
Quinn, P. J. and Dawson, R. M. C. (1969a). *Biochem. J.* **113**, 791–803.
Quinn, P. J. and Dawson, R. M. C. (1969b). *Biochem. J.* **115**, 65–75; (1970). *Biochem. J.* **116**, 671–680.
Quinn, J. R. and Pearson, A. M. (1964). *Nature* **201**, 928–929.
Reich, M. and Wainio, W. W. (1961a). *J. Biol. Chem.* **236**, 3058–3061.
Reich, M. and Wainio, W. W. (1961b). *J. Biol. Chem.* **236**, 3062–3065.
Reichlin, M., Fogel, S., Nisinoff, A. and Margoliash, E. (1966). *J. Biol. Chem.* **241**, 251–253.
Reichlin, M., Margoliash, E. and Nisinoff, A. (1968). *Fed. Proc.* **27**, Abstr. 34.
Reichlin, M., Nisinoff, A. and Margoliash, E. (1970). *J. Biol. Chem.* **245**, 947–954.
Robinson, A. B. and Kamen, M. D. (1968). "Osaka", pp. 383–387.
Rupley, J. A. (1964). *Biochemistry* **7**, 1648–1650.
Schejter, A. (1966). "Hemes and Hemoproteins", pp. 393–394.
Schejter, A. and George, P. (1965). *Nature* (London) **206**, 1150–1151.
Schejter, A., Aviram, I. and Sokolovsky, M. (1970a). *Biochemistry* **9**, 5118–5122.
Schejter, A., Aviram, I., Margalit, R. and Sokolowsky, M. (1970b). *8th Internat. Congress Biochem. Switzerland, Sympos.* **1**, 18.
Schejter, A., Glauser, S. C., George, P. and Margoliash, E. (1963). *Biochim. et Biophys. Acta* **73**, 614–643.
Shashoua, V. E. (1965). *Arch. Biochem. Biophys.* **111**, 550–558.
Shipley, G. G., Leslie, R. B. and Chapman, D. (1969a). *Biochim. et Biophys Acta* **173**, 1–10.
Shipley, G. G., Leslie, R. B. and Chapman, D. (1969b). *Nature* (London) **222**, 1–10.
Skov, K., Hofmann, T., and Williams, G. R. (1969). *Canad. J. Biochem.* **47**, 750–752.
Slater, E. C. (1949a). *Nature* (London) **163**, 532.
Slater, E. C. (1949b). *Biochem. J.* **44**, 305–318.
Smith, L. (1954). *Bacteriol. Rev.* **18**, 106–130.
Smith, L., Davies, H. D., Reichlin, M. and Margoliash, E. (1970). *Fed. Proc.* **29**, Abstr. 514.
Sokolovsky, M., Aviram, I. and Schejter, A. (1970). *Biochemistry* **9**, 5113–5118.
Stellwagen, E. (1964). *Biochemistry* **3**, 919–923.
Stellwagen, E. (1967). *J. Biol. Chem.* **242**, 602–606.
Stellwagen, E. (1968). *Biochemistry* **7**, 2893–2898.
Stellwagen, E. and Van Rooyan, S. (1967). *J. Biol. Chem.* **242**, 4801–4805.
Stewart, J. W. and Margoliash, E. (1965). *Canad. J. Biochem.* **43**, 1187–1206.
Storck, J., Tixier, R. and Uzan, A. (1964). *Nature* **201**, 835.
Sun, F. F. and Crane, F. L. (1969). *Biochim. et Biophys. Acta* **172**, 417–428.
Sutin, N. and Christman, D. R. (1961). *J. Am. Chem. Soc.* **83**, 1773–1774.
Takahashi, K., Titani, K., Furuno, K., Ishikura, H. and Minakami, S. (1958). *J. Biochem.* **45**, 375–378.
Takemori, S., Wada, K., Sekuzu, I. and Okunuki, K. (1962a). *Nature* (London) **195**, 456–457.
Takemori, S., Wada, K., Ando, M., Hosokawa, M., Sekuzu, I. and Okunuki, K. (1962b). *J. Biochem.* **52**, 28–37.
Taube, H. (1955). *J. Am. Chem. Soc.* **77**, 4481–4484.
Tsou, C. L. (1952). *Biochem. J.* **50**, 493–499.

Tsushima, K. and Miyajima, T. (1956). *J. Biochem.* **43**, 761–777.
Ulmer, D. D. (1965). *Biochemistry* **4**, 902–907.
Ulmer, D. D. (1966). *Biochemistry* **5**, 1886–1892.
Ulmer, D. D., Vallee, B. L., Gorchein, A. and Neuberger, A. (1965). *Nature* (London) **206**, 825–826.
Wada, K. and Okunuki, K. (1968). *J. Biochem.* **64**, 667–681.
Wada, K. and Okunuki, K. (1969). *J. Biochem.* **66**, 249–262.
Watanabe, S., Okada, Y. and Kitagawa, M. (1967). *J. Biochem.* **62**, 150–160.
Watt, G. D. and Sturtevant, J. M. (1969). *Biochemistry* **8**, 4567–4571.
Westcott, W. L. and Dickmann, S. R. (1954). *J. Biol. Chem.* **210**, 499–509.
Widmer, C. and Crane, F. L. (1958). *Biochim. et Biophys. Acta* **27**, 203–204.
Williams, G. R. (1963). *Canad. J. Biochem. Physiol.* **41**, 231–237.
Williams, J. N. Jr., and Jacobs, R. M. (1966). *Biochem. Biophys. Res. Commun.* **22**, 395–399.
Williams, J. N. Jr. and Jacobs, R. M. (1968). *Biochim. et Biophys. Acta* **154**, 323–331.
Wosilait, W. D. (1961). *Biochem. Pharmacol.* **7**, 221–225.
Yamanaka, T., Horio, T. and Okunuki, K. (1962). *J. Biochem.* **51**, 426–430.
Yonetani, T. (1968a). "Osaka", pp. 289–292.
Yonetani, T. (1968b). *Science* **159**, 654.
Yonetani, T. (1970). *Adv. Enzymol.* **33**, 309–335.
Yonetani, T. and Schleyer, H. (1968). "Osaka", pp. 535–550.

7. Cytochromes c in various tissues and cells

(a) *In animal organs, tissues and cell organelles.* Many of the cytochromes c have been discussed in preceding sections particularly in V, B1 and V, B2 dealing with their crystallization and amino acid sequence. We repeat here only that the acetyl glycine residue as N-terminal residue in vertebrates is replaced in invertebrates by several amino acids residues terminating in an unacetylated amino acid residue. The present section deals with their distribution in different animals, organs, tissues and cell organelles, with greater stress on their biological function than on their chemical structure, (see also Chapter VIII).

Of all organs, the heart has a higher cytochrome c (and other cytochrome) concentration than other organs; e.g. kidney and liver (Drabkin, 1949) and a much higher one than skeletal muscle (cf. Lemberg and Legge, 1949, p. 377; Akeson et al., 1969) except perhaps in the breast muscle of flying birds (Schollmeyer and Klingenberg, 1962). In pigeon breast muscle they found 60 μmoles/mg, while the white muscle of rabbit and smooth muscle contained only 2 μmoles/mg. Lack of vitamin B_{12} decreased the cytochrome c content of the rabbit myocardium (Nigro et al., 1958). González-Cadavid and Campbell (1967) give the concentration of cytochrome c in rat liver as 10 μmoles/g fresh (123 μg) which roughly corresponds to the 9–12 μmoles reported for cytochromes $(a + a_3)$ by Lemberg (1969). In the adrenal cortex, the cytochrome c concentration is only 15% of that of heart muscle (Spiro and Ball, 1961), that in the adrenal medulla only about 6%, cf. also

Tsou (1951). This is in harmony with the major use of NADPH for steroid hydroxylation by P-450 in the adrenals. The increase of c concentration in the brown adipose tissue of the hibernating hedgehog (Paléus and Johansson, 1968) is only due to the disappearance of water and lipid during hibernation. However, the increase of cytochrome c in the liver of rabbits (not of rats) on starvation appears to be a real increase (Beraud and Vanotti, 1955; Schmidt, 1956).

The intracellular distribution of cytochrome c in rat liver has been studied by González-Cadavid and Campbell (1967) and González-Cadavid et al. (1968). The greater part of cytochrome c was extracted by water at pH 4·0, the remainder by 0·15 M NaCl. This method appears to be preferable to the acid extraction used by Rosenthal and Drabkin (1943) which involves some denaturation. The spectrophotometric method for $a + a_3$, b, c and c_1 of Williams (1964) has not yet been tested sufficiently. González-Cadavid and Campbell found 57% of cytochrome c in the mitochondria, 24% in nuclei, 15% in microsomes, and 3% in the cell sap. However, the values for mitochondria must be considered as minimal, those of nuclei, microsomes and cell sap as maximal, on account of their contamination with mitochondria. In fact it is still doubtful whether rat liver nuclei contain any cytochrome c (Conover and Siebert, 1965; Conover, 1967, 1970; Yamagata and Sato, 1968), although its presence in the nuclei of the thymus of the rat and the calf has been established (Yamagata et al., 1966; Betel, 1967; Ueda et al., 1969). It is probably located in the inner nuclear membrane. As will be described in Chapter IX, cytochrome c is synthesized in the microsomes and later transferred into mitochondria.

The cytochrome c content of the diaphragmic muscle of rats is lowered by exposure of these animals for several months to an atmosphere containing only 10% oxygen (Duckworth, 1960). Castellino (1955) found a decrease of c in chronic CO-poisoning of man. In spite of the fact that the toxic action of CO is mainly due to its competition with oxygen for haemoglobin, the effect on cellular respiration is not entirely negligible and addition of cytochrome c has been found to increase oxygen consumption in brain slices, reversing the effect of CO (Dutkiewicz et al., 1960). Kidder et al. (1964) found cytochrome c 30–50% reduced in the steady state of the secreting bullfrog mucosa; anoxia caused complete reduction and abolition of proton secretion; on reoxygenation c was first completely oxidized before proton transport was re-established. Thiocyanate, which markedly inhibits proton secretion, also caused oxidation of cytochrome c.

The c content in the hearts of the turtle and the frog are low, about 1/5 of that in rat liver (Biörck and Paléus, 1961). In the Bonito fish, Takano et al. (1968) observed two different kinds of cytochrome c in young and old fish, but this appears exceptional and it is not excluded that the difference is due to one of species. The cytochrome c of the muscles of the primitive hagfish (*Myxine glutinosa*) is a typical vertebrate c although it has fewer lysines and thus a slightly lower isoelectric point, and only two histidines, but more arginine (Paléus *et al.*, 1969). The cytochrome c of the protochordate *Styela plicata* has been described by Yamanaka *et al.* (1964b). It is similar to mammalian c, but almost completely precipitable by ammonium sulphate, resembling arthropod cytochromes c.

Cytochromes c have been found in insects as well as in marine invertebrates, prawns, squids, molluscs and worms (Ghiretti *et al.*, 1959; Yamanaka *et al.*, 1964a,b; Yamanaka and Kamen, 1967). Insect and squid cytochrome c is precipitable by saturated ammonium sulphate, but not that of molluscs (Yamanaka *et al.*, 1964a). In insects, the nature of cytochrome c is not altered during metamorphosis, but the amount rises greatly in the last few days of pupal development in the somatic muscles and is thus much higher in the adult (Chan and Margoliash, 1966a,b; Yamanaka *et al.*, 1963b; Shappirio and Williams, 1957) both in the housefly and the *Samia* and *Platysamia* moths. It has also been found in the silkworm moth *Bombyx mori* (Tuppy, 1958; Tuppy and Dus, 1958; Ueda, 1959). In low temperature studies, Estabrook and Sacktor (1958) found besides typical cytochrome c (maxima at 548·5 and 545 nm) a cytochrome c (or $c_{1(?)}$) with maximum 551 nm which could be partially reduced in the presence of antimycin A; Goldin and Farnsworth (1966) also found a cytochrome with maximum at 551 nm, which is assumed to correspond to cytochrome c_1 of mammals, both in larval mitochondria and adult flight muscle sacrosomes of the fruitfly *Dropsophila melanogaster*.

Cytochrome c has also been found in the protozoa *Amoeba proteus*, *Chaos chaos* and *Tetrahymena geleii* (Møller, 1955) and *Orithidia fasciculata* (Kusel *et al.*, 1969), but Yamanaka *et al.* (1968) reported in *Tetrahymena pyroforme* a cytochrome c of $E_0' + 250$ mV, the α-band of which was situated at 553–555 nm. In this organism Turner *et al.* (1969) found an oxidase with Δ ferrous minus ferric maximum at 620, not 605 nm, of high O_2 affinity. Its oxidation, k_1 (first order) = 180/sec. was faster than that of c (120/sec.) and of b (55/sec.).

(b) *In fungi*. The amino acid sequence of *Saccharomyces oviformis* found by Narita *et al.* has been given in Table XB of Chapter V, B 2c (Narita and Titani, 1968). This can be considered as established in spite of the fact (Yaoi *et al.*, 1966) that the order of 6 large peptide sequences has not been rigorously proved (cf. Hiramatsu, 1967); however, the lysine in position 72 of animal cytochromes c is replaced by trimethyl lysine (De Lange *et al.*, 1970) in yeast c as well as that of other fungi. In *Neurospora crassa*, Scott and Mitchell (1969) found a precursor with lysine, not trimethyl lysine in position 72. Yaoi (1967) found no difference in the amino acid sequence of *S. oviformis*, *S. cerevisiae* and some respiration deficient mutants. It should also be noted that this is the structure of the isocytochrome-1 of yeast which forms 95% of the cytochrome c of yeast (Slonimski et al., 1965; Sherman et al., 1965, 1968b), and that most of the data on yeast cytochrome c refer to this isoenzyme. While the properties of yeast cytochrome c greatly resemble those of vertebrate cytochrome c, certain differences have been observed. Yeast c shows wider bands in chromatograms on Amberlite IRC 50 and more diffuse boundaries in paper electrophoresis (Minakami, 1955) and is eluted by lower NaCl concentrations (0·3 g. $Na^+/1$) from Amberlite XE 64 than horse c (Minakami et al., 1956). The redox potentials of yeast and heart c were almost identical in spite of previous claims of differences. The low temperature spectrum shows, in contrast to mammalian c only a single α-band at 546·5 nm (Estabrook, 1956). Armstrong et al. (1961) found the yeast c more easily denatured and with a slightly lower isoelectric point (9·85). The fact that the yeast cytochrome c contains a free cysteine SH-group near the C-terminal allows dimerizations by formation of a disulphide bridge, and the dimer is digested more readily by proteinases (Motonaga et al., 1965a,b). At pH 10, the MORD α-band of yeast cytochrome c is higher in comparison with the absorption band than in other cytochromes (Shashoua and Estabrook, 1966).

In anaerobically grown yeast, cytochrome c disappears during the exponential growth (Heyman-Blanchet and Chaix, 1959; Chaix, 1961; Ohaniance and Chaix, 1964, 1970; Heyman-Blanchet et al., 1964) and only a weak band at 552·5 nm (at $-190°C$) can be discovered, possibly c_1. Conversely, aeration of resting brewers' yeast produces the cytochrome spectrum of bakers' yeast (Chin, 1950). Yeast cytoplasmic mutants (ρ^-; petite vegetative) have a deficiency of all the cytochromes except c (Sherman and Slonimski, 1964); but see Mackler et al., (1965) who find c, c_1 and b only less decreased than aa_3; in other mutants (p and qy_1) the cytochrome deficiency appears to be more variable and less severe. These contain both isocytochromes-1 and -2 in the same proportion (Sels, 1966b), but synthesis of iso-2 in the presence of oxygen depends on the presence of its apoenzyme before oxygenation (Sels et al., 1966). In contrast to these are mutants of the CY_1 genetic locus coding for isocytochrome-1 which are deficient only in this, not in other cytochromes, not even iso-2 c. Iso-2 cytochrome c is of importance for the regulation of the synthesis of iso-1 c (Slonimski, 1953, 1960; Sels, 1964; Slonimski et al., 1965; Sherman, 1964, 1967; Margoliash, 1966; Sherman et al., 1966, 1968a,b; Clavilier et al., 1969).

The two isocytochromes c can be separated by chromatography on Amberlite XE64 (Sels et al., 1965; Sels, 1966a). They differ in their primary structure in the position of 20 amino acid residues, and are therefore quite different from the "isocytochromes c" of Flatmark, the existence of which in mammalian cells without secondary alterations has not yet been established (see V, B6b). The primary structure of iso-2-cytochrome c is known except for four residues (Stewart et al., 1966). Mirsky and George (1967) found iso-1 c^{2+} not as stable at pH 10 as horse c^{2+} or iso-2 c^{2+}; the differences between c^{2+} and c^{3+} observed in mammalian cytochrome c are also found in iso-2, but not in iso-1 c. Iso-2 c lacks the His-26 which is replaced by Asn but this does not affect the ORD spectrum (see also Stewart et al., 1966).

The differences in the amino acid composition and sequence of the cytochrome c from the fungi *Saccharomyces*, *Debaromyces*, *Candida* and *Neurospora* have been discussed in sections V, B2b and c (cf. also Yamanaka et al., 1964c). A large number (104) of different strains of yeast from 14 different genera have been classified by Haneishi and Shirasaka (1968) according to their affinity to Amberlite CG 50, their immunological reactivity with *Saccharomyces oviformis* and *Candida krusei* antisera, and their activities with yeast (*S. oviformis*) (see also Motonaga and Nakanishi, 1965), bovine and *Pseudomonas* oxidases. Group 4, with the weakest affinity to the cationic resin and the slowest activity with the yeast oxidase, resembled some algal and bacterial cytochromes c. The cytochrome c of *Candida krusei* (Shirasaka et al., 1968) was oxidized by yeast (*S. oviformis*) aa_3 oxidase much faster than horse cytochrome c, whereas both cytochromes were oxidized at about the same rate by mammalian aa_3 oxidase. *Pichia polymorpha c* has a redox potential of + 293 mV (Levchuk et al., 1967). Cytochrome c of *Torula utilis* has been studied by Lang and Yonetani (1968). A small magnetic field changed its typical ferric cytochrome spectrum to one similar to ferrihaemoglobin by decoupling nuclear and electronic spins. *Aspergillus oryzae c* (Yamanaka et al., 1963a) which does not react with aa_3 oxidase and only slowly with the oxidase of *Pseudomonas*, is thus rather different from yeast c. It has an acidic isoelectric point and is not adsorbed by Amberlite CG 50 at pH 7. At this pH, near its isoelectric point, the somewhat similar cytochrome c from *Ustilago sphaerogena*, the rust fungus, is still adsorbed to the resin (Neilands, 1952). This cytochrome can be extracted from the fungus by dilute alkali (pH 10–11) and is precipitated by saturated ammonium sulphate. Zviagilskaya et al. (1965) describe the cytochromes of *Endomyces magnusii*. The slime mould, *Physarum polycephalum* contains a cytochrome c (Yamanaka et al., 1962) which is rapidly oxidized by aa_3, but not by *Pseudonomas* oxidase, and which is also easily precipitated by saturated ammonium sulphate. Margolit and Schejter (1970) have compared the thermodynamics of the redox-reaction of two fungal cytochromes c (of bakers' yeast and *Candida*) with those of horse, turkey and tuna cytochromes c.

(c) *Mitochondrial. Cytochromes c in plants*. The cytochromes c of plant mitochondria apparently closely resemble those found in animals and yeast. The earlier literature is well assembled by Hill and Hartree (1953) and James (1957). There is also some, though not yet quite definitive, evidence for the presence of cytochrome c_1 (see below) (Lundegardh 1958, 1959, 1962a in wheat roots; James and Leech, 1960; Wiskich et al., 1960; Lance and Bonner, 1968). Its presence may cause the fact that the α-band has often been found in room temperature studies at a slightly longer wavelength (e.g. 552 nm in barley roots or skunk cabbage mitochondria) than that of purified cytochrome c. In low temperature studies the α-band is found at 547–548 nm (Bonner and Voss, 1961; Bonner, 1962; Lance and Bonner, 1968) while a second band at 552 nm is probably that of cytochrome c_1. In whole cells, however, a low temperature maximum at 552 nm may be caused by the presence of a cytochrome b (Bonner and Plesnicar, 1967).

The primary structure of some plant cytochromes c has been described in Chapter V, B2, their crystallization in V, B1 (see also Wasserman et al., 1963). Apparently all higher plant cytochromes c differ from animal ones by having two trimethyl lysine residues instead of lysine in position 72 and 86, whereas all fungal cytochromes c have only one trimethyl lysine residue in 72 (De Lange et al., 1970).

Wheat cytochrome c interacts with bovine aa_3 oxidase somewhat less rapidly than does bovine c, and reacts only slowly with *Pseudomonas* oxidase (Yamanaka and Okunuki, 1964). Algal cytochromes of type c, however, do not react with aa_3 and also only slowly with *Pseudomonas* oxidase, so that it appears that the algal cytochromes closely resemble cytochromes f of higher plants and that their function is photosynthetic rather than respiratory (Yamanaka and Okunuki, 1968).

Respiratory cytochromes also occur in algae, but at present the data are scanty. To them probably belong α absorption bands at 549 to 551 nm often accompanying bands at 552–555 nm (see below) in several algae. It cannot, however, always be decided whether such a cytochrome is a respiratory cytochrome c or a photosynthetic type cytochrome (see Chapter V, C). To this class belong the cytochrome c-549·5 of the green alga *Ulva*, a c-551·5 of the red alga *Polysiphonia* (Sugimura et al., 1968) and c-550 in a *Chlamydomonas* mutant which was rapidly oxidized by cytochrome aa_3 (Smillie and Levine, 1963). Other cytochromes c absorbing in this region, appear to be unusual. The diatom *Navicula pelliculosa* has, in addition to a type f cytochrome c-554, a c-550 of M.W. 34,000 which was autoxidizable and combined with CO and CN (Yamanaka and Kamen, 1965, 1967; Yamanaka et al., 1967). Still more unusual and perhaps closer to bacterial c_3 cytochromes, was c-549 of the blue-green alga *Anacystis nidulans*. This had a low E_0' of -260 mV, was highly autoxidizable and combined with CO, but not with CN (Holton and Myers, 1967a,b; Yamanaka et al., 1969). It had a low α-peak and a high γ/α ratio (7·7) unusual for respiratory cytochromes. It acted as NADH oxidase with a terminal oxidase probably of type o (see Chapter VI) not type aa_3, and was little inhibited by azide (Horton, 1968). This cytochrome formed a CO compound but did not appear to belong to c' cytochromes (see Chapter VI). It was more sensitive to high than to low pH and heat-labile, denatured at 60°.

The cytochrome c-556 of dark-grown *Euglena gracilis* with absorption bands at 556, 525 and 421 nm which accompanied the c-552 (A-type) (Nishimura, 1959; Colmano and Wolken, 1963; Perini *et al.*, 1964) had been considered as an f-type cytochrome closely related to c-552 by Ikegami *et al.* (1963) who postulated that it reacted close to c-552 in the photosynthetic electron chain. It has a somewhat lower redox potential (E_0' + 307 mV) according to these authors but + 370 mV according to Wolken and Gross, (1963), than that of c-552 (+ 380 mV). Perini *et al.* (1964) found this cytochrome autoxidizable and combining with CO. As its pyridine haemochrome had its α-band at 552·5 nm not at 550 nm, these authors suggested that one vinyl group of the haem may be unsaturated and uncombined and only one covalently bound by a thioether linkage. The pyridine haemochrome α-band of c-552·2 was also found at 553 nm (Nishimura, 1959). In more recent experiments, Raison and Smillie (1969) found it not autoxidizable and unable to combine with CO. They consider it as a respiratory cytochrome c which conveys electrons from cytochrome b to aa_3; in the steady state it is 22% reduced while cytochrome a is only 5% and cytochrome b 92% reduced.

C. CYTOCHROMES F AND SIMILAR ALGAL CYTOCHROMES

1. Cytochrome f.

Cytochrome f was isolated in a state approaching purity by Hill and Scarisbrick (1951) and by Davenport and Hill (1951/2) from parsley leaves, and named f from frons (Latin) for leaf. It is found only in the thylakoids of chloroplast to which it is firmly bound to their structure so that treatment with organic solvents (e.g. 50% ethanol) is required for its release. Its close relationship to cytochrome c was recognized by these authors (cf. Chapter II). In fact it has been suggested that cytochrome f may also be named c_6 (Nomenclature Commission Reports 1961 and 1965). Since, however, some doubts have arisen as to the justification of the names cytochrome c_4 and c_5 from *Azotobacter* (see the IUPAC-IUB report and Chapter VI), we retain the name cytochrome f at present. Chemically cytochrome f undoubtedly belongs to the family of cytochromes c. It contains the covalently linked prosthetic group, is not autoxidizable and does not react with CO.

Its preparation by the method of Davenport and Hill has been slightly modified by Forti *et al.* (1963) who use 1% Triton X-100 in the extraction medium. The principles of the method are, however, little altered and rely on extraction with about 50% ethanol, precipitation with cold (− 20°C) acetone, adsorption to Kieselguhr, elution with aqueous Tris-phosphate pH 8, and precipitation with ammonium sulphate at about 45% saturation.

There are some notable differences between cytochromes f and c, although it cannot yet be entirely excluded that some of them may be due to the mode of isolation. Cytochrome f is less heat-stable and is denatured at 58°C. Its pH stability range is also smaller; at pH 5 it begins to show a charge-transfer band and to react with CO. As f^{2+} it shows absorption bands at 555–554, 524, 422 and 430, as f^{3+} at 530, 410, 350 (weak) and 280 nm. The low temperature spectrum (Hill and Bonner, 1961) shows a double-peak at 552 and 548 nm. The ratio $E f^{2+} 442/E f^{3+} 280$ was found 2·0 by Hill and Davenport. Forti et al. (1965) give 2·80 and find $\Delta\epsilon$mM $(f^{2+} - f^{3+})$ 19·7. Although the α-band of f^{2+} was found by Hill et al. to be narrower than that of c^{2+}, it has a shoulder at about 552 nm and slopes more gradually towards shorter wavelengths. This may explain its slightly lower molar extinction.

The isoelectric point at 4·7 is quite different from that of most cytochromes c; in contrast to them it is an acidic protein. The molecular weight was found by Davenport and Hill (1952) 110,000 on the basis of sedimentation studies, with two prosthetic haem groups per molecule. More recently, Forti et al. (1965) have found that chromatography on Sephadex G 200 columns gives a M.W. of 245,000 with 4 haem groups (minimum M.W. per haem 63,000). Some of its properties may be due to this polymerization, but the shape of the α-band and the redox potential appears to be the same in cytochrome f and similar monomer algal cytochromes.

The E_0' at pH 7 (+ 365 mV) is higher than that of cytochrome c. The E_0' is independent of pH near neutrality, but shows a pK of 8·4 when the redox potential changes with pH $(E_0'/\Delta pH - 0·06)$. In pea and barley chloroplasts Bendall (1968) reported $E_0' + 350$ mV. Cytochrome f and the similar algal cytochromes are not oxidized by cytochrome aa_3 without the addition of cytochrome c.

2. Algal cytochromes of type f.

Cytochrome f was found by Davenport and Hill also in the four algal species (*Ulva*, *Vaucheria*, *Fucus serratus* and *Euglena*). However, in spite of similarities, the algal cytochromes of this type are extractable with aqueous solvents (Katoh and San Pietro, 1967; Ikegami et al., 1968), and there are some other differences. A large number of these algal cytochromes (17 from red algae, 10 from green, 4 from brown), some of which have been isolated well crystallized (see Chapter V, B1) have been described by Sugimura et al. (1968). The position of their absorption bands at 552·5–554·5 and γ-bands 415–418 nm, their high redox potential + 340 to 365, in some green algae as high as 390 mV) and their inability to react with cytochrome

aa_3 (Yamanaka and Okunuki, 1963, 1968) show that they are algal cytochromes f or closely related to them. Often the α-band shows the short wavelength shoulder also found in cytochrome f, but in some algal cytochromes (*Caulerpa, Euglena*) this may be missing. Readiness to crystallize appears once the purification has proceeded to E c^{2+}/E c^{3+} 280 of near 1.

Other algae in which cytochromes of this type have been found are *Vaucheria* (Sugimura et al., 1968), *Chlorella vulgaris* (Hill et al., 1953; Duysens, 1954), *Chlamydomonas* (Chance and Sager, 1957; Smillie and Levine, 1963; Gorman and Levine, 1965, 1966; Schuldiner and Okad, 1969; Hildreth, 1968; Hiyama et al., 1969), *Scenedesmus* (Kok, 1957; Powls et al., 1969), *Euglena* (Nishimura, 1959; Gross and Wolken, 1960; Colmano and Wolken, 1963; Olson and Smillie, 1963; Perini et al., 1964; Katoh and San Pietro, 1967; Ikegami et al., 1968; Mitsui and Tsushima, 1968; Raison and Smillie, 1969); of diatoms *Navicula pelliculosa* (Yamanaka and Kamen, 1965, 1966; Yamanaka et al., 1967); of brown algae, *Petalonia fascia* (Sugimura and Yakushiji, 1968); of red algae, *Porphyra* sp. (Yakushiji, 1935; Yakushiji et al., 1959; Duysens, 1955; Katoh, 1959a,b (crystalline), 1960; Nishimura and Takamiya, 1966; Nishimura, 1968; Mitsui and Tsushima, 1968), *Porphyridium cruentum* (Duysens, 1955; Duysens et al., 1961, 1962; Duysens and Amesz, 1962; Nishimura, 1966, 1968; Amesz et al., 1970); of blue-green algae *Anabaena cylindrica* or *variabilis* (Fujita and Myers, 1965a,b, 1966, 1967; Susor and Krogmann, 1966; Lightbody and Krogmann, 1967; Horton, 1968; Leach and Carr, 1969; Lee et al., 1969), *Phormidium luridum* (Biggins, 1967); *Anacystis nidulans* (Amesz and Duysens, 1962; Vredenberg and Duysens, 1965; Yamanaka et al., 1967, 1969; Holton and Myers, 1967a; Horton, 1968). Molecular weights of 10,500–13,500 and for *Anacystis* f 23,000 have been reported but it is not always clear whether these are minimum molecular weights as for *Anacystis* or true molecular weights. The γ/α ratio is frequently high (about 7) but occasionally lower (Sugimura et al., 1968).

Some of the algae show more than one cytochrome c α-band e.g. Phormidium (c-554 and c-549) and *Euglena* 558–556 besides 552 nm. *Euglena* c-552 has the typically high E_0' of cytochromes f (+ 300 mV), but has a lower γ/α band ratio, 5·6 than most of them; its isoelectric point is 5·5. The c-556 cytochrome has been discussed in V, B. There are three α-bands (552, 551, 549 nm) in *Scenedesmus* and three (554, 552, 549) in *Anacystis*. Whereas c-554 of *Anacystis* is a rather typical cytochrome of type f, the c-549 of low E_0' is perhaps a bacterial type respiratory one (see above). Cytochrome f is present even in dark-grown cells of *Chlamydomonas* but is not coupled to the photosystems; it becomes, however, rapidly coupled to them in 2 hrs of greening (Schuldiner and Okad, 1958). A similar observation on Mung bean leaves has been made by Hill and Bonner (1961).

Cytochrome f and plastocyanin are loosely bound to the lamellar structure in green plants and green algae (Fujita and Myers, 1966) but also in blue-green algae (Holton and Myers, 1967a).

The role of cytochromes f in the photosynthesis of higher plants and algae. Much of the early work on photosynthesis and the role of cytochromes in photosynthesis was carried out with algae, e.g. *Chlorella*, e.g. the work of Emerson (1957) on the two photosystems. In the red alga *Porphyridium cruentum*, Duysens, *et al.* (1961) demonstrated the oxidation of cytochrome f by light of photosystem I and its reduction by photosystem II. Chlorophyll-deficient mutants of *Chlamydomonas* and *Scenedesmus* have been of importance for the study of the role of cytochromes in photosynthesis. Photooxidation of cytochrome f in *Chlorella* was demonstrated by Witt *et al.* (1961) and Rumberg *et al.* (1962) when photosystem I alone was excited, or when photosystem II was separated from photosystem I by low temperature ($-150°C$) (but see Boardman *et al.*, 1971) or by the use of aged chloroplasts.

In the hypotheses discussed in Chapter IV, F3, cytochrome f was usually postulated as closest reductant of photooxidized P-700 in system I and as involved in the cyclic electron transport of this system, but it was also assumed to form the bridge between this and the electron transfer chain connecting it with photosystem II. Indeed rapid oxidation of cytochrome f at low temperature was found by Avron and Chance (1966), Hildreth *et al.*, (1966), Hildreth (1967) and Chance *et al.* (1967). However, the laser-light used (660 nm) stimulates both photosystems and it has now become uncertain whether some observations refer in fact to cytochrome f or to cytochrome b-559 (Floyd *et al.*, 1971). In the more recent theory of Knaff and Arnon (1969a), cytochrome f is assumed to act only in the cyclic electron system I, while plastocyanin is assumed to bridge the two photosystems IIa and IIb (see however Avron and Shneyour, 1971). The findings of Levine and Gorman (Gorman and Levine 1965, 1966; Levine and Gorman, 1966; Levine *et al.*, 1967; Hind, 1968; Levine, 1969) support that plastocyanin rather than cytochrome f is directly oxidized by photosystem I. This is largely based on experiments with pale green *Chlamydomonas* mutants some of which lack either plastocyanin or cytochrome f. Weaver (1969) found, however, the electron transport chain linking photosystems II and I interrupted in a mutant which lacked cytochrome f. In *Anabaena variabilis* Lightbody and Krogman (1967) found cytochrome f able to replace plastocyanin and found the photooxidation of f^{2+} not to depend on the presence of plastocyanin. They suggest that both act in parallel as did Kok *et al.* (1964) and Elstner *et al.* (1968). Without a role of cytochrome f in connecting the two photosystems it would be more difficult to explain the frequently observed photoreduction of cytochrome f by short λ red light (system II) counteracting the photooxidation by long λ red light

(Amesz and Duysens 1962; Duysens et al., 1961; Duysens and Amesz, 1962; Goedheer, 1963; Kelly and Sauer, 1965; Lundegårdh, 1964, 1965; Avron and Chance, 1966; Cramer and Butler, 1967; Nishimura, 1967, 1968), also in mutants of *Oenothera*, lacking photosystem II (Fork and Heber, 1968) with system II. Some connection between cytochrome f is further supported by observations of Fujita and Myers (1965a), Hind and Olson (1966), Katoh and San Pietro (1967a), Anderson and McCarty (1969) and Lee et al. (1969), whereas Raurainski et al. (1970) prefer a more indirect relationship, at least in whole algal cells. Thus cytochrome f is, together with plastocyanin, still left in the electron transfer chain combining the two photosystems in the latest scheme of Boardman et al. (1971) (see Chapter IV, F3). The role of plastocyanin, a copper protein discovered by Katoh (1960) has been more fully discussed by Kok and Rurainski (1965), Boardman (1968, p. 38), Hind (1968), Levine (1969), Anderson and McCarty (1969) and Marsho and Kok (1970). Its role as a copper-protein has been compared with the role of copper in cytochrome oxidase, but to the writers this comparison appears to have no more than possibly an evolutionary significance.

Yamamoto and Vernon (1969) have isolated a partially purified photosynthetic reaction centre I containing one P-700 per 30 moles of chlorophyll a and small amounts of cytochromes f and b_6. The major part of the cytochromes could be removed by detergent.

3. Models of photosynthesis.

Although the cytochromes of type c active in photosynthesis differ from respiratory cytochrome c, the latter has frequently been used in model experiments (Niemann and Vennesland, 1957, 1959; Niemann et al., 1959; Kok et al., 1963, 1964; Ke et al., 1965; Fujita and Myers, 1965b, 1966, 1967; Fujita and Murano, 1967; Krasnovskii and Mikhailova, 1969; Kelly and Porter, 1970). Digitonin extracts of chloroplasts contain a cytochrome c photooxidase (Niemann and Vennesland, 1957, 1959) which is heat-sensitive, but insensitive to cyanide and similar to the bacterial cytochrome c photooxidase of bacteria of Kamen and Vernon (Chapter VI); it consists of a chlorophyll bound to lipoprotein, soluble in digitonin, but not in water (photosystem I) and a water-soluble cytochrome c-reducing factor (CRF) which appears to be neither ferredoxin nor any of the cytochromes. It resembles plastoquinone, but differs from it by being not extractable by petroleum ether and tightly bound to the lamellae. Its amount is about one fifth of the chlorophyll. Galactolipids stimulate the photoreduction of c^{3+} which is inhibited by DCMU, an inhibitor of photosystem II (Chang and Lundin, 1965). Trimethyl-p-benzoquinone is able to photoreduce c^{3+} by photosystem I (Vernon and Shaw, 1965; Kelly and Sauer, 1965, 1968). This differs from the Hill reaction of chloroplasts which is stimulated by photosystem II (Sauer and Park, 1965). Vernon et al. (1965) and Tu and Wang (1969) observed a chlorophyllin-sensitized photoreduction of NADP⁺ by cytochrome c. For each molecule of NADP⁺ reduced, two c^{2+} were oxidized with an average quantum efficiency of about 5%.

Somewhat more distant from the reactions in the chloroplasts are chlorophyll-sensitized oxidation and reduction of haematin in pyridine (Brody et al., 1968). The photochemical reduction stimulated by red and green light is faster than the dark reduction. In the presence of air, chlorophyll also sensitizes the photooxidation of pyridine haemochrome. Insufficient attention has perhaps been paid to the water content of the system (cf. Lemberg, 1963 and Lemberg and Velins, 1965) which affects the stabilization in the ferrous or ferric state. Krasnovskii (1960) has reviewed the oxidation and reduction of chlorophyll and chlorophyll derivatives by a great variety of reactants.

Finally Kassner and Kamen (1967) and Eisenstein and Wang (1969) report a haematoporphyrin-sensitized photoreduction of ferredoxin by hydrogen donors, e.g. reduced glutathione. After illumination the system slowly returns to its original state in the dark. Spectrophotometric and EPR data show that an electron from GSH is first captured by excited haematoporphyrin to form a porphyrin-radical which then transfers its extra electron to ferredoxin.

REFERENCES

Åkeson, Å., Biörck, G. and Simon, R. (1969). *Acta Med. Scand.* **185**, 287–292.
Amesz, J. and Duysens, L. N. M. (1962). *Biochim. et Biophys. Acta* **64**, 261–278.
Amesz, J., Ven den Bos, P. and Dirks, M. P. (1970). *Biochim. et Biophys. Acta* **197**, 324–327.
Anderson, M. M. and McCarty, R. E. (1969). *Biochim. et Biophys. Acta* **189**, 193–206.
Armstrong, J. McD., Coates, J. H. and Morton, R. K. (1961). "Haematin Enzymes", pp. 385–388.
Avron, M. and Chance, B. (1966). 2nd W.-European Conference on Photosynthesis, 1965, Donker, Rotterdam.
Avron, M. and Shneyour, A. (1971). *Biochim. et Biophys. Acta* **226**, 498–500.
Barrett, J. and Sinclair, P. (1967). *Biochim. et Biophys. Acta* **143**, 279–281.
Bendall, D. S. (1968). *Biochem. J.* **109**, 46P.
Beraud, T. and Vannotti, A. (1955). *Schweiz. med. Wochschr.* **85**, 174–178.
Betel, T. (1967). *Biochim. et Biophys. Acta* **143**, 62–69.
Biggins, J. (1967). *Plant Physiol.* **42**, 1442–1446.
Biörck, G. and Paléus, S. (1961). *Nature* (London) **191**, 712–713.
Boardman, N. K. (1968). *Advances in Enzymol.* **30**, 1–79.
Boardman, N. K., Anderson, J. M. and Hiller, R. G. (1971). *Biochim. et Biophys. Acta* **234**, 126–136.
Bonner, W. D. (1962). *Fed. Proc.* **21**, p. 50.
Bonner, W. D. and Plesnicar, M. (1967). *Nature* **214**, 616–617.
Bonner, W. D. and Voss, D. O. (1961). *Nature* (London) **191**, 682–684.

Brody, M., Broyde, S. B., Yeh, C. C. and Brody, S. S. (1968). *Biochemistry* **7**, 3007–3015.
Castellino, N. (1955). *Folia Med.* (Naples) **38**, 838–850.
Chaix, P. (1961). "Haematin Enzymes", Vol. I, pp. 225–234.
Chan, S. K. and Margoliash, E. (1966a). *J. Biol. Chem.* **241**, 335–348.
Chan, S. K. and Margoliash, E. (1966b). *J. Biol. Chem.* **241**, 2252–2255.
Chance, B. and De Vault, D. (1964). *Ber. Bunsenges, Phys. Chem.* **68**, 722–726.
Chance, B. and Nishimura, M. (1960). *Proc. Nat. Acad. Sci.* **46**, 19–24.
Chance, B. and Sager, R. (1957). *Plant Physiol.* **32**, 548–561.
Chance, B., Schleyer, H. and Legallais, V. (1963). Plant and Cell Physiol. (Special Issue) pp. 337–346.
Chance, B., De Vault, D., Hildreth, W. W., Parsons, W. W. and Nishimura, M. (1966). *Brookhaven Sympos. Biol.* **19**, 115–125.
Chang, S. B. and Lundin, K. (1965). *Biochem. Biophys. Res. Commun.* **21**, 424–431.
Chin, H. C. (1950). *Nature* (London) **165**, 926–927.
Clavilier, L., Péré, G. and Slominski, P. P. (1969). *Molec. Genetics* **104**, 195–218.
Colmano, G. and Wolken, J. J. (1963). *Nature* (London) **198**, 783–784.
Conover, T. E. (1967). Abstracts *V*, 7th Internat. Congress, Tokyo, H–73.
Conover, T. E. (1970). *Arch. Biochem. Biophys.* **136**, 541–550.
Conover, T. E. and Siebert, G. (1965). *Biochim. et Biophys. Acta* **99**, 1–12.
Cramer, W. A. and Butler, W. L. (1967). *Biochim. et Biophys. Acta* **143**, 332–339.
Davenport, H. E., Hill, R. and Whatley, F. R. (1952). *Proc. Roy. Soc. London B* **139**, 327–345, 346–358.
De Lange, R. J., Glazer, N. and Smith, E. L. (1970). *J. Biol. Chem.* **245**, 3325–3327.
Discussions on Photosynthesis (H. K. Porter and R. Hill, eds.) (1963). *Proc. Roy. Soc. London B* **157**, 291–382.
Drabkin, D. L. (1949). In "Haemoglobins" (F. J. W. Roughton, and J. C. Kendrew, eds.) Butterworths Scientific Publications, London, pp. 47–51.
Duckworth, M. W. (1960). *J. Physiol.* **152**, 55–56P.
Dutkiewicz, J. S., Gwodz, B., Spett, K. and Spiock, F. (1960). *J. Physiol.* **152**, 482–486.
Duysens, L. N. M. (1954). *Science* **120**, 353–354.
Duysens, L. N. M. (1955). *Science* **121**, 210–211.
Duysens, L. N. M. (1963). *Proc. Roy. Soc. London B* **157**, 301–373.
Duysens, L. N. M. (1966). *Brookhaven Sympos. Biol.* **19**, 71–80.
Duysens, L. N. M. and Amesz, J. (1962). *Biochim. et Biophys. Acta* **64**, 243–260.
Duysens, L. N. M., Amesz, J. and Kamp, B. M. (1961). *Nature* (London) **190**, 510–511.
Eisenstein, K. K. and Wang, J. K. (1969). *J. Biol. Chem.* **244**, 1720–1728.
Elstner, E., Pistorius, E., Broeger, P. and Trebst, A. (1968). *Planta* **79**, 146–161.
Emerson, R., Chalmers, R. and Cederstrand, C. (1957). *Proc. Nat. Acad. Sci.* **43**, 133–143.
Epel, B. L. and Butler, W. L. (1970). *Plant Physiol.* **45**, 728–734.
Estabrook, R. W. (1956). *J. Biol. Chem.* **223**, 781–794.
Estabrook, R. and Sacktor, B. (1958). *Arch. Biochem. Biophys.* **76**, 532–545.
Floyd, R. A., Chance, B. and Devault, D. (1971). *Biochim. et Biophys. Acta* **226**, 103–112.
Fork, D. C. and Heber, U. W. (1968). *Plant Physiol.* **43**, 606–612.
Forti, G., Bertolè, M. L. and Parisi, B. (1963). *Biochem. Biophys. Res. Commun.* **10**, 384–389.
Forti, G., Bertolè, M. L. and Zanetti, G. (1965). *Biochim. et Biophys. Acta* **109**, 33–40.

Fujita, Y. and Murano, F. (1967). *Plant Cell Physiol.* (Tokyo) **8**, 269–281. CA **67**(5), 41046.
Fujita, Y. and Myers, J. (1965a). *Biochem. Biophys. Res. Commun.* **19**, 604–608.
Fujita, Y. and Myers, J. (1965b). *Arch. Biochem. Biophys.* **112**, 519–523.
Fujita, Y. and Myers, J. (1966). *Arch. Biochem. Biophys.* **113**, 730–737, 738–741; *Plant Cell Physiol.* **7**, 599–606.
Fujita, Y. and Myers, J. (1967). *Arch. Biochem. Biophys.* **119**, 8–15.
Ghiretti, F., Ghitertti-Magaldi, A. and Tosi, L. (1959). *J. Gen. Physiol.* **42**, 1185–1205.
Goedheer, J. C. (1963). *Biochim. et Biophys. Acta* **66**, 61–71.
Goldin, H. H. and Farnsworth, M. W. (1966). *J. Biol. Chem.* **241**, 3590–3594.
González-Cadavid, N. F. and Campbell, P. N. (1967). *Biochem. J.* **105**, 427–442.
González-Cadavid, N. F., Bravo, M. and Campbell, P. N. (1968). *Biochem. J.* **107**, 523–529.
Gorman, D. S. and Levine, R. P. (1965). *Proc. Nat. Acad. Sci.* **54**, 1665–1669.
Gorman, D. S. and Levine, R. P. (1966). *Plant Physiol.* **41**, 1643–1647, 1648–1656.
Gross, J. A. and Wolken, J. J. (1960). *Science* **132**, 537–538.
Haneishi, T. and Shirasaka, M. (1968). "Osaka", p. 404–413.
Heyman-Blanchet, T. and Chaix, P. (1959). *Biochim. et Biophys. Acta* **35**, 85–93.
Heyman-Blanchet, T., Ohaniance, L. and Chaix, P. (1964). *Biochim. et Biophys. Acta* **81**, 462–472.
Hildreth, W. W. (1967). *Fed. Proc.* **26**, Abstr. 2648.
Hildreth, W. W. (1968). *Plant Physiol.* **43**, 308–312.
Hildreth, W. W., Avron, M. and Chance, B. (1966). *Plant Physiol.* **41**, 983–991.
Hill, R. and Bonner, W. D. (1961). In (McElroy and Glass, eds.) "Light and Life", Johns Hopkins Press, Baltimore, p. 424.
Hill, R. and Hartree, E. F. (1953). *Ann. Rev. Plant. Physiol.* **4**, 115–150.
Hill, R. and Scarisbrick, R. (1951). *The New Phytologist* **50**, 98–111.
Hill, R., Northcote, D. W. and Davenport, H. E. (1953). *Nature* (London) **172**, 948.
Hind, G. (1968). *Biochim. et Biophys. Acta* **153**, 235–240.
Hind, G. and Olson, J. M. (1966). *Brookhaven Sympos. Biol.* **19**, 188–194.
Hiramatsu, A. (1967). *J. Biochem.* **62**, 364–372.
Hiyama, T., Nishimura, M. and Chance, B. (1969). *Plant Physiol.* **44**, 527–534.
Hiyama, T., Nishimura, M. and Chance, B. (1970). *Plant Physiol.* **46**, 163–168.
Holton, R. W. and Myers, J. (1967a). *Biochim. et Biophys. Acta* **131**, 362–374.
Holton, R. W. and Myers, J. (1967b). *Biochim. et Biophys. Acta* **131**, 375–384.
Horton, A. A. H. (1968). *Biochem. Biophys. Res. Commun.* **32**, 839–845.
Ikegami, I., Katoh, S. and Takamiya, A. (1968). *Biochim. et Biophys. Acta* **162**, 604–606.
Jagendorf, A. T. and Hind, G. (1963). In "Photosynthetic Mechanism of Green Plants" (B. Kok and A. T. Jagendorf, eds.) Nat. Acad. Sci.—Nat. Res. Council Publ., Washington, D.C., Publ. 1145.
James, W. O. (1957). *Advances in Enzymol.* **18**, 218–318.
James, W. O. and Leech, R. M. (1960). *Endeavour* **19**, 108–114.
Kassner, R. and Kamen, M. D. (1967). *Proc. Nat. Acad. Sci.* **58**, 2445–2449.
Katoh, S. (1959a). *J. Biochem.* **46**, 629–632.
Katoh, S. (1959b). *Plant Cell Physiol.* **1**, 29–38, 91–98.
Katoh, S. (1960). *Nature* (London) **186**, 138–139.
Katoh, S. and San Pietro, A. (1967a). *Arch. Biochem. Biophys.* **118**, 488–496.
Katoh, S. and San Pietro, A. (1967b). *Arch. Biochem. Biophys.* **121**, 211–219.
Ke, B., Vernon, L. P. and Shaw, E. R. (1965). *Biochemistry* **4**, 137–144.
Kelly, A. B. and Porter, G. (1970). *Proc. Roy. Soc.* (London) A **315**, 149–161.

Kelly, J. and Sauer, K. (1965). *Biochemistry* **4**, 2798–2802.
Keylock, M. J., Kirk, J. T. P. and Rogers, L. J. (1971). *Biochem. J.* **121**, 14P.
Kidder, G. W. III, Curran, P. F. and Rehm, W. S. (1964). *Fed. Proc.* **23**, p. 80.
Knaff, D. B. and Arnon, D. I. (1969a). *Proc. Nat. Acad. Sci. U.S.* **63**, 956–962.
Knaff, D. B. and Arnon, D. I. (1969b). *Proc. Nat. Acad. Sci. U.S.* **64**, 715–722.
Kok, B. (1957). *Nature* (London) **179**, 583–584.
Kok, B. and Rurainski, H. J. (1965). *Biochim. et Biophys. Acta* **94**, 588–590.
Kok, B., Hoch, G. and Cooper, B. (1963). *Plant Physiol.* **38**, 274–279.
Kok, B., Rurainski, H. J. and Harmon, E. A. (1964). *Plant Physiol.* **39**, 513–520.
Krasnovskii, A. A. (1960). *Am. Rev. Plant Physiol.* **11**, 313–410.
Krasnovskii, A. A. and Mikhailova, E. S. (1969). *Doklady* **185**, 938–941.
Kusel, J. P., Suriano, J. R. and Weber, M. W. (1969). *Arch. Biochem. Biophys.* **133**, 293–304.
Lance, C. and Bonner, W. D. (1968). *Plant Physiol.* **43**, 756–766.
Lang, G. and Yonetani, T. (1968). "Osaka", p. 551.
Leach, K. C. and Carr, N. G. (1969). *Biochem. J.* **112**, 125–126.
Lee, S. S., Young, A. M. and Krogmann, D. W. (1969). *Biochim. et Biophys. Acta* **180**, 130–136.
Lemberg, R. (1963). *Biochem. Z.* **338**, 97–105.
Lemberg, R. (1969). *Physiol. Revs.* **49**, p. 94.
Lemberg, R. and Legge, J. W. (1949). "Hematin Compounds and Bile Pigments", p. 377
Lemberg, R. and Velins, A. (1965). *Biochim. et Biophys. Acta* **104**, 487–495.
Levine, R. P. (1969). *Ann. Rev. Plant. Physiol* **20**, 523–540.
Levchuk, T. P., Kolesnik, L. V. and Losinov, A. B. (1967). *Mikrobiologiya* **36**, 233–239.
Levine, R. P. and Gorman, D. S. (1966). *Plant Physiol.* **41**, 1293–1300.
Levine, R. P., Gorman, D. S., Avron, M. and Butler, W. L. (1966). *Brookhaven Sympos. Biol.* **19**, 143–147.
Lightbody, J. J. and Krogmann, D. W. (1967). *Biochim. et Biophys. Acta* **131**, 508–515.
Lundegårdh, H. (1959a). *Biochim. et Biophys. Acta* **27**, 355–365.
Lundegårdh, H. (1959b). *Biochim. et Biophys. Acta* **35**, 340–353.
Lundegårdh, H. (1962). *Biochim. et Biophys. Acta* **57**, 352–358.
Lundegårdh, H. (1964). *Physiol. Plantarum* **17**, 202–206.
Lundegårdh, H. (1965). *Physiol. Plantarum* **18**, 269–274.
Mackler, B., Douglas, H. C., Will, S., Hawthorne, D. C. and Mahler, H. R. (1965). *Biochemistry* **4**, 2016–2020.
Margalit, R. and Schejter, A. (1970). *FEBS Letts.* **6**, 278–280.
Margoliash, E. (1966). "Hemes and Hemoproteins", p. 371–391.
Marsho, T. V. and Kok, B. (1970). *Biochim. et Biophys. Acta* **223**, 240–250.
Minakami, S. (1955). *J. Biochem.* **42**, 749–756.
Minakami, S., Ishikura, H. and Satake, K. (1956). *J. Biochem.* **43**, 575–577.
Mirsky, R. and George, P. (1967). *Biochemistry* **6**, 3671–3675.
Mitsui, A. and Tsushima, K. (1968). "Osaka", p. 459–466.
Møller, K. M. (1955). *Exptl. Cell Res.* **9**, 375–377.
Motonaga, K., Katano, H. and Nakanishi, K. (1965a). *J. Biochem.* **57**, 29–33.
Motonaga, K., Misaka, E., Nakajima, E., Ueda, S. and Nakanishi, K. (1965b). *J. Biochem.* **57**, 22–28.
Motonaga, K. and Nakanishi, K. (1965). *J. Biochem.* **58**, 98–100.
Narita, K. and Titani, K. (1968). *J. Biochem.* **63**, 226–241.
Neilands, J. B. (1952). *J. Biol. Chem.* **197**, 701–708.

Niemann, R. H. and Vennesland, B. (1957). *Science* **125**, 353–354.
Niemann, R. H. and Vennesland, B. (1959). *Plant Physiol.* **34**, 255–262.
Niemann, R. H., Nakamura, H. and Vennesland, B. (1959). *Plant Physiol.* **34**, 262–267.
Nigro, G., Romano, V. and Gravante, G. (1958). *Boll. soc. ital. biol. sper.* **34**, 1244–1246.
Nishimura, M. (1959). *J. Biochem.* **46**, 219–223.
Nishimura, M. (1966). *Brookhaven Sympos. in Biology* **19**, 132–142.
Nishimura, M. (1968). *Biochim. et Biophys. Acta* **153**, 838–847.
Nishimura, M. and Takamiya, A. (1966). *Biochim. et Biophys. Acta* **120**, 45–56.
Ohaniance, L. and Chaix, P. (1964). *Biochim. et Biophys. Acta* **90**, 221–227.
Ohaniance, L. and Chaix, P. (1970). *Bull. soc. chim. biol.* **52**, 1105–1109.
Olson, J. M. and Smillie, R. M. (1963). "Photosynthetic Mechanism of Green Plants". *Nat. Acad. Sci.—Nat. Res. Council Publ.* 1145, p. 56.
Paléus, S. and Johansson, B. W. (1968). *Acta. Chem. Scand.* **22**, 342–344.
Paléus, S., Tota, B. and Liljequist, G. (1969). *Compar. Biochem. Physiol.* **31**, 813–817.
Perini, F., Kamen, M. D. and Schiff, J. A. (1964). *Biochim. et Biophys. Acta* **88**, 74–90, 91–98.
Powls, R., Wong, J. and Bishop, N. I. (1969). *Biochim. et Biophys. Acta* **180**, 490–499.
Raison, J. K. and Smillie, R. M. (1969). *Biochim. et Biophys. Acta* **180**, 500–508.
Report of the Commission on Enzymes of the I.U.B. (1961). Symposium Series 20, Vol. 20, Pergamon Press, Oxford (1961). See also: Enzyme Nomenclature Recommendations of the I.U.B. (1964) by the Editors of Biochemical Journals, London, Elsevier, (1965).
Richmond, R. C. (1970). *Nature* (London) **225**, 1025–1028.
Rosenthal, O. and Drabkin, D. L. (1943). *J. Biol. Chem.* **149**, 437–450.
Rumberg, B., Müller, A. and Witt, H. T. (1962). *Nature* (London) **194**, 854–856.
Rurainski, H. J., Randles, J. and Hoch, G. E. (1970). *Biochim. et Biophys. Acta* **205**, 254–262.
Sauer, K. and Park, R. B. (1965). *Biochemistry* **4**, 2791–2798.
Schmidt, C. G. (1956). *Klin. Wochschr.* **34**, 457–461.
Schollmeyer, P. and Klingenberg, M. (1962). *Biochem. Z.* **335**, 426–439.
Schuldiner, S. and Okad, I. (1969). *Biochim. et Biophys. Acta* **180**, 165–177.
Scott, W. A. and Mitchell, H. K. (1969). *Biochemistry* **8**, 4282–4289.
Sels, A. A. (1964). *Arch. Internat. Physiol. et Biochim.* **72**, 529–530.
Sels, A. A. (1966a). *Arch. Internat. Physiol. et Biochim.* **74**, 729–730.
Sels, A. A. (1966b). *Arch. Internat. Physiol. et Biochem.* **74**, 730–731.
Sels, A. A., Fukuhara, H., Péré, G. and Slonimski, P. P. (1965). *Biochim. et Biophys. Acta* **85**, 486–502.
Sels, A. A., Zanen, J. and Marque, C. (1966). *Arch. Internat. Physiol. et Biochim.* **74**, 731–732.
Shappirio, D. C. and Williams, C. M. (1957). *Proc. Roy. Soc. London B* **147**, 218–232, 233–246.
Shashoua, V. E. and Estabrook, R. W. (1966). "Hemes and Hemoproteins", 427–430.
Sherman, F. (1964). *Genetics* **49**, 39–48.
Sherman, F. (1967). "Methods in Enzymol", Vol. **10**, 610–616.
Sherman, F. and Slonimski, P. P. (1964). *Biochim. et Biophys. Acta* **90**, 1–15.
Sherman, F., Stewart, J. W., Margoliash, E., Parker, J. and Campbell, W. (1966). *Proc. Nat. Acad. Sci.* **55**, 1498–1505.

Sherman, F., Stewart, J. W., Parker, J. H., Inhaber, E., Shipman, N., Putterman, G. J., Gardisky, R. L. and Margoliash, E. (1968a). *J. Biol. Chem.* **243**, 5446–5456.
Sherman, F., Stewart, J. W., Parker, J. H., Putterman, C. G. and Margoliash, E. (1968b). "Osaka", p. 257–268.
Sherman, F., Taber, H. and Campbell, W. (1965). *J. Mol. Biol.* **13**, 21–39.
Shirasaka, M., Nakayama, N., Endo, A., Haneishi, T. and Okazaki, H. (1968). *J. Biochem.* **63**, 417–424.
Sironval, C. (1962). *Physiologia Plantarum* **15**, 263–272.
Slonimski, P. P. (1953). "La Formation des Enzymes Respiratoires chez la Levure", Masson, Paris.
Slonimski, P. P. (1969). *Fed. Proc.* **28**, Abstr. 881.
Slonimski, P. P., Archer, R., Péré, A., Sels, A. and Somlo, M. (1965). "Mecanismes de la Regulation des Activities Cellulaires chez les Microorganismes". Centre Nat. de la Rech. Scient., Paris, p. 435.
Smillie, R. M. and Levine, R. P. (1963). *J. Biol. Chem.* **238**, 4058–4062.
Smith, L. and Chance, B. (1958). *Ann. Rev. Plant Physiol.* **9**, 449–482.
Spiro, M. J. and Ball, E. G. (1961). *J. Biol. Chem.* **236**, 225–230, 231–235.
Stevens, F. C., Glazer, A. N. and Smith, E. L. (1967). *J. Biol. Chem.* **242**, 2764–2779.
Stewart, J. W., Margoliash, E. and Sherman, F. (1966). *Fed. Proc.* **25**, Abstr. 2587.
Storey, B. T. (1968). *Arch. Biochem. Biophys.* **126**, 585–592.
Storey, B. T. (1970). *Plant Physiol.* **45**, 447–454.
Sugimura, Y. and Yakushiji, E. (1968). *J. Biochem.* **63**, 261–269.
Sugimura, Y., Toda, F., Murata, R. and Yakushiji, E. (1968). "Osaka", p. 452–458.
Susor, W. A. and Krogmann, D. W. (1966). *Biochim. et Biophys. Acta* **120**, 65–72.
Takano, T., Sugihara, A., Ando, O., Ashida, T., Kakudo, M., Horio, T., Sasada, Y. and Okunuki, K. (1968). *J. Biochem.* **63**, 808–810.
Tsou, C. (1951). *Biochem. J.* **49**, 658–662.
Tu, Shu-I. and Wang, J. H. (1969). *Biochemistry* **8**, 2970–2974.
Tuppy, H. (1958). *Z. Naturfschg.* **12b**, 784–788.
Tuppy, H. and Dus. K. (1958). *Monatshefte d. Chem.* **89**, 407–417.
Turner, G., Lloyd, D. and Chance, B. (1969). *Biochem. J.* **114**, 91P.
Ueda, K. (1959). *Compt. rend. soc. biol.* **153**, 1666–1669.
Vernon, L. P. and Avron, M. (1965). *Ann. Rev. Biochem.* **34**, 269–296.
Vernon, L. P. and Shaw, E. (1965). *Biochemistry* **4**, 132–136.
Vernon, L. P., San Pietro, A. and Limbach, D. A. (1965). *Arch. Biochem. Biophys.* **109**, 92–97.
Vredenberg, W. J. (1970). *Biochim. et Biophys. Acta* **223**, 230–239.
Vredenberg, W. J. and Duysens, L. N. M. (1965). *Biochim. et Biophys. Acta* **94**, 355–370.
Vredenberg, W. J. and Slooten, L. (1967). *Biochim. et Biophys. Acta* **143**, 583–594.
Wasserman, A. R., Garver, J. C. and Burris, R. H. (1963). *Phytochemistry* **2**, 7–14.
Weaver, E. C. (1969). *Plant Physiol.* **44**, 1538–1541.
Webster, D. A. and Hackett, D. P. (1966). *Plant Physiol.* **41**, 599–605.
Williams, J. N. Jr. (1964). *Arch. Biochem. Biophys.* **107**, 537–543.
Wiskich, J. T. and Morton, R. K. (1960). *Nature* (London) **188**, 659–660.
Wiskich, J. T., Morton, R. K. and Robertson, R. N. (1960). *Austral. J. Biol. Sci.* **13**, 109–122.
Witt, H. T., Müller, A. and Rumberg, B. (1961). *Nature* (London) **191**, 194.
Wolken, J. J. and Gross, J. A. (1963). *J. Protozool.* **10**, 189.
Yakushiji, E. (1935). *Acta Phytochim.* **8**, 325–329.

Yakushiji, E., Sugimura, Y., Sekuzu, I., Morikawa, I., and Okunuki, K. (1960). *Nature* (London) **185**, 105–106.
Yamagata, S. and Sato, R. (1968). *J. Biochem.* **64**, 549–556.
Yamagata, S., Ueda, K. and Sato, R. (1966). *J. Biochem.* **60**, 160–174.
Yamamoto, H. and Vernon, L. P. (1969). *Biochemistry* **8**, 4131–4137.
Yamanaka, T. and Kamen, M. D. (1965). *Biochem. Biophys. Res. Commun.* **19**, 751–754.
Yamanaka, T. and Kamen, M. D. (1966). *Biochim. et Biophys. Acta* **112**, 436–447.
Yamanaka, T. and Kamen, M. D. (1967). *Biochim. et Biophys. Acta* **143**, 425–426.
Yamanaka, T. and Okunuki, K. (1963). *Biochim. et Biophys. Acta* **67**, 379–393.
Yamanaka, T. and Okunuki, K. (1964). *J. Biol. Chem.* **239**, 1813–1817.
Yamanaka, T. and Okunuki, K. (1968). "Osaka", pp. 390–403.
Yamanaka, T., De Klerk, H. and Kamen, M. D. (1967). *Biochim. et Biophys. Acta* **143**, 416–424.
Yamanaka, T., Mizushima, H. and Okunuki, K. (1964a). *Biochim. et Biophys. Acta* **81**, 223–228.
Yamanaka, T., Mizushima, H., Miki, K. and Okunuki, K. (1964b). *Biochim. et Biophys. Acta* **81**, 386–388.
Yamanaka, T., Mizushima, H., Katano, H. and Okunuki, K. (1964c). *Biochim. et Biophys. Acta* **85**, 11–17.
Yamanaka, T., Nagata, Y. and Okunuki, K. (1968). *J. Biochem.* **63**, 753–760.
Yamanaka, T., Nakajima, H. and Okunuki, K. (1962). *Biochim. et Biophys. Acta* **63**, 510–512.
Yamanaka, T., Nishimura, I. and Okunuki, K. (1963a). *J. Biochem.* **54**, 161–165.
Yamanaka, T., Takemori, S. and Okunuki, K. (1969). *Biochim. et Biophys. Acta* **180**, 193–195.
Yamanaka, T., Tokugama, S. and Okunuki, K. (1963b). *Biochim. et Biophys. Acta* **77**, 592–601.
Yaoi, Y. (1967). *J. Biochem.* **61**, 54–58.
Yaoi, Y., Titani, K. and Narita, K. (1964). *J. Biochem.* **56**, 222–229.
Yaoi, Y., Titani, K. and Narita, K. (1966). *J. Biochem.* **59**, 247–256.
Zviagilskaya, P. H. and Karapetian, N. V. (1965). *Doklady* **163**, 497.

D. CYTOCHROME C_1

1. Introduction.

Cytochrome c_1 was first distinguished from cytochrome c and isolated by ammonium sulphate fractionation of a cytochrome oxidase preparation in cholate (Yakushiji and Okunuki, 1940, 1941). At first, the position of the α-band at 554 nm was thought to be explainable on the basis of admixture of a denatured protein protohaemochrome band to cytochrome c (Slater, 1949). Chance and Pappenheimer (1954) erroneously identified it with cytochrome b_5. In an admirable paper, Keilin and Hartree (1955) confirmed the existence of cytochrome c_1 and showed that it was this substance which had caused the "e"-band at 551–552 nm at $-190°C$ accompanying the cytochrome c band at 548 nm (Keilin and Hartree, 1949a,b, 1950). They found cytochrome c_1 like cytochrome c to be a haemoprotein with firmly linked "mesotype" haem prosthetic group and to be non-autoxidizable and not reacting with CO, but less thermo-stable than cytochrome c. Cytochrome c_1 occurs in yeast (Okunuki, 1966) but whether a similar band in *Acetobacter pasteurianum* is that of c_1 has not yet been established. A similar band in the red alga *Porphyra* is almost certainly that of an f-type cytochrome (cf. Chapter V, C), while a similar band in mung bean mitochondria will be discussed below. Stotz and co-workers (Clark *et al.*, 1954; Widmer *et al.*, 1954) had also observed an α-band at 554 nm in a factor linking "succinic" dehydrogenase with cytochrome c, and had tentatively identified it with Keilin's band "e", i.e. with cytochrome c_1. Ball and Cooper (1957) observed the simultaneous presence of cytochromes c and c_1 in heart muscle preparations and showed that c_1 cannot replace c (see also Glaze and Morrison, 1960) but worked in series with it. This function of cytochrome c_1 as cytochrome c-reductase was confirmed by Estabrook (1958). He could not confirm the claims of Yakushiji and Okunuki (cf. also Sekuzu *et al.*, 1960b) that c_1 reacted directly with cytochrome a.

2. Preparation of cytochrome c_1.

The starting material is either the heart muscle preparation (King, 1967) in the form of "green brei" and a cholate extract (Yakushiji and Okunuki, 1941; Okunuki *et al.*, 1958; Sekuzu *et al.*, 1960a; Orii *et al.*, 1969) or ox heart mitochondria (Crane *et al.*, 1956; Green *et al.*, 1959; Bomstein *et al.*, 1960, 1961; Rieske and Tisdale, 1967) as in the preparation of cytochrome aa_3 (see Chapter III). Cytochrome aa_3 is removed by ammonium sulphate precipitation in cholate or deoxycholate (cf. also Bernstein and Wainio, 1960) with or without addition of sodium lauryl sulphate (Duponol) while cytochrome b must be denatured at 30–40°C in 21% saturated ammonium sulphate. At 45–60% ammonium sulphate saturation, cytochrome c_1 is precipitated. Bile salts can be removed, but not without great loss of yield, by calcium phosphate gel absorption and elution with pH 7·4 phosphate. The purest preparations were claimed to contain 25–27 μmoles per mg. of protein. This would correspond to the minimum molecular weight per haem of 37,000 (Bomstein *et al.*, 1961; Orii and Okunuki, 1969). However, Rieske and Tisdale (1967) found by sedimentation data and haem content a minimum M.W. of 52,000 for the monomer in 0·5 mM sodium dodecylsulphate and a hexamer of this in solution (cf. also Okunuki, 1966, p. 240).

3. Properties.

The prosthetic group cannot be removed by acetone-HCl but by the silver sulphate method as haematohaem, like that of cytochrome c (Sekuzu *et al.*, 1960a). Green *et al.* (1959, 1960) found the redox potential to be + 223 mV at pH 7. Cytochrome c_1 is an acidic cytochrome (Orii *et al.*, 1962). The absorption bands of c_1^{2+} lie at 554, 524 and 418 nm, with a small broad band at 472 nm. The Soret band of c^{3+} is at 411 nm. The millimolar extinctions are 24·1, 11·6 and 116 at 554, 524 and 418 nm resp. (Green et al., 1959), with α/β 1·47 and γ/α 6·8 є mM (554–540 nm) of c_1^{2+} is 17·5, $\epsilon_{mM}^{554}(c_1^{2+} - c_1^{3+})$ 17·1,

In contrast to earlier statements, cytochrome c_1, like cytochrome c, does react with cyanide (Schejter and Berke, 1968). Cyanide shifts the Soret band causing a maximum increase at 427 nm. K equals $2\cdot3 \times 10^2$ as compared to $4\cdot3 \times 10^3$ for cytochrome c. C_1^{2+} CN is stable in contrast to c^{2+}CN. Cytochrome c_1^{2+}CN dissociates more slowly than c^{2+}CN but c_1^{3+}CN dissociates faster than c^{3+}CN. Whereas the CO-compound formed by modified cytochrome c at pH 13 is dissociated at lower pH, that of c_1 formed at pH 13 is somewhat more stable on neutralization. Like cytochrome c, c_1 has a closed crevice structure, but this is of less strength. A haemipeptide of cytochrome c_1 has been isolated by Wada et al., 1968.

4. Occurrence and biological role.

Storey (1969) found two cytochromes (with low temperature α-bands at 547 and 549 nm, corresponding to room temperature positions at 550 and 552 nm, in mung bean mitochondria. In spite of the similarity of the second α-band to that of c_1, it is doubtful whether the c-552 of the plant mitochondria plays a role similar to that of cytochrome c_1 in mammalian mitochondria. Cytochrome c-552 was faster oxidized than c-550, thus its place in the plant respiratory chain appears to be between cytochrome oxidase and cytochrome c rather than between cytochromes b and c. Both plant cytochromes c appear to be membrane-bound.

Cytochrome c_1 is found in a particle together with cytochrome b from which it can be isolated only after modification of cytochrome b, but apparently without modification of cytochrome c_1 (Goldberger and Green, 1963). The preparation of the bc_1 complex (complex III of the respiratory chain, coenzyme Q – cytochrome c reductase) has been described by Basford et al. (1957), by Takemori and King (1962), by King (1967b) and Rieske (1967a). The preparation of Rieske (see also Baum et al., 1967) contained 6·8 μmoles cytochrome b, 3·4 μmoles c_1, 6·2 μmoles non-haem iron (i.e. 3·1 μmoles NHI-protein with 2 atoms of Fe) and a little coenzyme Q besides phospholipids. The preparation of King contained the same amount of non-haem iron but less cytochromes b and $c(b/c = 1\cdot7)$. The complex will be discussed with regard to its role in the respiratory chain in Chapter VII.

Determinations of cytochrome c_1 in mitochondria have been described by Williams (1964) and by Rieske (1967b). While the method of Williams is entirely spectrophotometric, measuring absorption at several wavelengths, Rieske's method depends on physical separation of cytochromes c_1 and b. The formation of a ternary complex of cytochromes a, c_1 and c has been claimed by Orii et al. (1962), but there is no satisfactory evidence for it.

REFERENCES

Ball, E. G. and Cooper, O. (1957). J. Biol. Chem. **226**, 755–768.
Basford, R. E., Tisdale, H. D., Glenn, J. L. and Green, D. E. (1957). Biochim. et Biophys. Acta **24**, 107–115.
Baum, H., Rieske, J. S., Silman, H. I. and Lipton, S. H. (1967). Proc. Nat. Acad. Sci. **57**, 798–805.
Bernstein, E. H. and Wainio, W. W. (1960). Arch. Biochem. Biophys. **91**, 138–143.
Bomstein, R. A., Goldberger, R. and Tisdale, H. (1960). Biochem. Biophys. Res. Commun. **3**, 479–483.
Bomstein, R., Goldberger, R. and Tisdale, H. (1961). Biochim. et Biophys. Acta **50**, 527–543.
Chance, B. and Pappenheimer, A. M. (1954). J. Biol. Chem. **209**, 931–943.
Clark, H. W., Neufeld, H. A., Widmer, C. and Stotz, E. (1954). J. Biol. Chem. **210**, 851–861.
Crane, F. L., Glenn, J. L. and Green, D. E. (1956). Biochim. et Biophys. Acta **22**, 475–487.
Estabrook, R. W. (1958). J. Biol. Chem. **230**, 735–750.
Glaze, R. P. and Morrison, M. (1960). Fed. Proc. **19**, p. 34.
Goldberger, R. and Green D. E. (1963). In "The Enzymes", 2nd. ed. (Boyer, Lardy and Myrbäck, eds.), vol. **8**, pp. 81–96.
Green, D. E., Järnefelt, J. and Tisdale, H. D. (1959). Biochim. et Biophys. Acta **31**, 34–46.
Green, D. E., Järnefelt, J. and Tisdale, H. D. (1960). Biochim. et Biophys. Acta **38**, 160.
Keilin, D. and Hartree, E. F. (1949a). Nature (London). **164** 254–259.
Keilin, D. and Hartree, E. F. (1949b). Nature (London) **165**, 504.
Keilin, D. and Hartree, E. F. (1955). Nature (London) **176**, 200–206.
King, T. E. (1967a). "Methods in Enzymol". Vol **10**, 222–223.
King, T. E. (1967b). "Methods in Enzymol". Vol **10**, 202–208.
Oda, T. (1968). "Osaka", p. 500–515.
Okunuki, K. (1962). In "Oxygenases", p. 439.
Okunuki, K. (1966). In "Comprehensive Biochemistry" (M. Florkin and E. H. Stotz, eds.) vol. **14**, Chapter V, pp. 232–308.
Okunuki, K., Sekuzu, I., Yonetani, T. and Takemori, S. (1958). J. Biochem. **45**, 847–854.
Orii, Y., Sekuzu, I. and Okunuki, K. (1962). J. Biochem. **51**, 204–215.
Orii, Y. and Okunuki, K. (1969). Ann. Reports. Biol. Works Fac. Sci., Osaka Univ. **17**, 1–14.
Rieske, J. S. (1967a). "Methods in Enzymol". Vol **10**, 239–245.
Rieske, J. S. (1967b). "Methods in Enzymol". Vol **10**, 488–493.
Rieske, J. S. and Tisdale, H. S. (1967). "Methods in Enzymol". Vol. **10**, 349–352.
Schejter, A. and Berke, G. (1968). Biochim. et Biophys. Acta **162**, 459–461.
Sekuzu, I., Orii, Y. and Okunuki, K. (1960a). J. Biochem. **48**, 214–225.
Sekuzu, I., Takemori, S., Orii, Y. and Okunuki, K. (1960b). Biochim. et Biophys. Acta **37**, 64–71.
Slater, E. C. (1949). Nature (London) **163**, 532.
Storey, B. T. (1969). Plant Physiol. **44**, 413–421.

Wada, K., Matsubara, H. and Okunuki, K. (1968). "Osaka", p. 309–310.
Widmer, C., Clark, H. W., Neufeld, H. A. and Stotz, E. (1954). *J. Biol. Chem.* **210**, 861–867.
Williams, J. N. Jnr. (1964). *Arch. Biochem. Biophys.* **107**, 532–534.
Yakushiji, E. and Okunuki, K. (1940). *Proc. Imper. Acad. Tokyo* **16**, 299–302.
Yakushiji, E. and Okunuki, K. (1941). *Proc. Imper. Acad. Japan* **17**, 38–40.

Chapter VI

Bacterial cytochromes and cytochrome oxidases

A. General Introduction.
B. Cytochrome Oxidases and Nitrite Reductases
   1. Cytochrome a_1.
      a. Introduction.
      b. Cytochrome a_1 in yeast.
      c. Bacterial particulate preparations.
      d. Cytochrome a_1 in NO_3 and NO_2 respiration.
      e. Carbon monoxide and cyanide reactivity.
   2. Cytochrome o.
      a. Introduction.
      b. Properties of particulate and soluble preparations.
      c. Carbon monoxide and cyanide reactivity.
   3. Cytochrome d (a_2).
      a. Introduction.
      b. Spectral properties.
      c. Particulate preparation of cytochrome d.
      d. Occurrence and role as oxidase.
      e. Function as nitrate reductase.
      f. Reaction with cyanide.
   4. *Pseudomonas* and *Micrococcus* cytochrome d_1c.
      a. Introduction.
      b. Preparation.
      c. Physical properties.
      d. Cytochrome d_1c complexes with ligands.
      e. The haems of cytochrome d_1c.
      f. Cytochrome d_1c: a complex of two haemoproteins.
      g. Enzymic activity of cytochrome d_1c.
      h. Oxidase activity of *Pseudomonas* cytochrome d_1c with cytochromes c.
C. Cytochromes b.
   1. Introduction.
   2. Cytochrome b_1.
      a. Cytochrome b_1 of *E. coli*.
      b. Cytochrome b_1 from other organisms.

3. Cytochrome b-562.
   a. Introduction.
   b. Preparation and physical properties.
   c. Amino acid sequence and structure.
   d. Function of cytochrome b-562.
4. Other microbial cytochromes b.
5. Bacterial P-450.
   a. Introduction.
   b. *Pseudomonas* P-450$_{cam}$.
   c. *Rhizobium* P-450.
   d. P-450 from other bacteria.
6. Appendix.

D. Cytochromes c.
1. *Bacillus* cytochrome c.
   a. General properties.
   b. Biochemical aspects.
2. Cytochromes c of denitrifying organisms.
   a. Purification and general properties.
   b. Primary and tertiary structure.
   c. Other properties.
   d. Cytochromes c-552.
   e. *P. denitrificans* cytochrome c-553.
   f. High redox potential cytochromes c from other bacteria.
   g. *Halobacteria* cytochromes c.
   h. *Thiobacilli* cytochromes c.
3. Cytochromes c_4 and c_5 (c-555) of *Azotobacter*.
   a. Preparation.
   b. Spectral and other physical properties.
   c. Biochemical activity and role of cytochromes c_4 and c_5.
4. Cytochrome c_3 and other cytochromes of low oxidation-reduction potential.
   a. General properties of cytochrome c_3.
   b. Primary structure.
   c. Immunological reactions.
   d. Function of cytochrome c_3.
   e. Cytochrome c-553.
   f. Cytochrome cc_3.
   g. Cytochrome c-552 of *E. Coli*.
      (i) Occurrence and general properties.
      (ii) Function and induction of cytochrome c-552.
5. Cytochromes cc' and c'.
   a. Introduction.
   b. Preparation.
   c. Absorption spectra conversions and CD spectra.
   d. Magnetic susceptibility, ESR and Mössbauer studies.
   e. Reaction with ligands.
   f. Amino acid sequences and linkage of haems.
   g. Function of cytochrome cc'.
   h. Cytochrome c'.
   i. Cryptocytochrome c.
   j. Cytochrome c-552, 558.

6. Photosynthetic cytochromes c.
   a. Cytochrome c_2 (*Rhodospirillum rubrum*).
   b. Other cytochromes c_2.
   c. Flavin-containing cytochromes from *Chromatium* and *Chlorobium*.
   d. Other cytochromes c of photosynthetic bacteria.
   e. Bacterial photosynthesis.
E. Structural organization and coupled phosphorylation.
   1. Localization and organisation of the cytochrome complex.
   2. Coupled phosphorylation.
   References

A. GENERAL INTRODUCTION

Micro-organisms as a class in contrast to higher animals or plants exhibit a multiplicity of cytochrome oxidases—cytochrome aa_3 and d (a_2), d_1c, o, and there is a strong likelihood that these groups may yet have to be subdivided. Moreover various combinations of the cytochromes occur throughout the vast range of micro-organisms and in differing amount depending on the substrate being metabolised, the physiological state of the organism and the genetic composition of the particular strain.

A corollary of the plurality of oxidases is the development in these cells of alternate pathways for the electron flow from substrate to terminal electron acceptor. These pathways may become complex and branched, giving a flexible network for passage of the electron flux. The general availability of split-beam spectrophotometers capable of detecting small differences of spectral absorption in dense suspensions of micro-organisms and increasing access to double-beam spectrophotometers, a powerful tool for the study of the kinetics of respiratory components, have produced an upsurge of activity in this field, reviewed by White and Sinclair (1970).

The initial observations of Keilin (1927) and of Yaoi and Tamiya (1928) that the absorption bands of many bacteria were not at the same wavelengths as those of aerobic yeast and higher organisms have been followed by a considerable number of studies on a wide variety of bacteria, fungi and lower algae. Many of these investigations have been characterized by relatively superficial observations and there has been little of the intense study of particular cytochromes that has been a feature of the work on mammalian cytochromes. Problems associated with the production of large enough amounts of the organism and the mechanical breakage of the small bacterial cell, which often has a tough cell wall, e.g. *Staphylococcus aureus*, or is encapsulated e.g. *Aerobacter aerogenes* have contributed to the paucity of chemical knowledge of the bacterial cytochromes. Moreover the multitude of biochemical activities contained within a single bacterial cell in contrast to the specialisation of eukaryotic cells is frequently reflected in the presence in the cell of a variety of cytochromes and cytochrome oxidases. Except for the photosynthetic organisms these cytochromes are not partitioned between discrete organelles but are organised in a single membrane. It is characteristic that the haemoproteins of bacteria are usually more firmly bound to the embedding membrane than are the analogous haemoproteins in organelles. Consequently it has rarely been possible to release the component cytochromes from the membranes in sufficient amounts and without denaturation, often subtle, for adequate study. Nonetheless several of the more robust cytochromes c and a cytochrome b have now had their primary structure revealed.

The bacterial cytochromes participate in a diversity of interesting biochemical reactions and, as if in compensation for the difficulties enumerated, the ability of micro-organisms to adapt to modification of their environment facilitates experimental study of the physiological function of particular cytochromes *in situ*. Furthermore bacterial mutants provide not only an elegant means of obtaining knowledge of the inter-relationship of cytochromes and the other components in the electron transfer chain, but also a means to study the biosynthesis of cytochromes (Chapter IX).

B. CYTOCHROME OXIDASES AND NITRITE REDUCTASES

1. Cytochrome a_1

(a) *Introduction.* The demonstration by Kubowitz and Haas (1932) that *Acetobacter pasteurianum* contained a respiratory pigment which could combine with CO and had a photochemical absorption spectrum which generally resembled that obtained for the respiratory ferment of *Candida utilis*, was followed by an investigation in which Warburg et al. (1933), by the astute use of carbon monoxide and cyanide, were able to differentiate between an autoxidizable pigment having an absorption band at 589 nm in the reduced form and the other cytochromes, now designated b, c, and c_1. The 589 nm pigment was referred to by these workers as an oxygen transporting ferment. Their choice of organism was fortunate as in other strains cytochrome a_1 occurs together with cytochromes o and a_2 (cf. Keilin, 1966, Castor and Chance, 1959) so that the unequivocal demonstration that cytochrome a_1 has the role of an oxidase in these strains has not been possible. Latterly evidence has accumulated which implicates cytochrome a_1 as the terminal electron acceptor in nitrate reduction (White and Sinclair, 1970).

The introduction by Chance of double beam spectrophotometry enabled him to obtain refined CO-difference spectra and photo-dissociation spectra of the CO-compounds of cytochrome oxidases (Chance 1953a,b) and thereby to show that the CO-complex of cytochrome a_1 differed from that of cytochrome a_3 and that the CO-photodissociation spectrum of cytochrome a_1 agreed closely with the photoaction spectrum obtained by Kubowitz and

Haas (1932) for the respiratory enzyme of *A. pasteurianum*. Castor and Chance (1955, 1959) used a polarographic method for determining the concentration of oxygen in solution, greatly improving the accuracy and sensitivity of the determination of the photochemical action spectra of CO-inhibited respiration and thereby removed uncertainties of the earlier investigations and confirmed the identification of the oxidase of *A. pasteurianum* as cytochrome a_1.

(b) *Cytochrome a_1 in yeast.* Euler et al. (1927) using a spectrographic technique demonstrated that brewers top yeast contained cytochromes *a*, *b* and *c* in proportion and amounts found in mammalian cells. Brewer's bottom yeast which has a low respiration, exhibited weak cytochrome absorption bands, although the haematin content of both strains of yeast was similar. Further spectroscopic studies by Fink (1932) of different species and strains of yeast led to the identification of cytochrome a_1 in brewers yeast and yeast of vinous fermentation. The low concentration of this pigment and the diffuseness and position of its α-absorption band at 585–588 nm, has in anaerobic or quasi-anaerobic yeast resulted in confusion of identity with oxyhaemoglobin which has an absorption band at 578 nm. Indeed, spectroscopic studies have provided evidence for a haemoglobin-type pigment in strains of normal yeast (Keilin, 1953; Mok et al., 1969), petite mutants of yeast (Keilin and Tissières, 1954) and in *S. cerevisiae* grown aerobically in the presence of antimycin *A* (Yças, 1956). Ishidate et al. (1969a,b) have ascribed the absorption at 585–8 nm to cytochrome *c* peroxidase. Difficulty of recognition of cytochrome a_1 may be caused by the presence in the yeast of coproporphyrin or its zinc-complex, both having absorption bands in this region of the spectrum (cf. Sugimura et al., 1966; Pretlow and Sherman, 1967).

The appearance of the 588 nm band in the presence of dithionite, the CO-difference spectrum, and especially the sharpening of the 588 nm band in the presence of cyanide should distinguish cytochrome a_1 from other haemoproteins which can occur in anaerobic yeast (Lindenmayer and Smith, 1964) and lack of red or orange fluorescence from free porphyrins or zinc-porphyrin complex. Haem *a* has been isolated from anaerobic *S. cerevisiae* using an acetonitrile methylethylketone-HCl solvent mixture, and the prosthetic group has been further characterized by conversion of it to porphyrin *a* (Barrett, 1969). As only 100 µg porphyrin *a* was obtained from 100 g dry weight of yeast, identification of the 585–8 nm band with cytochrome a_1 in anaerobic yeast requires skill, but this procedure does provide conclusive evidence. Uncertainty exists as to the function of cytochrome a_1 in anaerobic yeast. Lindenmayer and Smith (1964) conclude that although the apparent cytochrome oxidase activity in anaerobic *S. cerevisiae* (LK 2912) was only 0·5% that of the same yeast grown aerobically the cytochrome a_1 present acted as a terminal oxidase. As mammalian cytochrome c was used in the cytochrome assay (Smith, 1954) rather than the probable natural electron donor cytochrome b_1, the real oxidase activity may have been greater. Ishidate et al. (1969a) on the other hand related the respiration in anaerobic cells in their strains of *S. cerevisiae* to the presence of P-450.

(c) *Bacterial particulate preparations.* Tissières (1952b) detected cytochrome a_1 in a succinic oxidase preparation from *Aerobacter aerogenes*. Cytochrome a_2 originally present in the whole cells had disappeared from the particles but cytochrome o, not then known, may have contributed to the observed oxidase activity.

From a strain of *Azotobacter vinelandii* in which photoaction spectra (Castor and Chance, 1959; Jones and Redfearn, 1967a) had indicated that cytochrome a, d and o were potential oxidases, Jones and Redfearn, (1967b) obtained small respiratory particles and fractionated them with the aid of sodium desoxycholate and salt into green particles which contained most of the cytochrome a_1, and also the cytochrome d, and into red particles which were rich in cytochrome o. The succinic-oxidase activity of the green particles however was low and they failed to oxidize ascorbate via a variety of electron mediators. The failure of blue light to substantially relieve CO-inhibited NADH oxidase in the original small particles, suggested that the involvement of cytochrome a_1 as an oxidase in the highly active db_1 complex was negligible.

Jurtshuk et al. (1969) noted that while there was substantial reduction of cytochromes $c_4 + c_5$ in an ETP fraction from *A. vinelandii*, neither the cytochrome a_1 nor cytochrome d present could be reduced by ascorbate in the absence of TMPD, suggesting that the c-type cytochromes may not serve as electron donor to the two oxidases.

A cytochrome oxidase preparation from *Ferrobacillus ferroxidans* which had activity against horse cytochrome c contained cytochrome a_1 (α-maximum 595 nm, Soret maximum 440 nm), but the presence of cytochrome o was not excluded (Blaylock and Nason, 1963).

A particulate electron transfer system from *Halobacterium cutirubrum* harvested at the early stationary phase of growth, contained cytochrome a_1 (α-maximum 592 nm) as the only oxidase for the main branch of electron transport from NADH and also a subsidiary branch from succinate, both streams of electrons going through c-type cytochromes which were the donors to cytochrome a_1 (Lanyi, 1969). CO-difference spectra had not revealed the presence of any cytochrome o in the particles (Lanyi, 1968). In contrast Cheah (1969) found cytochrome a_3 in his strain of *H. cutirubrum*. Cytochrome a_1 was present in slightly lower concentration than cytochrome o in membranes of *H. halobium* (Cheah, 1970b); the concentration of both cytochrome oxidases in this organism declined in the later stationary phase of growth. The cytochrome a_1 was readily reducible with ascorbate and

TMPD at 22° and − 196°C. The particulate formate oxidase from *Nitrobacter agilis* contained cytochromes a_1 and a_3; no b type cytochromes were present in the particles, only a c-550 (O'Kelley and Nason, 1970).

An oxidase capable of oxidizing horse cytochrome c was obtained from sonicates of *Pseudomonas aeruginosa* grown aerobically with peptone as sole carbon source and without nitrate (Azoulay and Couchoud-Beaumont, 1965). Glucose represses the biosynthesis of cytochrome a_1 (see below) in this organism and cytochrome d_1c is not formed in the absence of nitrate (Azoulay, 1964). The oxidase was partially purified by passage through Sephadex G100 and by the absorption of some impurities onto CM-cellulose at pH 8·0, resulting in a twenty fold increase of activity. The absorption of the α-bands of cytochrome a_1-592 and cytochrome c-552 was intensified but their relative intensities remained constant during purification of the oxidase. The absorption due to the b-type cytochromes had disappeared, so that it is unlikely that cytochrome o contributed to the oxidase activity. The Km for the oxidation of horse ferrocytochrome c was $1·5 \times 10^{-4}$ M, the pH optimum 8·0–8·5; the oxidase was not active with hydroquinone or ascorbate. Complete inhibition, reversed by light, of the oxidase activity was obtained with 80% CO and CN^-. Azide and hydroxylamine were also effective inhibitors.

A preparation was obtained from *Escherichia coli* str. McElroy, grown in highly aerobic and microaerophilic conditions, which was purified by chromatography on Sephadex, ammonium sulphate fractionation and chromatography on DEAE-cellulose. A cytochrome b component was still present in the preparation (Barrett and Sinclair, 1967a,b). This cytochrome "a_1b" preparation exhibited high peroxidase activity towards 2,3,6-trichloroindophenol.

(d) *Cytochrome a_1 in NO_3 and NO_2 respiration.* Evidence accumulated over the last decade points to the participation of cytochrome a_1 in dissimilatory nitrate reduction, or nitrate respiration, in certain bacteria. Addition of nitrate to intact cells of *Haemophilus parainfluenzae* under anaerobic conditions oxidized cytochrome a_1 more than it did the other cytochromes (White and Smith, 1962). However, the rate of electron transport with nitrate as acceptor is lower than oxygen. White (1962) noted that the biosynthesis of cytochrome a_1 in *H. parainfluenzae* is stimulated by growth in the presence of nitrate. Cytochrome a_1 in particulate fractions from *Nitrobacter winogradskyi* has similarly been oxidized with nitrate under anaerobic conditions in the presence of endogenous substrate (Kiesow, 1964) and NO_3^- can compete with O_2 in the particulate formate oxidase which contains cytochrome a_1 (O'Kelley and Nason, 1970).

Cytochrome a_1 in certain obligate chemoautotrophic bacteria has been implicated in the oxidation of NO_2^- to NO_3^-, a single 2e step. In *Nitrobacter agilis* cells and particulate preparations a_1 transfers electrons from

NO_2^- to a cytochrome c-550 (547, 519, 415 nm, 77°K) (Lees and Simpson, 1957; Aleem and Nason, 1959; Aleem, 1967). The thermodynamically unfavourable E_0' for the NO_2^-/NO_3^- ($E_0' = -430$ mV, pH 7·0) and cytochrome c-550 ($E_0' = 282$ mV, pH 7·0) couples, requires an energy input for the electron transfer from NO_2^- to cytochrome c to occur. A value of 15 kcal for the apparent activation energy of reduction of *Nitrobacter* cytochrome c by NO_2^- has been reported by Van Gool and Laudelout (1967); see also Van Gool and Laudelout, (1966).

Kiesow (1967) found that the reduction of cytochrome c by cytochrome a_1 was ATP dependent and could be inhibited by 2–4 dibromophenol, an uncoupler of oxidative phosphorylation. Evidence of reversed electron flow from cytochrome a_1 to cytochrome c-550 during NO_2^- oxidation in cell free preparations of *N. winogradskyi* was obtained by Van Gool and Laudelot (1967). Cytochrome a_1 in preparations treated with uncoupling agents or aged by heat is reduced by NO_2^- without concomitant reduction of cytochrome c (Aleem, 1971). The reversed electron flow from NO_2^- to NAD^+ has been postulated to be driven by the ATP generated in the terminal segments of the electron transport chain in which cytochrome a_3 is the actual oxidase (Aleem, 1968, 1970; Sewell and Aleem, 1969). An alternative, cyclic scheme involving ATP synthesis coupled to NADH oxidation (Kiesow, 1964) is difficult to reconcile with the fact that in *Nitrobacter* oxidation of nitrite is the sole source of energy for chemosynthesis. On the other hand, the presence in *Nitrobacter* nitrite oxidase particles of nitrite-cytochrome c reductase activity, which proceeds readily in the absence of added ATP, is not concordant with the hypothesis of reversed electron flow in nitrite oxidation in *Nitrobacter*, and Nason and co-workers (O'Kelley et al., 1970) suggest that the entry site of NO_2^- is not at cytochrome a_1, but at, or prior to, cytochrome c.

The spectral evidence (Butt and Lees, 1958; Aleem, 1969) for the presence of cytochrome a_3 in *Nitrobacter* particles in addition to a_1 is supported by the resolution at 77°K of the asymmetric peak of the a_1a_3 complex reduced with $Na_2S_2O_4$ or NO_2^- into two maxima, at 604 and 586 nm with Soret shoulders at 450 and 437 nm (O'Kelley et al., 1970). Although Hill and Taylor (1969) concluded from a photochemical action spectrum that cytochrome a_1 is the oxidase in *N. agilis* ATCC 14123 their data do not exclude the presence of either cytochromes a_3 or o. From digitonin-treated *N. agilis* particles Straat and Nason (1965) have separated cytochrome a_1, soluble at 144,000 g, maximum at 587 nm, from an a type cytochrome (α maximum 605 nm) which did not appear to combine with CO.

For an extended discussion of the energetics of the oxidation of NO_2^- in *Nitrobacter* and *Nitrosomonas* the reader is referred to the reviews of Aleem (1970), Wallace and Nicholas (1969) and Peck (1968).

(e) *Carbon monoxide and cyanide reactivity.* The classical CO-difference spectrum of cytochrome a_1 has been obtained with *Acetobacter pasteurianum* (Chance, 1953a). Small differences in the position of the maxima and the relative peak to rough absorption values have been observed in several bacteria where the CO-difference spectrum is complicated by the presence of other CO-combining pigments, notably cytochrome *o*. None of the cell-free preparations of cytochrome a_1 has been sufficiently free of other haemoproteins to ensure a single CO-spectrum.

With his new spectrophotometric methods and using the photodissociation of myoglobin-CO as a model, Chance (1953c) obtained molecular extinction coefficients of the α (589 nm) and Soret (427 nm) bands of the CO-compound of cytochromes a_1 of *A. pasteurianum*. The respective values of 12 and 87 cm^{-1}mM^{-1} compared well with those of 13 and 89cm^{-1} mM^{-1} which he calculated from the photoaction spectra obtained by the manometric procedure of Warburg and Negelein (1929).

Cytochrome a_1 may be estimated from the CO-reduced minus reduced difference spectrum using a value of 60 for the millimolar extinction co-efficients for wavelength pairs, E_{427}–E_{440} (Smith, 1955); for the reduced-oxidized difference spectrum Chance (1953c) calculated ϵ_{mM}^{427} as 120.

Accurate assay of cytochrome a_1 is difficult because of the contribution in most bacteria of other CO-combining pigments.

Cyanide complex. Cyanide reacts with cytochrome a_1 to give a sharpening of the α-maximum and displacement of this band to the red so that it occupies the peak of the α-maximum of cytochrome a_3. Lindenmayer and Smith (1964) noted the appearance of a trough at 450 nm in the cyanide-reduced difference spectrum of anaerobic yeast. Cheah (1970b) observed maxima at 592 and 440 nm in the CN-oxidized difference spectra of membrane fractions of *H. halobium*. As with cytochrome a_3-CO carbon monoxide was displaced by cyanide from its complex with cytochrome a_1 (Cheah, 1970a,b; see cytochrome *o*). In the presence of 0·5 mM cyanide CN-difference spectra of the nitrite reductase complex containing a_1 from *N. agilis* had peaks at 590 and 435 nm. Addition of nitrate to the enzyme preparation prior to the addition of cyanide effectively prevented inhibition of the nitrite reductase activity (Straat and Nason, 1965). Nitrate was competitive with cyanide at concentrations as high as 1 mM CN$^-$.

2. Cytochrome *o*

(a) *Introduction.* The existence of a cytochrome oxidase which has features of a *b*-type cytochrome was revealed by Chance (Chance, 1953a; Chance *et al.*, 1953) as the result of his elegant spectrophotometric studies two decades after the discovery of the *a*-type cytochrome oxidases. Although cytochtome *o* is present in many of the bacteria examined by the pioneers in the cytochrome field the spectroscopic techniques then available were inadequate to distinguish between this cytochrome b-type oxidase and the cytochromes b, from which cytochrome o differs principally in its rapid autoxidizability and its property of combining with CO and with CN in its native state. Even today few of the cytochromes o which have been provisionally identified have been rigorously characterized by the criterion of the photochemical action spectrum defined by Castor and Chance (1955, 1959).

Cytochrome o has been assumed to be present in intact micro-organisms or sub-cellular preparations which have exhibited in their CO-reduced difference spectra peaks at 578–565, 539–535, 410–419 and a trough at 560–547 nm. The position of the α-peak in the reduced-oxidized difference spectra ranges from 565 nm in *Staph. albus* and *Acetobacter suboxydans* (Smith, 1954) to 555 in *Staph. aureus* (Taber and Morrison, 1964). This might suggest at first that there may be classes of cytochrome o. Respiratory particles from *A. suboxydans* (Daniel, 1970) showed peaks at 557 and 565 nm in the CO difference spectrum while the reduced-oxidized spectrum had maxima at 568 and 588 nm. However there is no general correlation between the position of the α-peak of the reduced form and the near-red CO-difference peak. (cf. Table XI). Moreover some cytochromes b may be denatured readily, particularly in disrupted cells allowing CO to combine with the haem of a previously non-reactive cytochrome b.

Cytochrome o is usually estimated from the dithionite reduced + CO minus dithionite spectrum using a value of 80–90 for the ϵmM for the peak-plateau 417–460 (Chance, 1961). The ratio of absorption of the Soret/α peak [(417–47 nm)/(551–568 nm)] was 14·7. The photo-chemical data gave a larger Soret/α ratio. Daniel (1970) found that a more reproducible estimate of cytochrome o was obtained from peak minus trough measurements using ϵmM 417–432 nm = 170 and Soret/α peak [(417–432 nm)/(551–568 nm)] = 26.

The prosthetic group of cytochrome o has been characterized as protohaem by the formation of the pyridine haemochrome in purified electron transport particles (Taber and Morrison, 1964) or by the splitting-off of protohaem with methylethylketone-HCl from such a preparation (Cheah, 1969). These preparations may sometimes contain a-type cytochromes resulting in a mixture of haem a and protohaem.

(b) *Properties of particulate and soluble preparations.* Taber and Morrison (1964) found the absorption peak at 557 nm of a particle preparation from exponentially growing *Staph. aureus* could be resolved into 555 and 557 nm peaks at low temperature. The cytochrome-555 could be reduced by ascorbic acid-DCIP and thus be distinguished at room temperature from cytochrome b-557 which had a lower oxidation-reduction potential. Exposed to CO, the preparation which had been reduced with ascorbic

TABLE XI

Difference spectra maxima and minima of o-type cytochromes

Organism	Co-reduced minus reduced peaks and troughs nm (tr)			Reduced − oxidized α-peak nm	Ref.
Staphylococcus albus	567–568 tr 550	535–537	416–418	565	Castor and Chance (1955)
Acetobacter suboxydans	567–568 tr 550	535–537	416–418	565	Chance (1953a)
Achromobacter	572 tr 554	536	418	563	Arima and Oka (1965)
Rhodospirillum rubrum	576 tr 560	540	419–421	564	Taniguchi and Kamen (1965)
Rhizobium japonicum free cells	574 tr 556	539	415	562 (559·2, 77°K)	Appleby (1969b)
Myxococcus Xanthas	575	—	415	562	Dworkin and Niederpruem (1964)
Acetobacter suboxydans	568 + 560	536	417	557 + 565	Daniel (1970)
Vitreoscilla	570 tr 550	553	416	555 + 565	Webster and Hackett (1966a,b)
Leucothrix mucor	566 tr 550	533	416	558	Biggins and Dietrich (1968)
Leptospira sp.	575	540	417	—	Baseman and Cox (1969)
Bacillus megaterium KM	—	—	410–415	557	Broberg and Smith (1967)
Acetobacter suboxydans (IAM 1828)	567	535	417	558	Iwasaki (1966)
Staphylococcus aureus	568	533	415	555–557	Taber and Morrison (1964)
Halobacterium cutirubrum	578 tr 560	539	416	559	Cheah (1969)
Staphylococcus epidermis	573	537	417	558	Jacobs and Conti (1965)
Escherichia coli	567 tr 558	538	416	—	Revsin and Brodie (1969)
Haemophilus parainfluenzae	565 tr 547	532	416	—	White and Smith (1962)
Mycobacterium phlei	567 tr 557	538	416	—	Revsin and Brodie (1969)
Rhodopseudomonas spheroides	570	540	415	—	Sasaki et al. (1970)
Tetrahymena pyriformis	572 tr 557	—	418	—	Perlish and Eichel (1971)

acid and DCIP showed a spectrum characteristic of the cytochrome o-CO complex. The carbon monoxide action spectrum of the succinate-reduced preparation showed that cytochrome o was the principal oxidase of this *Staph. aureus*. Although cyanide (5×10^{-2} M) inhibited the cytochrome system, formation of a specific complex could not be detected spectroscopically.

Taniguchi and Kamen (1965) associated the succinate oxidase activity (180 mμ moles O_2 per min. per mg protein) in purified membrane fractions from dark-grown *Rhodospirillum* with a b-type cytochrome which at low temperature had α and β maxima at 564 and 540 nm respectively. This finding could be correlated with the appearance in the ascorbate-DCIP reduced minus oxidized spectrum of a peak at 564 nm. A second cytochrome b present of low-oxidation reduction potential was associated with the succinate-dehydrogenase activity.

The CO-difference and CO-action spectra obtained were characteristic for cytochromes o. The CO-complex is found to the same extent in the ascorbate-DCIP reduced membrane fractions as when succinate is the reductant. When dithionite was used the amount of CO-complex formed was 25% greater. No cytochrome a or cc' could be detected in these preparations. This membrane-bound cytochrome reacted with externally added *R. rubrum* cytochrome c_2. The succinate oxidase activity was inhibited 40% and the cytochrome c_2 oxidase activity 90% by cyanide (0·01 M). With azide (0·01 M) succinate oxidase activity was inhibited 80% and that of the cytochrome c_2 oxidase 90%.

Cytochrome o was the only detectable oxidase in particles from anaerobic-light grown *Rhodopseudomonas spheroides* but from aerobic-dark grown cells an a type cytochrome was the principal oxidase and cytochrome o was a minor component (Sasaki et al., 1970): incubation under low O_2 concentration of the light grown cells induced cytochrome o and repressed formation of the cytochrome a.

The cytochrome o components of *R. rubrum* (Taniguchi and Kamen, 1965) or *Staph. aureus* (Taber and Morrison, 1964) remained oxidized in the presence of 2-n-heptyl-4-hydroxyquinoline N-oxide and thereby can be distinguished from other b-type cytochromes in low-termperature difference spectra. Appleby (1969b) however found that the cytochrome o of *Rhizobium* free living cells and particles from them, characterized by photoaction spectra, was completely reducible by succinate in the presence of HQNO.

The photoaction spectrum of *R. rubrum* was also determined by Horio and Taylor (1965) who drew attention to the similarities between the spectrum of cytochrome o and that of cc'. The differences they concluded were no greater than the variations found amongst the cytochromes o and cc' cytochromes from different organisms. Their tentative identification of cytochrome cc' with cytochrome o cannot be sustained in the light of the evidence of the differences between the prosthetic groups of cytochrome cc' and the cytochrome o of other micro-organisms. Furthermore, while the rate of oxidation of cytochrome o is slower than that of yeast cytochrome oxidase, cytochrome cc' is very much less reactive in this respect, and the kinetic studies by Chance et al., (1966) appear to exclude an oxidase action for cytochrome cc'.

Cytochrome o is bound to the membrane in these bacteria and though various finely dispersed preparations have been obtained these do not appear to be truly soluble. Of the so-called soluble preparations none fulfill the criteria of high oxidase activity, a photoaction spectrum confirming that any respiration is related to the soluble cytochrome o, reducibility with ascorbic DCIP and ascorbic-TMPD and unaltered position of α and Soret maxima in the CO difference spectrum.

Three *Cyanophyta*, colourless algae, studied by Webster and Hackett (1966a) contained no a-type cytochromes but possessed CO-bonding pigments having absorption peaks at 570–535 and 416 nm. The photoaction spectrum for the relief of CO-inhibited respiration of *Vitreoscilla*, one of the *Cyanophyta*, confirmed that this pigment was the terminal oxidase. Two separate b-type cytochromes were obtained in highly purified form from *Vitreoscilla* (Webster and Hackett, 1966b). The cytochrome designated by fraction I had CO-difference spectra maxima at 570, 534 and 419 nm, was slowly autoxidizable and had a E_0' of $+ 1$ mV: the molecular weight was 27,000. Two α and two Soret bands were apparent in the liquid nitrogen difference spectra. Fraction II had CO-difference peaks at 566, 532 and 418 nm, a molecular weight of 22,500, an E_0' of $- 90$ mV volt and was rapidly autoxidizable. Both pigments combined with cyanide. Absolute spectra in 10^{-3} M KCN have absorption maxima at 540 (broad) and 416 nm. Cyanide difference spectra of both pigments are likewise similar and have maxima at 555 and 421 nm. Fraction II because of its extreme autoxidizability was considered to be a soluble form of the membrane-bound cytochrome o. The major objection that the E_0' of this fraction was too low for an oxidase was countered by the argument that the E_0' of this cytochrome could be different in the lipoid environment of the membrane. This argument could equally be applied to explain the low autoxidizability of Fraction I which has a higher E_0'. An alternative possibility that both these CO-reactive components combine in the membrane to form physiological cytochrome o was also suggested by Webster and Hackett (1969b).

Following analysis and concentration on Sephadex a cytochrome o preparation from *H. cutirubrum* (Cheah, 1969) had its CO-Soret peak displaced from 418 to 413 nm. A cytochrome b-558, identified as the probable effective cytochrome o of a strain of *Acetobacter suboxydans* (Iwasaki, 1966) grown on lactate was isolated and crystallized (Iwasaki, 1967). Oxidation of lactate in membrane fragments was stimulated when partially purified cytochrome b-558 was added to *A. suboxydans* membrane fragments which contained cytochrome a and another cytochrome b, but had no lactate oxidase activity without the added cytochrome. However, there had been no report that the crystallized b-558 retains cytochrome oxidase activity.

Broberg and Smith (1967) removed the cytochrome o of *Bacillus megaterium* KM (which also contains a^3) from the cell membrane by combining the latter with pancreatic lipase. In contrast to the behaviour of cytochrome o of the untreated membrane the cytochrome released into supernatant from the lipase-treated membrane was only slowly reduced by dithionite and the Soret peak of the dithionite reduced − oxidized difference spectra was shifted from 427 to 425 nm.

Cytochrome o and aa_3 occur together in *Mycobacterium phlei* and appear to be located in the segment of the respiratory chain containing the third site of phosphorylation (Orme et al., 1969; Revsin and Brodie, 1969). The complex $o + aa_3$ was solubilized with Triton X-100, partially digested with trypsin and fractionated with ammonium sulphate; a sevenfold purification in terms of the aa_3 content was achieved (Revsin et al., 1970a, 1970b). Over a pH range of 6·0–8·0 both the o and aa_3 components were reducible by ascorbate-TMPD or by NADH plus *M. phlei* electron transport particles. As observed with intact particles low concentrations of inorganic phosphate over the pH range 6·0–7·5 stimulated the rate of reduction of cytochromes aa_3 in the purified complex by ascorbate-TMPD, but at pH 8·0 similar phosphate concentrations enhanced the appearance of cytochrome o-CO in the CO-difference spectrum (Revsin et al., 1970a,b: Revsin and Brodie, 1967). The cytochrome o, but not aa_3 was refractory to reduction with dithionite unless the complex had had prior treatment with CO or cyanide (Revsin et al., 1970b). The evidence is not adequate to permit a correct explanation of this observation to be made.

Other reports have been made of cytochrome o and aa_3 occurring in the same organism, the relative amounts usually depending on the growth phase. With one exception none of the cytochromes o have been verified by photoaction spectra. Cytochrome aa_3 (α maximum of reduced aa_3 =607 nm) is the only detectable functional oxidase in membranes from exponential phase *Microccoccus denitrificans*, but the CO-difference spectrum of stationary phase cell had a shoulder at 415 nm (Scholes and Smith, 1968). As the intensity of absorption of the band at 415 nm is increased by treatment of the membranes with Triton X-100 or desoxycholate, the original 415 nm band may have arisen from a labile cytochrome b. A CO-combining pigment with a CO-difference peak at 422 nm in particles from both aerobic and anaerobic plus nitrate grown *M. denitrificans* had its CO-difference peak shifted to 418 nm on solubilization with Triton X-100

(Porra and Lascelles, 1965). More doubtful is the occurrence of cytochrome aa_3 in *Leptospira sp.* in which peaks were observed at 615, 553 and 425 nm in the oxidized-reduced difference spectrum of particles (Baseman and Cox, 1969). The Soret band of neither the oxidized-reduced nor the CO-reduced difference spectra showed evidence of either aa_3, or a_1. In contrast the presence of cytochrome aa_3 and o in free living *Rhizobium japonicum* is well established by spectrophotometric and inhibitor studies and by photoaction spectra (Appleby, 1969b). The bacteroids of N_2-fixing soyabean root nodules (Appleby, 1969a) and yellow lupin (Melik-Savkisyan et al., 1969) lack cytochrome oxidases.

Carbon monoxide difference spectra indicate the presence of cytochrome o in *Thiobacillus neapolitanus* and the oxygen affinity of the respiratory system is much lower than that for aa_3 (Ikuma and Hempfling, 1965). Cytochrome o and aa_3 occur together in *Thiobacillus denitrificans* (Peeters and Aleem 1970; Milhaud et al., 1958) and in *Hydrogenomonas eutropha* (Ishaque and Aleem, 1970).

Cytochrome o can be formed in large amounts under anaerobic conditions. Anaerobic cultures of *Haemophilus parainfluenzae* grown in the presence of nitrate or pyruvate contained three times as much cytochrome o as aerobic cultures (Sinclair and White, 1970). Anaerobically grown *Staphylococcus epidermidis* AT2 grown in a complex medium containing protohaemin formed nearly as much cytochrome o as did aerobically grown cells without added protohaemin (Jacobs and Conti, 1965). Cytochrome a was also formed in the *Staphylococcus* in both instances.

The relative activity of cytochrome o and the other cytochrome oxidases present in the same organism has been little explored. Studies with membranes, rather than whole cells should be assessed with caution. In whole cells of *H. parainfluenzae* Smith et al. (1970) found that cytochrome o reacts as rapidly with O_2 as does mammalian oxidase in heart muscle. The special topic of branched chain electron transfers in bacteria has been reviewed by White and Sinclair (1970) and factors affecting the different electron transfer systems in bacteria have also been discussed extensively by Smith (1968).

Although there have been reports of cytochrome o in organisms other than bacteria and the *Cyanophyta*, in none of these instances has the cytochrome o been rigorously identified. A haemoprotein with a cytochrome b-type reduced-oxidised spectrum found in a preparation from *S. cerevisiae* (Mok et al., 1969) had the Soret peak of its CO-difference spectrum at 408 nm, well outside the range for bacterial cytochromes o. The flavohaemoprotein, sulphate reductase of baker's yeast had a CO-difference spectrum for the dithionite reduced enzyme characteristic of cytochrome o (Soret peak at 416 nm) (Prabhakararao and Nicholas, 1969). A CO-reactive haemoprotein resembling cytochrome o has been reported in a parasitic worm *Moniezia expansa* (Cheah, 1968). In none of these instances has a photo-action spectrum of the presumptive cytochrome o been obtained.

(c) *Carbon monoxide and cyanide reactivity*. Respiratory systems in which cytochrome o is the terminal oxidase differ in their sensitivity to CO. A particulate preparation from dark grown *Rhodopseudomonas capsulata* (Klemme and Schlegel, 1969) containing a cytochrome b, maxima at 562, 525 and 426 nm but no a-type cytochromes, reacted with horse cytochrome c, DCIP and TMPD. Although the oxidase was inhibited by KCN and NaN$_3$ (half inhibition at 10 μM) it was active in the presence of a gas mixture of CO 90% + O$_2$ 10%. Almost no inhibition by CO was observed for the cytochrome o of dark-aerobic grown *R. rubrum* and in *Rhodopseudomonas spheroides* where an a-type cytochrome was the principal oxidase (Kikuchi and Motakawa, 1968; Kikuchi *et al.*, 1965). These preparations were solubilized with Triton X-100 and sodium cholate. One fourth or less of the respiration of membrane fragments from *B. megaterium* KM (Broberg and Smith, 1967) is relatively insensitive to CO, 13% of the respiration remained uninhibited in the presence of CO 95% O$_2$ 5%. The CO-inhibited oxidase was presumed to be cytochrome a_3 which in *B. megaterium* appears to be considerably more sensitive to CO than does the cytochrome a_3 of bakers' yeast (Castor and Chance, 1959).

A consequence of the differing sensitivity of the cytochromes o and other oxidases to CO is that the photo-action spectrum of a mixture of oxidases operating in parallel may not reflect accurately the relative activities of the oxidase (cf. Castor and Chance, 1959; Horio and Taylor, 1965).

Cytochrome o may combine with CO more slowly than does cytochrome a_3. Broberg and Smith (1967) found that for the full formation of the CO-complex of cytochrome o in membrane fragments of *B. megaterium* KM 15 minutes bubbling with CO was required, in contrast to cytochrome a_3 which had fully combined with CO in 30 seconds. This imposes difficulties on the estimation of both oxidases where the cytochrome content is high. Similar relative reactivities towards CO were found for the cytochromes o and a_3 of *H. cutirubrum* (Cheah, 1969) electron transport particles. In contrast the ascorbate-TMPD reduced cytochrome o of membrane fragments of *Mycobacterium phlei* (Revsin and Brodie, 1969) had fully combined with CO, as had cytochrome a_3, after 30 seconds of bubbling with CO. Solubilisation of the cytochrome o with pancreatic lipase did not diminish its affinity for CO. Exposure to urea, but not to guanidine, caused a large increase in the amounts of apparent CO-cytochrome o component in the CO-difference spectrum of the cytochrome a_3 complex from *M. phlei* (Revsin *et al.*, 1970a). Daniel (1970) has investigated the CO-affinity of the cytochrome o of *A. suboxydans*. The cytochrome which had a CO-peak at 560 nm was half-combined with CO at a CO concentration of 3·3 μM. The slope of the Hill plot was n = 1·9 and Daniel concluded that this indicated at least two co-operative CO-reactive sites. On standing at 4° the cytochrome changed sufficiently for the Hill plot value to fall to one. The other cytochrome which had a CO-peak at 568 nm was half-combined with CO at a level of 2 μM and the slope of the Hill plot did not significantly differ from $n = 1$.

Cytochrome o appears to be less sensitive to cyanide than is cytochrome a_3 and the degree of inhibition varies with the system in which cytochrome o is the terminal electron acceptor. The *A. suboxydans* NADH-oxidase was inhibited 65% by 0·75 mM KCN and the same system in aerobically-dark grown, *R. rubrum* (Taniguchi and Kamen, 1965) was not significantly affected by 0·3 mM KCN unless there was a period of pre-incubation. The ascorbate-DCIP system of *Acetobacter vinelandii* (Jones and Redfearn, 1967a) was 95% inhibited by 40 μM KCN. Cheah (1969) reported that the ascorbate-TMPD oxidase of *H. cutirubrum* was inhibited 84% by 0·5 μM CN^-. The cytochrome c oxidase system of aerobically-dark grown *R. rubrum* was 50% inhibited by 1·5 μM CN^- (Sasaki et al., 1970). In contrast to these, the succinate oxidase of free living cells of *Rhizobium japonicum* is very sensitive to CN^-, being completely inhibited at a level of 10 μM KCN (Appleby, 1969a,b). Significant spectral changes due to formation of a cyanide complex have not been observed for cytochrome o in active preparations; at the levels of CN^- used some of the inhibition observed may have been due to its action at a site other than the haem of cytochrome o.

CO can displace cyanide from its complex with cytochrome o whilst cyanide bound to haem a of the a-type cytochromes is not thus displaced. This difference in stability of the CN-complex of these two types of cytochrome oxidases has been used by Cheah (1970a,b) for quantitative determination of cytochrome o in the presence of cytochrome a_1 or a_3. However where the cytochrome is very sensitive to CN, as in *Rhizobium japonicum* the strength of the CN-protohaem bond does not permit displacement of the CN^- by CO.

3. Cytochrome d (a_2)

(a) *Introduction.* Cells of *E. coli* and *Shigella dysenteriae* grown aerobically were shown by Yaoi and Tamiya (1928) to possess an absorption band which differed from those of previously described cytochromes in that it lay well within the red region of the spectrum, at 630 nm. Keilin (1933) attributed this band to a component, designated a_2, of the cytochrome system of those bacteria. Negelein and Gerischer (1934) Tamiya and Yamaguchi (1933) and Fujita and Kodama (1934) independently found that cytochrome a_2 was autoxidizable and could combine with CO and cyanide. Unlike other cytochromes the oxidized form exhibited a strong band in the visible region of the spectrum at 645 nm. Fujita and Kodama (1934) also found that this cytochrome was widely distributed amongst

TABLE XII
Cytochromes d, d_1c and their derivatives

Cytochrome	Absorption maxima in nm; emM in brackets					Eγ/Eα	Authors
Cytochrome d* (*Escherichia coli*)							Negelein & Gerischer (1934)
Oxidized				647–643			Fujita & Kodama (1934)
Reduced				632–638			
Reduced plus CO cyanide				635–637			
Haemin-d	740			603			Barrett (1956)
Haematin-d				664			
Haem				618			
Haem-d plus CO				620–618			
Pyridine haemochrome-d				613			
Cyanide haemochrome-d				618			
Cytochrome d_1c (*Pseudomonas*)							
Oxidized,† pH 7·0				635	525	412	Yamanaka & Okunuki (1936b)
Reduced, pH 7·0			554, 549	652, 629	523 (460)‡	418	2·5
Reduced, pH 7·6			554, 549	655 (620)	523 (460)	418	2·7
Reduced, pH 5·6			554, 549	625	523 (460)	418	1·9
CO-reduced pH 7·0				658, 622			
CN-reduced pH 7·0				627	(472)	443–418	2·3
NO-reduced			554, 549	665	523		
Cytochrome d_1c (*Micrococcus*)							Newton (1969)
Oxidized, pH 6·0	702		(560)	640	525	435	(408)
Reduced, pH 6·0			553, 547	(655), 625	521	460	(418)
Reduced, pH 8·2			553, 547	655 (620)	520	460	(418)
Reduced, pH 3·9			553, 548	620	521	460	(418)
CO-reduced (pH 6·0)			553, 547	660–630	520		415
Pyridine haemochrome			549	618	520 (460)		412
Imidazole haemochrome			553, 547	625	520 (460)		418
CN-reduced (pH 7·5)			553, 547	625	520	475–445	418

NO$_2$-reduced (pH 6.0)						Newton (1969)
Initial	660–615	553, 547	520			
Final	630	560	520	455	411	
Cytochrome d_1c						Iwasaki and Matsubara (1971)
(*Alcaligenes faecalis*)						
Oxidized pH, 7.0	640				412	
ε mM	[21.2]	556	525	460	[151]	
	625	28.7	26.5	44.5	418	
Reduced, pH 7.0	18.5				189	
Haemin d_1	606	569	529		432	Yamanaka and Okunuki (1963c)
Haematin d_1	684			480	405	3.5
	[2.7]					3.0
Haem d_1	626				409	
	[14]					1.8
Pyridine-haemochrome d_1	620			462	436	408	119
	[24]						
Cyanide-haemochrome-d_1	632			480	449	(414)	
	[20]						2.0
NO-haem d_1	625			462	436	408	
	[20]						2.8
NO-haematin d_1	645					457	
	[13]						2.0
CO-haem d_1	644 (594)				(440)	413	
	[20]						

* By analogy with the *Pseudomonas* cytochrome d_1c εmM = 8·5 has been used in the reduced-oxidized spectrum for the wavelength pairs 630–615 nm (Jones and Redfern, 1966).
† Molar extinction values for the oxidized and reduced cytochrome d_1c for a less pure preparation are given by Horio et al. (1961).
‡ Indicates minor bands.

other bacteria, e.g. *Azobacter chroococcum, Proteus vulgaris, Acetobacter pasteurianum, Erbethella typhosum* and *Salmonella paratyphi*, and Moss and Tchan (1958) found it in several other *Azotobacter* sp., in *Azomonas insigne* and *Beijerinckia* species. Cytochrome a_2, with a_1, has been found in *Pseudomonas syringae*, and other closely related plant pathogens, which lack cytochrome c (Sands et al., 1967).

As the prosthetic group of cytochrome a_2 was shown to be an iron-chlorin (Barrett, 1956), this cytochrome was reclassified as cytochrome d (Commission on Enzymes, 1961), but much current literature still retains the original designation, a_2. Because of similarities in the red region of the absorption spectra of the oxidized and reduced forms of d and d_1c, and of their CO and cyanide complexes, confusion of identity may occur if examination is restricted to whole cells. Cytochrome d is characteristically tightly bound to the cytoplasmic membrane and thus differs from the soluble cytochromes of d_1c type which occur in *Pseudonomas* sp. and other bacteria.

(b) *Spectral properties*. No cell-free preparation of cytochrome d has been obtained which does not also contain cytochrome b_1c and a_1, and in one instance cytochrome aa_3 (Arima and Oka, 1965). Consequently the spectrophotometric data pertaining to cytochrome d are limited to the red end of the spectrum. The extinction of the Soret band can be expected, because of the chlorin-type haem, to be lower than that of the other cytochromes and masked by the cytochrome b_1 absorption peak in the 400–430 nm region. The properties of the haem complex do not suggest that the Soret peak of cytochrome d is likely to be displaced far from that of cytochrome b_1.

Table XII gives the position of the α band of cytochrome d and related compounds and derivatives.

(c) *Particulate preparation of cytochrome d*. Keilin and Harpley (1941) noted that cytochrome d in crushed *E. coli* could not oxidize mammalian cytochrome c. Moss (1954) obtained a particulate preparation from *A. aerogenes* by sonication at 9 KC which contained cytochrome d and a_1. Attempts to solubilize cytochrome d in deoxycholate, cetyl ammonium bromide and Teepol caused destruction of the cytochrome. A preparation obtained by grinding *A. aerogenes* with alumina contained cytochrome a_1 and succinate oxidase activity but the cytochrome d had been destroyed (Tissières, 1952b). This lability of cytochrome d may arise from the greater susceptibility of chlorin haems to oxidation by lipoperoxides.

Tissières (1956) found that the membrane fraction from *A. vinelandii* showed a relative intensification of the cytochrome d and b_1 and a decrease in cytochromes c. Membrane preparation obtained by lysozyme-EDTA treatment of *E. coli* at pH 9·0 contained cytochrome d (Tissières, 1961). Cytochrome d in respiratory particles from *H. parainfluenzae* could be fully reduced with succinate and NADH (Smith and White, 1962).

A significant advance was made by Jones and Redfearn (1967a,b) who achieved a physical separation of the branched pathways of electron flow in *A. vinelandii*. Small particles termed oxidosomes derived from a membrane preparation, were further fractionated by treatment with deoxycholate followed by density gradient centrifugation. The lighter red fraction was enriched in cytochromes o and $c_4 + c_5$ while the denser green particles contained a larger amount of cytochrome d and a_1. The cytochrome b_1 content of the red particles was twice as great as that of the green particles. The molar ratio $(c_4 + c_5)/d$ was 8·13 for the red particles and 0·87 for the green particles. Succinate oxidase and dehydrogenase activity and ascorbate-DCIP oxidase activity were much greater in the cytochrome o enriched particles, but the NADH oxidase activity had been totally destroyed by the deoxycholate treatment.

Barrett and Sinclair (1967a) obtained a cytochrome d enriched preparation from *A. aerogenes* by digesting particulate material with lysozyme and ribonuclease; the preparation after ammonium sulphate fractionation had succinoxidase activity. Digestion with trypsin at 4° in 1% cholate assisted the removal of some cytochrome b_1 and a_1. Lysozyme and ribonuclease treatment of particles obtained by high speed mixing of *E. coli* str. McElroy gave a preparation in which cytochrome d stayed in solution at 200,000 × g and was retarded on a G 100 Sephadex column. However, the cytochrome d and b_1 remained closely associated and are probably linked with membrane components. Cytochrome d was considerably more sensitive to destruction by alkali (pH 9·0) than was cytochrome b_1.

(d) *Occurrence and role as oxidase.* With few exceptions cytochrome d has been found only in organisms which also contain the cytochrome oxidases a_1 and o. Arima and Oka (1965) obtained spectroscopic evidence for the presence of cytochrome aa_3 occuring together with cytochrome d in a particle preparation from log phase cells of *Pseudomonas riboflavina* grown aerobically, and Barrett (1956) extracted haem d from cells of *Bacillus subtilis* and *Bacillus mycoides* known to contain cytochrome aa_3. Moss (1971) observed a band at 628–630 nm in reflectance spectra of *Bacillus* species containing cytochrome aa_3.

After the initial assumption that cytochrome d was an oxidase, certain doubts arose particularly amongst the early workers in the field, before the concepts of multiple oxidases and branched respiratory pathways became current, about the certainty of the function of cytochrome d. There is little correlation between the cytochrome d of bacteria and their respiratory activity (Tissières, 1952a; Moss, 1952; Castor and Chance, (1959). Cytochrome d in *E. coli* (Moss, 1952) and *A. aerogenes* (Moss, 1956) is formed only at low oxygen tension; for *A. aerogenes* the optimal environmental O_2 concentration was 10^{-6} M. In *Haemophilus parainfluenzae* the cytochrome d content was over a 100 fold greater in cells grown with poor aeration than in cells grown with vigorous aeration (White, 1967): there was no significant difference in the levels of cytochrome a_1 and o in organisms grown under the two aeration states.

The earlier literature (cf. Castor and Chance, 1959) contains conflicting evidence on the relief by light of the CO inhibition of respiration in cells containing cytochrome d. Castor and Chance (1959) obtained photoaction spectra of the relief of CO inhibition in several bacteria which provided *prima facie* evidence that cytochrome d is an oxidase. They, however, pointed out that respiration in the presence of CO is affected by the cytochrome with the least affinity for CO but that the relative importance of the individual oxidases may be different in the absence of CO. Furthermore, although photoaction spectra may demonstrate the existence of more than one pigment, no simple relationship exists between the oxidase activity of a pigment and height of the bands in the photoaction spectrum.

Metal deficient cells of *A. aerogenes* have a high O_2 uptake, although the absorption band of cytochrome d is hardly detectable (Tissières, 1952a). The absorption bands of cytochrome a_1 and b_1, however, were only reduced to one half their usual intensity. On the other hand Mizushima *et al.* (1960) noted a parallelism between the cytochrome d content and the cyanide resistant respiratory activity of *Achromobacter*. Smith *et al.* (1970) from a study of the kinetics of oxidation of the cytochromes of *H. parainfluenzae* using a rapid flow device found that cytochrome d reacted very rapidly with oxygen, while cytochrome a_1 reacted sluggishly.

The relative insensitivity of the CO-inhibited NADH oxidase of *A. vinelandii* to white light, the ability of red light, but not blue light to relieve substantially this inhibition, the failure of low concentrations of KCN to abolish this light relief of CO-inhibition and the relative insensitivity of the NADH oxidase to KCN and azide, point to cytochrome d as being the major terminal oxidase of the NADH pathway (Jones and Redfearn, 1967a).

Forty years after the discovery of cytochrome d it still cannot be said that the evidence for its role as an oxidase and its relative importance as such is conclusive. Clarification of its function must await the obtaining of a purified cytochrome d or the discovery of a mutant devoid of other oxidases. A mutant strain of *E. coli* (Cox *et al.*, 1970) had a ratio of cytochrome d to cytochrome o content of 1 : 0·8 instead of 1 : 2·7 of the parent strain.

Cytochrome d (and a_1) formation in streptomycin-dependent strains of *E. coli* required the presence in the medium of dehydrostreptomycin (Bragg and Polglase, 1963). The cytochrome b_1 and ubiquinone-8 content of this strain was normal and not streptomycin-dependent. Another streptomycin-dependent strain of *E. coli* did not produce cytochrome d even when streptomycin was added to the medium (Engelberg and Artman,

1961). In this strain of *E. coli* cytochrome *o* was assumed to account for the CO-sensitive respiration, although spectral evidence was not given.

Cytochrome *d* was reduced by D(−)-lactate in particles from *A. vinelandii* (Jurtshuk and Harper, 1968) and was present in small particles also containing cytochromes a_1 and *o*, which had high NADH, succinate and TMPD-oxidase activity (Jurtshuk et al., 1969). Ageing of these particles increased threefold their rate of oxidation of c_4 and c_5 and mammalian cytochrome *c* (Jurtshuk and Old, 1968). With NADH, malate, succinate and ascorbate-DCIP as substrates cytochrome *d* in *A. vinelandii* particles remained oxidized in the aerobic steady state, while cytochromes a_1, b_1 and c_4 and c_5 were substantially reduced (Jones and Redfearn, 1967a). The highest O_2 consumption was with NADH as substrate. The synthesis of cytochrome *d*, but not c_4 and c_5 which are associated with the cytochrome *o* pathway is greatly diminished in cells grown with urea as nitrogen source (Knowles and Redfearn, 1968); the NADH oxidase activity is markedly lowered while that of the succinate oxidase is little affected.

The occurrence of cytochrome *d* in a particle preparation from an obligate anaerobe *Desulphovibrio africanus* (Jones, 1971a) confirmed by identification of the prosthetic group as haem *d* (Jones, 1971b) implies that in this organism sulphate may be the ultimate acceptor of electrons from the cytochrome *d*.

(*e*) *Function as nitrate reductase.* Low concentrations of nitrate (10 mM) stimulate biosynthesis of cytochrome *d* in *E. coli* strains grown anaerobically under continuous culture conditions (Wimpenny and Cole, 1968; Cole and Wimpenny, 1968), but this is probably coincidental to the much greater stimulation of cytochromes *c*-552 and b_1 synthesis.

In general nitrate represses the synthesis of cytochrome *d*. Potassium nitrate at a level of 2·0% almost completely repressed cytochrome *d* synthesis in *E. coli* str. McElroy and *A. aerogenes* in aerated cultures; associated with this was a marked decrease in the amount of membrane fraction obtained from broken cells (Barrett and Sinclair, 1967b). The concentration of cytochrome *d* in cells of *H. parainfluenzae* grown anaerobically with nitrate is less than a sixth of that in cells grown with fumarate as the terminal acceptor (Sinclair and White, 1970). A similar finding was obtained with wild and mutant strains of *Hafnia* (Chippaux and Pichinoty, 1968).

Exceptions occur: nitrate does not repress cytochrome *d* synthesis in the cyanide-resistant strain of *Achromobacter* (Arima and Oka, 1965), nor in a low cytochrome *c* mutant of *H. parainfluenzae*. Cytochrome *d* synthesis in the wild strain of *H. parainfluenzae* is not repressed by nitrate in the presence of oxygen, and Sinclair and White (1970) have suggested that the oxygen effect is due to repression of nitrate reductase synthesis. However, in the *H. parainfluenzae* mutant there is ample nitrate reductase activity and cytochrome d synthesis is not repressed.

Cytochrome d in the cyanide-resistant *Achromobacter* is rapidly oxidized by nitrite, but not by nitrate (Arima and Oka, 1965). Nitrite appears to act as a competitive inhibitor of the oxidase for oxygen but is not used as an electron acceptor in this organism. Nitrate, but not nitrite, has been reported to oxidize cytochrome d of *H. parainfluenzae* in particles and in whole cells, but not as rapidly as it oxidizes the cytochrome c_1 (White and Smith, 1962). In the presence of formate the cytochrome d of *A. vinelandii* and other cytochromes were oxidized by nitrate and this oxidation was azide-sensitive (Cole and Wimpenny, 1968). Nitrite did not oxidize cytochrome d or a_1.

(*f*) *Reaction with cyanide.* Cytochrome d forms a complex with cyanide with an absorption maxima at 645 nm. No studies have been made of the affinity of cytochrome d or haem d for CN^- but there have been a number of observations with whole bacteria or particulate preparations which indicate that the cytochrome oxidase activity of cytochrome d is less sensitive to inhibition with cyanide than that of other oxidases. Jones and Redfearn (1967a) noted the relative insensitivity of the cytochrome $d + b_1$ electron pathway of *A. vinelandii* to cyanide, and to azide.

The cytochrome d of a particle preparation from cyanide resistant *Achromobacter* was slowly and only partially reduced in the presence of CN^- (10^{-3} M) with succinate or NADH as electron donor (Mizushima and Arima, 1960). Cytochrome d was more readily oxidized than cytochromes a_1 or b_1 in this cyanide-resistant strain in the presence of cyanide (Arima and Oka, 1965). The oxidase activity of cytochrome d in *E. coli*, *A. aerogenes* and *P. pseudomallei* was less sensitive to inhibition with cyanide than was the cytochrome a_1 oxidase activity. The cytochrome d content of aerated cultures of cyanide resistant *Achromobacter* was increased three-fold when KCN (10 mM) was present in the growth medium (Arima and Oka, 1965).

Finally photoaction spectra show that cytochrome d is responsible for the cyanide-resistant respiration in cyanide-resistant cells (Oka and Arima, 1965) although cytochrome o is the active oxidase in cyanide-sensitive *Achromobacter*.

4. Pseudomonas and Micrococcus cytochrome d_1c.

(*a*) *Introduction.* The finding in *P. aeruginosa* of an unusual greenish-brown soluble cytochrome (Horio, 1958a,b) which has a d-type and c-type haem in one protein molecule revealed a new class of cytochromes. A second cytochrome of this class has been obtained from *M. denitrificans* (Newton, 1967, 1969; Lam and Nicholas, 1969) from *Alcaligenes faecalis* (Iwasaki and Matsubara, 1971) and tentatively identified in a *Spirillum itersonii* (Clark-Walker, 1970). The first cytochrome of this type discovered has been referred to as *Pseudomonas* oxidase, although in the absence of nitrate it is not formed in aerated cultures of *P. aeruginosa* and its nitrite reductase activity is more pronounced than its oxidase activity. In *Micrococcus* the induction of the d_1c-type cytochrome is totally repressed in aerated cultures even in the presence of nitrate (Newton, 1969; Lam and Nicholas, 1969) and the oxidase activity of the purified cytochrome is very low.

(b) *Preparation*. Cytochrome d_1c can readily be obtained in soluble form by sonication of the cells, or for larger preparations by extraction of acetone-dried, or lyophilized cells, with 0·1 M Na citrate (Horio *et al.*, 1961a). Soaking bacteria in 50% saturated $(NH_4)_2SO_4$ overnight improved the extraction of the cytochrome from the cells (Kijimoto, 1968a). Rivanol, a substituted acridine, has been used by the Japanese workers to remove viscous material, but can have a deleterious effect on the enzyme (Kijimoto, 1968a). Streptomycin (1%) was equally as effective without harmful effect (Barrett and Collins, 1968). After precipitation of the enzyme with ammonium sulphate and dialysis against pH 6·0 buffer the enzyme was chromatographed on Amberlite (CG 50 type II) resin columns at that pH, and thereafter crystallized by addition of ammonium sulphate to the eluate. Diamond-shaped crystals appeared after 30 minutes standing at 30° (Yamanaka and Okunuki, 1963a); their E_{412}/E_{280} ratio was 1·2. *Pseudomonas* cytochrome d_1c could be freed from a brown-red impurity, probably a cytochrome c, by chromatography on Sephadex G-75 at pH 7·0 prior to crystallization (Kijimoto, 1968a). *Pseudomonas* cytochrome d_1c was not adsorbed on DEAE-cellulose (Newton, 1969).

Release of cytochrome d_1c from *M. denitrificans* required the use of a French Press followed by the removal of viscid material by deoxyribonuclease (Newton, 1969). Chromatography on DEAE-cellulose at pH 6·0 and on hydroxyapatite columns separated a major cytochrome d_1c component from two minor components which differed in spectral detail from the major component. Lam and Nicholas (1969) used an essentially similar procedure to purify the nitrite reductase (cytochrome d_1c of *M. denitrificans*, str. AM.W. NC1B 8944.

(c) *Physical properties*. A M.W. of 94,000 for the *Pseudomonas* protein was obtained from sedimentation and diffusion studies (Horio *et al.*, 1961a) and a value of 85,000 was obtained from studies using Sephadex gel (Newton, 1969). By the latter technique a molecular weight of 120,000 was estimated for the cytochrome d_1c of *M. denitrificans* (Newton, 1969) and similarly a M.W. of 90,000 was found for the crystallized cytochrome d_1c of *A. faecalis* (Iwasaki and Matsubara, 1971). Amino-acid analyses have recently shown (Nagata *et al.*, 1970) that the molecular weight of the *Pseudomonas* cytochrome is 67,325 (597 amino acid residues and two haems). The larger molecular weight of the *Micrococcus* enzyme may then be that of a dimer. By extrapolation to zero mobility an isoelectric point at pH 5·8 was estimated for the *Pseudomonas* cytochrome d_1c (Horio et al., 1961a) but the pH focussing technique gave an isoelectric point at pH 7·1 for *Pseudomonas* cytochrome and pH 3·85 for that from *Micrococcus* (Newton, 1969). The higher isoelectric point of the *Pseudomonas* cytochrome is surprising as it has a considerable excess (35) of acid residues over basic residues.

After the haem d_1 has been removed by acetone-HCl the resulting haem *c*-protein moiety is insoluble at neutral pH (Yamanaka and Okunuki, 1963c). The small excess (7) of hydrophilic amino acid residues may explain this insolubility (Nagata et al., 1970). Only two cysteines are present in this large molecule, and these are bound to the haem *c* so that no -SH group is necessary for either of the enzymic activities of *Pseudomonas* cytochrome d_1c. Two iron atoms were found corresponding to the two haems. No copper was detected in the *Pseudomonas* cytochrome, but a separate copper protein (azurin) is present in this organism (Horio, 1958b; Yamanaka et al., 1963a). A blue protein, probably a copper-protein has also been isolated from *M. denitrificans* (Newton, 1967).

The two haems are so situated in the proteins, or protein complex, that they maintain independent oxidation potentials. Electrochemical titration of the *Pseudomonas* cytochrome (70% pure) showed that the haem *d* component has an E_0' (pH 7·0, 24°) 70 mV more negative (Horio et al., 1961b) than that of the cytochrome *c*-550 component which has an E_0' (pH 7·0, 20°) of + 288 mV (Kamen and Horio, 1970).

The spectra of cytochrome d_1c from both bacteria (Table XII) is a composite of the spectra of a cytochrome *c* which has a split α-band and a cytochrome with a chlorin type haem. Although the band of the reduced form of the cytochrome *c* component places it in the class of aberrant cytochromes *c* having an unusual environment in the immediate vicinity of the haem *c*, most of the interest in the spectra has centred on the contributions of the green haem.

In both *Pseudomonas* and *Micrococcus* ferrous cytochrome d_1c the position and shape of the absorption band in the red region of the spectrum is pH-dependent. In whole cells and crude preparations this maximum of absorption is generally observed at 625–630 nm. A small shoulder or peak (at 465 nm) appears on the side of the large Soret absorption band and is enhanced in the reduced minus oxidized difference spectrum. This peak has been described as the Soret peak of the haem d_1 component of cytochrome d_1c, but is probably a satellite band of the main Soret band of that component. The pyridine haemochrome of haem d_1 which has a peak at 432 nm has satellite bands at 454 and 408 nm (Yamanaka, 1963). The spectrum of the cytochrome *c*-550 component is pH-independent.

Low temperature spectra (Newton, 1969) of the reduced *Micrococcus* cytochrome d_1c show enhanced splitting of the double α-band of the cytochrome c component, while the β-band shows a major and three minor peaks; the weak band at 700 nm of the oxidized form is less apparent at 77°K.

(d) *Cytochrome d_1c complexes with ligands.* In pyridine the cytochrome d_1c gives a typical haemochrome spectrum with a sharp band at 549 nm and an Eα/Eβ ratio of 1·66, instead of the ratio of 1·0 in the native cytochrome. The green haem is displaced from the protein to give a green pyridine haemochrome without added reductant, a feature characteristic of chlorin haems. This auto-reduction of the green haem in alkaline pyridine contrasts with the very slow reduction of the d component of the cytochrome. The effect of other ligands is summarized in Table XII.

The spectrum of the CO complex differs little in the visible region from that of the ferro-cytochrome d_1c except that the absorption maximum in the red region is shifted a few nm to the red and the 460 nm band is absent, thus strengthening the assignment of the origin of this band to the haem d_1 component.

In the presence of CN^- (10^{-3} M) the 460 nm peak of the reduced cytochrome disappears and is replaced by one at 443 nm with a faint shoulder at 472 nm. Dithionite-reduced cytochrome d_1c formed a compound with NO_2^- and NO which had a maximum of absorption at 665 nm (Yamanaka and Okunuki, 1963b). Newton (1969) observed that NO_2^- caused the disappearance of the 435 nm shoulder and the 700 nm band of the oxidized cytochrome d_1c, with a sharpening of the 640 nm band. Under these conditions the haem c was reduced but after a few minutes became oxidized and strong bands developed at 455 and 630 nm. Cyclic oxidation and reduction could be maintained by discrete additions of dithionite and Newton considered these changes to be a possible model for nitrite reductase.

(e) *The haems of cytochrome d_1c.* The green haem is removed from the cytochrome by acetone-HCl or by methylethylketone-pH 3·5 buffer. The spectral properties of the *Pseudomonas* green haem and its derivatives have been described by Yamanaka and Okunuki (1963a), and the spectra of the haemin from the *Micrococcus* nitrite reductase are similar (Newton, 1967). Distinct spectral differences exist between these two haems and the haem d of cytochrome d (Barrett, 1956) as shown in Table XII.

These differences are emphasized in the spectra of the acid-haemin: in haem d the band in the red-green region is simple while that of haem d_1 is complex. The spectral characteristics of the "chlorin" and its Cu and Zn derivatives are markedly different to those of chlorin d (Barrett and Newton, 1968). Haem d_1 is hydrophilic in contrast to the lipophilic nature of haem d. The properties of haem d_1 suggest that this prosthetic group is not the iron complex of a true chlorin, but one in which the conjugation of the tetrapyrrole ring is modified by the introduction of hydroxyl groups.

(*f*) *Cytochrome d_1c: a complex of two haemoproteins.* Kijimoto (1968b) has suggested that *Pseudomonas* cytochrome d_1c may be a tight complex of two protein components, each carrying one haem. *Pseudomonas* cytochrome d_1c was treated with 1% sodium dodecyl sulphate prior to centrifugation in a sucrose density gradient. By this technique fractions of different M.W. having a low to high ratio of haem d_1 to haem c were obtained. The addition of the cytochrome c enriched component to the cytochrome d_1 enriched fractions increased the cytochrome oxidase activity of the latter (cytochrome c-551 as electron donor), but did not increase the nitrite reductase activity. End-group analysis has not confirmed whether or not cytochrome d_1c is a complex of individual haemoproteins, as the yield of di-(2,4-dinitrophenyl)-lysine was very low in the N-end group determinations of Nagata *et al.* (1970).

(*g*) *Enzyme activity of cytochrome d_1c.* The *Pseudomonas* cytochrome d_1c has both cytochrome oxidase and nitrite reductase activity, the latter appearing to be of greater physiological importance (Yamanaka and Okunuki, 1963a; Kijimoto, 1968a). In the presence of air one mole of the crystallized cytochrome d_1c oxidizes 154 and 600 moles of *Pseudomonas* ferrocytochrome c-551 per min at 16° and 27° respectively, while anaerobically 250 moles of the ferrocytochrome c-551 were oxidized per min per mole of the enzyme by nitrite at 19°. The turnover number at 37° was calculated to be 2,400 for the oxidase activity and 4,000 for the nitrite reductase (Yamanaka and Okunuki, 1963a). The Km for O_2 is 28 μM at pH 6·0 and 30° and the Km for nitrite is 53 μM at pH 6·0 and 27°. The pH optimum for the cytochrome oxidase is 5·1 and for the nitrite-reductase activity, 6·5; while nitrite-reductase activity is not inhibited by CO, the oxidase activity of the enzyme with various electron donors is inhibited by nitrite, (Yamanaka *et al.*, 1961). Cyanide on the other hand is a strong inhibitor of both the oxidase and nitrite reductase activity.

Kijimoto (1968a) has studied the oxidase kinetics of oxidation of *Pseudomonas* d_1c and the inhibition of the oxidase activity by nitrite with ascorbate as electron donor and has concluded that both ligands react with the same site, haem d, but that differences in the steric structures of the ligands cause a different degree of conformational change in the enzyme protein.

Pseudomonas cytochrome d_1c can also catalyse the oxidation by O_2 of the *Pseudomonas* blue copper-protein, E_0' at pH 6·4 = + 328 mV (Horio *et al.*, 1961a). The turnover number is very similar to that for the oxidation of reduced *Pseudomonas* cytochrome c-551. The nitrite-reductase preparation of *P. aeruginosa* (Walker and Nicholas, 1961) gave spectra which indicated the presence of cytochrome and copper-protein. Addition of copper stimulated the nitrite reductase activity of this preparation.

Enzymically active cytochrome d_1c has been reconstituted (Yamanaka and Okunuki, 1962, 1963d) from its haem d_1 and the protein moiety, which still retains the c type haem after the acetone-HCl split. The nitrite reductase and cytochrome oxidase activities of the reconstituted enzyme are 54% and 37% respectively of the native enzyme.

The cytochrome oxidase activity of the *Micrococcus* cytochrome d_1c is weak, particularly compared with that of the cytochrome aa_3 present in both aerobically and anaerobically grown cells. The turnovers of *Micrococcus* cytochrome c were 50 and 250 μmoles per μmole of haem d per minute for the oxidase and nitrite reductase activity respectively (Newton, 1969). Lam and Nicholas (1969) found that the Km for O_2 was about 27 μM and for NO_2^-, 46 μM, with horse heart cytochrome c as electron donor, and as for the *Pseudomonas* enzyme the oxidase activity was inhibited by NO_2^-. The pH optimum of the nitrite-reductase enzyme was 6·7 but for the oxidase it was 8·0. The cytochrome d_1c of *A. faecalis* also has strong nitrite reductase activity (Matsubara and Iwasaki, 1971).

(h) *Oxidase activity of Pseudomonas cytochrome d_1c with cytochromes c.* In addition to *P. aeruginosa* cytochrome c-551 *Pseudomonas* d_1c generally reacts with acidic cytochromes c. There are exceptions, however; *Chlorobium thiosulphatophilum* cytochrome c, although it has an isoelectric point at about pH 10·0, reacts fairly rapidly with *Pseudomonas* oxidase but poorly with cow oxidase (Yamanaka and Okunuki, 1968a). Reactivity of the cytochromes with *Pseudomonas* d_1c is not greatly affected by differences of some + 200 mV in oxidation-reduction potential. *Micrococcus* cytochrome c-554 ($E_0' = +$ 180 mV) reacted very rapidly with *Pseudomonas* oxidase and the cytochromes c of both *Navicula pelliculosa* ($E_0' = +$ 340 mV) and *C. thiosulphatophilum* ($E_0' = +$ 140 mV) also reacted fairly rapidly with this oxidase. However, *R. rubrum* cytochrome c_2 reacts poorly with the *Pseudomonas* oxidase even though it has a high E_0' value and is slightly acidic (Yamanaka et al., 1963b). The cytochromes c which have $E\gamma/E\alpha$ ratios higher than 6 reacted rapidly with *Pseudomonas* oxidase in contrast to those which have ratios of $E\gamma/E\alpha$ of 4·5 and react rapidly with cow oxidase (Yamanaka and Okunuki, 1968a).

C. CYTOCHROMES *B*

1. Introduction

The bacterial cytochromes b like their counterparts in the higher organisms have protohaemin as prosthetic group, and consequently the position of the α-maxima of their absorption spectra range from 557 to 564 nm. Those cytochromes which have their α-peak situated at 560–563 nm are usually designated cytochrome b and those with α-peak at 557–560 nm are referred to as cytochrome b_1. However this demarcation is by no means clear cut. The position of the α-maxima used to characterise the cytochrome b may have been taken from a difference and not an absorption spectrum; difference maxima are often displaced from those observed in absorption spectra. Unlike the situation with the bacterial cytochromes c very few soluble preparations of b-type cytochromes have been obtained, so that characterisation of the bacterial cytochromes b has been limited to spectroscopic observations and biochemical studies of cytochromes b in particulate complexes, in association with oxidases, other cytochromes or a variety of reductases, or even, as in the photosynthetic bacteria, with chlorophyll.

If a broader statement may be made, it is that the b type cytochrome associated with the classical type oxidase cytochrome aa_3 is spectrally similar to, and its function and position in the electron transfer chain of the bacteria is analogous to those of mitochondrial cytochrome b. The cytochromes b associated with the more specifically bacterial cytochrome oxidases, cytochromes a_1, d (a_2) and o appear to be of b_1 type. The small soluble cytochrome b-562 found in many members of the *Enterobacteriaceae* is another sub-class, while the cytochromes b that are intimately associated with the photosynthetic complex in the photosynthetic bacteria may here be regarded as a separate subgroup until they can be defined on a chemical basis. The function of the latter cytochromes b in photosynthesis has been especially reviewed by Kamen and Horio (1970). As knowledge of the physical parameters and structure of the cytochromes becomes available the sub-classification of the cytochromes b may have to be further extended.

Cytochrome o, because it has protohaem as prosthetic group and because of the position of the alpha band of the reduced form deduced from preparations in which a true cytochrome b is invariably present, is sometimes classified as a cytochrome b (Bartsch, 1968, Kamen and Horio, 1970). This is primarily because of numerous observations of b-type cytochromes which may react sluggishly with oxygen or with CO (cf. Kamen and Horio 1970) and has led to the corollary that there may be a gradation of oxidase activity between the classical non-autoxidable cytochrome b and the rapidly autoxidizable cytochrome o. It seems unlikely however that an efficient oxidase would not have a structure peculiar to that particular function of the haemoprotein.

2. Cytochrome b_1

(a) *Cytochrome b_1 of E. coli.* The early spectroscopic observations of Keilin (1927) demonstrated the existence of a b-type cytochrome in *E. coli*. Keilin and Harpley (1941) observed oxidation of succinate and lactate by crushed cells of aerobically grown *E. coli*, which possessed cytochrome b_1, d and a_1, but no detectable cytochrome c. Cytochromes b_1 and a_1 were shown to be present in a succinic-oxidase particulate preparation of *A. aerogenes* (Tissières, 1952b). Wrigley and Linnane (1961) achieved partial purification of the cytochrome b_1 solubilized from a formate dehydrogenase-b_1 complex of *E. coli*. By trypsin digestion cytochrome b_1 was released from a particulate pyruvate-oxidase system of an acetate-requiring mutant rich in this oxidase (Williams and Hager, 1960); the purified preparation in the reduced form had absorption maxima at 556, 523 and 427 nm. From *E. coli* grown anaerobically in the presence of nitrate Taniguchi and Itagaki (1960) isolated a nitrate reductase system rich in cytochrome b_1; by digestion with snake venom at pH 8·0 in the presence of EDTA and 0·1% sodium deoxycholate about 40–60% of the particle-bound cytochrome b could be obtained in a soluble form. Fractionation of this with ammonium sulphate at pH 5·3 and 8·0, followed by chromatography on hydroxyapatite yielded a cytochrome b_1 which was of 50% purity (Fujita, *et al.*, 1963). The apparent molecular weight of this preparation was 6–800,000 and the E_0' (pH 7·0) was $-$ 20 mV. This preparation was strongly autoxidisable but the autoxidation was insensitive to 10 mM CN^- or azide. Neither in the ferrous nor ferric form could the cytochrome bind CO, cyanide or ethyl isocyanide, unless previously denatured at 60°.

These preparations of cytochrome b_1 though showing evidence of some alteration of the native cytochrome b_1 are of importance as they reveal some organisational relationship of cytochrome b_1 to certain important dehydrogenases and reductases and the sensitivity of this haemoprotein to disturbance of its structure by hydrolytic enzymes.

A different approach to the purification of cytochrome b_1 was taken by Deeb and Hager (1964) who released the cytochrome by sonication at 10 KC from particles obtained by disrupting aerobically grown *E. coli* with glass beads by high speed blending. Purification was achieved by chromatography on calcium-phosphate-cellulose columns, followed by crystallization of the cytochrome b_1 by dialysing it against ammonium sulphate solution of increasing strength. The yield of cytochrome b_1 was high: 38 mgs from 700 mgs of *E. coli* cell paste representing 27% of the cytochrome b_1 in the particulate fraction (Hager and Deeb, 1967).

The molecular weight at neutral pH, obtained from sedimentation data, is 500,000. Eight haems are present so that the minimum molecular weight is 60,000. At high pH values cytochrome b_1 is dissociated into a monomeric species of molecular weight 60,000. Octomeric b_1 has an E_0' (pH 7·0) of $-$ 340 mV in contrast to the monomeric form which has an E_0' of 0 mV. Less pure preparations of cytochrome b_1 had an E_0' (pH 7·0) ranging from $-$ 10 mV to $-$ 16 mV, similar to the oxidation-potential of the highly purified cytochrome b_1 obtained by Fujita *et al.*, (1963). A colourless protein (potential-modifying protein) which separates out at lower stages of purification, results on re-addition to crystalline cytochrome b_1 in a more positive value of the E_0'. Spectral data for crystallized cytochrome b_1 are given in Table XIII.

TABLE XIII
Absorption and difference spectra of cytochromes b purified or solubilized

Cytochrome b	Wavelength in nm (ϵmM) in parentheses (s) shoulder			References
Cytochrome b_1 of *Escherichia coli*				Hager and Deeb (1967)
Reduced	564 (22·1)	527·5 (15·2)	425 (12·7)	
Oxidized	564 (83)	532 (11)	418 (107)	365 (42·5)
Reduced − oxidized				
Maxima	557·5 (16)	527·5 (6)	427·5 (60)	
Minima			455 (−5·3)	409 (−25)
Cytochrome b-562 of *Escherichia coli*				Hager and Itagaki (1967)
Reduced	562 (31·6)	531·5 (17·4)	427 (180·1)	324
Oxidized	564 (9·7)	530 (10·6)	418 (117·4)	365
Reduced − oxidized		562 (−24·6)		
Cytochrome b-558 of *Streptomyces griseus*				Inoue and Kubo (1965)
Reduced	558	527	424	
Oxidized		527 (s) 510 (s)	410	
CO-reduced-reduced	575 (s) 558	528	425	376 (s) 276
Cytochrome b-558 of *Bacillus amitratum*				Hauge (1961)
Reduced	562	532	428	300
Oxidized	565	530	419	365

Acetobacter suboxydans				
Reduced	558		437	Iwasaki (1966)
Oxidized	645		493, 402	
CO-reduced	574	544	423	
Pseudomonas b-567 (denitrifying species)				
Reduced	567	540	435	Gray and Ely (1971)
Rhodospirillum rubrum				
Reduced*	557.5	526	428	Kakuno et al. (1971)
Oxidized	565	530	417 365	
Rhodopseudomonas spheroides				
Reduced	559	526	428	Orlando and Horio (163)
Oxidized	565	533	419 365	
Rhodopseudomonas palustris				
Reduced	558.5	528	425	Kamen and Horio (1970)
Oxidized	565	532	419 365	

* $E\gamma/E\alpha = 4\cdot 7$.

Ruiz-Herra and De Moss (1969) have concluded from spectroscopic and enzymatic studies with normal and mutant strains of *E. coli* K-12 that the cytochrome b_1 induced by nitrate is different from the cytochrome b_1 produced in its absence by both aeroboic and anaerobic cells. Moreover, the inhibitory effect of HQNO on the oxidation of the nitrate-specific cytochrome b_1 by nitrate indicated that this b_1 consists of two components of different redox-potential.

(b) *Cytochrome b_1 from other organisms.* There have been numerous reports of cytochromes b_1 in a variety of bacteria (cf. Smith, 1954; Smith, 1961). Only those which are of special interest or have been partly purified will be considered here.

The denitrifiers *M. denitrificans* and *Ps. denitrificans* grown anaerobically in the presence of nitrate each have a cytochrome *b* which is relatively soluble. The partially purified preparations of cytochrome b_1 (Vernon, 1956a) (maxima of absorption of the reduced form 559, 528 and 426 nm) were rapidly oxidized by nitrate. The preparations of cytochrome b_1 reacted with oxygen but not cyanide, and neither were reducible, with ferrocyanide or ascorbic acid. Reduction of the cytochrome b_1 with succinate was slow, but NADH rapidly reduced the cytochrome in the presence of extracts from the corresponding bacteria. A soluble cytochrome *b*-558 has also been obtained from an unidentified strain of a *Pseudomonad*, grown anaerobically with nitrate (Vernon, 1956b). A cytochrome *b*-558 is present in particles obtained from *P. stutzeri* (Kodama and Shidara, 1969) in amounts about one third that of the cytochrome *c*-552 in the overall cell-free extract. A thermolabile cytochrome *b*-559 together with the oxidases a_2 and a_1 was present in a particulate preparation from *Pasteurella tularensis* (Fellman and Mills, 1960). This succinate-oxidase preparation was not sensitive to antimycin A, as might be expected in the absence of a cytochrome *c*.

A cytochrome *b*-560 present in a halotolerant *Micrococcus* (Hori, 1963) is common to both nitrate and oxygen terminal oxidase pathways. Transfer of electrons from cytochrome *b*-560 to the two species of cytochrome *c*-554 (I and II) is antimycin-sensitive. Cytochrome *b*-560 is not directly associated with the nitrate reductase.

A. pasteurianum and *A. peroxydans* each contain a *b*-type cytochrome with a maximum of absorption at 558–557 nm. Since cytochrome aa_3 is absent it is concluded that the cytochrome *b* is of the b_1 type. These organisms also contain cytochrome *o*, although, minimally in *A. pasteurianum*, some of the absorption in the reduced spectrum of the cytochrome b_1 is probably contributed by cytochrome *o*. In *Acetobacter suboxydans* (ATCC 621) which contains no *a* type cytochromes (Chance, 1953a; Daniel, 1970) a second cytochrome *b* is revealed at 77°K with α-peak at 564 nm (Daniel, 1970). Cytochrome-564 (77°K) shows only a very low percentage reduction in the aerobic steady state, indicative of it being an oxidase.

Kusaka et al. (1964) report that $M.$ $smegmatis$ and $M.$ $paratuberculosis$ have cytochrome b_1 with peaks in the oxidized-reduced difference spectrum at 560 and 430 nm; these organisms have also a and o type cytochrome oxidases. This contrasts with the presence of cytochrome b (peaks in difference spectra at 565 and 432 nm) in $M.$ $avium$ and $M.$ $tuberculosis$ H37 RA. The cytochrome b of $M.$ $phlei$ exhibited a peak at 562 nm in the succinate-reduced minus oxidized difference spectrum of particles; in the reduced form at 77°K the maximum was at 561 nm. Spectroscopy at 77°K revealed that a cytochrome b-556, as well as a b-560, was present in $Micrococcus$ $lysodeikticus$; the b-556 was reduced by NADH and L-malate and was coupled to the dehydrogenases of the electron transport chain (Lukoyanova and Taptykova, 1968).

Aerobically-grown $P.$ $syringae$ exhibit absorption maxima at 559, 530 and 429 nm at 22°C, but at 77°K distinct peaks at 563, 557, 536 nm and 528 are observed; the Soret peak is not split (Sands et al., 1967). No cytochrome c was detected. Since the CO-reduced minus reduced spectra at 77°K has peaks at 571, 538, 416 nm cytochrome o may have contributed to the absorption spectrum of the reduced cytochromes.

3. Cytochrome b-562

(a) *Introduction.* Although multiple bacterial cytochromes c have been known for some time, the plurality of cytochrome b in bacteria was only revealed by Fujita and Sato (1963) by their discovery of a soluble b-type cytochrome in anaerobic $E.$ $coli$ grown on nitrate. This cytochrome b of low molecular weight also occurs in the aerobic cell where it is present in concentration of less than one third of the concentration of cytochrome b_1 (Itagaki and Hager, 1965). Usually cytochrome b-562 is difficult to detect in intact cells although at liquid air 77°K Hochster and Nozzolio (1960) observed that the α-band attributed to cytochrome b_1 in $Xanthomonas$ $phaseoli$ was greatly intensified and divided into two peaks, 557 and 564 nm, indicating the presence of cytochrome b-562. By fractionation of the soluble cytochromes on DEAE-cellulose Fujita (1966b) was able to show that cytochrome b-562 was present in a number of strains of $E.$ $coli$, $Erwinia$ $aroideae$, $Erwinia$ $carotovora$, $Proteus$ $vulgaris$ and $Serratia$ $marcescens$.

(b) *Preparation and physical properties.* Cytochromes b-562 may be obtained from $E.$ $coli$ by disintegration of the cells, or by the extraction of acetone powders (Hager and Itagaki, 1967). A highly purified preparation of the cytochrome was obtained by Fujita (1966a) from $E.$ $coli$ str. Yamagutchi. Itagaki and Hager (1966) have crystallized cytochrome b-562 from $E.$ $coli$ str. B. Purification involved chromatography on calcium phosphate gel and fractionation with ammonium sulphate (Hager and Itagaki, 1967).

The molecular weight, from ultracentrifugal data, of b-562 is between 11,900 and 12,700. A value of 12,000 is given by haem anaylsis and 11,954 from the amino acid analysis. The isoelectric point lies between pH 7 and 8.

Ferricytochrome b-562 gives reddish-amber crystals of average length 0·06 mm, and the ferrous form light-red crystals of 0·1 to 0·2 mm in size. From 1 kg of *E. coli* cell paste, 14–15 mg of recrystallized cytochrome b-562 are obtained, an overall yield of 36%.

Spectra of b-562 are presented in Table XII. The twice recrystallized b-562 has a value of 1·5 for the ratio Eα/E 280 nm. and the apoprotein has an absorption band of weak intensity at 277 nm, a reflection of the low content of aromatic amino acids.

Cytochrome b-562 does not combine with CO in the pH range 3·0 to 10·5, but at higher pH values CO causes a shift of the Soret maximum from 427 to 425 nm. Neither cyanide nor azide react with the haem group at neutral pH. Ascorbic acid, cysteine and $NaBH_4$, as well as $Na_2S_2O_4$, readily reduce cytochrome b-562.

(c) *Amino acid Sequence and Structure.* The amino acid sequence of cytochrome b-562 (*E. coli*) has been determined by Itagaki and Hager, (1968). No cysteine, cystine or tryptophan residues were amongst the 103 amino-acid residues. The amino terminal residue is alanine and arginine was found at the carboxyl terminal; two histidines and two methionines are present in the peptide chain.

Although it has been suggested that both ligands to the haem iron of cytochrome b-562 are histidine (Itagaki and Hager, 1968) there is no direct evidence that both histidines are coordinated to the iron. One histidine is near the centre of the molecule in similar fashion to one of the co-ordinating histidines of myoglobin (Kendrew, 1961) but the second histidine is separated by only three residues from the carboxyl terminus. If this histidine was the second ligand to the haem this would be unique amongst the haemoproteins which so far have had their primary structures determined (see Tables VIII, IX and X). A lysine residue, several of which are scattered fairly evenly throughout the chain, or methionine 35 may be considered as the possible sixth ligand. On the other hand it should be stated that methionine is not present in microsomal cytochrome b_5 and evidence has been presented against lysine being the sixth ligand (Ozols and Strittmatter, 1969).

Warme and Hager (1970a,b) have studied the combination of mesohaem and mesohaem anhydride (see also Chapter V, B3a) with apocytochrome b-562 in order to ascertain which lysines are in proximity to the haem carboxyls. The major sites of reaction with mesohaem anhydride are lysine 50 and secondly lysine 108.

Kendrew (1961) has shown by x-ray crystallography that arginine-45 in whale myoglobin is the site of attachment of the haem propionic carboxyl. In other myoglobins lysine is substituted for arginine at residue 45: Warme and Hager (1970b) by aligning the histidine 66 and the histidine 106 with myoglobin distal haem-linked histidine 64 and histidine 97, rather than histidine 93 (the proximal haem linkage), achieve a better appearance of homology between cytochrome b-562 and myoglobin. The number of amino acid residues between the histidines and lysine concerned with the propionic carboxyl binding in each haemoprotein is similar, 15 and 18, for cytochrome b-562 and myoglobin respectively. Although there are only three amino-acids common to both haemoproteins in this partial sequence the substitutions are generally conservative. However any similarity of the tertiary structure can only be slight, otherwise such a profound difference in function and reactivity with molecular oxygen and CO would not be observed. Further there are 43 (approximately 40%) more amino acid residues in myoglobin and these can be expected to have an effect on the tertiary structure around the haem group of myoglobin.

(d) *Function of Cytochrome b-562.* The concentration of cytochrome b-562 in aerobic *E. coli* although lower than that of cytochrome b_1 (4 mg/100 g and 20 mg/100 g of cell paste respectively) (Hager and Itagaki, 1967; Hager and Deeb, 1967) indicates that it has a significant function in the electron transfer processes of the cell. There appears to be little difference in the b-562 content of cells grown aerobically or anaerobically in synthetic media or in the presence of nitrates in complex media (Fujita, 1966b). Though not "membrane bound" cytochrome b-562 may be a component of a system located in the membrane, its small size facilitating release from the site of its function. Hager and Itagaki (1967) have noted that cytochrome b-562 can serve as an electron acceptor for various flavoprotein-linked oxidations.

4. Other Microbial Cytochromes b

Cytochrome b-562 of *M. lysodeikticus* has been released from particles obtained by lysis of cells with lysozyme and ribonuclease (Jackson and Lawton, 1959). After solubilization of the cytochrome with Cetrimide the maximum had shifted to 558 nm, and a further shift to 552–554 nm occurred on prolonged storage of the cytochrome b. This is suggestive of alteration of one of the vinyl groups of the haem. The separated cytochrome b was autoxidizable even in the presence of cyanide (5×10^{-2} M). Ferrocyanide did not reduce the cytochrome b-558, but with ascorbate partial reduction was obtained.

A soluble cytochrome b-562 (Table XIII) occurs in *Bacillus anitratum* grown in aerated cultures and has been considerably purified (Hauge, 1960, 1961). The E_0' of this preparation (M.W. 290,000) was + 120 to 140 mV.

This cytochrome b-562 could be reduced by a glucose dehydrogenase purified from *B. anitratum*, provided that a non-dialyzable factor was present. The b-562 and the glucose-dehydrogenase system contrasts with that in *Bacillus megaterium* where the oxidation of glucose required the coupling of a soluble NAD-dehydrogenase with the entire membrane-bound cytochrome oxidases and a cytochrome b (oxidized−difference maxima at 557 and 427 nm) (Broberg *et al.*, 1969). The E_0' of the b-557 *in situ* is $+8$ mV and menaquinone present in the membrane in 5–10 fold molar excess over cytochrome-557, appears to function between cytochrome b-557 and the dehydrogenases (Kröger and Dadak, 1969).

In *Bacillus subtilis* (Smith, 1954; Chaix and Petit, 1956; Miki *et al.*, 1967) *Mycobacterium phlei* (Revsin *et al.*, 1970a,b) *Mycobacterium tuberculosis* (Kearney and Goldman, 1970) *Corynebacterium diptheriae* (Kufe and Howland, 1968) and other bacteria where cytochrome aa_3 is a principal cytochrome oxidase the cytochrome b spectrally resembles mitochondrial b and has a similar role to it in the electron transport chain. None of these bacterial cytochromes b has yet been purified to any extent.

A b-type cytochrome has been reported to be the dominant cytochrome in *Actinoplanacae* (Domnas and Grant, 1970). The α and Soret absorption maxima were at 563–560 nm and 430–435 nm respectively. Cytochrome aa_3 was present in some strains, but no cytochrome c. In the absence of CO-difference spectra the presence of cytochrome o is not excluded. Heim *et al.* (1957) found that of thirteen species of *Streptomycetaceae* five had only a cytochrome b (peaks in the oxidation-reduction difference spectra 563–558 and 430–424 nm); the others also had a cytochrome c but aa_3 was not detected in any. The extracts were however prepared by sonication at pH 10·5 and this treatment may have destroyed the aa_3. The cytochrome b extracted by the alkaline method and partially purified (Birk *et al.*, 1957) was readily autoxidizable and this autoxidation was not inhibited by CN^-, azide, CO, antimycin A or HQNO. It could be reduced with NADH, NADPH, $FMNH_2$ and slowly with succinate but not with ascorbate. The cytochrome b could readily reduce beef heart ferricytochrome c or ferricyanide. Niederpruem and Hackett (1961) on the contrary found cytochrome aa_3 as well as cytochrome c associated with the particles obtained by sonication of the mycelia of four other species of *Streptomyces*. The cytochrome b was not fully reducible with succinate or NADH. The succinic-oxidase activity was cyanide-sensitive; the endogenous respiration of the cells was inhibited 70% by 10^{-3} M CN^-. Antimycin A and HQNO were only slightly inhibitory. A CO-binding pigment (peak at 418–420 nm in difference spectra) accumulated in old cultures of the *Streptomyces*.

The properties of these cytochromes suggest that they should be classified as cytochrome b rather than b_1 at least until they have been further characterized. The properties of the cytochromes b that have been isolated from the bacteria suggest that they have been partially modified during extraction.

Cytochrome b-558 of *Streptomyces griseus* has been crystallized (Inoue and Kubo, 1965). The cytochrome was released from the cells by incubation of the mycelium with lysozyme and steapsin, followed by extraction with deoxycholate (1%) at pH 8·5. Purification was achieved by repeated fractionation with ammonium sulphate. The crystallized cytochrome b_1 extracted with ethyl alcohol yielded a yellow substance. The acidic cytochrome was homogenous on centrifugation and had a sedimentation co-efficient of 16·4 S. Spectrally (Table XIII) the cytochrome was similar to that of the cytochrome b_1 preparation of *S. fradiae* (Birk et al., 1957). CO reacts with the crystallized *S. griseus* cytochrome b_1 at alkaline pH (maximum 425 nm). The cytochrome is autoxidizable but its autoxidation is not inhibited by CN^-. Cytochrome b-558 is reduced by reduced vitamin K_3 and by cysteine but is only slightly reduced by ascorbic acid: it is not reduced by NADH, succinate or lactate. The cytochrome had no peroxidase or catalase activities.

Halobacteria. Two cytochromes b have been observed in the cytoplasmic membranes of several species of *Halobacterium*. Cytochrome b-559 (557 nm at 77°K) of *H. cutirubrum* (Lanyi, 1968) transfers electrons from succinate, NADPH and α-glycerophosphate to cytochrome c-555 and thence to cytochrome a-592. Cytochrome b-563 participates in an alternative electron transfer pathway, which includes cytochrome c-550 and is operative at pH 9·4. Succinate does not reduce cytochrome b-563.

Cytochrome b-561 (559 nm, 77°K) of *H. halobium* can be reduced by either ascorbate or ascorbate-TMPD (Cheah, 1970a) and is thus different from cytochrome b_1 which is not known to be reduced by these reductants. Cheah (1970a) suggests that b-561 transfers electrons to cytochrome o directly or to cytochrome a_1, through a cytochrome c_1, present in low levels in this organism. The transfer of electrons from cytochrome b-561 to the cytochrome c is antimycin A and HQNO sensitive. Cytochrome b-564 of *H. halobium*, present at only a third of the concentration of b-561 was not reducible by ascorbate-TMPD (Cheah, 1970a).

By differential reduction with various substrates at 77°K Cheah (1970c) demonstrated the presence of three cytochromes b, b-562, b-557 and b-559 (77°K spectra) in isolated membranes of *H. salinarium*. Reduced minus oxidized spectra at 22°C show peaks at 557 nm with succinate as reductant, at 558 nm with α-glycerophosphate and at 560 nm with ascorbic acid. Cytochrome b-562 (77°K) was not identified in the spectra at 22°C. Ascorbate and ascorbate-TMPD reduce cytochrome b-557 (77°K) in the membranes and this reduction is sensitive to HQNO. Kinetic and inhibitor studies and the difference spectra obtained at 22°C and 77°K have led Cheah (1970b) to conclude that electrons may transfer from the cytochrome

b complex either through cytochrome c-553 (549, 77°K) to cytochrome a_3, or directly from cytochrome b-557 (77°K) to cytochrome o. Cytochrome b-562 (77°K) may be involved in the oxidation of α-glycerophosphate.

Rhizobium. A particle-bound cytochrome b which exhibits an α peak at 559 nm in the oxidized − reduced difference spectrum is found associated with a cytochrome c-551 and cytochromes aa_3 and o in free-lving Rhizobium aponicum. (Appleby and Bergersen, 1958; Tuzimura and Watanabe, 1964; Appleby, 1969a). At 77°K the α-peak of this cytochrome b shifts to 556 nm and if succinate is the reductant a peak at 559·5 nm is also obtained, and this peak has been attributed to cytochrome o. In the presence of cyanide and with succinate as a reductant only 30% of the apparent cytochrome b-559 is reduced, in contrast to the cytochrome-551 which is 100% reduced. The cyanide-insensitive component is the cytochrome b-556 (77°K) whilst the cyanide-sensitive component is either cytochrome o or Rhizobium haemoglobin. From the steady state levels of reduction of the cytochromes aa_3, c and b in the particles it has been suggested (Appleby, 1969b) that the sequence of electron transfer is the same as in mitochondria. However cytochrome b-556 is not maintained in the reduced form in the presence of antimycin when NADH is added to oxidized particles. The physical properties of the cytochrome b and c may be sufficiently different for the antimycin not to be able to exert its inhibitory action. The cytochrome b may in any case transfer electrons to cytochrome o.

Romanoff et al., (1970) by spectroscopy at 77°K have also detected two b-type cytochromes as well as aa_3 and c-549 (77°K) in Rhizobium leguminosarum and Rhizobium trifolii; nitrogen-fixing strains had 40% more cytochrome than non-effective strains.

A cytochrome b-559 is found in Rh. japonicum bacteroids (Appleby, 1969a). The absence of low temperature spectra and the lack of the oxidases aa_3 and o in the bacteroids make identification with the cytochrome b-559 of the free living Rhizobium tenuous. Cytochrome c-550 is present in the bacteroids as are also the CO-reactive pigments c-552 and Rhizobium P-450.

In bacteroids from N_2-fixing nodules of yellow lupin the cytochrome c to cytochrome b ratio is rather greater than in the membrane fraction of the nodules, but in ineffective nodules it is about equal (Melik-Sarkisyan et al., 1969).

Chemoautotrophs. A soluble cytochrome b (α-557 nm; E_0' + 150 mV) has been purified from Thiobacillus neopolitanus. The b-557 is reduced by thiosulphate in the presence of thiosulphate-oxidizing enzyme (Trudinger, 1961a,b). A minor amount of a cytochrome b-562 was detected in the intact bacteria and this was also reducible by thiosulphate. The cytochrome b (α-band 563 nm) of Thiobacillus concretivorus participates in both the sulphate and sulphide oxidation pathways (Moriarty and Nicholas, 1969;

1970). Aleem (1965) found that the cytochrome b in *Thiobacillus novellus* grown autotrophically with thiosulphate does not mediate electron flow from that substrate to O_2. A cytochrome b does however participate in these bacteria in the succinate-oxidase pathway and is a site of generation of ATP used in the reduction of NAD by succinate, involving reversal of electron transfer from cytochrome c (Aleem 1966a). Evidence for the position of cytochrome b in the complex electron pathway involving the oxidation of sulphur compounds in the obligate chemautotroph *Thiobacillus neopolitanus*, which utilizes O_2, and *Thiobacillus denitrificans* which can use either O_2 or NO_3, has been presented by Peeters and Aleem (1970).

During the oxyhydrogen reaction in the chemoautotroph *Hydrogenomonas eutropha* the ATP-dependent reversal of electron transfer from succinate to NAD^+ is effected from the level of cytochrome b as well as c (Ishaque and Aleem, 1970); Bongers (1967) concluded from inhibitor studies that phosphorylation with hydrogen as electron donor occurs exclusively at a site in an abbreviated chain between H_2 and cytochrome b.

Photosynthetic organisms. The facultative photoheterotrophic non-sulphur purple bacteria, e.g. *Rhodospirillum rubrum*, *Rhodopseudomonas spheroides*, grown both in the dark-aerobic (Taniguchi and Kamen, 1963; Whale and Jones, 1970) or the light-anaerobic mode, contain appreciable amounts of b-type cytochromes. Cytochromes which have their α-peak at 557·5–559 nm have been purified from several bacteria of this group (Orlando and Horio, 1961; Bartsch, 1970; Kakuno, 1970; latter two cited by Kamen and Horio, 1970). Like cytochrome b, they form aggregates of M.W. 500,000. The R. rubrum b-557·5 has an E_0', pH 7·0 of -210 mV; the chromatophore membrane exhibits a peak at 430 nm on differential reduction with succinate and $NADH_2$ and this peak is attributed to two cytochromes which have E_0', pH 7·0, of $+20$ mV and -200 mV respectively.

No cytochromes b have been detected in the photoautotrophic purple sulphur bacterium *Chromatium*.

5. Bacterial P-450

(a) *Introduction.* Cytochrome P-450 has been obtained from three widely different bacterial sources; from *Rhizobium japonicum* bacteroids (Appleby, 1967, 1968), from *Pseudomonas putida* grown on (+)-camphor as component of an inducible 5-exomethylene hydroxylase, (Katagiri et al., 1968) and from *Corynebacterium* sp. 7E1C (Cardini and Jurtshuk, 1970). A CO-binding pigment occurring in *Nitrosomonas europea* (Rees and Nason, 1965) was probably a cytochrome of d_1 type, not P-450, as the Soret peak of the reduced form was at 460 nm, not 410–415 nm. Free-living air-grown *Rhizobium japonicum* contain a very small amount of soluble P-450 (Daniel and Appleby, 1971) but *Rhizobia* grown anaerobically with nitrate contain large amounts of P-450. The P-450 content of *P. putida*

TABLE XIV

Absorption spectra of cytochrome p-450$_{cam}$ and Rhizobium p-450c

		Absorption Peaks in nm emM in parentheses		Reference
P-450$_{cam}$ plus D-camphor				Gunsalus et al. (1971)
Oxidized	646 (4·5)		391 (87)	
Reduced		540 (10)	408 (73)	
P-450$_{cam}$ without substrate				
Oxidized		571 (10·5)	417 (105)	
Reduced		535 (10) 540 (14)	411 (71)	
P-450$_{cam}$ plus D-camphor reduced				
+ O$_2$		552 (14)	418 (62)	
+ CO		550 (12)	447 (106)	
+ Ethyl isocyanide	576	549	453 (104)	Griffin and Peterson (1971)

Δ spectra (max − min) oxidized + substrate − oxidized without substrate

Substrate				
D-Camphor (0·5 μmole/l)			388–419 (170)	Gunsalus (1968)
Laurate (0·2 μmole/l)			388–419 (6)	
Aniline (5 μmole/l)			429–390 (70)	
Imadazole (10 μmole/l)			427–408 (170)	
KCN (10 μmole/l)			440–413 (245)	
Rhizobium P-450$_c$ *				
Oxidized	640 (2·5)	568 (11·0)	535 (11·6)	417 (105)	360 (37)	Appleby and Daniel (1971)
Reduced			545 (14·8)	408 (86)	—	
Reduced + CO			550 (14·3)	446 (s) (105)	422 (43)	

* Cytochrome P-450$_c$ is the major component from *R. japonicum* bacteroides. The oxidized pigment as isolated is predominantly low spin. Rhizobium P-450 components a, b, b_1 have variously low to high spin structures. (Appleby and Daniel, 1971.)

increased from practically none to 0·15 n moles/mg dry weight when this bacterium was grown in media containing (+)-camphor (Peterson, 1970).

The *Rhizobium* and *Pseudomonas* P-450 differ from microsomal P-450 in the relative ease with which they are released from the cytoplasmic membrane, providing soluble preparations which have led to the crystallization of P-450$_{cam}$ (Yu and Gunsalus, 1970a) and to highly purified preparations of the *Rhizobium* P-450 (Appleby and Daniel, 1971).

(b) *Pseudomonas* P-450$_{cam}$. The methylene hydroxylase system of *P. putida* (Hedegaard and Gunsalus, 1965) was resolved into three components: an iron-sulphide protein, putidaredoxin; a flavoprotein, putidaredoxin reductase; and P-450$_{cam}$ by Katagiri et al. (1968) who stabilized the P-450$_{cam}$ by carrying out the separations in the presence of (+)-camphor. Subsequently the haemoprotein was obtained in pure form by the Kanazawa group and by the Urbana group (Yu and Gunsalus, 1970b; Dus et al., 1970).

The P-450$_{cam}$-camphor complex crystallized as needles, but in the absence of substrate P-450$_{cam}$ crystallized as long plates (E 392/E 280 = 1·47). Mercaptosuccinate was used as a protective agent during crystallization of P-450$_{cam}$; mercaptoethanol and dithiothreitol were not suitable as they both could react with P-450$_{cam}$ (Yu and Gunsalus, 1970a). Spectral data for P-450$_{cam}$ are given in Table XIV.

The P-450$_{cam}$ molecule contains one haem group and one glucosamine residue. On the basis of the amino-acid analysis the molecular weight is 45,070 (Dus et al., 1970). This compares well with the value of 46,000 calculated from sedimentation data (Yu and Gunsalus, 1970); a value of 49,000 was obtained by the Katagawa group. The isoelectric point of P-450$_{cam}$ determined by the electrofocussing technique is pH 4·55, that of the P-450 camphor complex, 4·67. This difference is ascribed to the release of a weakly basic group or the masking of an acid (Dus et al., 1970).

P-450$_{cam}$ protein consists of a single polypeptide chain which has valine at the carboxyl terminal and asparagine as the amino terminal residue (Dus et al., 1970). The amino terminal is poorly accessible to chemical and enzymatic reagents. The presence of six cysteine residues is of particular interest in the amino acid analyses, in view of the evidence that one ligand of the haem-iron is a mercaptide-sulphur; histidines are numerous, but there is only one tryptophan residue. The locus and mode of attachment of the glucosamine to the polypeptide chain are not known.

The EPR spectrum of P-450$_{cam}$ was studied over the temperature range 4·2–80°K (Tsai et al., 1970). Substrate-free P-450$_{cam}$ gives an anisotropic EPR spectrum which is typical in gross appearance to that given by the low spin components of ferric haem. The locations of the g values, 2·45, 2·26 and 1·91 are similar to those of the thiol-treated ferric haemoglobin and myoglobin (Blumberg and Peisach, 1970) and also mixtures of haem compounds made with thiols and substituted pyridines (Bayer et al., 1969).

The crystal field components of P-450$_{cam}$ are similar to those of compounds prepared from haems and thiols irrespective of the nature of the other non-mercaptide ligand and whether or not the haem is incorporated into a protein. A mercaptide-sulphur iron-linkage is thus established in P-450$_{cam}$.

A minor signal attributable to high spin ferric haem (g = 8, 4, 1·8) and equivalent to less than 7% of the haem was observed at temperatures below 20°K (Tsai et al., 1970). This high spin signal increased, to represent 60% of the haem, on addition of (+)-camphor to P-450$_{cam}$. The 40% of the haem left in the low spin form could not be converted to the high spin form by (−)-camphor. The high spin signal has a rhombic character stronger than that observed for any type of haem compound (E = 0·33 cm^{-1}; D = 3·8 cm^{-1}; E/D = 0·087) (cf. Peisach and Blumberg, 1970).

The EPR evidence is consistent with the view that camphor causes a shift of the mercaptide sulphur away from the haem iron (Yu et al., 1969). This appears not to be due to direct displacement of the sulphur ligand by the substrate, as N-phenylimidazole, a competitive inhibitor of camphor, does not abolish the low spin character of P-450$_{cam}$. A significant change in the oxidation-reduction potential of P-450$_{cam}$ on combination with substrate has been noted by Peterson and Dutton (1971).

Ethyl isocyanide binds more firmly to the ferrous state (KS 9 × 10^{-6} M) than the ferric state of P-450$_{cam}$. The finding that camphor competes very effectively with the binding of ethyl isocyanide to ferrous P-450 indicates that the complex binding site is near the haem iron (Griffin and Peterson, 1971).

Muller-Eberhard et al. (1969) have studied the transfer of P-450$_{cam}$ haem to apomyoglobin and haemopexin. Incubation of P-450$_{cam}$, in the 416 nm form (P-420) at 25° with apomyoglobin led to loss of the 416 nm and 360 nm absorption bands. The reaction is relatively slow, half exchange occurring in 2–5 minutes and equilibrium being approached in 20 minutes, at pH 7·4. There is a concomitant loss of hydroxylase activity but partial restoration is obtained if protohaemin is added. The binding of the haem was followed using ^{59}Fe or ^{14}C haem. Haemopexin, a larger molecule (77,000) was not as efficient as apomyoglobin in removing the haem.

Only the specific substrates (+) and (−)-camphor and their 1,2 lactones induce the type I spectral shift with difference maximum at 388 nm and minimum at 419 nm (Gunsalus, 1968). Other substrates which induce a type II spectral shift in liver and adrenal microsomal and mitochondrial P-450 are ineffective with P-450$_{cam}$. Aniline causes, as with the mammalian P-450, a type II shift with minima and maxima near 393 and 430 nm respectively. The crystallized enzyme has a specific activity of 26 in the complete methylene hydroxylase assay, which requires NADH (Yu and Gunsalas, 1970b). EPR, CD and Mössbauer measurements have been used to study formation of products by equimolar P-450$_{cam}$-putidaredoxin (Gunsalus et al., 1970). Excess putidaredoxin increases the catalytic activity per P-450 molecule. In the reconstituted system the rate-limiting step appears to be the transfer of the second electron to the oxygenated ferrous P-450.

An important development towards the understanding of the mechanism of action of P-450$_{cam}$ was the discovery of a new spectral species (absorption maxima at 555, 418 and 355 nm), attributed to an oxygen complex of P-450$_{cam}$ (Ishimura et al., 1971). The spectral characteristics are different to those of the presumed oxygenated species of microsomal P-450 (Estabrook et al., 1971; cf. Chapter IV). The oxygenated species is obtained only with ferrous P-450$_{cam}$ in the presence of excess O_2. It is not obtained with H_2O_2 nor can ferric P-450$_{cam}$ be oxygenated. CO converts the new species to CO-P-450$_{cam}$ and this transformation is reversed by oxygenation. The EPR signal at $g = 2\cdot26$ disappears on formation of the oxygenated complex. The oxygenated form is stable at 4–8°, but slowly converts to the high spin form at 20°. This conversion is greatly accelerated in the presence of stoichiometric amounts of putidaredoxin. Ishimura et al. (1971) have proposed a reaction mechanism based on involvement of the oxygenated complex in the reaction cycle.

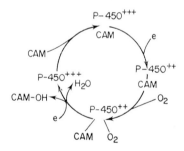

FIG. 8. Reaction mechanism of P-450$_{cam}$ according to Ishimura et al. (1971).

Although O_2 forms a complex with ferrous cytochrome P-450$_{cam}$ the state of the iron may not be strictly analogous to that in HbO_2 but may be closer to that of oxygenated cytochrome a_3 (cf. Chapters III B5f and IV D7).

(c) *Rhizobium* P-450. The cytochrome P-450 present in *Rhizobium* bacteroids was readily released by disruption of the bacteroids in a French Press, in the presence of deoxyribonuclease. The P-450 preparation after sequential chromatography on Sephadex G-75 and G-100 gels and DE 52 anion exchange cellulose was 95% pure and very stable (Appleby and Daniel, 1968). A molecular weight of 40–45,000 for *Rhizobium* P-450 was calculated from gel studies. Absorption and difference spectra for the CO,

CN⁻ and isocyanide complexes are given in Table XIV. At low concentrations of isocyanide (at pH 6·8) an aberrant complex (peak at 455 nm) is obtained, while at high isocyanide concentrations a typical haemochrome type spectrum (peak at 430 nm) is formed (Appleby 1967, 1968). Quantitative conversion of *Rhizobium* P-450 to P-420 is obtained with CO at pH 5·3.

No breakdown of *Rhizobium* P-450 occurs on incubation of the hydroxylase with *Trimeresuras flavoviridis* venom or with a highly purified phospholipase A from the Australian black snake. Incubation of the P-450 preparation with steapsin or deoxycholate caused disappearance of P-450 without the appearance of the P-420 form (Appleby, 1968). Appleby and Daniel (1971) on the basis of induction, chromatography and EPR studies have concluded that *Rhizobium* P-450 may be a family of separable haemoproteins, thereby strengthening the "many separate enzyme" hypothesis for P-450 activity. The M.W. and CO-affinity of the P-450 of free living *Rh. japonicum* are different to those of the bacteroids (Daniel and Appleby, 1971). Two additional CO-combining pigments, P-420 and P-428, have been reported to occur in *Rh. japonicum* bacteroids (Appleby, 1969a); so far they have not been further characterized.

(d) P-450 *from other bacteria*. *Corynebacterium* P-450 is bound to particles or at least associated with lipids (Cardini and Jurtshuk, 1970). Inhibition, spectral and induction studies confirm the participation of P-450 in n-octane hydroxylation in *Corynebacterium* (7 EIc). Mass spectrometry showed that molecular oxygen is incorporated into the substrate during hydroxylation.

A P-420 type cytochrome has been detected in the mixed function dimethylamine oxidase system of *Pseudomonas aminovorans* (Jarman *et al.*, 1970). Its spectral properties generally resemble the 420 nm microsomal pigment (Omura and Sato, 1964). A mixed function oxidase was reported to be present in *P. oleovorans* by Peterson *et al.* (1966); a peak was observed at 419 nm, but not at 450 nm. However, Peterson (1970) who noted a five-fold increase in the cytochrome *b* component of the cells when *P. oleovorans* was grown on hexane, concluded that no P-450 was present, but rather cytochrome *o*. The partially purified hydroxylase of *P. oleovorans* gave EPR signals (g = 1·94, g = 2·03) consistent with the presence of a labile sulphide iron-protein and confirmed the absence of haem-iron (McKenna and Coon, 1970). Moreover the enzyme was inhibited by cyanide ($K_i = 3·1 \times 10^{-4}$ M with laurate) but not by CO. Known specific organic inhibitors of the P-450 mixed function oxidase system (cf. Chapter IV) inhibited the aerobic hydroxylation of β-carotene in *Staphylococcus aureus* U-71 (Hammond and White, 1970). A CO-binding pigment (peak at 448–460 nm) is induced in anaerobically grown cells of this strain when they are aerated.

6. Appendix

An unusual haemoprotein, originally called cytochrome b_4 by Egami et al. (1953) which occurs in a halotolerant *Micrococcus* no. 203, a denitrifier, has been studied by Mori and Hirai (1968) and designated by them as "cytochrome B-574". Nitrate added to the medium depressed formation of this haemoprotein, while promoting the formation of cytochrome b-560; acriflavine (2 pts per million) enhanced the yield of "cytochrome B-574". In the reduced form absorption maxima were at 574, 537 and 418 nm; the E_0' (pH 7·0) was + 50 mV. "Cytochrome B-574" was convertible, in intact cells and in extracts (in these by simple dilution) to a c-type cytochrome with absorption maxima at 552 and 548 nm and E_0' (pH 7·0) = 190 mV. The haem could not be extracted from the cytochrome c form whereas it was extractable from the "B-574" form. An intermediate form which had an α-band at 554 nm was also identified. The haem could be split from this by acetone-HCl to give a pyridine haemochrome with an α-peak at 553 nm. The conversion of "B-574" to the 552, 548 form was reversible, the reversal occurring on allowing the "cytochrome c" to stand *in vacuo* in the presence of $Na_2S_2O_4$.

The position of the Soret peak (418 nm) of reduced "cytochrome B-574" suggests that the haem is not protohaem but has the vinyl groups saturated. Unlike the absorption bands in the visible region the position of the Soret maximum is much less sensitive to the effects of structural changes in the protein.

The haemochrome spectrum of the haem extracted from the intermediate form suggests that only one vinyl group is altered. It is interesting in this context that a haem which gives mesohaemochrome-type spectra has been extracted from cells of *Desulphovibrio africanus* (Jones, 1971a). Hori (1963) investigating the same strain of *Micrococcus* mentioned above obtained two c-type cytochromes c-554 and c-548, 552 but not "B-574": this may have been due to the presence of excess nitrate in the medium.

While a conversion of a protohaem cytochrome b into a cytochrome c with covalent linkage between prosthetic group and protein by thioether linkage has been observed (Morton and Shepley, 1965) such a reaction would not be expected to be so readily achieved nor to be reversible. Another unexplained point is the position of the α-band; one explanation may be that a strongly electrophilic side chain of the haem is unstable and readily loses its electrophilic character. Such reactions are known in the Schiffs base formation of haem a (Lemberg, 1962) or in the ketoenol transformations of cryptoporphyrins p (Clezy et al., 1964); however, the ease of conversion as well as the position of the Soret band at 418 nm appears to exclude this explanation. Certainly these experiments require further study.

D. CYTOCHROMES C

1. Bacillus cytochrome c

(a) *General properties*. The early visual observations of Keilin (1925), Yaoi and Tamiya (1928) and Tamiya and Yamaguchi (1933) on intact *B. subtilis* and other bacilli revealed the presence of a cytochrome system similar to that of muscle and yeast, including the presence of a cytochrome c-550. Spores of *B. subtilis* also contained cytochromes but at about 6% of the level found in the vegetative forms (Keilin and Hartree, 1949). At liquid air temperature Keilin and Hartree (1949) detected in *B. subtilis* and *Bacillus licheniformis* a second c-type cytochrome (α-maximum at K 78° –552 nm), generally assumed later to be analagous to mitochondrial cytochrome c_1. The ingenious application of the spectrographic technique to the low temperature studies of the cytochromes of the strict aerobe *B. subtilis* str. 3610 (Chaix and Petit, 1956, 1957) showed that the relative amounts of the two cytochromes c varied depending on the age and growth rate of the culture. In bacilli collected after 7 hours of growth the absorption at 552 nm was greater than that at 548 nm in the low temperature spectra. In some strains of these bacteria (cf. Miki *et al.*, 1967) the cytochrome with the 552 nm α-maximum of absorption could not be detected at low temperature.

The cytochromes in the *Bacilli* are membrane-bound (Weibull, 1953; Weibull and Bergstrom, 1958). Vernon and Mangum (1960) were able to release some cytochrome c by the use of lipase from membrane fragments obtained by sonication of *Bacillus megaterium* and *B. subtilis*.

The partially purified cytochrome had absorption peaks at 550, 520 and 413 nm in the reduced form; the oxidized-reduced difference spectrum of the particles exhibited a peak at 552 nm. The E_0' (pH 7·0) of the cytochrome was + 250 mV. The ready autoxidizability of the cytochrome may have been due to the lipase treatment. Yamagutchi *et al.* (1966) released a cytochrome c-550 from protoplasts of *B. megaterium* KM and *B. subtilis* IFO 326 with 0·02 M NaOH (pH 8·0). A molecular weight of 12,000 was indicated by the centrifugation data. Spectrally the soluble cytochrome resembled mammalian cytochrome c.

Miki and Okunuki (1969a) have obtained stable preparations of both the cytochrome c-550 and cytochrome c-554 (α-maximum, 552 nm at 78°K) by passage of *B. subtilis* K_1 wild strain, suspended in 0·89% KCl + 5 mM EDTA through a Sorvall-Ribi cell fractionator, at 14,000 Kg/cm². Repeated chromatography on DEAE cellulose at pH 6·0 separated the c-550 from the c-554, and the former was further purified by chromatography on Amberlite CG-50. Purification was about 40-fold and the yield from 1·8 kg wet weight of bacilli was 10·2 mg of cytochrome c-554.

The minimum molecular weight determined from the haem content was 13,000 for cytochrome c-550 and 14,300 for cytochrome c-554 (Miki and Okunuki, 1969b). Gel filtration gave molecular weights of 12,500 and 14,000 respectively for c-550 and c-554 (Miki and Okunuki, 1969a). The E_0' (pH 7·0) of cytochrome c-550 was + 210 mV, and that of cytochrome c-554 − 80 mV. Neither cytochrome combined with CO or oxygen at neutral pH. The absorption maxima of cytochrome c-550 were 550 (25·5), 520 (13·1), 414 (127·5), 316 (31·4), and 279 (39·4), in the reduced form; 528 (11·8), 407 (92·5) and 279 (30·6) in the oxidized form.

The absorption maxima of the second cytochrome c were: reduced form 554 (20·0), 521 (14·5), 417 (14·36), 316 (40·0), 280 (33·9); oxidized form 523 (10·2), 409 (106·7), and 280 (24·0). Not only is the absorption of the α-maximum low, but a pronounced shoulder at 550 nm (14·5) is present. At 79°K the α-maximum was resolved into two clearly separated peaks 552 and 546 nm; the β peak was also split into multiple distinct peaks, main peaks 420 and 527 nm. Cytochrome c-550 had a single α-peak at 548 nm.

Neither urea, guanidine nor sodium dodecyl sulphate alter the shape of the α-peak of cytochrome c-554. The reduction of cytochrome c-554 with ferrous sulphate was not biphasic.

The high extinction of the Soret and γ peaks has been associated with asymmetric α-peaks which have a shoulder on the shorter wavelength side of the peak (cf. Yakushiji et al., 1960; Yamanaka and Okunuki, 1968b; Sugimura and Yakushiji, 1968; Meyer et al., 1968).

CD spectra provided evidence for the existence in cytochrome c-552 of a stronger haem-ligand interaction than is found in mammalian cytochrome c. The chromophore appears to be in an anisotropic environment and a large change in ellipticity occurs in the Soret region of the CD spectrum on oxidation and reduction of cytochrome c-554 (Miki and Okunuki, 1969a). Cytochrome c-550 is slightly basic (isoel. point at pH 8·65) but cytochrome c-554 differs considerably from mitochondrial cytochromes c in having its isoelectric point at pH 4·44. This acid character of the cytochrome is largely due to the high content of aspartic and glutamic acid.

(b) *Biochemical aspects.* The localization of cytochromes c-550 and c-554 in electron transport particles of *B. subtilis* was shown by their low temperature spectroscopy (Miki and Okunuki, 1969b). Cytochrome a-601 and c-550 were fully reduced, but cytochrome c-554 only partially. Addition of NADH to the electron transport particles caused rapid reduction of cytochrome c-550 and cytochrome a-601. *Bacillus* cytochrome a-601 oxidized cytochrome c-550 and c-554 at similar rates despite the difference in the isoelectric point of the two cytochromes. However, the rates are much lower than those obtained when yeast cytochrome c was the electron donor. Both cytochrome c-550 and c-554 were reduced at the same rate by

NADH in the presence of a small amount of *Bacillus* electron transport particle. Antimycin inhibits the transfer of electrons from b-560 to the cytochromes c although as the reduction of c-550 is not complete in the presence of this inhibitor there may be an alternative pathway from NADH to the cytochrome a-601 in *B. subtilis*.

Cytochrome c-550 and c-554 reacted rapidly with *Pseudomonas* cytochrome oxidase, the rates being comparable with those obtained with *P. stutzeri* cytochrome c-552, (turnover number: c-550, 110·4, c-554, 72·0, *P. stutzeri* c-522, 148·0). Neither beef heart nor yeast cytochrome oxidase could oxidize ferrocytochrome c-550 or c-554 at significant rates.

Although the function of cytochrome c-550 is, in virtue of the physical properties and biochemical reactivity of the cytochrome, considered to be analagous to that of mitochondrial cytochrome c, that of cytochrome c-554 is less certain. Miki and Okunuki (1969b) have proposed a scheme in which cytochrome c-554 is an electron carrier between cytochrome c-550 and cytochrome b, analagous to cytochrome c_1. This is somewhat doubtful in view of the low oxidation-potential and acidic character of c-554. The occurrence of cytochrome o in the electron transport particles has been demonstrated by Broberg and Smith (1967) and interaction of cytochrome c-554 with the electron chain of this oxidase is a possibility.

2. Cytochromes c of denitrifying organisms

(a) Purification and general properties. Many species of *Pseudomonads*, facultative anaerobes, grown under anaerobic conditions in the presence of nitrate produce c-type cytochromes of low molecular weight which have spectra and oxidation reduction potentials similar to animal cytochrome c. Cytochromes c of larger molecular weight and of dihaem character have also been obtained from these bacteria and their oxidation-reduction potential is similar to the cytochrome b.

Cytochromes c-551. The small cytochrome c found in *P. fluorescens* is historically important as it was the first soluble cytochrome studied in the non-photosynthetic bacteria (Kamen and Vernon, 1955; Kamen and Tokeda, 1956) and the first bacterial cytochrome to have its complete amino-acid sequence established (Ambler, 1963b). Knowledge of this structure had important implications for the haem-ligand studies of mammalian cytochrome c.

Lenhoff and Kaplan (1953) obtained cell free extracts from *P. fluorescens*, grown in the denitrifying mode and noted the acid nature of the cytochrome c and the presence of a cytochrome peroxidase which accompanied the cytochrome c fraction. Considerable separation of the cytochrome c from the peroxidase, which reacted specifically with the *Pseudomonas* cytochrome c (Lenhoff and Kaplan, 1956) was achieved. Formation of the cytochrome and the peroxidase was repressed by oxygen (Lenhoff et al., 1956).

Investigating the respiratory components of *Ps. aeruginosa*, Horio (1958a) and Horio *et al.* (1958) partially purified two soluble cytochromes c, c-551 and c-554. An elaborate purification procedure which included chromatography on polymethylacrylic resin, aluminium oxide and zone electrophoresis in starch, yielded a crystalline preparation (needles) of cytochrome c-551 (Horio *et al.*, 1960). The thrice recrystallized cytochrome had a purity index ($E^{red}_{551} - E^{red}_{570}/E_{280}$ nm) of 1·27. The iron content (0·69% Fe) indicated a molecular weight of 8,100 and ultracentrifugation data 7,600. On electrophoresis in the Tiselius apparatus the cytochrome c-551 behaved as a single protein, with an isoelectric point of 4·7. The E_0' (pH 6·5) was + 0·286 mV. The reduced cytochrome had maxima of absorption at 551 (ϵmM 28·3), 521, 416 (ϵmM) 316, 290, 280; the oxidized at 409, 360, 290, 280 nm.

A simpler purification procedure applied to the cytochrome c-551 of *P. fluorescens* (str. 005/1 Kaplan) employing chromatography on CM-cellulose and DEAE-cellulose, enabled Ambler (1963a) to obtain cytochrome c-551 in high purity ($E^{red}_{551} - E^{red}_{570}$ nm/E_{280} nm = 1·17) and sufficient amounts (7·5 mg/100 g wet weight of cells) for amino acid sequence studies.

The iso-electric point of the *P. fluorescens* cytochrome c-551 was 4·7. The molecular weight of the cytochrome calculated on the tyrosine, histidine and arginine content was 9,000. The 82 amino acid residues (5 analyses) were in accord with this value. The number of residues obtained by Coval *et al.*, (1961) for *P. aeruginosa* c-551 was only 65 (one analysis). The difference noted in the number of amino acid residues and molecular weight determination of the two cytochromes c-551 may be due to a real difference in their size, albeit the spectral and other data of these two cytochromes agree. In contrast the amino acid analyses of the blue copper-protein of the two bacteria are closely similar (Ambler, 1963a).

(b) *Primary and tertiary structure.* The amino-acid sequence of *P. fluorescens* cytochrome c-551 is given in table IX (Chapter V). The haem, as in the animal cytochrome c, is at the amino acid end of the peptide chain. The sequence immediately around the haem is His-Gly-Cys-Val-Ala-Cys-His (16) (Ambler, 1962). The NH_2 terminal amino-acid is glutamic acid and the COOH terminal is lysine (Ambler, 1963a,b). Only one histidine is present and the demonstration of this played an important part in casting doubt on the necessity of histidine occupying both co-ordination positions in mammalian cytochrome c (Hettinger *et al.*, 1966; Fanger *et al.*, 1967). The elegant experiment of Harbury and co-workers strongly implicated methionine 61 as the second ligand to haem iron. Trifluoracetylation of the α- and ϵ-amino groups of *P. fluorescens* cytochrome c-551 resulted in a derivative which still had haemochrome character at pH 7·0 and when the lysine residues are guanidated and the ϵ-amino groups acetylated the modified cytochrome c-551 still gives haemochrome type spectra at pH 1·6 (Vinogradov and Harbury, 1967). Consequently lysine was ruled out as the second ligand of the haem iron. That this ligand is the methionine is stongly indicated by the fact that treatment of cytochrome c-551 with bromoacetate at pH 7·0 in the absence of cyanide alkylated only the methionine at position 22 whereas in the presence of cyanide, which combined with the cytochrome b, methionine 61 was alkylated and the spectrum was altered. Furthermore, only the methionine at position 22 reacted with iodoacetate at pH 5·4 while at pH 3·0 both methionines were alkylated resulting in a spectral change. A three-dimensional structure has been proposed (Dickerson, 1971) for cytochrome c-551 based on the x-ray analysis of mammalian cytochromes c and amino-acid sequences.

(c) *Other properties.* Like mammalian ferricytochrome c *Pseudomonas* c-551 in the ferric form has a weak absorption band ($\epsilon_M = 810$ at 25°) at 695 nm. Added imidazole (0·11 M) at pH 7·0 abolishes the 695 nm band, but it is fully regenerated at pH 6·0 and below in the presence of imidazole (Vinogradov, 1970). Without added imidazole the 695 nm band disappears at acid pH. Between pH 7·0 and 10·0 the 695 nm band persists, suggesting that the His-haem-Met complex remains stable in this pH region.

An interesting difference in the chemical reactivity of this cytochrome was manifested in the splitting of the haem from the protein. The silver method (Paul, 1951) was ineffectual; a combination of performic acid and mercuric acetate was necessary to cleave the thio-ether bond, whereas performic acid alone released the haem from cytochrome c (Margolisah, 1962).

The kinetics of electron transfer between the copper-containing protein azurin and cytochrome c-551 has been studied (Antonini *et al.*, 1970). The reaction in both directions is fast. The second order rate constant at 20° is 3×10^{-6} M^{-1} sec^{-1} for the reaction between Fe^{2+} and Cu^{2+} and $1·4 \times 10^6$ M^{-1} sec^{-1} for the reaction between Fe^{3+} and Cu^+. Mammalian cytochrome c could not adequately replace cytochrome c-551. The kinetics of the reduction and oxidation of cytochrome c-551 with dithionite and ferricyanide are much slower than the rates of reaction between cytochrome c-551 and azurin.

The cytochromes c-554 partially purified by Horio *et al.* (1960) from *P. aeruginosa* and by Ambler (1963a) from *P. fluorescens* have not been further described.

(d) *Cytochromes c-552.* The formation of cytochromes c of the denitrifying *Pseudomonas* is usually repressed in these bacteria when they are grown in the presence of oxygen. To the contrary Kodama and Shidara (1969) found a considerable amount of two cytochromes c-552 present in about equal amounts, in *P. stutzeri* str. van Niel when the organism was grown in aerated culture in the absence of nitrate. Cytochrome c-552 (I) was found in the soluble fraction of cell free extract and cytochrome c-552 (II) in the particulate fraction; the latter cytochrome could be released from the particles by acetone treatment.

The spectra of cytochrome c-552 (I) are similar to that of the Pseudomonas c-551; reduced form: 552 (ϵmM 31·0), 523, 418 nm (ϵmM 150·8); oxidized 530, 410 (ϵmM 100·4), 284 nm. There are however, significant differences in the relative strength of the maxima of absorption of the $P.$ $stutzeri$ cytochromes c, e.g., cytochrome c-552 (I) $E_{552\ nm}/E_{523\ nm}$ = 1·90, $E_{418\ nm}/E_{552\ nm}$ = 4·95; cytochrome c-552 (II) $E_{552\ nm}/E_{523\ nm}$ = 1·18, $E_{418\ nm}/E_{552\ nm}$ = 7·89. The ϵmM of the band of reduced c-552 (II) is only 19·5.

A minimum molecular weight of 8,100–8,600 was established for c-552 (I) which appears to exist as a monomer. Cytochrome c-552 (II) has a minimum molecular weight of 11,000 whilst centrifugation data indicated 20,000. The E_0' (pH 7·0) for c-552 (I) was + 277 mV and for c-552 (II) + 283 mV. The latter cytochrome is slightly autoxidizable but neither combine with CO at neutral pH. The absorbancy of the α-peak of c-552 (II) is not increased in the presence of various protein denaturants. Cytochrome c-552 (II) has been crystallized in the form of needles. As yields of 25 mg of c-552 (I) and 10 mg (II) can be obtained from 100 g wet weight of cells a structure determination should not be difficult to obtain. A particle-bound cytochrome c-552 E_0' (pH 7·0) ca. + 300 mV has been reported by Iwasaki (1960) to be present in $P.$ $denitrificans$ grown anaerobically.

The role of these cytochromes in the aerobic cells has not been established. However, a cytochrome o-type pigment and a cytochrome b (α max. 558 nm) is present in aerobic $P.$ $stutzeri$ (Kodama and Shidara, 1969) and the cytochromes c-552 may be linked to these.

The two cytochromes c-552 have been obtained from anaerobic cells of $P.$ $stutzeri$ (Kodama and Shidara, 1969). In anaerobically grown $P.$ $stutzeri$ cytochrome c-552 (I and II) appears to be associated with the nitrite reductase system; considerable increase of cytochrome c-552 relative to cytochrome b was obtained at the terminal phase of growth in cultures containing nitrate or when the organism was grown entirely in the presence of nitrite (Kodama et al., 1969). Cytochrome d was also induced in the presence of nitrite but not nitrate. Miyata and Mori (1968) observed that cytochrome c-552 of $P.$ $stutzeri$ could be substituted for the cytochrome c-552 of $P.$ $denitrificans$ in the $P.$ $denitrificans$ denitrifying enzyme-TMPD system. When lactate + yeast lactate dehydrogenase was the electron donating system in the conversion of nitrite to NO, $P.$ $denitrificans$ cytochrome c-552 (E_0' (pH 7·0) + 300 mV) was as efficient in transferring electrons to the denitrifying enzyme as was $P.$ $denitrificans$ cytochrome c-553 (E_0' (pH 7·0) + 93 mV) but only a fourth as effective as cytochrome c-553 in the more highly active ascorbate-TMPD-NO$_2$ system. Yamanaka et al. (1963c) obtained a cytochrome c-552 from *Pseudomonas saccharophila* grown aerobically in the presence of nitrate. This had an E_0' (pH 7·0) + 237 mV. This cytochrome reacted rapidly with *Pseudomonas* cytochrome oxidase but not with cow heart cytochrome oxidase. Starch-gel electrophoresis (pH 7·0) gave a main component and a second faster running component. No cytochrome c-554 was detected in the organism. A different cytochrome c (maxima 550, 523 and 417 nm reduced form) was found in a hydrogen-utilizing strain of *P. saccharophila* grown autotrophically (Bone, 1963). The E_0' at pH 7·8 was + 90 mV. About 23% of the cytochrome in the cell was associated with particles. A cytochrome b was also present.

(e) *P. denitrificans cytochrome c-553.* *P. denitrificans*, when grown aerobically produces a soluble CO-combining and autoxidizable cytochrome c of large molecular weight as well as the atypical cryptocytochrome c (Iwasaki and Shidara, 1969a). Partial reduction of the cytochrome with dithionite, in the presence of nile blue, reveals that the cytochrome has two haems, one contributing to a maximum of absorption at 555 nm and the other at 551 nm (Iwasaki and Shidara, 1969a). The E_0' (pH 7·0) was estimated to be − 90 mV for the 551 nm and between 0 and − 90 mV for the 555 nm component. The iron content (0·245%) and the haem content gave a molecular weight of 23,000 but the sedimentary coefficient and Sephadex chromatography pointed to a molecular weight of 45,000. The crystallized cytochrome had maxima of absorption in the reduced form at 553 (42·8), 524 (288) and 419 (359). The ratio $E_{409\ ox}/E_{280\ ox} = 5·5$.

The crystallized cytochrome c-553 could combine with CO at pH 7·0 ($Fe^{2+}CO$ γ max. 416 nm (445)) and with cyanide ($Fe^{2+}CN$ γ max. 420 nm (338); $Fe^{3+}CN$ γ max. 412 nm (247)). Combination of cytochrome c-553 with CO could be detected in the intact bacterial cell.

Despite the low oxidation-reduction potentials the purified cytochrome c-553 could be substituted for cytochrome c-552 in the denitrifying system of *P. denitrificans* (Miyata and Mori, 1969).

(f) *Other cytochromes c-550 of high redox potential.* *M. denitrificans cytochrome c-550.* The first bacterial cytochrome to be isolated which closely resembled mammalian cytochrome c in its spectra, E_0' (250 mV, pH 7·0) and size was obtained from *M. denitrificans* by Kamen and Vernon (1955). Its acidic character distinguished it from cytochrome c. Cytochrome c-550 is bound to the bacterial membrane primarily by electrostatic linkages, and by extraction with KCl and chromatography on DEAE and Sephadex-75 pure c-550 has been obtained in high yields and crystallized in the form of prisms (Scholes et al., 1971). At 550 nm the ϵmM is 26·8, based on an iron content of 0·39%, and the ratio $E_{550\ nm}/E_{280\ nm}$ is 1·2. The M.W. calculated from the amino acid analyses (135 residues) is

14,861, but lower values are obtained by iron analysis (14,287) and equilibration centrifugation (14,119). Like the *Pseudomonas* c-551 the *M. denitrificans* c-550 contains only one histidine so that the second ligand to the haem iron may be one of the four methionines present. Tyrosine (3) and phenylalanine (4) are present in c-550 but no tryptophan was found. An excess of dicarboxylic over dibasic residues account for the acidity of the cytochrome.

Cytochrome c-550 is rapidly oxidized by *M. denitrificans* oxidase (aa_3 type) and more slowly by beef heart oxidase (Kamen and Vernon, 1955; Smith *et al.*, 1966), but deoxycholate treatment of the heart oxidase increased the reactivity with c-550 ten-fold. Smith *et al.* conclude that the charge on a specific area rather than the overall protein charge is of greater importance for the interaction of cytochrome c with oxidases.

Spirillum itersonii cytochrome c-550. A cytochrome c-550 (550 (27·6), 522 (16·1), 416 (146) reduced form; oxidized 412 (119)) has been obtained from *S. itersonii* grown aerobically and anaerobically (Clark-Walker and Lascelles, 1970). The molecular weight was 10,411, calculated from the amino acid analysis and 10,800 from sedimentation data. The E_0' (pH 7·0) was +297 mV, and the isoelectric point 9·6. Only one methionine and histidine were present in this cytochrome and in this respect it is similar to the *Pseudomonas* cytochrome c-551.

About 85% of the cytochrome, previously called c-552 is membrane-bound in aerobic cultures. The concentration of cytochrome c-552 is four times as great in cells grown under low aeration and the cytochrome c can be fully reduced with succinate, implying that it is all integrated with the respiratory electron chain (Clark-Walker *et al.*, 1967). This higher concentration of cytochrome c-552 is not accompanied by an increased respiratory rate in the cells. Cytochrome c_1 is absent from these bacteria but presumptive evidence for the presence of cytochrome o was obtained (Clark-Walker *et al.*, 1967). Nitrate reductase activity was found in these cells even when nitrate was absent from the cultures.

When *S. itersonii* is grown anaerobically in the presence of nitrate the proportion of soluble cytochrome c-550 found is increased to 50%. A functional relationship between nitrate reductase (Gouthier *et al.*, 1970) and the soluble cytochrome c-550 has not been demonstrated, but nitrite was found to be more effective than nitrate in stimulating synthesis of soluble cytochtome c-550.

Cytochromes c from Other Organisms. *Bordatella pertussis* and other species contain a cytochrome c which has its maxima of absorption in the reduced form at 550, 520 and 418 nm and another c with maxima at 553, 520 and 418 nm (Sutherland, 1963). The cytochromes c present in about equal amounts in these bacteria have been partially purified. The E_0' (pH 7·0) of c-550 is + 259 mV and that of c-553 is + 192 mV; c-550 is slowly autoxidizable. There is some evidence that they participate in the oxidation-reduction of the copper protein, azurin, present in *Bordatella* sp. (Sutherland and Wilkinson, 1963).

(g) *Cytochromes c from halotolerant Micrococcus.* A typical cytochrome c, maxima of absorption at 551, 521 and 416 nm in the reduced form, of small molecular weight and E_0' + 249 mV (pH 7·0) has been isolated from a halotolerant *Micrococcus* strain 203 (Hori, 1961a). A cytochrome c-554 (554, 521 and 418 nm in the reduced form) is also present in aerobic cells. The M.W. of c-554 is about 18,000 and the E_0' is + 180 mV (Hori, 1961b). A third c-type cytochrome has a distinctly split α-peak in the reduced form, maxima of absorption at 554–548, 525 and 418 nm. The ϵmM values are roughly twice those of normal cytochromes c, and two haems c may be present in this cytochrome c. While ultracentrifugal data indicates a M.W. of 18,000, a minimum M.W. of 9,000 is given by iron analysis. The E_0' of + 113 mV is the lowest of the three cytochromes. Anaerobic nitrate-grown cells contain six times as much cytochrome c-554,548 as c-551: in aerobic cells the three cytochromes are present in equal amounts. Hori (1963) has concluded from the effect of inhibitors on electron transport in solubilized oxidase and nitrate reductase preparations that c-551 is the electron donor to the oxidase and c-554,548 to the nitrate reductase in the *Micrococcus*.

(h) *Thiobacilli.* Cytochromes of medium to high redox potential occur in the *Thiobacilli*. Milhaud et al. (1958) isolated a cytochrome c-551 with E_0' (pH 7·0) of + 140 to + 150 mV from *T. thioparus* and reported that the cytochrome c-552 of *T. denitrificans* had an E_0' of + 270 mV, the highest amongst the *Thiobacilli* cytochromes c. Three cytochromes c were obtained from *Thiobacillus* by Trudinger (1961a): c-550 (γ-band reduced form 416 nm, oxidized 409·5 nm) had an E_0' of + 200 mV, c-553·5 (γ-band reduced form 418 nm, oxidized 410·5 nm) had an E_0' of + 210 mV, and a third with an α-band displaced well to the red region c-557 (γ-band reduced form 419 nm, oxidized 409 nm) E_0' + 155 mV. All three were autoxidizable at pH 7·0, but c-557 only slowly. None combined with CO in the native form, but all were much more readily denatured than mammalian cytochrome c. Cytochromes c-550 and c-557 were basic proteins while c-553·5 was acidic. Cytochrome c-553·5 was shown to participate in the oxidation of thiosulphate to tetrathionate by thiosulphate oxidase (Trudinger, 1961b); c-550 and c-557 required the presence of c-553·5 as intermediary in order to react rapidly with the oxidase.

Aleem (1969) has demonstrated a cytochrome c complex (absorption peaks at 557, 555, 553 and 550 nm reduced form) in *T. neapolitanus*. Electrons from thiosulphate and sulphate enter the respiratory chain of *T. neapolitanus* at the level of cytochrome c in a process whereby the enzymic oxidation of these substrates catalyzed by aa_3 is coupled to the generation of energy. Evidence for the coupling of NADH to endogenous cytochrome c in *T. neapolitanus* was obtained by Trudinger and Kelly (1968). A cytochrome c-550 which has been purified 150-fold from *T. concretivoras* (Moriarty and Nicholas, 1969) is involved in the formation of polysulphur intermediates. Carbon monoxide can bind to this cytochrome, thus inhibiting the production of polysulphur compounds. Similarly the oxidation of inorganic sulphur substrates is coupled to the reduction of c-551 in *T. thiooxidans* (London, 1963) which was also found to have a high coenzyme Q content (Cook and Umbreit, 1963). A c-550 was also found in *T. thioparus* by Cook and Umbreit and the properties of a purified sulphite: cytochrome c oxidoreductase from this bacterium has been studied by Lyric and Suzuki (1970).

Cytochrome c-550 of *T. novellus* was partly purified and shown to be the electron acceptor for a sulphite : oxidoreductase which had also been purified from this organism (Charles and Suzuki, 1966); this cytochrome was not autoxidizable. The oxidation of thiosulphite by *T. novellus* involves aa_3 and c-550 (Aleem, 1966a,b). Cytochrome c-550 of *T. novellus* was subsequently highly purified by Yamanaka et al. (1971). Its absorption maxima in the reduced form are at 550 nm (ϵmM $= 25\cdot 8$), 520 and 414·5 nm: oxidized form, 410 nm. The E_0' (pH 7·0) is $+$ 276 mV. Based on 118 amino acid residues the M.W. is 13,270, but rather larger values were given by physical methods. The isoelectric point was at pH 7·5. The capacity of c-550 to donate electrons to various oxidative enzymes was investigated by Yamanaka et al. An acidic cytochrome c was also obtained in a purified form from *T. novellus*. The absorption maxima in the reduced form were at 551 nm (ϵmM $= 19\cdot 6$), 522 and 416 nm; in liquid nitrogen (551 sh) 548, 528, 520·8; oxidized form, 410·5 nm. The E_0' (pH 7·0) was $+$ 260 mV and the isoel. point pH 5·2. A cytochrome c-sulphite:oxidoreductase remained associated with the purified preparation and the c-551 was rapidly reduced by sulphite.

An unusual hemoprotein with haem c-type prosthetic group has been isolated from *T. novellus* (Yamanaka and Okunuki, 1970). The absorption maxima of the reduced form are at 550 (ϵmM $= 16\cdot 8$), 520 and 415 nm; oxidized form 398 nm. The extinction of the γ-band of the reduced form differs from that of typical cytochrome c by being less than that of the γ-band of the oxidized form. This haemoprotein has peroxidase activity using a variety of cytochromes c as electron donors, but surprisingly reacts poorly with *T. novellus* c-550. It is not clear yet whether the preparation is that of an impure cytochrome c or is actually that of a peroxidase which unlike other known peroxidases has haem c as its prosthetic group.

The wider aspects of energy metabolism of *Thiobacilli* and other microorganisms which metabolise inorganic sulphur compounds have been reviewed by Trudinger (1967, 1969) and Peck, (1968).

3. Cytochrome c_4 and c_5 (c-555) of Azotobacter

(a) *Preparation.* Tissières and Burris (1956) applied the designation c_4 and c_5 to two cytochromes present in *Azotobacter vinelandii* which resembled cytochrome c in several important properties. n-Butanol was used to release the cytochromes from the cells and this method has been the basis of all later preparations. Although highly purified preparations were obtained by Tissières (1956) and the two cytochromes were crystallized by Neumann and Burris (1959), the inclusion of several dialysis stages, an empirical basic lead acetate precipitation and the detrimental effect of acidic chromatography resulted in uncertain and low yields. A modified method using DEAE and CM cellulose gave highly purified cytochrome c_4 for CD studies (Van Gelder et al. 1968). An improved procedure gave double the yields obtained by the previous methods (Swank and Burris, 1969). This included chromatography on DEAE cellulose and passage through Sephadex and electrofocussing. The yields of purified cytochrome from cells of *A. vinelandii* grown with air as nitrogen source were 32–52 mg cytochrome c_4 and 13–16 mg c_5. The purity ratios $E\alpha/E_{276}$ of the reduced form were 1·30 for c_4 and 1·16 for cytochrome c_5, and were similar to the values reported for the crystalline preparation (Neuman and Burris, 1959).

(b) *Spectral and other physical properties.* On the basis of haem analysis the minimum molecular weight of the crystallized cytochrome c_4 was 11,200 and of cytochrome c_5 11,500. As isolated cytochrome c_4 and c_5 have apparent molecular weights, as determined by passage through Sephadex G-100 of 25,000 ± 900 and 24,480 ± 1,000 respectively (Swank and Burris, 1969). Van Gelder et al. (1968) reported a value of 23,500 obtained similarly for cytochrome c_4; the same values were obtained in gel chromatography in the presence of mercaptoethanol (0·1 M). Cytochrome c_4 behaved as a monomer in the presence of denaturing agents on gel chromatography and polyacrylamide gel electrophoresis, whereas cytochrome c_5 was shown to be a dimer of a monohaemoprotein of molecular weight 12,100 (Swank and Burris, 1969). Neumann and Burris (1959) noted the tendency of cytochrome c_5 on ultracentrifugation to aggregate, to give several molecular species.

The absorption maxima of cytochrome c_4 ferrous form are at 551, 522, 414 and 314 nm ($E \frac{\gamma red}{\alpha red} = 6·5$, $E \frac{\alpha red}{\beta red} = 1·3$); γ maximum of ferric form 409 nm (Neumann and Burris, 1959). On the basis of an iron content of 0·46% the ϵmM for cytochrome c_4 ferrous form are α-band 23·8, β-band 17·6, γ-band 157·2; ϵmM of γ-band ferric form, 115·8; ϵmM 270 nm 20·5 (Tissières, 1956). The absorption maxima for cytochrome c-555 are reduced from 555, 524, 416, 318 nm; oxidized 414 nm ($E \frac{\gamma red}{\alpha red} = 6·0$; $E \frac{\alpha red}{\beta red} = 1·4$) (Neumann and Burris, 1959). Calculated on a M.W. of 12,100 the ϵmM of cytochrome c_5 at 554 nm (ferrous) is 26·2 and at 554 (ferric) is 9·6 (Swank and Burris, 1969). There is no splitting of the α-maximum of reduced cytochrome c_4 at 77°K (Jones and Redfearn, 1967a).

The isoelectric points of oxidized and reduced cytochrome c_4 were 4·40 and 4·69 and for oxidized and reduced cytochrome c_5 4·20 and 4·44. These values are in accord with similarity of the aspartic acid, glutamic acid and lysine content of the two cytochromes (per haem). Cytochrome c_5 is low in aromatic residues; no phenylalanine is present and only one tyrosine and tryptophan. Cytochrome c_4 has six tyrosines and five phenylalanines, but no tryptophan, and roughly 3 times as many aromatic residues on a haem basis (Swank and Burris, 1969). Cytochrome c_5 like *P. fluorescens* cytochrome *c*-551 and *Chromatium* cytochrome *c'* has only one histidine, but has two methionines and, interestingly, two extra cysteines. Cytochrome c_4 has four histidines and 8 methionines, but only the requisite number of cysteines for the 4 thioether linkages to its two haems.

The E_0' over the pH range 6·0–7·5 is + 300 mV for cytochrome c_4 and + 320 mV for cytochrome c_5 (Tissières, 1956). Neither cytochrome was appreciably autoxidizable, and while cytochrome c_5 showed a slight tendency to combine with CO this is possibly due to denaturation which occurs during the acidic isoelectric focussing (Swank & Burris, 1969).

The CD spectra of cytochrome c_4 (Van Gelder *et al.* 1968) more strongly resemble those of cytochrome *c* undecapeptide (Urry, 1967) than of cytochrome *c*. The CD band of the Soret region is of simple form and it shifts to longer wave-length and magnitude corresponding with changes observed in the absorption spectrum. The CD extremum at 314 nm (a haemochrome band) is positive in both cytochrome c_4 and the monomeric haem-undecapeptide, in contrast to the CD spectra of cytochrome *c*. Such differences that are observed between the spectra of the oxidized and reduced forms of cytochrome c_4, over the whole wavelength range, are attributed to the haem chromophore which appears not to be subject to gross change in its environment or change of valency of the iron.

A third cytochrome, minor c_4, representing about 5% of the cytochrome *c* present in the broken cells of *A. vinelandii* str. O.P. was partially purified by Swank and Burris (1969). Spectrally it is identical with cytochrome c_4, but has half the molecular weight.

Apart from the position of its α-peak at 555 nm, there is little to justify the retention of the term c_5 for the monohaem cytochrome *c* of *Azotobacter*. The continued use of this term obscures the essential similarity of cytochrome *c*-555 with the other bacterial cytochromes of similar molecular weight and high oxidation-reduction potential.

(c) *Biochemical activity and role of cytochromes c_4 and c_5*. In assays with oxidase containing particles from *A. vinelandii* the K_m value for cytochrome c_5 was 3 fold lower than that for cytochrome c_4 and the first order rate constant k (sec^{-1} mg protein^{-1}) for cytochrome c_5 (0·12) was three times that of cytochrome c_4 (0·038). Cytochrome c_5 (and minor c_4) was oxidized by *Azotobacter* small particles at double the rate of cytochrome c_4, but did not react with beef heart oxidase, whereas cytochrome c_4 was oxidized at 20% the rate of beef cytochrome c. Oxidation of cytochromes c_4 and c_5 was very sensitive to inhibition with CN^- (50% at 10^{-5} M KCN) indicating that the oxidation is mediated by cytochrome c_1 or o, not cytochrome a_2. Swank and Burris (1969) consider that cytochromes c_5 and minor c_4 interact directly with the oxidase. However, cytochrome c_5 is present in lesser amounts in the cells than cytochrome c_4 and Jones and Redfearn (1967a) could not detect it in their strain of *Azotobacter* (NCIB 8660). Ascorbate-DCIP can donate electrons to the cytochrome c_4 and c_5 pathway at the cytochrome c_4 and c_5 level (Jones and Redfearn, 1967a).

Optimal yields of cytochromes c_4 and c_5 are obtained when N_2 rather than ammonia is the source of nitrogen (Swank and Burris, 1969). The amount of cytochromes c_4 and c_5, and cytochrome c_1, was not greatly diminished in cells grown on urea whereas the amount of cytochrome a_2 formed was very low (Knowles and Redfearn, 1968). A direct role for these cytochromes c in nitrogen fixation, as suggested by Ivanov *et al.* (1967) is theoretically unlikely and negative evidence for such a role has been presented by Gvozdev *et al.* (1968). The increase of cytochrome c_4 and c_5 content in cells using N_2 as nitrogen source is probably related to the increased demand for ATP formation, by oxidative phosphorylation, during nitrogen fixation.

The acetylene-reducing activity of a nitrogenase preparation from *Azotobacter chroococcum* was stimulated by the presence of cytochrome c_4 (Yates, 1970b).

4. Cytochrome c_3 and other cytochromes of low oxidation-reduction potential

(a) *General properties of cytochrome c_3.* The observation that the sulphate-reducer *Desulphovibrio*, an exacting anaerobe, contained a cytochrome c (Butlin and Postgate, 1953; Postgate, 1954; Ishimoto *et al.*, 1954), was important as it demonstrated that contrary to the current beliefs (cf. Schaeffer and Nisman, 1952; Keilin and Slater, 1953) cytochromes could occur in obligate anaerobes. Postgate's (1956) finding that the cytochrome from *D. desulphuricans* str. Hildenborough (now renamed *D. vulgaris*) has an oxidation-reduction potential some 500 mV lower than that of mammalian cytochrome c revealed the versatility of the c-type cytochromes. The cytochrome of *D. vulgaris* was named cytochrome c_3 by Postgate (1956), and subsequently several other *Desulphovibrio* species have been shown to contain a cytochrome c of this type (Le Gall *et al.*, 1965; Drucker and Campbell, 1969).

A procedure for the purification of cytochrome c_3 of *D. vulgaris* which
K

gives yields of 60–80 mg of cytochrome of purity index 2·78 ($E^{red}_{552\ nm} - E^{red}_{570\ nm}/E^{oxid}_{280\ nm}$) is described by Drucker and Campbell (1969). Lower yields of cytochrome c_0 were obtained by these workers from *D. desulphuricans*. Storage of the cells at $-20°C$ for several weeks before extraction facilitated purification of the cytochrome c_3. The cytochrome from *D. salexigens* behaved as an acidic protein rather than a basic protein, requiring an additional chromatography, when cells had been stored for less than six months (Drucker et al., 1970b). To obtain good yields it was necessary to maintain the cytochrome in the reduced form during certain stages of the purification. Ferricytochrome c_3 of *D. vulgaris* has been crystallized as micro-needles, (purity index 2·74) by Horio and Kamen (1961). This crystallized cytochrome c_3 had absorption peaks at 552, 522 and 418 nm at pH 7·0, in the reduced form; 409 and 349 nm in the oxidized form.

The cytochrome purified by Postagte (1956) had an E_0' (pH 7·0) of -205 mV; there is no peak at 695 nm in the ferric form, nor a peak at 280 nm because of the low aromatic amino acid content. Centrifugal analysis indicated a molecular weight of 13,000. The iron content was 0·92% pointing to the presence of at least two haems in cytochrome c_3. Horio and Kamen (1961) drew the same conclusion from a comparison of the extinction of the band of reduced crystalline cytochrome c_3 with that of cytochrome c.

Drucker et al. (1970c) have redetermined the iron content of the cytochromes c_3, using a more sensitive assay and an improved procedure for desalting the haemoprotein prior to iron analysis. The iron content obtained was 1·21% for the cytochrome c_3 of *D. vulgaris* and 1·17% and 1·24% for *D. desulphuricans* and *D. gigas* respectively. These values are consistent with the presence of three rather than two haems in each of the cytochromes c_3 and are in accord with the extinctions of the band of the reduced cytochromes (e.g. *D. vulgaris* $\epsilon^{red}_{552\ nm} = 84$). The molar extinction at 550 nm of the reduced pyridine haemochrome of *D. vulgaris* and *D. desulphuricans* (88·3 and 78·9 respectively) strengthens the conviction that three haems are present in cytochrome c_3. On the basis of similar data the cytochrome c_3 of *D. salexigens* appears to have three haems (Drucker et al., 1970c).

Eight cysteines are present in the cytochromes c_3 of the four *Desulphovibrio* species so far examined, so that there are sufficient sites for attachment of three haems (cf. Drucker et al., 1970b; Ambler et al., 1970).

Drucker et al. (1970b,c) have determined the molecular weight of cytochromes c_3 of *D. vulgaris* (14,700), *D. sulphuricans* (13,500) and *D. salexigens* (13,870) from sedimentation velocity data. These values compare well with those calculated from amino acid analyses, including three haems (13,466, 13,676 and 13,901 respectively). The cytochrome c_3 of *D. vulgaris* and *D. salexigens* are basic cytochromes, isoelectric points 9·9 and

10·8 respectively, the greater content of glutamic acid and fewer lysine residues in *D. desulphuricans* results in this cytochrome c_3 having an isoelectric point of 8·0. The experimentally determined values were very close to those calculated from the amino acid composition (Drucker *et al.*, 1970b).

The isoelectric point of *D. gigas* c_3 is at pH 5·2 and this is reflected in the very different amino acid composition of this cytochrome (Bruschi-Heriaud and Le Gall, 1967).

(*b*) *Primary structure*. The amino acid sequence of cytochromes c_3 of *D. vulgaris* and *D. gigas* (Ambler, 1968; Ambler *et al.*, 1969) show that the peptide chain has only two sites for normal cytochrome *c* type haem attachments, so that the third haem presumably is attached to one of the two sequences Cys-A-B-C-D-Cys-His, which occur in each protein (see Table IX). One of these unusual sequences is within a residue or two of the carboxyl terminal amino acid. One of the free normal cytochrome *c* linkages is spaced 15 residues further away from the amino terminal compared to cytochrome *c*; the other cytochrome *c* linkage is about the same distance in from the carboxyl end of the peptide chain. These two haems are thus somewhat further apart than are the two haems in cytochromes *cc'* (see 5f).

In view of the present emphasis on the importance of homologies in the three-dimensional structure of the cytochromes *c* (Margoliash *et al.*, 1968; Dickerson, 1971) it is of particluar interest that though there is little homology between the cytochromes c_3 and the other cytochromes for which amino acid sequences are available, the evolutionary invariants tyrosine 48 and 67 have their counterparts in the cytochrome c_3 of *D. vulgaris* (45, 67) and *D. gigas* (46, 68).

The size of the peptide loop (9 residues) between the site of attachment of the amino terminal haem and the first invariant tyrosine is similar to that of the equivalent peptide loop (10 residues) in *Pseudomonas* cytochrome *c* (Dickerson, 1971).

The dissimilarities between the amino acid sequence of the two cytochromes c_3 is considerable. Only 49 of the two cytochrome c_3 residues of *D. gigas* match residues in the *D. vulgaris* cytochrome c_3 when the haem binding sites are aligned. The other histidines and aromatic residues are mostly conserved, but not the hydrophobic residues. Fifty residues are not identically matched. Despite the differences, DNA base ratios are similar (60 + 61% guanine and cytosine), consequently greater discrepancies in the amino acid sequences are expected of *D. desulphuricans* and *D. salexigens* where the DNA base ratio is 55 and 46% guanine and cytosine (Postgate and Campbell, 1966).

Methionine is absent from the cytochrome c_3 of *D. gigas* and only in *D. vulgaris* cytochrome c_3 are there enough methionine residues to provide a ligand to each haem (Drucker et al., 1970b). The very low redox potential of cytochromes c_3 may in part be attributed to a different amino acid at the sixth coordination position of the haem iron. As the cytochromes c_3 so far discovered have the same oxidation-reduction potential and the same spectrum, it is probable that this ligand is identical in each case. At least six histidines are present in the cytochromes c_3 (Coval et al., 1961, Horio and Kamen, 1961; Drucker et al., 1970b,c), enough to provide both ligands to the haem. However, the position that the histidines occupy in the peptide chain almost certainly excludes histidine from being the second ligand to more than one of the cytochrome c_3 haem irons. A probable candidate for this role is lysine (cf. Margoliash and Schejter, 1966).

The similarity of the absorption and optical rotary properties of the cytochromes c_3 of the three species of *Desulphovibrio* indicate that the haem environment in all these proteins is similar, despite substantial difference in their amino acid composition (Drucker et al., 1970a). Cytochromes c_3 have circular dichroism bands which are of an order of magnitude larger than those of mammalian cytochrome *c*, and their Soret band in the CD spectra is split into two components of opposite sign in both oxidized and reduced states. There is no inversion in sign of the two components on reduction, unlike that observed for the Soret band of mammalian cytochrome *c*.

(c) *Immunological reactions.* Drucker and Campbell (1966) prepared antisera against cytochrome c_3, and showed that there is no immunological cross-reaction between the cytochromes of *D. desulphuricans* and *D. vulgaris* (Drucker and Campbell, 1969). Neither the cytochrome c_3 of *D. salexigens* (Drucker et al., 1970b) nor that of *D. gigas* (Drucker and Campbell, 1969) shares a common precipitating antigenic determinant with the cytochromes c_3 of *D. desulphuricans* or *D. vulgaris*. This immunological specificity contrasts with the cross reactions that occur amongst the mammalian *c*-type cytochromes.

(d) *Function of cytochrome c_3.* The cytochrome c_3 of *D. vulgaris* (*D. desulphuricans* str. Hildenborough) has been implicated in the reduction of sulphate, sulphite and thiosulphate (Postgate, 1956). Earlier work in this area has been reviewed by Ishimoto et al. (1958), Postgate (1961) and Newton and Kamen (1961). The involvement of cytochrome c_3 in the sulphate reductase complex has been demonstrated by Ishimoto and Yagi (1961). The participation of cytochrome c_3 in the formic hydrogenase system has been shown by Williams et al. (1964). Cytochrome c_3 mediated the reduction of FMN and FAD by a purified hydrogenase preparation from *D. vulgaris* (NCIB 8303) (Sadana and Morey, 1961). A particulate hydrogenase, solubilized and highly purified catalyzes both the evolution of H_2 from $Na_2S_2O_4$ in the presence of cytochrome c_3 ($pK_m = 4.15$) and the reduction of H_2 by ferricytochrome c_3, (Yagi et al., 1968, Yagi, 1970).

Ferredoxin could not replace cytochrome c_3. The optimum pH for the cytochrome c_3 dependent H_2 evolution was 5–6; for cytochrome c_3 reduction by H_2, pH 8–9.

The hydrogenase from *D. vulgaris* reduced ferricytochrome c_3 of *D. salexigens* at a rate 1,000 times slower than it did the c_3 from *D. vulgaris* or *D. desulphuricans* (Haschke and Campbell, 1971). Akagi (1967) studied the phosphoroclastic reaction of *D. desulphuricans* (str. 8303) and showed that cytochrome *c* and ferredoxin are closely associated during the process of molecular H_2 evolution from pyruvate. He proposed a sequential scheme in which ferredoxin transferred electrons from pyruvate to cytochrome c_3, the latter then donating the electrons to hydrogenase. Cytochrome c_3 was not essential for the phosphoroclastic reaction, but the rate of hydrogen formation was faster in the presence of the cytochrome. An enzyme system from *D. vulgaris* separable into two major components required cytochrome c_3, ferredoxin and hydrogenase to form thiosulphate from sulphite (Suh and Akagi, 1969).

(*e*) *Cytochrome c-553*. Cytochrome *c*-553 which has a higher redox potential, E_0' (pH 7·0) = 0 to $-$ 100 mV, was obtained from *D. vulgaris* (Hildenborough) by Le Gall and Bruschi-Heriaud (1968), and from *D. desulphuricans* NCIB 8380 and *D. gigas* NCIB 933 (Ambler *et al.*, 1970). Cytochrome *c*-553 has a molecular weight of 9,000 and a single haem; its isoelectric point is 8·6. The 80 amino acid residues contain two cysteines, only one histidine but an unusually large number of methionines (5) (Bruschi-Heriaud *et al.*, 1970). As in cytochrome c_3 there are no tryptophan residues and phenyl-alanine is also absent.

The absorption spectrum of *c*-553 is similar to that of c_3 except that it has a peak at 280 nm also, and in the ferric form a small peak at 685 nm. The amino terminal residue, alanine, is common to *c*-553 and c_3, but *c*-553 has leucine as a terminal carboxyl residue instead of glutamic acid.

Yet another cytochrome *c* with a molecular weight of about 26,000 has been detected in *D. vulgaris*, *D. gigas* and *D. sulphuricans*. The physiological function of cytochrome *c*-553 and of the larger cytochrome is unknown, (Bruschi-Heriaud *et al.*, 1970).

(*f*) *Cytochrome cc_3*. A cytochrome *c* of larger M.W. (26,000) has recently been obtained from *D. vulgaris*, *D. gigas* and *D. desulphuricans*. Cytochrome cc_3 as it has been designated is deduced by analogy with the spectra of cytochrome c_3 to have 4–6 haems. Amino acid analyses of *D. vulgaris* cc_3 show that it contains 16 cysteines and 16 histidines, but no methionine (Ambler *et al.*, 1970). Comparison of the amino acid composition of the three cytochromes isolated from *D. vulgaris* shows that cc_3 is not a dimer of c_3 or *c*-553. Cytochrome cc_3 has a low redox potential as it is not reduced by ascorbic acid. The spectrum of the ferric form does not exhibit a peak at 695 nm.

Cytochrome cc_3 or *D. gigas* participated in the reduction of thiosulphate by molecular hydrogen with an activity in the reaction comparable to the ferredoxins. NADPH + H$^+$, rubredoxin oxido-reductase and formic dehydrogenase are known to reduce c_3-553, but the physiological role of this cytochrome remains obscure.

(g) *Cytochrome c-552 of E. coli.* (i) *Occurrence and general properties.* Although Yamagutchi (1937) had obtained spectroscopic evidence for the presence of a cytochrome of c-type in *Escherichia coli*, the extremely low level in aerobic cells and the obscuring of the α-band (552 nm) by that of cytochrome b_1 (557 nm) present in much greater amounts contributed to the delay in the general acceptance of Yamagutchi's findings (cf. Keilin and Harpley, 1941). Lazzarini and Atkinson (1961) observed a c-type cytochrome in a purified preparation of NADP-specific nitrite reductase from *E. coli* grown in the presence of 0·01 M nitrite. A survey of several strains of *E. coli* and other *Enterobacteriaceae* grown anaerobically revealed the presence of a soluble cytochrome (Gray et al., 1963) in these organisms. *Escherichia aurescens* contained the greatest amount of cytochrome c-552, the next richest source was an *E. coli* strain K-12 containing a fourth as much. Fujita and Sato (1963a, 1966b) observed that the presence of 0·1% NaNO$_3$ in the growth medium increased the yield of cytochrome c and that the Yamagutchi strain was superior to *E. coli* K-12 and other strains as a source of cytochrome c-552. Very small amounts of another c-type cytochrome having its α-band at 550 nm was also obtained from the Yamagutchi strain.

Cytochrome c-552 of *E. coli* has presented difficulties in purification, partly because of its tendency to exist as a polymer. The purest preparation has been obtained by Fujita (1966a), who employed the Yamagutchi strain, releasing the cytochrome from the cell with butanol and then used ammonium sulphate precipitation, and chromatography on hydroxyapatite and DEAE-cellulose followed by passage through Sephadex G-75. Attempts to crystallize the cytochrome from ammonium sulphate solutions produced a sphaerulite-like precipitate.

Sedimentation analysis of this preparation gave a single well defined peak. The molecular weight estimated by the Archibald method was 136,000. The minimum molecular weight calculated from the haem and iron contents were 12,300 and 11,100 respectively. Gray et al. (1963) noted that cytochrome c-552 appeared to exist in the cell as a large polymer. Cytochrome c-552 behaved as an acidic protein in the Tiselius apparatus.

The reduced cytochrome has maxima at 552 (31·1), 523 (176), 420 (148), 280 nm (30·2) in the ferrous form; oxidized at 562 (6·9), 532 (9·7), 409 nm (106). This cytochrome c-552 did not combine with CO at pH 7·0, unless previously heated at 100°. Cytochrome c-552 was autoxidizable and had an E_0' (pH 7·0) between − 194 and − 220 mV (using phenosafranin).

A highly autoxidizable cytochrome c-552 was obtained (Barrett & Sinclair, 1967b) from *E. coli* str. McElroy which differed from the cytochrome c-552 from the Yamagutchi strain in the relative ease whereby it could denature to combine with CO at 50°; intact cells contained variable amounts of the CO-combining form. The spectra were similar to those obtained for cytochrome c-552 Yamagutchi and the oxidation-reduction potential was slightly less negative, $-$ 150 mV (pH 7·0) (using phenosafranine). This cytochrome c was extremely difficult to purify; the best preparation was 20% pure and behaved in ultrafiltration and on Sephadex as though the molecular weight was 50–70,000. Starch gel electrophoresis of the purified cytochrome gave two cytochrome bands. The major component had an isoelectric point of about pH 4·4 and the minor component migrated more slowly towards the anode. The relative amounts of the two components were the same whether the organism was grown under aerobic or micro-aerophilic conditions.

(*ii*) *Function and induction of cytochrome c-552.* The low oxidation-reduction potential of cytochrome c-552 and the repression by oxygen of its formation in a number of the *Enterobacteriaceae* (Wimpenny et al., 1963) point to a role for it in anaerobic electron transport. Gray and Wimpenny (1965) noted the simultaneous induction by anaerobiosis of the cytochrome and of the hydrogenase and hydrogenlyase activities in various strains of *E. coli* and suggested that cytochrome c-552 is involved in the formic-hydrogenlyase reaction of *E. coli*.

The preponderance of evidence indicates however that this cytochrome c-552 participates in a nitrite reductase system. Reduced cytochrome c-552 was reoxidized rapidly by nitrite and hydroxylamine but not by nitrate. These oxidations were inhibited by cyanide. On the basis of growth experiments Fujita and Sato (1966b) concluded that nitrite is the true inducer of cytochrome c-552 and that induction by nitrate is an indirect effect. Low levels of nitrite and nitrate stimulate the biosynthesis of cytochrome c-552 preferentially without causing an increase of cytochrome b_1. The presence of 0·5% nitrate in the growth medium represses the formation of cytochrome c-552 but not that of cytochrome b_1 (Fujita and Sato, 1966b). Cole (1968) showed that nitrite added to the growth media of *E. coli* (K-12 and Yamaguchi strains) caused an increase of nitrite reductase activity concomitantly with the increase of cytochrome c-552 content. Conversely high concentrations of nitrate stimulated nitrate reductase and cytochrome b_1 formation (Cole and Wimpenny, 1968). The concentration of cytochrome c-552 in various *Enterobacteriaceae* grown anaerobically in the presence of 0·1% nitrate was 5–10% that of cytochrome b_1 (Fujita, 1966a).

Intact cells or spheroblasts of *E. coli* catalyse the reduction of nitrite to ammonia when glucose is the electron donor and the activity is proportional to the cytochrome c-552 present (Fujita and Sato, 1967). As cytochrome c-552 is located either in the cell-wall or outer cytoplasmic membrane (Fujita and Sato, 1966a) and the reduced cytochrome can be reoxidized directly by nitrate, Fujita and Sato (1966a) suggested that cytochrome c-552 might detoxify the nitrite, the cell then utilizing the products for growth.

The function of the c-552 of *E. coli* McElroy is more obscure though it appears to be involved in sulphite metabolism and ability to synthesize the cytochrome is lost by this organism if it is not maintained on sulphite-rich growth media (Barrett & Sinclair, 1967b). In the presence of a particulate preparation from *E. coli* str. McElroy sulphite caused cyclic oxidation of reduced cytochrome c-552. Cyanide (10 mM) inhibited this reaction. Neither nitrite, nitrate or sulphate could replace sulphite. Oxygen stimulates the formation of this c-552 and under microaerophilic conditions cytochrome c-552 McElroy is present in the bacterial cell in greater concentration than cytochrome b_1. Nitrate and anaerobiosis suppress its formation.

A related bacterium *Salmonella typhimurium* produced abundant cytochrome c (551, 525 and 416 nm, reduced form) when L-glutamate was the carbon source in growth media but was totally repressed by glucose (Richmond and Kjeldgaard, 1961).

5. Cytochromes cc' and c'

(a) *Introduction.* A haemoprotein which resembled myoglobin spectroscopically in having a strong band at 640 nm in the oxidized form and a broad band at 550–560 nm when reduced and combined with CO, was discovered in trichloroacetic acid extracts of *R. rubrum* by Elsden *et al.* (1953). The pigment, however, lacked oxygen-carrying capacity and the pyridine and cyanide haemochromes were the same as those prepared from cytochrome c; consequently Vernon and Kamen (1954) tentatively called this pigment pseudohaemoglobin. The presence of this haemoprotein in large amounts in representative species of other photosynthetic bacteria (Vernon and Kamen, 1954) led Hill and Kamen to propose that the term RHP (*Rhodopseudomonas* haemoprotein) be used for this class of cytochromes, and this term remained current until it became apparent that these haemoproteins were essentially anomalous cytochromes c, or cytochromoids (Kamen and Bartsch, 1961). The designation, cytochrome cc' is now used for these haemoproteins of this sub-class of cytochrome c which have two haems covalently bound by thioether linkages to the protein, and cytochrome c' for those with only one haem (cf. International Commission for Nomenclature of Enzymes, 1966). However the term cc' suggests a difference of environment of the two haems, which had previously been assumed, but which seems to have become more uncertain recently.

(b) *Preparation.* Trichloroacetic acid, though used for extracting both cytochromes cc' and cytochromes c from acetone powders of photosynthetic bacteria, gave low yields of native haemoproteins (Vernon and Kamen, 1954). A preliminary extraction with sulphuric acid at pH 4·0 was used to remove cytochrome from lyophilised cells of *R. rubrum* prior to extraction from the cells of the cytochrome cc' with 1% citric acid. This procedure has been replaced by mechanical rupture or sonication of lyophilized *R. rubrum* cells in Tris buffer at pH 8·0, or phosphate buffer at pH 7·5 for *Chromatium sp.* (Horio and Kamen, 1961; Bartsch and Kamen, 1960). The extracts are chromatographed on Sephadex G-25 and on DEAE cellulose equilibrated with Tris at pH 7·8 for *R. rubrum* cytochrome cc' or with phosphate at pH 7·0 for that of *Chromatium sp.*, thereby removing ferredoxin. Rechromatography on DEAE cellulose and fractional precipitation with ammonium sulphate leads to preparations of very high purity. Horio and Kamen (1961) obtained crystalline *R. rubrum* cytochrome cc' in a procedure using aluminium oxide equilibrated at pH 7·1, instead of ion-exchange celluloses. Preparations suitable for the large amounts of the cytochromes cc' required for amino acid sequence studies have been described by Kamen *et al.*, (1963).

(c) *Absorption spectra conversions and CD spectra.* At pH 7·0 oxidized cytochrome cc' has a broad Soret peak at 390 nm and the reduced form has a peak at 423 nm with a prominent shoulder at 430 nm. At pH 10 the Soret peak assumes a new shape having a maximum at 402 nm and a shoulder at 370 nm whereas that of the reduced form remains unchanged. On further increase of pH to 12·5 both the oxidized and reduced cytochrome cc' exhibit spectra typical of c-type cytochrome, accompanied by considerable intensification of the Soret bands. This spectral conversion is reversible. The spectra of the oxidized cytochrome cc' at pH 7·0, 10 and 12·5 have been called Type I, II and III respectively; those of the reduced form at pH 7–10 and 12·5 as Type a and n respectively (Imai *et al.*, 1969a). In the presence of alcohol, ketones and phenols immediate conversion of Type I spectrum of the oxidized pigment to Type III occurs at pH 7·0 without the appearance of the Type II spectrum. Type II spectrum can similarly be converted to Type III at pH 10·0. The effectiveness of a series of solvents is a function of their hydrophobicity. Higher concentrations of solvents are necessary to induce the spectral conversion of the reduced form from Type a to n. Urea and guanidine induce the same spectral conversions but at a lower rate.

Increase of ionic strength has a reverse effect; at pH 12·5 the spectral conversion in the presence of 2 M KCl proceeds only as far as Type IV with oxidized cytochrome cc', while the spectral conversion of the reduced form from Type a to n is suppressed. Type III spectra are converted to Type II spectra at pH 12·5 on adding 2 M KCl to the oxidized cytochrome.

TABLE XV
Cytochromes of type c'. Absorption maxima and difference spectra. Wavelength in nm

Cytochrome	α	β	γ	δ	
			εmM in parentheses†		
Rhodospirillum rubrum cc'*					Bartsch (1963)
Reduced, free	563 (7)	550 (22)	426–433 (106–165)		
Reduced, associated with chromatophore	562 (20)	532 (2)	429		
Oxidized, free	638 (5.9)		390 (112)		
			(159)		
Reduced plus CO	564 (24)	534 (25.6)	416.5 (480)		
Reduced plus CO – reduced					
Maxima	569 (2.3)	534 (6.2)	416.5 (338)		
	590	553	434		
Minima	(– 2.8)	(– 0.3)	(– 130)		
Rhodopseudomonas palustris c'					Bartsch (1963)
Reduced	570 (10)	552 (10.4)	435 (9.2)	426 (192)	
Oxidized	642 (2.75)	500 (10.4)	398 (85)	283 (21)	
Reduced plus CO	570 (11)	535 (12.6)	418 (262)		
Reduced plus CO – reduced					
Maxima	572 (1.7)	534 (0.4)	417 (180)		
	592	555	436		
Minima	(– 2.4)	(– 1)	(– 25.5)		

					Reference
Pseudomonas denitrificans cc'					Cusanovich et al. (1970)
Reduced	550 (7·1)			434 (426) (87·5) (97·0)	
Oxidized	635 (3·7)	495 (10·2)		400 (80·0)	280 (30·8)
Reduced plus CO	564 (10·5)		534 (11·8)	418 (210·0)	
Reduced in 3N NaOH	550 (24·9)		522 (13·2)	416 (156·2)	
Oxidized in 3N NaOH	575 (6·3)		538 (9·4)	413 348 (99·0) (25·1)	
Pseudomonas stutzeri c-552, 558					Kodama and Mori (1969b)
Reduced	558 (18·3)	526 (10·5)	552 (18·8)	421 (51)	
Oxidized	530			409 360 (s) (136)	
Reduced plus CO	558 (19·1)	527	552 (s)	415 (148)	
Reduced, 0·1 NaOH	552 (27)		524	418 (213)	

* Spectrum of Chromatium cc' shows little variance from that of R. rubrum cc'. Bartsch (1963).
† emM. Recalculated for single haem.

To a lesser extent glycerol had the same effect as 2 M KCl. This effect may arise from intensification of helical structure, as observed for polylysine (Blauer, 1964), which protects the intra molecular hydrophobic interactions essential for the maintenance of the abnormal spectra.

Further insight into the complexity of this type of haemoprotein is provided by the CD study of *R. rubrum* cytochrome cc' and *R. palustris* cytochrome c' by Imai et al. (1969b). The CD spectra of both cytochromes in the oxidized form had a large positive extremum at 406 nm and a broad negative extremum at 325 nm at pH 7·0 and 10·3. The strength of these bands was much larger than for any other haemoproteins so far reported. At pH 12·3 the CD bands were greatly decreased.

The helical content of both the oxidized cytochromes calculated from the value of θR at 222 nm was 63%, assuming no contribution from the haem to that band, but was much diminished at pH 12·3. The high helical content in these cytochromes is ascribed to the large number of alanine residues in their structure.

The CD spectra of the reduced form of the two cytochromes exhibited a large and complex positive band at 430–435 nm, which, as it occurs also in the CD spectra of the monohaem cytochrome c, cannot have its origin in haem–haem interaction (Urry, 1967). A positive band at 263 nm characterized as being due to a porphyrin-iron-histidine complex in other haemoproteins (Flatmark and Robinson, 1967) increased in strength when the pH was raised from 7·0 to 12·4. The helical content of the reduced cytochrome was similar at both pH 7·0 and 12·3 to that of the oxidized form.

In general it appears that the atypical spectra of this sub-class of cytochromes reflect the abnormally large optical asymmetry in the vicinity of the haem and that changes in the absorption spectra with pH and other agencies reflect destruction of a hydrophobic environment around the haem. Alterations in the gross helical configuration of the protein are probably not involved in the spectral conversions.

(*d*) *Magnetic susceptibility, ESR and Mössbauer studies.* Magnetic susceptibility studies confirm the deduction made from optical studies that in the ferrous state cytochromes cc' and c' are fully high spin compounds (Ehrenberg and Kamen, 1965). However, a mixture of high and low spin forms was indicated for the ferric cytochromes by the findings of values of 4·35–5·30 BM instead of the expected 5·92 BM for high spin iron with 5 unpaired electrons. Temperature optical difference spectra provided support for this interpretation of their data, but contrary to the effect obtained on alkaline ferrimyoglobin and ferrihaemoglobin the high spin content is increased at lower temperatures. Tasaki et al. (1967) measured the magnetic susceptibility of *R. rubrum* cytochrome cc' down to 4·2°K. The BM values at 150°K and at 4·2°K were similar to those obtained for myoglobin for both the Fe^{2+} and Fe^{3+} states suggesting that there may be an alternative interpretation for the data of Ehrenberg and Kamen (1965).

Tasaki et al. (1967) noted that at alkaline pH the BM value for haemoglobin was markedly lowered, attributing this to the complexity of the haemoprotein. In this context the observations of Ehrenberg and Kamen (1965) on the ESR of these haemoproteins are of interest. Strong ESR absorption at higher magnetic fields (lower g values) compared to that for myoglobin (g = 6·0) were obtained for cytochromes cc', e.g. that of R. rubrum g = 4·7, for R. palustris cytochrome c' g = 4·9. The ESR peaks of the cytochromes were very broad and their breadth increased with temperature. This feature of the ESR spectra cannot be due to interaction of the haems as it is exhibited by the monohaem cytochrome c'.

Mössbauer spectroscopy studies (Moss et al., 1968) confirm that oxidized cytochromes cc' have a highly distorted high spin electronic configuration, as found in methaemoglobin. However, the signs of electric field gradients are opposite for methaemoglobin and the cytochrome cc', indicating that the iron co-ordination is very different in the two types of haemoprotein. The distinctly high spin state of the iron in ferrous cytochrome cc' indicates a weak ligand field environment (Orgel, 1966) contrasting with the strong ligand field of the iron in Chromatium cytochrome c-552 and R. rubrum cytochrome c_2. The Mössbauer spectra of all forms of the cytochrome cc' displayed a single pair of Mössbauer absorption lines indicating that the ligand pair of each haem is identical and also implying that in Chromatium cytochrome cc', which has only 2 histidines (Bartsch et al., 1961) that this amino acid cannot be the sixth ligand to either haem. Substantial differences in the quadrupole splitting in the Mössbauer spectra of R. rubrum and of Chromatium cytochrome cc' are considered to be due not to differences in the ligands but to more subtle variations in the environment of the haem iron.

(e) *Reaction with ligands.* Taniguchi and Kamen (1963) studied the spectroscopic responses of R. rubrum cytochrome cc' to exposure to a variety of ligands which can react with the accessible haem group of the haemoglobins and myoglobins. Only NO and CO are able to penetrate to the prosthetic haem groups of the protein. The anionic ligands, azide, fluoride and cyanide did not react with the haem and neither do the larger uncharged ligands, nitrosobenzene and 4-methylimidazole. Chromatium cytochrome cc' behaved similarly towards these ligands. Cytochrome cc' has a much lower affinity for NO than for CO, which is contrary to the universal experience that NO is a more powerful ligand for haematin compounds than is CO (Gibson and Roughton, 1955).

The NO-ferricytochrome cc' compound exhibits spectra similar to the NO complexes of free haematins and haemoproteins of the open type (Keilin, 1955) and are distinguishable from NO-compounds of ferrocytochrome cc'. Certain facts led Taniguchi and Kamen (1963) to doubt that NO reacts as a true ligand. These were the obtaining of the ferric form on alkalinization of the NO-complex of cytochrome cc' instead of the ferrous form, contrary to the experience of Ehrenberg and Szczepkowski (1960) with the NO adduct of ferricytochrome c; the photostability of NO products of both ferri- and ferro-cytochrome cc'; and the anomalous order of affinity of NO and CO.

The affinity of cytochrome cc' for CO is much lower than for most other haemoproteins; the half saturation value for the formation of the CO complex of *R. rubrum* cytochrome cc' is c. 0·015 atm. at all pH values. Unlike the CO compounds of other haemoproteins, that of *R. rubrum* cytochrome cc' has a lower rate of photodissociation at acidic pH (Bartsch and Kamen, 1958; Taniguchi and Kamen, 1963).

In a kinetic analysis of the reaction of cytochrome cc' with CO, Gibson and Kamen (1966) found wide differences in the behaviour of the *Chromatium* and *R. rubrum* cytochromes cc' to CO. *Chromatium* cytochrome cc' was half saturated by 0·2 mM CO while that of *R. rubrum* requires a concentration of 400 mM CO. In stopped-flow experiments the rate of combination of CO with *R. rubrum* cytochrome cc' was independent of CO concentration over the range 1 mM to 500 μM CO whereas with *Chromatium* cytochrome cc' the rate was proportional to CO concentration over the range 38 μM to 500 μM CO.

Flash photochemical experiments gave biphasic recombination reactions with rates independent of CO concentrations in both cases. In *R. rubrum* cytochrome cc' the photochemical behaviour was analogous to that of myoglobin, the rate of dissociation of CO being dependent on light intensity. For the *Chromatium* haemoprotein the rate was independent of light intensity.

Anomalous behaviour was obtained when the flash dissociation of the CO-complex was carried out in the presence of oxygen. Instead of the expected fast oxidation of the cytochromes there was a rapid reformation of the CO-enzymes. In contrast, when ferricyanide was used as an oxidant the cytochromes were rapidly oxidized on flash dissociation of the CO. For *Chromatium* cytochrome cc' the rate of oxidation with 15 μM ferricyanide was about 0·1 $\sec^{-1}$ as compared to 0·001 $\sec^{-1}$ for oxidation by 150 μM ferricyanide. The behaviour of *R. rubrum* cytochrome cc' was quite similar. As an explanation of their findings Gibson and Kamen (1966) suggested that the protein structure envelops the two haems to form a cage, thus constraining diffusion of the ligands.

This hypothesis was extended to explain the sluggish oxidation of cytochrome cc' by oxygen compared with the rapid oxidation obtained with ferricyanide and rapid reduction of this cytochrome by dithionite even in the presence of oxygen (Chance et al., 1966).

(f) Amino-acid sequences and linkage of haems. In spite of considerable efforts and the experience of determining the amino acid sequence of *R. rubrum* cytochrome c_2 in Kamen's laboratory, only fragmentary sequences have been determined for *R. rubrum* and *Chromatium* cytochromes *cc'*, although a haem-peptide has been obtained from the interesting monohaem cytochrome *c'* of *R. palustris* no sequence for this is yet available (Dus *et al.*, 1967). Elucidation of the structure of this cytochrome would be extremely valuable for the theoretical spectroscopist as the peculiar spectral features of this haemoprotein have their origin solely in the interaction of a single haem with the protein.

The largest sequence established is that of the *Chromatium* 27 amino acid dihaem-peptide (Dus *et al.*, 1962). The presence of two haems was established by spectral analysis of the pyridine haemochrome. The sequence Lys-Cys-Ala-Gln-Cys-His accounted for the attachment of one haem, but the presence of only one other cysteine implies that either the covalent binding of the haem to the peptide chain was through one thioether linkage leaving a free vinyl side chain or as suggested by the grouping Ser-Ala-Lys-Cys-His through a mono-oxyether adduct to serine. So far it has not been possible to split the dihaem-peptide into two separate haem peptides.

A sequence containing 13 amino-acid residues, including the grouping Lys-Cys-Leu-Ala-Cys-, thus accounting for the site of one haem, has been established for *R. rubrum* cytochrome *cc'* (Dus and Sletten, 1968). The location of the other two cysteines of this cytochrome *cc'* is not known, and even though there is one more cysteine than in *Chromatium* cytochrome *cc'*, it is not certain that the second haem is bound to the protein through a two-thioether link. That one of the haems in cytochrome *cc'* has a different covalent linkage might be an explanation of the low yields of haematohaem or haematoporphyrin from *Chromatium* cytochrome *cc'* (Barrett and Kamen, 1961).

Partial sequence data (Dus and Kamen, 1963) of the N-terminal portion of *Chromatium* cytochrome *cc'* locates the cysteine pair at positions 24 and 27, and the third cysteine at position 39. The location of the haem nearest to the N-terminal amino acid in *R. rubrum* cytochrome *cc'* is at positions 26 and 29. In both cytochromes the N-terminal amino acid is alanine.

(g) Function of cytochrome cc'. Early researches indicated that cytochrome *cc'* might have the function of an oxidase and was identifiable with cytochrome *o* (Kamen and Bartsch, 1961; Horio and Taylor, 1965). Such identification is contraindicated by the differences in binding of the haem to the protein in the two cytochromes. Furthermore kinetic studies in whole *R. rubrum* cells in which the reactions of the oxidase with CO and oxygen have been examined reveal no trace of a pigment which has the kinetic characteristics of cytochrome *cc'* and the absence of any significant oxidase activity in *Chromatium* which has only mesohaem-proteins (Cusanovich,

1966). The E_0' (pH 7·0) values of the isolated cytochromes cc' are below those expected for a cytochrome oxidase. However, the observation of Cusanovich and Kamen (1968) that *Chromatium* cytochrome cc' embedded in the chromophore does not bind CO and that *R. rubrum* cc' in the membrane has only 15% of the CO-binding capacity of crystallized cc' (Kakuno et al., 1970) raises the possibility that the autoxidizability and E_0' (pH 7·0) may also be different for cytochrome cc' *in situ*.

General experience with the photosynthetic heterotrophic bacteria has indicated that in these organisms too, cytochrome cc' is unlikely to have the role of an oxidase; Taniguchi and Kamen (1965) found cytochrome cc' in dark-grown *R. rubrum* was undetectable even by immunological methods, in membranes, and only trace amounts were detectable in cytoplasmic fluid; Geller (1962) found that the cytochrome cc' content of the same organism similarly grown was less than a tenth of that in anaerobic-light-grown cells; Kikuchi et al. (1965) could only detect minute amounts of cytochrome cc' in dark-grown *Rps. spheroides*.

On the other hand evidence, though not explicit, is pointing to a function for cytochrome cc' in the complex electron pathways of bacterial photosynthesis. Cytochrome cc' was identified spectrally as one of the four functional cytochromes oxidized when *Chromatium* cells were illuminated anaerobically with light of 800–900 nm (Olson and Chance, 1960a,b). Preferential oxidation of cytochrome cc' was observed when *Chromatium* cells were irradiated with light at 850 nm (Morita, 1968) while Cusanovitch et al. (1968) have associated cytochrome cc' with the bacteriochlorophyll-haemoprotein complex containing the 890 nm active centre.

(h) *Cytochrome c'*. A monohaem variant of this sub-class of cytochromes c occurs in *R. palustris* str. van Neil 2137 (De Klerk et al., 1965), in *Rhodopseudomonas gelatinosa* and in *Rhodopseudomonas molischianum* (cf. Bartsch, 1968). The spectral characteristics of *R. palustris* cytochrome c' differ little from those reported for the various double haem-cytochromes cc' (Bartsch, 1963; De Klerk et al., 1965). The main point of distinction at neutral pH is that the splitting of the Soret band is more pronounced, suggesting that the haem environment is even more assymetric than in the cytochrome cc', and the absence of tryptophan absorption in the UV region. The characteristic transition to cytochrome c-type spectra occurs at pH 14·0, higher than is found for the cytochromes cc'. The spectra of the NO and CO complexes at pH 7·0 resemble those of cytochrome cc' at pH 5·0, rather than those prepared at pH 7·0. The E_0' of *R. palustris* c' (pH 7·0 + 105 mV) is notably higher than that of the cytochromes cc' but is comparable to that of the monohaem variant of *P. denitrificans*.

The minimal molecular weight of *R. palustris* cytochrome c' based on the presence of 137 amino acid residues and one protohaem is 14,820 (Dus et al., 1967). Ultracentrifuge studies indicated that this cytochrome c' has a molecular weight of 13,400 and a similar value was obtained by chromatography on Sephadex. These results exclude the possibility that the purified cytochrome c' exists as a dimer.

Cytochrome c' is soluble over a wider pH range than are the cytochromes cc' and in contrast to these has an isoelectric point at pH 9·7 (Henderson and Nankiville, 1966). The high isoelectric point was attributed by Dus et al. (1967) to the presence of at least 8 amide groups and an excess of three basic residues (19 lysine and 2 arginine) over the acidic (18 aspartic plus glutamic acid) amino acid residues.

A feature of the amino acid composition is the low content of aromatic residues, 6 (1 tyrosine, 5 phenylalanine, no tryptophan) against 17 (7 tyrosine, 8 phenylalanine, 2 tryptophan) of R. rubrum cytochrome cc' (M.W. 28,840). As only one histidine is present in R. palustris cytochrome c' it is interesting to note in connection with the identity of the sixth ligand to the iron that the amino acid analyses account for two methionines (Henderson and Nankiville, 1966; Dus et al., 1967). As in the cytochromes cc' a high alanine content, 29 residues, was found.

A haempeptide containing 24 amino acid residues including the single histidine and both the cysteines has been obtained by tryptic digestion of cytochrome c'; the haem is split off in 80% yield by $HgCl_2$ in acid (Dus et al., 1967). This evidence is compatible with the haem being linked by two thioether bonds to the protein.

(i) *Cryptocytochrome c*. A cytochrome which could combine with CO and had spectral features common to the cytochromes cc' (Table XV) was discovered in the denitrifier *Pseudomonas denitrificans* by Iwasaki (1960) who named it cryptocytochrome c when attempts to split off the haem with acetone-HCl revealed that the cytochrome was an anomalous cytochrome c. Cryptocytochrome c has been isolated from anaerobically grown *P. denitrificans* and crystallized in the form of needles of 1 mm length in the ferric state, and rectangular prisms for the ferrous state (Suzuki and Iwasaki, 1962). Ultracentrifuge analysis of the recrystallized cytochrome gave a molecular weight of 20,400. The ϵmM of the pyridine haemochrome at 550 nm was 40 indicating the presence of two haems in each molecule, a conclusion supported by an iron content of 0·46%. Cryptocytochrome c has also been prepared in crystalline form from aerobically grown cells (Iwasaki and Shidara, 1969b); a molecular weight of 15,500 was given by the iron content of 0·36%, whereas gel filtration indicated a molecular weight of 28,000. The purity index confirmed by ultracentrifugal studies (Cusanovich et al., 1970) (E400 nm (ox))/(E280 nm) was 2·9 for the crystallized cytochrome from both aerobic and anaerobic *Ps. denitrificans*. 6M guanadine-HCl effects dissociation of the cytochrome into subunits of 14,000 molecular weight, a value consistent with that obtained from amino acid analyses (13,624 per haem) (Cusanovich et al., 1970).

Ferrocryptocytochrome c gives a photodissociable CO-compound (Iwasaki and Shidara, 1969b) and the ability of the cytochrome to combine with CO is found in the crude cell extracts (Iwasaki, 1970). The dissociation constant, K_d, is $1 \cdot 18 \times 10^{-4}$ at 20° and is much less than that for CO-myoglobin (Cusanovich et al., 1970); photodissociation studies indicate a complex situation typical of the cytochromes cc' (Cusanovich and Gibson, 1970). A reversible complex is obtained with NO, but a compound is not formed with nitrite.

In the presence of 50% propanol both reduced and oxidized crypto-cytochrome c are reversibly converted to a form which exhibits a spectrum of "cytochrome c-553" and which still combines with CO (oxidized 528 and 409 nm; reduced 553, 524 and 419 nm; CO-compound 570, 536 and 419 nm). Sodium dodecylsulphate and sodium dodecylbenzene sulfonate, 0·01%, and 4 M guanidine also effected this reversible spectral conversion. Sodium deoxycholate, 1%, sodium salicylate, 1 M, and sodium benzoate, 1·5 M, only effected the conversion of the oxidized cryptocytochrome c to "cytochrome c-553". Cusanovich et al. (1970) found that the ferrous form showed no spectral changes in the pH range 5–10, but in the ferric forms this spectral change occurred at pH 7·1, 2·3 pH units lower than found for the cytochrome cc' of the photosynthetic bacteria. With 2 N NaOH both ferrous and ferric forms undergo the spectral conversion (Iwasaki and Shidara, 1969a,b). A water molecule may be complexed to the haem in ferric cc'. Even at the high pH obtained the structural conversion was still reversible. A tighter structure of cryptocytochrome c around the haem than is found in cytochrome cc' of R. rubrum and Chromatium is suggested by the inability of 6 M urea to effect the spectral conversion. A haemochrome could not be formed by imidazole 1 mM.

The redox potential (E_0' pH 7·0 = + 132 mV) is markedly pH-dependent above pH 7·0; the slope is close to the theoretical (60 mV/pH) for the involvement of a proton in the redox process and thus consistent with the interpretation of the spectral studies in this pH region (Cusanovich et al., 1970).

123 amino acid residues are present per haem; one fifth of the total residues are alanine, a feature typical of cytochromes cc'. The N-terminal residue is apparently blocked; the release of the C-terminus by carboxy-peptidase A and B was poor, so that neither terminal amino-acid has yet been identified. Two histidines and two methionines are present in each sub-unit; the aromatic content is rather low. Cusanovich et al. (1970) have proposed the structure of a 23 amino acid sequence bearing the haem, based on determination of the structure of the peptides obtained by cyanogen bromide cleavage of the enzyme and digestion with proteolytic enzymes. The sequence is NH_2-Ala-Ala-Phe-Gly-Asp-Val-Gly-Ala-Ser-Glu-Ala-Asp-Cys-Lys-Ala-Cys-His-Ser-Gly-Asn-Ala-Tyr-Arg-COOH,

presenting no unusual features around the haem, and consequently the typical cc' type spectra must be determined by the secondary and tertiary structure. Since these are not yet known, the chemical basis of the spectral difference between cytochromes c' and cytochrome c remains to be elucidated.

The biological function of cryptocytochrome c is not yet known. The denitrifying enzyme in *P. denitrificans* is a copper-enzyme (Iwasaki et al., 1963) but cryptocytochrome c has not been excluded as an electron carrier participating in the system. Anaerobic cells contain twice as much of this cytochrome as do aerobic cells (Iwaski and Shidara, 1969b). The E_0' suggests that cryptocytochrome c is not directly linked with the oxidase of the aerobic bacteria. Cryptocytochrome c is rapidly reduced by NADH in the presence of a soluble fraction from broken cells and this reductase activity is equally high in anaerobic cells.

(j) *Cytochrome c*-552, 558. A CO-combining cytochrome which occurs in the denitrifier *Pseudomonas stutzeri* (Kodama and Shidara, 1969) is grouped with the cytochromes cc' because of the marked electronic distortion of the haem reflected in the spectrum of the reduced cytochrome. The α band has a maximum at 558 nm, with a shoulder at 552 nm; the ϵmM 18·8 at 558 nm is unusually low. The β band has double maxima, at 533 nm and 527 nm, while the Soret band exhibits only a single peak at 421 nm. The oxidized cytochrome has a very broad absorption band, maximum 529 nm, and a Soret peak at 409 nm. A typical bacterial cytochrome c-552 spectrum (Soret band at 418 nm), with considerable intensification of the absorbancy of each maximum, can be obtained in 0·1 M NaOH, in 6 M urea at pH 7·0 (but not with 4 M guanidine) and in the presence of aliphatic alcohols (Kodama and Mori, 1969b). On removal of these agents the spectrum reverted to the original form.

The presence of two haems in cytochrome c-552, 558 is indicated by a minimum molecular weight of 37,000, based on haem analysis, and a molecular weight of 70–74,000 derived from ultracentrifugation and gel filtration. The haems are covalently bound to the protein and from the evidence of the pyridine haemochrome of the cytochrome do not possess a free vinyl group. The absence of a free vinyl group together with the position of the Soret band, typical for cytochromes c, point to a disturbance of the iron-porphyrin electronic field as a source of the spectral anomalies.

Cytochrome c-552, 558 probably has a low E_0' as it is reduced by dithionite, but not by borohydride, ascorbate or cysteine, at pH 7·0. The capacity of cytochrome c-552, 558 to combine with CO is present in intact cells and no increase in its CO-combining capacity is observed during purification of the cytochrome. The electron transfer system in which cytochrome c-552, 558 functions has not been identified. A cytochrome c with an asymmetrical α-peak has been crystallized from *Alcaligenes faecalis*

(Iwasaki and Matsubara, 1971). Cytochrome c-557 (551) has two haems c and has an M.W. of 65,000 determined by gel filtration c-557 (551) combines with CO at pH 7·0, and the oxidized form combines with CN (0·01 M), the Soret peak shifting 4 nm towards the red.

These cytochromes and the cytochromes cc' and c' may be variants which have arisen during the evolution of the cytochromes b.

6. Photosynthetic cytochromes c

(a) *Cytochrome c_2 (Rhodospirillum rubrum)*. Cytochromes c_2 occur in the non-sulphur purple photosynthetic bacteria. They are monohaem-proteins of molecular weight 12–14,000, have E_0' (pH 7·0) of + 280 mV or higher and their absorption spectra are very similar to those of cytochrome c. Cytochrome c_2 does not combine with CO at neutral pH and is not autoxidizable. It is considered to be analogous to the type f cytochromes of algae (cf. Chapter V, C) and the equivalent of mitochondrial cytochrome c, though its function is primarily as an electron carrier in the photosynthetic process.

R. rubrum cytochrome c_2 which is taken as a representative of this group was the first bacterial cytochrome isolated and was named cytochrome c_2 at the suggestion of Hill and Keilin (Elsden *et al.*, 1953). The *R. rubrum* cytochrome c_2 has been more widely investigated than other members of this group and is the only one in which the primary structure has been established.

Preparation. Cytochrome c_2 is a soluble cytochrome and is easily released by lyophilization or by sonication of the bacteria. A preparation which is greater than 98% pure by spectral and electrophoretic criteria is described by Kamen *et al.* (1963). The homogeneity of the pigment was established by isoelectric focussing for the purpose of amino acid sequence studies (Dus and Sletten, 1968). Amberlite CG50 was used in the chromatographic procedure for convenience in the large scale preparations, but alumina was effectively employed to obtain crystalline cytochrome c_2 (Horio and Kamen, 1961). More recently DEAE cellulose has been used in the chromatographic procedure to obtain material suitable for Mössbauer spectroscopy (Moss *et al.*, 1968).

Spectral properties. Compared to cytochrome c the band of reduced cytochrome c_2 has a higher extinction; several other small differences are observed in the absorption spectra at room temperature (Horio and Kamen, 1961; Bartsch, 1968), and at low temperature (Estabrook, 1956), noticeably the absence of splitting of the Soret peak. Differences are more apparent in the CD spectra (Flatmark and Robinson, 1968). The spectrum of reduced cytochrome c_2 is similar to that of cytochrome c. However, comparison of the spectra of the oxidized forms revealed marked qualitative differences. In the region 240–300 nm cytochrome c_2 exhibits one positive ellipticity band at 251 nm and three negative bands (at about 276, 279 and 289 nm) of which only that at 251 nm is identical in frequency, sign and amplitude to one observed in bovine heart cytochrome c. The most prominent difference however, is the absence in the CD spectra of ferrocytochrome c_2 of the very strong ellipticity band centred at 273 nm, attributed in cytochrome c to the porphyrin-iron-histidine complex (Flatmark and Robinson, 1968; see Chapter V, B4b).

The molecular weight of $R.$ $rubrum$ c_2 is 12,940, calculated from its amino acid composition and haem content (Dus and Sletten, 1968) and a monomeric state is indicated by the molecular weight of 12,750 ($\pm$ 1,250) given by ultracentrifugal analysis (Horio and Kamen, 1961). The E_0' of cytochrome is pH-dependent, varying from + 380 mV at pH 5·0 to + 300 mV at pH 8·0. Crystallized cytochrome c_2 had an E_0' of + 320 mV at pH 7·0 (Horio and Kamen, 1961). The isoelectric point at 0° of the ferrous form is 6·11 and the ferric form 6·44.

The rhombic crystals of the oxidized and reduced forms of cytochrome c_2 are isomorphous (Kraut et al., 1968) indicating that only small conformational changes may accompany the change in oxidation state of the haemoprotein. Mössbauer studies (Moss et al., 1968) show that the quadrupole splittings of the reduced cytochromes are typical of the low spin Fe^{2+} haem coordination (cf. Bearden et al., 1968) and that a greater degree of asymmetry in the environment of the iron is found in the ferrous than in the ferric form.

Amino acid sequence. Paléus and Tuppy (1959) determined the structure of a 14 amino acid haempeptide obtained by tryptic digestion of cytochrome c_2, establishing that the haem was bound by two cysteines in the sequence Cys-Leu-Ala-Cys-His-Thr, and Levin (1961) using peptic digestion confirmed part of this structure and identified several of the amino acid residues on the amino terminal side of the haem. These studies were an important step in showing the essential similarity of the structure in the haem region of the mammalian and bacterial cytochromes c. The complete amino acid sequence of 112 residues (Dus and Sletten, 1968; Dus et al., 1968; Sletten et al., 1968) reveals that the two cysteines are at positions 14 and 17, respectively and that the second of the two histidines is at position 41. The methionine corresponding to that at position 80 in mammalian cytochrome c (cf. Table VIII) is displaced to position 91, largely due to an insert of eight amino acids (residues 81–88). The probability that methionine is the sixth ligand to the iron is supported by the presence of only one histidine (and one methionine) in *Rhodopseudomonas capsulatus* cytochrome c_2. Interestingly this cytochrome c_2 also has 112 amino acid residues, but a molecular weight of only 13,090 due to the replacement of some of the aspartic acid, glutamic acid and leucine residues by smaller residues (Dus and Sletten, 1968). The clustering of hydrophobic residues in *R. rubrum* cytochrome c_2 is less pronounced than in cytochrome c, while the clustering of basic residues is limited to a single lysyl-lysine. Dus et al. (1968) have concluded that the high content of homologous sequences indicate that cytochrome c_2 is even older on an evolutionary scale than the *P. fluorescens* cytochrome c.

As mentioned above, c_2 of *R. rubrum* and *Rps. capsulatus* contain an additional sequence of 8 residues; these are located between positions 77 and 78 of the eukaryotic cytochromes, a region which is absolutely consistent in the eukaryotic cytochromes c. This extra sequence probably blocks the left channel which in cytochrome c allows access of small ions to the haem (Krejcarek et al., 1971). This probably explains the lack of reactivity of cytochrome c_2 with KCN.

The 220 MH_3 proton magnetic resonance spectra of cytochromes c_2 of *R. rubrum* and *Rps. capsulatus* show a resonance signal at $-3\cdot1$ ppm corresponding to that found for the methionine S-methyl-iron linkage in cytochrome c ($-3\cdot4$ ppm) (see Chapter V). An additional resonance between $-1\cdot9$ to $-2\cdot4$ ppm is assigned to the alkyl side chain of the extra amino acid sequence (Turner et al., 1971).

Preliminary 5 Å x-ray analysis (Kraut et al., 1968) indicates that the cytochrome has a large central core of low electron density, probably packed hydrophobic side chains. The haem is embedded well into the molecule on one side and there may be two short segments of α helix of two to three turns. The sixth ligand to the iron could be methionine.

Function of cytochrome c_2. Cytochrome c_2 is the most abundant of the c-type cytochromes in cells cultured photosynthetically. Unlike cytochrome cc', significant amounts are present in dark-grown cells; Geller (1962) found that the cytochrome c_2 content in dark-aerobically grown cells was a third that present in light-grown cells, and was only slightly less than the cytochrome b content.

The photo-oxidation of cytochrome c_2 by cell free extracts of light-anaerobic *R. rubrum* cells was demonstrated by Vernon and Kamen (1953), but some evidence that it might have a role in a respiratory chain was obtained by Chance and Smith (1955) and by Smith (1957).

The cytochrome c_2 content of membrane fragments from dark-grown *R. rubrum* was about that of the cytochrome b, which included the cytochrome o fraction (Taniguchi and Kamen, 1965). The oxidation of equine cytochromes c by membrane fragments from dark-grown *R. rubrum* was 4 and 1·5 times respectively as rapid as that of cytochrome c_2. A similar lower rate of oxidation of cytochrome c_2 compared with cytochrome c was obtained with the oxidase system of dark-grown *Rps. spheroides* (Kikuchi et al., 1965).

The lower effectiveness of cytochrome c_2 as an electron donor in the cytochrome oxidase system was attributed by these workers to its high E_0'

(+ 320 mV at pH 7·0). A c-type cytochrome (c-549, $E_0{}' + 300$ mV) is the main c component of respiratory particles prepared from aerobically grown *Rh. spheroides* (Whale and Jones, 1970). A small amount of photoactive chlorophyll was present in the particles.

Greater interest, however, resides in the place of cytochrome c_2 in the photosynthetic electron transport system, though this has been difficult to demonstrate. On the basis of comparable oxidized and reduced difference spectra, and other properties, it has been assumed that cytochrome c_2 was responsible for the light-induced changes in the cytochrome of *R. rubrum* and related bacteria (Bartsch, 1963; Smith and Ramirez, 1959; Chance and Smith, 1955). Cytochrome c_2 of cell-free extracts was oxidized when the extracts were illuminated under conditions where phosphorylation of ADP occurred (Smith and Baltscheffsky, 1959).

(*b*) *Other cytochromes c_2.* Other cytochromes c_2 have been investigated in some detail. *Rhodospirillum molischianum* produces two cytochromes of c_2 type; one of molecular weight 13,443 and a smaller one of 10,221 (112 and 93 amino acid residues respectively) (Dus *et al.*, 1970). The smaller cytochrome c_2 is predominant in young cultures.

The absorption spectra of the two cytochromes are very similar to those of mammalian cytochrome c. The ϵmM of the α-band is 29·76 for the larger cytochrome c_2 and 29·15 for the smaller cytochrome. At low temperature the α-bands are split into three satellite bands and the intensity of the $c_{\alpha 1}$ and $c_{\alpha 2}$ bands is lower in the larger cytochrome c_2.

CD spectra (Flatmark *et al.*, 1970) exhibit a marked similarity, though differences exist in the 245–300 nm region due to disparity of amino acid content. An apparent α-helix content was calculated for both haemoproteins. A difference in the relative magnitude of the ellipticity bands (412, 415 nm) arising from the Soret transition in the ferrous form of the two cytochromes c_2 is of interest in respect to the difference in the $E_0{}'$ (pH 7·0) of cytochrome c_2 and the smaller variant (+ 288 mV and + 381 mV).

Both cytochromes are highly basic and this can be attributed to the low aspartic acid and glutamic acid content. Two histidines are present in each cytochrome, but surprisingly while cytochrome c_2 has only one methionine the smaller cytochrome c_2 which has 29 fewer residues has three methionines. The N-terminal sequence of both cytochromes is Ala-Asp-Ala-Pro-Pro-, but there was little homology in the region of the C-terminus (large M.W. cytochrome c_2, leucine; small M.W. cytochrome c_2, valine). Peptide maps of tryptic digests show noticeable homologies.

Despite the physical similarity with mammalian cytochrome c neither of the two *R. molischianum* cytochromes c_2 could restore any succinate oxidation to rat kidney mitochondria depleted of endogenous cytochrome c (Flatmark *et al.*, 1970).

Cytochrome c_2 of *Rps. palustris* str. Van Niel 2137 has been crystallized in the oxidized form (Henderson and Nankiville, 1966). The minimum molecular weight 17,000 is based on an iron content of 0·33%. The ϵmM at 552 nm of the reduced form at pH 7·0 is 23·2. The isoelectric point is 10·6 and the E_0' (pH 7·0) is + 310 mV (Kamen and Vernon, 1955). The cytochrome c_2 crystallized by Morita (1960) from an unidentified strain of *Rps. palustris* had an isoelectric point of 7·7, an E_0' (pH 7·0) of + 330 mV and an M.W. of 15,600 estimated by ultracentrifugation. The cytochrome c_2 from both strains had the α peak of the reduced form at 552 and the Soret peak correspondingly displaced to 418 nm.

A cytochrome c_2 with peaks at 550, 522 and 414 nm in the reduced form has been obtained from *Rhodomicrobium vannieli* (Morita and Conti, 1963) and partially purified. The E_0' (pH 7·0) was + 304 mV and the isoelectric point was close to neutrality. In membrane fractions the distribution of cytochrome c_2 paralleled that of the bacteriochlorophyll. A second cytochrome c with maxima in the reduced form at 553, 521 and 423 nm and which had a lower, but not specified, oxidation-reduction potential was also present in the membranes.

Cytochrome c-553 (550) which has some properties resembling the algal cytochromes f has been extracted from *Chromatium* chromatophores, using 50% acetone-Tris buffer at pH 7·0 (Cusanovich and Bartsch, 1969). The α-peak of the reduced form is asymmetrical and is of low intensity 553, 550 nm, ϵmM = 16·3, 15·7; 522 and 417·5 nm, ϵmM = 14·3 and 115). The E_0' (pH 7·0) is 330 mV, similar to that of the membrane bound cytochrome c-555 which may be the primary reductant of the reaction centre bacteriochlorophyll (Cusanovich *et al.*, 1968). The minimum molecular weight from amino acid analyses is approximately 13,000. A marked heterogeneity in size is shown by ultracentrifugal analyses of c-553 (550), the M.W. ranging from 13,000 to 30,000. This cytochrome has two histidine residues and four methionines. The N-terminal sequence is Ala-Glu-Glu-Leu-. Evidence available at present does not allow for identification of the solubilized c-553 (550) with c-550.

(c) *Flavin-containing cytochromes from Chromatium and Chlorobium.* Cytochromes which have a flavin tightly bound to the cytochrome have been obtained from the obligate purple bacterium *Chromatium D* (Bartsch, 1961) and from the green photosynthetic bacterium *Chlorobium thiosulphatophilum* str. NCIB 8346 and 8327.

Chromatium cytochrome c-552, with the flavin still attached, was crystallized from 70% saturated ammonium sulphate (Bartsch *et al.*, 1968) following purification of the cytochrome by chromatography on DEAE-cellulose (Bartsch and Kamen, 1960; Kamen *et al.*, 1963). Cytochrome c-552 with identical spectroscopic and chromatographic properties has also been isolated from *Chromatium* strains Van Niel 9P813 and C20. The acid and alkali lability of cytochrome c-552 appears to account for the differences in the cytochromes c-552 purified by Newton and Kamen (1956) and the later preparations.

The molecular weight estimated by the rapid equilibrium procedure of Van Holde and Baldwin (1958) is 72,000 ± 6,000. Some inhomogeneity was evident in the sample but this value is considered to be more reliable than the earlier one of Bartsch and Kamen (1960) because of a probable error in the diffusion constant used. *Chlorobium* cytochrome c-553 is smaller; a molecular weight of 50,000 was given by Sephadex chromatography and by sedimentation (Meyer et al., 1968). A notable difference between the *Chromatium* cytochrome and that of *Chlorobium* is that the former has two haems and can combine with one CO while cytochrome c-553 has one haem and does not combine with CO.

Cytochrome c-552 has an E_0' (pH 7·0) of + 10 mV and is readily autoxidizable while cytochrome c-553 has an E_0' (pH 7·0) of + 98 mV and is only slowly autoxidized in air. The isoelectric point of the *Chromatium* cytochrome is 5·1 and that of *Chlorobium* is 6·7.

The flavin moiety in each cytochrome may be covalently bound to the protein. Treatment with cold saturated urea or 6 M guanidine at pH 8·0, or an excess of organomercurial at neutral pH was necessary to release the flavin (Bartsch et al., 1968; Meyer et al., 1968). The flavin material obtained did not fluoresce, but treatment with a protease yielded a flavin having some of the properties of a flavin mononucleotide. Hendricks and Cronin (1971) have however characterised the flavin separated from *Chromatium* c-552 by the urea treatment as an adenine dinucleotide on which the dimethyl isoalloxazine ring is substituted in the 8-methyl position; the substitution does not appear to be histidine.

The effective redox potential of the flavin *in situ* is more reducing than that of the haems. A similar situation is found in cytochrome b_2 (Horomi and Sturtevant, 1965).

The extinction coefficients of *Chromatium* cytochrome c-552 are twice as great as those of the *Chlorobium* cytochrome c-553 and horse cytochrome c and although the α bands of the reduced forms are displaced to the red, the position of the Soret band is the same for all three (Bartsch et al., 1968). Extinction values reported earlier for cytochrome c-552 (Bartsch, 1963) are a third too high. The only significant deviation from normal cytochrome spectra in these flavocytochromes is observed in the oxidized state.

The ORD spectrum of cytochrome c-552 is quite different from that of other c-type cytochromes. The optical rotation of the visible and aromatic regions is featureless but large and sharp Cotton effects show within the Soret bands and these are particularly marked in the CO-complex (Bartsch et al., 1968; Yong and King, 1970). The existence of haem-haem interaction in this cytochrome (Urry, 1966) as a source of the optical rotation was proposed by Bartsch (1968). The Cotton effects within the Soret band are closer together in the reduced form and have a greater amplitude than in the oxidized form. This is attributed to a more symmetrical alignment of the haem planes in the reduced form. As only three transitions are observed, instead of the expected four within the Soret band of the oxidized form and the CO complex of cytochrome c-552, Yong and King (1970) suggest that the environment of one haem is different from the other, with degeneration of one haem, and that the CO is not equally shared by the two haems, as was proposed by Bartsch *et al.*

The ORD spectra of *Chlorobium* cytochrome c-553 show no unusual features (Bartsch *et al.*, 1968).

Chromatium cytochrome c-552 shows no anomaly in its magnetic properties. The effective magnetic moment of the ferric form, 2·7 ($\pm$ 0·3) Bohr magnetons, is only slightly higher than the value expected for such iron (1·73–2·6 B.M.) (Ehrenberg and Kamen, 1965). No significant ESR signal was detected. The Mössbauer spectra of oxidized and reduced *Chromatium* c-552 are similar to those of *R. rubrum* cytochrome c_2, though the E_0' (pH 7·0) differs by + 280 mV. The Mössbauer spectra of the CO-complex resemble those of the cytochrome cc'-CO complex. No difference in the environment of the two haems of cytochrome c-552 could be detected by Mössbauer spectroscopy.

(*d*) *Other cytochromes c of photosynthetic bacteria.* Gibson (1961) isolated from the photosynthetic green bacterium *Chlorobium* (strains Lascelles and Van Niel) two cytochromes c. These were identified as cytochromes c-554, a basic protein with a molecular weight of 12,000 estimated by haem analysis and E_0' (pH 7·0) + 140 mV; the other, an acidic protein, cytochrome c-554, has since been characterized as a flavocytochrome c (cf. above). Neither cytochrome was autoxidizable or reacted with carbon monoxide or cyanide. Together they accounted for 0·5% of the dry weight of the bacteria.

Improved extraction and purification procedures enabled Meyers *et al.* (1968) to isolate yet another cytochrome c-551 from both the strains of *Chlorobium thiosulphatophilum* and to recharacterize the two described by Gibson (1961). Cytochrome c-551 has two haems and a molecular weight of 45,000 estimated by ultracentrifugation by the method of Ehrenberg (1957), but Sephadex-chromatography gives a value of 60,000. The isoelectric point of the oxidized haemoprotein is 6·0 at 28° and the oxidation-reduction potential over the range pH 6–8 is + 135 mV. The absorption spectrum is typical of a c type cytochrome. A spectrally similar cytochrome c-551 occurs in *Chloropseudomonas ethylicum* (Olson and Shaw, 1969).

Cytochrome c-555 (Gibson's c-554) has its isoelectric point at pH 10·5 (25°). The oxidation-reduction potential is pH-dependent, decreasing from + 145 mV at pH 6·0 to + 114 mV at 8·0. The molecular weight of cytochrome c-555 estimated by Sephadex chromatography agrees with that obtained by Gibson (1961) by haem analysis, but a lower value, 9970, was calculated by Meyer et al. (1968) from the amino acid analysis. An unusual feature of the amino acid composition is the presence of 8 methionine residues and the absence of arginine and phenylalanine.

The absorption spectrum of cytochrome c-555 shows a markedly asymetrical peak and a low α/β (1·2) and α/γ (0·14) ratio. At liquid N_2 temperature the α-peak is split to give a major peak at 553 nm and a lesser peak at 548 nm. Similar low α/β and α/γ ratios are found with the cytochrome c-556 of *R. palustris* (Horio and Bartsch, 1968). At liquid nitrogen temperature the α-peak of cytochrome c-556 exhibits a splitting of 2·5 nm. Some acidic algal cytochromes c-553 also exhibit an asymmetric α-peak (Yamanaka and Okunuki, 1968b). Cytochrome c-555 has also been purified from *C. ethylicum* (Olson and Shaw, 1969).

Chlorobium c-553 reacts fairly rapidly with *Pseudomonas* oxidase, in contrast to the cytochrome c_2 of the purple non-sulphur bacteria, where the activity is either low or absent. A significant reactivity is also found towards cow cytochrome oxidase; this finding is contradictory to the conclusion that *Chlorobium* is more primitive than the purple nonsulphur bacteria (Yamanaka and Okunuki, 1968b).

The cytochromes c of photosynthetic bacteria are liable to alteration during extraction from the cell, particularly if warm citrate at pH 5·4 is used. Orlando and Maes (1967) used acrylamide disc electrophoresis to analyse extracts from *Rps. spheroides*, *Rps. palustris*, *R. rubrum* and *Chromatium*. As many as 5 electrophoretically distinct components were obtained with similar spectral properties and numerous satellite bands were present. Some of the distinct electrophoretic species appeared to be artefacts and not multiple forms of natural cytochrome c_2. Electrophoretically similar cytochromes c_2 were obtained from *R. rubrum* and *Rps. spheroides*. Cytochrome c-553 was the predominant cytochrome in the extracts of aerobically grown *Rps. spheroides*. The multiple components of the cytoplasmic cytochromes of R. rubrum have been further investigated by Bartsch et al. (1971) and Kakuno et al. (1971) using chromatography and electro-focussing.

(e) *Bacterial photosynthesis*. Photosynthesis in bacteria differs from that of plants in two important ways: Anaerobic conditions are essential and the electron donor used is not water but an inorganic compound of hydrogen or sulphur; oxygen is not involved in this process. A further distinction is that the non-sulphur bacteria typically use an organic compound as carbon source, whereas the purple and green sulphur-bacteria, like plants, use CO_2 as their sole carbon source.

Several reviews have appeared in recent years on aspects of bacterial photosynthesis and these are often complementary in their survey of this subject, as they reflect the active interest of the authors' laboratory (Vernon, 1968; Lascelles, 1968; Hind and Olson, 1968; Horio and Kamen, 1970). A general review (Pfennig, 1967) on the photosynthetic bacteria, including their morphology and physiology, provides a valuable introduction to the biology of this group of organisms.

Light-induced oxidation of cytochromes was shown to occur in *R. rubrum* (Duysens, 1954; Chance and Smith, 1955; Smith and Ramirez, 1959). A quantum requirement of two for each electron was found for the oxidation of a *c*-type cytochrome by light of low intensity asborbed by bacteriochlorophyll *a* (Olson and Chance, 1960b). The introduction of low temperature (80–77°K) spectroscopy to the study of the light-induced oxidation of cytochromes (Chance and Nishimura, 1960) was a major step towards the elucidation of the relation of the cytochromes to the primary reaction of photosynthesis. The primary reaction was defined as the reaction occurring at high quantum efficiency and with no change of rate at very low temperature. Vredenberg and Duysens (1963, 1964) using temperatures down to 53°K found that the photoxidation of cytochromes in *Chromatium* D and *R. rubrum* is temperature-dependent and concluded, contrary to the view of Chance and Nishimura (1960) and Nishimura *et al.* (1964) that cytochromes were not the primary oxidant in the photoreaction but were close to the site of the primary photoreaction. This was confirmed later by Kihara and Chance (1969). Photo-oxidation studies at 77°K with the chromatophores from several species of photosynthetic bacteria show that the ambient redox potential influences the relative degree of oxidation of *c*-type cytochromes (cf. Dutton *et al.*, 1970; Kihara and Dutton, 1970).

Olson and Chance (1960a,b) proposed that there were two different pathways of electron transfer from the primary oxidant and that cytochromes c-552 was of principal importance in photosynthetic electron transport. Later work with *Chromatium* D (Cusanovich *et al.*, 1968; Morita *et al.*, 1965), *R. rubrum* (Sybesma, 1967; Sybesma and Fowler, 1968) and *Chloropseudomonas ethylicum* (Sybesma, 1967) has shown that the high redox potential cytochromes *c* are components of a cyclic electron transport system and that the low potential cytochromes *c* are components of substrate-linked non-cyclic electron transport. A similar role for cytochromes c-558 and c-553 is indicated in the bacteriochlorophyll *b* containing *Rhodopseudomonas* strain (NHTC-133) by Olson and Nadler (1965), and for the *c*-type cytochromes in *Thiococcus sp.*, a bacteriochlorophyll *b* containing sulphur bacterium (Olson *et al.*, 1969).

Light-induced absorbance changes in *Chromatium* D have been interpreted by Morita (1968) and Cusanovich *et al.* (1968) as indicating that the high redox potential c-555 (c-556) and the low potential c-552 (c-553) are oxidized by two different photochemical centres. Similar evidence for two distinct light reactions in *R. rubrum* has been presented (Sybesma and

Fowler, 1968; Sybesma, 1969; Fowler and Sybesma, 1970) and supported by the results of flashing light experiments (Sybesma and Kok, 1969); the differential reduction of pyridine nucleotides by light in *R. rubrum* has been interpreted in support of the two photocentre concept (Govindjee and Sybesma, 1970). The observations of Morita that 890 nm light was more effective for c-555 oxidation and 800 nm more effective for c-552 oxidation has been confirmed by other workers (see Parson and Case, 1970). However, a fluorescence action spectra had earlier indicated that 880 nm light is more effective than is 800 nm light for inducing the oxidation of c-552 and that this oxidation is linked to the B890 bacteriochlorophyll complex (Amesz and Vredenberg, 1966), so that an explanation which takes into account the changes in the efficiency of energy transfer at high levels of absorbed light has been preferred for the differences in light reactivity observed by Morita and other investigators. Action spectra for the oxidation of c-555 in *Chromatium* chromatophores shows a light of 800 nm or 900 nm is equally effective if the data are examined on the basis of absorbed light (Suzuki *et al.*, 1969).

The laser-flash studies (Parson, 1968, 1969b; Parson and Case, 1970; Seibert and Devault, 1970) confirm the view that a single photochemical centre in which the active species is $P883^+$, oxidizes the high and low redox potential cytochromes of *Chromatium* D, and similarly in *Rps. viridis*, in which $P960^+$ is the photoreactive form of the bacteriochlorophyll *b* (Case *et al.*, 1970).

Subchromatophore fractions from several bacteria have been prepared in which the bulk of the light-harvesting bacteriochlorophyll *a* has been removed by detergents, but which retain a specialized photoreactive form of bacteriochlorophyll *a* now designated P883 (formerly, P870 or P890; cf. Parson, 1968; Dutton, 1971). The P883 upon illumination undergoes reversible bleaching to $P883^+$. Another component P800 which shows a reversible blue shift, but no bleaching, upon exposure to light is also present. This P800 can transmit harvested light energy more efficiently to P883 than can the bulk chlorophyll, B890 (Clayton and Sistrom, 1966). A similar relationship exists between the P830 form of bacteriochlorophyll *b* and the photoactive P985 (or P960) of the reaction centre of *Rhodopseudomonas* sp. (Olson and Clayton, 1966). A reaction centre obtained from *Rps. spheroides* using the cationic detergent cetylmethyl ammonium bromide contained P883, P800, c-type cytochrome and ubiquinone in the molar ratio of 2·4 : 2 : 1 : 10 (Jolchine *et al.*, 1969). That cytochrome c might not be an essential component of the reaction centre was suggested by the finding of a much lower ratio of cytochrome c to the other components in a mutant strain of *Rh. spheroides*. By using Triton X-100 a reaction centre was obtained from *Rps. spheroides* which contained P883, c-552, b-562 and ubiquinone in the molar ratio of 1 : 1 : 2 : 13 (Reed, 1969;

Reed et al., 1970). From *Rps. viridis*, which has bacteriochlorophyll *b* as its photopigment, a reaction centre was obtained which contained P958, P830, *c*-558, *c*-553 in the ratio of 1 : 2 : 2 : 5 (Thornber et al., 1969). A complex solubilized by dodecysulphate (anionic) with a M.W. of 500,000 contained both cyclic and non-cyclic photoelectric transport systems of *Chromatium* D (Thornber, 1970). The ratio of P883, P + B800, B890, *c*-556, *c*-552 was 1 : 4 : 40 : 2 : 7. Carotenoids were also present. Using this preparation to study the spectral changes induced in the reaction centre by light (868 nm) from a liquid dye laser, Dutton et al., (1971) claim to identify an additional photooxidant P600$^+$ which is more effective than P883$^+$ in oxidizing *c*-553. However, this finding has not yet been corroborated with intact cells. Supporting evidence has come from Ke and Chaney (1971) who found that in subchromatophore fractions from a carotenoidless mutant of *Chromatium*, *c*-553 and *c*-555 are oxidized only in the fraction containing the P883 (P890) form of bacteriochlorophyll, and that the oxidation is dependent on the redox state of the fraction. The quantum efficiency of the oxidation of *c*-553 and *c*-555 in *Chromatium* D by the single photocentre is close to 1 (Seibert and Devault, 1970) and a similar value was obtained for the photooxidation of cytochrome c_2 (Loach and Sekura, 1968).

The photoreaction centre acts as a reducer for the primary electron acceptor X which has not been identified, and then serves as the oxidant for the cytochromes *c* of the cyclic and non-cyclic electron pathways (cf. Parson, 1969a). In *Chromatium* the primary acceptor has an E_0' of $-$ 135mV or $-$ 160 mV (Dutton, 1971; Cramer, 1969) and the bacteriochlorophyll an E_0' of $+$ 415 mV. A low potential light reaction involving *c*-553 but not *c*-555 has been detected in this organism in which the primary acceptor has an E_0' of 318 mV (Seibert et al., 1971). A low potential system with an E_0' of $-$ 350 mV for the primary acceptor has been detected in *Rhodopseudomonas gelatinosa* (Dutton, 1971). It is not certain whether there is a new reaction centre or a different spectral form of P883 operating in these systems. Okayama et al., (1970) have reported that in *R. rubrum* chromatophores there are two different pigments which undergo light absorbance changes, Liac-860 and Liac-890. The midpoint oxidation-reduction potentials for the spectral changes associated with Liac-860 and Liac-890 are $+$ 450 mV and a value more negative than $-$ 100 mV respectively.

E. STRUCTURAL ORGANIZATION AND COUPLED PHOSPHORYLATION

1. Localization and organization of the cytochrome complex

In bacteria the cytochromes are located not in a separate organelle, but in the plasma or cytoplasmic membranes, attached with varying degrees of firmness to the interior of the cell wall. There may be specialized areas which cleave to form vesicles or particles of defined size on disruption of the membranes, e.g. the chromatophores of photosynthetic bacteria or fragments in *Azotobacter*. These specialized areas may be in the form of partition-like membranes joined in an array at their margins to resemble grana, as in *Rhodopseudomonas viridis* or the chemoautotroph *Nitrocystis*, or not joined at their margin as in *Rh. molischianum* (Drews and Giesebrecht, 1965; Giesebrecht and Drews, 1966; Murray and Watson, 1965, 1966). The chemistry and structure of bacterial cell walls and internal membranes has been extensively discussed in monographs by Salton (1964) and Gelman *et al.* (1967) and in a review by Salton (1967; cf. also Mirsky, 1969). The topography in particular of the cell wall is reviewed by Glauert and Thornley (1969) and the structure and organization of the membranes and cell wall of *M. lysodeikticus* specifically by Salton and Nachbar (1971). The cytoplasmic membrane is the limiting surface of protoplasts, or spheroblasts (Heppel, 1967) which have been used extensively in studies of the organization of cytochrome *c*. Miura and Mizushima (1968) separated spheroblast membranes of *E. coli* K-12 into a supposedly outer membrane, containing most of the carbohydrate, and another fraction containing the cytochromes and 70% of the phospholipid. Lysozyme-EDTA treatment of *E. coli* reveals the presence of a barrier between the plasma-membrane and the cell wall, the periplasmic region (cf. Glauert and Thornley, 1969). Cytochromes have been found attached to isolated plasma membranes of gram-negative bacteria (Norton *et al.*, 1963; Salton and Ehtisham-ud-din, 1965). The succinoxidase system was active in membranes obtained by lysozyme treatment of *E. coli* (Nagata *et al.*, 1967). The surface structure of the plasma membrane of *M. lysodeikticus* is granular but after extraction of enzymes with deoxycholate the surface of the membrane is smooth and cytochromes and succinoxidase remain embedded in the matrix (Ellar *et al.*, 1971; Salton *et al.*, 1968) and Munoz *et al.* (1968a,b) conclude that the ATPase of the membrane is located in stalked particles. Spherical structures of c. 80 Å diameter have been observed attached to vesicular membrane fragments of *A. vinelandii* (Jurtshuk *et al.*, 1969; Jones and Redfearn, 1967b).

On the other hand, Simakova *et al.* (1968, 1969) consider that ATPase is embedded in the stroma of the plasma membrane. Infra-red spectroscopy reveals that the lipids in *M. lysodeikticus* membranes have little trans character, the configuration of lowest energy and this may have a bearing on ion transport across the plasma membrane (Green and Salton, 1970).

Cytochromes of *B. subtilis* (Ferrandes *et al.*, 1966; Sedar and Burde, 1965) were claimed to be located in the mesosomes, invaginations of the cytoplasmic membrane (Fitz-James, 1960, 1968; Kaye and Chapman, 1963; Cota-Robles, 1966), and several workers have reported the presence of haemoproteins in mesosomal preparations (cf. Salton, 1967). However, recent reports indicate that the plasma membrane is the main location of the cytochromes. From respiratory inhibitor studies Reavley and Rogers (1969) concluded that in *Bacillus licheniformis* the cytochromes are located in the plasma membrane. The mesosomal membranes of *M. lysodeikticus* are deficient in cytochromes, succinate dehydrogenase and ATPase (Ellar *et al.*, 1971). Succinate dehydrogenase and NADH oxidase activities were 30 times greater in pure plasma membranes from *B. subtilis* than in the mesosome fraction (Ferrandes *et al.*, 1970).

The cytochromes of *M. lysodeikticus* are bound to the plasma membrane by hydrophobic forces and the ATPase by ionic forces. The respiratory chain could be disrupted by detergents into two fractions, one containing b-556 and the malate and NADH-dehydrogenases and the other cytochromes b-560 and c-550 (Gelman *et al.*, 1970; Tikhonova *et al.*, 1970). The substrate portion of the electron transport chain is solubilized more readily than the terminal one. Cytochromes and succinic dehydrogenase remain attached to the membrane of *B. megaterium* while the NADH dehydrogenase and ATPase were readily removed (Mizushima *et al.*, 1966, 1967).

The sequential relationship of the mitochondrial cytochromes and the linking of the electron transport with oxidative phosphorylation is discussed in Chapter VII. The pattern of arrangement of the electron transport chains seen in the mitochondria though not identical appear to be paralleled in the bacterial membrane system (cf. Nachbar and Salton, 1970; Rothfield *et al.*, 1969) although multiple electron pathways create greater complexity in some bacteria (Sinclair and White, 1970; Ackrell and Jones, 1971a,b).

A homogeneous particle enriched in lipid and NADH dehydrogenase has been obtained from *M. lysodeikticus* (Nachbar and Salton, 1970) and it is significant that a similar particle has been obtained from liver microsomal preparations (MacLennan *et al.*, 1967). From the fluorescence quenching observed using 8-anilino-1-napthalene sulphonic acid as a probe with membrane fragments from *M. lysodeikticus* it is probable that the microenvironment and the membrane conformation that characterizes the energized state are similar in some bacterial and mitochondrial membranes (Matsubara, 1970). The quenching of the fluorescence of atebrin by membrane fragments from *A. vinelandii* when oxidizing malate (Eilermann, 1970) shows that conformational changes occur in the membrane proteins similar to those observed in chloroplasts (Kraayenhof, 1970).

There are some differences between the interaction or structure of the cytochrome complexes of bacterial and mitochondrial membranes and these are exemplified in the response of the electron transport systems to inhibitors (cf. Gelman *et al.*, 1967; Smith, 1968). The electron transport system of *M. lutea* (a-598, b-562, b-557, c-554 and c-549) was insensitive to antimycin A and had low sensitivity to rotenone (Erickson and Parker, 1969). The succinoxidase system of *A. vinelandii* was insensitive to antimycin A and thenoyl-trifluoro-acetone and had low sensitivity to HQNO (Jurtshuk *et al.*, 1969). The malate oxidase system of *E. coli* particles was not inhibited by antimycin A (Cox *et al.*, 1968). Phosphorylation coupled to oxidation of hydrogen with fumarate in the chemoautotroph *D. gigas* is not inhibited by oligomycin (Barton *et al.*, 1970). The mode of preparation may affect the sensitivity of the membrane fraction to inhibitors; the NADH and succinate oxidase systems in particulate fractions obtained by lysis of spheroblasts of *E. coli* by the non-ionic detergent Brij-58 were sensitive to antimycin A and HOQNO, probably at the cytochrome b_1 level (Birdsell and Cota-Robles, 1970). The entire-energy transfer system of *M. denitrificans* is firmly attached to membrane fragments and the response to uncouplers of oxidative phosphorylation and inhibitors of energy transfer is generally similar to mitochondrial membranes (Imai *et al.*, 1967).

The structure and composition of the differentiated areas of membranes in which are localized the bacterial photosynthesis systems, and in some members of the *Athiorhodaceae*, a respiratory chain have been reviewed by Lascelles (1968). Holt and Marr (1965a,b; see also Hickman and Frenkel, 1965) presented convincing evidence that isolated chromatophores arise from fragmentation of a tubular network of intracytoplasmic membrane and Gibson's (1965a,b) findings that chromatophores have both vesicular and reticular characteristics is concordant with their conclusions. Further evidence that chromatophores are part of a continuous membrane system is provided by analysis of their lipid composition by Gorchein *et al.* (1968), protein studies (Oelze *et al.*, 1969; Oelze and Drews, 1969a) and from a kinetic study of the labelling of membranes with [2-^{14}C] acetate (Oelze and Drews, 1969b). The vesicular nature of this membrane particularly in relation to its ability to sustain a pH-gradient has been studied by several workers (cf. Ketchum and Holt, 1970).

Boll (1970a,b) has attempted to elucidate the architecture of electron transport systems in *R. rubrum* membranes by a study of the activation-inactivation response of their enzyme properties to different detergents.

2. Coupled phosphorylation

The topic of phosphorylation coupled to oxygen and nitrate respiration in bacteria has been extensively reviewed by Gelman *et al.* (1967) and specialized aspects by Smith (1968) and the following discussion is primarily concerned with the efficiency of coupled phosphorylation. Much interest recently has been directed to the inter-relationship of electron transfer activity, efficiency of energy conservation and the environmental oxygen concentration (Ackrell and Jones, 1971b).

The observed efficiency of oxidative phosphorylation by subcellular particles from bacteria is usually considerably less (P : O = 1·0) than that of mitochondria (see Gelman et al., 1967) and this in part is due to damage sustained by the membrane fragments during isolation. The membranes of *M. denitrificans* are more stable (Imai et al., 1967) and osmotically disrupted *M. denitrificans* yields particles which give a P : O ratio of 1·5 and a P : NO_3 of 0·9 with NADH as electron donor (John and Whatley, 1970); with succinate lower values were obtained. The chemoautotrophs *N. agilis* and *N. winogradskyi* yield membrane preparations which have efficient oxidative phosphorylation systems (P : O ratio of up to 2·8 with NADH) which do not require coupling factors (Aleem, 1968; Kiesow, 1964). A P : O ratio with succinate as substrate of 1·9 was obtained with a small particle fraction from *Thiobacillus novellus* (Cole and Aleem, 1970). The looser coupling of respiration to phosphorylation (Yates, 1970a,b; Eilermann et al., 1970; Ackrell and Jones, 1971a,b) contribute to the low P : O ratios. The number of phosphorylation sites may be fewer than in mitochondria or there may be a phosphorylation pathway with two or three sites and a non-phosphorylation pathway together in the intact cell or membrane fraction. Ackrell and Jones (1971a) have detected three phosphorylation sites in membrane preparations (P : O, NADH = 1·0) from *A. vinelandii*, site I at the level of NADH, site II at the level of ubiquinone, site III on the $c_4 + c_5 \to a_1 + o$ pathway. As the P : O ratio of intact *A. vinelandii* is 2 when oxidizing NAD^+ linked substrates (Knowles and Smith, 1970) it is assumed that coupling is either weak or that oxidative phosphorylation is lacking from the $b_1 \to a_2$ pathway, or is of very low efficiency.

Heterotrophically grown *Pseudomonas saccharophila* which has both cytochromes *o* and *a*, has three phosphorylation sites coupled to the oxidation of generated NADH with an overall P : O ratio of 0·92, while only two sites are functional in the autotrophically (H_2 and CO_2) grown cells, these having cytochrome *o*, but not *a* (Ishaque et al., 1971). However, in *Hydrogenomonas eutropha* (Ishaque and Aleem, 1970) and *M. denitrificans* (Knobloch et al., 1970) grown autotrophically under H_2, three phosphorylation sites are involved in the coupled oxidation of NADH or H_2.

Early attempts to estimate the P : O ratio directly in intact cells gave low values. However, Ramirez and Smith (1968) obtained a P : O ratio of 2·8 for the coupled oxidation of β-hydroxybutyrate by O_2 in *R. rubrum* and *Rps. spheroides* using the luciferin-luciferase assay for ATP. The efficiency of phosphorylation in whole bacterial cells has been estimated indirectly by the molar growth yield of ATP method of Bauchop and Elsden (1960) who showed that under anaerobiosis several organisms yield 10·5 g dry weight of cells per mole of ATP generated from the catabolism of glucose. Hadjipetrou et al. (1964, 1965) found that in oxygen or nitrate respiration

30 g of cells of *A. aerogenes* are formed per pair of reducing equivalents passed to O_2 or NO_3^- during growth on glucose and these workers concluded that the P : O ratio for growth under O_2 is 3 for coliform bacteria and this is confirmed by Harrison and Maitra (1969) for *A. aerogenes*. Factors which contribute to uncertainties of the P : O estimate by the method are discussed in a review by Forrest (1969).

Using a more sophisticated and direct approach employing stopped-flow techniques Hempfling (1970) also finds that in *E. coli* B and *A. aerogenes* oxidative phosphorylation is as efficient as in higher organisms.

The low P : O ratio of bacterial membrane particles can be due to the loss of soluble coupling factors. Two isolated coupling factors, one heat labile and of low molecular weight, restore oxidative phosphorylation to *M. lysodeikticus* membrane fragments (Ishikawa, 1970). The reconstituted system was sensitive to Gramicidin A. A heat labile and a heat stable soluble factor could restore oxidative phosphorylation to washed membrane fragments of *Alcaligenes faecalis* (Adolfsen and Moudrianakis, 1971b). Scocca and Pinchot (1968) found that coupling factors could act as inhibitors of electron transport in *Alcaligenes faecalis* and proposed that these could act as regulators of respiration. Coupled phosphorylation could be restored to light-inactivated extracts of *M. phlei* by the trans-isomer of menaquinone (Phillips *et al.*, 1970); the cis-isomer only restored the oxidation. Menaquinone has been implicated in the coupling process which joins electron flux and ATP synthesis (Kufe and Howland, 1968). The role of metal ions in the binding of coupling factors to the membrane and the activation of phosphorylation has been studied by Adolfsen and Moudrianakis (1971a). Electron transport particles from *M. phlei* require either soluble factors to restore their coupled phosphorylation or treatment of the particles with trypsin (Orme *et al.*, 1969; Bogin *et al.*, 1970a,b). The two treatments are additive in effect but act differently. Trypsin unmasks latent ATPase and the treated particle requires coupling factors for maximal phosphorylation. The restoration of oxidative phosphorylation by heating which is also obtained with *M. lysodeikticus* particles but not by mitochondria or submitochondrial particles, may be due to structural changes or removal of a regulator of coupled phosphorylation. Photophosphorylation was restored by soluble factors to *R. capsulatus* after this had been diminished by sonication (Baccarini-Melandri *et al.*, 1970). The cyclic photosynthetic transport chain in *R. rubrum* is coupled to phosphorylation at two points, one localized between cytochromes *b* and *c* and the other between bacteriochlorophyll and cytochrome *c* (see Vernon, 1968). Evidence has been presented for the existence of a coupling site located between NAD^+ photoreduction (Jones and Vernon, 1969; Keister and Minton, 1969; Yamashita and Kamen, 1969). This photoreduction occurs via an energy-dependent reversed electron transfer pathway. Isaev *et al.*

(1970) from a study of the accumulation of ions in the chromatophore membrane confirm the localization of the three energy generating sites, and identify a fourth in the NADH-dehydrogenase segment of the chromatophore redox chain.

Reversed electron transfer, which has been extensively studied in mitochondria (see Chapter VII) has already been cited in connection with nitrite and sulphite oxidation in chemoautotrophs. Its occurence in bacterial photosynthesis has been investigated by Baltscheffsky (1967, 1969a,b) and by Yamashita et al. (1969).

REFERENCES

Ackrell, B. A. C. and Jones, C. W. (1971a,b). *Europ. J. Biochem.* **20**, 22–28, 29–35.
Adolfsen, R. and Moudrianakis, E. N. (1971a,b). *Biochemistry* **10**, 434–440, 440–446.
Akagi, J. M. (1967). *J. Biol. Chem.* **242**, 2478–2483.
Aleem, M. I. H. (1965). *J. Bacteriol.* **90**, 95–101.
Aleem, M. I. H. (1966a). *Biochim. et Biophys. Acta* **128**, 1–12.
Aleem, M. I. H. (1966b). *J. Bacteriol.* **91**, 729–736.
Aleem, M. I. H. (1967). *Bacteriol. Proc.* **67**, 112.
Aleem, M. I. H. (1968). *Biochim. et Biophys. Acta* **162**, 338–347.
Aleem, M. I. H. (1969). *Anton van Leeuwenhoek* **35**, 379–391.
Aleem, M. I. H. (1970). *Ann. Rev. Plant Physiol.* **21**, 67–90.
Aleem, M. I. H. and Nason, A. (1959). *Biochem. Biophys. Res. Commun.* **1**, 323–327.
Aleem, M. I. H. (1971). Personal communication.
Ambler, R. P. (1962). *Biochem. J.* **82**, 30P.
Ambler, R. P. (1963a). *Biochem. J.* **89**, 341–349.
Ambler, R. P. (1963b). *Biochem. J.* **89**, 349–378.
Ambler, R. P. (1968). *Biochem. J.* **109**, 47P.
Ambler, R. P., Bruschi-Heriaud, M. and Le Gall, J. (1968). *FEBS Lett.* **5**, 115–117.
Ambler, R. P., Bruschi-Heriaud, M. and Le Gall, J. (1970). Communication 10th Int. Congress of Microbiology. In "Recent Advances in Microbiology" pp. 25–39.
Amesz, J. and Vredenberg, W. J. (1966). *Biochim. et Biophys. Acta* **126**, 254–261.
Antonini, E., Finazzi-Agro, A., Avigliano, L., Guerrieri, P., Rotilio, G. and Mondovi, G. (1970). *J. Biol. Chem.* **18**, 4847–4856.
Appleby, C. A. (1967). *Biochim. et Biophys. Acta* **147**, 399–402.
Appleby, C. A. (1968). "Osaka", pp. 666–679.
Appleby, C. A. (1969a). *Biochim. et Biophys. Acta* **172**, 71–87.
Appleby, C. A. (1969b). *Biochim. et Biophys. Acta* **172**, 88–105.
Appleby, C. A. and Daniel, R. M. (1971). In "Proceedings of the Second International Symposium on Oxidases and Related Oxidation-Reduction Systems" (T. E. King, H. S. Mason, M. Morrison, eds.).
Appleby, C. A. and Bergersen, F. J. (1958). *Nature* **182**, 1174.
Arima, K. and Oka, T. (1965). *J. Bacteriol.* **90**, 734–743.
Azoulay, E. (1964). *Biochim. et Biophys. Acta* **92**, 458–464.
Azoulay, E. and Couchoud-Beaumont, P. (1965). *Biochim. et Biophys. Acta* **110**, 301–311.

Baccarini-Melandri, A., Gest, H. and San Pietro, L. (1970). *J. Biol. Chem.* **245**, 1224–1226.
Baltscheffsky, M. (1967). *Biochem. Biophys. Res. Commun.* **28**, 270–276.
Baltscheffsky, M. (1969a,b). *Arch. Biochem. Biophys.* **130**, 646–652; **133**, 46–53.
Barrett, J. (1956). *Biochem. J.* **64**, 626–639.
Barrett, J. (1969). *Biochim. et Biophys. Acta* **177**, 442–455.
Barrett, J. and Collins, S. (1968). Unpublished results.
Barrett, J. and Kamen, M. D. (1961). *Biochim. et Biophys. Acta* **50**, 573–575.
Barrett, J. and Newton, N. (1968). Unpublished results.
Barrett, J. and Sinclair, P. R. (1967a). *Abstr. 7th Int. Congress of Biochem.*, Tokyo H-107.
Barrett, J. and Sinclair, P. R. (1967b). *Biochim. et Biophys. Acta* **143**, 279–281.
Barton, L. L., Le Gall, J. and Peck, H. D. (1970). *Biochem. Biophys. Res. Commun.* **41**, 1036–1042.
Bartsch, R. G. (1961). *Fed. Proc.* **20**, 43.
Bartsch, R. G. (1963). In "Bacterial Photosynthesis" (H. Gest, A. San Pietro and L. P. Vernon, eds.) p. 475–489, Antioch Press, Yellow Springs.
Bartsch, R. G. (1968). *Ann. Rev. Microbiol.* **37**, pp. 181–250.
Bartsch, R. G. and Horio, T. (1968). Cited in Meyer *et al.* (1968).
Bartsch, R. G. and Kamen, M. D. (1958). *J. Biol. Chem.* **230**, 41–63.
Bartsch, R. G. and Kamen, M. D. (1960). *J. Biol. Chem.* **235**, 825–831.
Bartsch, R. S., Coval, M. L. and Kamen, M. D. (1961). *Biochim. et Biophys. Acta* **51**, 241–245.
Bartsch, R. G., Meyer, T. and Robinson, A. B. (1968). "Osaka", pp. 443–451.
Bartsch, R. G., Kakuno, T., Horio, T. and Kamen, M. D. (1971). Cited in Kamen and Horio (1970).
Baseman, J. B. and Cox, C. D. (1969). *J. Bact.* **97**, 1001–1004.
Bauchop, T. and Elsden, S. R. (1960). *J. Gen. Microbiol.* **23**, 457–469.
Bayer, E., Hill, H. A. O., Roder, A. and Williams, R. J. P. (1969). *Chem. Commun.* **1969**, 109.
Bearden, A. J., Moss, T. H., Caughey, W. S. and Beaudreau, C. A. (1956). *Proc. Nat. Acad. Sci. U.S.* **53**, 1246–1253.
Biggins, J. and Dietrich, W. E. (1968). *Arch. Biochem. Biophys.* **128**, 40–50.
Birdsell, D. C. and Cota-Robles, E. H. (1970). *Biochim. et Biophys. Acta* **216**, 250–261.
Birk, Y., Silver, A. W. S. and Heim, A. H. (1957). *Biochim. et Biophys. Acta* **25**, 227–228.
Blauer, G. (1964). *Biochim. et Biophys. Acta* **79**, 547–562.
Blaylock, B. A. and Nason, A. (1963). *J. Biol. Chem.* **238**, 3453–3465.
Blumberg, W. and Peisach, J. (1970). In "Structure and Function of Macromolecules and Membranes" (B. Chance, C. P. Lee and T. Youctaw, eds.). Academic Press, New York.
Bogin, E., Higashi, T., and Brodie, A. F. (1970a). *Biochem. Biophys. Res. Commun.* **41**, 995–1001.
Bogin, E., Higashi, T., and Brodie, A. F. (1970b). *Proc. Nat. Acad. Sci.* **67**, 1–6.
Boll, M. (1970a). *Arch. Mikrobiol.* **71**, 1–8.
Boll, M. (1970b). *Experientia* **26**, 956–957.
Bone, D. H. (1963). *Nature* **197**, 517.
Bongers, L. (1967). *J. Bacteriol.* **93**, 1615–1623.
Bragg, P. I. and Polglase, W. J. (1963). *J. Bacteriol.* **86**, 544–547.
Broberg, P. L. and Smith, L. (1967). *Biochim. et Biophys. Acta* **131**, 479–489.

Broberg, P. L., Welsch, M. and Smith, L. (1969). *Biochim. et Biophys. Acta* **172**, 205–216.
Bruschi-Heriaud, M. and Le Gall, J. (1967). *Bull. Soc. Chim. Biol.* **49**, 753–758.
Bruschi-Heriaud, M., Le Gall, J., Le Guen, M. and Dus, K. (1970). *Biochem. Biophys. Res. Commun.* **38**, 607–616.
Butlin, K. R. and Postgate, J. R. (1953). In "Microbial Metabolism". Symp. 6th Inter. Congr. Microbiol. pp. 126–143.
Butt, D. and Lees, H. (1958). *Nature* **182**, 723–733.
Cardini, C. and Jurtshuk, P. (1970). *J. Biol. Chem.* **245**, 1789–1796.
Case, G. D., Parson, W. W. and Thornber, P. J. (1970). *Biochim. et Biophys. Acta* **223**, 122–128.
Castor, L. N. and Chance, B. (1955). *J. Biol. Chem.* **217**, 453–465.
Castor, L. N. and Chance, B. (1959). *J. Biol. Chem.* **234**, 1587–1592.
Chaix, P. and Petit, J. F. (1957). *Biochim. et Biophys. Acta* **22**, 66–71.
Chaix, P. and Petit, J. F. (1959). *Biochim. et Biophys. Acta* **25**, 481–486.
Chance, B. (1953a). *J. Biol. Chem.* **202**, 383–396.
Chance, B. (1953b). *J. Biol. Chem.* **202**, 397–406.
Chance, B. (1953c). *J. Biol. Chem.* **202**, 407–416.
Chance, B. (1961). "Haematin Enzymes", p. 433.
Chance, B. and Nishimura, M. (1960). *Proc. Nat. Acad. Sci.* **46**, 19–24.
Chance, B. and Smith, L. (1955). *Nature* **175**, 803–806.
Chance, B., Smith, L. and Castor, L. N. (1953). *Biochim. et Biophys. Acta* **12**, 289–298.
Chance, B., Horio, T., Kamen, M. D. and Taniguchi, S. (1966). *Biochim. et Biophys. Acta* **112**, 1–7.
Charles, A. M. and Suzuki, I. (1966). *Biochim. et Biophys. Acta* **128**, 522–534.
Cheah, K. S. (1968). *Biochim. et Biophys. Acta* **153**, 718–720.
Cheah, K. S. (1969). *Biochim. et Biophys. Acta* **180**, 320–333.
Cheah, K. S. (1970a). *Biochim. et Biophys. Acta* **197**, 84–86.
Cheah, K. S. (1970b). *Biochim. et Biophys. Acta* **205**, 148–160.
Cheah, K. S. (1970c). *Biochim. et Biophys. Acta* **216**, 43–53.
Chippaux, M. and Pichinoty, F. (1968). *Europ. J. Biochem.* **6**, 131–134.
Clark-Walker, G. D. (1970). Personal communication.
Clark-Walker, G. D. and Lascelles, J. (1970). *Arch. Biochem. Biophys.* **136**, 153–159.
Clark-Walker, G. D., Rittenberg, B., and Lascelles, J. (1967). *J. Bacteriol.* **94**, 1648–1655.
Clayton, R. K. and Sistrom, W. R. (1966). *Photochem. Photobiol.* **5**, 661–668.
Clezy, P. S., Parker, M. J., Barrett, J. and Lemberg, R. (1964). *Biochim. et Biophys. Acta* **82**, 361–372.
Cole, J. A. (1968). *Biochim. et Biophys. Acta* **162**, 356–368.
Cole, J. A. and Wimpenny, J. W. T. (1968). *Biochim. et Biophys. Acta* **162**, 39–48.
Cole, J. S. and Aleem, M. I. H. (1970). *Biochem. Biophys. Res. Commun.* **38**, 736–743.
Cook, T. M. and Umbreit, W. W. (1963). *Biochemistry* **2**, 194–196.
Cota-Robles, E. H. (1966). *J. Ultrastructure Res.* **16**, 626–639.
Coval, M. L., Horio, T. and Kamen, M. D. (1961). *Biochim. et Biophys. Acta* **51**, 246–260.
Cox, G. B., Newton, N. A., Gibson, F., Snoswell, A. M. and Hamilton, J. A. (1970). *Biochem. J.* **117**, 551–562.

Cox, G. B., Snoswell, A. M. and Gibson, F. (1968). *Biochim. et Biophys. Acta* **153**, 1–12.
Cramer, W. A. (1969). *Biochim. et Biophys. Acta* **189**, 54–59.
Cusanovich, M. A. (1966). Cited by Kamen, M.D. in "Hemes and Hemoproteins", p. 521–526.
Cusanovich, M. A. and Bartsch, R. G. (1969). *Biochim. et Biophys. Acta* **189**, 245–255.
Cusanovich, M. A. and Gibson, Q. H. (1970). Cited in Cusanovich *et al.* (1970).
Cusanovich, M. A. and Kamen, M. D. (1968). *Biochim. et Biophys. Acta* **153**, 376–396.
Cusanovitch, M. A., Bartsch, R. G. and Kamen, M. D. (1968). *Biochim. et Biophys. Acta* **153**, 397–417.
Cusanovich, M. A., Tedro, S. and Kamen, M. D. (1970). *Arch. Biochem. Biophys.* **141**, 557–570.
Daniel, R. M. (1970). *Biochim. et Biophys. Acta* **216**, 328–341.
Daniel, R. M. and Appleby, C. A. (1971). *Proc. Austral. Biochem. Soc.* **4**, 72.
Deeb, S. S. and Hager, L. P. (1964). *J. Biol. Chem.* **239**, 1024–1031.
De Klerk, H., Bartsch, R. G., Kamen, M. D. (1965). *Biochim. et Biophys. Acta* **97**, 275–280.
Dickerson, R. E. (1971). *J. Mol. Biol.* **57**, 1–15.
Domnas, A. and Grant, N. G. (1970). *J. Bacteriol.* **101**, 652–653.
Drews, G. and Giesebrecht, P. (1965). *Arch. Mikrobiol.* **52**, 242–250.
Drucker, H. and Campbell, L. L. (1966). *Bacteriol. Proc.* **66**, 71.
Drucker, H. and Campbell, L. L. (1969). *J. Bacteriol.* **100**, 358–364.
Drucker, H., Campbell, L. L. and Wordy, R. W. (1970a). *Biochemistry* **9**, 1519–1527.
Drucker, H., Trousil, E. B. and Campbell, L. L. (1970b). *Biochemistry* **9**, 3395–3400.
Drucker, H., Trousil, E. B., Campbell, L. L., Barlow, G. H. and Margoliash, E. (1970c). *Biochemistry* **9**, 1515–1518.
Dus, K. and Kamen, M. D. (1963). *Biochem. Z.* **338**, 364–375.
Dus, K. and Sletten, K. (1968). "Osaka", pp. 293–303.
Dus, K., Bartsch, R. G. and Kamen, M. D. (1962). *J. Biol. Chem.* **237**, 3083–3093.
Dus, K., de Klerk, A., Bartsch, R. G., Horio, T. and Kamen, M. D. (1967). *Proc. Nat. Acad. Sci.* **57**, 367–370.
Dus, K., Katagiri, M., Yu, C. A., Erbes, D. L. and Gunsulus, I. C. (1970). *Biochem. Biophys. Res. Commun.* **40**, 1423–1430.
Dus, K., Sletten, K. and Kamen, M. D. (1968). *J. Biol. Chem.* **243**, 5507–5518.
Dutton, P. L. (1971). *Biochim. et Biophys. Acta* **226**, 65–80.
Dutton, P. L., Kihara, T. and Chance, B. (1970). *Arch. Biochem. Biophys.* **139**, 236–240.
Dutton, P. L., Kihara, T., McCray, J. A., Thornber, J. P. (1971). *Biochim. et Biophys. Acta* **226**, 81–87.
Duysens, L. N. M. (1954). *Nature* **173**, 692–693.
Dworkin, M. and Niederpruem, D. J. (1964). *J. Bacteriol.* **87**, 316–322.
Egami, F., Itahishi, M., Sato, R. and Mori, T. (1953). *J. Biochem.* **40**, 527–534.
Ehrenberg, A. (1957). *Acta Chem. Scand.* **11**, 1257–1270.
Ehrenberg, A. and Kamen, M. D. (1965). *Biochim. et Biophys. Acta* **102**, 333–340.
Ehrenberg, A. and Szczepkowski, T. W. (1960). *Acta Chem. Scand.* **14**, 1684–1692.
Eilermann, L. J. M. (1970). *Biochim. et Biophys. Acta* **216**, 231–238.

Eilermann, L. J. M., Pandit Hovenkamp, H. G. and Kolk, A. H. J. (1970). *Biochim. et Biophys. Acta* **197**, 25–30.
Ellar, D. J., Munoz, E. and Salton, M. R. J. (1970). *Biochim. et Biophys. Acta* **225**, 140–150.
Ellar, D. J., Thomas, T. D. and Postgate, J. A. (1971). *Biochem. J.* **122**, 44–45P.
Elsden, S. R., Kamen, M. D. and Vernon, L. P. (1953). *J. Am. Chem. Soc.* **75**, 6347–6348.
Engelberg, H. and Artman, M. (1961). *Biochim. et Biophys. Acta* **47**, 553–560.
Erickson, S. K. and Parker, G. L. (1969). *Biochim. et Biophys. Acta* **180**, 56–62.
Estabrook, R. W. (1956). *J. Biol. Chem.* **223**, 781–794.
Estabrook, R. W., Hildebrandt, A. G., Baron, J., Netter, K. J. and Leibmann, K. (1971). *Biochem. Biophys. Res. Commun.* **42**, 132–139.
Euler, H. V., Fink, H. and Hellström, H. (1927). *Z. Physiol. Chem.* **169**, 10–51.
Fanger, M. W., Hettinger, T. P. and Harbury, H. A. (1967). *Biochemistry* **6**, 713–720.
Fellman, J. H. and Mills, R. C. (1960). *J. Bact.* **79**, 800–806.
Ferrandes, B., Chaix, P. and Ryter, A. (1966). *C. R. Acad. Sci. Paris* **263**, 1623–1635.
Ferrandes, B., Grehel, C. and Chaix, P. (1970). *Biochim. et Biophys. Acta* **223**, 292–308.
Fink, H. (1932). *Z. Physiol. Chem.* **210**, 197–219.
Fitz-James, P. C. (1960). *J. Biochem. Biophys. Cytol.* **8**, 507–528.
Fitz-James, P. C. (1968). In "Microbial Protoplasts, Spheroblasts and L-forms" (L. B. Guze, ed.) p. 124, Williams and Wilkins, Baltimore.
Flatmark, T., Dus, K., de Klerk, H. and Kamen, M. D. (1970). *Biochemistry* **9**, 1991–1996.
Flatmark, T. and Robinson, A. B. (1968). "Osaka", pp. 318–327.
Forrest, W. W. (1969). "Microbial Growth", Symposium **19**, *Soc. Gen. Microbiol.*, p. 65–85.
Fowler, C. F. and Sybesma, C. (1970). *Biochim. et Biophys. Acta* **197**, 276–283.
Fujita, T. (1966a). *J. Biochem.* **60**, 204–215.
Fujita, T. (1966b). *J. Biochem.* **60**, 329–334.
Fujita, A. and Kodama, T. (1934). *Biochem. Z.* **273**, 186–197.
Fujita, T. and Sato, R. (1963). *Biochim. et Biophys. Acta* **77**, 690–693.
Fujita, T. and Sato, R. (1966a). *J. Biochem.* **60**, 568–577.
Fujita, T. and Sato, R. (1966b). *J. Biochem.* **60**, 691–700.
Fujita, T. and Sato, R. (1967). *J. Biochem.* **62**, 230–238.
Fujita, T., Itagaki, E. and Sato, R. (1963). *Biochem. J.* **53**, 282–289.
Geller, D. M. (1962). *J. Biol. Chem.* **237**, 2947–2954.
Gel'man, N. S., Lukoyanova, M. A. and Ostrovskii, D. N. (1967). "Respiration and Phosphorylation of Bacteria". Plenum Press, New York. (A. I. Oparin, ed.)
Gel'man, N. S., Tikhonova, G. V., Simokova, J. M., Lukoyanova, M. A., Taptykova, S. D. and Mikelsaar, H. M. (1970). *Biochim. et Biophys. Acta* **223**, 321–331.
Gibson, J. (1961). *Biochem. J.* **79**, 151–158.
Gibson, K. D. (1965a). *Biochemistry* **4**, 2042–2051.
Gibson, K. D. (1965b). *J. Bacteriol.* **90**, 1059–1072.
Gibson, Q. H. and Kamen, M. D. (1966). *J. Biol. Chem.* **241**, 1969–1976.
Gibson, Q. H. and Roughton, F. J. W. (1955). *Proc. Roy. Soc.* (London). Ser. B **143**, 310–334.
Giesbrecht, P. and Drews, G. (1966). *Arch. Mikrobiol.* **54**, 297–330.

Glauert, A. M. and Thornley, M. J. (1969). *Ann. Rev. Microbiol.* **23**, 159–198.
Gorchein, A., Neuberger, A. and Tait, G. H. (1968). *Proc. Roy. Soc.* (London) Ser. B. **170**, 311–318.
Gouthier, D. K., Clark-Walker, G. D., Garrard, W. T. Jr. and Lascelles, J. (1970). *J. Bacteriol.* **102**, 797–803.
Govindjee, R. and Sybesma, C. (1970). *Biochim. et Biophys. Acta* **223**, 251–260.
Gray, C. T. and Wimpenny, J. W. T. (1965). "Mechanisms de Regulation des Activities cellulaires chez les Microorganisms" (ed. by C.N.R.S.), Paris, No. 124, pp. 523.
Gray, C. T. and Ely, S. (1971). *Fed. Proc.* **30**, Abstr. 534.
Gray, C. T., Wimpenny, J. W. T., Hughes, D. E. and Ranlett, M. (1963). *Biochim. et Biophys. Acta* **67**, 157–160.
Green, D. H. and Salton, M. R. J. (1970). *Biochim. et Biophys. Acta* **211**, 139–147.
Griffin, B. and Paterson, J. A. (1971). *Arch. Biochem. Biophys.* **145**, 220–229.
Gunsalus, I. C. (1968). *H. S. Zeit. für Physiol. Chem.* **249**, 1610–1613.
Gunsalus, I. C., Tsai, R. L., Tyson, C. A., Msu, M-C., Yu, C-A. (1970). *Abstract. Proc. Magnetic Resonance Conf.*, Oxford.
Gunsalus, I. C., Tyson, C. A. and Lipscomb, J. D. (1971). Second Internat. Symposium on Oxidases, Memphis.
Gvozdev, R. I., Sadkov, A. P., Yokolev, V. A. and Alfimova, E. Ya. (1968). *Izv. Akad. Nauk. SSSR, Ser. Biol.* **6**, 838.
Hadjipetrou, L. P., Gerrits, J. P., Teulings, F. A. G. and Stouthamer, A. H. (1964). *J. Gen. Microbiol.* **36**, 39–50.
Hadjipetrou, L. P. and Stouthamer, A. H. (1965). *J. Gen. Microbiol.* **38**, 29–34.
Hager, L. P. and Deeb, S. S. (1967). "Methods in Enzymology" **10**, pp. 367–372.
Hager, L. P. and Itagaki, E. (1967). Methods in Enzymology" **10**, pp. 373–378.
Hammond, R. K. and White, D. C. (1970). *J. Bacteriol.* **103**, 607–610.
Harrison, D. E. F. and Maitra, P. K. C. (1969). *Biochem. J.* **112**, 647–656.
Haske, R. and Campbell, L. (1971). *J. Bacteriol.* **105**, 249–258.
Hauge, J. G. (1960). *Biochim. et Biophys. Acta* **45**, 250–262.
Hauge, J. G. (1961). *Arch. Biochem. Biophys.* **94**, 308–318.
Hedegaard, J. and Gunsalus, I. C. (1965). *J. Biol. Chem.* **240**, 4038–4043.
Heim, A. H., Silver, W. S. and Birk, Y. (1957). *Nature* **180**, 608–609.
Hempfling, W. P. (1970). *Biochim. et Biophys. Acta* **205**, 169–182.
Henderson, R. W. and Nankiville, D. D. (1966). *Biochem. J.* **98**, 587–593.
Hendricks, R. and Cronin, J. R. (1971). *Biochem. Biophys. Res. Commun.* **44**, 313–318.
Heppel, L. A. (1967). *Science* **156**, 1451–1455.
Hettinger, T. P., Fanger, M. W., Vinogradov, S. N. and Harbury, H. A. (1966). *Fed. Proc.* **25**, Abstr. 2588.
Hickman, D. D. and Frenkel, A. W. (1965). *J. Cell. Biol.* **25**, 279–291.
Hill, J. E. and Taylor, C. P. S. (1969). *Canad. J. Biochem.* **47**, 507–509.
Hind, G. and Olson, J. M. (1968). *Ann. Rev. Plant Physiol.* **19**, 249–282.
Hiromi, K. and Sturtevant, J. M. (1965). *J. Biol. Chem.* **240**, 4662–4668.
Hochster, R. M. and Nozzolio, C. G. (1960). *Canad. J. Biochem. Physiol.* **38**, 79–93.
Holt, S. C. and Marr, A. G. (1965a,b). *J. Bacteriol.* **89**, 1402–1412, 1413–1420.
Hori, K. (1961a). *J. Biochem.* **50**, 481–485.
Hori, K. (1961b). *J. Biochem.* **50**, 440–449.
Hori, K. (1963). *J. Biochem.* **53**, 354–363.
Horio, T. (1958a). *Biochem. J.* **45**, 195–205.

Horio, T. (1958b). *J. Biochem.* **45**, 267–279.
Horio, T. and Bartsch, R. G. (1968). Cited in Meyer et al. (1968).
Horio, T. and Kamen, M. D. (1961). *Biochim. et Biophys. Acta* **48**, 266–286.
Horio, T. and Kamen, M. D. (1970). *Ann. Rev. Biochem.* **39**, 673–700.
Horio, T. and Taylor, G. P. S. (1965). *J. Biol. Chem.* **240**, 1772–1775.
Horio, T., Higashi, T., Sasagawa, M., Kusai, K., Nakai, M. and Okunuki, K. (1960). *Biochem. J.* **77**, 194–201.
Horio, R., Higashi, T., Matsubara, H., Kusai, K., Nakai, M. and Okunuki, K. (1958). *Biochim. et Biophys. Acta* **29**, 297–302.
Horio, T., Higashi, T., Yamanaka, T., Matsubara, H., and Okunuki, K. (1961a). *J. Biol. Chem.* **236**, 944–951.
Horio, T., Kamen, M. D. and de Klerk, H. (1961b). *J. Biol. Chem.* **236**, 2783–2787.
Ikuma, H. and Hempfling, W. P. (1965). In Trudinger, P. A. (1967).
Imai, K., Asano, A. and Sato, R. (1967). *Biochim. et Biophys. Acta* **143**, 462–476.
Imai, Y., Imai, K., Sato, R. and Horio, T. (1969a). *J. Biochem.*, (Tokyo) **65**, 225–237.
Imai, Y., Imai, K., Ikeda, K., Hamaguchi, K. and Horio, T. (1969b). *J. Biochem.*, (Tokyo) **65**, 629–637.
Inoue, Y. and Kubo, H. (1965). *Biochim. et Biophys. Acta* **110**, 57–65.
Isaev, P. I., Liberman, E. A., Samuilov, V. D., Skulachev, V. P. and Tsofina, L. M. (1970). *Biochim. et Biophys. Acta* **216**, 22–29.
Ishaque, M. and Aleem, M. I. H. (1970). *Biochim. et Biophys. Acta* **223**, 388–397.
Ishaque, H., Donowa, A. and Aleem, M. I. H. (1971). *Biochem. Biophys. Res. Commun.* **44**, 245–251.
Ishidate, K., Kawaguchi, K., Tawaga, K. and Hagihara, B. (1969a). *J. Biochem.* (Tokyo) **65**, 375–383.
Ishidate, K., Kawaguchi, K. and Tagawa, K. (1969b). *J. Biochem.* (Tokyo) **65**, 385–392.
Ishikawa, W. (1970). *J. Biochem.* **67**, 297–312.
Ishimoto, M. and Yagi, T. (1961). *J. Biochem.* **49**, 103–109.
Ishimoto, M., Koyaka, J. and Nagai, J. (1954). *Bull. Chem. Soc., Japan* **27**, 564–565.
Ishimoto, M., Kondo, J., Kamdydind, T., Yagi, T. and Shivaki, M. (1958). *Proc. Int. Symp. Enzyme Chem.* (Tokyo) p. 229.
Ishimura, Y., Ullrich, V. and Peterson, J. A. (1971). *Biochem. Biophys. Res. Commun.* **42**, 140–146.
Itagaki, E. and Hager, L. P. (1965). *Fed. Proc.* **24**, 545. Abstr. 2301.
Itagaki, E. and Hager, L. P. (1966). *J. Biol. Chem.* **241**, 3687–3695.
Itagaki, E. and Hager, L. P. (1968). *Biochem. Biophys. Res. Commun.* **32**, 1013–1019.
Ivanov, I. D., Matkhanov, G. I., Belov, M. Yu, and Gogleva, T. V. (1967). *Mikrobiol.* **36**, 205–209.
Iwasaki, H. (1960). *J. Biochem.* (Tokyo) **47**, 174–184.
Iwasaki, H. (1966). *Plant Cell Physiol.* (Tokyo) **7**, 199.
Iwasaki, H. (1967). 7th Intern. Congress Biochem., Tokyo. Abstract H-4: also cf. R. G. Bartsch (1968) in *Ann. Rev. Microbiol.* **22**, 191–192.
Iwasaki, H. (1970). Private communication.
Iwasaki, II. and Matsubara, T. (1971). *J. Biochem.* **69**, 847–848.
Iwasaki, H. and Shidara, S. (1969a). *J. Biochem.* **66**, 775–781.
Iwasaki, H. and Shidara, S. (1969b). *Plant Cell Physiol.* **10**, 291–305.
Iwasaki, H., Shidara, S., Suzuki, H. and Mori, T. (1963). *J. Biochem.* (Tokyo) **53**, 299–303.

Jackson, F. L. and Lawton, V. D. (1959). *Biochim. et Biophys. Acta* **35**, 76–84.
Jacobs, N. J. and Conti, S. F. (1965). *J. Bact.* **89**, 675–679.
Jarman, T. R., Eady, R. R. and Lange, P. J. (1970). *Biochem. J.* **119**, p. 55.
John, P. and Whatley, F. R. (1970). *Biochim. et Biophys. Acta* **216**, 342–352.
Jolchine, G., Reiss-Husson, F. and Kamen, M. D. (1969). *Proc. Nat. Acad. Sci. U.S.* **64**, 650–653.
Jones, C. W. and Redfearn, E. R. (1966). *Biochim. et Biophys. Acta* **113**, 467–481.
Jones, C. W. and Redfearn, E. R. (1967a). *Biochim. et Biophys. Acta* **143**, 340–353.
Jones, C. W. and Redfearn, E. R. (1967b). *Biochim. et Biophys. Acta* **143**, 354–362.
Jones, C. W. and Vernon, L. P. (1969). *Biochim. et Biophys. Acta* **180**, 149–164.
Jones, H. E. (1971a). *J. Bacteriol.* **106**, 339–346.
Jones, H. E. (1971b). *Arch Mikrobiol.* **80**, 78–86.
Jurtshuk, P. and Harper, L. (1968). *J. Bacteriol.* **96**, 678–686.
Jurtshuk, P. and Old, F. (1968). *J. Bacteriol.* **95**, 1790–1797.
Jurtshuk, P., May, A. K., Pope, L. M. and Aston, P. R. (1969). *Canad. J. Microbiol.* **15**, 797–807.
Kakuno, T., Bartsch, R., Horio, T. and Kamen, M. D. (1970). Cited in Kamen and Horio (1970).
Kakuno, T., Bartsch, R., Nishikawa, K. and Horio, T. (1971). *J. Biochem.* **70**, 79–94.
Kamen, M. D. and Bartsch, R. G. (1961). "Haematin Enzymes" pp. 419–435.
Kamen, M. D. and Horio, T. (1970). *Ann. Rev. Biochem.* **39**, 673–700.
Kamen, M. D. and Takeda, Y. (1956). *Biochim. et Biophys. Acta* **21**, 518–523.
Kamen, M. D. and Vernon, P. (1955). *Biochim. et Biophys. Acta* **17**, 10–22.
Kamen, M. D., Bartsch, R. G., Horio, T. and de Klerk, H. (1963). "Methods in Enzymology" (S. P. Colowick and N. O. Kaplan, eds.) Academic Press. Vol. 6, pp. 391–404.
Katagiri, M., Ganguli, B. N. and Gunsalus, I. C. (1968). *J. Biol. Chem.* **243**, 3543–3546.
Kaye, J. J. and Chapman, G. B. (1963). *J. Bacteriol.* **86**, 536–543.
Ke, B. and Chaney, T. H. (1971). *Biochim. et Biophys. Acta* **226**, 341–345.
Kearney, E. B. and Goldman, D. S. (1970). *Biochim. et Biophys. Acta* **197**, 197–205.
Keilin, D. (1925). *Proc. Roy. Soc.* (London) B **98**, 312–339.
Keilin, D. (1927). *Compt. Rend. Soc. Biol. Paris* **96**, Sp39–Sp68.
Keilin, D. (1933). *Nature* **132**, 783.
Keilin, D. (1953). *Nature* **172**, 390–393.
Keilin, D. (1955). *Biochem. J.* **59**, 571–579.
Keilin, D. (1966). "The History of Cell Respiration and Cytochrome". pp. 276–280. Cambridge University Press, London.
Keilin, D. and Harpley, C. H. (1941). *Biochem. J.* **35**, 688–692.
Keilin, D. and Hartree, E. F. (1949). *Nature* **164**, 254–259.
Keilin, D. and Slater, E. C. (1953). *Brit. Med. Bull.* **9**, 89–97.
Keilin, D. and Tissières, A. (1954). *Biochem. J.* **57**, xxix–xxx.
Keister, D. L. and Minton, N. Y. (1969). *Biochemistry*, **8**, 167–173.
Kendrew, J. C. (1961). *Sci. Amer.* **205**, 96–110.
Ketchem, P. A. and Holt, S. C. (1970). *Biochim. et Biophys. Acta* **196**, 141–161.
Kiesow, L. (1964). *Proc. Nat. Acad. Sci. U.S.* **52**, 980–988.
Kiesow, L. (1967). In "Current Topics in Bioenergetics". Vol. 2 (Sanadi, D. R., ed.). New York, Academic Press, pp. 195–233.
Kihara, T. and Chance, B. (1969). *Biochim. et Biophys. Acta* **189**, 116–124.
Kihara, T. and Dutton, P. L. (1970). *Biochim. et Biophys. Acta* **205**, 196–204.

Kijimoto, S. (1968a). *Ann. Rep. Biol. Works, Osaka Univ.* **16**, 1–18.
Kijimoto, S. (1968b). *Ann. Rep. Biol. Works Osaka Univ.* **16**, 19–34.
Kikuchi, G. and Motokawa, Y. (1968). "Osaka", pp. 174–181.
Kikuchi, G., Saito, Y. and Motokawa, Y. (1965). *Biochim. et Biophys. Acta* **94**, 1–14.
Klemme, J. H. and Schlegel, H. G. (1969). *Arch. Mikrobiol.* **68**, 326–354.
Knobloch, K., Ishaque, M. and Aleem, M. I. H. (1970). *Arch. Mikrobiol.* **67**, 114–125.
Knowles, C. J. and Redfearn, E. R. (1968). *Biochim. et Biophys. Acta* **162**, 348–355.
Knowles, C. and Smith, L. (1970). *Biochim. et Biophys. Acta* **197**, 152–160.
Kodama, T. and Mori, T. (1969b). *J. Biochem.* (Tokyo) **65**, 621–628.
Kodama, T. and Shidara, S. (1969). *J. Biochem.* (Tokyo) **65**, 351–360.
Kodama, T., Shimoda, K. and Mori, T. (1969). *Plant Cell Physiol.* **10**, 855–865.
Kraayenhof, R. (1970). *Fed. Europ. Biochem. Soc. Lett.* **6**, 161–165.
Kraut, J., Singh, S. and Alden, R. A. (1968). "Osaka", pp. 252–254.
Krejcarek, G. E., Turner, L. and Dus, K. (1971). *Biochem. Biophys. Res. Commun.* **42**, 983–991.
Kröger, A. and Dadák, V. (1969). *Europ. J. Biochem.* **11**, 328–340.
Kubowitz, F. and Haas, E. (1932). *Biochem. Z.* **255**, 247–277.
Kufe, D. W. and Howland, J. L. (1968). *Biochim. et Biophys. Acta* **153**, 291–293.
Kusaka, T., Sato, R. and Shoji, K. (1964). *J. Bacteriol.* **87**, 1383–1388.
Lam, Y. and Nicholas, D. J. D. (1969). *Biochim. et Biophys. Acta* **180**, 459–472.
Lanyi, J. L. (1968). *Arch. Biochem. Biophys.* **128**, 716–724.
Lanyi, J. K. (1969). *J. Biol. Chem.* **244**, 2864–2869.
Lascelles, J. (1968). *Adv. in Microbiol. Physiol.* **2**, 1–39.
Lazzarini, R. A. and Atkinson, D. E. (1961). *J. Biol. Chem.* **236**, 3330–3335.
Lees, H. and Simpson, J. R. (1957). *Biochem. J.* **65**, 297–305.
Le Gall, J. and Bruschi-Heriaud, M. (1968). "Osaka", pp. 467–472.
Le Gall, J., Mazza, G. and Dragoni, N. (1965). *Biochim. et Biophys. Acta* **99**, 385–387.
Lemberg, R. (1962). *Nature* **193**, 373–374.
Lenhoff, H. M. and Kaplan, N. O. (1953). *Nature* **172**, 730–731.
Lenhoff, H. M. and Kaplan, N. O. (1956). *J. Biol. Chem.* **220**, 967–982.
Lenhoff, H. M., Nicholas, D. J. D. and Kaplan, N. O. (1956). *J. Biol. Chem.* **220**, 983–995.
Levin, K. (1961). *Acta Chem. Scand.* **15**, 1739–1746.
Lindenmayer, A. and Smith, L. (1964). *Biochim. et Biophys. Acta* **93**, 445–461.
Loach, P. A. and Sekura, D. L. (1968). *Biochemistry* **7**, 2642–2649.
London, J. (1963). *Science* **140**, 409–410.
Lukoyanova, M. A. and Taptykova, S. D. (1968). *Biokhimiya* **33**, 888–894.
Lyric, R. M. and Suzuki, I. (1970). *Canad. J. Biochem.* **48**, 334–343.
McKenna, E. J. and Coon, J. (1970). *J. Biol. Chem.* **245**, 3882–3889.
MacLennan, D. H., Tzagoloff, A. and McConnell, D. G. (1967). *Biochim. et Biophys. Acta* **131**, 59–80.
Margoliash, E. (1962). *J. Biol. Chem.* **237**, 2161–2174.
Margoliash, E. and Schejter, A. (1966). *Adv. in Protein Chemistry* **21**, 113–286.
Margoliash, E., Fitch, W. M. and Dickerson, R. E. (1968). *Brookhaven Sympos. in Biology* **21**, 259–305.
Matsubara, J. K. (1970). *Fed. Proc.* **29**, Abstr. 3739.
Matsubara, T. and Iwasaki, H. (1971). *J. Biochem.* **69**, 859–868.

Melik-Sarkisyan, S. S., Karapetyan, N. Y. and Kretovich, V. L. (1969). *Dokl. Akad. Nauk. SSSR* **188**, 930–933. *Chem. Abstr.* **72**(6) 51775.
Meyer, T. E., Bartsch, R. G., Cusanovich, M. A. and Mathewson, H. J. (1968). *Biochim. et Biophys. Acta* **153**, 854–861.
Miki, K. and Okunuki, K. (1969a). *J. Biochem.* **66**, 831–843.
Miki, K. and Okunuki, K. (1969b). *J. Biochem.* **66**, 845–854.
Miki, K., Sekuzu, I. and Okunuki, K. (1967). *Ann. Rep. Biol. Works Fac. Sci., Osaka Univ.* **15**, 33–58.
Milhaud, G., Aubert, J. P. and Millet, J. (1958). *C.R. Acad. Sci. Paris* **246**, 1766–1769.
Mirsky, R. (1969). *Biochemistry* **8**, 1164–1169.
Miura, T. and Mizushima, S. (1968). *Biochim. et Biophys. Acta* **150**, 159–161.
Miyata, M. and Mori, T. (1968). *J. Biochem.* **64**, 849–861.
Miyata, M. and Mori, T. (1969). *J. Biochem.* **66**, 463–471.
Mizushima, S. and Arima, K. (1960c). *J. Biochem.* **48**, 205–213.
Mizushima, S., Miura, T. and Ishida, M. (1966). *J. Biochem.* **60**, 256–261.
Mizushima, S., Mitral, T. and Ishida, M. (1967). *J. Biochem.* **61**, 146–148.
Mizushima, S., Oka, T. and Arima, K. (1960). *J. Biochem.* **48**, 205–213.
Mok, T. C. K., Rickard, P. A. D. and Moss, F. J. (1969). *Biochim. et Biophys. Acta* **172**, 438–449.
Mori, T. and Hirai, K. (1968). "Osaka", pp. 681–690.
Moriarty, D. J. W. and Nicholas, D. J. D. (1969). *Biochim. et Biophys. Acta* **184**, 114–123.
Moriarty, D. J. W. and Nicholas, D. J. D. (1970). *Biochim. et Biophys. Acta* **216**, 130–138.
Morita, S. (1960). *J. Biochem.* (Tokyo) **48**, 870–873.
Morita, S. (1968). *Biochim. et Biophys. Acta* **153**, 241–247.
Morita, S. and Conti, S. F. (1963). *Arch. Biochem. Biophys.* **100**, 302–307.
Morita, S., Edwards, M. and Gibson, J. (1965). *Biochim. et Biophys. Acta* **109**, 45–58.
Morton, R. K. and Shepley, K. (1965). *Biochim. et Biophys. Acta* **96**, 349–356.
Moss, F. (1952). *Aust. J. Expt. Biol. Med. Sci.* **30**, 531–540.
Moss, F. (1954). *Aust. J. Expt. Biol. Med. Sci.* **32**, 571–575.
Moss, F. (1956). *Aust. J. Expt. Biol. Med. Sci.* **34**, 395–405.
Moss, F. (1971). Private communication.
Moss, F. and Tchan, Y. T. (1958). *Proc. Linnean Soc. N.S.W.* **83**, 161–164.
Moss, T. H., Bearden, A. J., Bartsch, R. G. and Cusanovich, M. A. (1968). *Biochemistry* **7**, 1583–1590.
Muller-Eberhard, Y., Liem, H. H., Yu, C-A. and Gunsalus, J. C. (1969). *Biochem. Biophys. Res. Commun.* **35**, 229–235.
Munoz, E., Freer, J. H., Ellar, D. J. and Salton, M. R. J. (1968a). *Biochim. et Biophys. Acta* **150**, 531–533.
Munoz, E., Nachbar, M. S., Schor, M. R. and Salton, M. R. J. (1968b). *Biochem. Biophys. Res. Commun.* **32**, 539–546.
Murray, R. G. E. and Watson, S. W. (1965). *J. Bacteriol.* **89**, 1594–1609.
Nachbar, M. S. and Salton, M. R. J. (1970). *Biochim. et Biophys. Acta* **223**, 309–320.
Nagata, J., Yamanaka, T. and Okunuki, K. (1970). *Biochim. et Biophys. Acta* **221**, 668–671.
Nagata, Y., Shibiya, I. and Maruo, B. (1967). *J. Biochem.* **61**, 623–632.
Negelein, E. and Gerischer, W. (1934). *Biochem. Z.* **268**, 1–7.

Neumann, N. P. and Burris, R. H. (1959). *J. Biol. Chem.* **12**, 3286–3290.
Newton, J. W. and Kamen, M. D. (1956). *Biochim. et Biophys. Acta* **21**, 71–80.
Newton, J. W. and Kamen, M. D. (1961). In "The Bacteria" (I. C. Gunsalus and R. J. Stanier, eds.) Vol. II pp. 397–423. Academic Press, New York.
Newton, N. (1967). *Biochem. J.* **105**, 21C.
Newton, N. (1969). *Biochim. et Biophys. Acta* **185**, 316–331.
Niederpruem, D. J. and Hackett, D. P. (1961). *J. Bacteriol.* **81**, 557–563.
Nishimura, M., Roy, S. B., Schleyer, H. and Chance, B. (1964). *Biochim. et Biophys. Acta* **88**, 251–266.
Norton, J. E., Bulmer, G. S. and Sokatch, J. R. (1963). *Biochim. et Biophys. Acta* **78**, 136–147.
Oelze, J. and Drews, G. (1969a). *Arch. Mikrobiol.* **69**, 12–19.
Oelze, J. and Drews, G. (1969b). *Biochim. et Biophys. Acta* **173**, 448–455.
Oelze, J., Biedermann, M. and Drews, G. (1969). *Biochim. et Biophys. Acta* **173**, 436–447.
Okayama, S., Kakuno, T. and Horio, T. (1970). *J. Biochem.* **68**, 19–29.
O'Kelley, J. C. and Nason, A. (1970). *Biochim. et Biophys. Acta* **205**, 426–436.
O'Kelley, J. C., Becker, G. E. and Nason, A. (1970). *Biochim. et Biophys. Acta* **205**, 409–425.
Olson, J. M. and Chance, B. (1960a,b). *Arch. Biochem. Biophys.* **88**, 26–39; 40–53.
Olson, J. M. and Nadler, K. D. (1965). *Photochem. Photobiol.* **4**, 783–791.
Olson, J. M. and Clayton, R. K. (1966). *Photochem. Photobiol.* **5**, 655–660.
Olson, J. M. and Shaw, E. K. (1969). *Photosynthetica* **3**, 288–290.
Olson, J. M., Caroll, J. W., Clayton, M. L., Gardner, G. M., Linkins, A. E. and Moreth, C. M. C. (1969). *Biochim. et Biophys. Acta* **172**, 338–339.
Omura, T. and Sato, R. (1964). *J. Biol. Chem.* **239**, 2379–2385.
Orgel, L. E. (1966). In "An Introduction to Transition Metal Chemistry". Methuen, London: John Wiley, N.Y.
Orlando, J. A. and Maes, A. A. (1967). *Biochim. et Biophys. Acta* **140**, 459–467.
Orlando, J. A. and Horio, T. (1963). *Biochim. et Biophys. Acta* **50**, 367–369.
Orme, T. W., Revsin, B. and Brodie, A. F. (1969). *Arch. Biochem. Biophys.* **134**, 172–179.
Ozols, J. and Strittmatter, P. (1969). *J. Biol. Chem.* **244**, 6611–6619.
Paléus, S. and Tuppy, H. (1959). *Acta Chem. Scand.* **13**, 641–646.
Parson, W. W. (1968). *Biochim. et Biophys. Acta* **153**, 248–259.
Parson, W. W. (1969a). *Biochim. et Biophys. Acta* **189**, 384–396.
Parson, W. W. (1969b). *Biochim. et Biophys. Acta* **189**, 397–403.
Parson, W. W. and Case, G. D. (1970). *Biochim. et Biophys. Acta* **205**, 232–245.
Paul, K. G. (1951). *Acta Chem. Scand.* **5**, 389–405.
Peck, H. D. Jr. (1968). *Ann. Rev. Microbiol.* **22**, 489–518.
Peeters, T. and Aleem, M. I. H. (1970). *Arch. Mikrobiol.* **71**, 319–330.
Peisach, J. and Blumberg, W. (1970). In "Structure and Function of Macromolecules and Membranes" (B. Chance, C. P. Lee and T. Yonetani, eds.), Academic Press, New York.
Perlish, J. S. and Eichel, H. J. (1971). *Biochem. Biophys. Res. Commun.* **44**, 759–1014.
Peterson, J. A. (1970). *J. Bacteriol.* **103**, 714–721.
Peterson, J. A., Basa, D. and Coon, M. J. (1966). *J. Biol. Chem.* **241**, 5162–5164.
Peterson, J. A. and Dutton, L. (1971). Cited by Ishimura *et al.* (1971).
Pfennig, N. (1967). *Ann. Rev. Microbiol.* **21**, 285–324.
Phillips, P. G., Revsin, B., Drell, E. C. and Brodie, A. F. (1970). *Arch. Biochem. Biophys.* **139**, 59–66.

Porra, R. J. and Lascelles, J. (1965). *Biochem. J.* **94**, 120–126.
Postgate, J. R. (1954). *Biochem. J.* **56**, P xi.
Postgate, J. R. (1956). *J. gen. Microbiol.* **14**, 545–572.
Postgate, J. R. (1961). "Haematin Enzymes", pp. 407–418.
Postgate, J. R. and Campbell, L. L. (1966). *Bacteriol. Rev.* **30**, 732–738.
Prabhakararao, K. and Nicholas, D. J. D. (1969). *Biochim. et Biophys. Acta* **180**, 253–263.
Pretlow, T. P. and Sherman, F. (1967). *Biochim. et Biophys. Acta* **148**, 629–644.
Ramirez, J. and Smith, L. (1968). *Biochim. et Biophys. Acta* **153**, 466–475.
Reaveley, D. A. and Rogers, H. J. (1969). *Biochem. J.* **113**, 67–79.
Reed, D. W. (1969). *J. Biol. Chem.* **244**, 4936–4941.
Reed, D. W., Rareed, D. and Israel, H. W. (1970). *Biochim. et Biophys. Acta* **223**, 281–291.
Rees, M. and Nason, A. (1965). *Biochem. Biophys. Res. Commun.* **21**, 248–256.
Revsin, B. and Brodie, A. F. (1967). *Biochem. Biophys. Res. Commun.* **28**, 635–640.
Revsin, B. and Brodie, A. F. (1969). *J. Biol. Chem.* **244**, 3101–3104.
Revsin, B., Marquez, E. D. and Brodie, A. F. (1970a). *Arch. Biochem. Biophys.* **136**, 563–573.
Revsin, B., Marquez, E. D. and Brodie, A. F. (1970b). *Arch. Biochem. Biophys.* **139**, 121–129.
Richmond, M. and Kjeldgaard, N. O. (1961). *Acta Chem. Scand.* **15**, 226.
Romanoff, V. L., Taptykova, S. D. and Kretovich, V. L. (1970). *Dokl. Akad. Nauk. SSSR* **193**, 1189–1191.
Rothfield, C., Weiser, M. and Endo, A. (1969). In "Membrane Proteins", p. 27. Little, Brown & Co., Boston.
Ruiz-Herrera, J. and De Moss, J. A. (1969). *J. Bacteriol.* **99**, 720–729.
Sadana, J. C. and Morey, A. V. (1961). *Biochim. et Biophys. Acta* **50**, 153–163.
Salton, M. R. J. (1964). "The Bacterial Cell Wall", Elsevier, Amsterdam.
Salton, M. R. J. (1967). *Ann. Rev. Microbiol.* **21**, 417–442.
Salton, M. R. J. and Ehtisham-Ud-Din, A.F.M. (1965). *Aust. J. Exptl. Biol. Med. Sci.* **43**, 255–264.
Salton, M. R. J. and Nachbar, M. S. (1971). In "Anatomy and Biogenesis of Mitochondria and Chloroplasts" (N. K. Boardman, A. W. Linnane and R. M. Smillie, eds.) pp. 42–52. North Holland Publishing Co., Amsterdam.
Salton, M. R. J., Freer, J. H. and Ellar, D. J. (1968). *Biochem. Biophys. Res. Commun.* **33**, 909–915.
Sands, D. C., Gleason, F. H. and Hildebrand, D. C. (1967). *J. Bacteriol.* **94**, 1785–1786.
Sasaki, T., Motokawa, Y. and Kikuchi, G. (1970). *Biochim. et Biophys. Acta* **197**, 284–291.
Schaeffer, P. and Nisman, B. (1952). *Ann. Inst. Pasteur* **82**, 109–110.
Scholes, P. B. and Smith, L. (1968). *Biochim. et Biophys. Acta* **153**, 363–375.
Scholes, P. B., McLain, G. and Smith, L. (1971). *Biochemistry* **10**, 2072–2076.
Scocca, J. J. and Pinchot, G. B. (1968). *Arch. Biochem. Biophys.* **124**, 206–217.
Sedar, A. W. and Burde, R. M. (1965). *J. Cell. Biol.* **27**, 53–66.
Seibert, M. and Devault, D. (1970). *Biochim. et Biophys. Acta* **205**, 220–231.
Seibert, M., Dutton, P. L. and Devault, D. (1971). *Biochim. et Biophys. Acta* **226**, 189–192.
Sewell, D. L. and Aleem, M. I. H. (1969). *Biochim. et Biophys. Acta* **172**, 467–475.
Simakova, I. M., Lukoyanova, M. A., Birjuzova, V. I. and Gelman, N. S. (1968). *Biokhimiya* **33**, 1047–1052.

Simakova, I. M., Lukoyanova, M. A., Birjuzova, V. I. and Gelman, N. S. (1969). *Biokhimiya* **34**, 1271–1278.
Sinclair, P. R. and White, D. C. (1970). *J. Bacteriol.* **101**, 365–372.
Sletten, K., Dus, K., de Klerk, H. and Kamen, M. D. (1968). *J. Biol. Chem.* **243**, 5492–5506.
Smith, L. (1954). *Bacteriol. Revs.* **18**, 106–130.
Smith, L. (1955). "Methods in Enzymology", Vol. 2 (S. P. Colowick and N. O. Kaplan, eds.) pp. 732–740. Academic Press, New York.
Smith, L. (1957). In "Research in Photosynthesis" (H. Gafron, ed.) pp. 179–183. Interscience, New York.
Smith, L. (1961). In "The Bacteria". (I. C. Gunsalus and R. J. Stanier, eds.) Vol. II, pp. 365–396. Academic Press, N. Y. and London.
Smith, L. (1968). In "Biological Oxidations" (T. P. Singer, ed.) pp. 55–122. Interscience, New York.
Smith, L. and Baltscheffsky, J. (1959). *J. Biol. Chem.* **234**, 1575–1579.
Smith, L. and Ramirez, J. (1959). *Arch. Biochem. Biophys.* **79**, 233–244.
Smith, L. and White, D. C. (1962). *J. Biol. Chem.* **237**, 1337–1341.
Smith, L., Newton, N. and Scholes, P. (1966). "Hemes and Hemoproteins", pp. 395–404.
Smith, L., White, D. C., Sinclair, P. R. and Chance, B. (1970). *J. Biol. Chem.* **245**, 5096–5100.
Straat, P. A. and Nason, A. (1965). *J. Biol. Chem.* **240**, 1412–1426.
Sugimura, T., Okabe, K., Nagoa, M. and Gunge, N. (1966). *Biochim. et Biophys. Acta* **115**, 267–275.
Sugimura, Y. and Yakushiji, E. (1968). *J. Biochem.* **63**, 261–269.
Suh, B. and Akagi, J. M. (1969). *J. Bact.* **99**, 210–215.
Sutherland, I. W. (1963). *Biochim. et Biophys. Acta* **73**, 162–164.
Sutherland, I. W. and Wilkinson, J. F. (1963). *J. Gen. Microbiol.* **30**, 105–112.
Suzuki, H. and Iwasaki, H. (1962). *J. Biochem.* (Tokyo) **52**, 193–199.
Suzuki, Y., Morita, S. and Takamiya, A. (1969). *Biochim. et Biophys. Acta* **180**, 114–122.
Swank, R. T. and Burris, R. H. (1969). *Biochim. et Biophys. Acta* **180**, 473–489.
Sybesma, C. (1967). *Photochem. Photobiol.* **6**, 261–267.
Sybesma, C. (1969). *Biochim. et Biophys. Acta* **172**, 177–179.
Sybesma, C. and Fowler, C. E. (1968). *Proc. Nat. Acad. Sci. U.S.* **61**, 1343–1348.
Sybesma, C. (1969). *Biochim. et Biophys. Acta* **172**, 177–179.
Sybesma, C. and Kok, B. (1969). *Biochim. et Biophys. Acta* **180**, 410–413.
Taber, H. and Morrison, M. (1964). *Arch. Biochem. Biophys.* **105**, 367–379.
Tamiya, H. and Yamaguchi, S. (1933). *Acta Photochim.* (Japan) **17**, 233–244.
Taniguchi, S. and Itagaki, E. (1960). *Biochim. et Biophys. Acta* **44**, 263–279.
Taniguchi, S. and Kamen, M. D. (1963). *Biochim. et Biophys. Acta* **74**, 438–455.
Taniguchi, S. and Kamen, M. D. (1965). *Biochim. et Biophys. Acta* **96**, 395–428.
Tasaki, A., Otsuka, J. and Kotani, M. (1967). *Biochim. et Biophys. Acta* **140**, 284–290.
Thornber, J. P. (1970). *Biochemistry* **9**, 2688–2698.
Thornber, J. P., Olson, J. M., Williams, D. M., Clayton, M. L. (1969). *Biochim. et Biophys. Acta* **172**, 351–354.
Tikhonova, G. V., Simakova, I. M., Lukoyanova, M. A., Taptykova, S. D., Mikelsaar, K. L. N. and Gelman, N. S. (1970). *Biokhimiya* **35**, 1123–1130.
Tissières, A. (1952a). *Biochem. J.* **50**, 279–288.
Tissières, A. (1952b). *Nature* **169**, 880–881.

Tissières, A. (1956). *Biochem. J.* **64**, 582–589.
Tissières, A. (1961). "Haematin Enzymes", pp. 218–223.
Tissières, A. and Burris, R. H. (1956). *Biochim. et Biophys. Acta* **20**, 436–437.
Trudinger, P. A. (1961a,b) *Biochem. J.* **78**, 573–680, 680–686.
Trudinger, P. A. (1967). *Rev. Pure Appl. Chem.* **17**, 1–24.
Trudinger, P. A. (1969). *Adv. Microbiol. Physiol.* **3**, 111–158.
Trudinger, P. A. and Kelly, D. P. (1968). *J. Bacteriol.* **95**, 1962–1963.
Tsai, R., Yu, C. A., Gunsalus, I. C., Peisach, J., Blumberg, W., Orme-Johnson, W. H. and Beinert, H. (1970). *Proc. Nat. Acad. Sci. U.S.* **66**, 1157–1163.
Turner, L., Kennel, S. and Dus, K. (1971). Cited in Krejcarek *et al.* (1970).
Tuzimura, K. and Watanabe, I. (1964). *Plant Cell Physiol.* (Tokyo) **5**, 157–170.
Urry, D. W. (1966). "Hemes and Hemoproteins', pp. 435–442.
Urry, D. W. (1967). *J. Biol. Chem.* **242**, 4441–4448.
Van Gelder, B. F., Urry, D. W. and Beinert, H. (1968). "Osaka", pp. 335–339.
Van Gool, A. and Laudelut, H. (1966). *Biochim. et Biophys. Acta* **113**, 41–50.
Van Gool, H. and Laudelut, H. (1967). *J. Bacteriol.* **93**, 215–220.
Van Holde, K. E. and Baldwin, R. L. (1958). *J. Phys. Chem.* **62**, 734–743.
Vernon, L. P. (1956a). *J. Biol. Chem.* **222**, 1035–1044.
Vernon, L. P. (1956b). *J. Biol. Chem.* **222**, 1045–1049.
Vernon, L. P. (1968). *Bacteriol. Rev.* **32**, 243–261.
Vernon, L. P. and Kamen, M. D. (1953). *Arch. Biochem. Biophys.* **44**, 298–311.
Vernon, L. P. and Kamen, M. D. (1954). *J. Biol. Chem.* **211**, 643–662.
Vernon, L. P. and Mangum, J. H. (1960). *Arch. Biochem. Biophys.* **90**, 103–104.
Vinogradov, S. N. (1970). *Biopolymers* **9**, 507–509.
Vinogradov, S. N. and Harbury, A. H. (1967). *Biochemistry* **6**, 709–712.
Vredenberg, W. J. and Duysens, L. N. M. (1963). *Nature* **197**, 355–357.
Vredenberg, W. J. and Duysens, L. N. M. (1964). *Biochim. et Biophys. Acta* **79**, 456–463.
Walker, G. C. and Nicholas, D. J. D. (1961). *Biochim. et Biophys. Acta* **49**, 350–360.
Wallace, W. and Nicholas, D. J. D. (1969). *Biol. Rev.* **44**, 359–392.
Warburg, O. and Negelein, E. (1929). *Biochem. Z.* **214**, 64–100.
Warburg, O., Negelein, E. and Haas, E. (1933). *Biochem. Z.* **227**, 171–183.
Warme, P. K. and Hager, L. H. (1970a). *Biochemistry* **9**, 4237–4244.
Warme, P. K. and Hager, L. H. (1970b). *Biochemistry* **9**, 4244–4251.
Webster, D. A. and Hackett, D. P. (1966a). *Plant Physiol.* **41**, 599–605.
Webster, D. A. and Hackett, D. P. (1966b). *J. Biol. Chem.* **241**, 3308–3315.
Weibull, C. (1953). *J. Bacteriol.* **66**, 696–702.
Weibull, C. and Bergström, L. (1958). *Biochim. et Biophys. Acta* **30**, 340–351.
Whale, F. R. and Jones, O. T. G. (1970). *Biochim. et Biophys. Acta* **223**, 146–157.
White, D. C. (1962). *J. Bacteriol.* **83**, 851–859.
White, D. C. (1967). *J. Bacteriol.* **93**, 567–573.
White, D. C. and Smith, L. (1962). *J. Biol. Chem.* **237**, 1032–1036.
White, D. C. and Sinclair, P. R. (1970). *Advances in Microbiol. Physiol.* **5**, 173–211.
Williams, F. R. and Hager, L. P. (1960). *Biochim. et Biophys. Acta* **38**, 566–567.
Williams, J. P., Davidson, J. T. and Peak, H. D. (1964). *Bacteriol. Proc.* **1964**, 110–111.
Wimpenny, J. W. T., Ranlett, M. and Gray, C. T. (1963). *Biochim. et Biophys. Acta* **73**, 170–172.
Wimpenny, J. W. T. and Cole, J. A. (1968). *Biochim. et Biophys. Acta* **148**, 233–242.
Wrigley, C. W. and Linnane, A. W. (1961). *Biochem. Biophys. Res. Commun.* **4**, 66–70.

Yagi, T. (1970). *J. Biochem.* **68**, 649–657.
Yagi, T., Honya, M. and Tamiya, N. (1968). *Biochim. et Biophys. Acta* **153**, 699–705.
Yakushiji, E., Sugimura, Y., Sekuzu, I., Morikawa, I. and Okunuki, K. (1960). *Nature* **185**, 105–106.
Yamagutchi, S. (1937). *Botan. Mag.* (Tokyo) **51**, 457.
Yamagutchi, T., Tamura, G. and Arima, K. (1966). *Biochim. et Biophys. Acta* **124**, 413–414.
Yamanaka, T. (1963). *Ann. Rep. Sci. Works Fac. Sci., Osaka Univ.* **11**, 77–115.
Yamanaka, T. and Okunuki, K. (1962). *Biochim. et Biophys. Acta* **59**, 755–756.
Yamanaka, T. and Okunuki, K. (1963a). *Biochim. et Biophys. Acta* **67**, 379–393.
Yamanaka, T. and Okunuki, K. (1963b). *Biochim. et Biophys. Acta* **67**, 394–406.
Yamanaka, T. and Okunuki, K. (1963c). *Biochim. et Biophys. Acta* **67**, 407–416.
Yamanaka, T. and Okunuki, K. (1963d). *Biochem. Z.* **338**, 62–72.
Yamanaka, T. and Okunuki, K. (1968a). "Osaka", pp. 390–403.
Yamanaka, T. and Okunuki, K. (1968b). *J. Biochem.* **63**, 341–346.
Yamanaka, T. and Okunuki, K. (1970). *Biochim. et Biophys. Acta* **220**, 354–356.
Yamanaka, T., Ota, A. and Okunuki, K. (1961). *Biochim. et Biophys. Acta* **53**, 294–308.
Yamanaka, T., Kijimoto, S. and Okunuki, K. (1963a). *J. Biochem.* **53**, 256–259.
Yamanaka, T., Okunuki, K. and Horio, T. (1963b). *Biochim. et Biophys. Acta* **73**, 165–167.
Yamanaka, T., Miki, T. and Okunuki, K. (1963). *Biochim. et Biophys. Acta* **77**, 654–656.
Yamanaka, T., Takenami, S., Akiyama, N. and Okunuki, K. (1971). *J. Biochem.* **70**, 349–358.
Yamashita, J. and Kamen, M. D. (1969). *Biochem. Biophys. Res. Commun.* **34**, 418–425.
Yamashita, J., Kamen, M. D. and Horio, T. (1969). *Arch. Microbiol.* **66**, 304–314.
Yaoi, H. and Tamiya, H. (1928). *Proc. Imp. Acad. Japan* **4**, 436–439.
Yates, M. G. (1970a). *FEBS Letters* **8**, 271–285.
Yates, M. G. (1970b). *J. Gen. Microbiol.* **60**, 393–401.
Yças, M. (1956). *Expt. Cell Res.* **11**, 1–6.
Yong, F. C. and King, T. E. (1970). *J. Biol. Chem.* **245**, 1331–1335.
Yu, C-A. and Gunsalus, I. C. (1970a). *Bacteriol. Proc.* **1970**, 140.
Yu, C-A. and Gunsalus, I. C. (1970b). *Biochem. Biophys. Res. Commun.* **40**, 1431–1436.
Yu, C-A., Ganguli, B. N., Bartholomaus, R. C. and Gunsalus, I. C. (1969). *Bact. Proc.* 1969, 127.

Chapter VII

The cytochromes in the respiratory chain

A. Introduction.
   1. General considerations.
   2. The respiratory chain.
   3. The position of the cytochromes in the mitochondrion.
   4. The respiratory control.
B. The stoichiometry of cytochromes in the respiratory chain.
C. The respiratory chain as a whole.
   1. Influence of water and some inhibitors.
   2. Phospholipids.
   3. Structural protein.
   4. Kinetics of the respiratory chain and percentage reduction of cytochromes.
D. The respiratory chain between cytochrome c and oxygen.
   1. Interaction of cytochrome c with cytochrome oxidase.
   2. Cytochrome c-cytochrome oxidase complexes.
   3. Interaction of cytochromes c and a.
   4. The specificity of cytochrome c structure for its interaction with cytochrome oxidase.
   5. Inhibitors of cytochrome c oxidase.
   6. Oxidative phosphorylation on site 3.
E. The respiratory chain between cytochromes b and c.
   1. The bc_1 complex.
   2. Inhibitors of the respiratory chain between cytochromes b and c_1.
   3. NADH- and succinate cytochrome c reductases and oxidases.
F. The antimycin-insensitive cytochrome reductases.
   1. In mitochondria.
   2. Microsomal cytochrome reductases.
   3. The role of coenzyme Q (ubiquinone) and of non-haem-iron protein.
   4. Inhibitors of cytochrome reductases.
G. Indirect inhibition of the respiratory chain.
H. Mechanism of electron transfer and energy conservation.
   References.

A. INTRODUCTION

1. General considerations

So far in this book our attention has been centred at the chemical structure of the individual cytochromes and their classification according to chemical and physical criteria, with occasional glimpses at their biological significance and connections between structure and function. Only with regard to the problem of photosynthesis (Chapters IVF, VC and VI) and to the role of cytochromes in anaerobic bacterial reactions (Chapter VI) has an attempt at integration into biologically functioning systems been made. An attempt at such an integration of the cytochromes in cell respiration has been left to this chapter. It should be realized that it requires a far more complex integration than that attempted in the previous chapters. No attempt is being made to write a complete monograph on cellular respiration, and as the title indicates the centre point of vision remains the cytochromes. Yet it is impossible to neglect the whole complexity of that subject. We must be concerned with the whole interaction of the cytochromes with one another, their interaction with oxygen on the one hand and with substrates on the other through cytochrome reductases, and while we draw a line between these and the dehydrogenases which do not interact with cytochromes, this limitation is in itself somewhat arbitrary. All this, however, is only a small part of the complexity. Firstly the biological problem is not only one of the biological catalysis, but one of multienzyme system and their cybernetic control. Thus no consideration of cytochrome activities in cellular respiration can be considered without their role in energy conservation and oxidative phosphorylation. In intact mitochondria respiration and oxidative phosphorylation are coupled processes. It is also impossible to exclude the consideration of the structure of mitochondria and their membranes, and finally problems of ion transport through these membranes and also the fact that some of the substrates are anions. All this together forms a multidimensional problem of immense complexity, a complexity with which even our most refined biophysical methods cannot deal adequately, nor are computers so far of great help when we know all too well that the data which we can feed into them are pitifully inadequate. A shrewd study of the preceding chapters will reveal this inadequacy.

This is not written to underestimate the great advances which have been made in this field, but to set them in the right perspective of the immensity of the task. This in fact, is a substantial part of biology as a whole. The modern overstress on the genetic problems themselves—which cannot of course be omitted entirely from the picture—has perhaps tended to overshadow the real significance of the problems of biochemistry proper, to establish an integrated relationship between genetic control, intracellular microscopic and submicroscopic structure, and the structure of the individual catalysts, not only as static complex haemoprotein structures, but in their conformational fluidity in interaction within the whole system. Obviously we stand at the very beginning of fully realizing this complexity and this chapter is an attempt to document this.

What has been said about the phenomena is equally true with regard to theories which attempt to bring the great variety of these phenomena under a single aspect. At the very best they are gross oversimplifications of the true complexity. Thus the single respiratory chain of Chance (see subchapter 4) with kinetics based on the collision of individual cytochromes in that chain, had soon to be modified by the realization that the ratio of cytochromes is not as stoichiometric as this picture foresaw, and that the kinetics of cytochrome interactions demand a more complex picture of "branching" chains. Serious doubts have also arisen as to the adequacy of a theory of cytochromes colliding by rotating at fixed points in the chain. On the other hand, the attempts of Green and his co-workers who originally assumed four "complexes" in simple stoichiometry, connected by "mobile" carriers such as cytochrome c and coenzyme Q has undergone so important qualification and modification that today its theoretical importance is far less than its value for preparative purposes. Not only is there no invariable stoichiometric relation of the complexes, there is little reliable evidence for the existence of these complexes as subunits of the respiratory chain, and it is doubtful whether the "mobile" carriers are truly mobile. At present no theory of electron transport is adequate to explain all the phenomena, and the same can be said about the theories of oxidative phosphorylation and energy conversion. It is not unexpected that we stand only at the beginning of the correlation between the structure of individual haemoproteins and their relation to submicroscopic and microscopic structure of membranes and cellular organelles. The problems connected with the biosynthesis of the cytochromes in the cell and in organelles has been relegated to a separate chapter (Chapter IX).

2. The respiratory chain

The mitochondrial chain can be depicted with some simplification and particular stress on the role of the cytochromes in it as depicted in Fig. 9.

The field has been reviewed by Slater (1958, 1966b), Green (1959), Klingenberg and Bücher (1960), Nicholls (1963), Ernster and Lee (1964), Racker (1965), Mahler and Cordes (1966), Green and Silman (1967), Chance (1965a) and Chance et al. (1967a, 1968a). The Keilin-Hartree heart muscle preparation (King, 1967a) is a particulate suspension of physically disintegrated mitochondrial membranes. The four complexes I-IV, which can be separated will be discussed below. Only complexes III and IV are part of the cytochrome system, whereas I and II function as

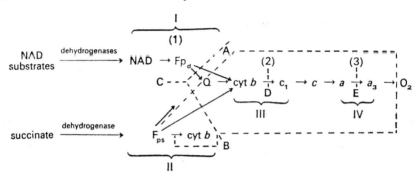

FIG. 9. The respiratory chain.
Cytochromes underlined and in space surrounded by interrupted line.
I, II, III, IV the four complexes that can be isolated (1), (2), (3) the three sites of phosphorylation.
A–E points of attack by inhibitors.
A: barbiturates, rotenone, piericidin, etc.
B: thenoyltrifluoroacetone, etc.
C: connected indirectly through oxidative phosphorylation, only in coupled mitochondria; oligomycin, atractylate, HN_3.
D: antimycin A. quinoline-N-oxides, substituted naphthoquinones, BAL.
E: HCN, HN_3.
Q: ubiquinone Q_{10}.

cytochrome reductases. Under certain conditions, however, cytochrome b can be found an integral part of the succinic-dehydrogenase system, but this is probably due to an alteration of the usual attachment of cytochrome b, the role of which in the respiratory chain has not yet been fully elucidated. It is not clear whether cytochrome c acts as a mobile carrier, or only when attached to complexes III or IV. There appears little if any sound evidence for the complex scheme of Jacobs and co-workers (Jacobs and Andrews, 1965, 1967; Jacobs et al., 1968) which is based on unphysiological reductants (tetrachloroquinols) and postulates two oxidases, one reaching directly with cytochrome c_1, the other with cytochrome c. The role of ubiquinone (Q) is not yet fully established. It is not clear whether it is attached to complex III or to complex I, and it can apparently be by-passed (see below). Still less certain is whether non haem iron-proteins are essential in complexes I, II and particularly in III. The figure also shows the three sites at which oxidative phosphorylation ($ADP + P_i \rightarrow ATP$) coupled to electron transport occurs (see A4). Much information on the respiratory chain has been derived from the study of inhibitors, although as will be seen below, there are few if any inhibitors the action of which is entirely restricted to a single site (subchapters D5, E5 and F5). Barbiturates, rotenone and piericidin mainly inhibit the oxidation of complex I, thenoyltrifluoroacetone that of complex II (thus cytochrome reductases).

The interaction between cytochromes b and c_1 is the main site of inhibition by antimycin A, 2-n-heptyl-4-hydroxyquinoline N-oxide, some substituted naphthoquinones and dimercaptopropanol plus oxygen. Cyanide inhibits the interaction of cytochromes a and a_3, although the mechanism of this inhibition is less simple than had previously been assumed. The situation is still more complex with other inhibitors of the oxidase, notably hydrazoic acid (HN_3) which inhibits phosphorylation-coupled cellular respiration far more than respiration in the absence of respiratory control. Azide inhibition thus partially resembles that by such inhibitors as oligomycin which act on cellular respiration in an indirect way through oxidative phosphorylation. Other substances, called uncouplers, e.g. dinitrophenols, or ionophoretic antibiotics, such as valinomycin, the primary action of which is on cation transport through the mitochondrial membrane, can accelerate the rate of respiration.

3. The position of the cytochromes in the mitochondrion

It has been recognized for a considerable time that the mitochondrion, an organelle of animal, plant and fungal cells, is the "powerhouse of the cell" and the site of cellular respiration. The history of this discovery is given by Keilin and Slater (1953), Cleland and Slater (1953), Palade (1952, 1956) and Sjöstrand (1953). In bacteria, the cytoplasmic cell membrane fulfils the same function. The large amount of literature on the ultrastructure of the mitochondria cannot be quoted here and the reader must be referred to a number of reviews (Lehninger, 1962, 1964, 1967; Chance et al., 1967b; Parsons et al., 1967; Ernster and Drahota (eds., 1969); Racker, 1970; see also Burnstein and Racker, 1970; Hendler, 1971). In this difficult field between molecular and microscopic structure, finality has not yet been reached, and serious errors have been committed by premature assignment of structures seen in the electron microscope to certain mitochondrial functions. Several methods of preparation of mitochondria have been collected in Methods in Enzymology, vol. 10 (1967) for mitochondria of animals (pp. 3–7, 75–122), plants (pp. 123–135) and fungi (pp. 135–147), (see also Schneider and Hogeboom, 1950; Hogeboom, 1955; Ohnishi et al., 1966; Sharp et al., 1967 and Weiss et al., 1970). Yeast mitochondrial inner membrane contains cytochrome c oxidase, cytochrome b, cytochrome c reductases and the oligomycin-sensitive ATPase (Tzagoloff, 1969; Tzagoloff et al., 1968b). By various methods (e.g. sonication and treatment with digitonin) smaller "submitochondrial" particles have been prepared (Methods in Enzymology **10**, 179–212). Mitochondria have an outer and an inner mitochondrial membrane, the latter being invaginated to form the cristae which surround the matrix. The respiratory chain is known to be present in this inner membrane and cytochromes form 5–10% of the inner membrane surface (Klingenberg,

1967) with one cytochrome a per 50,000 Å^2. Methods for the separation of the two membranes have been described by Parsons et al. (1966), Parsons and Yano (1967), Parsons and Williams (1967), Kopaczyk (1967), Sottocasa (1967), Sottocasa et al. (1967a) and Craven et al. (1969), see also Hoppel and Cooper (1969), Brunner and Bygrave (1969) and Parkes and Thompson (1970). The "elementary particles" observed in the electron microscope and at first assumed to be the respiratory subunits (Fernandez-Moran et al., 1964), later called "headpieces", inner membrane subunits, or simply "knobs", do not in fact contain cytochromes, but the oligomycin-sensitive ATPase which is of importance for energy conservation (Chance et al., 1964; Stasny and Crane, 1964; Racker et al., 1965; Kagawa and Racker, 1966; Oda, 1968). Green et al. later extended the concept of "elementary particle" to include the base piece, a part of the inner membrane, the stalk and the headpiece as a "tripartite particle" and essential subunit of the membrane (Green et al., 1965; Green and Hechter, 1965; Green and Perdue, 1966; McConnell et al., 1966; Tzagoloff et al., 1967; Penniston and Green, 1968). Stoeckenius (1970, p. 64) however, concludes that there is little justification for this subunit theory, that the picture of an assembly of globular lipoprotein complexes may not be correct, and may have to be replaced by that of small segments of a somewhat modified (Danielli and Davson, 1935) type of phospholipid membrane with a low degree of order in the plane of the membrane; only in chloroplasts is there at present more reliable evidence for definite subunits. Wallach and Gordon (1968) develop a picture of a protein lattice penetrated by cylinders of phospholipid with contact both to polar and non-polar regions of the protein. The inner mitochondrial membrane also contains phosphorylating enzymes partly resolved by Linnane (1958), Linnane and Titchener (1960), Pullman et al. (1960), Kagawa and Racker (1966), cf. also Chance et al. (1967b) and Methods in Enzymology **10**, pp. 522–537, (1967).

In submitochondrial particles obtained by sonication Estabrook and Holowinski (1961), Ernster (1965) and Lee and Ernster (1966) hypothesized that they were turned "inside-out", i.e. that the side of the inner membrane facing the matrix was now facing outside (see also Kornberg and McConnell, 1971). This does not occur in digitonin treatment (Malviya et al., 1968). The matter is, however, not entirely clear (Stoeckenius, p. 70f) and recently, Chance et al. (1970c) concluded that sonicated submitochondrial particles have their respiratory chains partly outside and partly inside. This would be easy to understand if only some vesicles are turned inside-out on sonication and formation of new vesicles, but the authors conclude that all are equal in having partly inside and partly outside respiratory chains. It is not easy to understand how this could have arisen.

There has been a great deal of speculation with regard to the position of the cytochromes on the inner or outer side of the inner mitochondrial membrane, and whether all of them are on the same side of this membrane. On the basis of the chemiosmotic theory of Mitchell (1966a,b), electrons are transported through the membrane (see the next subchapter) so that some cytochromes may be on the inner, some on the outer side of the inner membrane. Prezbindowski et al. (1968) find cytochrome c_1 on the inside of cytochrome aa_3 on the outside of a fibrous layer. Nicholls et al. (1969) place cytochrome a_3 near the matrix side of the inner membrane and cytochrome c on the intracristal side, acting as a mobile carrier between inner and outer membranes, not between single assemblies of the inner membrane. Klingenberg (1970) comes to similar conclusions with regard to cytochromes a_3 and c, but also places cytochromes b and c_1 on the matrix side. The two-sided arrangement of respiratory carriers is said to be in agreement with the vectorial transfer of electrons and protons through the membrane postulated by Lundegårdh (1960), Robertson (1960) and Mitchell (1966b). Ferricyanide experiments indicated low accessibility of c_1 and b. These results differ from those of Tyler (1970) who concluded from experiments with ferricyanide that both cytochromes c and c_1 are on the outer cristae surface, accessible to ferricyanide, but that cytochromes aa_3 are on the inner surface so that electrons must pass the membrane twice from cytochrome aa_3 to cytochrome b. The concept raises, however, difficult problems (see e.g. the formulation of the Mitchell hypothesis in Slater, 1966b) which places cytochromes aa_3 and b on opposite sides of the membrane, and in later alterations of the theory (cf. Mitchell, 1970a), the findings do not agree and the evidence is not convincing (see also Chance et al., 1970c). Another unsolved problem is that the thinness of the inner membrane does not allow the active dimer of aa_3 unless they sit side by side horizontally in the membrane; a similar difficulty exists for the cytochrome bc_1 complex. This would require a horizontal mosaic with a certain degree of definite two-dimensional order allowing for aa_3 and bc_1 horizontal coupling. Finally, in certain, e.g. adrenal mitochondria, a P-450-dependent system (see Chapter IVD) and the cytochrome aa_3 system can compete with one another (see e.g. Sauer, 1970) and also for high energy intermediates (see Chapter VIIH). The nature of the EPR signals found in mitochondria and submitochondrial particles is not yet fully elucidated (Sands and Beinert, 1960).

4. The respiratory control

Again, an attempt to discuss this subject is beyond the purport of this monograph. The reader must be referred to textbooks (e.g. Mahler and Cordes, 1966) and a few essential papers and reviews such as that of Chance and Baltscheffsky (1958), Chance and Williams (1956a), Chance (1961b),

Lehninger and Wadkins (1962), Slater (1966b), Boyer (1967) and of Lardy and Ferguson (1969). In the remaining short paragraph we shall consider these phenomena only from the one-sided aspect of cellular respiration and the electron (or hydrogen) carriers themselves, the cytochromes. Oxidative phosphorylation and energy conservation is a phenomenon not restricted to the respiratory chain but has also been found in glycolytic and photosynthetic reactions; these will not be discussed here. The development of the subject and its state were excellently summed up by Slater (1966a,b; 1967b; Tager et al., 1966) with fair consideration being given to "chemiosmotic" and membrane-potential theory of Mitchell and the theory of energy-rich intermediates developed by Slater and Chance. Chance and Williams (1955a-d, 1956a) made the important discovery that respiration is greatly increased by the presence of ADP and Pi. The "coupled respiration" of state 3 greatly exceeds that in their absence (state 4). They called this phenomenon "respiratory control". This makes it necessary to study for each inhibitor (or accelerator) of respiration whether its effect is one directly exerted on the components of the respiratory chain, or indirectly through their effect on oxidative phosphorylation. There are finally substances called "uncouplers" which sometimes increase the rate of respiration even more than "coupling" with ADP + Pi does, e.g. in submitochondrial particles obtained by sonication (Vallin, 1968); the respiration in the submitochondrial particles is thus still under respiratory control (Lee and Ernster, 1965). Under other conditions uncouplers can also decrease succinate oxidation (Papa et al., 1969). The difference between the effect of these substances (e.g. dinitrophenols which "uncouple" and substances, like oligomycin, which inhibit oxidative phosphorylation) is clearly set out by Slater on pp. 358–360 of his review (see Fig. 10). Uncouplers like dinitrophenol and e.g. salicylic acid (Penniall, 1958) release cellular respiration from respiratory control and cause a hydrolysis of ATP by stimulating ATPase (Chance et al., 1963); oligomycin (see below) inhibits ATP formation and coupled respiration but has no effect on uncoupled respiration. The stoichiometry of uncouplers has been studied by Kurup and Sanadi

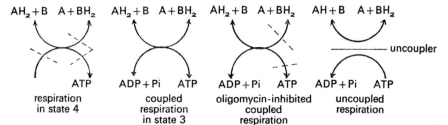

Fig. 10. Coupled, coupled-inhibited and uncoupled cellular respiration. (According to Slater, 1966b).

(1968). This fails, however, to explain the inhibition of respiration by uncouplers sometimes found.

B. THE STOICHIOMETRY OF CYTOCHROMES IN THE RESPIRATORY CHAIN

According to the theory of a linear respiratory chain by Chance and Williams (1955, 1956a) the ratio of cytochromes $a_3 : a : c : c_1 : b$ in the respiratory chain should be and was claimed to be $1 : 1 : 1 : 1 : 1$, making the ratios $(a + a_3)/c$ two and both the ratios $(a + a_3)/(c + c_1)$ and c_1/c unity. It has been pointed out by Lemberg (1969, p. 93 see also Lemberg et al., 1956) that the actual findings do not always agree with these ratios, if corrected for the presence of cytochromes in other parts of the cell and the use of more recent extinction coefficients. The ratios c_1/c is 0·5 rather than 1, and the ratio $(a + a_3)/c$ is not 2, but usually between 1 and 2. Two more recent papers confirm the lack of strict stoichiometry. Vanneste (1966a,b) finds the ratio $(a + a_3) : c : c_1 : b$ to be $4 : 2 : 1 : 2$ in ox heart mitochondria. Williams (1968) finds the ratio $(a + a_3) : c : c_1 : b$ $2 : 2 : 1 : 2$ in rat liver and ox heart but quite different ratios in other organs, e.g. in rat intestines $(a + a_3)/(c + c_1)$ 3·8 not 0·67 and in guinea pig liver 1·45 not 0·87 as found by Estabrook (1958). He found, however, more c_1 than previous workers, probably because his method is based on spectrophotometric data, not on separation (see also Koller, 1968). Klingenberg (1970b) reports a rather constant ratio of $a_3 : a : c_1 : b$ of $1 : 1 : 0·5 : 1$ in animal mitochondria with a greater variation of cytochrome c concentration. In fungal mitochondria e.g. in yeast cytochrome c was much higher, cytochromes b and particularly cytochromes a and a_3 lower, with a particularly high c/a ratio, > 10, in Neurospora. The ratio aa_3/b greatly varies on aeration (Biggs and Linnane, 1963). There appears also to be no simple stoichiometry between dehydrogenases and cytochromes.

In mitochondria of the skeletal muscle ratios of $(a + a_3) : (c + c_1) : b$ have been found to be $1 : 1·13 : 1·14$ for rat leg and $1 : 1·23 : 0·8$ for pigeon breast muscle, indicating a relatively low content of type a cytochromes (see also Evtodienko and Mokhova, 1969). Although the removal of cytochrome c in preparation of the "inner mitochondrial membrane" is expected the ratios of $(a + a_3) : (c + c_1) : b$ of $1 : 0·17 : 0·23$ to $0·4$ found by Kopaczyk (1967) indicate a surprisingly small amount of cytochromes b and c_1.

C. RESPIRATORY CHAIN AS A WHOLE

1. Influence of water and some inhibitors

Water is an essential milieu of the activity of the respiratory chain in spite of the fact that many of its reactions are of hydrophobic nature and phospholipid is also essential. Laser and Slater (1960) found 40–50% of inhibition of cellular respiration in heart muscle and bacteria and about 75% in yeast when water was replaced by 95–98% D_2O. The inhibition by heavy water on NADH- and succinate oxidation was confirmed by Estabrook et al. (1964) and Tyler and Estabrook (1966). This is perhaps of great importance in so far as it provides some counterbalance to the present stress on the hydrophobic nature of the inside of respiratory enzymes. It appears possible that chains of water molecules in between the hydrophobic parts may connect distant parts of the enzyme molecule by H-transporting water chains (cf. e.g. Bücher, 1953; Klotz et al., 1958). Chance and Spencer (1959) had previously found inhibition by glycerol, and the fact that the steady state reduction of the cytochromes remained unaltered suggested a non-specific effect on their interaction. D_2O increases the P/O ratio, but mainly or entirely owing to its effect on respiration, not on oxidative phosphorylation (Muraoka and Slater, 1968), perhaps by slowing down the hydrolysis of a hypothetical "energy-rich" intermediate $A \sim C^*$. Hyperosmolarity may decrease respiration by impeding the entry of substrate to their respective dehydrogenases (Atsmon and Davis, 1967). Zinc inhibits non-phosphorylating respiration (Nicholls and Malviya, 1968; see also Chistyakov et al., 1969).

2. Phospholipids

The presence of phospholipids is, however, also essential for cellular respiration. Their effects on cytochrome oxidase have been discussed in Chapter III A3 (see also Lemberg, 1969, pp. 59–60) and Chapter V B9. A good recent review is that of Chapman and Leslie (1970), particularly with regard to the role of phospholipids in mitochondrial membranes. The fact that the mitochondrial membranes are rich in phospholipids and that phospholipids are essential for cellular respiration has been recognized for many years (see Ball and Cooper, 1949, Edwards and Ball, 1954, Polonosky and Gay, 1956; Ball and Barrnett, 1957) but Ball and Cooper (1957) could remove 90% of the phospholipids by repeated treatment with deoxycholate without disturbing the binding of aa_3 and b. All mitochondria appear to contain about 27% total lipid. Of this more than 90% is phospholipid and of this P-choline and P-ethanolamine form about 77%, cardiolipin 20% and the others, particularly P-inositol, the remainder. The outer mitochondrial membrane is particularly rich in phospholipids (about 45% dry weight) and contains more P-inositol and far less cardiolipin than the inner membrane (see Ernster and Kuylenstierna, 1970). The various parts

---

* The symbol $\sim$ often called "squiggle", means a so-called "energy-rich" linkage releasing energy on hydrolysis.

of the respiratory chain have similar phospholipid concentrations (Brierley et al., 1962; Green and Fleischer, 1963). Microsomes contain 0·39 mg phospholipid per mg of protein of which P-choline and P-ethanolamine together form 80–90%, P-inositol and P-serine the remainder.

A protohaem-proteolipid containing 35% lipid has been extracted from heart microsomes (Shichi et al., 1965). Snake venom (Minakami et al., 1964) and lysolecithin (Honjo and Ozawa, 1968) inhibit mitochondrial respiration. Phospholipids also play a role in the reactivation of ATPase in submitochondrial particles (Bulos and Racker, 1968) and in the activity of cytochrome c reductases in the heart muscle (Badano et al., 1967).

The results of extraction of lipids with isooctane (Igo et al., 1959) cannot be taken as a reliable guide for measuring the effects of phospholipids on respiratory activity. Not only is ubiquinone removed by this procedure (see F3), but the solvent remaining in the preparation has itself an inhibitory effect (Pollard and Bieri, 1958) and there is also some irreversible effect on the organization of the respiratory chain (Redfearn et al., 1960). Extraction with 90% acetone (Lester and Fleischer, 1961; Brierley and Merola, 1962) similarly causes a requirement for added cytochrome c although cytochrome c is still present after the extraction. Richardson et al. (1963) found that up to 90% of the phospholipids could be removed without effect on the structural integrity, which was assumed to be maintained by the "structural protein".

3. Structural protein

In 1961 a colourless insoluble protein was isolated by Green and co-workers (Green et al., 1961; Criddle et al., 1961, 1962) from mitochondria. This represents about 40% of the total mitochondrial protein from ox heart and kidney and 65% from ox liver. Several methods for the preparation of structural protein have been described. Whereas those of Criddle et al. (1962) and Richardson et al. (1964) disagree in some properties, further purifications (Allman et al., 1967b; Criddle, 1967) have given more consistent results. Since it combines with cytochromes and some other constituents of the mitochondrial membrane to form soluble complexes it has been considered by Green and co-workers as "structural protein" (Silman et al., 1967; Rieske et al., 1967b).

Doubts have, however, been expressed by Conover and Barany (1966) and by Senior and MacLennan, (1970). Zalkin and Racker (1965) have found the coupling factor 4 which couples cellular respiration with oxidative phosphorylation very similar in its properties to structural protein. Bruni and Racker (1968) have shown that cytochrome b itself can play a structural role in the organization of the respiratory chain. Zahler et al. (1969) reported that the amount of structural protein correlates with matrix space rather than with mitochondrial membrane space and consists mainly of matrix protein. It is certainly not required for the activity of the respiratory chain (Kopaczyk et al., 1966; Green and Tzagoloff, 1966) nor does it seem essential for the maintenance of the tripartite membrane structure. Stoeckenius (1970, p. 55) concludes: "The existence of a layer of structural protein between the lipid and the enzyme proteins that would have specific binding sites for the enzyme proteins and assure their proper spatial arrangement is an attractive idea, but so far, remains unproven". Racker (1970, p. 167) states: "Possibly all proteins of the membrane have a structural as well as a catalytic role and the capabilities of self-assembly". It is possible, however, that a minor component of the membrane, possibly a gene product of mitochondrial DNA plays an important role in the initiation of membrane formation.

4. Kinetics of the respiratory chain and percentage reduction of cytochromes

The ingenious spectrophotometric methods of Chance and co-workers (Chance, 1952a,b, 1954, 1957, 1964, 1965c, 1967; Chance and Hess, 1959a; Chance and Sacktor, 1958; Chance et al., 1965; Chance and Schindler, 1965) have made it possible to measure the rate of individual cytochrome interactions in the respiratory chain and the percentage reduction of the cytochromes. There are, however, some difficulties in these estimations which have not yet been fully overcome. Calculating the state of reduction of cytochromes a and a_3 separately, while now possible in principle, has not yet been widely used and most workers rely on the estimation of ΔA 445-455 nm as measuring cytochrome a_3 and ΔA 605-590 nm as measuring cytochrome a. The need for correcting such measurements has been discussed in Chapter III B5 and more fully by Lemberg (1969, p. 95). The similarity of the absorption spectra of cytochromes c and c_1 raises another difficulty and most workers have been content to measure the degree of reduction of the sum of both (cf. e.g. Vázquez-Colón and King, 1967). A separation of c and c_1 now appears possible (cf. Chapter V, D), but has not found sufficient application. Finally difficulties with regard to cytochrome b (see Chapter IV and subchapter E5 below) have not yet been solved. It thus appears wise to accept the present data on the kinetics of the respiratory chain worked out by Chance et al. (1965), Chance and Schindler (1965) and Chance et al. (1967b) as provisional, see also Lee et al. (1965a) and Mokhova (1965) and, for the yeast Candida utilis, Wohlrab (1969a). The values given for the pseudo-first-order constants of the reductions of cytochromes a_3, a, c and c_1 in the phosphorylation-coupled respiratory chain ("state 3") in liver mitochondria are approximately >490 (probably 800), 360, 200 and 150 sec^{-1} respectively, with half-times of reduction of 1 msec., 2 msec. and 4 msec. respectively for a_3, a and c. The respiratory control by ADP increases the total respiratory oxidation greatly, but that of individual parts of the chain differentially, and "uncoupling" increases the rate by a further 10–30%. The rate of respiration after all ADP has become converted to ATP ("state 4") becomes so slow that Muraoka and Slater (1969) find the redox state of individual cytochromes now practically dictated by the thermodynamic equilibrium and no longer by a kinetic steady state. As expected, the percentage of reduction of the cytochromes nearest to oxygen is then lowest, that of the cytochromes nearest the substrate highest, i.e. in the order $a_3 > a > c > c_1 > b$. This is not necessarily true for the chain in phosphorylating mitochondria. Chance and Williams (1955c, 1956a) and Chance and Hess (1959a,b) have postulated that the points of the chain at which oxidative phosphorylation occurs correspond to the "crossover points", i.e. the points at which on addition of ADP a component on the oxygen side of the crossover point becomes more reduced, one on the substrate side more oxidized. They have supported this theory by computer solutions of differential equations (Chance et al., 1958; Chance, 1961a; Garfinkel et al., 1970). Several such crossover points have been recorded, e.g. between cytochromes a and c and c and b (Chance and Williams, 1955c, 1956a; Klingenberg and Kröger, 1967) and between NADH and flavoprotein. However, Muraoka and Slater (1969) have also found a crossover point between a and a_3 and under other conditions no crossover point between cytochromes (i.e. a phosphorylation site between a_3 and O_2) on ascorbate-TMPD reduction. They conclude that a crossover point does not necessarily identify a phosphorylation site (see below). Lemberg (1969, p. 96) had pointed out (see also Chance, 1967 and Lee et al., 1965a) that there was no definite evidence that cytochrome a was under all conditions more oxidized than cytochrome c so that under these conditions the chain appeared to be

$$c_1 \to c \to a_3 \to O_2$$
$$\updownarrow$$
$$a$$

(cf. also Yakushiji and Okunuki, 1940) rather than $c_1 \to c \to a \to a_3 \to O_2$. This is in agreement with recent findings of the redox potential for cytochrome a close to that of cytochrome c (Dutton et al., 1970).

Chance and Spencer (1959) discovered that the "steady state" of the respiratory chain could be "frozen in" by sudden cooling in liquid nitrogen and this method has been frequently used in studies by Chance, Bonner, Estabrook, Wilson and others. Physical alteration of the absorption, however, raises some problems for quantitative estimations, which have been partially overcome in the method of Elliott and Doebbler (1966).

Certain electron carriers, e.g. TMPD (tetramethyl-p-phenylene diamine), PMS (phenazine methosulphate) and tetrazolium salts can short-circuit parts of the respiratory chain and thus by-pass the sites at which some inhibitors react with the chain (see e.g. Mustafa et al., 1968 and Nachlas et al., 1960). This will be discussed below (D5, E5, F5).

D. THE RESPIRATORY CHAIN BETWEEN CYTOCHROME c AND OXYGEN

1. Interaction of cytochrome c with cytochrome oxidase

Reviews on this subject have been written by Slater (1958), Nicholls (1963), Yonetani and Ray (1965) and by Lemberg in the review on cytochrome oxidase (1969). In this book the subject has been touched upon in Chapter III, B3 in respect to the assay of the oxidase. Nowadays purified solubilized preparations of the oxidase and pure cytochrome c are available, and their interaction has been studied in great detail (see Chapter III B5, c,e,f). What remains uncertain, however, is whether this interaction is a good model for what happens in the respiratory chain reaction in the intact mitochondrion. This is said to contain an "endogenous" cytochrome c with some unique properties (Tsou, 1952). A distinction has been made between "closed" succinoxidase (or NADH oxidase) and "open succinoxidase (or NADH oxidase). In the "closed" system sufficient endogenous c has been left in the mitochondrion so that added, exogenous, cytochrome c stimulates oxidative activity little or not at all. Closed succinoxidase has been described by Keilin and Hartree (1938, 1940), Tsou (1952), Keilin and King (1960) and by Lee et al. (1965a); closed NADH-oxidase by Slater (1950a,b), Tsou (1952), Green et al. (1954), Estabrook and Mackler (1957), Smith and Camerino (1963) and Camerino and Smith (1964). "Open" oxidases require the addition of soluble cytochrome c and according to Smith and Camerino soluble cytochrome c does not react with endogenous c. It has been shown, however, that under certain conditions the rate of oxidation of exogenous c is as high as that of endogenous (Clark et al., 1954; Mackler and Penn, 1957; Estabrook, 1959). Cytochrome c can be extracted from the mitochondrion by so mild a reagent as 0·15 M KCl (Jacobs and Sanadi, 1960; Low and Vallin, 1963; Edwards and Criddle, 1966) and reincorporated without substantial loss of activity. A closed system is also converted to an open one by small amounts of bile salts, e.g. 0·1% of deoxycholate (Smith and Conrad, 1961; Estabrook, 1961; Smith and Camerino, 1963) or by 0·5 M guanidine (Orii and Okunuki, 1967). Phospholipase A releases cytochrome a from mitochondria (Ambe and Crane, 1958). Machinist et al. (1962) have proposed that cytochrome c is bound in a negatively charged pore in the mitochondrial membrane with the negatively charged phospholipids predominating at the surface of the pore.

The turnover rate of cytochrome c per mole haem a approaches values of the maximal rate of interaction between cytochromes a and a_3 (see Chapter

III, B3). It therefore appears likely that the observed differences between exogenous and endogenous cytochrome *c* may be not genuine differences between two types of compounds, but may be due to differences in membrane permeability (cf. Lee *et al.*, 1969a,b,c). The phenomena observed after sonication, and in liver mitochondria also after mechanical fragmentation (Muscatello and Carafoli, 1969) and inversion of the membranes of submitochondrial particles, lend support to this hypothesis. That cytochrome *c* does not readily penetrate into mitochondria is long known (Lehninger *et al.*, 1954). Wojtzak and Zaluska (1969) have found the outer mitochondrial membrane impenetrable to cytochrome *c*. Salt releases cytochrome *c* from mitochondria only after hypotonic swelling and disruption of the outer membrane (Streichman and Avi-Dor, 1967); this explains the results of Jacobs and Sanadi (1960) and agrees with the findings of Parsons *et al.* (1966) and Caplan and Greenawalt (1966). Lee and Carlson (1968) found that only cytochrome *c* present at the outer surfaces of the inner mitochondrial membrane (or on the inner surface of inverted sonic submitochondrial particles) can participate in integrated respiratory chain function (see also Arion and Wright, 1970). In sonicated submitochondrial particles the cytochrome *c*/haem *a* ratio can be increased to 17, by incubation with cytochrome *c* in a sucrose-Tris pH 8·7 medium and the cytochrome *c* is removable by high salt concentration. In contrast with non-sonicated submitochondrial particles the cytochrome *c*/haem *a* ratio reaches only 5 and this cytochrome *c*, like endogenous *c*, is not removed by high salt concentrations (Lenaz and MacLennan, 1966). Cytochrome *c* is also bound by reconstituted succinoxidase and NADH-oxidase complexes (Richardson and Fowler, 1963; King, 1966) and exogenous *c* is bound to untreated mitochondria by protein, by both high and low affinity (Williams and Thorp, 1970) and stabilized by phospholipid. Under certain conditions it can be incorporated into a heart muscle preparation (Cheng and Tu, 1965). The phosphorylating electron transport can be reconstituted in cytochrome *c*-deficient yeast by both yeast iso-cytochromes *c* as well as by mammalian *c* (Mattoon and Sherman 1966).

Nevertheless, it remains uncertain whether the relation between cytochrome *c* and the oxidase can be reconstituted exactly once it has been broken. Baba *et al.* (1967) showed that antigen-antibody type of interactions inhibiting oxidase activity occur between the oxidase and various peptides of cytochrome *c*, particularly those containing the residues 2–9, 39–45 and 81–103, while no such interactions occur with the haem-carrying peptides and peptides containing the long invariable part (residues 70–80) present in many cytochrome species. It is possible that such inactivating linkages between oxidase and cytochrome *c* are absent in mitochondria, but may occur in reactions of the oxidase with soluble cytochrome *c*. Nicholls (1965) assumed that endogenous cytochrome *c* is bound at a high affinity side of the oxidase whereas exogenous cytochrome c reacts largely with the low affinity side of the oxidase. Abnormal linkage of either type may explain the "inhibition" of the oxidase by increasing concentrations of cytochrome c found by L. Smith and co-workers (Smith, 1955; Smith and Conrad, 1961; Davies et al., 1964; Smith and Minnaert, 1965; see also Chapter III, B3). These authors, and more recently Mochan and Kabel (1969) found that both c^{2+} and c^{3+} inhibit owing to their interaction as basic protein with the acidic oxidase, which they compare with the similar inhibition by polylysine. Other workers (McGuiness and Wainio, 1962; Yonetani and Ray, 1965) found, however, only c^{3+} inhibitory and this is left undecided by Hollocher (1962, 1964). There are many studies on the kinetics of the oxidation of cytochrome c by solubilized oxidase (Yonetani, 1960, 1962; Gibson and Greenwood, 1963; Greenwood et al., 1964; Gibson et al., 1965; Kuboyama et al., 1962, 1963; Minnaert, 1965; Slater et al., 1965; Okunuki, 1962; Sekuzu et al., 1967; Horio and Ohkawa, 1968; Lemberg and Cutler, 1970), while studies of the kinetics with particulate oxidase have been carried out by Stotz et al. (1938), Slater (1949c), Minnaert (1961a,b; 1965), King and Takemori (1964), Nicholls et al. (1969) as well as by Smith et al., Hollocher et al. and McGuiness and Wainio (see above). King and Takemori found electron transport from reconstituted succinate-cytochrome c reductase to soluble cytochrome c slower than that to oxygen in the intact mitochondrial respiratory chain, but the difference disappears on the addition of a small amount of cholate to the cytochrome c-reductase system.

2. Cytochrome c-cytochrome oxidase complexes

The formation of a complex between cytochrome c and cytochrome oxidase has been postulated by many authors, whereas Keilin (1930), Slater (1949c), Kiese and Reinwein (1953) and Yonetani (1962) assumed only a labile enzyme-substrate type of complex. There is no certainty either, whether the complex, if it exists, is stimulatory for the oxidase activity or inhibitory. The most convincing evidence for complex formation comes from the work of King et al. and of Orii et al. King et al. (1965) and Kuboyama et al. (1962, 1963) isolated a complex of molar rate cytochrome c/haem a of 1 by Sephadex G-200 chromatography. This complex differed from a mixture of the two components by a strong Soret maximum of the CO-compound at 415 nm, whereas that of a_3CO lies at 430 nm and c^{2+} does not react with CO. It also differed by infra-red bands at 950 and 1050 cm^{-1} and by higher activity with the bc_1 complex and with succinic dehydrogenase. It is readily dissociated in an Amberlite IRC 50 column or by media of high ionic strength, but not by centrifugation. It appears doubtful that this complex can play a role in mitochondria, since their photochemical action spectrum does not indicate the presence of a

415 nm maximum. Perhaps this maximum is formed by a modification of cytochrome c to a CO-reacting species, although we have been unable to find evidence for this to occur on mixing an oxidase solution in Emasol-phosphate with cytochrome c. The evidence of Orii et al. (1962) is based on the formation of a single spot of a R_F between that of cytochrome c and of the oxidase, also with a ratio of cytochrome c/haem a of 1. This complex also appears to be a stimulatory one. According to Okunuki (1962) complex formation may occur through the reaction of an ϵ-NH_2 group of cytochrome c-lysine with the formyl group in haem a. While the spectrum shows that this cannot be due to aldimine formation (which does not occur at physiological pH) a type of more labile binding is perhaps not excluded. In contrast to these observations, Nicholls (1964) reports that a complex of the composition 2 haem a/cytochrome c is precipitated at low ionic strength and is dissociated at higher ionic strength; this is supposed to be an inhibitory complex. Later evidence of Nicholls et al. (1969) for the existence of an oxidase-cytochrome c complex formed on addition of cytochrome c to a Keilin-Hartree preparation does not support its activity *in vivo* since cytochrome a_3 and c are supposed to be placed on opposite sites of the inner membrane (see above). Gibson and Greenwood (1963), Greenwood et al. (1964) and Gibson et al. (1965) interpreted their kinetic results with the assumption of an inhibitory complex between cytochrome c and cytochrome a. Their findings have been differently interpreted by Lemberg (1969), (see Chapter III, B5f,g), i.e. by assuming different forms of oxidized cytochrome oxidase. Antonini et al. (1970) while also rejecting the assumption of an inhibitory complex between cytochromes a and c, have suggested another explanation.

3. Interaction of cytochromes c and a

The anaerobic interaction between cytochromes c and a has been studied by Minnaert (1965), Slater et al. (1965), Horio and Ohkawa (1968) and Lemberg and Cutler (1969, 1970) with some contradictory conclusions. Slater and Minnaert concluded from their observations that one equivalent of cytochrome c^{2+} is able to reduce two equivalents of cytochrome a^{3+} (cf. also Tzagoloff and MacLennan, 1966), whereas Lemberg and Cutler find that one molecule of c^{2+} is required to reduce one a^{3+} and a second is required to reduce simultaneously one of the Cu^{2+} atoms of the oxidase. With "ferric oxidase" the interaction of c^{2+} with a^{3+} is faster than that with a_3^{3+} and again two molecules of c^{2+} are required, one for the reduction of the a_3^{3+} iron and one for that of the second Cu^{2+}. The results with "oxygenated" oxidase do not show the large difference in the reaction rates of a and a_3 (see Chapter III, B5f,g). Whereas Minnaert (1961a,b), Chance (1964) and Chance et al. (1965) find cytochrome a always more oxidized in the steady state than cytochrome c and thus postulate a chain

$c_1 \to c \to a \to a_3 \to O_2$, Yakushiji and Okunuki (1940) assumed the chain

$$c_1 \to c \to a_3 \to O_2$$
$$\uparrow$$
$$a$$

This has been discussed above. However, the claim of Okunuki et al. (1965) that cytochrome c is required for the autoxidation of cytochrome oxidase must be rejected since the reaction is extremely rapid in the complete absence of cytochrome c (Gilmour et al., 1969) and does not lead to a reversibly oxygenated oxidase.

4. The specificity of cytochrome c structure for the interaction with cytochrome oxidase

Studies by the Okunuki school (Takemori et al. 1962a,b; Wada, 1964; Okunuki et al., 1965) have shown that acetylation, succinylation, treatment with trinitrobenzene sulphonate and trichloracetic acid decrease the activity of cytochrome c in the cytochrome oxidase system (see Chapter V, B6b). Since succinylation inactivated more completely than acetylation, decreasing the basic charge more, and since guanidinylation did not diminish the activity, the destruction of a specific positively charged site of cytochrome c was the decisive factor. The overall basic character of the cytochrome c cannot, however, be decisive, since the acidic cytochrome c of *Thiobacillus* reacts even faster with the mammalian aa_3 oxidase than the basic cytochrome c of *Chlorobium* (Yamanaka and Okunuki, 1968), and since the acidic cytochrome c of *Micrococcus denitrificans* reacts rapidly with the aa_3 oxidase of the organism (Smith et al., 1966). The specific basic site of cytochrome c required for the interaction with the oxidase is possibly one of the two lysines 72 or 73.

5. Inhibitors of cytochrome c oxidation

These have been mainly discussed in the chapter on cytochrome oxidase (III, B6). The inhibition of azide is only partly due to a reaction with cytochrome oxidase and partly to its effect on oxidative phosphorylation (Bogucka and Wojtczak, 1966; Palmieri and Klingenberg, 1967; Muraoka and Slater, 1969). This may explain why Yonetani and Ray (1965) found the inhibition by azide non-competitive with cytochrome c while that by cyanide was uncompetitive. Alkylhydroxynaphthoquinones inhibit the mitochondrial oxidation of TMPD (Howland, 1967) as well as the electron transfer between cytochromes b and c_1 (see below).

6. Oxidative phosphorylation on site 3

After an initial failure to find phosphorylation in the oxidation of ferrocytochrome c (Slater, 1955), Lehninger et al. (1954), Nielsen and

Lehninger (1954) and Maley and Lardy (1954) established these phosphorylations (site III) (cf. also Judah, 1951 and Slater, 1954, 1961b, 1966b). The P/O ratio was found to be 1 (Sanadi and Jacobs, 1967) although Ramirez and Mujica (1964) found a somewhat higher value. It is, however, not yet definitely established at which point of the respiratory chain between cytochrome c and oxygen energy conservation with ATP formation occurs. Based on the position of the "crossover point", Chance and Williams (1956a,b) and Chance et al. (1958) concluded that site III was between cytochromes c and a. However, this has been made unlikely by the findings of Wilson (1970), Wilson and Dutton (1970a) and Dutton et al. (1970) that in mitochondria the redox potential of cytochrome a (190–205 mV) was even slightly below that of cytochrome c. Even before this, Wilson and Chance (Wilson, 1967a,b; Wilson and Chance, 1966, 1967; Wilson and Gilmour, 1967) had suggested a position between cytochromes a and a_3, and Ramirez and Mujica (1964) two phosphorylation sites, one between a and a_3, the second between a_3 and oxygen, while Slater had pointed out that for thermodynamic reasons a position between a_3 and oxygen is preferable (Slater, 1969; Storey, 1970a,b). Recent investigations of Muraoka and Slater (1969) support this assumption. Depending on the state of the mitochondria, the substrate and other conditions "crossover points" can be found in different postions of the chain, including none (i.e. one between a_3 and oxygen) (Slater, 1967a). Storey (1970a,b) postulates a contribution of the oxidation of cytochrome a_3 by oxygen to oxidative phosphorylation at sites I and particularly site II, although his rather involved theory of the mechanism of this reaction appears unattractive. The recent determinations of redox potentials of cytochromes a and a_3 inside mitochondria (Wilson and Dutton, 1970a) indicate that ATP lowers the potential of a_3 from about 390 mV, to about 300 mV, indicating a phosphorylation site at a_3^{2+}, whereas ATP has no effect on the potential of cytochrome a; uncouplers restore the high potential of a_3. Wilson assumes that the 200 mV difference between a_3 and a is a sufficient source of energy for phosphorylation, but his findings can also be interpreted as supporting a site between a_3^{2+} and oxygen. It should also be noted that these potentials in intact mitochondria are based on a resolution of a sigmoid type curve obtained by plotting E_0' terms log (red/ox) by simultaneous spectrophotometric ($\Delta 445$-455) and potentiometric anaerobic measurements, into two linear curves of n $= 1$ by a process of extrapolation. This probably does not give results of great accuracy. Nevertheless, it is an important attempt (cf. also Horio and Ohkawa, 1968) to discover the redox potentials of components in a mixture; it may also help to explain contradictory findings (see above D3) on the interaction of cytochrome c and cytochrome oxidase. However, the role of the oxidase copper in these reactions still remains obscure.

The phosphorylation site III is not inhibited by low concentrations of phenazin-methosulphate (Löw et al., 1964). According to Chance et al. (1967a) the phosphorylation site III cannot be explained by the chemiosmotic theory of Mitchell, since there is no evidence here for the role of a hydrogen carrier, such as hydroquinone, being oxidized. Brzhevskaya and Nedelina (1969) found the EPR signal accompanying the oxidation of exogenous cytochrome c increased by additions of a phosphate acceptor, decreased by uncouplers.

Theories of oxidative phosphorylation will be discussed below (subchapter H). The one theory of Brinigar and Wang (1964) specific for site III which assumed a role of the formyl side chain of haem a has later been abandoned by Wang in favour of the general theory assuming the formation of an energy-rich 1-phosphoimidazole (Brinigar et al., 1967; Wang, 1970; Tu and Wang, 1970), although the role of a binding of the haem iron between two imidazole nitrogens (Cooper et al., 1968) can no longer be maintained for cytochrome c. Huang and Wang (1967) found phosphorylation when ferroprotohaemochrome was autoxidized in the presence of solutions containing AMP, P_i and imidazole; in the absence of imidazole the $O_2^{\frac{1}{2}}$ radical was formed; this was prevented in the presence of imidazole and an imidazole radical bound to haem iron was formed instead.

E. THE RESPIRATORY CHAIN BETWEEN CYTOCHROMES b AND c

1. The bc_1-complex

For several reasons this part of the respiratory chain between cytochromes b and c is perhaps the most important one from the point of view of the interaction of the cytochromes in the respiratory chain. It is the region where two branches of the relatively slow di-hydrogen transport of the non-cytochrome part of the respiratory chain, one coming from NADH, the other from succinate, meet the rapid one-electron (or one-hydrogen) transport between the cytochromes. It contains the phosphorylation site II and also the site of reversed electron transport (Hinkle et al., 1967) and can be by-passed by TMPD (Lee et al., 1967). It has been particularly difficult to elucidate, probably because of the close involvement of cytochrome b with particles and its instability outside them. At first even the function of cytochrome b was in some doubt (Chance, 1958) and while more cautious preparations of the particles have now ensured that cytochrome b always plays an important role, its structure and uniformity is by no means established. Recent evidence indicates that two cytochromes b are involved in animal mitochondria (IV, B and below) while their number is greater in plant mitochondria (IV, F2c).

The isolation of a complex containing cytochromes b and c_1 has been described in Chapter IV, B, since it was used in the attempts of preparing cytochrome b. It can be readily separated from cytochrome oxidase by its greater resistance to ammonium sulphate precipitation in cholate solution. It is remarkable that there is little real difference between the bc_1 complex, isolated by King (1966, 1967b) and the so-called coenzyme Q-cytochrome c reductase, the complex III of the Green school (Baum et al., 1967a,b; Rieske, 1967). Both contain about two molecules of cytochrome b per one molecule of cytochrome c_1, in addition to non-haem-iron and phospholipid. The only differences are the higher cytochrome content and slightly higher, but variable contents of phospholipid and of ubiquinone (coenzyme Q_{10}) in the preparation of Green's co-workers; the M.W. was found to be about 300,000. Kopaczyk et al. (1968a,b) assume this complex to be a highly purified stable segment of the respiratory membrane, freed of phosphorylation factor CF_0 which is left in the "stalk". It is this complex which is the site II of oxidative phosphorylation and also the site of the inhibition by antimycin and 2-n-heptyl-8-hydroxyquinoline-N-oxide (HOQNO).

Much has been learned recently on the nature of the bc_1 complex and on cytochrome b by the use of antimycin. Antimycin reacts with the cytochrome b moiety, not with the c_1 moiety (Yamashita, 1968; Yamashita and Racker, 1969; Berden and Slater, 1970) or, more correctly with one of the two cytochrome b molecules (Berden and Slater, 1970). While the cytochrome c_1 can be isolated from the complex in apparently unaltered form (see Chapter V, D), this does not appear so to be with regard to cytochrome b. Ohnishi (1966) obtained a cytochrome b of M.W. 21,300 by glycerol-2 M guanidine precipitation and washing with 4 M urea. To judge from the mode of preparation and its ability to react with carbon monoxide this can hardly be native cytochrome b (see also Berden and Slater, 1970), although the last-named authors find that it may contain 5–28% of native cytochrome b. Possibly to this is due the fact that Yamashita and Racker (1968, 1969) have been able to reconstitute an active bc_1 complex by the addition of cytochrome c_1. Racker and co-worker (cf. also Baum et al., 1967b) suggest that cytochrome b may not only act as an electron transporting cytochrome, but may also be necessary as a constitutive protein necessary for the assembly of the other constituents.

Antimycin prevents the separation of cytochromes b and c_1 by an effect on the conformation of the bc_1 complex (Rieske et al., 1967a; Baum et al., 1967a); it also prevents its digestion by trypsin and makes SH groups less accessible to mercurials. It inhibits the electron transfer between cytochromes b and c_1 causing discontinuity in their reduction inside the complex, and increases the amount of cytochrome b that can be reduced by succinate or NADH. It also causes a slight shift of the absorption maximum, presumably of ferrocytochrome b (Slater et al., 1970b). There appear to be additional actions of antimycin with the particulate complex in the Keilin-Hartree preparation. Bryla and Kaniuga (1968), Bryla et al. (1969a,b), Slater (1970b), Slater et al. (1970a,b,c) and Bryla et al. (1971) have presented evidence that in the intact chain, different from the solubilized bc_1 complex, the bc_1 complex is present in two polymer (or oligomer) states R and T with which antimycin acts as an allosteric inhibitor (cf. Monod et al., 1965). The sigmoidal curve of antimycin binding is explained by assuming that it combines with high affinity to the low energy polymer R present in low concentration before combining with low affinity with the high energy state polymer T present in high concentration. The isolated bc_1 complex is assumed to be the R monomer. It is suggested that only one of the two cytochrome b molecules (b and b') reacts with antimycin, that one which is reducible by dithionite, but not by NADH or succinate. According to Slater et al. both b and b' (or b_i) form energy-rich forms.

These results differ from those of Wilson and Dutton (1970b) according to whom only one of the two forms of cytochrome b can form an energy-rich form. (See also the criticism of the results of Wilson and Dutton by Devault, 1971). From the fact that the redox potential of this is raised by ATP, they conclude that it is the ferric, not the ferrous form of this cytochrome b which is energized by ATP. The potential of the form which undergoes energizing by ATP was changed from $+35$ mV to $+245$ mV. This is ascribed to a change of ligand binding from $b^{3+} - L_1$ to $b^{3+} - L_3$, in which the original energy comes from the oxidation of b^{2+} to b^{3+} by c_1^{3+}:

$$b^{2+} - L_1 + c_1^{3+} \rightarrow b^{3+} - L_2 + c_1^{2+}$$
$$b^{3+} - L_2 + ADP + P_i \rightarrow b^{3+} - L_1 + ATP$$
$$b^{2+} - L_1 + c_1^{3+} + ADP + P_i \rightarrow b^{3+} - L_1 + c^{2+} - ATP$$

which is reversible only if

$$b^{2+} - L_1 \rightarrow b^{3+} - L_1 + e^- \text{ is about } 0 \text{ mV and}$$
$$b^{2+} - L_2 \rightarrow b^{3+} - L_2 + e^- \text{ is about } 250 \text{ mV}.$$

An alternative, but less likely explanation, would require that the source of energy is the oxidation of another constituent of the respiratory chain by $b^{3+} - L_1$ leading to an energized $b^{2+} - L_2$. Slater adduced evidence for an energized $b^{2+} \sim X$, based on a small difference (1–2 nm) of the α-maximum of the spectrum caused by ATP which is ascribed to energizing b^{2+}. The Chance and the Slater schools now take a stand opposite to theirs in 1961 (Chance, 1961a; Slater, 1961) when Chance argued for $b^{2+} \sim X$, Slater for $b^{3+} \sim X$. In later papers, Slater et al. (1970a,b;

Wegdam et al., 1970) suggest that oxidative phosphorylation is accompanied by a reaction $b^{2+} \sim X\ b_i^{3+} \rightleftharpoons b^{2+}b_i^{3+} \sim X$, with energy conserved by both the oxidation of b_1^{2+} and the reduction of b^{3+}, with $b^{2+} \sim X$ and $b_1^{3+} \sim X$ possibly differing from b^{2+} and b^{3+} in protein conformation rather than in ligand binding. This hypothesis appears to demand energy-rich forms for both types of cytochrome b, but so far only one has been definitely established.

Ultrasonic treatment is claimed to inhibit respiration in heart mitochondria; the respiration is restored by ATP, Fe^{2+} and chelators (Skulachev et al., 1966). Since antimycin inhibits the restoration, the authors postulate that the block is between cytochromes b and c_1. Coenzyme Q_{10} and NHI are present in the bc_1 complex but it is not yet certain whether they play an essential role in it. The role of NHI protein and coenzyme Q in cellular respiration will be discussed in subsection F3.

2. Inhibitors of the respiratory chain between cytochromes b and c_1

(a) *Antimycin*. A mixture of closely related complicated depsopeptides, in which an acyl- and alkyl-substituted dilactone ring is combined by a peptide bond to 3-formamido-salicylic acid (Strong et al., 1960; Farley et al., 1965) is found as an antibiotic in some *Streptomyces* species (Ahmad et al., 1965). It has been used as a specific inhibitor of electron transfer between cytochromes b and c by Potter and Reif (1952), Chance (1952a), Keilin and Hartree (1955) and Estabrook (1958, 1962). Although under certain conditions high concentrations of uncouplers have been shown to release the antimycin inhibition (Howland, 1968; Kovac et al., 1970a) its main action is directly on electron transfer rather than on oxidative phosphorylation. It also inhibits the energy-requiring reduction of NAD^+ by succinate (Chance and Hollunger, 1961) and, at even lower concentrations, that coupled with oxidation of TMPD- ascorbate (Low et al., 1963). Baum et al. (1967a) assumed antimycin to act on a non-haem iron protein between cytochromes b and c_1 but its action appears to be primarily on a cytochrome b, the absorption spectrum of which is altered by antimycin (Chance, 1958 and above; Kovac et al., 1970a,b). Antimycin reacts with isolated (though altered) cytochrome b, not with isolated cytochrome c_1 (Yamashita and Racker, 1969) although it is not clear why the inhibition for succinoxidase should be greater than that for succinate-cytochrome c reductase. According to Rieske and Zaugg (1962) one mole of antimycin is required per one of the two moles of cytochrome b for complete inhibition. The apparently much greater inhibition of isolated cytochrome "b" is at least partly due to the fact that only a part of this is active.

As has been discussed above, antimycin is an allosteric inhibitor (see subchapter H), bound firmly to one polymer form of the bc_1 complex (Bryła et al., 1969a,b; 1971; Kaniuga et al., 1969a,b; Slater, 1970b;

Slater et al., 1970a). This is in addition to the conformational changes which it causes in the monomer bc_1 complex (Rieske and Zaugg, 1962; Baum et al., 1967a). Antimycin is removed by extraction with ether (Bryta and Kaniuga, 1968) or by freezing in 0 2 M guanidine (Rieske et al., 1967a), all from the bc_1 complex, but only partly from the Keilin-Hartree preparation though with complete abolition of its inhibitory effect. Antimycin can dissociate cytochrome b from the bc_1 complex and cause it to combine with succinic dehydrogenase (Ernster et al., 1969).

After kinetic isolation of the terminal portion of the respiratory chain of uncoupled rat liver mitochondria by antimycin A and extraction of cytochrome c, four reducing equivalents are found on the oxygen side of the antimycin block, probably two of haems a and two of copper (Wohlrab, 1969b). Without removal of cytochrome c, three more reducing equivalents are present, two due to cytochrome c, one to cytochrome c_1. This, of course, assumes that the antimycin inhibition is complete.

The antimycin-sensitive site is by-passed by tetramethyl-p-phenylene diamine (Lee et al., 1965b; Packer and Mustafa, 1966). Brown et al. (1965) found a small by-pass in heart mitochondria, inhibited by hydroxyalkylnaphthoquinones (but not by quinoline-N-oxide) but this may not be present in intact mitochondria (Okui et al., 1967). Short chain ubiquinones, but not Q_{10} can also by-pass the antimycin site (Fynn, 1969a,b), but Q_{10} appears only to compete with antimycin.

In the protozoon *Tetrahymena pyriformis* antimycin A has been found to stimulate growth and metabolism in a well-aerated medium (Shug et al., 1968).

(b) *Other inhibitors of this section of the respiratory chain.*

2-alkyl-4-hydroxyquinoline-N-oxides. These were first known as antagonists of dihydrostreptomycin. They were found to inhibit succinoxidase, as well as the oxidation of bacterial cytochrome b, which latter is not inhibited by antimycin (Lightbown and Jackson, 1956). The most frequently used respiratory inhibitor of this class is the 2-n-heptyl compound (HOQNO), but compounds with other 2-alkyl-groups are equally or even more strongly effective. The action on the heart muscle preparation was studied by Chance (1958) and was found to resemble closely that of antimycin and to cause the same spectroscopic shift of the α-band of cytochrome b. Both succinoxidase, NADH-oxidase, and the respective cytochrome c reductases are inhibited (Hatefi et al., 1961c; Howland, 1963), and the effects of HOQNO and antimycin are additive (Nijs, 1967). HOQNO inhibition is more sensitive to uncouplers, e.g. 2,4-dinitrophenol, which causes a partial release of the inhibition, than that by antimycin, but the difference is quantitative rather than qualitative (Kaniuga et al., 1969b).

2-hydroxy-3-alkyl-1,4-naphthoquinones. Some naphthoquinones, in particular SN5949 (Ball et al., 1947; Reif and Potter, 1953) have very similar inhibitory properties as HOQNO. Inhibition can be abolished by serum albumin (Potter and Reif, 1953) or by a large excess of lipophilic substances, e.g. coenzyme Q (Hendlin and Cook, 1960). Whereas Widmer *et al.* (1954) had concluded that these inhibitors inhibit the reaction of cytochrome c_1 with c, their inhibition of the reaction of cytochrome b with c_1 was established by Estabrook (1958, 1962). In some members of this groups, uncouplers relieve the inhibition (Howland, 1963, 1965) while other members act as inhibitors of oxidative phosphorylation.

2,3-Dimercaptopropanol (BAL). BAL ("British Anti-Lewisite") was found by Slater (1948, 1949a) to inhibit succinoxidase and NADH-oxidase (1950a,b) in the presence of oxygen. He assumed the presence of a labile heamatin compound active between cytochromes b and c which was destroyed by coupled oxidation. The coupled oxidation affects, however, a myoglobin admixture (Deul and Thorn, 1962) rather than the factor itself. In spite of many studies (Widmer *et al.*, 1954; Thorn, 1956; Stoppani and Brignone, 1957; Wheeldon, 1958; Baksby and Gershbein, 1961; Deul and Thorn, 1962; Kirschbaum and Wainio, 1965) the BAL reaction is still not satisfactorily explained (Slater, 1958), and some of the observations are contradictory. When cytochrome c_1 became known, it was an obvious candidate but Slater (1949b) found no destruction of cytochrome c_1 by BAL treatment, and Slater and Colpa-Boonstra (1961) no effect on the absorption spectrum. While this possibility can perhaps not yet be safely excluded, it appears likely that BAL reacts with essential SH groups of either cytochrome c_1 or a cytochrome b. The latter possibility cannot be safely excluded on the ground that the BAL-treated heart-muscle preparation can still bind antimycin, and should be re-investigated. Recently, Baksky and Gershbein (1971) found that BAL and thioglycolate depressed succinic and NADH-cytochrome c reductases and assume that the BAL-sensitive factor acts as a link between the two systems.

Phenylhydrazine. Phenylhydrazine, or more correctly a product of its oxidation, phenyldiimide, inhibits respiration most strongly at the bc_1 complex but also destroys cytochromes aa_3, c and c_1 (Asami, 1968). Amytal, though primarily an inhibitor of dehydrogenases, can at high concentrations also block the reaction between cytochromes b and c_1 (Pumphrey and Redfearn, 1963).

3. NADH- and succinate-cytochrome c reductases and oxidases

This is an extension of the reaction between cytochromes b and c_1 described above towards both sides; towards the substrate, including thus the action of their dehydrogenases as cytochrome b reductases containing flavoproteins (Slater, 1966b,c; Singer, 1968); and towards cytochrome c, including the interaction of cytochrome c_1 with cytochrome c, and towards oxygen including this as well as the action of cytochrome oxidase.

The literature in this field is very large and cannot be fully quoted, partly because the papers are of greater interest for the structure and properties of the flavin dehydrogenases which no longer fall into the ambit of this book, partly because it was not clear in many of the earlier papers to which part of the overall system a special observation referred. How intricate these problems often are is illustrated by the observation of the Singer school and others (Cremona et al., 1963; Kaniuga, 1963; Mersmann et al., 1966; Watari et al., 1963; Singer and Sanadi, 1966) that a secondary alteration of a succinic dehydrogenase could cause it to become active as a succinic-cytochrome c reductase. The NADH-dehydrogenase of Mahler et al. (1952) is similarly convertible into an artificial NADH-cytochrome c reductase (King and Howard, 1962a,b, 1967; Ziegler et al., 1959; Ringler et al., 1960; Kaniuga and Veeger, 1962, 1963; Kaniuga, 1963; Minakami et al., 1963).

Succinate-cytochrome c-reductase, e.g. that of the heart muscle, is antimycin A-dependent and requires the presence of phospholipids (Badano et al., 1967). It is inactivated by phospholipases, but activity can be restored by the addition of intact phospholipids. It can be restored from soluble preparations (see King et al., below).

It is, however, not yet certain that any of the soluble NADH-cytochrome c reductases isolated from mitochondria are unaltered parts of the cytochrome c reductase present in the particle (King and Howard, 1962a, 1967; Mackler et al., 1968), although these preparations appear less altered than the Mahler enzyme (Mahler et al., 1952; Mahler and Elowe, 1954). Such an enzyme has also been obtained from yeast (Mackler, 1967a). Their interaction with cytochrome c is, e.g. not inhibited by antimycin or amytal. It is also noteworthy that snake venom treatment at 37° yields a NADH-cytochrome c reductase, but at 30° a NADH-dehydrogenase able to react with ferricyanide but not with cytochrome c (Singer et al., 1956; Singer, 1968). Chance et al. (1967a), Hassinen and Chance (1968) and Scholz et al. (1969) postulate two different flavoproteins, Fp_{D1} and Fp_{D2} between NADH and cytochrome b. While these findings may explain some of the divergencies found with NADH-cytochrome c reductase, they fail to explain the similar divergencies found with succinate-cytochrome c reductase. Our knowledge of cytochrome b reductases in mitochondria is more limited (Ernster et al., 1969) but especially important. Two such flavoproteins, a non-fluorescent Fp_{ha} and a fluorescent Fp_{hf} have been reported in plant mitochondria by Storey (1970b) and Erecinska and Storey (1970); of these Fp_{ha} appears to react with cytochrome b-553, cytochromes c and the aa_3 oxidase, while Fp_{hf} interacts with the unknown cyanide-insensitive oxidase. Albracht and Slater (1969) could reconstitute a rotenone-sensitive but antimycin insensitive NADH-dehydrogenase from a Q-free complex III preparation on addition of coenzyme Q and doubt the need for two different flavoproteins (see also Ragan and Garland, 1969). NADH-oxidase has not yet been fully restored by the addition of cytochrome c (Camerino and King, 1965; see also Cunningham et al., 1967). NADH-cytochrome c reductase and NADH-oxidase have also been found in the protozoon *Tetrahymena pyriformis* (Eichel, 1956).

The presence of NADH and succinate oxidases in the mitochondrial membrane was first demonstrated by Siekevitz and Watson in 1952 (see 1956) and by Ball and Barrnett (1957). Estabrook (1957) found the NADH-cytochrome c reductase inhibited by antimycin and by amytal (see also Baltscheffsky and Baltscheffsky, 1958). Complexes I (NADH-dehydrogenase), II (succinate dehydrogenase), Complex III (coenzyme Q-cytochrome c reductase which includes the bc_1 complex) and the green Complex IV (cytochrome oxidase; see Chapter III and Seki, 1969) were isolated by the schools of Green (Green and Silman, 1967) and of King (1966). Reconstitution of succinoxidase could be achieved provided that succinate was present (King, 1962) from the extract of an acetone powder of heart muscle by tris buffer pH 8·9 (Singer et al., 1956) by the method of Keilin and King (1958, 1960), by phosphate-borate extraction at pH 9·5 (King, 1961, Wilson and King, 1967) or by thiourea (King and Howard, 1967), see also Kuboyama et al. (1962) and Arion and Racker (1970). In this reconstitution respiratory control and energy-dependent reversed electron transport are also re-established (Lee et al., 1969b). NADH-cytochrome c reductase can be reconstituted from I + III, succinate-cytochrome c reductase from II + III (Tisdale, 1967), NADH-oxidase from I + III + IV, and succinate oxidase from II + III + IV. This can be done by e.g. incorporation of a soluble dehydrogenase in a particulate cytochrome system (King, 1967b) or by mixing separate membraneous preparations under certain conditions which are not yet entirely clear; e.g. the earlier data of Hatefi et al. (1961a,e) and Green and Wharton (1963) found combination if the individual membranes were mixed at high concentration, whereas in later methods co-precipitation after a dispersion of the membranes in detergents was used (Fowler and Richardson, 1963; Green and Hechter, 1965; Tzagoloff and McConnell, 1967; Tzagoloff et al., 1967). According to Jasaitis (1969) ATP accelerated the formation of the cytochrome c–cytochrome oxidase complex (c + IV) and decreased its cytochrome c content and also accelerated the formation of succinate-cytochrome c reductase.

Again, Hatefi et al., Fowler and Richardson and Green and Wharton postulated stoichiometric reactions, at least for Complexes I, II and III, but this was later abandoned (Green and Silman, 1967; Tzagoloff et al., 1967) and replaced by a "statistical" interpretation. There is certainly no stoichiometric ratio between Complexes III and IV in different tissues. Initially cytochrome b had been assumed to be present in Complex II, but this was later abandoned (Green and Tzagoloff, 1966). Cytochrome c and coenzyme Q were initially considered as mobile carriers between the four complexes but it was later found that these must be taken up by the membrane system (Tzagoloff et al., 1967; Oda, 1968; Penniston and Green, 1968). It is difficult to decide whether the individual complexes are present as such in the inner mitochondrial membrane, forming a kind of mosaic, or are only formed during the isolation procedures. The role of coenzyme Q as the central point of the system where the two dehydrogenase branches meet the bc_1 complex was postulated by Hatefi et al. 1961, by Green and Brierly (1965), Green and Tzagoloff (1966) and by Baginsky and Hatefi (1969), but its role, or at least its central role, is still open to doubt. While non-haematin iron is found in most complexes, its specific role is also not yet decided (see below). By structural damage to mitochondria cytochrome b may be separated from cytochrome c_1 and transferred to succinic dehydrogenase (Lee et al., 1965a):

$$\begin{array}{c} Fp_s \\ \diagdown \\ Q \to b \to c_1 \to c \text{ giving} \\ \diagup \\ Fp_d \end{array}$$

$$\begin{array}{c} Fp_s - b \\ \diagdown \\ Q \text{ separated from } c_1 \to c. \\ \diagup \\ Fp_d \end{array}$$

Antimycin similarly interrupts the chain between b and c_1 (Bruni and Racker, 1968; Bryla and Kaniuga, 1968). Zinc ions appear also to inhibit electron transport between cytochromes b and c_1 (Chistyakov et al., 1969). For succinoxidase, Hultin et al. (1969a,b) have developed a bioelectrode conception as a more physical explanation of its activity.

Finally, the large part (about 80%) of mitochondrial flavoproteins not reducible by NADH or succinate, but by $Na_2S_2O_4$, is active as fatty acid-cytochrome b reductase system (Garland et al., 1967).

F. THE ANTIMYCIN-INSENSITIVE CYTOCHROME REDUCTASES

1. In mitochondria

In addition to the NADH-cytochrome c reductases which are probably artifacts derived by secondary alterations of flavin-type NADH-dehydrogenases, there are other such enzymes present in the outer, not the inner, mitochondrial membrane which have been studied by Mahler et al. (1958), Raw et al. (1958), Sottocasa et al. (1967a,b), Parsons et al. (1967), Brunner and Bygrave (1969), Takesue and Omura (1970a,b) and Kuylenstierna et al. (1970) which are identical with or closely related to the microsomal cytochrome b_5 reductase but not derived from microsomal contaminations (Landriscina et al., 1970).

2. Microsomal cytochrome reductases

(a) *NADH-cytochrome b_5 reductases.* The FAD-enzyme EC 1.6.2.2 was isolated by Strittmatter and Velick (1956, 1957), Penn and Mackler (1958), Mahler et al. (1958), Omura and Sato (1964), Strittmatter (1965) and Mackler (1967b) from liver microsomes and by Kersten et al. (1958) from adrenal microsomes. It is a NADH-cytochrome b_5 reductase which does not directly react with cytochrome c but reduces it via cytochrome b_5. Its isolation has been described by Strittmatter (1967b). The reaction mechanism has been studied by Strittmatter (1965, 1967a), Loverde and Strittmatter (1968) and Hara and Minakami (1971). The most reactive of its lysyl residues is essential for the reaction of the enzyme with cytochrome b_5. Mackler (1967b) isolated a complex of high M.W. containing cytochrome b_5 in addition to flavin and non-haem-iron protein. The importance of this reductase for the desaturation of stearyl coenzyme A has been studied by Holloway and Wakil (1970). It is also, together with cytochrome b_5, probably involved in the oxidation of 7 α-hydroxycholesterol (Wada and Shimakawa, 1969). Phospholipid is required for the activity of microsomal NADH-cytochrome c reductase, but not for the NADHPH-cytochrome c reductase (Jones and Wakil, 1967). Iyanagi and Yamazaki (1969) found a similar enzyme in microsomes which can reduce cytochrome c in the absence of cytochrome b_5 but in the presence of quinones by a monoelectron transfer. Rapoport and co-workers (Rapoport and Wagenknecht, 1957; Wagenknecht et al., 1959; Wagenknecht and Rapoport, 1964) have studied the cooperation of this enzyme with cytosol NADH-oxidation systems. This enzyme is much less sensitive to trypsin than the NADPH-cytochrome c reductase (Orrenius et al., 1969). In addition to this enzyme, another enzyme or enzyme system EC 1.6.99.3) reduces cytochrome c; according to Schulze et al. (1970) they reduce 70–90% of cytochrome c. As with other microsomal cytochrome c reductases it may function as such *in vivo.*

(b) *NADPH-cytochrome c reductase.* This enzyme is another metal-free FAD-protein of M.W. 68,000, (EC 1.6.2.3) which was studied by Horecker (1950), Lang and Nason (1959), Williams and Kamin (1962), Phillips and Langdon (1962), Nishibayashi et al. (1963) and Masters et al. (1965, 1967) and in yeast by Haas et al. (1940, 1942); however the latter is said to contain FMN. It acts directly with cytochrome c with NADPH as better hydrogen donor than NADH and also acts as a transhydrogenase (Ichikawa and Yamano, 1969). The isolation of NADPH-cytochrome c reductase from liver microsomes and separation from cytochrome b_5 by Sephadex chromatography has been described by Omura and Takesue (1970). In the presence of naphthoquinones, e.g. vitamin K_3, it can also catalyse the one-electron reduction of cytochrome b_5 (Iyanagi and Yamazaki, 1969; cf. also Nilsson, 1969) but more slowly (Nishibayashi et al., 1970). It is inhibited by an endogenous NADPH-pyrophosphatase of rat liver (Sasame and Gillette, 1970). Probably the same enzyme has been found in fungi, yeast, *Neurospora* and *Aspergillus* (Sorger, 1963; Cove, 1967; Cove and Coddington, 1965) and differs from a similar bacterial enzyme. With molybdenum it can act as nitrate reductase. It reacts with exogenous rather than with the endogenous cytochrome c of the mitochondrial respiratory chain. An NADPH-cytochrome f reductase is found in chloroplasts (Forti and Sturani, 1968) which is probably identical with the NADPH-cytochrome c_{554} reductase found by Yamanaka and Kamen (1965) in the diatom alga *Navicula pelliculosa* (see also Forti, 1966).

The NADPH-cytochrome c P-450 reductase (EC 1.6.99.3) and its role in the hydroxylation reactions by P-450 has been discussed in Chapter IV, D7. Cytochrome c inhibits the P-450 catalysed hydroxylation of heptane and laurate (Das et al., 1968). In non-solubilized form it also acts as azoreductase (Hernandez et al., 1967).

(c) *Lactate-cytochrome c reductases of yeast*. Some of the lactate-cytochrome c reductases of yeast have been discussed in connection with cytochrome b_2 in Chapter IV, G. In addition to this haemoprotein, however, which acts as L(+)-lactate-cytochrome c reductase, several non-haem containing flavoproteins acting as D(−)-lactate-cytochrome c reductases have been described by Nygaard (1961a,b; see also Chapter IV, G), Graae and Nygaard (1962); Somlo (1962), Gregolin and Singer (1962a,b), Roy (1964), Gregolin and d'Alberton (1964) and Gregolin et al. (1964). Anaerobic yeast contains D(−)-lactate dehydrogenase, but no lactate-cytochrome c reductase. On induction this forms first the D(−)-lactate-cytochrome c reductase, and after a lag phase the L(+)-lactate-cytochrome c reductase. The D(−)-lactate enzyme is highly specific for cytochrome c. It is inhibited by other carboxylic acids substrate-competitively, and by positively charged protein acceptor-competitively. The D(−)-lactate enzyme is a Zn-FAD enzyme. Neither D(−) nor L(+)-lactate cytochrome c reductase is diminished in aerated "petite yeast" (Mahler et al., 1964).

3. The role of coenzyme Q (ubiquinone) and non-haem-iron protein

Both will be discussed together since recent evidence indicates that their reaction with the respiratory chain may be more intimately connected than had previously been assumed (Cox et al. 1970). While the presence of coenzyme Q_{10} (ubiquinone), discovered by Crane et al. (1957) and Crane and Glenn (1957) in both the bc_1 complex (see above) and in mitochondrial cytochrome reductase has long been established (Crane 1960, Crane and Low, 1966; Morton, 1965; Crane, 1968), its role in the electron transfer as link between b and c_1 and also in the cytochrome reductase has not been firmly established (see also Ogamo et al., 1968), although the school of Green put coenzyme Q in the centre of the interactions (see e.g. Hatefi et al., 1959, 1962; Green and Brierley, 1965). Redfearn and Pumphrey (1960) find the rate of reduction of Q not equivalent to the steady-state reduction of electron transfer and conclude that it lies on a branch pathway of the respiratory chain. Part of the reason which led Chance and Redfearn (1961, see also Chance, 1961a; 1965b; Redfearn, 1961) to question the role of Q in the electron transport was due to the fact that the apparently straightforward evidence from a removal of the lipophilic Q by extraction of particles with isooctane and reactivation by re-addition of Q was in fact far more complex, isooctane not only acting in the extraction of Q but as inhibitor and dislocator of the respiratory chain (Redfearn and Pumphrey, 1958; Redfearn et al., 1960; Chance and Redfearn, 1961), but later Redfearn and Whittaker (1966) leave the question open, and Szarkowska (1966) using pentane extraction of freeze-dried beef heart mitochondria provided better evidence for reversibility by re-addition of Q for both NADH- and succinoxidase (see also Albracht and Slater, 1969 and Gawron et al., 1969). Kröger and Klingenberg (1967) concluded from their experiments that Q functions in the main electron pathway between the flavoproteins and the cytochromes. The experiments of Storey and Chance (1967) and Storey (1968) while being interpreted as evidence against a role of coenzyme Q in intact mitochondria, in fact support that at least in submitochondrial particles two parallel pathways exist, one through Q, the other through cytochrome b and that at least for succinate oxidation the pathway through Q is more important. At this time too, no distinction was made between coenzyme Q as a direct link in the electron carrier in the respiratory chain and its more complex connection with it through non-haem-iron (NHI) proteins. This proposition was made in fact first by Redfearn and King (1964) and has been now demonstrated for the function of Q in $E.$ $coli$ (Cox et al., 1970). Finally at a state of isolation of mitochondrial cytochrome c reductases which included the bc_1 system, it was difficult to decide whether Q or NHI affected the chain between say, NADH and cytochrome b, or between cytochrome b and c_1. An essential role for NHI-protein in the bc_1 reaction was postulated by Rieske et al. (1964a,b) and Baum et al. (1967b). However, NHI was not present in isolated cytochromes c and c_1 which could be re-united to an active bc_1

complex (Yamashita and Racker, 1969). Kawakita and Ogura (1969) concluded from o-phenanthroline inhibition experiments that NHI was essential for the bc_1 electron transport rather than for that of the flavoprotein-cytochrome reductases. The coenzyme Q content of the bc_1 complex was only one-fifth of that of the heart muscle, making Takemori and King (1964a) doubt whether Q played an essential role in the bc_1 part of the chain. Takemori and King (1964b) found that Q competitively reversed the antimycin inhibition, but the great excess of Q necessary for the reversal (Brown et al., 1965) diminished the value of this evidence while Sanadi et al. (1967) found NADH-coenzyme Q reductase to react more rapidly with cytochrome c than with coenzyme Q_{10}. Evidence for a role of Q in the reduction of cytochrome b was brought by Ernster et al. (1969) and Rossi et al. (1970) whose evidence indicates that Q has some indirect effect on cytochrome b itself and a regulatory function and modifies the action of inhibitors on succinic dehydrogenase as well as the action of antimycin on the bc_1 complex.

There is far less satisfactory evidence for the action of tocopherols and vitamins K as essential factors in the main respiratory chain (Slater, 1958; 1960, 1961a; Crane et al., 1959; Crane, 1960, 1968; Slater et al. 1961) but see Caswell and Pressman (1968) and Krogstad and Howland (1966) with reference to menadione.

While the presence of NHI-proteins or iron sulphur proteins (Hall and Evans, 1969) in the flavoprotein dehydrogenases and cytochrome c reductases of mitochondria is beyond doubt, their role in the reductase mechanism is still not definitely established. Doeg (1961) found NHI essential for the succinic-Q reductase. It was assumed to be concerned with energy conservation rather than with electron transfer (Butow and Racker, 1965; Vallin and Löw, 1968; in yeast, Chance et al., 1967a; Schatz et al., 1966). However, Chance et al. (1967a) established the presence of NHI in the Fp_{D1} part of NADH-cytochrome b reductase, and also either for Fp_{D2} or Fp_s. Beinert and Lee (1961) and Beinert et al. (1962, 1965) found a g = 1·93 EPR-maximum characteristic for NHI-proteins. This was also found by King (1962) in his NADH-cytochrome c reductase, but was different from that found by Hollocher and Commons (1960, 1961) in the NADP-dehydrogenase of Singer et al. (1956). Beinert considers NHI-protein an essential part of the flavin dehydrogenases. Tyler et al. (1965) found evidence for the function of highly reactive SH groups in the NHI-protein part of NADH-dehydrogenase. Bois and Estabrook (1969) established it as the site of the rotenone inhibition of NADH oxidase. Rieske et al. (1964a,b) found it active in coenzyme Q-cytochrome c reductase, although it is present in smaller concentration (2 Fe/1 cytochrome c_1) than in the cytochrome reductases. In E. coli, Cox et al. (1970) found NHI forming a bridge between coenzyme Q and the respiratory chain in two different places, between Fp_D and cytochrome b_1 and between cytochrome b_1 and the terminal oxidases. The EPR signals of NHI-protein were eliminated by piericidin. In yeast Ohnishi (1970) found that site I of oxidative phosphorylation, missing in anaerobic yeast, could be induced by aeration without, however, the g = 1·94 signal appearing (cf. also Garland et al., 1970).

It has been shown in Chapter V, B6c that superoxide radical ion formed by autoxidation of flavins, reduces cytochrome c^{3+}. This $O_{\overline{2}}$ interaction with the respiratory chain is strongly inhibited by catechols (Miller, 1970).

4. Inhibitors of cytochrome reductases

The inhibition of cellular respiration by narcotics, e.g. urethane, was one of the earliest phenomena observed in cellular respiration (Battelli and Stern, 1911; Warburg, 1949; Keilin, 1925, 1930). At that time a general physicochemical mechanism of this inhibition was assumed, although it was soon recognized that the inhibition concerned the dehydrogenase systems rather than the cytochrome chain. These inhibitors, particularly amytal and other barbiturates, rotenone, piericidin A (Chance and Hollunger, 1963b; Hall, et al., 1966; Hatefi, 1968), some guanidines, and also an anthraquinone from *Cassia*, rhein (Kean et al., 1970), were at low concentrations specific for NADH-dehydrogenase and bound at the same site (Coles et al., 1968; Horgan et al., 968), whereas thenoyltrifluoroacetone, a strong iron chelator (Tappel, 1960) like o-phenanthroline (cf. Yang et al., 1958) specifically inhibited succinyldehydrogenase (Redfearn et al., 1965; King, 1966). The inhibitors of NADH-dehydrogenase appear all bound to the same site and differences in their action assumed at first (cf. e.g. Redfearn et al., 1965) have become doubtful. At higher concentration multiple inhibition occurs (Chance and Hollunger, 1963a; Ernster, 1963; Jeng et al., 1968; in bacteria, Kosaka and Ishikawa, 1968; Snoswell and Cox, 1968; Teeter et al., 1969). Curious differences in reduction of cytochromes b and c_1 have been reported by Jeng and Crane (1968). The site of the rotenone inhibition was placed by Chance et al. (1967a) and Scholz et al. (1969) between two different flavoproteins, Fp_{D1} and Fp_{D2}, while Hatefi et al. (1969) placed it between a flavoprotein and the non-haem-iron protein with a typical EPR band at g = 1.93 (see above) but neither evidence is satisfactory (Albracht and Slater, 1969). It appears possible that NHI-protein is the real site of rotenone inhibition (Bois and Estabrook, 1969) and that the difference of succinate- and NADH-dehydrogenase is due to the nature and accessibility of the NHI-protein in them to inhibitors. The action of these inhibitors appears to be less concerned, however, with electron transport per se than with energy coupling (Redfearn et al., 1965; Gómez-Pyou, 1969) with the possible exception of amytal (Jalling et al., 1957; Pumphrey and Redfearn, 1963, but see Schatz and Racker,

1966). Gutman et al. (1970) place phosphorylation site I between the rotenone-binding site and the NHI of $g = 1\cdot93$. Under certain conditions these substances can even stimulate oxygen uptake in avocado (Lips and Biale, 1966). Uncouplers such as dicumarol, inhibit succinate oxidation by inhibition of succinate dehydrogenase, while other uncouplers are far weaker inhibitors (Wilson and Merz, 1969).

G. INDIRECT INHIBITION OF THE RESPIRATORY CHAIN

Since respiration is under the control of oxidative phosphorylation reactions, it is possible that an observed inhibition of the respiratory chain is not due to the direct action of the inhibitor on electron transfer, but to its effect on oxidative phosphorylation. Several such instances have already been mentioned in passing, for azide in Chapter VII, D5, for inhibitors of the b–c_1 reaction in E2b, and for inhibitors of cytochrome reductase in F4. The fact that azide acts as inhibitor of oxidative phosphorylation was already known to Loomis and Lipman (1949), see Judah (1951) and Slater (1955). Recent papers (Bogucka and Wojtczak, 1966; Wilson and Chance, 1967; Wilson and Azzi, 1968; Vigers and Ziegler, 1968; Riemersma, 1968; Muroaka and Slater, 1969) have confirmed and extended these observations. In azide, in addition to inhibition of oxidative phosphorylation, Palmieri and Klingenberg (1967) and Palmieri and Quagliarielli (1970) assume active azide anion accumulation. The action of hydroxylamine appears to resemble that of azide in that it is a far more active inhibitor of respiration in state 3 of mitochondria than in the non-coupled state 4 (Wikstrom and Saris, 1969).

In this regard, azide resembles oligomycin (Lardy et al., 1958, 1964; Huijing and Slater, 1961; Pinna et al., 1967; Davies, 1963; Kovac et al., 1970b) or similar compounds like rutamycin in spite of the more specific action of azide on cytochrome oxidase. Huijing and Slater (1961) predicted that oligomycin would become a very useful reagent to distinguish between phosphorylating and non-phosphorylating pathways (see e.g. Currie and Gregg, 1965) and also to show whether an energy-requiring reaction required ATP, as e.g. in the beating of single heart cells (Harary and Slater, 1965) or whether ATP formation itself was not necessary, as e.g. for the reversed electron transport of the transhydrogenase reaction (Snoswell, 1962; Tager et al., 1963; Lee et al., 1964; Ernster, 1965; Lee and Ernster, 1965; Groot, 1970; Slater and Ter Welle, 1969; Wilson and Cascarino, 1970). Both oligomycin and atractyloside inhibit ATP formation but for quite different reasons. While oligomycin interferes with the last stage of ATP formation from an energy-rich non-phosphorylated precursor, atractyloside inhibits only its formation from externally added

ADP, preventing its passage across the mitochrondrial membrane. In sonic submitochondrial particles very low concentrations of oligomycin can even stimulate oxidative phosphorylation in the presence of factor F_1 (Fessenden et al., 1966; Lam et al., 1966; Racker and Horstman, 1967). In brain slices oligomycin prevents only the K^+ stimulation of respiration, but does not inhibit respiration otherwise (Minakami et al., 1963; Van Groningen and Slater, 1963; Tobin and McIlwain, 1965; Tobin and Slater, 1965).

It is now known that the action of oligomycin is the inhibition of a mitochondrial ATPase which requires a soluble factor (F_1) to become oligomycin-sensitive (Kagawa and Racker, 1966; MacLennan and Tzagoloff, 1968).

Some trialkyl- or triaryltin compounds have actions similar to those of oligomycin (Aldridge and Street, 1970; Stockdale et al., 1970) and probably act by mediating an anion-OH^- exchange across the mitochondrial membrane.

The structure of atractyloside is that of complex diterpenoid glycoside with sulphate ester groups which is said to resemble the molecular structure of ADP (Allmann et al., 1967a,b; cf. also Vignais et al., 1966). Atractyloside is not an inhibitor of oxidative phosphorylation per se but the site of its action is the mitochondrial outer membrane inhibiting a nucleoside monophosphokinase (Bruni and Azzone, 1964; Bruni et al., 1965a,b; Kemp and Slater, 1964; Chappell and Crofts, 1965; Heldt et al., 1965; Brierley and O'Brien, 1965; O'Brien and Brierley, 1965; Charles and Van den Bergh, 1967; Gangal and Bessman, 1968; Winkler and Lehninger, 1968; Weidemann et al., 1970; Mitra and Bernstein, 1970; Souverijn et al., 1970).

H. MECHANISM OF ELECTRON TRANSFER AND ENERGY CONVERSION IN CELLULAR RESPIRATION

Four main theories on the mechanism of electron transport in cytochromes have been discussed:

(1) The respiratory chain theory of Chance (Chance and Williams, 1956a; Chance, 1967; Chance et al., 1968b) which assumes electron transport in a linear assembly of cytochromes with thermal rotation of the cytochrome molecules around a fixed point at which they are attached in the membranes. This was later modified by evidence of branching in the chains (Gibson and Greenwood, 1964; Chance, 1965c; Chance et al., 1967b; Bogin et al., 1969; Wohlrab and Jacobs, 1967; Wohlrab, 1970; Chance, 1970b). At very low pO_2 the interaction between cytochrome c and the oxidase is slow. If the rates of oxidation of c^{2+} are determined as function of CO-inhibited oxidase molecules in the presence of antimycin (Wohlrab, 1970) evidence for branching of the respiratory chains is

obtained by the demonstration that very few CO-free oxidase molecules can maintain the oxidation of cytochrome c. The aerobic autoxidation of cytochrome in mitochondria from which cytochrome c had been extracted has a slow and a fast phase; by addition of cytochrome c, the slow phase can be titrated back to the fast phase found in normal mitochondria. The slow phase is interpreted as an interchain reaction and cytochrome c is required for rapid interaction between the chains. No such slow autoxidation had, however, been observed in the experiments of Gilmour et al. (1969) with soluble cytochrome c-free oxidase. It has also been shown that electron transport can be reversed and the energy acquired in the processes in one part of the chain used for energy-requiring processes in other parts of the chain (Chance (ed.) Symposium, 1963), as e.g. in the coupling of the energy-requiring reduction of NAD^+ with succinate oxidation (Klingenberg and Schollmeyer, 1960; Chance and Hollunger, 1961; Chance and Fugman, 1961; Tager et al., 1962; Penefsky, 1962; Hinkle et al., 1967) with TMPD + ascorbate oxidation (Löw et al., 1963) or the transhydrogenase reaction, the energy-requiring reduction of $NADP^+$ by NADH (Chance et al., 1967b; Wilson and Bonner, 1970); the latter is of importance for the formation of NADPH required in biosynthetic reactions (Pullman and Schatz, 1967). The theory has become less plausible by what is now known on the structure of cytochrome c established by x-ray diffraction methods (see Chapter V, 3c). There is no evidence for two electron channels at 180° angles which could conduct electrons to the haem iron either directly, or by conjugated double bonds; the only such channel for cytochrome c appears to be one leading from the periphery of the porphyrin, a methene bridge, to the iron (Dickerson et al., 1967; 1970; Castro and Davis, 1969), at least unless the third theory (see below) is assumed to allow electron movements through the "hydrophobic channel" packed with hydrophobic residues.

(2) The "complex" theory of Green and his school did not afford any real explanation of the interaction of the complexes, nor inside one complex, as e.g. the bc_1 complex. Both theories 1 and 2 now seem to be largely abandoned by Green (1970) (see also Young et al., 1970 and Chance, 1970a). Both the notions of stoichiometric ratios of cytochromes in the respiratory chain and of stoichiometric reaction between the "complexes" have become doubtful.

(3) The original conduction band theory of Szent-Gyorgyi (1941) and Evans and Gergely (1949) has been frequently discussed (Szent-Gyorgyi et al., 1961, 1968; Avery et al., 1967) and modified (Cope, 1963, 1965; Cope and Straub, 1969; Rosenberg and Postow, 1969) or criticized (Taylor, 1960; see also discussion Amherst Symposium, 1965, p. 67f). Activation energy for electronic conduction in cytochrome oxidase is lower than that in serum albumin (Straub and Lynn, 1967). There appears,

however, to be no satisfactory mechanism for electron transfer through aliphatic regions of the protein, unless the theory is replaced by the theory of conformational change or a theory close to it like that of Fröhlich (1970) who speaks of long-range forces based on dipole moments of interacting metastable enzyme conformations.

(4) This theory, first applied by Boyer (1965) to a non-cytochrome component of the respiratory chain, suffered at first from a lack of definite evidence for conformational changes in a cytochrome. This evidence has meanwhile been found, and in 1969 it could be predicted that the conformation theory would become of increasing importance for the explanation of electron transfer in the respiratory chain (Lemberg, 1969).

Before discussing this in detail, it should be mentioned that the two main current theories of oxidative phosphorylation both of Chance and Slater and of Mitchell, involve cytochromes. The chemical theory (Slater, 1953; Slater and Colpa-Boonstra, 1961; Slater, 1966b, Chance and Williams, 1956a; Chance, 1961a) did this implicitly by leaving open the possibility that some of the non-phosphorylated energy-rich intermediates ($I \sim X$ or $A \sim C$) might be energy-rich cytochromes. The "chemiosmotic" theory of Mitchell (Mitchell, 1961, 1966, 1967, 1970b; Mitchell and Moyle, 1965, 1967a,b, 1968, 1969a) did it by assuming that cytochromes acted as recipients of electrons and protons transported through the mitochondrial inner membrane or, e.g. in the cytochrome oxidase reaction (Mitchell, 1970a). However, the lack of connection in Mitchell's theory between the electron transfer in the respiratory chain and the membrane alterations caused by them is the present greatest weakness of the "chemiosmotic" or better "electrochemical theory" (cf. Pullman and Schatz, 1967). The theory was initially devised to explain ATP formation, but it is now known that the formation of ATP is only a last step, ATP forming in many instances the exchangeable energy currency, necessary for some reactions (e.g. the beating of single heart cells, Harary and Slater, 1965) but not for others, as for some reactions of reversed electron transport, e.g. the transhydrogenase reaction (Danielson and Ernster, 1963). Moreover, the recent experiments of Painter and Hunter (1970a,b and Froede and Hunter, 1970) have shown that oxidative phosphorylation and ATP formation can be carried out in solution (see also Brinigar et al., 1967 and Wang, 1970) provided that reduced and oxidized glutathione are present. Since both SH and SS-groups are present in proteins and particularly in non-haem-iron proteins it is possible that these play an important role in oxidative phosphorylation (see also Klotz et al., 1958; Cross and Wang, 1970; and Gautheron, 1970).

(5) The theory which is probably the most important one, at least for the cytochromes, is that postulating conformational changes in the tertiary structure of cytochromes, and possibly changes in the quaternary structure

of some individual cytochromes and of cytochrome complexes, or complexes of cytochromes with other proteins (Lemberg, 1971). The whole of the complicated energy interactions described in subchapter A, 1 can be written:

$$\text{Redox energy} \rightleftharpoons \begin{array}{c}\text{conformational}\\\text{energy in}\\\text{cytochromes}\end{array} \rightleftharpoons \begin{array}{c}\text{conformational}\\\text{energy in other}\\\text{proteins and in}\\\text{membranes}\end{array} \rightleftharpoons \begin{array}{c}\text{ATP-bound}\\\text{energy}\end{array}$$

I   II   III   IV
↓
energy of
ion gradients
IV

A similar scheme has recently been given by Green (1970) in a Symposium of the National Academy of Science, U.S.A. He omitted II from his scheme and particularly stressed the membrane changes, but mentioned also the prior conformational changes in cytochromes. Chance (1970a) in the same Symposium developed a similar approach. A similar scheme was developed by Slater (1970a) who used the less definite term "energy pressure" for II. Earlier Slater (1966a) had stated: "I believe that the mechanism of oxidative phosphorylation is locked away in the intimate details of the oxidoreduction between adjacent members of the respiratory chain".

In this book devoted to cytochromes, stress will be on II, the conformational changes and their energy relations of cytochromes and, to a smaller extent, on the relations of II and III. Not only is the energy first provided by the cytochrome interactions, but as now has been demonstrated, the conformational changes, e.g. of cytochrome b precede the slower membrane changes (Chance, 1970a; Chance et al., 1970d).

There is now ample evidence for conformational changes in cytochromes. Such changes in cytochrome a_3 have been found by Lemberg and coworkers (see Chapter III, B5f, B5g, B7 and Chapter VII, D1 and Williams, 1970) while Nicholls (1968) had assumed allosteric changes in cytochrome a. They have been confirmed by redox potential evidence for an "energized" form of cytochrome a_3 by Wilson (1970), Wilson and Dutton (1970a) and by Wohlrab and Ogunmole (1971). The latter authors showed how energization affects a_3^{2+} and lowers the affinity of it to carbon monoxide.

Conformational changes which occur in the oxidation of ferro- to ferricytochrome c have been firmly established by x-ray diffraction methods by Dickerson and others (Chapter V, B3b,c). It is of interest that these changes may perhaps not involve a ligand change of the haem iron and only a minor change of the spin character of the iron, but nevertheless a definite conformational change of the protein. The claim of Webster (1962, 1963;

Webster et al., 1963; Webster and Green, 1964) of an energized cytochrome c has later been withdrawn (see Sanadi, 1965, p. 42). Perhaps too much significance has been attached, however, to the existence of metastable energized products. It is known from photochemical reactions that an unstable singlet form can be involved in other reactions. If c_A and c_B mean two conformationally different forms, c_A stable in c^{2+}, c_B in c^{3+}, the unstable form c_A^{3+} will probably be formed first on oxidation of c_A^{2+}. This has a higher energy content than c_B^{3+} and may rapidly interact with other cytochromes or protein molecules as energy-rich compound, like the active conformational form of reduced haemoglobin obtained by flash photolysis from CO-haemoglobin (Gibson, 1959). In fact the thermodynamic difference between the O_2/H_2O couple and the c_2^+/c^{3+} couple is unnecessarily large for the formation of a single ATP molecule on phosphorylation site III, while the c^{2+}/c^{3+} and b^{2+}/b^{3+} difference is too small for it. Storey (1970a,b) and Chance et al. (1970d) therefore assume that part of the energy of the oxygen-cytochrome c interaction may be available for the cytochrome c-cytochrome b reaction.

There is definite evidence for the existence of at least one, possibly two forms of energized cytochrome b (see Chapter IV, B2, and Chance et al., 1970b) and of energized forms of the cytochrome bc_1 (Chapter VII, E1). There is also evidence for conformational changes in the hydroxylating activity of P-450 (Yong et al., 1970) and in bacterial cytochromes of *Chromatium* (De Vault, 1968). In a recent paper De Vault (1971) points out the thermodynamic relevance of the redox potential changes for the transfer of energy of electron transport of cytochromes to oxidative phosphorylation.

Of these conformational changes that involving cytochrome c is not an allosteric change, since it involves the active haem group itself, but is perhaps nearer to the group of conformational changes described by Koshland (1958, 1960) in his "induced fit" theory. The changes to energized forms of cytochromes a and b are probably allosteric changes in the sense of Monod et al. (1963), more so than in the more specific sense of Monod et al. (1965) worked out for haemoglobin. The importance of the conformational change for the function of haemoglobin is now clearly evident from the work of Perutz (1970). The conformational changes of complexes of different cytochromes may also be allosteric changes, if the term is not restricted to the influence of a small molecule on the conformation of a protein at a point away from the active centre. There appears to be no reason why such allosteric effects should not also be produced by the binding of a large protein molecule on such a site.

The slowest of the energization reactions appears to be the energization of b_1. This has a half time of 200 msec (Wilson and Dutton, 1970b; Chance, 1970a), still 10 times faster than the changes in membrane charge and structure, which by the technique of fluorescent probes (Azzi et al., 1969; Brocklehurst et al., 1969; Chance, 1970a; Waggoner and Stryer, 1970; Katsumata et al., 1970) have been found to have a half-time of 2–3 secs. These mainly reside in the bc_1 region (Nordenbrand and Ernster, 1970). Chance (1970a) finds with ethidium bromide a fast initial fluorescence increase of the same rate as that of the cytochrome b_1-energization, and postulates that this is due to an interaction of the fluorescent probe with a cytochrome on the outside of the membrane. This is partially based on observations of alterations of proton resonance NMR spectra which indicate that the interacting molecules are paramagnetic, ferric cytochromes or ferrous NHI-proteins, and partly on the need of oligomycin for the fluorescence increase in sonicated submitochondrial particles—in these electron flow does not result in energy coupling in the absence of oligomycin (Lee et al., 1969a). Both Chance and Green (see also Young et al., 1970) now assume that conformational changes in cytochromes precede the membrane changes (see also Bronk and Jasper, 1970). The latter would be the membrane changes observed in the electron microscope (Palade, 1964; Hackenbrock, 1968; Harris et al., 1968; Green et al., 1968; Penniston et al., 1968; Vanderkooi and Green, 1970), as well as the membrane changes accompanying oxidative phosphorylation and proton and cation transport in the mitochondrial membrane. According to Margoliash et al. (1970) cytochrome c plays a direct role in ion transport (cf. also Robertson, 1960, 1969). With regard to the role of ionophorous antibiotics for ion transport the reader must be referred to reviews of Pressman (1968, 1970), see also Cockrell et al. (1967).

REFERENCES

Ahmad, K., Schneider, H. G. and Strong, F. M. (1950). *Arch. Biochem. Biophys.* **28**, 281–294.
Albracht, S. P. J. and Slater, E. C. (1969). *Biochem. et Biophys. Acta* **189**, 308–310.
Aldridge, W. N. and Street, B. W. (1970). *Biochem. J.* **118**, 171–179.
Allmann, D. W., Harris, R. H. and Green, D. E. (1967a). *Arch. Biochem. Biophys.* **122**, 766–782.
Allmann, D. W., Lauwers, A. and Lenaz, G. (1967b). "Methods in Enzymol". **10**, 433–437.
Ambe, K. S. and Crane, F. L. (1958). *Science* **129**, 98–99.
Antonini, E., Brunori, M., Greenwood, M. and Malmström, B. G. (1970). *Nature* **228**, 936–937.
Arion, W. J. and Racker, E. (1970). *J. Biol. Chem.* **245**, 5186–5194.
Arion, W. J. and Wright, B. V. (1970). *Biochem. Biophys. Res. Commun.* **40**, 594–599.
Asami, K. (1968). *J. Biochem.* **63**, 425–433.
Atsmon, A. and Davis, R. P. (1967). *Biochem. et Biophys. Acta* **131**, 221–233.
Avery, J., Bay, Z. and Szent-Gyorgyi, A. (1967). *Proc. Nat. Acad. Soc. U.S.* **47**, 1742–1744.

Azzi, A., Chance, B., Radda, G. K. and Lee, C. P. (1969). *Proc. Nat. Acad. Sci. U.S.* **62**, 612–619.
Baba, Y., Mizushima, H., Ito, A. and Watanabe, H. (1967). *Biochem. Biophys. Res. Commun.* **26**, 505–509.
Badano, B. N., Boveris, A., Stoppani, A. O. M. and Vidal, J. C. (1967). *C. r. Soc. Biol.* **167**, 2083–2085.
Baginsky, M. L. and Hatefi, Y. (1969). *J. Biol. Chem.* **244**, 5313–5319.
Baksky, S. and Gershbein, L. L. (1961). *Fed. Proc.* **20**, I 237.
Baksky, S. and Gershbein, L. L. (1971). *Biochem. Biophys. Res. Commun.* **42**, 1765–1771.
Ball, E. G. and Barrnett, R. J. (1957). *J. Biophys. Biochem. Cytol.* **25**, 1023–1036.
Ball, E. G. and Cooper, O. (1949). *J. Biol. Chem.* **180**, 113–124.
Ball, E. G. and Cooper, O. (1957). *J. Biol. Chem.* **226**, 755–763.
Ball, E. G., Anfinsen, G. B. and Cooper, O. (1947). *J. Biol. Chem.* **168**, 257–270.
Baltscheffsky, H. and Baltcheffsky, M. (1958). *Biochem. et Biophys. Acta* **29**, 220–221.
Batelli, F. and Stern, L. (1911). *Biochem. Z.* **30**, 172–194.
Baum, H., Rieske, J. S., Silman, H. I. and Lipton, S. H. (1967a). *Proc. Nat. Acad. Sci.* **57**, 798–805.
Baum, H., Silman, H. I., Rieske, J. S. and Lipton, S. H. (1967b). *J. Biol. Chem.* **242**, 4876–4887.
Beinert, H. and Lee, W. (1961). *Biochem. Biophys. Res. Commun.* **5**, 40–45.
Beinert, H., Heinen, W. and Palmer, G. (1962). *Brookhaven Sympos.* **15**, 229–263.
Beinert, H., Palmer, G., Cremona, T. and Singer, T. P. (1965). *J. Biol. Chem.* **240**, 475–480.
Berden, J. A. and Slater, E. C. (1970). *Biochim. et Biophys. Acta* **216**, 237–249.
Biggs, D. R. and Linnane, A. W. (1963). *Biochim. et Biophys. Acta* **78**, 785–788.
Bogin, E., Higashi, T. and Brodie, A. F. (1969). *Science* **165**, 1364–1367.
Bogucka, K. and Wojtczak, L. (1966). *Biochim. et Biophys. Acta* **122**, 381–392.
Bois, R. and Estabrook, R. W. (1969). *Arch. Biochem. Biophys.* **129**, 362–369.
Boyer, P. D. (1965). "Amherst", II, 1007.
Boyer, P. D. (1967). In (T. P. Singer ed.) "Biological Oxidation", Wiley, N.Y. pp. 193–235.
Brierley, G. P. and Merola, A. J. (1962). *Biochim. et Biophys. Acta* **64**, 205–217.
Brierley, G. P., Merola, A. J. and Fleischer, S. (1962). *Biochim et Biophys. Acta* **64**, 218–228.
Brierley, G. and O'Brien, R. L. O. (1965). *J. Biol. Chem.* **240**, 4532–4539.
Brinigar, W. S. and Wang, J. H. (1964). *Proc. Nat. Acad. Sci.* **52**, 699–704.
Brinigar, W. S., Knaff, D. B. and Wang, J. H. (1967). *Biochemistry* **6**, 36–42.
Brocklehurst, J. R., Freedman, R. B., Hancock, D. J. and Radda, G. K. (1970). *Biochem. J.* **116**, 721–731.
Bronk, J. R. and Jasper, D. K. (1970). *Biochem. J.* **116**, 3P.
Brown, C. B., Russell, J. R. and Howland, J. L. (1965). *Biochim. et Biophys. Acta* **110**, 640–642.
Bruni, A. and Azzone, G. F. (1964). *Biochim. et Biophys. Acta* **93**, 462–474.
Bruni, A. and Racker, E. (1968). *J. Biol. Chem.* **243**, 962–971.
Bruni, A., Contessa, A. R. and Scallela, P. (1965a). *Biochim. et Biophys. Acta* **100**, 1–12.
Bruni, A., Luciani, S. and Bortigon, C. (1965b). *Biochim. et Biophys. Acta* **97**, 434–441.
Brunner, G. and Bygrave, F. L. (1969). *Europ. J. Biochem.* **8**, 530–534.
Bryła, J. and Kaniuga, Z. (1968). *Biochim. et Biophys. Acta* **153**, 910–913.

Bryła, J., Kaniuga, Z. and Slater, E. C. (1969a). *Biochim. et Biophys. Acta* **189**, 317–326.
Bryła, J., Kaniuga, Z. and Slater, E. C. (1969b). *Biochim. et Biophys. Acta* **189**, 327–336.
Bryła, J., Ksiezak, H., Rode, W. and Kaniuga, Z. (1971). *Biochim. et Biophys. Acta* **226**, 213–220.
Brzhevskaya, O. N. and Nedelina, O. S. (1969). *Biofizika* **14**, 647–652, *Chem. Abst.* **71**, (13) 119568.
Bücher, T. (1953). Advances in Enzymol. **14**, 1–48.
Bulos, B. and Racker, E. C. (1968). *J. Biol. Chem.* **243**, 3891–3900.
Burstein, C., and Racker, E. (1970). *8th Internat. Congress Biochem.*, Switzerland 1969. Sympos. **3**, 161–162.
Butow, R. A. and Racker, E. (1965). *Biochim. et Biophys. Acta* **96**, 18–27.
Camerino, P. W. and King, T. E. (1965). *Biochim. et Biophys. Acta* **96**, 18–27.
Camerino, P. W. and Smith, L. (1964). *J. Biol. Chem.* **239**, 2345–2350.
Caplan, A. I. and Greenawalt, J. W. (1966). *J. Cell. Biol.* **31**, 455–472.
Castro, C. E. and Davis, H. F. (1969). *J. Am. Chem. Soc.* **91**, 5405–5407.
Caswell, A. H. and Pressman, B. C. (1968). *Arch. Biochem. Biophys.* **125**, 318–325.
Chance, B. (1952a). *Nature* (London) **169**, 215.
Chance, B. (1952b). *J. Biol. Chem.* **197**, 567–576.
Chance, B. (1954). In (W. D. McElroy and B. Glass) "The Mechanism of Enzyme Actions", Johns Hopkins Press, Baltimore, pp. 399–452.
Chance, B. (1957). In (S. P. Colowick and N. O. Kaplan, eds.) "Methods of Enzymol". Vol. **4**, 273–329. Academic Press, New York.
Chance, B. (1958). *J. Biol. Chem.* **233**, 1223–1229.
Chance, B. (1961a). "Haematin Enzymes", pp. 597–624.
Chance, B. (1961b). *Cold Spring Harbor Sympos. Biol.* **26**, 289–299.
Chance, B. ed. (1963). "Energy-linked Functions of Mitochondria". Academic Press, New York, p. 282.
Chance, B. (1964). In "Oxygen in the Animal Organism" (London Sympos. 1963) p. 415. Pergamon, London.
Chance, B. (1965a). *J. Gen. Physiol.* **49**, 163–188.
Chance, B. (1965b). In (R. A. Morton ed.) "Biochemistry of Quinones", pp. 460 462, 489–498; Academic Press, New York.
Chance, B. (1965c). "Amherst" pp. 929–938.
Chance, B. (1967). *Keilin Memorial Lecture. Biochem. J.* **103**, 1–18.
Chance, B. (1970a). *Proc. Nat. Acad. Sci. U.S.* **67**, 560–571.
Chance, B. (1970b). *Fed. Proc.* **29**, Abstr. 2762.
Chance, B. and Baltscheffsky, M. (1958). *Biochem. J.* **68**, 283–295.
Chance, B. and Fugmann, U. (1961). *Biochem. Biophys. Res. Commun.* **4**, 317–322.
Chance, B. and Hess, B. (1959a). *J. Biol. Chem.* **234**, 2404–2412.
Chance, B. and Hess, B. (1959b). *J. Biol. Chem.* **234**, 2413–2415.
Chance, B. and Hollunger, G. (1961). *J. Biol. Chem.* **236**, 1534–1568.
Chance, B. and Hollunger, G. (1963a). *J. Biol. Chem.* **238**, 418–431.
Chance, B. and Hollunger, G. (1963b). *J. Biol. Chem.* **238**, 432–438.
Chance, B. and Redfearn, E. (1961). *Biochem. J.* **50**, 632–644.
Chance, B. and Sacktor, B. (1958). *Arch. Biochem. Biophys.* **76**, 509–531.
Chance, B. and Schindler, F. (1965). "Amherst" II, 921–929.
Chance, B. and Spencer, E. L. (Jr.). (1959). *Discuss. Faraday Soc.* **27**, 200–205.
Chance, B. and Williams, G. R. (1955a). *J. Biol. Chem.* **217**, 383–393.
Chance, B. and Williams, G. R. (1955b). *J. Biol. Chem.* **217**, 395–407.
Chance, B. and Williams, G. R. (1955c). *J. Biol. Chem.* **217**, 405–427.

Chance, B. and Williams, G. R. (1955d). *J. Biol. Chem.* **217**, 429–438.
Chance, B. and Williams, G. R. (1956a). *Adv. Enzymol* **17**, 65–134.
Chance, B. and Williams, G. R. (1956b). *J. Biol. Chem.* **221**, 477–489.
Chance, B., Bonner, W. D. and Storey, B. T. (1968a). *Ann. Rev. Plant Physiol.* **19**, 295–320.
Chance, B., Erecińska, M., Wilson, D., Dutton, P. L. and Devault, D. (1970b) *8th Internat. Congress. Biochem.*, Switzerland, 1969, *Sympos.* **4**, pp. 239–240.
Chance, B., Erecińska, M. and Lee, C-P. (1970c). *Proc. Nat. Acad. Sci.* **66**, 928–935.
Chance, B., Ernster, L., Garland, P. B., Lee, C-P., Light, P. A., Ohnishi, T., Ragan, C. I. and Wong, D. (1967a). *Proc. Nat. Acad. Sci. U.S.* **57**, 1498–1505.
Chance, B., Holmes, W., Higgins, J. and Connelly, C. M. (1958). *Nature* **182**, 1190–1193.
Chance, B., Lee, C. and Mela, L. (1967b). *Fed. Proc.* **26**, 1341–1354.
Chance, B., Lee, C-P., Mela, L. and Devault, D. (1968b). "Osaka", pp. 475–484.
Chance, B., Parsons, D. F. and Williams, G. R. (1964). *Science* **143**, 136–139.
Chance, B., Schoener, D. and Devault, D. (1965). "Amherst", pp. 907–942.
Chance, B., Williams, G. R. and Hollunger, G. (1963). *J. Biol. Chem.* **238**, 439–444.
Chance, B., Wilson, D. F., Dutton, P. L. and Erecińska, M. (1970d). *Proc. Nat. Acad. Sci. U.S.* **66**, 1175–1182.
Chapman, D. and Leslie, R. B. (1970). In (E. Racker ed.) "Membranes of Mitochondria and Chloroplasts". ACS Monograph. Van Nostrand Reinhold Co., 1970, pp. 91–126.
Chappell, J. B. and Crofts, A. R. (1965). *Biochem. J.* **95**, 707–716.
Charles, R. and Van den Bergh, S. G. (1967). *Biochim. et Biophys. Acta* **131**, 393–396.
Cheng, S. Y. and Tu, H-T. (1965). *Chem. Abstr.* **62**, 16568a.
Chistyakov, V. V., Smirnova, E. G., Jasaitis, A. and Skiladov, V. P. (1969). *Chem. Abstr.* **70** (2) 8120.
Clark, H. W., Neufeld, H. A., Widmer, C. and Stotz, E. (1954). *J. Biol. Chem.* **210**, 851–861.
Cleland, K. W. and Slater, E. C. (1953). *Biochem. J.* **53**, 547–556.
Cockrell, R. S., Harris, E. J. and Pressman, B. C. (1967). *Nature* **215**, 1487–1488.
Coles, C., Griffiths, D. E., Hutchinson, D. W. and Sweetman, A. J. (1968). *Biochem. Biophys. Res. Commun.* **31**, 983–989.
Conover, T. E. and Barany, M. (1966). *Biochim. et Biophys. Acta* **127**, 235–238.
Cooper, T. A., Brinigar, W. S. and Wang, J. H. (1968). *J. Biol. Chem.* **243**, 5854–5858.
Cope, F. W. (1963). *Arch. Biochem. Biophys.* **103**, 352–365.
Cope, F. W. (1965). "Amherst" I, pp. 51–71.
Cope, F. W. and Straub, K. D. (1969). *Bull. Math. Biophys.* **31**, 761–764. *Chem. Abstr.* **72**, (8) 75045.
Cove, D. J. (1967). *Biochem. J.* **104**, 1033–1039.
Cove, D. J. and Coddington, A. (1965). *Biochim. et Biophys. Acta* **110**, 312–318.
Cox, G. B., Newton, N. A., Gibson, F., Snoswell, A. M. and Hamilton, J. A. (1970). *J. Biochem.* **117**, 551–562.
Crane, F. L. (1960). *Arch. Biochem. Biophys.* **87**, 198–202.
Crane, F. L. (1968). In (T. P. Singer ed.) "Biological Oxidations", pp. 533–580. Interscience, New York.

Crane, F. L. and Glenn, J. L. (1957). *Biochim. et Biophys. Acta* **24**, 100–107.
Crane, F. L. and Low, H. (1966). *Physiol. Revs.* **46**, 662–695.
Crane, F. L., Hatefi, Y., Lester, R. L. and Widmer, C. (1957). *Biochim. et Biophys. Acta* **25**, 220–221.
Crane, F. L., Widmer, C., Lester, R. L., Hatefi, Y. and Fechner, W. (1959). *Biochim. et Biophys. Acta* **31**, 476–489, 490–501, 502–512.
Craven, P. A., Goldblatt, P. J. and Basford, R. E. (1969). *Biochemistry* **8**, 3525–3532.
Cremona, T., Kearney, E. B., Villavicencio, M. and Singer, T. P. (1963). *Biochem. Z.* **338**, 407–442.
Criddle, R. S. (1967). "Methods in Enzymol". **10**, 668–676.
Criddle, R. S., Bock, R. M., Green, D. E. and Tisdale, H. D. (1961). *Biochem. Biophys. Res. Commun.* **5**, 75–80.
Criddle, R. S., Bock, R. M., Green, D. E. and Tisdale, H. D. (1962). *Biochemistry* **1**, 827–842.
Cross, R. J. and Wang, J. H. (1970). *Biochem. Biophys. Res. Commun.* **38**, 848–854.
Currie, W. D. and Gregg, C. T. (1965). *Biochem. Biophys. Res. Commun.* **21**, 9–15.
Cunningham, W. P., Crane, F. L. and Sottocasa, G. L. (1965). *Biochim. et Biophys. Acta* **110**, 265–276.
Danielson, L. and Ernster, L. (1963). *Biochem. Z.* **338**, 188–205; *Biochem. Biophys. Res. Commun.* **10**, 91–96.
Das, M. L., Orrenius, S. and Ernster, L. (1968). *Europ. J. Biochem.* **4**, 519–523.
Davies, E. J. (1963). *Biochim. et Biophys. Acta* **96**, 528–529.
Davies, H. C., Smith, L. and Wasserman, A. R. (1964). *Biochim, et Biophys. Acta* **85**, 238–246.
Davson, H. and Danielli, J. F. (1943). "The permeability of natural membranes". Cambridge Univ. Press.
Deul, D. H. and Thorn, M. B. (1962). *Biochim. et Biophys. Acta* **59**, 426–436.
Devault, D. (1968). "Osaka", pp. 488–499.
Devault, D. (1971). *Biochim. et Biophys. Acta* **225**, 193–199.
Dickerson, R. E., Kopka, M. L., Borders, C. L., Varnum, J. C., Weinzierl, J. E. and Margoliash, E. (1967). *J. Mol. Biol.* **29**, 77–95.
Dickerson, E. E., Eisenberg, R. E., Takano, T., Battfay, O. and Samson, L. (1970). *8th Internat. Congress Biochem., Switzerland, 1969. Sympos.* **1**, p. 13.
Doeg, K. A. (1961). *Fed. Proc.* **20**, 44.
Dutton, P. L., Wilson, D. F. and Lee, C. P. (1970). *Biochemistry* **9**, 5077–5082.
Edwards, D. L. and Criddle, R. S. (1966). *Biochemistry* **5**, 583–588.
Edwards, S. W. and Ball, E. G. (1954). *J. Biol. Chem.* **209**, 619–633.
Eichel, H. J. (1956). *J. Biol. Chem.* **222**, 121–136.
Elliott, W. B. and Doebbler, G. F. (1966). *Anal. Biochem.* **15**, 463–469.
Erecińska, M. and Storey, B. T. (1970). *Plant Physiol.* **46**, 618–624.
Ernster, L. (1963). *Proc. 5th Internat. Congress Biochem.* **5**, 146–153.
Ernster, L. (1965). *Fed. Proc.* **24**, 1222–1236.
Ernster, L. and Drahota, Z. (1969) eds. "Mitochondria, structure and function". *Fed. of European Biochem. Soc., 5th Meeting, Prague*, 1968, Academic Press, London and New York.
Ernster, L. and Kuylenstierna, B. (1970). In (E. Racker ed.) "Membranes of Mitochondria and Chloroplasts" pp. 172–212. Van Nostrand Reinhold Co., N.Y.
Ernster, L. and Lee, C-P. (1964). *Ann. Rev. Biochem.* **33**, 729–788.
Ernster, L., Lee, I-Y., Norling, B. and Persson, B. (1969). *European J. Biochem.* **9**, 299–310.
Estabrook, R. W. (1957). *J. Biol. Chem.* **227**, 1093–1108.

Estabrook, R. W. (1958). *J. Biol. Chem.* **230**, 735–750.
Estabrook, R. W. (1959). *Fed. Proc.* **18**, 223.
Estabrook, R. W. (1961). "Haematin Enzymes" I, 276–278.
Estabrook, R. W. (1962). *Biochim. et Biophys. Acta* **60**, 249–258.
Estabrook, R. W. and Holowinski, A. (1961). *J. Biophys. Biochem. Cytol.* **9**, 19–28.
Estabrook, R. W. and Mackler, B. (1957). *J. Biol. Chem.* **229**, 1091–1103.
Estabrook, R. W., Gonze, J. and Tyler, D. D. (1964). *Fed. Proc.* **28**, II, 1326.
Evans, M. G. and Gergely, J. (1949). *Biochim. et Biophys. Acta* **3**, 198–204.
Evtodienko, Yu. V. and Mokhova, E. (1969). *Chem. Abstr.* **69**, (13) 103711.
Farley, T. M., Strong, F. M. and Bydalek, T. L. (1965). *J. Am. Chem. Soc.* **87**, 3501–3504.
Fernandez-Moran, H., Oda, T., Blair, P. V. and Green, D. E. (1964). *J. Cell. Biol.* **22**, 63–100.
Fessenden, J. M., Racker, E. and Dannenberg, A. (1966). *J. Biol. Chem.* **247**, 2483–2486.
Forti, G. (1966). *Brookhaven Sympos. Biol.* **19**, 195–201.
Forti, G. and Sturani, E. (1968). *Europ. J. Biochem.* **3**, 461–472.
Fowler, L. R. and Richardson, S. H. (1963). *J. Biol. Chem.* **238**, 456–463.
Froede, H. C. and Hunter, F. E. Jr. (1970). *Biochem. Biophys. Res. Commun.* **38**, 954.
Fröhlich, H. (1970). *Nature* **228**, 1093.
Fynn, G. H. (1969a). *Biochim. et Biophys. Acta* **180**, 244–252.
Fynn, G. H. (1969b). *Biochem. J.* **112**, 9p.
Galzigna, L. and Sartorelli, L. (1968). *Experientia* **24**, 19–20.
Gangal, S. W. and Bessman, S. P. (1968). *Biochem. Biophys. Res. Commun.* **33**, 675–681.
Garfinkel, D., Garfinkel, L., Pring, M., Green, S. B. and Chance, B. (1970). *Ann. Rev. Biochem.* **39**, 473–498.
Garland, P. B., Chance, B., Ernster, L., Lee, C. P. and Wong, D. (1967). *Proc. Nat. Acad. Sci.* **58**, 4696–4702.
Garland, P. B., Bray, R. C., Clegg, R. A., Haddock, B. A., Light, P. A., Ragan, C. I., Skyrme, J. E. and Swann, J. C. (1970). *8th Internat. Congress Biochem. Switzerland, Sympos.* **2**, p. 145.
Gautheron, D. C. (1970). *Bull. Soc. Chem. Biol.* **52**, 499–521.
Gawron, A., Mahagan, K. P., Limetti, M. and Glaid, III, A. J. (1969). *Arch. Biochem Biophys.* **129**, 461–467.
Gibson, Q. H. (1959). *Biochem. J.* **71**, 293–303.
Gibson, Q. H. and Greenwood, C. (1963). *Biochem. J.* **86**, 541–554.
Gibson, Q. H. and Greenwood, C. (1964). *J. Biol. Chem.* **239**, 586–603.
Gibson, Q. H., Greenwood, C., Wharton, C. and Palmer, G. (1965). *J. Biol. Chem.* **240**, 888–894.
Gilmour, M. V., Lemberg, M. R. and Chance, B. (1969). *Biochim. et Biophys. Acta* **172**, 37–51.
Gomez-Pyou, A., Sandoval, F., Peña, G., Chavez, E. and Tuena, M. (1969). *J. Biol. Chem.* **244**, 5339–5345.
Graae, J. and Nygaard, A. P. (1962). *J. Biol. Chem.* **237**, 3255–3258.
Green, D. E. (1959). *Adv. Enzymol.* **21**, 73–129.
Green, D. E. (1970). *Proc. Nat. Acad. Sci. U.S.* **67**, 544–549.
Green, D. E. and Brierley, G. P. (1965). (R. A. Morton ed.) In "Biochemistry of Quinones", pp. 405–432, Academic Press, New York.
Green, D. E. and Fleischer, S. (1963). *Biochim. et Biophys. Acta* **70**, 554–582.
Green, D. E. and Hechter, O. (1965). *Proc. Nat. Acad. Sci.* **53**, 318–325.
Green, D. and Perdue, J F. (1966). *Proc. Nat. Acad. Sci.* **55**, 1295–1302.

Green, D. E. and Silman, I. (1967). *Ann. Rev. Plant Physiol.* **18**, 147–178.
Green, D. E. and Tzagoloff, A. (1966). *Arch. Biochem. Biophys.* **116**, 293–304.
Green, D. E. and Wharton, D. C. (1963). *Biochem. Z.* **338**, 335–348.
Green, D. E., Asai, J. A., Harris, R. A. and Penniston, J. T. (1968). *Arch. Biochem. Biophys.* **125**, 684–705.
Green, D. E., Mackler, B., Repaske, R. and Mahler, H. R. (1954). *Biochim. et Biophys. Acta* **15**, 435–437.
Green, D. E., Tisdale, R. S., Criddle, R. S. and Bock, R. M. (1961). *Biochem. Biophys. Res. Commun.* **5**, 109–114.
Green, D. E., Wharton, D. C., Tzagoloff, A., Rieske, J. S. and Brierley, G. P. (1965). "Amherst", II, pp. 1032–1065.
Greenwood, C., Gibson, Q. H., Wharton, D. C., Palmer, G. and Beinert, H. (1964). *Fed. Proc.* **23**, Abstr. 1328.
Gregolin, C. and d'Alberton, A. (1964). *Biochem. Biophys. Res. Commun.* **14**, 103–108.
Gregolin, C. and Singer, T. P. (1962a). *Biochim. et Biophys. Acta* **57**, 410–412.
Gregolin, C. and Singer, T. P. (1962b). *Nature* **193**, 659.
Gregolin, C., Scallela, P. and d'Alberton, A. (1964). *Nature* **203**, 1202–1203.
Griffith, J. S. (1961). *Proc. Nat. Acad. Sci.* **47**, 1083–1094.
Groot, G. S. P. (1970). *Biochem. J.* **116**, 14p.
Gutman, M., Mayr, M., Oltzik, R. and Singer, T. P. (1970). *Biochem. Biophys. Res. Commun.* **41**, 40–44.
Haas, E., Horecker, B. L. and Hogness, T. R. (1940). *J. Biol. Chem.* **136**, 747–774.
Haas, E., Harrer, C. J. and Hogness, T. R. (1942). *J. Biol. Chem.* **143**, 341–349.
Hackenbrock, C. R. (1968). *Proc. Nat. Acad. Sci.* **61**, 598–605.
Hall, C., Wu, M., Crane, F. L., Takahashi, H., Tamura, S. and Folkers, K. (1966). *Biochem. Biophys. Res. Commun.* **25**, 373–377.
Hall, D. O. and Evans, M. C. W. (1969). *Nature* **223**, 1342–1348.
Hara, T. and Minakami, S. (1971). *J. Biochem.* **69**, 317–330.
Harary, I. and Slater, E. C. (1965). *Biochim. et Biophys. Acta* **99**, 227–233.
Harris, R. A., Penniston, J., Asai, J. and Green, D. E. (1968). *Proc. Nat. Acad. Sci.* **59**, 830–837.
Hassinen, I. and Chance, B. (1968). *Biochem. Biophys. Res. Commun.* **31**, 895–900.
Hatefi, Y. (1968). *Proc. Nat. Acad. Sci.* **60**, 733–740.
Hatefi, Y., Haavik, A. G., Fowler, L. R. and Griffiths, D. E. (1962). *J. Biol. Chem.* **237**, 2661–2669.
Hatefi, Y., Haavik, A. G. and Griffiths, D. E. (1961a). *Biochem. Biophys. Res. Commun.* **4**, 441–446.
Hatefi, Y., Haavik, A. G. and Griffiths, D. E. (1961b). *Biochem. Biophys. Res. Commun.* **4**, 447–453.
Hatefi, Y., Haavik, A. G. and Jurtshuk, P. (1961c). *Biochim. et Biophys. Acta* **52**, 106–108.
Hatefi, Y., Jurtshuk, P. and Haavik, A. G. (1961d). *Biochim. et Biophys. Acta* **52**, 119–129.
Hatefi, Y., Jurtshuk, P. and Haavik, A. G. (1961e). *Arch. Biochem. Biophys.* **94**, 148–155.
Hatefi, Y., Lester, R. L., Crane, F. L. and Widmer, C. (1959). *Biochim. et Biophys. Acta* **31**, 490–501.
Hatefi, Y., Stempel, K. E. and Haustein, W. G. (1969). *J. Biol. Chem.* **244**, 2358–2365.
Heldt, H. W., Jacobs, H. and Klingenberg, M. (1965). *Biochem. Biophys. Res. Commun.* **18**, 174–179.

Hendler, R. W. (1971). *Physiol. Revs.* **51**, 66–97.
Hendlin, D. and Cook, T. M. (1960). *J. Biol. Chem.* **235**, 1187–1191.
Hernandez, P. H., Gillette, J. R. and Mazel, P. (1967). *Biochem. Pharmacol.* **16**, 1859–1875.
Hinkle, P. C., Butow, R. A., Racker, E. and Chance, B. (1967). *J. Biol. Chem.* **242**, 5169–5173.
Hogeboom, G. H. (1955). Methods in Enzymol. **1**, 16–19.
Hollocher, T. C. (1962). *Arch. Biochem. Biophys.* **98**, 12–16.
Hollocher, T. C. (1964). *Nature* **202**, 1006–1007.
Hollocher, T. C. and Commoner, B. (1960). *Proc. Nat. Acad. Sci.* **46**, 416–427.
Hollocher, T. C. and Commoner, B. (1961). *Proc. Nat. Acad. Sci.* **47**, 1355–1374.
Holloway, P. W. and Wakil, S. J. (1970). *J. Biol. Chem.* **245**, 1862–1865.
Honjo, I. and Ozawa, K. (1968). *Biochim. et Biophys. Acta* **162**, 624–627.
Hoppel, C. and Cooper, C. (1969). *Arch. Biochem. Biophys.* **135**, 173–183.
Horecker, B. L. (1950). *J. Biol. Chem.* **183**, 593–605.
Horgan, D. J., Casida, J. E. and Singer, T. P. (1968). *Fed. Proc.* **27**, Abstr. 1744.
Horio, T. and Ohkawa, J. (1968). *J. Biochem.* **64**, 393–404.
Howland, J. L. (1963). *Biochim. et Biophys. Acta* **73**, 665–667.
Howland, J. L. (1965). *Biochim. et Biophys. Acta* **105**, 205–213.
Howland, J. L. (1967). *Biochim. et Biophys. Acta* **131**, 247–254.
Howland, J. L. (1968). *Biochim. et Biophys. Acta* **153**, 309–311.
Huang, K. and Wang, J. H. (1967). *7th Internat. Congress Biochem. Tokyo, Abstr.* V-H91.
Huijing, F. and Slater, E. C. (1961). *J. Biochem.* (Japan) **49**, 493–501.
Hultin, E., Liljeqvist, G., Lundblad, G., Paléus, S. and Stahl, G. (1969a). *Acta Chem. Scand.* **23**, 3426–3434.
Hultin, E., Paléus, S., Tota, B., Liljeqvist, G. (1969b). *Acta Chem. Scand.* **23**, 3417–3425.
Igo, R. P., Mackler, B. and Hanahan, D. V. (1959). *J. Biol. Chem.* **234**, 1312–1314.
Ichikawa, Y. and Yamano, T. (1969). *J. Biochem.* **66**, 351–360.
Iyanagi, F. and Yamazaki, I. (1969). *Biochim. et Biophys. Acta* **172**, 370–381.
Jacobs, E. E. and Andrews, E. C. (1965). "Amherst", II pp. 784–805.
Jacobs, E. E. and Andrews, E. C. (1967). *7th Internat. Congress Biochem. Tokyo Abstr. V*, H-100.
Jacobs, E. E., Andrews, E. C., Wohlrab, H. and Cunningham, W. (1968). "Osaka", pp. 114–128.
Jacobs, E. E. and Sanadi, D. R. (1960). *J. Biol. Chem.* **235**, 531–534.
Jalling, O., Löw, H., Ernster, L. and Lindberg, O. (1957). *Biochim. et Biophys. Acta* **26**, 231–232.
Jasaitis, A. (1969). *Chem. Abstr.* **71** (13), 119943.
Jeng, M. and Crane, F. L. (1968). *Biochem. Biophys. Res. Commun.* **30**, 465–470.
Jeng, M., Hall, C., Crane, F. L., Takahashi, N., Tamura, S. and Folkers, K. (1968). *Biochemistry* **7**, 1311–1322.
Jones, P. D. and Wakil, J. S. (1967). *J. Biol. Chem.* **242**, 5267–5273.
Judah, J. D. (1951). *Biochem. J.* **49**, 271–285.
Kagawa, Y. and Racker, E. (1966). *J. Biol. Chem.* **241**, 2461–2482.
Kaniuga, Z. (1963). *Biochim. et Biophys. Acta* **73**, 550–564.
Kaniuga, Z. and Veeger, C. (1962). *Biochim. et Biophys. Acta* **60**, 435–437.
Kaniuga, Z. and Veeger, C. (1963). *Biochim. et Biophys. Acta* **77**, 339–342.
Kaniuga, Z., Bryła, J. and Slater, E. C. (1969a). *FEBS Sympos.* **19**, 285–295.
Kaniuga, Z., Bryła, J. and Slater, E. C. (1969b). *20th Colloquium Ges. Biol. Chem. Mosbach*.

Katsumata, Y., Niyazi, A. and Ozawa, T. (1970). *J. Biochem.* **68**, 423–425.
Kawakita, A. and Ogura, Y. (1969). *J. Biochem.* **66**, 203–211.
Kean, E. A., Gutman, M. and Singer, T. P. (1970). *Biochem. Biophys. Res. Commun.* **40**, 1507–1513.
Keilin, D. (1925). *Proc. Roy. Soc. London* **13**, 98, 312–339.
Keilin, D. (1930). *Proc. Roy. Soc. London B* **106**, 418–444.
Keilin, D. and Hartree, E. F. (1938). *Proc. Roy. Soc. London B* **125**, 171–186.
Keilin, D. and Hartree, E. F. (1940). *Proc. Roy. Soc. London B* **129**, 277–306.
Keilin, D. and Hartree, E. F. (1955). *Nature* **176**, 200.
Keilin, D. and King, T. E. (1958). *Nature* **181**, 1520–1522.
Keilin, D. and King, T. E. (1960). *Proc. Roy. Soc. London B* **152**, 163–187.
Keilin, D. and Slater, E. C. (1953). *Brit. Med. Bulls.* **9**, 89–96.
Kemp, A. and Slater, E. C. (1964). *Biochim. et Biophys. Acta* **92**, 178–180.
Kersten, H., Kersten, W. and Staudinger, H. (1958). *Biochim. et Biophys. Acta* **27**, 598–608.
Kiese, M. and Reinwein, D. (1953). *Biochem. Z.* **324**, 51–59.
King, T. E. (1961). *J. Biol. Chem.* **236**, 2342–2344.
King, T. E. (1962). *Biochim et Biophys. Acta* **59**, 492–494.
King, T. E. (1966). *Advances in Enzymol.* **28**, 155–236.
King, T. E. (1967a). "Methods in Enzymol". **10**, 202–208.
King, T. E. (1967b). "Methods in Enzymol". **10**, 322–331.
King, T. E. and Howard, R. L. (1962a). *J. Biol. Chem.* **237**, 1686–1698.
King, T. E. and Howard, R. L. (1962b). *Biochim. et Biophys. Acta* **59**, 489–492.
King, T. E. and Howard, R. (1967). "Methods in Enzymol". **10**, 275–294.
King, T. E. and Takemori, S. (1964). *J. Biol. Chem.* **239**, 3559–3569.
King, T. E., Kuboyama, M. and Takemori, S. (1965). "Amherst" II, 707–735.
Kirschbaum, J. and Wainio, W. W. (1965). *J. Biol. Chem.* **240**, 462–466.
Klingenberg, M. (1970a). *8th Internat. Congress Biochem. Switzerland 1969, Sympos.* **2**, 154–155.
Klingenberg, M. (1970b). *8th Internat. Congress Biochem., Switzerland 1969, Sympos.* **3**, 240.
Klingenberg, M. and Bücher, T. (1960). *Ann. Rev. Biochem.* **29**, 669–708.
Klingenberg, M. and Kröger, A. (1967). In (E. C. Slater, Z. Kaniuga and L. Wojtczak, eds.) "Biochemistry of Mitochondria," Academic Press, London, Warsaw, p. 11.
Klingenberg, M. (1967). In (C. Quagliariello, ed.) "Mitochondrial Structural Compartmentation". Round Table Discuss. 1966, publ. 1967, pp. 124–162.
Klingenberg, M. and Schollmeyer, P. (1960). *Biochem. Z.* **335**, 231–242.
Klotz, I. M., Ayers, J., Holl, J. Y. C., Horowitz, M. G. and Heiney, R. E. (1958). *J. Amer. Chem. Soc.* **80**, 2132–2141.
Koller, K. (1968). *Chem. Abstr.* **69** (3) 16739.
Kopaczyk, K. C. (1967). "Methods in Enzymol". **10**, 253–258.
Kopaczyk, K. C., Asai, J., Allmann, D. W., Oda, T. and Green, D. E. (1968a). *Arch. Biochem. Biophys.* **123**, 602–621.
Kopaczyk, K. C., Asai, J. and Green, D. E. (1968b). *Arch. Biochem. Biophys.* **126**, 358–379.
Kopaczyk, K. C., Perdue, J. and Green, D. E. (1966). *Arch. Biochem. Biophys.* **115** 215–225.
Kornberg, R. D. and McConnell, H. M. (1971). *Biochemistry* **10**, 1111–1120.
Kosaka, T. and Ishikawa, Sh. (1968). *J. Biochem.* **63**, 506–513.
Koshland, D. E. (1958). *Proc. Nat. Acad. Sci.* **44**, 98–104.
Koshland, D. E. (1960). *Advances in Enzymol.* **22**, 45–97.

Kováč, L., Šmigaň, P., Hrušovska, E. and Hess, B. (1970a). *Arch. Biochem. Biophys.* **139**, 370–379.
Kováč, L., Hrušovska, E. and Šmigaň, P. (1970b). *Biochim. et Biophys. Acta* **205**, 520–522.
Kröger, A. and Klingenberg, M. (1966). *Biochem. Z.* **344**, 317–336.
Krogstad, D. J. and Howland, J. L. (1966). *Biochim. et Biophys. Acta* **118**, 189–191.
Kuboyama, M., Takemori, S. and King, T. E. (1962). *Biochem. Biophys. Res. Commun.* **9**, 534–539, 540–544.
Kuboyama, M., Takemori, S. and King, T. E. (1963). *Fed. Proc.* **22**, Abstr. 2542.
Kurup, C. K. R. and Sanadi, D. R. (1968). *Arch. Biochem. Biophys.* **126**, 722–724.
Kuylenstierna, B., Nicholls, D. G., Homoller, S. and Ernster, L. (1970). *Europ. J. Biochem.* **12**, 419–426.
Lam, K. W., Warshaw, J. B. and Sanadi, D. R. (1966). *Arch. Biochem. Biophys.* **117**, 594–598.
Landriscina, C., Papa, S., Coratelli, P., Mazzarella, L. and Quagliariello, E. (1970). *Biochim. et Biophys. Acta* **205**, 136–147.
Lang, C. A. and Nason, A. (1959). *J. Biol. Chem.* **234**, 1874–1877.
Lardy, H. A. and Ferguson, S. M. (1969). *Ann. Rev. Biochem.* **38**, 991–1034.
Lardy, H. A., Connelly, J. L. and Johnson, D. (1964). *Biochemistry* **3**, 1961–1968.
Lardy, H. A., Johnson, D. and McMurray, W. C. (1958). *Arch. Biochem. Biophys.* **78**, 587–597.
Laser, H. and Slater, E. C. (1960). *Nature* **187**, 1115–1117.
Lee, C-P., and Carlson, K. (1968). *Fed. Proc.* **27**, Abstr. 3460.
Lee, C-P. and Ernster, L. (1965). *Biochem. Biophys. Res. Commun.* **18**, 523–529.
Lee, C-P. and Ernster, L. (1966) in (J. M. Tager, S. Papa, E. Quagliariello and E. C. Slater eds.) "Regulation of metabolic processes in mitochondria" Elsevier, Amsterdam, vol. **7**, pp. 218–234.
Lee, C-P., Azzone, G. F. and Ernster, L. (1964). *Nature* **201**, 152–155.
Lee, C-P., Ernster, L. and Chance, B. (1969a) *Europ. J. Biochem.* **8**, 153–163.
Lee, C-P., Estabrook, R. W. and Chance, B. (1965a). *Biochim. et Biophys. Acta* **99**, 32–45.
Lee, C-P., Johansson, B. and King, T. E. (1969b). *Biochem. Biophys. Res. Commun.* **35**, 234–238.
Lee, C-P., Kimelberg, H. and Johansson, B. (1969c). *Fed. Proc.* **28**, Abstr. 2260.
Lee, C-P., Nordenbrand, K. and Ernster, L. (1965b). "Amherst" II, pp. 960–970.
Lee, C-P., Sottocasa, G. L. and Ernster, L. (1967) "Methods in Enzymol". **10**, 33–37.
Lehninger, A. L. (1962). *Physiol. Revs.* **42**, 467–517.
Lehninger, A. L. (1964). "The Mitochondrion", Benjamin, New York.
Lehninger, A. L. (1967). *Fed. Proc.* **26**, 1333.
Lehninger, A. L., Hassan, M. and Sudduth, H. C. (1954). *J. Biol. Chem.* **210**, 911–922.
Lehninger, A. L. and Wadkins, C. L. (1962). *Ann. Rev. Biochem.* **31**, 47–78.
Lemberg, R. (1969). *Physiol. Revs.* **49**, 59–60, 93–104.
Lemberg, R. (1971). John Falk Memorial Lecture. *Search (Australia)* **2**, 160–165.
Lemberg, R. and Cutler, M. E. (1969) in (H. C. Freeman, ed.) Proc. Internat. Conf. on Coordination Chemistry, Science Press, Sydney.
Lemberg, R. and Cutler, M. E. (1970). *Biochim. et Biophys. Acta* **197**, 1–10.
Lemberg, R., Morell, D. B., Lockwood, W. H. and Stewart, M. (1956). *Chem. Ber.* **89**, 309–374.

Lenaz, G. and MacLennan, D. H. (1966). *J. Biol. Chem.* **241**, 5260–5265.
Lester, R. L. and Fleischer, S. (1961). *Biochim. et Biophys. Acta* **47**, 358–377.
Lightbown, J. W. and Jackson, F. L. (1956). *Biochem. J.* **63**, 130–137.
Linnane, A. W. (1958). *Biochim. et Biophys. Acta* **30**, 221–222.
Linnane, A. W. and Titchener, E. B. (1960). *Biochim. et Biophys. Acta* **39**, 469–478.
Lips, S. H. and Biale, J. B. (1966). *Plant Physiol.* **41**, 797–802.
Löw, H. and Vallin, I. (1963). *Biochim. et Biophys. Acta* **69**, 361–374.
Löw, H., Alm, B. and Vallin, I. (1964). *Biochem. Biophys. Res. Commun.* **14**, 347–352.
Löw, H., Vallin, I. and Alm, B. (1963). In (B. Chance, ed.) "Energy-linked Functions in Mitochondria", Academic Press, New York, p. 5.
Loomis, W. I. and Lipmann, F. (1949). *J. Biol. Chem.* **179**, 503–504.
Loverde, A. and Strittmatter, P. (1968). *J. Biol. Chem.* **243**, 5777–5787.
Lundegårdh, H. (1960). *Biochim. et Biophys. Acta* **39**, 162–164.
McConnell, D. G., Tzagoloff, A., MacLennan, D. H. and Green, D. E. (1966). *J. Biol. Chem.* **241**, 2373–2382.
McGuiness, E. T. and Wainio, W. W. (1962). *J. Biol. Chem.* **237**, 3273–3277.
MacLennan, D. H. and Tzagoloff, A. (1968). *Biochemistry* **7**, 1603–1610.
Machinist, J. M., Das, M. L., Crane, F. L. and Jacobs, E. E. (1962). *Biochem. Biophys. Res. Commun.* **6**, 475–478.
Mackler, B. (1967a). "Methods in Enzymol". **10**, 294–296.
Mackler, B. (1967b). "Methods in Enzymol". **10**, 551–553.
Mackler, B. and Penn, N. (1957). *Biochim. et Biophys. Acta* **24**, 294–300.
Mackler, B., Erickson, R. J., Davis, S. D., Mehl, T. D., Sharp, Ch., Wedgwood, R. J., Palmer, G. and King, T. E. (1968). *Arch. Biochem. Biophys.* **125**, 40–45.
Mahler, H. R., Sarkar, N. K., Vernon, L. P. and Alberty, R. A. (1952). *J. Biol. Chem.* **199**, 585–606.
Mahler, H. R. and Elowe, D. (1954). *Biochim. et Biophys. Acta* **14**, 100–107.
Majler, H. R., Raw, I., Molinari, R. and Do Ameral, D. F. (1958). *J. Biol. Chem.* **233**, 230–239.
Mahler, H. R., Mackler, B., Grandchamp, S. and Slonimski, P. P. (1964). *Biochemistry* **3**, 668–682.
Mahler, H. R. and Cordes, E. H. (1966). "Biological Chemistry", Harper and Row and J. Weatherhill, New York and Tokyo, pp. 592–623.
Maley, G. F. and Lardy, H. A. (1954). *J. Biol. Chem.* **210**, 903–909.
Malviya, A. N., Parsa, B., Yokaiden, R. E. and Elliott, W. B. (1968). *Biochim. et Biophys. Acta* **162**, 195–209.
Margoliash, E., Barlow, G. H. and Byers, V. (1970). *Nature* **228**, 723–726.
Masters, B. S. S., Kamin, H., Gibson, Q. H. and Williams, C. H. Jr. (1965). *J. Biol. Chem.* **240**, 921–931.
Masters, B. S. S., Williams, C. H. Jr., and Kamin, H. (1967). "Methods in Enzymol". **10**, 565–573.
Mattoon, J. R. and Sherman, F. (1966). *J. Biol. Chem.* **241**, 4330–4338.
Mersmann, H., Luthy, L. and Singer, T. P. (1966). *Biochim. Biophys. Res. Commun.* **25**, 43–48.
Miginiac-Maslow, M. (1968). *C.r. soc. biol.* **162**, 12–16.
Miller, R. W. (1970). *Canad. J. Biochem.* **48**, 935–939.
Minakami, S., Cremona, T., Ringler, R. L. and Singer, T. P. (1963). *J. Biol. Chem.* **238**, 1529–1537.
Minakami, S., Kakinuma, K. and Yoshikawa, H. (1963). *Biochim. et Biophys. Acta* **78**, 808–811.

Minakami, S., Schindler, F. J. and Estabrook, R. W. (1964). *J. Biol. Chem.* **239**, 2042–2049, 2049–2054.
Minnaert, K. (1961a). *Biochim. et Biophys. Acta* **50**, 23–24.
Minnaert, K. (1961b). *Biochim. et Biophys. Acta* **54**, 26–41.
Minnaert, K. (1965). *Biochim. et Biophys. Acta* **110**, 42–56.
Mitchell, P. (1961). *Nature* (London) **191**, 144.
Mitchell, P. (1966a). *Biol. Rev. Cambridge Philos. Soc.* **41**, 445–502.
Mitchell, P. (1966b). "Chemiosmotic coupling in oxidative and photosynthetic phosphorylation," Glynn Res. Ltd., Bodwin.
Mitchell, P. (1967). *Nature* **214**, 1327–1328.
Mitchell, P. (1970a). *8th Internat. Congress Biochem. Switzerland, 1969. Sympos. 3*, p. 160.
Mitchell, P. (1970b). *Biochem. J.* **116**, 5–6P.
Mitchell, P. and Moyle, J. (1965). *Nature* (London) **208**, 147.
Mitchell, P. and Moyle, J. (1967a). *Nature* (London) **213**, 137–139.
Mitchell, P. and Moyle, J. (1967b). *Biochem. J.* **105**, 1147–1162.
Mitchell, P. and Moyle, J. (1968). *European J. Biochem.* **4**, 530–539.
Mitchell, P. and Moyle, J. (1969a). *European J. Biochem.* **7**, 471–484.
Mitra, R. S. and Bernstein, I. A. (1970). *Arch. Biochem. Biophys.* **141**, 519–524.
Mochan, E. and Kabel, B. S. (1969). *Fed. Proc.* **28**, Abstr. 3492.
Mokhova, E. N. (1965). *Biofizika* **10**, 571–577.
Monod, J., Changeux, J. P. and Jacob, F. (1963). *J. Mol. Biol.* **6**, 306–329.
Monod, J., Wyman, J. and Changeux, J. P. (1965). *J. Mol. Biol.* **12**, 88–118.
Morton, R. A. (1965). *Endeavour* **24**, 81–86.
Muraoka, S. and Slater, E. C. (1968). *Biochim. et Biophys. Acta* **162**, 170–174.
Muraoka, S. and Slater, E. C. (1969). *Biochim. et Biophys. Acta* **180**, 221–226; 227–236.
Muscatello, U. and Carafoli, E. (1969). *J. Cell. Biol.* **40**, 602–621.
Mustafa, M. G., Cowger, M. L., Labbe, R. F. and King, T. E. (1968). *J. Biol. Chem.* **243**, 1908–1918.
Nachlas, M. M., Margulis, S. I., and Seligman, A. M. (1960). *J. Biol. Chem.* **235**, 2739–2743.
Nicholls, P. (1963). In "The Enzymes" (P. D. Boyer, H. Lardy and K. Myrbäck, eds., 2nd ed.) Vol. **8**, 3–40.
Nicholls, P. (1964). *Arch. Biochem. Biophys.* **106**, 25–48.
Nicholls, P. (1965). "Amherst", II, pp. 764–779.
Nicholls, P. (1968). "Osaka", pp. 76–85.
Nicholls, P. and Malviya, A. N. (1968). *Biochemistry* **7**, 305–310.
Nicholls, P., Mochan, E. and Kimelberg, H. K. (1969). *FEBS Letters* **3**, 242–246.
Nielsen, S. O. and Lehninger, A. L. (1954). *J. Am. Chem. Soc.* **76**, 3860–3861.
Nijs, P. (1967). *Biochim. et Biophys. Acta* **143**, 454–461.
Nilsson, R. (1969). *Biochim. et Biophys. Acta* **184**, 237–251.
Nishibayashi, H., Omura, T. and Sato, R. (1963). *Biochim. et Biophys. Acta* **67**, 520–522.
Nishibayashi, H., Yamashita, H. and Sato, R. (1970). *J. Biochem.* **67**, 199–210.
Nordenbrand, K. and Ernster, L. (1970). *8th Internat. Congress Biochem., Sympos.* **3**, 153–154.
Nygaard, A. P. (1961a). *J. Biol. Chem.* **236**, 920–925.
Nygaard, A. P. (1961b). *J. Biol. Chem.* **236**, 2128–2132.
O'Brien, R. L. O. and Brierley, G. (1965). *J. Biol. Chem.* **240**, 4527–4531.
Oda, T. (1968). "Osaka", pp. 500–515.
Ogamo, A., Suzuki, Y. and Okui, S. (1968). *J. Biochem.* **63**, 582–590.

Ohnishi, K. (1966). *J. Biochem.* **59**, 1–23.
Ohnishi, T. (1970). *Biochem. Biophys. Res. Commun.* **41**, 344–352.
Ohnishi, T., Kawaguchi, K. and Hagihara, B. (1966). *J. Biol. Chem.* **241**, 1797–1806.
Okui, S., Suzuki, Y. and Momose, K. (1967). *J. Biochem.* **54**, 471–476.
Okunuki, K. (1962). In (O. Hayaishi, ed.). "Oxygenases", pp. 409–467.
Okunuki, K., Wada, H., Matsubara, H. and Takemori, S. (1965). "Amherst", II pp. 549–564.
Omura, T. and Sato, R. (1964). *J. Biol. Chem.* **239**, 2370–2378.
Omura, T. and Takesue, S. (1970). *J. Biochem.* **67**, 249–257.
Orii, Y. and Okunuki, K. (1967). *J. Biochem.* **61**, 388–400.
Orii Y., Sekuzu, I. and Okunuki, K. (1962). *J. Biochem.* **51**, 204–215.
Orrenius, S., Berg, A. and Ernster, L. (1969). *Europ. J. Biochem.* **11**, 193–200.
Packer, L. and Mustafa, M. G. (1966). *Biochim. et Biophys. Acta* **113**, 1–12.
Painter, A. A. and Hunter, F. E. Jr. (1970a). *Science* **170**, 552–553.
Painter, A. A. and Hunter, F. E. Jr. (1970b). *Biochem. Biophys. Res. Commun.* **40**, 360–395.
Palade, G. E. (1952). *Anat. Record* **114**, 427.
Palade, G. E. (1956). In (O. H. Gaebler, ed.). "Enzyme Units of Biological Structure and Function," p. 185. Academic Press, New York.
Palade, G. E. (1964). *Proc. Nat. Acad. Sci.* **52**, 613–643.
Palmieri, F. and Klingenberg, M. (1967). *Europ. J. Biochem.* **1**, 439–446.
Palmieri, F. and Quagliariello, E. (1970). *Boll. Soc. Ital. Biol. Sper.* **46**, 125–128.
Papa, S., Lofrumento, N. E., Paradies, G. and Quagliariello, E. (1969). *Biochim. et Biophys. Acta* **180**, 35–44.
Parkes, J. G. and Thompson, W. (1970). *Biochim. et Biophys. Acta* **196**, 162–169.
Parsons, D. F. and Williams, G. R. (1967). "Methods of Enzymol". **10**, 443–448.
Parsons, D. F. and Yano, Y. (1967). *Biochim. et Biophys. Acta* **135**, 362–364.
Parsons, D. F., Williams, G. R. and Chance, B. (1966). *Ann. New York Acad. Sci.* **137**, 643–666.
Parsons, D. F., Williams, G. R., Thompson, W., Wilson, D. and Chance, B. (1967) In "Mitochondrial Structure and Compartmentalisation", p. 29. Adriatica, Bari.
Penefsky, H. S. (1962). *Biochim. et Biophys. Acta* **58**, 619–621.
Penn, N. and Mackler, B. (1958). *Biochim. et Biophys. Acta* **27**, 539–543.
Penniall, R. (1958). *Biochim. et Biophys. Acta* **30**, 247–251.
Penniston, J. T. and Green, D. E. (1968). "Osaka", pp. 516–520.
Penniston, J. T., Harris, R. A., Asai, J. and Green, D. E. (1968). *Proc. Nat. Acad. Sci.* **59**, 624–631.
Perutz, M. T. (1970). *Nature* **228**, 726–739.
Phillips, A. H. and Langdon, R. G. (1962). *J. Biol. Chem.* **237**, 2652–2660.
Pinna, L. A., Lorin, M., Moret, V. and Siliprandi, N. (1967). *Biochim. et Biophys. Acta* **143**, 18–25.
Pollard, C. J. and Bieri, J. G. (1958). *Biochim. et Biophys. Acta* **30**, 658–659.
Polonovsky, J. and Gay, A. (1956). *Bull. Soc. Chim. Biol.* **38**, 475–480.
Potter, V. R. and Reif, A. E. (1952). *J. Biol. Chem.* **194**, 287–297.
Pressman, B. C. (1968). *Fed. Proc.* **27**, (6), 1283–1288.
Pressman, B. C. (1970). In (E. Racker, ed.) "Membranes of Mitochondria and Chloroplasts", pp. 213–244. Van Nostrand Reinhold Co., New York.
Prezbindowski, K. S., Ruzicka, F. J., Sun, F. F. and Crane, F. L. (1968). *Fed. Proc.* **27**, Abstr. 1377.
Pullman, M. E. and Schatz, G. (1967). *Ann. Rev. Biochem.* **36**, 539–610.

Pullman, M., Penefsky, H. S., Datta, A. and Racker, E. (1960). *J. Biol. Chem.* **235**, 3322–3330.
Pumphrey, A. M. and Redfearn, E. R. (1963). *Biochim. et Biophys. Acta* **74**, 317–327.
Racker, E. (1965). "Mechanisms in Bioenergetics" Academic Press, New York.
Racker, E. (1970). In "Membranes of Mitochondria and Chloroplasts (E. Racker, ed.) pp. 127–168. Van Nostrand Reinhold Co., New York.
Racker, E. and Horstman, L. L. (1967). *J. Biol. Chem.* **242**, 2547–2551.
Racker, E., Tyler, D. D., Estabrook, R. W., Conover, T. E., Parsons, D. F. and Chance, B. (1965). "Amherst", II, pp. 1077–1096.
Ragan, C. I. and Garland P. B. (1969). *Europ. J. Biochem.* **10**, 399–410.
Ramirez, J. and Mujica, A. (1964). *Biochim. et Biophys. Acta* **86**, 1–13.
Rapoport, S. and Wagenknecht, C. (1957). *Naturwiss.* **44**, 515–516.
Raw, I., Molinari, R., Do Amaral, D. F. and Mahler, H. R. (1958). *J. Biol. Chem.* **233**, 225–229.
Redfearn, E. R. (1961). In "Biological Structure and Function". London, New York: Academic Press.
Redfearn, E. R. and King, T. E. (1964). *Nature* **202**, 1313–1316.
Redfearn, E. R. and Pumphrey, A. M. (1958). *Biochim. et Biophys. Acta* **30**, 437–438.
Redfearn, E. R. and Pumphrey, A. M. (1960). *Biochem. J.* **76**, 64–71.
Redfearn, E. R. and Whittaker, P. A. (1966). *Biochim. et Biophys. Acta* **118**, 413–418.
Redfearn E. R., Pumphrey, A. M. and Fynn, G. H. (1960). *Biochim. et Biophys. Acta* **44**, 404–415.
Redfearn, E. R., Whittaker, P. A. and Burgos, J. (1965). "Amherst" II, pp. 943–959.
Reif, A. E. and Potter, V. R. (1953). *J. Biol. Chem.* **205**, 279–290.
Richardson, S. H. and Fowler, L. R. (1963). *Arch. Biochem. Biophys.* **100**, 547–553.
Richardson, S. H., Hultin, H. O. and Fleischer, S. (1964). *Arch. Biochem. Biophys.* **105**, 254–260.
Richardson, S. H., Hultin, H. O. and Green, D. E. (1963). *Proc. Nat. Acad. Sci.* **50**, 821–827.
Riemersma, J. C. (1968). *Biochim. et Biophys. Acta* **153**, 80–87.
Rieske, J. S. (1967). "Methods in Enzymol". **10**, 488–493.
Rieske, J. S. and Zaugg, W. S. (1962). *Biochem. Biophys. Res. Commun.* **8**, 421–426.
Rieske, J. S., Hansen, R. E. and Zaugg, W. S. (1964a). *J. Biol. Chem.* **239**, 3017–3022.
Rieske, J. S., Zaugg, W. S. and Hansen, R. E. (1964b). *J. Biol. Chem.* **239**, 3023–3030.
Rieske, J. S., Baum, H., Stoner, C. D. and Lipton, S. H. (1967a). *J. Biol. Chem.* **242**, 4854–4866.
Rieske, R. S., Lipton, S. H., Baum, H. and Silman, H. I. (1967b). *J. Biol. Chem.* **242**, 4888–4896.
Ringler, R. L., Minakami, S. and Singer, T. P. (1960). *Biochem. Biophys. Res. Commun.* **3**, 417–421.
Robertson, R. N. (1960). *Biol. Rev. Cambridge Philos. Society* **35**, 231–264.
Robertson, R. N. (1969). *Endeavour* **26**, 134–139.
Rosenberg, B. and Postow, E. (1969). *Ann. New York Acad. Sci* **158**, 161–190.
Rossi, E., Norling, B., Persson, B. and Ernster, L. (1970). *Europ. J. Biochem.* **16**, 508–513.

Roy, B. L. (1964). *Nature* **201**, 80–81.
Sanadi, D. R. (1965). *Ann. Rev. Biochem.* **34**, 21–43.
Sanadi, D. R. and Jacobs, E. E. (1967). "Methods in Enzymol". **10**, 38–41.
Sanadi, D. R., Pharo, R. L. and Sordahl, L. H. (1967). "Methods in Enzymol". **10**, 297–302.
Sands, R. H. and Beinert, H. (1960). *Biochem. Biophys. Res. Commun.* **3**, 47–52.
Sasame, H. A. and Gillette, J. R. (1970). *Arch. Biochem. Biophys.* **140**, 113–121.
Sauer, L. A. (1970). *Arch. Biochem. Biophys.* **139**, 340–350.
Schatz, G. and Racker, E. (1966). *J. Biol. Chem.* **241**, 1429–1438.
Schatz, G., Racker, E., Tyler, D. D., Gonze, J. and Estabrook, R. W. (1966). *Biochem. Biophys. Res. Commun.* **22**, 585–590.
Schneider, W. C. and Hogeboom, G. H. (1950). *J. Biol. Chem.* **183**, 123–128.
Scholz, R., Thurman, R. G., Williamson, J. R., Chance, B. and Bücher, T. (1969). *J. Biol. Chem.* **244**, 2317–2324.
Schulze, H. U., Gallenkamp, H. and Staudinger, Hj. (1970). *Z. physiol. Chem.* **351**, 809–817.
Seki, S. (1969). *Acta. Med. Okayama* **23**, 69–88; 175–200. *Chem. Abstr.* **72** (2) 9379.
Sekuzu, I., Mizushima, S., Hirota, T., Yubisui, T., Matsumura, Y. and Okunuki, K. (1967). *J. Biochem.* **62**, 710–718.
Senior, A. E. and MacLennan, D. H. (1970). *J. Biol. Chem.* **245**, 5086–5095.
Sharp, C. W., Mackler, B., Douglas, H. C., Palmer, G. and Felton, S. P. (1967). *Arch. Biochem. Biophys.* **122**, 810–812.
Shichi, H., Sugimura, Y. and Funahashi, S. (1965). *Biochem. et Biophys. Acta* **97**, 483–497.
Shug, A. L., Ferguson, Sh. and Sarago, E. (1968). *Biochem. Biophys. Res. Commun.* **32**, 81–85.
Siekevitz, P. and Watson, M. L. (1956). *J. Biophys. and Biochem. Cytol.* **2**, 653–669.
Silman, H. I., Rieske, J. S., Lipton, S. J. H. and Baum, H. (1967). *J. Biol. Chem.* **242**, 4876–4875.
Singer, T. P. (1968). In (T. P. Singer, ed.) "Biological Oxidations", Interscience, pp. 339–377.
Singer, T. P. and Sanadi, D. R. (1966). *J. Biol. Chem.* **242**, 4867–4875.
Singer, T. P., Kearney, E. B. and Bernath, P. (1956). *J. Biol. Chem.* **223**, 599–613.
Sjöstrand, F. S. (1953). *Nature* (London) **171**, 30.
Skulachev, V. P., Evtodienko, Yu. V., Smirnova, E. G., Chistyakov, V. G. and Yasaitis, A. A. (1966). *Chem. Abstr.* **66**, (1) 71; **66** (7) 52233; **66** (13) 112237.
Slater, E. C. (1948). *Nature* **161**, 405.
Slater, E. C. (1949a). *Biochem J.* **45**, 14–30.
Slater, E. C. (1949b). *Nature* **163**, 532.
Slater, E. C. (1949c). *Biochem. J.* **44**, 305–318.
Slater, E. C. (1950a). *Biochem. J.* **46**, 484–503.
Slater, E. C. (1950b). *Nature* **165**, 674.
Slater, E. C. (1953). *Nature* **172**, 975.
Slater, E. C. (1954). *Nature* **174**, 1143.
Slater, E. C. (1955). *Biochem. J.* **59**, 392–405.
Slater, E. C. (1958). *Advances Enzymol.* **20**, 147–199.
Slater, E. C. (1960). 4th Internat. Congress Biochem. 1958, XI Vitamin Metabolism. Permagon Press, 316–344.
Slater, E. C. (1961a). *Amer. J. Clin. Nutrition* **9**, 50–60.
Slater, E. C. (1961b). *Biochim. et Biophys. Acta* **48**, 117–131.

Slater, E. C. (1966a). (E. C. Slater, Z. Kaniuga and L. Wojtczak) "Biochemistry of Mitochondria", Academic Press, London, Warsaw 1–10.
Slater, E. C. (1966b). In (M. Florkin and E. H. Stotz, eds.) "Comprehensive Biochemistry" **14**, 327–396. Elsevier, Amsterdam.
Slater, E. C. (1966c). (E. C. Slater, ed.) "Flavins and Flavoproteins", Elsevier, Amsterdam.
Slater, E. C. (1967a). 7th Internat. Congress Biochem., Tokyo Sympos. Abstracts. VI, 2, 1.
Slater, E. C. (1967b). European J. Biochem. **1**, 317–326.
Slater, E. C. (1969). In (S. Papa, J. M. Tager, E. Quagliariello and E. C. Slater eds.) "The Energy Level and Metabolic Control in Mitochondria", Adriatica Ed., Bari. pp. 255–259.
Slater, E. C. (1970a). In (J. M. Tager, S. Papa, E. Quagliariello and E. C. Slater eds.) "Electron Transport and Energy Conservation" p. 1–3, pp. 363–369. Adriatica Ed., Bari.
Slater, E. C. (1970b). In (J. M. Tager, S. Papa, E. Quagliariello, and E. C. Slater, eds.) "Electron Transport and Energy Conservation" pp. 533–536, Adriatica Ed., Bari.
Slater, E. C. and Colpa-Boonstra, J. P. (1961). "Haematin Enzymes" II, pp. 575–596; 622.
Slater, E. C. and Terwelle, H. F. (1969). 20th Colloquium, Ges. Biol. Chem., Mosbach.
Slater, E. C., Berden, J. A. and Bertina, R. M. (1970a). 8th Internat. Congress. Biochem. Sympos. IV, pp. 157–159.
Slater, E. C., Lee, C. P., Berden, J. A. and Wegdam, H. J. (1970a,b). Nature **226**, 1248–1249.
Slater, E. C., Lee, C. P., Berden, J. H. and Wegdam, H. J. (1970c). Biochim. et Biophys. Acta **223**, 354–364.
Slater, E. C., Rudney, H., Bouman, H. and Links, J. (1961). Biochim. et Biophys. Acta **47**, 497–515.
Slater, E. C., Van Gelder, B. F. and Minnaert, K. (1965). "Amherst" II, pp. 667–700.
Smith, L. (1955). J. Biol. Chem. **215**, 833–846.
Smith, L. and Camerino, P. W. (1963). Biochemistry **2**, 1428–1432; 1432–1439.
Smith, L. and Conrad, H. (1961). "Haematin Enzymes" I, pp. 260–274.
Smith, L. and Minnaert, K. (1965). Biochim. et Biophys. Acta **105**, 1–14.
Smith, L., Newton, N. and Scholes, P. (1966). "Hemes and Hemoproteins", pp. 395–403.
Snoswell, A. M. (1962). Biochim. et Biophys. Acta **60**, 143–157.
Snoswell, A. M. and Cox, G. B. (1968). Biochim. et Biophys. Acta **162**, 455–458.
Somlo, M. (1962). Biochim. et Biophys. Acta **65**, 333–346.
Sorger, G. J. (1963). Biochem. Biophys. Res. Commun. **12**, 395–401.
Sottocasa, G. L. (1967). Biochem. J. **105**, 1P.
Sottocasa, G. L., Kuylenstierna, B., Ernster, L. and Bergstrand, A. (1967a). "Methods in Enzymol". **10**, 448–463.
Sottocasa, G. L., Kuylenstierna, B., Ernster, L. and Bergstrand, A. (1967b). J. Cell Biol. **32**, 415–438.
Souverijn, J. M., Weigers, P. J., Groot, G. S. and Kemp, A. Jr. (1970). Biochim. et Biophys. Acta **223**, 31–35.
Stasny, J. T. and Crane, F. L. (1964). J. Cell. Biol. **22**, 49–62.
Stockdale, M., Dawson, A. P. and Selwyn, M. J. (1970). European J. Biochem **15**, 342–351.

Stoeckenius, W. (1970). In (E. Racker, ed.) "Membranes of Mitochondria and Chloroplasts" pp. 53–90. S.C.S. Monograph; Van Nostrand Reinhold Co., New York.
Stoppani, A. O. M. and Brignone (1957). *Arch. Biochem. Biophys.* **66**, 91–99.
Storey, B. T. (1968). *Arch. Biochem. Biophys.* **126**, 585–592.
Storey, B. T. (1970a). *J. Theor. Biol.* **28**, 233–259.
Storey, B. T. (1970b). *Plant Physiol.* **46**, 625–630.
Storey, B. and Chance, B. (1967). *Arch. Biochem. Biophys.* **121**, 279–289.
Stotz, E., Altschul, A. M. and Hogness, T. R. (1938). *J. Biol. Chem.* **124**, 745–754.
Straub, K. D. and Lynn, W. D. (1967). *Fed. Proc.* **26**, Abstr. 2646.
Streichman, S. and Avi-Dor, Y. (1967). *Biochem. J.* **104**, 71–77.
Strickland, E. H. and Ackerman, E. (1965). *J. Theor. Biol.* **10**, 114–124.
Strittmatter, P. (1965). *J. Biol. Chem.* **240**, 4481–4487.
Strittmatter, P. (1967a). *J. Biol. Chem.* **242**, 4630–4636.
Strittmatter, P. (1967b). "Methods in Enzymol". **10**, 561–565.
Strittmatter, P. and Velick, S. F. (1956). *J. Biol. Chem.* **221**, 277–286.
Strittmatter, P. and Velick, S. F. (1957). *J. Biol. Chem.* **228**, 785–799.
Strong, F. M., Dickie, J. P., Loomans, M. E., Van Tamelen, E. E. and Dewey, R.S. (1960). *J. Am. Chem. Soc.* **82**, 1513–1514.
Szarkowska, L. (1966). *Arch. Biochem. Biophys.* **113**, 519–525.
Szent-Gyorgyi, A. (1941). *Science* **93**, 609–611.
Szent-Gyorgyi, A. E., Isenberg, I. and McLaughlin, J. (1961). *Proc. Nat. Acad. Sci.* **47**, 1089–1093.
Szent-Gyorgyi, A. (1968). *Proc. Nat. Acad. Sci* **58**, 2012–2014.
Tager, J. M., Howland, J. L. and Slater, E. C. (1962). *Biochim. et Biophys. Acta* **58**, 616–618.
Tager, J. M., Howland, C. J. L., Slater, E. C. and Snoswell, A. M. (1963). *Biochim. et Biophys. Acta* **77**, 266–275.
Tager, J. M., Veldsema-Currie, R. D. and Slater, E. C. (1966). *Nature* **212**, 376–379.
Takemori, S. and King, T. E. (1964a). *J. Biol. Chem.* **239**, 3546–3558.
Takemori, S. and King, T. E. (1964b). *Science* **144**, 852.
Takemori, S., Wada, K., Ando, M., Hosokawa, M., Sekuzu, I. and Okunuki, K. (1962a). *J. Biochem.* **52**, 28–37.
Takemori, S., Wada, K., Sekuzu, I. and Okunuki, K. (1962b). *Nature* **195**, 456–457.
Takesue, S. and Omura, T. (1970a). *Biochem. Biophys. Res. Commun.* **40**, 396–401.
Takesue, S. and Omura, T. (1970b). *J. Biochem.* **67**, 259–276.
Tappel, A. L. (1960). *Biochem. Pharmacol.* **3**, 289–296.
Taylor, C. P. S. (1960). *Dissertation Abstr.* **21**, No. 4, 150.
Teeter, M. E., Baginsky, M. L. and Hatefi, Y. (1969). *Biochim. et Biophys. Acta* **172**, 331–333.
Thorn, M. B. (1956). *Biochem. J.* **63**, 420–436.
Tisdale, H. D. (1967). "Methods in Enzymol" Vol. **10**, pp. 213–215.
Tobin, R. B. and McIlwain, H. (1965). *Biochim. et Biophys. Acta* **105**, 191–192.
Tobin, R. B. and Slater, E. C. (1965). *Biochim. et Biophys. Acta* **105**, 214–220.
Tsou, C. L. (1952). *Biochem. J.* **50**, 493–499.
Tu, Sh.-I. and Wang, J. H. (1970). *Biochemistry* **9**, 4505–4509.
Tyler, D. D. (1970). *Biochem. J.* **116**, 30–31P (1970).
Tyler, D. D. and Estabrook, R. W. (1966). *J. Biol. Chem.* **241**, 1672–1680.
Tyler, D. D., Butow, R. A., Gonze, J. and Estabrook, R. W. (1965). *Biochem. Biophys. Res. Commun.* **19**, 551–557.

Tzagoloff, A. (1969). *J. Biol. Chem.* **244**, 5020–5026.
Tzagoloff, A. and McConnell, D. G. (1967). *Fed. Proc.* **26**, Abstr. 153.
Tzagoloff, A. and MacLennan, J. H. (1966). In (J. Peisach, P. Aisen and W. E. Blumberg, eds.) "The Biochemistry of Copper," pp. 253–265. Academic Press, N.Y.
Tzagoloff, A., MacLennan, D. H. and Byington, K. H. (1968b). *Biochemistry* **7**, 1596–1602.
Tzagoloff, A., MacLennan, D. H., McConnell, D. G. and Green, D. E. (1967). *J. Biol. Chem.* **242**, 2051–2061.
Vallin, I. (1968). *Biochim. et Biophys. Acta* **162**, 477–486.
Vallin, I. and Löw, H. (1968). *Europ. J. Biochem.* **5**, 402–408.
Vanderkooi, G. and Green, D. E. (1970). *Proc. Nat. Acad. Sci. U.S.* **66**, 615–621.
Van Groningen, H. E. M. and Slater, E. C. (1963). *Biochim. et Biophys. Acta* **99**, 527–530.
Vanneste, W. H. (1966a). *Biochemistry* **5**, 838–848.
Vanneste, W. H. (1966b). *Biochim. et Biophys. Acta* **113**, 175–178.
Vázquez-Colón, L. and King, T. E. (1967). *Arch. Biochem. et Biophys.* **122**, 190–195.
Vigers, G. A. and Ziegler, F. D. (1968). *Biochem. Biophys. Res. Commun.* **30**, 83–88.
Vignais, P. V., Duce, E. D., Vignais, P. M. and Huet, J. (1966). *Biochim. et Biophys Acta* **118**, 465–483.
Wada, F. and Shimakawa, H. (1969). *J. Biochem.* **66**, 871–872.
Wada, K. (1964). *Ann. Reports, Sci. Works, Fac. Sci. Osaka Univ.* **12**, 19–54.
Wagenknecht, C. and Rappoport, S. (1964). *Acta Biol. Med. Ger.* **12**, 542–551.
Wagenknecht, C., Rappoport, S. and Nieradt-Hiebsch, Ch. (1959). *Acta Biol. Med. Ger.* **3**, 321–329.
Waggoner, A. S. and Stryer, L. (1970). *Proc. Nat. Acad. Sci.* **67**, 579–589.
Wallach, D. F. H. and Gordon, A. (1968). *Fed. Proc.* **27**, 1263–1268.
Wang, J. H. (1970). *Science* **167**, 25–30.
Warburg, O. (1949). "Heavy Metal Prosthetic Groups and Enzyme Action". Clarendon Press, Oxford.
Watari, H., Kearney, E. B. and Singer, T. P. (1963). *J. Biol. Chem.* **238**, 4063–4073.
Webster, G. (1962). *Biochem. Biophys. Res. Commun.* **7**, 245–249.
Webster, G. (1963). *Biochem. Biophys. Res. Commun.* **13**, 399–404.
Webster, G. and Green, D. E. (1964). *Proc. Nat. Acad. Sci. U.S.* **52**, 1170–1176.
Webster, G., Smith, A. L. and Hansen, M. (1963). *Proc. Nat. Acad. Sci. U.S.* **49**, 259–266.
Wegdam, H. J., Berden, J. A. and Slater, E. C. (1970). *Biochim. et Biophys. Acta* **223**, 365–373.
Weidemann, M. J., Erdelt, H. and Klingenberg, M. (1970). *Europ. J. Biochem.* **16**, 313–335.
Weiss, H., Jagow, G., Klingenberg, M. and Bücher, T. (1970). *Europ. J. Biochem.* **14**, 75–82.
Wheeldon, L. W. (1958). *Biochim. et Biophys. Acta* **29**, 321–332.
Widmer, C., Clark, H. W., Neufeld, H. A. and Stotz, E. (1954). *J. Biol. Chem.* **219**, 851–867.
Wikstrom, M. K. F. and Saris, N. E. L. (1969). *European J. Biochem.* **9**, 160–166.
Williams, C. H., Jr., and Kamin, H. (1962). *J. Biol. Chem.* **237**, 587–595.
Williams, J. N. Jr. (1968). *Biochim. et Biophys Acta* **162**, 175–181.
Williams, J. N. Jr. and Thorp, S. L. (1970). *Arch. Biochem. Biophys.* **141**, 622–631.

Williams, R. J. P. (1970). *8th Internat. Congress Biochem.*, *Switzerland*, *1969*, *Sympos.* 2, 126–128.
Wilson, D. F. (1967a). *7th Internat. Congress Biochem.*, *Tokyo*, V H-81.
Wilson, D. F. (1967b). *Biochim. et Biophys. Acta* **131**, 431–440.
Wilson, D. F. (1970). *Fed. Proc.* **29**, Abstr. 882.
Wilson, D. F. and Azzi, A. (1968). *Arch. Biochem. Biophys.* **126**, 724–726.
Wilson, D. F. and Chance, B. (1966). *Biochem. Biophys. Res. Commun.* **23**, 751–756.
Wilson, D. F. and Chance, B. (1967). *Biochim. et Biophys. Acta* **131**, 421–430.
Wilson, D. F. and Dutton, P. L. (1970a). *Arch. Biochem. Biophys.* **136**, 583–585.
Wilson, D. F. and Dutton, P. L. (1970b). *Biochem. Biophys. Res. Commun.* **39**, 59–64.
Wilson, D. F. and Gilmour, M. V. (1967). *Biochim. et Biophys. Acta* **143**, 52–61.
Wilson, D. F. and King, T. E. (1967). *Biochim. et Biophys. Acta* **131**, 265–279.
Wilson, D. F. and Merz, R. (1969). *Arch. Biochem. Biophys.* **129**, 79–85.
Wilson, M. and Cascarono, J. (1970). *Biochim. et Biophys. Acta* 216, 54–62.
Wilson, S. B. and Bonner, W. D. (1970). *Plant Physiol.* **46**, 25–30; 31–35.
Winkler, H. H. and Lehninger, A. (1968). *J. Biol. Chem.* **243**, 3000–3008.
Wohlrab, H. (1969a). *Fed Proc.* **28**, Abstr. 3673.
Wohlrab, H. (1969b). *Biochem. Biophys. Res. Commun.* **35**, 560–564.
Wohlrab, H. (1970). *Biochemistry* **9**, 474–479.
Wohlrab, H. and Jacobs, E. E. (1967). *Biochem. Biophys. Res. Commun.* **28**, 991–1002.
Wohlrab, H. and Ogunmole, G. B. (1971). *Biochemistry* **10**, 1103–1106.
Wojtczak, L. and Zaluska, H. (1969). *Biochim. et Biophys. Acta* **193**, 64–72.
Yakushiji, E. and Okunuki, K. (1940). *Proc. Imper. Acad. Japan* **16**, 299–302.
Yamanaka, T. and Kamen, M. D. (1965). *Biochem. Biophys. Res. Commun.* **19**, 751–754.
Yamanaka, T. and Okunuki, K. (1968). Osaka, pp. 390–403.
Yamashita, S. (1968). *Fed. Proc.* **27**, Abstr. 1735.
Yamashita, S. and Racker, E. (1968). *J. Biol. Chem.* **243**, 2446–2467.
Yamashita, S. and Racker, E. (1969). *J. Biol. Chem.* **244**, 1220–1227.
Yang, W. C., Yanasugondia, D. and Webb, J. L. (1958). *J. Biol. Chem.* **232**, 659–668.
Yonetani, T. (1960). *J. Biol. Chem.* **235**, 3138–3143.
Yonetani, T. (1962). *J. Biol. Chem.* **237**, 550–559.
Yonetani, T. and Ray, G. (1965). *J. Biol. Chem.* **240**, 3392–3398.
Yong, F. C., King, T. E., Oldham, S., Waterman, M. R. and Mason, H. S. (1970). *Arch. Biochem. Biophys.* **138**, 96–100.
Young, J. H., Blondin, G. A., Vanderkooi, G. and Green, D. E. (1970). *Proc. Nat. Acad. Sci.* **67**, 550–559.
Zahler, W. L., Ozawa, H. and Fleischer, S. (1969). *Fed. Proc.* **28**, Abstr. 3685.
Zalkin, H. and Racker, E. (1965). *J. Biol. Chem.* **240**, 4017–4022.
Ziegler, D. M., Green, D. E. and Doeg, K. (1959). *J. Biol. Chem.* **234**, 1916–1921.

Chapter VIII

Some physiological aspects of cytochromes

A. Introduction.
B. General biological aspects.
   1. Distribution of cytochromes in the cell.
   2. Anoxia and hyperoxia.
   3. Temperature regulation and acclimatization.
   4. Nutrition.
   5. Histochemistry.
C. Mammalian and other vertebrate organs.
   1. Heart and skeletal muscle.
   2. Liver, kidney, spleen.
      a. Liver mitochondria.
      b. Liver microsomes.
      c. Kidney and spleen.
   3. Nerves and brain.
   4. Other mammalian tissues.
      a. Gastric and intestinal mucosa.
      b. Arteries.
      c. Blood.
      d. Skin and glands.
      e. Spermatozoa.
      Effects of hormones on respiratory enzymes.
      a. Sexual hormones.
      b. Adrenals.
      c. Thyroid.
   6. Tumours.
D. Invertebrates.
E. Plants and fungi.
   References.

A. INTRODUCTION

In a postscript to a recent Symposium, Drabkin (1966) complained about the "forgotten cell", with mitochondria and microsomes becoming the major subjects of research. Yet there seem to be valid reasons for temporary setting aside of problems of the complexity of the whole cell. We have seen in the preceding chapters that even the slightly less complex problems on the level of e.g. mitochondria are not yet overcome. The difficulties at the level of the whole cell or still more so the total organism are still several magnitudes greater. There is the age-old contempt of the "holist" for the plodding analyser, and his criticism is justified if at no stage attempts are made to integrate the partial knowledge into a whole. Yet such attempts are bound to fail if they are made prematurely, and for this reason our present knowledge of the physiology of the cytochromes leaves much to be desired. There is much uncertain, and even contradictory evidence and a satisfactory integration often can not be achieved.

Several aspects of cytochrome physiology had to be discussed in preceding chapters since they were intimately connected with the chemistry and biochemistry of individual cytochromes, e.g. cytochrome aa_3 (Chapter III, B1), the physiology and pharmacology of cytochromes b_5 and P-450 (Chapter IV, C and IV, D) and of the enterochromes (Chapter IV, E), the role of cytochromes c and $c(f)$ in green plant and algal photosynthesis (Chapters IV, F and V, C) and in bacterial photosynthesis (Chapter VI) as well as the role of cytochrome c in mammalian organs (Chapter V, D) and of cytochromes in bacterial metabolism (Chapter VI) and in the respiratory chain (Chapter VII, see also IV, F). Thus it remains to discuss the more physiological aspects in which frequently more than one type of cytochrome is involved. This may give the chapter the unsatisfactory character of "left-overs", but its purpose is to integrate our knowledge of the chemistry and biochemistry of the cytochromes already discussed from the viewpoint of more physiological and general biological aspects.

B. GENERAL BIOLOGY

1. Distribution of cytochromes in the cell

Again, the occurrence of cytochromes aa_3, b, c and c_1 in the inner membrane of mitochondria, of these and related cytochromes in the bacterial cytoplasmic membrane, of b_5 and P-450 in adrenal mitochondria and in microsomes has been described previously. The occurrence of cytochrome c in some cell nuclei has also been briefly mentioned (Chapter V, B7). In spite of earlier findings of cytochrome oxidase but not cytochrome aa_3 (Bereznay et al., 1970a,b) and of NADH-oxidase in rat liver cell nuclei (Penniall et al., 1964; Currie et al., 1966), the presence of cytochrome oxidase in them is still in doubt (Klouwen et al., 1965; Conover, 1967a,b, 1968) although recently Bereznay and Crane (1971) have claimed that cytochrome aa_3 is present in the nuclear membrane in 30% of the mitochondrial concentration. 91–95% of the cytochrome aa_3 of rat liver is present in mitochondria, 4% in microsomes and the remainder in cell sap, pre-microsomes and nuclei (Davidian et al., 1968). It will be shown, however, in Chapter IX that the synthesis of apocytochrome c is carried out in the endoplasmic reticulum. In contrast nuclei from rat and calf thymus cells contain cytochrome aa_3, c, b and cytochrome c reductase, the nucleus containing no less than 40% of the total cytochrome aa_3 (Yamagata at al., 1966; Betel, 1967; Conover, 1967a,b; Ueda et al., 1969). This nuclear respiratory chain is linked with oxidative phosphorylation (Klouwen et al., 1965; Betel and Klouwen, 1967). The presence of cytochrome c is, however, still uncertain, although Tarshis et al. (1968) have reported effects of γ-radiation on cytochromes b and c in thymocyte nuclei.

By a highly sensitive method, Perry et al. (1959) studied the distribution of respiratory enzymes in single living cells, e.g. in the "Nebenkern", an accumulation of mitochondria of 5–10 μ diameter, in grasshopper spermatids. An absorption maximum at 420 nm with a shoulder at 434 nm (probably due to c^{2+} and b^{2+}) was found in an anaerobic glucose-containing suspension; a single band at 413 nm was found in the glucose-free suspension. NADH fluorescence at 443 nm was also concentrated in the "Nebenkern". The maxima of absorption found in single kidney and liver cells in mammals also agreed with those of cytochromes aa_3, b and c (Thorell et al., 1965).

2. Anoxia and hyperoxia

In moderate anoxia, corresponding to a height of up to 9,000 M above sea level, the well-known increase in haemoglobin and myoglobin is accompanied by an increase of cytochrome oxidase in brain, kidney and skeletal muscles, but not of heart and liver of rabbits and rats (Popov, 1966). Why cytochrome b should be more oxidized in chronic hypoxia of rats (Costa and Taquini, 1970) is not easy to understand. Mice acclimatized for oxygen for a fortnight at a height of 3,800 M above sea level survived small doses of cyanide better than mice acclimatized at 2,000 M or at sea level, probably on account of the changes in cytochrome oxidase (Barbashova et al., 1967). In severe hypoxia, corresponding to a height of more than 10,000 M, the cytochrome oxidase is generally found decreased (Popov, 1966; Ziegler, 1967). This decrease is also found in anaphylactic shock in guinea pigs, but ascribed to acidosis (Malyuk, 1965, 1966). Respiratory control was decreased in severe hypoxia, at less than 300 mM Hg pressure in rats (Ziegler, 1967).

Under conditions of severe hypoxia, injections of cytochrome c or "lipid cytochrome c" have been found to increase survival in rabbits (Ruiz et al., 1965), to abolish abnormalities of the electrocardiogram in human patients (Lee et al., 1966; Ziliotto and Curri, 1967) and in rabbits (Currie et al., 1966), to prevent ventricular fibrillation in cats after coronary ligature

(Brikker et al., 1969), to improve frog heart contraction after x-ray damage (Bryukhanov and Manoilov, 1967) and to improve cerebral atrophy in mice (Currie et al., 1966). While the cause of this effect is unknown and indeed difficult to understand, these findings are strong indications of the beneficial effects of cytochrome c injections and further studies appear desirable.

The regulation of pulmonary respiration is well known to be under partial control of chemoreceptors in the carotid body, a tiny organ of arterial capillary sinuses branching off the carotid artery and supplied by a sinus nerve (cf. Korner and Uther, 1970). In the cat the whole organ is about 1 mm in diameter. In spite of the smallness it has recently become possible to implicate reduced cytochrome a_3 as factor causing the firing discharge rate of the sinus nerve. Woods (1967) showed that cytochrome oxidase was present in the rabbit carotid. Carbon monoxide in the transfusion fluid increased the firing rate, while illumination (which releases CO and probably allows reoxidation of the oxidase) decreases the discharge. Purves (1970a,b) found that the decrease of chemoreceptor discharge bears a closer relation to the decrease in oxygen consumption than to the total blood flow. In contrast to earlier findings of Daly et al. (1954), it appears no longer possible to explain the great oxygen sensitivity of the carotid body by an exceptionally large oxygen consumption (Fay, 1970); the oxygen consumption is now found to be only one fifth of that reported by Daly et al. Mills and Frans (1970) have confirmed the close correlation between the reduction of cytochrome a_3 measured at 445 nm and the discharge rate of the receptor in the cat's carotid; they found, however, that the decrease of the discharge on oxygenation is far less sensitive to an increase of pO_2 than would be expected from the oxygen affinity of mitochondrial cytochrome oxidase. It would be premature to accept this as evidence for a different cytochrome a_3 in the carotid body. Apart from the fact that the absorption difference 445–465 nm measures cytochrome $a_3 + a$ rather than a_3 alone, and that reduced haemoglobin may also contribute to it, the differences found may be due to kinetic rather than thermodynamic factors, or to a special attachment of the cytochrome oxidase to the carotid membrane.

The toxic effect of hyperbaric oxygen is of special interest, both with regard to the clinical therapeutic effects of 100% and hyperbaric oxygen of 2–3 atm pressure, and of the breathing of pure oxygen in space travel. These aspects have been reviewed in the 1965 Symposium on "Hyperbaric Oxygen" in the New York Academy of Science and by Haugaard (1968). The toxic effects are fortunately less severe in man than in other animal species (Barber et al., 1970). Its biochemical explanation is still uncertain and is perhaps not the same in all conditions. It is agreed that the cytochrome chain and cytochrome oxidase in particular is not itself sensitive to hyperbaric oxygen (Riggs, 1945; Jamieson and Chance, 1966; Haugaard, 1968). The effect is on the substrate side of the cytochrome chain, either on SH-groups of flavin cytochrome-reductases (Barron, 1936; Dixon et al., 1960; see also below for algae) or on nicotinamide nucleotides; Dixon suggested that flavin autoxidation may lead to the formation of destructive radicals (see Chapter V, B6c). Chance et al. (1965, 1966) confirmed the oxidation of flavin enzymes in hyper-oxygenation, but found oxidized, not radical flavin. The oxidation was also too slow to account for the rapid onset of convulsions in hyperbaric oxygen. They postulate the rapid effect to be on the energy-requiring reduction of NADH and consequently of NADPH which appears to be very sensitive to hyperbaric oxygen. Oxidation of NADH could be demonstrated fluorometrically in pigeon and rat heart mitochondria, and also in rat kidney, liver and brain, in ascites tumour cells, and in yeast. Convulsions followed the NADH-oxidation in 5, death in 10 minutes. However, Straub (1967) found no decrease of NADH in 3 atm of oxygen in rats, although it produced toxic effects. Willis and Kratzing (1970) found hyperbaric oxygen to decrease the level of ascorbate but not of reduced glutathione.

In *Euglena*, in a low phosphate medium and generally in other euglenoid algae, e.g. *Astasia*, hyperbaric oxygen decreased the succinate- and NADH-cytochrome *c* reductases (Bégin-Heick and Blum, 1967; Blum and Bégin-Heick, 1967; Bégin-Heick, 1970). Toxic effects of oxygen on the annelid tube worm *Tubifex tubifex* have been described by Walker (1970).

3. Temperature regulation and acclimatization

The mechanism whereby warmblooded animals are able to keep their body temperature in narrow limits is still only partially elucidated. The source of the energy is the cytochrome system (see Chapter VII). Cytochrome oxidase is proportional to the body surface rather than to body weight, i.e. to (body weight)$^{0.67}$ and its activity varies less than 25% from its maximal activity. There is no increase of succinoxidase on cold acclimatization but it does decrease in hot acclimatization, at least in the liver but not in the muscle (Chaffee and Roberts, 1971). The percentage of the energy converted to heat is under hormonal control (see below), although other mechanisms such as shivering also play a role. Fat is the energetically richest substrate and adipose tissue (Smith and Horwitz, 1969), particularly brown adipose tissue abounds in the mammalian newborn and in cold-acclimatized and hibernating animals, at the same time playing a role as insulator. This tissue contains a large number of mitochondria which differ from those found in white adipose tissue (Rafael et al., 1970), so that direct conversion of one type to the other appears unlikely. The P/O ratio is low and only 25% of the energy is used for the production of ATP (Prusiner et al., 1968), in contrast to white adipose tissue. The usual cytochromes are, however, present in brown adipose tissue (Hook and Barron, 1941; Hagen and Ball, 1960; Joel and Ball, 1960, 1962; Eichel, 1959). The concentration of cytochrome c resembles that found in skeletal muscle but the concentration of the other respiratory enzymes is smaller. Paléus and Johannson (1968) found little difference in cytochrome c content between adipose tissue of hibernating and non-hibernating hedgehogs. Perhaps only the substrate site of phosphorylation (I) is active in brown adipose tissue (Kornacher and Ball, 1968; Aldridge and Street, 1968; Editorial in Nature, 1968) although the P/O ratio can be increased by serum albumin (Guillory and Racker, 1968). The uncoupling may well be caused by free fatty acids (Pressman and Lardy, 1956). Catecholamines strongly stimulate heat production (Prusiner et al., 1968; Kornacher and Ball, 1968; Reed and Fain, 1968). The mechanism of this stimulation and its relative role is not yet clear (see the recent review of Chaffee and Roberts, 1971). According to Prusiner (1970) the fatty acids released from the triglyceride stores by norepinephrine have the dual function of providing a heat-rich metabolite, and to decrease respiratory control; however, there is at least partial energy coupling even at maximal stimulation. A good correlation between the steady state redox levels of the cytochromes and the electron flux is indicated by trapped steady state spectra in liquid nitrogen. Reed and Fain (1968) suggest that catecholamines stimulate the formation of cyclic AMP thus initiating uncoupling either directly or through release of fatty acids. There also appears to be much desaturation of fatty acids in adipose tissue (Oshino and Sato, 1971).

Brown adipose tissue is increased by acclimatization to the cold (Jansky, 1961, 1963; Jansky et al., 1969). In the squirrel oxidative processes are distinctly depressed during hibernation in kidney and liver but not in the heart (Rusev and Stefanov, 1970). In the ground squirrel, *Citellus lateralis*, the number of mitochondria in the liver, cytochrome oxidase, cytochromes b and c, and oxygen uptake with succinate plus ADP were all decreased in hibernation (Shug et al., 1971). The sluggish oxygen uptake under phosphorylating conditions may be caused by a change in penetration of ADP and/or Pi into the mitochondria, since the uncoupler salicylanilide stimulated oxygen uptake to a rate even greater than in the normal animal. Depocas (1966) found a higher cytochrome c content in abdominal, but not in other muscles of the cold-acclimatized rat, whereas cytochrome c was increased in rat myocardium at an external temperature of 38° (D'Allessandro et al., 1966b). Heat acclimatization causes a decrease of cytochrome oxidase and succinoxidase in the liver, but a slight increase in heart and skeletal muscle (see also below with regard to the action of aldosterone).

The concentration of cytochrome oxidase and succinic dehydrogenase in the liver of poikilothermic vertebrates was found to be inversely proportional to the melanin content (Morgan and Singh, 1969; Morgan *et al.*, 1967; see also the earlier reviews of Bullock, 1955 and Prosser, 1955). In the eel, Malessa (1969) found the red muscles with much higher respiration, cytochrome oxidase and succinic dehydrogenase content and much more susceptible to temperature acclimatization than the slowly respiring and primarily glycolytic white muscles; the high respiration of cold-adapted eels may be biologically significant for their spawning migration. Temperature acclimatization in the goldfish has been studied by Freed (1965) and Caldwell (1969). Cytochrome oxidase, succinate- and NADH-cytochrome *c* reductases were all higher at 5–10° than at 30°. Maximum activity of cytochrome oxidase was found at 25°; two hours exposure to 45° inactivated 80% of the oxidase.

Similar effects were found in the nauplia of the brine shrimp *Artemia salina* (Engel and Angelovic, 1968). In the fiddler crab *Uca*, Vernberg and Vernberg (1968a) found increases of cytochrome oxidase by cold exposure in muscles, liver, brain and gills of species normally living in moderate climate, but not in tropical species; in a study on the effect of temperature acclimatization on the oxidative phosphorylation in blowfly muscle sarcosomes, Davison (1971) found the sarcosomes more sensitive *in vitro* than *in vivo*. The acclimatization pattern of the host snail was effected by larval trematode infestation (Vernberg and Vernberg, 1968b). Semenova (1967) reported an increase of granules in *Amoeba proteus* at 4°.

4. Nutrition and cytochromes

Minerals. Both copper and iron are essential constituents of cytochrome oxidase and iron is a constituent of all cytochromes. Copper has been for a long time recognized as an essential factor, the lack of which produces diseases in animals such as "enzootic ataxia" or "swayback" in lambs and "falling disease" in cattle (Bennets *et al.*, 1948; Howell and Davison, 1959; see also Lemberg *et al.*, 1962). The symptoms have been found to be due to a lack of cytochrome oxidase, in particular in the central nervous system (Barlow, 1963). Not only is copper an essential constituent of the oxidase but it also plays a role in the formation of haem *a*, its prosthetic group (see Chapters II and III, and Lemberg, 1969). It is also needed for the synthesis of phospholipids (Gallagher and Reeve, 1971). Cytochrome oxidase deficiency due to lack of Cu has also been found in rats (Dreosti, 1967) and in Wilson's disease in man (Shoheir and Schreffler, 1969). In Cu-deficiency in yeast, Wohlrab and Jacobs (1967) found only 5% of the usual acid-acetone extractable haem *a* in mitochondria which possibly still contain the apoprotein.

Iron deficiency lowers cytochrome oxidase in the kidney but not in the heart of rats (Beutler, 1959), but in long-lasting human iron-lack anaemia Korecky and Rakusan (1967) found a decrease in the heart. Petrov and

Kalinin (1969) found cytochrome oxidase decreased in the liver and other tissues and Dagg et al. (1966) in the buccal mucosa. The effect of iron on the biosynthesis of cytochromes is discussed in Chapter IX, A. In experimental ricketts, liver cytochrome oxidase was decreased but restored to normal by addition of Cu to the diet (Smolyar, 1968).

Prolonged protein deficiency decreases cytochromes b, c and c_1 in rat liver to about half the normal (Williams et al., 1966a,b, 1969 but see Pokrovskii et al. 1969, who found cytochrome oxidase but not the other cytochromes decreased in starvation) and phospholipid follows the same pattern. On restoration of protein feeding all rose to normal in 8 days. If DL-methionine was added to the diet only cytochrome c_1 was found decreased. Hepatic microsomal P-450 and the induction of its formation by phenobarbitone were decreased on a 15% casein diet which was otherwise nutritionally adequate (Marshall and McLean, 1967), and further decreased on a 3% casein diet.

5. Histochemistry

The most reliable methods of determining cytochrome oxidase and other cytochromes in tissues are spectrophotometric methods which are by now sufficiently sensitive (see e.g. Chapter VIII, B1 and Hess and Pope, 1953). By such methods the relative concentrations of cytochrome oxidase in rat tissue mitochondria, with rat heart mitochondria set as 100 were found to be 61 in kidney, 42 in brain cortex, 48 in cerebellum, 26 in liver, 32 in skeletal muscle and 10% in spleen; see also Lemberg et al. (1956) for rats and bovine cytochrome oxidase.

However, in some instances where exceptionally small amounts of cytochrome oxidase are present or a great excess of other haem pigments, histochemical methods relying on enzyme catalysts must be used. While most of the earlier methods are now somewhat suspect on account of lack of specificity—they may e.g. also measure peroxidative activity, more recent methods are more specific and sensitive, as e.g. the method of Nachlas et al. (1958) and the widely applied method of Burstone (1961). Seligman et al. (1967) use the osmophilic nature of indoanilines formed from N-benzyl-p-phenylenediamine for locating cytochromes by electron microscopy.

C. CYTOCHROME OXIDASE AND CYTOCHROMES IN MAMMALIAN AND OTHER VERTEBRATE ORGANS AND TISSUES

1. Heart and skeletal muscles

Heart. Although glycolytic metabolism is found in heart muscle cells and produces ATP, it and the phosphocreatine reserves are insufficient to provide the necessary energy for the work of the heart except for a very short period. The main part of the energy (98%) for contraction depends on aerobic metabolism and the cytochrome system (Keul et al., 1965a,b; Challoner, 1968b; Doll et al., 1965; Hatt et al., 1968). Non-phosphorylating respiration corresponds to only 13% of the total myocardial respiration in vivo, but to 85–90% of the basal oxygen consumption. The major part of energy is thus required for the work of contraction. Cyanide-insensitive respiration is only 3% of the total and about 20% of the oligomycin-insensitive respiration. In the human heart, of 100 ml O_2, 97 ml are utilized by mitochondria, 85 ml by aerobic formation of ATP, 2 ml are consumed by the membranes, 10 ml for mitochondrial non-phosphorylating oxidation and 3 ml for extramitochondrial, perhaps microsomal oxidation. Oxygen consumption by the human myocardium ($M\dot{V}O_2$) equals 4·8–8·4 μl/min/mg and 55–76% of the arterio-venous difference is extracted by the beating heart, depending primarily on the arterial oxygen pressure. In the quiescent papillary heart muscle of the cat, Cranefield and Greenspan (1960) found an oxygen consumption of 2·84 μl per mg wet tissue per hour, in the kitten 4·05 μl. The Q_{O_2} of rat heart slices was increased from 9·9 in air to 18·2 in 80% oxygen (Peschel and Georgiade, 1958). Dependence on pO_2 was greater in heart and kidney than in liver and spleen slices. Other data on oxygen consumption of hearts and on cytochrome c reductase have been reported by Lang et al. (1963), by Neely et al. (1967) and Gamble et al. (1970) in rats, and by Graham et al. (1968) in dogs. The succinic oxidase activity of the heart electric tissue (bundle of Sachs) is much lower than that of the heart muscle, and only as high as that of skeletal muscle (Eisenberg, 1958). Heart work depends on ATP and is inhibited by oligomycin which causes rapid arrest with decrease of ATP and of oxygen consumption. Recently it has been found possible to obtain single beating heart cells from embryonic chicken hearts and young rat hearts which continue to beat rhythmically for up to 4 days (Harary and Farley, 1963). They form synchrononously beating nests of cells, the beating of which depends on ATP and is arrested by oligomycin (Harary and Slater, see Chapter VII, H). Beating heart cells must have a mechanism for maintaining a high steady state of ATP, perhaps involving creatine phosphate (Seraydarian et al., 1968).

Experimental cardiac infarction, e.g. by ligation of coronary arteries or lymphatics decreased cytochromes (Cho and Symbas, 1965; Frolkis, 1968a,b; Golubev, 1968). A minimum of 20% of normal was reached after 4 days followed by a slow increase. Decrease was more rapid for cytochrome oxidase (in 2–3 days) and increase afterwards was also faster. The decrease was not restricted to the directly affected and necrotizing parts of the heart muscle. In some regions cytochrome oxidase was decreased more than the dehydrogenases. Cardiopulmonary by-pass in dogs greatly decreased cytochrome oxidase and NADH-oxidase (Cho, 1965). Bykov and Novikov (1968) found an extracorporeal shunt of the left ventricle in the dog to increase cytochrome oxidase activity in the left ventricle 2–3 times. In the foetus more cytochrome c is present in the right than in the left ventricle, but the difference disappears after birth (Dallman, 1966); in hypertrophied hearts with congenital lesions cytochrome c was higher in the more severely affected ventricle. Derhachev and Manoilov (1967) reported x-irradiation to increase cytochromes a and b in the heart, but not in the liver. In rat hearts hypertrophied by exercise (swimming), the mitochondrial mass was greatly increased (Arcos et al., 1968). There were, however, some foci with decreased succinic dehydrogenases even with optimal activity. Chronic cardiac denervation of the heart, in contrast to that of skeletal muscles, does not impair the functional integrity of the contractile apparatus or the regulation of oxygen consumption by the work load (Coleman, 1969, 1970).

In myocarditis, e.g. that produced by streptococcal toxin (Sokolov, 1966), cytochrome oxidase and succinic dehydrogenase were increased except when necrotic changes set in (Milanov and Dashev, 1965; Shakhnazorov, 1967), whereas ATPase was always decreased; Viti (1968) found a decrease of cytochrome c in heart muscle mitochondria (more than in other tissues) after administration of diphtheria toxin to guinea pigs. In amphibian hearts which beat slowly (cf. Wittenberg, 1970) cytochrome a_3 was 92–98% oxidized (Ramirez, 1959). During activity cytochromes a_3 and b were oxidized, cytochromes a and c reduced, indicating crossover points between a_3 and a and between c and b. The same was found in perfused lobster hearts (Ramirez, 1964) in which the concentration of cytochrome a_3 was higher than in frog or toad hearts. These findings seem to support a site of phosphorylation between cytochromes a and a_3 (but see Chapter VII, D6). It also is remarkable that the extinction coefficient at 445 nm is increased which does not harmonize with the assumed greater oxidation of cytochrome a_3, this band being to a larger extent due to a_3^{2+} than to a^{2+}. Digitalis (acetylstrophantidin) increased myocardial oxygen consumption but somewhat decreased cardial efficiency (Covell et al., 1966). Effects of histamine and endotoxins have been studied by Cho et al. (1965, 1966) in rabbits and dogs. The effects appear to be complex. Histamine added to animals in shock slightly increased respiration, but it decreased respiration if added to normal mitochondria in vitro. Plasmocid considerably decreased the cytochrome oxidase of the heart muscle (Balogh et al., 1967). Pyrogens increase NADH-cytochrome c reductase and succinic dehydrogenase in the cardiac muscle of young rats, but decrease them in old rats (Miyanishi et al., 1969). Serotonin decreased the rate of cytochrome reduction in rabbit heart muscle mitochondria (Zubovskaya, 1968). Ouabain, a cardioactive steroid, modifies the Na^+K^+

ATPase so that phosphorylation by inorganic orthophosphate becomes possible (Albers et al., 1968). Atherogenic diet increased oxygen uptake in heart muscle slices and cytochrome oxidase, with a correlation between the increased oxygen uptake and the degree of fatty degeneration of the liver (Pojda, 1966).

Skeletal muscles. Muscles which move rapidly whether in flight muscles of birds (Tucker, 1966) or in thoracic flight muscles of insects have a high content of cytochrome oxidase and cytochromes. In budgerigars flying in wind tunnels, the major energy source for the first 137 M in 15 sec was anaerobic. A 40 g bird in flight used as much as 2 litres of O_2, whereas a sitting bird used only 320 ml. Flight increased oxygen consumption by 2·7 ml/g muscle/min; in insect flight muscles similar values (1·4–7·3) have been reported.

Makinen and Lee (1968) compare the cytochrome contents of skeletal muscles of man, dog and rat. Mitochondria with satisfactory respiratory control have been prepared from rat and human skeletal muscles by Mockel and Dumont (1969). Data for the cytochrome c content of various muscles have been given by Nigro et al. (1969), see also Chapter V, B7. Most mammalian muscles contain both red and white muscle fibres and their red colour depends on the presence of myoglobin rather than on cytochromes. However, in general, red muscles contain five times more cytochrome oxidase than white muscles, and in tuna muscles the metabolic rates of red muscles are about six times those of white muscles (Gordon, 1968). In some instances a correlation between cytochrome oxidase and myoglobin content has been found, e.g. in horse muscles both increase with age (Lawrie, 1953). Often the red muscles are those of slow but repetitive activity. Cytochrome b is oxidized suddenly at the end of the muscle contraction (Jöbsis, 1959, 1963). The data support either adsorption of the high energy donor to the contractile protein during contraction or involvement of the donor in the relaxation phase. While in normal contractile activity cytochrome b and NADH are oxidized they, as well as cytochrome c, are reduced in tetanic contraction. Lactate and pyruvate reduced NAD, cytochrome b and occasionally also cytochrome c and flavoproteins more than did succinate. ATP is the energy donor in muscular contraction. The capacity for oxidation of pyruvate was doubled by treadmill exercise in the gastrocnemius muscle of the rat (Holloszy, 1967). Cytochrome oxidase, succinoxidase, NADH-cytochrome c reductase, cytochrome c and NADH- and succinic dehydrogenases (the latter only in severe exercise) were all increased; the total protein was also increased by 60%. Coupling and respiratory control remained high and ATP production thus rose concomitantly. Cytochrome oxidase decreased with age (Kunkel and Campbell, 1952; Petrasek, 1961; but cf. above in the horse). Intermittent, but not continuous, starvation also increased cytochrome oxidase in the skeletal muscles and the diaphragm. Uterine muscle contained fewer sacrosomes than heart muscle and the absorption spectrum of the 100,000 × g fraction (Gaudemer et al., 1963), also indicated qualitative differences in the cytochromes. Histochemically, the Burstone method showed an even distribution of the sarcosomes in slow twitching muscles whereas fast-twitch muscles showed an accumulation of sarcosomes on both sides of the Z-line of the fibres (Millington and Orr, 1966). Oxidative phosphorylation contributes a smaller fraction of energy to the contraction of smooth than of striated muscles, but the main difference is in the quantity of the mitochondria (Stephens and Wrogemann, 1970).

Denervation greatly decreased the cytochrome oxidase in skeletal muscles (Schmidt, 1952; Hearn, 1959; Honig et al., 1971) and also succinic oxidase, dehydrogenase and cytochrome c (Schmidt and Schlief, 1956). Cytochrome c reductases were even more decreased than cytochrome oxidase (Hearn, 1959). In small animals (e.g. in the rat *gracilis* muscle) oxygen extraction by the muscle remained constant and depended on oxygen transport, whereas in dog muscles with far smaller oxygen consumption, the resting metabolism was independent of oxygen transport.

Respiration and oxidative phosphorylation in the frog *sartorius* muscle have been investigated by Chance (1959) and Chance and Weber (1963). The respiration of fully aerobic muscles is limited by supply of substrate, of ADP (primarily) or Pi. With insufficient substrate, the cytochromes are largely oxidized and addition of ADP further oxidizes cytochrome b and NADH, half maximally at about 10 μM concentration. The cytochrome content is 2 μmoles per kg fresh weight, about one tenth of heart muscle. In the resting muscle cytochrome b is about 30% reduced, suggesting about half saturation with substrate; addition of pyruvate or lactate caused further reduction. Electric stimulation of the muscles increased the oxidation of cytochrome b parallel to the number of twitches. The effects disappeared spontaneously after stimulation. With low concentrations of azide, cytochromes a and c were reduced, this effect disappearing in 7–8 minutes.

The cytochrome oxidase and cytochrome content of some invertebrate muscles has been a matter of controversy for a long time. In contrast to earlier findings and those of Kmetec et al. (1963) and Lee and Chance (1968), muscles of the semi-anaerobic *Ascaris lumbricoides* were found to contain cytochrome oxidase, cyanide-sensitive respiration and the usual set of cytochromes by Kikuchi and Ban (1961). Cytochrome c was reduced by ascorbate, but not by cytochrome c. In a recent paper, Cheah and Chance (1970) confirm the findings of cytochromes a, b and c and also a small amount of cytochrome a_3 by low temperature spectroscopy. The respiration is CO- and CN-sensitive, and an antimycin A-sensitive TMPD oxidase is present. The flavin enzymes also are similar to those of mammals.

Succinoxidase was observed in oyster muscles as early as 1947 by Humphrey. The muscles of *Aplysia* have been studied by Wittenberg et al. (1965a,b), see also Chapter III, B1.

In contrast to these slowly operating muscles, the very rapidly contracting muscles of insects, particularly the thoracic flight muscles, contain high concentrations of cytochromes. Red and white muscles of the American cockroach *Periplaneta americana* have been studied by Nakatsugawa (1960). Its muscles contain the normal cytochromes in mitochondria, but red muscles about three times as much as white muscles and males more than females, particularly cytochrome a. The high cytochrome content and oxidative phosphorylation of mitochondria (sarcosomes) in insect flight muscles was simultaneously discovered by Sacktor (1953a,b, 1955) and by Lewis and Slater (1953, 1954). Chance and Sacktor (1958) and Sacktor et al. (1959) found α-glycerophosphate the major substrate in housefly sarcosomes; it raised cytochrome turnover to that found in the flying insect (see also Estabrook and Sacktor, 1958; Chapter IV, B). Mitochondria with high respiratory control and P/O ratios were obtained by Van den Bergh and Slater (1960, 1962). Honey bee pupae contain only cytochromes aa_3 and b_5, which latter disappears 2 days before the emergence of the bee (Herold and Borei, 1963). In later pupal and early adult life there is a progressive increase of cytochromes aa_3, $b + c_1$, c and of mitochondria which is complete 20 days after the emergence of the bee. Cytochrome b_5 is intimately connected with the protein synthesis for the myofibrils, the other cytochromes with the high energy demands of the flight muscles. In sheep blowfly *Lucilia cuprina*, cytochrome oxidase appeared first in small particles of less than 1 μ diameter, which later became aggregated to giant sarcosomes of 1–10 μ diameter. The highest oxygen uptake of the flight muscle was 1360 μl O_2/hour/mg protein (Lennie and Birt, 1967).

The indirect roles played by myoglobin, haemoglobins and other oxygen-carriers in the transport of oxygen to the mitochondrial membrane have been discussed by Jones (1963) in an extensive review and by Wittenberg (1970).

2. Liver, kidney and spleen

(a) *Liver mitochondria*. The method of separating rat liver mitochondria from lysosomes has been developed by Myers and Slater (1957) and improved by Loewenstein et al. (1970). Cytochrome oxidase in the liver has been reviewed by Drabkin (1954). A single rat liver mitochondrian contains about 17,000 respiratory chains based on the haem a content (Estabrook and Holowinski, 1961). In spite of marked differences in the respiratory enzyme content of different animals, the turnover number of cytochromes is nearly constant, and this is probably due to variations in their spatial arrangement. In the senile rat, cytochrome oxidase is decreased in the central and mid-zonal lobe areas, but increased in the periportal areas where activity is always higher (Kobayashi, 1965); giant nuclear and binucleate hepatic cells with high oxidase activity occur more frequently in the liver of the senile rat. X-ray radiations cause a transient decrease of cytochrome oxidase in the rat liver (Dancewicz et al., 1965). According to Minnaert (1960) endogenous substrate for rat liver mitochondria is continuously formed, probably by lipid breakdown. Q_{O_2} (about 13·5) rapidly decreases but respiration continues for hours. Aspirin was found to increase cytochrome oxidase and succinic dehydrogenase in old chicks (Nakane et al., 1967), Pinto and Bartley (1969) find a negative correlation between GSH oxidation and oxygen uptake in liver homogenates which is not easy to understand. In foetal rat liver the respiratory enzymes are only $\frac{1}{2}-\frac{1}{8}$ of those in adult liver in mitochondria, even less ($\frac{1}{4}-\frac{1}{20}$) in the homogenate (Jakovcic et al., 1971). Since the number of mitochondria in the postnatal period rose only by a factor of two there appears to be also an increase of the specific activity of the system. Mitochondrial cytochrome oxidase in the human liver increased during the first two days after birth, probably due to an effect of the birth process (Smith, 1970). Spectrophotometric studies on the redox state of cytochromes a, b, c and b_5 are reported by Drynov et al. (1968) in aerobic and anaerobic conditions with various substrates, as well as with amytal and with 2,4-DNP as uncoupler. Evtodienko et al. (1966) discuss the possibility of a role of cytochrome b_5 in non-phosphorylating liver mitochondria; they find a rapid reduction of cytochrome b_5 by 2,4-DNP and by NADH substrate, and oxidation by cytochrome c or TMPD. The method of Williams and Thorp (1969) gives somewhat higher values than those found by others (see Chapter V, D) in liver, brain and lung, but the same values in heart and kidney. The mitochondria and microsome-free plasma membranes (Vassiletz et al., 1967) contain only 0·08 μmoles cytochrome c per g protein, 0·03 of cytochrome b and 0·16 μmoles protohaem, but 5·7 μmoles of iron and 2·1 of copper, some NADH and NADPH-cytochrome c and cytochrome b_5 reductases. The Q_{O_2} was 6·0 and 9·0 with NADH and NADPH respectively (see also Fleischer et al., 1971).

Obstruction of the extrahepatic bile duct in patients increased cytochrome oxidase according to Maksimova et al. (1966) but Kartashova et al. (1966) found no change. Carbon tetrachloride decreased cytochrome oxidase and succinic dehydrogenase according to Zamfirescu-Gheorghiu et al. (1967) in chronic experimental hepatitis of rabbits, but in frog liver Hunter (1965) found a 17% decrease in cytochrome oxidase. In the early stages of the intoxication in mouse liver cytochrome c was decreased but later returned to normal (Squame et al., 1968). Endotoxin of E. coli greatly decreased tissue respiration and cytochrome oxidase, whereas no such decrease was found in a sterile turpentine abscess (Shkolovov, 1966). Infection by *Trypanosoma cruzi* had at first no effect on liver cytochrome oxidase or succinic dehydrogenase, but after seven days they decreased to about one third when breakdown of structure and loss of mitochondria occurred (Mercado, 1969).

(b) *Liver microsomes.* Cytochrome b_5 and P-450 have been discussed in Chapter IV, C and D, the microsomal reductases in Chapter VII, F. In the present chapter we report the more physiological findings together with a number of very recent observations. More P-450 was found in smooth-faced than in rough microsomes (Holtzman et al., 1968); this was confirmed for P-450 after phenobarbitone induction but not after methylcholanthrene induction (see also Murphy et al., 1969). Panchenko et al (1969) found the rough microsomes enzymically more active. Cytochromes b_5 and P-450 (at least after phenobarbitone-induction) were also found in Golgi membranes (Ichikawa and Yamano, 1970; Fleischer et al., 1971). There are marked species-differences in the cytochrome content of liver microsomes. Less P-450, b_5, hydroxylases and demethylases are found in the human liver than in rat or rabbit liver (Alvares et al., 1969; Ackerman, 1970). NADH-cytochrome b_5 and c reductases were also lower in human liver and K_m and V_{max} differed (but see Soyka, 1970, in human infant liver). The spectral change caused by aminopyrine in P-450 was greatest in the mouse, lowest in the guinea pig and was correlated to N-demethylation (Flynn et al., 1970). Aniline hydroxylation and type II spectral change were also greatest in the mouse, lowest in the guinea pig. Binding capacity for aniline did not differ in mice and rabbits, although aniline hydroxylation was higher in mice (Kato et al., 1970a,c). In contrast, hexobarbital hydroxylation was proportional to the amount bound to P-450.

Estabrook and co-workers (Cohen and Estabrook, 1971; Oshino et al., 1971; Hildebrandt and Estabrook, 1971) have recently adduced evidence that NADPH-cytochrome c reductase also functions in the reduction of cytochrome b_5 and that cytochrome b_5 plays a role in the mechanism of the function of P-450. This is supported by the stoichiometry of formaldehyde formation, aminopyrine demethylation and NADPH oxidation. Both NADH and NADPH are oxidized together and cytochrome b_5^{2+} provides one of the electrons required for the reduction of P-450. There is no evidence for a direct reduction of P-450 by NADH. A scheme is suggested in which NADPH reduces ferric cytochrome P-450, while b_5^{2+} reduces oxygenated cytochrome P-450. Soyka (1970) describes the isolation of cytochrome b_5 and of the NADH-cytochrome c reductase system from human infant liver. Oshino et al. (1971) have studied the function of cytochrome b_5 in fatty acid desaturation. A steady state of b_5 oxidation is reached by the balance of the NADPH reduction and its oxidation by a cyanide-sensitive factor which is the key enzyme combining stearyl-coenzyme A desaturation with the oxidation of cytochrome b_5. NADH reduces cytochrome b_5 much faster than NADPH so that no change of the steady redox state of b_5 is observed, unless the NADH-cytochrome b_5 reductase is inhibited by p-chloromercuriphenylsulphonate. The desaturation of stearyl-coenzyme A by ascorbate also depends on the action of cytochrome b_5. It is still unknown whether polyunsaturated fatty acids are formed by a similar mechanism; this does not appear to be influenced by the removal of cytochrome b_5 by proteolysis. A taurochenodeoxy-cholate-6β-hydroxylase in rat liver microsomes has been described (Voigt et al., 1970).

Phenobarbitone and methylcholanthrene induction of P-450 and of NADPH-cytochrome reductases in starved and glucose-fed rats has been studied by Kato (1967) and Kato et al. (1969). Starvation increased P-450 in males, but decreased it in females; glucose decreased it in both. Thyroxine and adrenalectomy decreased P-450, while morphine and alloxan had no effect. The spectral effect of hexobarbitone on P-450 was, however, decreased by morphine as well as by thyroxine, while aniline hydroxylation was increased by alloxan and starvation as well as by thyroxine. Androgens stimulate hexobarbitone hydroxylation, but not aniline hydroxylation. Hexobarbitone-induced, but not aniline-induced spectroscopic changes are markedly greater in males than in females (Kato et al., 1970a). Hanninen et al. (1969) found no memory effect on repeating phenobarbitone induction. Ageing in nitrogen for 90 hours after phenobarbitone induction causes a larger decrease of the type I than of type II spectral change; the latter was only parallel to the small loss of haem (Hewick and Fouts, 1970). No qualitative alteration of the CO-spectrum occurred on ageing in the perfused rat liver (see also Cinti and Schenkman, 1970). The type I substrate spectral alteration was found only when P-450 was either in the oxidized or in the ferrous state, not during the metabolism (Sies and Brauser, 1970). Hexobarbitone addition after phenobarbitone-induction decreased the rate of reduction of NADPH. The NADPH-cytochrome c reductase in the phenobarbitone-induced rats was less stable *in vivo* than the non-induced enzyme (Kuriyama and Omura, 1971). Dietary lipid is required for phenobarbitone-induction (Marshall and McLean, 1971). Enzymes catalysing aniline and nitroaniline oxidation parallel microsomal P-450 more closely than the enzymes catalysing hexobarbitone hydroxylation in liver regenerating after partial hepatectomy (Gram et al., 1968).

Some of these recent results still require confirmation and others cannot be readily understood on the basis of our present knowledge; an attempt at their rationalization at the present stage would appear to be premature. It should also be noted that the reaction of the so-called microsomal cytochrome c reductases probably does not occur *in vivo* directly since cytochrome c is not normally present in microsomes.

The ligation of the extrahepatic bile duct causes a decrease of aminopyrine demethylation and of binding of type I substrates, but little decrease of binding of type II substrates, of P-450 and of NADPH-P-450 reductase (Hutterer et al., 1970). The dye 3-methyl-4-dimethylaminoazobenzene increases endoplasmic reticulum, but on the contrary decreases microsomal electron transport and P-450 possibly by the effect of a dye metabolite (Arcasoy et al., 1968).

Contradictory findings have been published on the effect of carbon tetrachloride on P-450. A decrease has been reported by Archakov et al. (1969), Greene et al. (1969) and Sasame et al. (1968) while Barker et al. (1969) and Marshall and McLean (1969) found an initial decrease followed by an increase. An increase of NADH- and NADPH-cytochrome c reductases by CCl_4 has been reported by Cleveland and Smukler (1965). A decrease of P-450 has been ascribed by Archakov, but not by Sasame to the action of peroxides.

Cytochrome b_5 in rat liver is decreased by castration (Brentani and Raw, 1965) and starvation, and in guinea pig liver by scurvy (Degkwitz et al., 1968). The decrease of P-450 but not of cytochrome b_5 caused by starvation is restored by ascorbate. Orotic acid decreases P-450 only in the male rat (Holtzman and Gillette, 1966, 1969) and only aniline hydroxylase not ethylmorphine hydroxylase is affected. Fatty acid deficiency decreased P-450 and aryl-4-hydroxylation but left the phenobarbitone-induction intact (Kaschnitz, 1970). Laurate specifically inhibits type I ω-hydroxylation of high fatty acids and inhibits drug demethylation competitively (Orrenius and Thor, 1969). The effect of laurate is abolished by 3 mM higher fatty acids; in this concentration they do not convert P-450 to P-420. An increase of protein in the diet which greatly increased liver weight, failed to increase P-450 during the first 96 hours (Marshall and McLean, 1968).

(c) *Kidney and spleen.* The renal medulla, in contrast to the cortex, has few mitochondria and a mainly anaerobic metabolism, but small amounts of the usual cytochromes are present (Kean et al., 1962). The oxygen consumption of kidney slices remains steady on anaerobic storage for the first six hours at 4°, but decrease thereafter whether oxygen is present or not (Rockman et al., 1967a,b). At 38° there was a marked loss of activity even in anaerobiosis. In kidney perfusion with oxygenated Ringer-phosphate, cytochrome oxidase decreased in the absence of dinitrophenol (Mazarean et al., 1967). In the presence of 10^{-4}M DNP oxidase activity slightly increased during the first two hours, but later a decrease accompanied structural changes. In contrast to the cytochrome oxidase of the liver, that of the kidney was not increased by aspirin. Renal clearance of cytochrome c in the rat exceeds that of inulin; the decrease of the clearance with increasing concentration of cytochrome c in the plasma indicates that there is tubular secretion of cytochrome c in addition to glomerular filtration (Gemba, 1966; Gemba et al., 1966). Cytochrome b_5 of pig kidney is found in soluble form in the homogenate (Mangum et al., 1970). Heterogenic blood transfusion decreases the cytochrome oxidase of the kidney (Federov et al., 1968). Q_{O_2} in the dog kidney was stimulated by low concentrations of K^+ and Ca^{2+} and inhibited by ouabain. K^+ competitively diminished the ouabain inhibition (De Jairala et al., 1969). Lethal radiation doses affected the mitochondria of spleen and thymus, lowering their cytochrome c content but did not affect other mitochondria (Scaife, 1966).

3. Nerves and brain

Energy for ion transport particularly the "sodium pump" plays an essential role in the mechanism of nerve tissue and in the transmission of action potentials along nerve fibres (see e.g. McIlwain and Gore, 1951; Kety, 1955; Elliott, 1955; Moore and Strasberg, 1969). Although brain forms only 2·5% of the human body's weight, its oxygen consumption is about 25% of the total.

It has been shown by Tobin and McIlwain (1965) using oligomycin that the energy of K^+, or of electrical stimulation, is mediated by ATP, not immediately derived from other energy-rich intermediates of oxidative phosphorylation. Oxygen consumption in the confused, compressed or comatose brain is decreased (Sacktor and Packer, 1962). Stimulation causes an increase in succinoxidase and a redistribution of mitochondria in the nerve cell as well as a swelling of the mitochondria (Titova et al., 1962). NAD is reduced by energy-requiring reversal of electron transport as well as by NAD-linked substrates. Osmotic regulations through energy-rich compounds are of special importance for nerve cells on account of their volume restrictions. In stimulation of the superior cervical ganglion Brauser et al. (1970) found a 50% decrease of oxygen and glucose uptake and a 170% increase of lactate production. The respiration was as tightly controlled as in perfused liver mitochondria. The percentage of oxidation of cytochromes in the steady state was measured by the technique of Chance et al. (1962a), see also Brauser (1968). In the resting state cytochrome a_3 was more than 97% oxidized, c 85% and b 37%. Measurement of 605–630 nm of cytochrome a indicated sufficient aerobiosis during stimulation. Cytochrome b was measured at 564–575 nm, NADH by fluorescence. Ungar et al. (1956) observed spectrophotometric changes on stimulation indicating changes of nerve proteins and alterations of tyrosine and cysteine hydrogens which somewhat resemble those produced by urea denaturation but are reversed by rest.

Brain mitochondria are similar in composition to those of heart and liver (Parson and Basford, 1967) but somewhat heterogeneous in cytochrome composition (Pigareva et al., 1965) and still more so with regard to substrate utilization (Blockhuis and Veldstra, 1970). α-glycerophosphate is the predominant substrate rather than other NAD-dependent substrates or succinate (Ringler and Singer, 1958; Sacktor et al., 1959; Sacktor and Packer, 1962); in this they resemble insect flight muscle sarcosomes. Mitochondria can be concentrated around the nucleus of the sympathetic ganglion, or more dispersed in the cytoplasm (Sharma, 1967) or in dendrites, e.g. the olfactory bulb dendrites. In the rat cerebral cortex, Hess and Pope (1960) found mitochondria in neurons and their processes rather than in the neuroglia, the non-nervous supporting tissue elements. In cat adrenergic hypogastric nerves mitochondria move away from a state of constriction thus decreasing cytochrome oxidase in the zone of compression in the damaged axons (Banks et al., 1969). Cotman et al. (1968) have isolated from rat brain, by zonal centrifugation in 30% sucrose, a membrane fraction enriched in nerve endings which may, however, contain some plasma membranes. The Na^+K^+-stimulated ATPase content was high, but cytochrome oxidase was only one-seventh of that of mitochondria and antimycin-insensitive NADH-oxidase only one-tenth.

The distribution of cytochromes in the cortex of the large hemispheres of the brain differed considerably from that in the brain stem (Pigareva, 1963; Pigareva and Kamysheva, 1963) both in relative and absolute concentrations; the brain stem contained roughly twice as much of cytochromes as the cortex and even more of cytochromes c and a. The distribution of cytochrome oxidase (Hess and Pope, 1960) and of NADH-cytochrome c reductase (Busnyuk, 1967) varied in different layers of the cortex; except for the top layer the upper layers showed higher enzyme concentration. In animals born in a rather immature state, like rats, cytochrome oxidase and succinic dehydrogenase of the brain were lower at birth than in animals born in a more mature state (Yamada, 1961; Buikis, 1966). The oxidase content of the adult guinea pig brain exceeded that of rat liver (Lindall and Frantz, 1967). The gray matter of brain is far richer in cytochrome oxidase than the white so that the content of the cortex exceeds that of the white brain (Hess and Pope, 1954; Shimizu et al., 1957; Manocha and Bourne, 1966; Gulidova, 1967). Cytochrome oxidase and succinic-cytochrome c reductase of the rat brain cortex were present in two subfractions of sucrose gradient fractionation (Bogolepov and Dovedova, 1969). In injured sensory neurons of the rabbit subjected to a pinching of the sciatic nerve at the pelvis succinate dehydrogenase and cytochrome oxidase were decreased (Baranov and Prokopenko, 1969). In the sympathetic ganglion of the rat, succinic dehydrogenase was more active at the periphery of the ganglion than in the nerve fibres between the ganglia; cytochrome oxidase was stronger in these than in the ganglion (Sharma, 1967).

In anoxia of the rat brain caused by ischaemia of ligation or by nitrogen after ligation of the carotid artery, decrease of cytochrome oxidase and succinic oxidase develops more slowly than of monoamine oxidase, but faster than that of α-glycerophosphate dehydrogenase (Spector, 1963; Macdonald and Spector, 1963). However, Barbashova and Grigoreva (1969) found an increase of cytochrome oxidase in adaptation to hypoxia which occurred only in adult, not in newborn rat brain. Smialek and Hamberger (1970) found also an increase of cytochrome oxidase in rabbit brain in moderate hypoxia caused by six hours in an atmosphere of 92% nitrogen, 8% oxygen, or 24 hours after section of both common carotid arteries. Incorporation of labelled leucine into the mitochondrial protein of the cortex was also increased. The effects of hypoxia and hypercapnia (increased p_{CO_2}) as stimulants of lung respiration by phrenic nerve activity are complex owing to a stimulatory effect at the peripheral chemoreceptor and a depressing central effect (Cherniak et al., 1970). Tranquilizing drugs, e.g. chloropromazine, inhibited cytochrome oxidase far more than brain succinoxidase which in turn is less sensitive to them than liver succinoxidase (Helper et al., 1958). Both forebrain and cerebellum cytochrome oxidase were strongly inhibited (Moraczewski and Anderson, 1966). Sera of psychiatric patients were, however, found to increase both cytochrome oxidase and succinic dehydrogenase (Sorokina, et al., 1966); antisera from rabbits immunized with rat brain mitochondria had the same effect. Succinic dehydrogenase and cytochrome oxidase were decreased by hydrocortisone after bilateral adrenalectomy (Cherkasova et al., 1968). Oestradiol markedly increased the cytochrome oxidase of the hypothalamus in newborn rats (Heim, 1969). Castration significantly decreased cytochrome oxidase in the anterior and posterior but not in the middle hypothalamus without altering succinic dehydrogenase, thus affecting the cytochrome c-oxygen part of the respiratory chain; testosterone had the opposite effect (Moguilevsky, 1966, 1969). There is no very effective counter-agent against the inhibition of cytochrome oxidase by cyanide in the brain, but sodium thiosulphate which penetrates into the mitochondria and combines with cyanide to form thiocyanate appears to be more effective than nitrite which acts on haemoglobin to form cyanide-bonding methaemoglobin (Estler, 1965; Schubert and Brill, 1968).

The localization of respiratory enzymes in different parts of the brain has been studied by Hess and Pope (1954, 1960) Shimizu et al. (1957), Hyden et al. (1958), Friede (1961), Titova et al. (1962), Tolani and Talwar (1963), Nandy and Bourne (1964), Ridge (1964, 1967), Manocha and Bourne (1966, 1967), Gulidova (1967), Busnyuk (1967) and Sharma (1968). The content of cytochrome oxidase and succinic dehydrogenase, a measure of metabolic activity, is highest in the phylogenetically recent structures of the cerebral and cerebellar cortex, caudate nucleus, also in brain stem and thalamus. Low concentration of free bilirubin (20 μM) inhibit the respiration of brain mitochondria, oxidative phosphorylation and respiratory control and cause swelling of mitochondria (Mustafa et al., 1969) causing the pathology of "Kernicterus". Bilirubin differs, however, from uncouplers and appears to act on ion transport rather in the way of gramicidin, valinomycin and nonactin.

Mammalian brain microsomes as well as those of fowl, toad and carp, contain cytochrome b_5 and NADH and NADPH cytochrome b and c reductases, chiefly in the smooth-surfaced endoplasmic reticulum of the cortex, less in the brain stem (Inouye and Shinagawa, 1965; Inouye et al., 1966; Kamino and Inouye, 1970; see also Guiditta and Aloj, 1965). The presence of cytochrome P-450 has not yet been ascertained, but van Dyke (1969) suggests that anaesthetics are bound to P-450 in brain membranes.

The electric organs of the fishes *Torpedo* and *Electrophorus* have been studied by Chalazonitis and Arvanitaki (1956) and by Aubert et al. (1964). In the *Torpedo* there was some evidence for aerobic catalysis by cytochromes at the ventral surface of the electroplex, with electron and proton transfer dorso-ventrally. In *Electrophorus* a decrease of absorption at 420 nm was found on stimulation, "just conceivably due to a haemoprotein".

Retina. Cytochrome oxidase and cytochrome c are present in the retina of the eye, though in rather small concentration (De Berardinis, 1960); the outer retinal layer contains most of the cytochrome oxidase and succinic dehydrogenase (Shukolyukov, 1967). Retinal microsomes have been studied by Heath and Fiddick (1965) and Shichi (1969). Cytochrome b_5 appears to be present only in minor amounts but P-450 has been demonstrated. NADH and NADPH are oxidized even in the absence of ascorbate, which increases their oxidation. Cytochrome oxidase is also present in the lens of the eye (Korkonen and Korkonen, 1966). In the cornea, however, a cyanide-sensitive, but formalin-resistant oxidase has been found only in wounds (Weimar and Haraguchi, 1965).

Cochlea. In the cochlea of the inner ear, Conti and Borgo (1965) found cytochrome oxidase histochemically in mitochondria of the organ of Corti, and most abundantly in the stria vascularis, where groups of cells have an intense metabolic activity.

4. Cytochromes in other mammalian tissues

(*a*) *Gastric mucosa.* In an early theory Robertson (1960) postulated a direct role of cytochromes and their valency change in the production of hydrochloric acid by the gastric mucosa. This was later replaced by the theory of Davies (Forte and Savies, 1964; Forte et al., 1967) and Kasbekar and Durbin (1965) according to which ATP hydrolysis provides the necessary energy. In the resting state when proton excretion is inhibited, e.g.

by thiocyanate, the mitochondria would be in state 4, but stimulation of gastric secretion, e.g. by histamine, would cause a shift to state 3 by respiratory control. More recently, however, Hersey and Jöbsis (1969) have found it impossible to describe their results on the basis of this theory. In the bullfrog mucosa, the Δ spectrum anoxic minus oxygenated showed the maxima of the cytochromes with a remarkably high aa_3 maximum at 603 nm. A large reduction of all components accompanied the stimulation of gastric secretion by histamine whereas thiocyanate caused their oxidation. The reduction caused by histamine was not due to hypoxia and oxidation by thiocyanate brought the mitochondria closer to the maximally respiring state, perhaps by an uncoupling effect. A more direct relation between cytochromes and gastric secretion than in the ATP theory is therefore postulated, but the nature of the cytochrome activity remains unexplained. An involvement of cytochrome oxidase in gastric HCl-production had also been supported by Vitale et al. (1956), as histamine was found to increase the oxidase and succinic dehydrogenase. In studying the thiocyanate inhibition of gastric secretion, Kidder et al. (1966) and Kidder (1970) found redox changes only in cytochrome c not in cytochrome aa_3, b or c_1 and assume the formation of a cytochrome c-thiocyanate complex. Here proton secretion is considered an alternative to rather than a consequence of ATP formation as it is in the Mitchell hypothesis. Increased proton excretion is connected with reduction of c^{3+} which they ascribe to extramitochondrial cytochrome c present in the mucosa. Thus the pendulum has swung back to postulating a more direct role of cytochromes in gastric acid secretion (see also Heinz, 1967). In iron-deficiency gastric biopsies show strong activity of cytochrome oxidase in the parietal cells (Stone, 1968).

Intestinal mucosa. Cytochrome oxidase, cytochromes, and Na^+K^+-stimulated ATPase in villi of intestinal mucosa have been studied by Oda and Seki (1966), Oda et al. (1969), Quigley and Gotterer (1969), Strelina (1970) and Bostelmann et al. (1970).

Pancreas. Honjo et al. (1968a,b) have isolated mitochondria with high respiratory control from guinea pig pancreas. An endogenous phospholipase produces lysolecithin and albumin is required for its inhibition.

(b) *Arteries.* Cytochrome oxidase and succinic dehydrogenase are localized in the intima and media in the aorta and adjoining arteries of several mammals (Sotonyi et al., 1965). In spite of the high rate of aerobic glycolysis oxidative phosphorylation with high P/O ratio has been found in the guinea pig aorta (Ritz and Kirk, 1967). In early arteriosclerotic changes in man, dog and rabbit, oxidative capacity of the intima, and less in the media is increased but later it is decreased (Maier and Haimovici, 1965; Matova and Sukasova, 1967). The early increase in the intima possibly precedes lesions in the aorta found in the region of lipid spots and may be an adaptation to increased energy requirements. The later decrease may be due to fibrosis. Yamada *et al.* (1966) found cytochrome *c* to depress hypercholesteraemia and atherosclerosis in rabbits.

(*c*) *Blood.* The cytochrome oxidase content of normal erythrocytes is very small, if indeed it is cytochrome oxidase (see Chapter III). However, cytochrome oxidase has been found in red cells formed rapidly after haemorrhage (Fedorov and Pekus, 1970). Oxygen consumption in young chicken erythrocytes is six to eight times higher than in mature red cells (Augustin and Rapoport, 1959). The decrease of respiration of rabbit reticulocytes in their maturation is caused by a factor found in the haemolysate (Altenbrunn *et al.*, 1959; Rapoport *et al.*, 1961) and loss of succinate oxidase in maturation in faster than other changes. Analytical methods for the measurement of cytochrome oxidase in red cells of children have been described by Nartsissov (1968).

In leucocytes small amounts of cytochrome *c*, less than 4% of those in rat heart, have been found by Evans (1962) to accompany myeloperoxidase; in leukaemic leukocytes oxygen uptake is only partly mediated by cytochrome aa_3, the remainder by systems of lower oxygen affinity. Cytochrome oxidase has been demonstrated in normal leukocytes by the Burstone method (Quaglino *et al.*, 1965; Ronin, 1967); it is probably present in mitochondria. It is highest in granuloblasts and granulocytes, moderate in monocytes, least in lymphocytes. Succinic dehydrogenase is weakest in neutrophils, somewhat stronger in eosinophils, lymphocytes and monocytes. Minakami and Mokhova (1968) found a cytochrome a/c ratio of one in leukocytes and the concentration of cytochrome *a* less than one-tenth of that of myeloperoxidase. Whether the higher oxidative activity in leukaemia is due to cytochrome oxidase or to the peroxidase is still uncertain. Higher activity of succinate-cytochrome *c* reductase and of transhydrogenase has been found in homogenates of white cells in chronic lymphocytic or acute leukaemia than in normal leukocytes of cells in chronic myeloic leukaemia (Evans and Getz, 1968), indicating a location in mitochondria. Leukaemia made no difference to the respiratory enzymes in mouse spleen mitochondria (Sacktor, 1964). Yunusova and Bolobonkin (1967) suggest that the study of changes in cytochrome oxidase in leukocytes may be useful for the diagnosis of hepatitis in children.

(*d*) *Skin and glands.* Cytochrome oxidase and succinic dehydrogenase were found in the ciliated epithelian cells of the respiratory tract; the beating of the cilia was stopped by ATPase and by smoking (Cress *et al.*, 1965).

Using the Burstone method, Montagna and Soon Yun (1961) found moderate cytochrome oxidase activity associated with mitochondria throughout the Malpighian layer, strong in the hair follicles and pilary canal, in sebaceous glands and eccrine sweat glands; see also Loewenthal and Hins (1963). Apocrine glands were reactive in secretory cells, but not in their duct. Meissner corpuscles and myelinated nerves emerging from them were moderately active. Cytochrome oxidase content is high in parotid, submaxillary and sublingual glands of man and animals (Mori and Mizushima, 1965). Ohlin (1965b) found it decreased by sympathetic and para-sympathetic denervation of the submaxillary gland of the rat; it was not significantly affected by hypophysectomy, thyroxine or testosterone, although the weight of the gland and its secretory response to pilocarpine decreased (Ohlin, 1965a). Lacrimal glands have been studied by Kuehnel (1968).

In the mammary gland cytochrome b_5 has been found in microsomes (Bailie and Morton, 1955). The mitochondria of the mammary gland of the guinea pig showed respiratory control and were able to synthesize protein (Jones and Gutfreund, 1959). The effects of inhibitors and uncouplers on the respiration of slices of the avian salt gland have been studied by Chance et al. (1964) and by Van Rossum (1965a,b). Amytal caused little change on the level of reduction of cytochromes, while dicoumarol showed a cross-over point between cytochromes b and $(c + c_1)$. Evidence was obtained for the reversal of electron transport.

The respiratory enzymes of the prostate of the rat have been studied by Harding and Samuels (1961) and those of the adenomatous prostate in man by Hopkinson and Jackson (1961). They are similar, though at smaller concentrations than those of the rat liver. A specific NADPH-cytochrome c reductase is absent and NADPH is oxidized by the NADPH-NAD hydrogen transferase. This may explain the high citric acid excretion of the prostate found by Humphrey and Mann (1948). Endogenous respiration in the human adenomatous prostate is low and also its cytochrome c content, possibly due to hypoxia.

(e) *Spermatozoa*. Cytochrome oxidase and the usual mitochondrial cytochromes (Mann, 1945) and succinic dehydrogenase (Lardy and Phillips, 1945) are present in mammalian sperm cells (Gonze, 1959). They are concentrated in the flagellae on which the motility of the sperms depend (Zittle and Zitin, 1942; Nelson, 1955; Semakov, 1961; Iosinov, 1966; Higashi and Kawai, 1970). The effects of freezing on sperm vitality have been studied by Semakov (1961) and Khavinzon (1968); it is essential to preserve succinic dehydrogenase and cytochrome oxidase. The localization of cytochrome oxidase during the mitochondrial specialization in the spermatogenesis of prosobranch snails has been studied by Anderson (1970). Initially a smaller part of the oxidase was consistently found in the matrix, but in mature spermatozoa it is only found in the inner membrane and the cristae of the mitochondria, while the enzyme disappears from the matrix. Sea urchin spermatozoa contain cytochrome oxidase (Mohri, 1956). Cytochromes a_3, a, b, c and c_1 were found in the sperms of two species of sea urchin in the ratio 1 : 1 : 1 : 2 : 1, one unit being 0·05

μmoles/g protein (Wilson and Epel, 1968). Antimycin A caused the usual shift of the α-band of cytochrome b from 560 to 565 nm, and an increase of substrate-reducible b which may be only apparent and due to the higher molar extinction of the antimycin-combined form of b. The identification of cytochrome c_1 requires further proof; it throws some doubt on the claim of Gonze (1962) of a special cytochrome-557 in the sperm of the clam *Spisula*. Results similar to those of Wilson and Epel with sea urchin sperms have been obtained with the spermatozoa of the fresh-water mussel by Higashi and Kawai (1970). No cytochrome c_1 was found, however, in the low temperature spectrum.

5. Effect of hormones on respiratory enzymes

(a) *Sexual hormones.* Ovaries, like adrenals have a dual cytochrome system, although their P-450 content is lower (Cammer and Estabrook, 1967; Cooper and Thomas, 1970). Cytochrome oxidase is present in the sheath cells and highest in mature follicles (Aratei et al., 1967). Human chorionic gonadotrophin doubled both cytochrome oxidase and P-450 content of ovaries (Cooper and Thomas, 1970). Sheep corpora lutea were rich in cytochrome oxidase which varied with oestrus and pregnancy (Arvy and Mauleon, 1964). Cytochrome P-450 is present in the mitochondria of bovine corpora lutea in about 25–30% of the concentration found in adrenal cortex mitochondria (Yohro and Horie, 1967). In the rat uterus cytochrome oxidase and succinic dehydrogenase were also very active in oestrus but low in ovariectomized rats (Okla, 1965).

In the rat uterus, 17 β-oestradiol (2 μg/kg) injected daily for 3 days, trebled the NADH-cytochrome c reduction (Christiane and Klitgaard, 1968). Stilboestrol, more than hexoestrol, inhibited respiration and succinate oxidation in state 3, but stimulated it in state 4 (Vallejos and Stoppani, 1967a,b). The main action of stilboestrol is on the oxidation of reduced NADH-flavoprotein and cytochrome b (De Otamendi and Stoppani, 1967). Rat pregnancy causes no significant effect on V_{max} and K_m of P-450 hydroxylation and demethylation activities for 20 days, but later V_{max} was decreased (Guarino et al., 1969). The human amnion contains cytochrome c (Brame and Overby, 1970).

P-450 is also present in testicular microsomes and plays a role in the biogenesis of androgens (Machino et al., 1969). CO decreases the activity of 17-α-hydroxylase and of steroid C_{17}–C_{20} lyase. Androgens decrease P-450, 11-β- and 18-β-hydroxylases and corticosteroids while they increase cytochrome aa_3 in adrenal mitochondria by 200% (Brownie et al., 1968, 1970). Testosterone is a competitive substrate for deoxycorticosterone hydroxylation in the bovine adrenal cortex (Johnson et al., 1970); both are type I substrates of P-450 with half-saturation at about 3 μM concentration and probably react at the same site of P-450.

A single dose of aldosterone increased cytochrome oxidase and succinoxidase in the liver, but repeated doses had the opposite effect. Aldosterone restored cytochrome aa_3 in rat kidney mitochondria after decrease by adrenalectomy, but had no effect on cytochrome b_1, c or c_1 (Kirsten et al., 1970). The molar ratio c/haem a, normally 1·1 was increased to 1·8 by adrenalectomy and restored to normal by aldosterone.

(b) *Adrenals.* Normal adrenal mitochondria contain almost seven times as much cytochrome P-450 than cytochrome aa_3 (Harding et al., 1966; Cammer and Estabrook, 1967). In guinea pig strains with high adrenal hydroxylation, the P-450 concentration was also higher (Burstein, 1967). In the adrenal cortex P-450 is mostly found in the fascicular zone and in microsomes more than in mitochondria (Ichikawa et al., 1970). Cytochrome b_5 was highest in the centre of the adrenal cortex. In the adrenals, transhydrogenase plays an important role in the generation of intramitochondrial NADPH for the hydroxylation reactions of P-450, and forms a link between it and the respiratory chain serving as a regulation mechanism for corticosteroid formation (Oldham et al., 1968; Wilson et al., 1968). Following adrenal enucleation in rats, cytochrome P-450 was completely restored in 8 weeks and to 78% in five weeks (Gallant and Brownie, 1969). The NADPH generation may be the major factor in 11β-hydroxylation.

Small amounts of adrenocorticotrophic hormone (ACTH) caused a large increase of the oxidation of nicotinamide nucleotides, probably NADH, in the adrenal gland of hypophysectomized rats; this was studied by the fluorescence method of Chance et al. (1962a). ACTH and whole body x-irradiation of rats increased cytochrome b_5 and NADH-cytochrome reductase but did not affect NADPH-cytochrome c reductase or P-450 (Yago et al., 1967). This field has been reviewed by Simpson et al. (1969).

(c) *Thyroid gland.* The predominant importance of the thyroid gland for the regulation of oxygen uptake of the human and mammalian body and for its basal metabolism was recognized early. The mechanism of action of the thyroidal hormones, however, has still not yet been elucidated. Until recently two theories have been in the foreground of studies.

(1) A reaction on the energy conservation process involving an uncoupling effect of the hormones and mitochondrial swelling (Lardy and Feldott, 1951; Maley and Lardy, 1955; Tapley and Cooper, 1956; Niemeyer et al., 1951; Martius and Hess, 1957; Lehninger, 1962). It is, however, now clear that these phenomena require much higher than physiological concentrations of the thyroid hormones (Tata et al., 1963; Roodyn et al., 1965; Bronk, 1965; Glick and Bronk, 1965; Ernster, 1965; Tata, 1966; Kadenbach, 1966a,b; Werner and Nauman, 1968). If they are involved at all, they are not of significance for the anabolic action, but only

for the catabolic and calorigenic effects of thyrotoxicosis (cf. Hülsmann, 1970; Challoner, 1968a). The action of thyroid hormones considerably differs from that of uncouplers (Michel and Leblanc, 1969; Michel and Pairault, 1969), e.g. thyroxine or triodothyronine activate succinate oxidation only in very small concentrations and even 2–5 μM inhibit in state 4, whereas the uncouplers stimulate at all except extremely high concentrations. The authors assume that the hormones produce uncouplers as well as inhibitors of electron transport (see also Hoch, 1968). This does not, however, agree with the findings of other workers. Skelton et al. (1971) found added thyroxine to decrease the efficiency of conversion of cellular energy into work energy in cat muscles, but there was no increase of efficiency in the hypothyroid state. Undamaged mitochondria have been obtained from the thyroid (Gonze and Tyler, 1965) and the respiratory pathway is very similar to that in other mammalian cells.

(2) A direct effect of the thyroid hormones on the cytochromes and other respiratory enzymes of mitochondria has frequently been observed (see the review of Ernster and Lee, 1964 and Werner and Nauman, 1968). Barker (1955) found that thyroid hormones increased the Q_{O_2} of liver, heart, skeletal muscle, kidney and salivary gland, but not of brain, spleen, testes, ovary, prostate, thymus and lymph nodes. While cytochrome oxidase is undoubtedly increased by thyroid hormones, its concentration is usually too high and not rate-determining, so that changes in its concentration cannot play a great role. Also stimulation of electron transport does not always lead to an increased BMR (Bronk, 1963, 1966). Increases of cytochrome oxidase, succinoxidase and cytochrome c by thyroid hormones, and inversely their decrease by thyroidectomy have been noted frequently (Katsura, 1954; Maley, 1955, 1957; Ball and Cooper, 1957; Schole, 1959; Bronk, 1960a; Bronk and Bronk, 1962; Hoch, 1965; Kadenbach, 1964, 1966a; Suzuki et al., 1967; Michel et al., 1968; Hoch and Motta, 1968), although some deviating results have been reported (Katsura, 1954) with large amounts of hormones acting on the dog heart (Suzuki, 1954; Klitgaard, 1966). Polacov and Cilento (1970) found thyroxine to increase the reduction of cytochrome c^{3+}. Different effects of thyroxine on NADH-cytochrome c reductase have been observed by Christiane and Klitgaard (1968). Ball and Cooper (1957) found thyroxine to inhibit transhydrogenase; one would suppose this to cause an increase of NADH oxidation which has not been found. Suppression of oxaloacetate formation by the thyroid hormone and of its inhibitory effect on succinoxidase have been found by Clarke and Ball (1955) and Wolff and Ball (1957). Doubts on stimulation of the terminal oxidase being a sufficient explanation of the increased metabolic rate have been expressed by Bronk (1963) who found dehydrogenases not increased by thyroxine. Some of these doubts have been allayed by the finding of Kadenbach

(1966a) that "pacemaker" dehydrogenases were indeed increased. An increase in oxidation of cytochromes, but a decrease of oxidation by flavoproteins was found by Michel et al. (1968) and by Raoul et al. (1970). Raw and Da Silva (1965) found thyroxine to increase P-450 and NADPH-P-450 reductase, but to decrease considerably cytochrome b_5 and NADH-cytochrome b_5 reductase.

(3) While these observations support an action of the thyroid hormones on the oxidative enzymes, e.g. in the muscle, recent observations indicate that the thyroid hormones do not act on untreated mitochondria *in vitro*, and this and a lag period in the increase *in vivo* has been taken as evidence that the hormones stimulate the biosynthesis of the respiratory enzymes (see Chapter IX; Tata et al., 1963; Moury et al., 1964; Roodyn et al., 1965; Ernster, 1965; Tata, 1963, 1966). Lucas et al. (1970) found a rapid binding of ^{131}I-labelled thyroid hormones to rat liver mitochondria which had, however, no relation to the metabolic changes. Observations on electron-microscopic changes of muscle mitochondria in hyper- and hypo-thyroid states have been made by Ernster (1965) and Tata (1966).

The Askanazy-Hürthle cells of the thyroid itself are rich in cytochrome oxidase (Tremblay, 1962; De Groot and Dunn, 1964, 1968). Calf thyroid contains more cytochrome oxidase and antimycin-sensitive succinate-cytochrome c reductase than NADH-oxidase. Catalytic amounts of cytochrome c tripled the NADH-oxidase in an antimycin-resistant pathway. NADH-cytochrome c reductase (EC.1.6.2.1) and NADPH-cytochrome c reductase (EC.1.6.2.3) accompanied cytochrome oxidase in the thyroids of man, rat and calf. Cytochrome oxidase is increased in thyrotoxicosis (Graves disease) but not in adenoma. De Groot and Davis (1962) found evidence that it is not cytochrome oxidase, but a peroxidase which is responsible for the iodination process.

The effects of the parathyroid glands on mitochondrial ion transport have been studied by Kimmich and Rasmussen (1966). The P/O ratio of about 1 of the ascorbate-TMPD reaction (phosphorylation site 3) was not altered. Accumulation of Mg^{2+}, K^+ and phosphate was found, accompanied by uncoupling and stimulation of mitochondrial respiration.

6. Cytochromes in malignant tumours

In the well-known theory of Warburg on tumours (1930, 1956) the lack of cellular respiration has been considered the cause of the high aerobic glycolysis which is indeed a feature of tumour cells. The reason for this high aerobic glycolysis still remains unexplained (Weinhouse, 1956; Chance and Hess, 1959a, b; Borst, 1960; Boxer and Devlin, 1961; Chance, 1964; Bickis and Quastel, 1965; Nigam, 1966; Gordon et al., 1967;

Terranova et al., 1967; Galeotti et al., 1970). It is not due to the uncoupling of oxidative phosphorylation. Mitochondrial compartmentation possibly plays a role, but more probably lack of interaction of cytoplasmic and possibily microsomal reducing systems with mitochondria which is under genetic control.

Overwhelming evidence has come to light which shows that lack of mitochondrial respiration per se cannot be the cause. L-cells are considered sarcosomal cells of low malignancy except for susceptible 3CH mice. Their cytochrome and cytochome oxidase is markedly higher than that of Ehrlich ascites tumour cells (King and King, 1968a,b). Adebonjo et al. (1961) exposed these cells for up to three months to 95% N_2, 5% CO_2; this decreased their cytochrome redox content to the level of Ehrlich ascites tumour cells, and increased aerobic glycolysis, but failed to produce increased malignancy.

Cytochrome oxidase, while low in some malignant cells is not consistently lower than in normal tissues of the same kind (see e.g. Haven et al., 1967; Jones, 1965) and certainly not lower than in other types of non-malignant tissues. Drabkin and Rosenthal (1943) excluded lack of cytochrome c as a factor in causing malignancy (cf. also Schmidt and Schlief, 1955; Monier et al., 1959).

Leukaemia virus (Friend) increased the cytochrome c content of heart and kidney, though not in liver, brain and skeletal muscle of infected mice (D'Allessandro et al., 1966a). Using an improved method of low temperature spectrophotometry and Nagase as proteolytic agent, Ritter and Elkin (1967) found high concentrations of cytochrome a_3, a, b and c in both hyperdiploid (ELD) and tretraploid (EL_2) mouse ascites tumour cell, with a cytochrome c concentration not below that of normal rat liver, although perhaps not all were reactive in the electron transport. The cytochrome content of the mitochondria for the tumour cells showed them to be very competent to perform aerobic metabolism. In ascites hepatomas of rat and mice Sato and Hagihara (1970) found cytochrome c normal though only half as much of cytochrome a and b as in normal cells.

Squamous carcinoma cells of the skin had a higher than normal cytochrome oxidase concentration (Carruthers et al., 1959) and Woernley et al. (1959) find a larger concentration of it than previously discovered in Ehrlich ascites cells. In carcinoma (Walker-256) cytochrome oxidase was also increased in the tumour (Greene and Haven, 1957), but it was decreased in mice breast carcinomas and Yoshida sarcoma (Nagata, 1957). A lowering of cytochrome oxidase with increase of peroxidase was reported by Neufeld et al. (1958) in Walker carcinoma; however, some normal tissues (intestine, stomach, spleen, lung) showed similar high peroxidase/oxidase action. A high level of cytochrome oxidase and succinic dehydrogenase was found in adenocarcinoma (Suzuki, 1966). In transplantable rat hepatoma cells, cytochrome oxidase was decreased but the distribution of the cytochromes was the same as in normal liver (Wattiaux and Wattiaux-de Conmink, 1968). Cytochrome oxidase was increased in a malignant human blue naevus although its endogenous respiration was low (Rilcy and Pack, 1966, 1967). Lack of cytochrome oxidase and of mitochondrial cytochromes is therefore not the cause of malignancy.

With regard to the microsomal enzymes, however, there is far better evidence for their depression in malignancy. Ikeda et al. (1965) found sarcoma to depress cytochromes b_5 and P-450. Walker-256 carcinoma showed decreased drug hydroxylation, P-450 and binding capacity in males which could be increased by testosterone (Kato et al., 1970c). Sato and Hagihara (1970) found neither cytochrome b_5 nor P-450 in ascites hepatomas of rat and mice. Sugimura et al. (1966) found microsomal cytochromes as well as NADH-cytochrome c reductase absent in Yoshida hepatoma, though not in all hepatomas. Both NADPH- and NADH-cytochrome c reductase were decreased in liver microsomes of tumour-bearing rats (Kato et al., 1968). The suppression of the microsomal liver cytochrome is ascribed by Takahashi and Kato (1969) to a "toxohormone." The possibility must be borne in mind, however, that the microsomal cytochromes might be used up in detoxification reactions with carcinogens which make them less stable. Borukaeva et al. (1969) have studied EPR changes caused by the interaction of carcinogens with microsomal NO-haemo-proteins.

D. INVERTEBRATES

The cytochrome systems of invertebrates frequently resembles that of the vertebrates closely, the differences being more often quantitative rather than qualitative ones. Some observations on invertebrates muscles, from the very rapidly moving thoracic flight muscles to slow-moving muscles of parasitic worms have been discussed above (C1). More sensitive methods of examination for cytochromes (e.g. low temperature spectra) and cytochrome oxidase have further decreased the differences assumed to exist. Thus Humphrey (1947) did not find cytochrome oxidase in the oyster and Jodrey and Wilbur (1955) erroneously believed that evidence of inhibition by dithiocarbamate made it unlikely that cytochrome oxidase was the terminal oxidase in oysters. The presence of it in oysters has now been established by Hoshi (1958), Kawai (1958, 1959) and Mengebier and Wood (1967). Rogers (1949, 1962) found no evidence for aerobic metabolism and the cytochrome system in parasitic worms, e.g. *Ascaris*, which live in the intestine in an environment poorly supplied with oxygen, yet it is now clear (Smith, 1969 and see above) that their aerobic metabolism with cytochrome oxidase as terminal oxidase suffices for their energy supply. In contrast to previous findings they also contain mitochondria, though they are rather unstable.

Table XVI summarizes a large number of findings on the cytochrome system in invertebrates, from Ascidia and Echinodermata to Arthropods, molluscs and worms. On the whole, the more highly developed an invertebrate is, the more closely does its cytochrome system appear to resemble that of the vertebrates, but even some negative findings for more primitive organisms should be accepted with caution, for their study with more sensitive methods may well reveal the presence of cytochromes, though perhaps at rather low concentration.

There appears to be a certain resemblance of the cytochrome systems of some primitive invertebrates to those of bacteria, e.g. the presence of autoxidizable type b cytochromes (cytochrome o). It is, however, doubtful whether in the presence of cytochrome aa_3 with its high oxygen affinity the presence of other terminal oxidases of low oxygen affinity, e.g. cytochrome o could play any role, particularly in an environment of constantly low oxygen pressure. One interesting difference between some low arthropods and vertebrates is that the cytochromes of the former, like haemoglobins, can be increased under conditions of low oxygen content of the aqueous environment (Fox, 1955).

The realm of invertebrates also provides some interesting specific problems in the development of their eggs, the phenomenon of egg diapause, and in insects in their metamorphosis and the pupal diapause. These alterations, too, have been revealed as quantitative rather than qualitative, in contrast to earlier assumptions based on observations of cyanide and azide inhibition. Some of these observations have also been included in the table.

E. CYTOCHROMES IN PLANT AND FUNGI

The role of cytochromes in higher plant and algal respiration and photosynthesis has been discussed in preceding chapters with reference to the various chemical classes of cytochromes, in particular in Chapters III, B1, IV, F and V, 7c and VC, that of fungi in III, B1, IV, G and V, 7c. In the present subchapter a few more physiological aspects of plant and fungal cytochromes are discussed.

The distribution of cytochrome oxidase in the cells of shoots of *Rauwolffia* has been studied by Mia and Pathak (1963). The enzyme is richest in the roots of the sugar cane (Alexander, 1966), its K_m for cytochrome c^{2+} being 4.5×10^{-5} M. While cytochrome oxidase is cyanide-sensitive, the succinoxidase is only 50% inhibited by 10^{-3} M cyanide and also by antimycin A (Simon, 1957). Wilson (1970) found oxidative phosphorylation in the *Arum spadix* and in cultured sycamore cells, but only one

TABLE XVI
Cytochromes in invertebrates

Phylum	Order	Species	Part	Cytochromes	Reference
Tunicata	Ascidiacea		Eggs and larvae	Formation of cyt. oxidase in larval development; special effect on tail formation	Reverberi, 1957
Echinodermata	Echinoidea		Eggs	Cyt. oxidase. Development blocked by puromycin	Dontsova and Neifakh, 1966
	Echinoidea	Paracentrotus lividus	Eggs	Mitochondria, cyt. oxidase, NADH-cyt. c reductase, succinate dehydrogenase. Microsomes, cyts. aa_3, b, c	Rapoport et al. 1958
	Echinoidea	Paracentrotus lividus	Eggs	Cyt. oxidase increased 30% on fertilization	Maggio and Ghiretti-Magaldi 1958 Maggio, 1959 Maggio and Monroy 1959
	Echinoidea	Paracentrotus lividus	Eggs	Inhibitor decreased on fertilization	Maggio et al., 1960
	Echinoidea	Dendraster (Sanddollar)		Cyt. oxidase in early cleavage state	Berg, 1958
	Echinoidea	Lythechinus pictus	Sperm	Room and low temp. aa_3, b, c, c_1	Wilson and Epel, 1968
	Echinoidea	Strongylocentrotus purpuratus			
	Holothuroidea	Parastidiopus tremulus	Muscle	No cytochromes	Mattison, 1962
Arthropoda	Merostomata	Limulus polyphemus	Gills	Cyt. oxidase, succinoxidase; CN- and azide inhibition	Person and Fine, 1959

Group	Species	Tissue	Notes	References
Crustacea, Decapoda Brachyurea (crabs)	Carcinas maenas	Hepatopancreas	Mitoch., oxid. phosphoryl., cyt. oxidase, succinoxidase, CO-inhibition, amytal inhibition, aa_3, b, c, fp, NAD $\geqslant$ (cyts.). Basically as in vertebrates	Beechey, 1961a,b Beechey et al., 1963 Burrin and Beechey, 1962, 1963, 1964
Crustacea, Decapoda Brachyurea (crabs)	Brackish water crabs	Gill mitoch.	Cyt. oxidase, effect of salinity	King, 1966
Anostraca	Artemia salina	Nauplia	Cyt. oxidase, effect of salinity and temperature	Engel and Angelovic, 1968; Roels, 1970
Decapoda	Crab Callinectus sapidus Lobster Homarus americanus Shrimp Pandalus sp.	Muscles	Low temp. spectra aa_3, b, c; c_1 except lobster	Tappel, 1960 Pablo and Tappel, 1961
Decapoda, Astacura	Astacus leptodactylus	Muscles	Specially high aa_3 (40 μmoles/kg wet) aa_3, little b	Mattison, 1962
		Hepatopancreas	Cyt. oxidase, succinoxidase, CN-inhibition; NAD(P)H-cyt. c reductases	Obuchowicz, 1962
		Nerves	Cyt. oxidase, correlated with functional state	Zaguskin et al. 1967
Crustaceans etc.	Astacus leptodactylus	Nerves	NAD(P)H-cyt. c reductases sporadically	Giuditta and Aloj, 1965
Crustacea, Branchiopoda Conchostraca	Leptestheria mayeti Caenestheria inopinata Limnadia lenticularis Daphnia magna	Mandible muscles	Increase of cytochromes (as well as of haemoglobin) on O_2-lack	Fox, 1955
Crustacea, Cirripedia			More cyt. oxidase in barnacles adapted to longer air exposure in intratidal zone	Augenfeld, 1967

TABLE XVI (continued)

Phylum	Order	Species	Part	Cytochromes	References
Arthropoda	Insecta Lepidoptera	Galleria melonella		Cyt. oxidase, terminal oxidase in larvae, succinoxidase CN-inhibition. Cyt. oxidase decreasing in starvation and in pupae	Wojtczak, 1952
	Insecta Lepidoptera	Galleria melonella	Pupae	Succinoxidase decreases in late larvae and pupa, rises from 3rd day of pupation	Sulkowski and Wojtczak, 1958
	Insecta Lepidoptera	Galleria melonella	Intracellular location	Cyt. oxidase mostly in mitoch., 10% cytoplasmic. Succinate-cyt. c reductase only in mitochondria	Wojtczak et al., 1958
	Insecta Lepidoptera	Hyalophora cecropia	Pupae, heart	Pupal hearts inhibited only slowly by CN, adult hearts strongly, also by CO	Harvey and Williams, 1958
	Insecta Lepidoptera	Hyalophora cecropia	Pupae	In pupal hearts cyt. c, not cyt. oxidase limiting factor	Kurland and Schneiderman, 1959
	Insecta Lepidoptera	Hyalophora cecropia	Pupae	Injury slightly increases CO metabolism	Harvey, 1961
	Insecta Lepidoptera	Hyalophora cecropia	Larvae Pupae	Moderately high aa_3, b, c, b_5 Low aa_3, b_5, no b and c in diapause	Shappirio and Williams, 1957a
			Adult	Normal cyt. system induced by prothoracic gland hormone. Low temp. c_1	
	Insecta Lepidoptera	Hyalophora cecropia	Wing epithelium	Cyt. oxidase, NADH oxidase, succinate-c reductase disappear in pupal diapause, reappear on termination	Shappirio and Williams, 1957b

Insecta Lepidoptera	*Hyalophora cecropia*	Pupae	Injury in diapause increases respiration, cyt. oxidase, succinate-*c* reductase, NADH oxidase, CO-sensitivity. No evidence for alternative oxidase	Shappirio and Harvey, 1965
Insecta Lepidoptera	*Bombyx mori*	Eggs and larvae	Cyt. oxidase decreased in ovaries in egg-laying, increased in developing egg after diapause. In midgut max. at 3rd day of larvae: aa_3, c_1, b_5, fp	Ueda, 1961
Insecta Lepidoptera	*Bombyx mori*	Eggs	Cyt. oxidase and b_5, but not *c* or NADH-oxidase in early stages including diapause. Mitoch. and "lipid-rich particles". Antimycin-insensitive NAD(P)H reductases in these and microsomes; no succinate-*c* reductase. NADH-oxidase by combining mitoch. and soluble fraction. Later typical cyt.-oxidase system	Chino, 1963
Insecta Lepidoptera	*Bombyx mori*	Hibernating eggs	No *c* in early stages of hibernating eggs, but in newly hatched larvae	Osanai, 1967
Insecta Coleoptera	*Sitophilus granarius* and *Tenebroides mauritanicus*		Cyanide poisoning, anaer. and aerob.	Bond, 1962

TABLE XVI (continued)

Phylum	Order	Species	Part	Cytochromes	References
Arthropoda	Insecta Coleoptera	*Leptinotarsa decemlineata* (Colorado beetle)	Metamorphosis	Cyt. oxidase and succinic dehydrogenase in early prediapause declines with retirement into soil (minimum oxidase activity and low efficiency of insecticides)	Ushatinskaya, 1959
	Insecta Hymenoptera	*Apis mellifera* (Honey bee)	Eggs	In early development c deficiency in both worker and queen. Cyt. oxidase and O_2 consumption low but in queen $2 \times$ as many mitoch. and cyt. oxidase as in worker	Osanai and Rembold 1968
	Insecta Diptera (flies)	*Musca domestica*	Pupae	DDT-resistant pupae have less cyt. oxidase, more CN-resistance	Sacktor, 1951
	Insecta Diptera (flies)	*Musca domestica*	Thoracic muscles	Cyt. oxidase in pupae correlated with metabolic activity during oocyte development	Lachinova, 1967
	Insecta Diptera (flies)	*Musca domestica*	Pupae	P-450 and dieldrin resistance not parallel	Matthews and Casida, 1970
	Insecta Diptera (flies)	*Drosophila melanogaster*	Pupae	Effect of density on cyt. oxidase	Ward and Bird, 1969
	Insecta Diptera (flies)	*Drosophila melanogaster*	Larvae	Cyt. oxidase, not always lowered in periods of retarded activity, only one mutant lacking cyt. oxidase	Farnsworth, 1964, 1965
				Low temp. spectra aa_3, b, c, no c_1 but cyts.-551 and c-555	Goldin and Farnsworth, 1966

Some physiological aspects of cytochromes

	Insecta Blattodea (cockroaches)	*Periplaneta americana*	Muscles, heart, gut, nerves, brain	Usually more in female	Sacktor and Bodenstein, 1952
	Insecta Blattodea (cockroaches)	*Periplaneta americana*	Muscle	Normal cyt. system, more c than in rat muscle	Harvey and Beck, 1953
	Insecta Blattodea (cockroaches)	*Periplaneta americanus*		Respiratory control	Cochran, 1963
Mollusca	Cephalopoda	*Loligo subulata*	Muscle	aa_3, not in other molluscs	Pablo and Tappel, 1961
	Cephalopoda	*Sepia officinalis*	Neurons	Cytochrome pattern possibly as complicated as in vertebrates. Antimycin-insensitive NAD(P)H-c reductases	Giuditta and Aloj, 1965
	Cephalopoda	*Octopus vulgaris*	Muscle	aa_3, b, c, cyt. oxidase, succinoxidase, CO-inhibition	Ghiretti-Magaldi, et al., 1957
	Cephalopoda	*Octopus vulgaris* and *Eldone moschata*	Muscle and hepatopancreas	Same, also cyt. h	Ghiretti-Magaldi, 1958a,b
	Bivalva	Oyster	Gills, mantle	Cyt. oxidase, cyt. b, low temp. spectra aa_3, b, no c. Active cyt. system though conc. low	Kawai 1956, 1958
	Bivalva	Oyster	Gills, mantle	aa_3, b, c	Hoshi, 1958
	Bivalva	*Crassostrea gigas*, *Pinctada martensii*, *Mytilus crassitesta*	Visceral cells	Normal cyts., in heart c, b; b predominant in other tissues. Cyt. oxidase	Kawai, 1959
	Bivalva	*Crassostrea virginica*	Mantle	Much cyt. oxidase, activity depressed by parasites (dinoflagellate *Cochlidium*, *Minchiana nelsonii*)	Mengebier and Wood, 1967
	Bivalva	*Mytilus edulis*	Muscle	aa_3, $c + c_1$, b	Ryan and King, 1962
	Bivalva	*Mytilus galloprovincialis*	Liver, gills, muscle	CN- and azide inhibition only partial	Shapiro, 1967

TABLE XVI (continued)

Phylum	Order	Species	Part	Cytochromes	References
Mollusca	Bivalva	Cristaria plicata Hyriopsis schlegelii	Heart, sperm, unfertlized eggs, gills, adductor muscle	c, b predominant, autoxidizable: b and c in gills. Succinate-c reductase (antimycin-insensitive) in mitoch. and microsomes, only b in microsomes	Kawai, 1961
	Bivalva	Spisula solidissima (clam)	Nerves, ganglia	Cyt. b, myoglobin-like, M.W. 20,000	Strittmatter and Burch, 1963
	Ophistobranchia, Aplysiomorpha	Aplysia punctata, depilans	Neurons	Haemoprotein	Chalazonitis and Arvanitaki 1951
	Ophistobranchia, Aplysiomorpha	Aplysia depilans, limacina	Buccal muscle, gizzard	Cyt. oxidase, CO-inhibition, NADH-c reductase, aa_3, b, c, c_1	Ghiretti et al., 1959b
	Ophistobranchia, Aplysiomorpha	Aplysia punctata, depilans	Hepatopancreas	Cyt. h	Tosi and Ghiretti, 1960
	Ophistobranchia, Aplysiomorpha	Aplysia californica	Neurons, muscle	Myohaemoglobin-like 570, 548, 423 nm	Wittenberg et al., 1965a,b
	Gastropoda Neogastropoda	Buscyon caniculatum	Radula muscle	Myohaemoglobin-like 570, 548, 422 nm	Wittenberg et al., 1965a,b
	Gastropoda Neogastropoda	Pila globosa	Stomach	Cyt. b Soret peak	Reddy and Swami, 1964
	Gastropoda Pulmonate	Viviparus viviparina	Hepatopancreas	Cyt. oxidase, CN-inhibited	Obuchowicz and Zerbe, 1962
	Gastropoda Pulmonate	Viviparus viviparina	Hepatopancreas	NADH-c reductase more than cyt. oxidase	Obuchowicz and Klepke, 1963
	Gastropoda Stylammatophora	Helix pomatia	Heart	Cyt. oxidase and cyt. c	Cardot, 1969
	Gastropoda Stylammatophora	Helix pomatia	Neurons	Distribution of cyt. h	Chalazonitis and Gola, 1964

Phylum/Class	Subgroup	Species	Tissue/Organ	Notes	Reference
	Gastropoda Stylammatophora	Helix pomatia	Spermatozoa	Mitochondria	Anderson and Personne, 1970
	Gastropoda Prosobranch	Neptunea despecta	Giant muscle cells and radula	Cyt. oxidase (histochem)	Justensen, 1966
	Gastropoda Neogastropoda	Urosalpinx cinareus	Boring organ	Cyts. including c. No boring in nitrogen	Person et al., 1967
Annelida	Polychaeta Aphroditidae	Aphrodite aculeata	Nerve ganglia	Myoglobin-like cpts.	Wittenberg et al. 1965a,b,c
		Halodysna. Also in Nereis, Arenicola, Lumbricus, Tubifex			Ghiretti and Ghiretti-Magaldi, 1959; Ghiretti et al., 1959a; Ghiretti-Magaldi et al., 1958b
	Polychaetas Sabellidae	Spirographis spallanzani		No aa_3, but a cyt. b	Petrucci, 1955
	Oligochaeta	Eisenia foetida. Peloscolex relutinus	Cyt. oxidase,	partial CN-inhibition and CO-inhibition	Urich, 1965
	Oligochaeta Lumbriculidae	Lumbricus terrestris	Chlorogogous tissue. Body wall	Mitoch., cyt. oxidase, succinoxidase	
Echiduroidea		Urechis caupo		Mitoch., more oxidase, succinoxidase	Rothschild and Tyler, 1958
Sipunculoidea		Sipunculus nudus		Cyt. system more like bacterial than yeast system	Ghiretti and Ghiretti-Magaldi, 1959
Acantocephala		Amphiporus sp.		,, ,,	Wittenberg et al., 1965c

TABLE XVI (continued)

Phylum	Order	Species	Part	Cytochromes	References
Aschelminthes	Nematoda	Trichinella spiralis	Adult	aa_3, b, c. Less active in adult worms than in larvae. Cyt. oxidase, CN- and CO-inhibited, succinoxidase	Goldberg, 1957
	Nematoda	Haemonchus contortus	Larvae	Cyt. oxidase, CN-inhibited. NADH-c reductase antimycin-inhibited. Branched chain, one to aa_3, the other probably to cyt. b.	Moon and Schofield, 1968
	Nematoda	Ascaris lumbricoides	Muscle	Cyt. oxidase, antimycin-insensitive NAD(P)H-c reductases, antimycin-sensitive succinate-c reductase, autoxidizable cyt. b, probably of secondary importance; cytochrome c	Kikuchi et al., 1959. Kikuchi and Ban, 1961
	Nematoda	Ascaris lumbricoides	Eggs	Cyt. oxidase, succinoxidase, CN-sensitive. Adults muscle, see Cl	Kimetec et al., 1963
	Nematoda	Ascaris lumbricoides	Eggs	No cyt. oxidase (but cyt. c) in unembryonated eggs, but in embryonated eggs. Succinate-c reductase antimycin-sensitive	Costello et al., 1963
	Nematoda	Ascaris lumbricoides	Larvae	Third stage homogenates contained cyt. oxidase, 4th stage not, anaerobic to aerobic metabolism on 3rd	Sylk, 1969

Some physiological aspects of cytochromes 425

Group	Subgroup	Species	Tissue	Description	Reference
Nematina				moult; in 4th stage terminal flavoprotein-oxidase. In guinea pigs arrest of larval development at stage 3	Poluhovich, 1970
Platyhelminthes	Trematoda (Digenea, flukes)	*Prostoma rubrum* *Fasciola hepatica*		Cyt. oxidase in free-living worm but not in digestive system of liver fluke, possible peroxidase	Humiczewska, 1966
	Cestoda	*Dendrocoelum lacteum* *Moniezia expansa*		Low temp. spectra aa_3, b, c, autoxidizable b (o)	Cheah, 1968
	Eucestoda (tapeworm)	*Taenia hydatigena*		Two terminal oxidases a_3 and o. Cyt. b-555 with small antimycin sensitivity. In particular fraction c and c_1 predominate; succinoxidase	Cheah, 1967
	Trematoda Eucestoda	*Hymenoplepsis diminuta*	Gonads and embryos	NADH-c reductase and probably two pathways. No oxidase during embryogenesis. Enzyme decrease towards end of development	Rybicka, 1967
Coelenterata	Hydrazoa	*Hydra littoralis*		Nematocysts contain an inhibitor of succinoxidase	Kline and Waravdekar, 1960
Protozoa	Flagellates Coccidia (Malaria parasites)	*Polytonella caeca*		Mitoch., aa_3, $c + c_1$, b, fp. Inhibited by antimycin A, rotenone, piericidin	Lloyd and Chance, 1968
	Flagellates Coccidia (malaria parasites)	*Plasmodium berghei gallinaceum*		Mitoch. and whorls of cystoplasmic membranes, cyt. oxidase histochem.	Theakston *et al.*, 1969
		Plasmodium knowlesii		Cyt. oxidase in platelet-free prep.	Scheitel and Miller, 1969

TABLE XVI (continued)

Phylum	Order	Species	Part	Cytochromes	References
Protozoa	Trypanosomatids	*Crithidia fasciculata*		Cyt. aa_3, c, b_{555}, CO-inhibition, oxid. phosphorylation. Acriflavin depresses cyt. oxidase, less b and c, c-555	Hill and White, 1968a,b Chan and Hill, 1969
	Ciliata, Hymenostomatida	*Tetrahymena pyriformis*		Cytochrome d? (620 nm) terminal oxidase, b, c_1; rotenone-sensitive NADH-c reductase. Antimycin inhibits b and c reoxidation succinate reduces d and c, only partly b	Kobayashi, 1965 Turner et al., 1969
	Amoeba	*Dictyostelium discoideum* *Harmanella castellani*		Mitoch., low a/c ratios	Erickson and Ashworth, 1969

phosphorylation site appears to be connected with CN-insensitive respiration. The phosphorylation requires a P_{O_2} of 70 μM. It is suggested that the mechanism controls heat production in the spadix. The differences of the Mung bean mitochondria from both animal and yeast mitochondria have been described by Ikuma and Bonner (1967; see also Chapter IV, F2). In lupine roots, mitochondria are particularly found in fine root cells and their cyanide- and azide-sensitive cytochrome oxidase and the parallel content of succinic dehydrogenase indicate an active role in root absorption (Potapov et al., 1964). In the mitochondria of the Jerusalem artichoke tubers, tetrazolium salts even in lower concentration than usually used for histochemical assays, inhibit the oxidation of NAD-linked substrates between NADH and cytochrome b. They also uncouple oxidative phosphorylation (Palmer and Kalina, 1966). Sweet potato (*Ipomoea batata*) roots, intact or wounded, contain several kinds of particles differing in their succinate dehydrogenase/cytochrome oxidase, but not in their succinoxidase/cytochrome oxidase ratio (Sakano et al., 1968). They show considerable state 4 respiration, at least partially due to the respiratory chain but this is less sensitive to 0·3 μM chloropromazine (Carmeli and Biale, 1970). Mycorrhizal as well as uninfected roots of *Fagus sylvatica* showed cyanide-, azide- and CO-sensitive cytochrome oxidase activity (Harley and Rees, 1959). The herbicide Paraquat stimulates the slow cytochrome c reduction by NADH in crude spinach leaf extracts, and also in pea roots as well as in light and dark-grown *Chlorella* (Black and Myers, 1966) with NADH and NADPH.

The adaptive synthesis of cytochromes in yeast will be discussed in Chapter IX. Tightly coupled mitochondria from yeast have been isolated by Ohnishi and Hagihara (1964, 1965) and Mattoon and Balcavaga (1967), from *Neurospora* by Greenawalt et al. (1967). The uncoupler 2,4-dinitrophenol inhibits yeast respiration at the second phosphorylation site (Stoppani et al., 1969). Fungal cytochromes similar to those in yeast have also been found in higher fungi, e.g. *Polystictus versicolor*, and in other fungi, e.g. *Neurospora crassa* (Boulter, 1955; Boulter and Derbyshire, 1957), *Aspergillus niger* (Higgins and Friend, 1970) and in *Rhodotorula mucilaginosa* (Kitsutani et al., 1970). Fungal respiration is specifically inhibited by 3-(p-chlorophenyl)-1,1-dimethyl urea which prolonged reduction of cytochromes c and a (Mukasa et al., 1966).

The distribution of mitochondria from a vegetative "petite" yeast mutant in zonal centrifugation did not differ from that in wild type yeast, and in spite of enzymic differences the distribution of both cytochrome oxidase and NADH-cytochrome c reductase was uniform in the different zones (Avers et al., 1965). The distribution of cytochrome oxidase and cytochrome c reductases in various fractions obtained by zonal centrifugation from *S. carlsbergensis*, anaerobically grown and after various times of adaptation to air, has been studied by Lloyd et al. (1970). The mutant did not contain cytochrome aa_3 or b, but the cytochrome c oxidation was blocked by cyanide (Kazakova et al., 1969). The full respiratory function of cytochrome c-deficient yeast mutants could be restored by the addition of extrinsic cytochrome c from yeast (iso-1 and iso-2) as well as from horse heart. The cytochrome c was firmly bound, and with it the uptake of P_i and O_2, as well as the P/O ratio increased (Mattoon and Sherman, 1966). The stoichiometry of the binding of c was not altered by its type, although horse heart was slightly inferior in reconstituting succinoxidase. Exogenous c was taken up by water suspensions of *Candida* (*Torulopsis utilis*) as well as by *S. cerevisiae* and decreased the viable count. The cytoplasmic membrane lost its capacity to retain intracellular constituents which were released into the medium, but phosphate protected the spheroblasts against cytochrome c. Mother cells were more susceptible than buds (Svihla et al., 1969). Low temperature studies on some yeasts by Zviagilskaya and Karapetyan (1965) have been reported above (Chapter V, B6c), see also Terui and Sugimoto (1969). *Candida lipolytica* can produce cytochrome c with n-alkanes, higher fatty acids or higher alcohols as sole carbon and energy source (Tanaka and Fukui, 1970; Volland and Chaix, 1970).

In *Neurospora crassa* Edwards and Woodward (1969) found much cytochrome c and cytochrome oxidase activity, but while they observed an α-band of type a at 583 nm as well as one at 557 nm, they found only one γ-band at 422 nm, possibly due to an aldimine bond between the formyl group of haem a and the protein (see Chapter III, B4b).

In contrast to earlier findings of Nord (1945), Kikuchi and Barron (1959) found the cytochrome system in *Fusarium lini*, succinate- and NADH-oxidase as well as aa_3, b and c; in contrast to Nord, nitrate or hydroxylamine reductase was absent unless the organism had grown on these substrates. b_3 (?) was present in mitochondria as well as in microsomes (cf. Chapter IV, F2). Cytochrome oxidase, aa_3, CN-sensitive with ascorbate-TMPD, but antimycin A-insensitive oxygen uptake with Krebs cycle intermediates and a low P/O ratio were found in young mycelia of *Aspergillus oryzae* (Kawakita, 1971). Cytochrome b was largely reduced in aerobic suspensions and hardly oxidized even by ferricyanide. Gleason (1968) and Gleason and Unestam (1968) have found cytochromes a, b and c, but no c_1 in the aquatic fungi *Leptomitus lacteus* and *Apodachlya punctata* by normal and low temperature spectra, as well as cytochromes of type A. The α-bands of *Blastocladiella* (Cytriomycetes) were at 604, 559, 556 and 548 nm, those of *Monoblepharioides* at 608, 653, 559, 555 and 548 nm, while those of *Oomycetes* were at 602, 562, 555 and 549 nm. Horgan and Griffin (1969) found the mitochondria in *Blastocladiella emersonii* in zoospores and sporangia and in salt-induced sporangia as active or more active than in ordinary zoospores and sporangia.

In *Phellinus tremulae*, cytochrome oxidase as well as ascorbate and tyrosine oxidase were present (Fedorov and Staichenko, 1969). Sporangiophores of *Phycomyces blakesleeanus* contained cytochrome c in wild type and in a carotenoid-deficient albino mutant (Wolken, 1969).

Spores of actinomycetes (*Streptomyces fluorescens*, *Micropolyspora rectiovirgata* and *vulgaris*, *Actinobiphida dichotoma* contained cytochromes in young submerged cultures decreasing with age (Taptykova *et al.*, 1969). Except in the last species, spores also contained cytochromes. Cytochrome oxidase was present in the first two organisms.

REFERENCES

Ackerman, E. (1970). *Biochem. Pharmacol.* **19**, 1955–1973.
Adebonjo, F. O., Bensch, K. G. and King, D. W. (1961). *Cancer Res.* **21**, 252–256.
Albers, R. W., Koval, G. V. and Siegel, G. V. (1968). *Mol Pharmacol.* **4**, 324–336.
Aldridge, W. N. and Street, B. W. (1968). *Biochem. J.* **107**, 315–317.
Alexander, A. G. (1966). *J. Agric. Univ. Puerto Rico* **50**, 131–145.
Altenbrunn, H. J., Rapoport, S. and Hahn, C. (1959). *Acta Biol. Med. Germ.* **2**, 599–620.
Alvares, A. O., Schilling, G. R., Levin, W., Kuntzman, R., Brand, L. and Mark, L. C. (1969). *Clin. Pharmacol. Ther.* **10**, 655–659.
Anderson, W. A. (1970). *J. Histochem. Cytochem.* **18**, 201–210.
Anderson, W. A. and Personne, P. (1970). *J. Histochem. Cytochem.* **18**, 783–793.
Aratei, H., Oprisor, M. and Leporda, G. (1967). *Morfol. Norm. Patol.* (Bucharest) **12**, 131–170. *Chem. Abstr.* **68** (3) 20138.
Arcasoy, M., Smuckler, E. A. and Benditt, E. P. (1968). *Amer. J. Pathol.* **52**, 841–868.
Archakov, A. I., Panchenko, L. F., Karuzina, I. I. and Aleksandrova, T. A. (1969). *Biokhimiya* **34**, 604–609. *Chem. Abstr.* **71** (6) 47363.
Arcos, J. C., Sokal, R. S., Sun, S. C., Argus, M. F. and Burch, G. E. (1968). *Exptl. Mol. Pathol.* **8**, 49–65.
Arvy, L. and Mauleon, P. (1964). *Compt. rend. soc. biol.* **158**, 453–457.
Aubert, X., Chance, B. and Keynes, R. D. (1964). *Proc. Roy. Soc.* **160**, 211–245.
Augenfeld, J. M. (1967). *Physiol. Zool.* **40**, 92–96. *Chem. Abstr.* **66** (3) 110437.
Augustin, H. W. and Rapoport, S. (1959). *Acta Biol. et Med. Germ.* **3**, 433–449.
Avers, C., Rancourt, U. and Lin, F. (1965). *Proc. Nat. Acad. Sci. U.S.* **54**, 527–534.
Bailie, N. and Morton, R. K. (1955). *Nature* **176**, 111–113.
Ball, E. G. and Cooper, O. (1957). *Proc. Nat. Acad. Sci. U.S.* **43**, 357–364.
Balogh, U. Jr., Hoyt, R. F. Jr. and Pragay, D. A. (1967). *Lab. Invest.* **16**, 211–219. *Chem. Abstr.* **66** (8) 63942.
Bank, P., Mangnall, D. and Mayer, D. (1969). *J. Physiol.* **200**, 745–762.
Baranov, V. F. and Prokopenko, S. M. (1969). *Doklady* **187**, 443–445. *Chem. Abstr.* **71** (12) 110859.
Barbashova, Z. I. and Grigoreva, G. I. (1968). *Chem. Abstr.* **71** (6) 47333.
Barbashova, Z. I., Vasilev, P. V. and Uglova, N. N. (1967). *Chem. Abstr.* **66** (12) 103656.
Barber, R. E., Lee, J. and Hamilton, W. K. (1970). *New Engl. J. Med.* **283**, 1478–1784.

Barker, E. A., Arcasoy, M. and Smukler, E. A. (1969). *Chem. Abstr.* **172** (11) 109199.
Barker, S. B. (1951). *Physiol. Revs.* **31**, 205–231.
Barker, S. B. (1955). *Endocrinology* **57**, 414–418.
Barlow, R. M. (1963). *J. Comp. Pathol. Therop.* **73**, 61–67.
Barron, E. S. G. (1936). *J. Biol. Chem.* **113**, 695–697.
Bedrak, E. and Samoiloff, M. R. (1967). *Canad. J. Physiol. Pharmacol.* **45**, 717–722.
Beechey, R. B. (1961a). *Nature* **192**, 975–976.
Beechey, R. B. (1961b). *Comp. Biochem. Physiol.* **3**, 161–174.
Beechey, R. B., Burrin, D. H. and Baxter, M. I. (1963). *Nature* **198**, 1277–1279.
Bégin-Heick, N. (1970). *Canad. J. Biochem.* **43**, 257–258.
Bégin-Heick, N. and Blum, J. J. (1967). *Biochem. J.* **105**, 813–820.
Bennets, H. W., Beck, A. B. and Harley, R. (1948). *Austral. J. Vet. Sci.* **24**, 237–244.
Berezney, R. and Crane, F. L. (1971). *Biochem. Biophys. Res. Commun.* **43**, 1017–1023.
Berezney, R., Funk, L. K. and Crane, F. L. (1970a). *Biochim. et Biophys. Acta* **223**, 61–70.
Berezney, R., Funk, L. K. and Crane, F. L. (1970b). *Biochem. Biophys. Res. Commun.* **38**, 93–98.
Berg, W. E. (1958). *Exptl. Cell Res.* **14**, 398–400.
Betel, I. (1967). *Biochim. et Biophys. Acta* **143**, 62–69.
Betel, I. and Klouwen, H. M. (1967). *Biochim. et Biophys. Acta* **137**, 453–467.
Beutler, E. (1959). *Acta Haematol.* **21**, 371–377.
Bickis, I. J. and Quastel, J. H. (1965). *Nature* **205**, 44–46.
Black, C. C. and Myers, L. (1966). *Chem. Abstr.* **66** (2) 10126.
Blockhuis, G. G. D. and Veldstra, H. (1970). *FEBS Letters* **11**, 197–199.
Blum, J. J. and Bégin-Heick, N. (1967). *Biochem. J.* **105**, 821–829.
Bogolepov, N. and Dovedova, E. L. (1969). *Tsitologiya* **11**, 189–200. *Chem. Abstr.* **70** (12) 102968.
Bond, E. J. (1962). *Nature* **193**, 1002–1003.
Borst, P. (1960). *J. Biophys. Biochem. Cytol.* **7**, 381–383.
Borukaeva, M. R., Raikhman, L. M. and Shabalkin, V. A. (1969). *Doklady* **189**, 651–654 (*Proc. Acad. Sci. USSR*).
Bostelmann, W., Engelmann, B., Ernst, B. and Nagel, K. H. (1970). *Acta Biol. Med. Germ.* **25**, 169–176.
Boulter, D. (1955). *Biochem. J.* **60**, XXI.
Boulter, D. and Derbyshire, E. (1957). *J. Exptl. Bot.* **8**, 313–318.
Boxer, G. E. and Devlin, J. D. (1961). *Science* **134**, 1495–1501.
Brame, R. G. and Overby, J. R. (1970). *Obstet. Gynec.* **36**, 425–428.
Brauser, B. (1968). *Z. Analyt. Chem.* **237**, 8–17.
Brauser, B., Bücher, Th. and Dolivo, M. (1970). *FEBS Letters* **8**, 297–300.
Brentani, R. and Raw, I. (1965). *Exptl. Cell Res.* **38**, 213–215.
Brikker, V. N., Volpert, E. E., Ganelina, I. E., Issakyan, L. A. and Krioskaya, V. (1969). *Kardiologiya* **10**, 95–97.
Bronk, J. R. (1960a). *Biochim. et Biophys. Acta* **37**, 327–336.
Bronk, J. R. (1963). *Science* **141**, 816–818.
Bronk, J. R. (1965). *Biochim. et Biophys. Acta* **97**, 9–15.
Bronk, J. R. (1966). *Science* **153**, 638–639.
Bronk, J. R. and Bronk, M. S. (1962). *J. Biol. Chem.* **237**, 897–903.

Brownie, A. C., Colby, H. D., Gallant, S. and Skelton, F. R. (1970). *Endocrinol.* **86**, 1085–1092.
Brownie, A. C., Skelton, F. R., Gallant, S., Nicholls, P. and Elliott, W. B. (1968). *Life Sci.* **7**, 765–771.
Brykhanov, O. A. and Manoilov, S. E. (1967). *Chem. Abstr.* **69** (8) 57327.
Buikis, I. (1966). *Chem. Abstr.* **66** (5) 35856.
Bullock, T. H. (1955). *Biol. Rev.* **30**, 342–377.
Burrin, D. H. and Beechey, R. B. (1962). *Biochem. J.* **83**, 1P.
Burrin, D. H. and Beechey, R. B. (1963). *Biochem. J.* **87**, 48–53.
Burrin, D. H. and Beechey, R. B. (1964). *Comp. Biochem. Physiol.* **12**, 245–258.
Burstein, S. (1967). *Biochem. Biophys. Res. Commun.* **26**, 697–703.
Burstone, M. S. (1961). *J. Histochem. Cytochem.* **9**, 59–65.
Busnyuk, M. M. (1967). *Chem. Abstr.* **68** (1) 1295.
Bykov, E. G. and Novikov, Yu G. (1968). *Chem. Abstr.* **70** (8) 65901.
Caldwell, R. S. (1969). *Comp. Biochem. Physiol.* **31**, 79–93.
Cammer, W. and Estabrook, R. E. (1967). *Arch. Biochem. Biophys.* **172**, 721–747.
Cardot, J. (1969). *Compt. rend. soc. biol.* **163**, 873–877.
Carmeli, C. and Biale, J. B. (1970). *Chem. Abstr.* **72** (10) 96647.
Carruthers, C., Woernley, D. L., Baumier, A. and Davis, B. (1959). *Cancer Res.* **19**, 330–333.
Chaffee, R. R. J. and Roberts, J. C. (1971). *Ann. Rev. Physiol.* **33**, 155–202.
Chalazonitis, N. and Arvanitaki, A. (1951). *Bull. Inst. Oceanogr.* **48**, No. 996. *Chem. Abstr.* **46**, 5214.
Chalazonitis, N. and Arvanitaki, A. (1956). *J. Physiol.* (Paris) **48**, 430–435.
Chalazonitis, N. and Gola, N. (1964). *Compt. red. soc. biol.* **158**, 1908–1914.
Challoner, D. R. (1968a). *Amer. J. Physiol.* **214**, 365–369.
Challoner, D. R. (1968b). *Nature* **217**, 78–79.
Chan, S. K. and Hill, G. C. (1969). *Bact. Proc.*, pp. 142–162.
Chance, B. (1959). *Ann. New York Acad. Sci.* **81**, 477–489.
Chance, B. (1964). *Acta Union Intern. Contre le Cancer* **20**, 1028–1032.
Chance, B. and Hess, B. (1959a). *J. Biol. Chem.* **234**, 2421–2427.
Chance, B. and Hess, B. (1959b). *Science* **129**, 700–708.
Chance, B. and Sacktor, B. (1958). *Arch. Biochem. Biophys.* **76**, 509–531.
Chance, B. and Weber, A. (1963). *J. Physiol.* **169**, 263–277.
Chance, B., Jamieson, D. and Coles, H. (1965). *Nature* **206**, 257–263.
Chance, B., Jamieson, D. and Williamson, J. R. (1966). *8th Internat. Conf. on Hyperbaric Medicine. Nat. Acad. Sci. Washington, D.C.*
Chance, B., Jöbsis, F. and Schoener, B. (1962a). *Science* **137**, 499–508.
Chance, B., Lee, C-P., Oshino, R. and Van Rossum, G. D. V. (1964). *Am. J. Physiol.* **206**, 461–468.
Cheah, K. S. (1967). *Comp. Biochem. Physiol.* **20**, 867–875.
Cheah, K. S. (1968). *Biochem. et Biophys. Acta* **153**, 718–720.
Cheah, K. S. and Chance, B. (1970). *Biochim. et Biophys. Acta* **223**, 55–60.
Cherkasova, L. S., Taits, M. Yu, Tyrtyshina, G. F. and Filimonov, M. M. (1968). *Chem. Abstr.* **71** (9) 27827.
Cherniak, N. S., Edelman, N. H. and Lahiri, S. (1970). *Respir. Physiol.* **11**, 113–126.
Chin, C. H. (1950). *Nature* **165**, 926–927.
Chino, H. (1963). *Arch. Biochem. Biophys.* **102**, 400–415.
Cho, Y. W. (1965). *Experientia* **21**, 440–441.
Cho, Y. W. and Symbas, P. N. (1965). *Life Sci.* **4**, 1461–1466.
Cho, Y. W., Akbari-Ford and Dugge, E. J. (1966). *Experientia* **22**, 242–244.

Cho, Y. W., Theogaraj, T., Aviado, D. M. and Bellet, S. (1965). *Arch. Intern. Pharmacodyn.* **158**, 314–323.
Christiane, M. I. and Klitgaard, H. M. (1968). *Proc. Soc. Exp. Biol.* **128**, 107–109.
Cinti, D. L. and Schenkman, J. B. (1970). *Fed. Proc.* **23**, Abstr. 573.
Clarke, E. C. and Ball, E. G. (1955). *Fed. Proc.* **14**, Abstr. 623.
Clemmons, J. J. (1964). *Fed. Proc.* **23**, Abstr. 1737.
Cleveland, P. S. and Smuckler, E. A. (1965). *Proc. Soc. Exp. Biol. Med.* **120**, 808–810.
Cochran, D. G. (1963). *Biochim. et Biophys. Acta* **78**, 393–403.
Cohen, B. S. and Estabrook, R. W. (1971). *Arch. Biochem. Biophys.* **143**, 37–75.
Coleman, H. N., Dempsey, P. J. and Cooper, T. (1970). *Amer. J. Physiol.* **218**, 475–478.
Coleman, H. N., Sonnenblick, E. H. and Braunwald, E. (1969). *Amer. J. Physiol.* **217**, 291–296.
Conover, T. E. (1967a). *Curr. Top. Bioenerg.* **2**, 235–267. *Chem. Abstr.* **68** (4) 26846.
Conover, T. E. (1967b). *Abstr. 7th Internat. Conf. Biochem., Tokyo* V, H73.
Conover, T. E. (1968). *Fed. Proc.* **27**, Abstr. 1738.
Conti, A. and Borgo, M. (1965). *Laryngoscope* **75**, 830–839.
Cooper, J. M. and Thomas, P. (1970). *Biochem. J.* **117**, 24P.
Corwin, L. M. and Schwarz, K. (1959). *J. Biol. Chem.* **234**, 191–197.
Costa, L. E. and Taquini, A. C. (1970). *Acta Physiol. Latin-Amer.* **20**, 103–109. *Chem. Abstr.* **73** (9) 85623.
Costello, L. C., Oya, H., Smith, W. (1963). *Arch. Biochem. Biophys.* **103**, 345–351.
Cotman, C., Mahler, H. R. and Anderson, N. G. (1968). *Biochim. et Biophys. Acta* **163**, 272–275.
Covell, J. W., Braunwald, E., Ross, J. Jr. and Sonnenblick, E. H. (1966). *J. Clin. Invest.* **45**, 1535–1542.
Cranefield, P. F. and Greenspan, K. (1960). *J. Gen. Physiol.* **44**, 235–249.
Cress, H. R., Spots, A. and Heatherington, D. C. (1965). *J. Histochem. Cytochem.* **13**, 677–683.
Currie, W. D., Davidian, N. M., Elliott, W. B., Rodman, N. F. and Penniall, R. (1966). *Arch. Biochem. Biophys.* **113**, 156–166.
Dagg, J. H., Jackson, J. M., Curry, B. and Goldberg, A. (1966). *J. Haematol.* **12**, 331–333.
D'Alessandro, L., Conti, F., Stangherlin, P. and Catalano, G. (1966a). *Chem. Abstr.* **65** (13) 20639. *Biochim. Appl.* **13**, 109–144.
D'Allessandro, L., Conti, L. I. and Egidio, G. (1966b). *Chem. Abstr.* **65** (13) 20630. *Biochim. Appl.* **13**, 131–138.
Dallman, P. R. (1966). *Nature* **212**, 608–609.
Daly, M. de B., Lambertson, C. I. and Schweitzer, A. (1954). *J. Physiol.* **125**, 67–89.
Dancewicz, A. M., Mazanowska, A. and Panfil, B. (1965). *Chem. Abstr.* **65** (3) 4231.
Davidian, N. M., Penniall, R. and Elliott, W. B. (1968). *FEBS Lett.* **2**, 105–108.
Davison, T. F. (1971). *Comp. Biochem. Physiol.* **38**, 21–34.
De Berardinis, E. (1960). *Acta Opthalmol.* **38**, 488–507.
Degkwitz, E., Luft, D., Pfeiffer, U. and Staudinger, Hj. (1968). *Z. physiol. Chem.* **349**, 465–471.
De Groot, L. J. and Davis (1962) *Endocrinol.* **70**, 492–504.
De Groot, L. J. and Dunn, A. D. (1964). *Biochim. et Biophys Acta* **92**, 205–222.

De Groot, L. J. and Dunn, A. D. (1968). *J. Lab. Clin. Med.* **71**, 984–988.
De Jairala, S. W., Kayra, A., Garcia, A. P. and Rasia, M. L. (1969). *Biochim. et Biophys. Acta* **183**, 137–143.
De Otamendi, M. E. and Stoppani, A. O. M. (1967). *Rev. Soc. Argent. Biol.* **43**, 140–151.
Depocas, F. (1966). *Canad. J. Physiol. Pharmacol.* **44**, 875–880.
Derhachev, E. V. and Manoilov, S. E. (1967). *Chem. Abstr.* **69** (8) 57339.
Dixon, M., Maynard, J. M. and Morrow, P. F. W. (1960). *Nature* **186**, 1022.
Doll, E., Keul, J., Steim, H., Maiwald, C. and Reindell, H. (1965). *Pflügers. Arch. Ges. Physiol.* **282**, 28–42.
Dontsova, G. V. and Neifakh, A. A. (1966). *Doklady* **167**, 215–218.
Drabkin, D. L. (1954). *Physiol. Revs.* **31**, 345–431.
Drabkin, D. L. (1966). "Hemes and Hemoproteins", pp. 599–604.
Drabkin, D. L. and Rosenthal, O. (1943). *J. Biol. Chem.* **150**, 131–141.
Dreosti, I. E. (1967). *Chem. Abstr.* **67** (12) 106402.
Drynov, I. D., Mokhova, E. N. and Petukhov, E. V. (1968). *Chem. Abstr.* **71** (13) 119818.
Editorial (1968). *Nature* **218**, 321–322.
Edwards, D. L. and Woodward, D. P. (1969). *FEBS Lett.* **4**, 193–196.
Eichel, H. J. (1959). *Biochem. Biophys. Res. Commun.* **1**, 923–927.
Eisenberg, M. A. (1958). *Arch. Biochem. Biophys.* **74**, 372–389.
Elliott, K. A. C. (1955). In (K. A. C. Elliott, I. H. Page and J. H. Quastel, eds.) "Neurochemistry", Thomas Springfield, p. 53.
Engel, D. W. and Angelovic, J. W. (1968). *Comp. Biochem. Physiol.* **26**, 749–752.
Erickson, S. K. and Ashworth, J. M. (1969). *Biochem. J.* **113**, 567–568.
Ernster, L. (1965). *Fed. Proc.* **24**, 1222–1236.
Ernster, L. and Lee, C. P. (1964). *Ann. Rev. Biochem.* **33**, 729–788.
Estabrook, R. W. and Sacktor, B. (1958). *J. Biol. Chem.* **233**, 1014–1019.
Estabrook, R. W. and Holowinski, A. (1961). *J. Biophys. Biochem. Cytol.* **9**, 19–28.
Estler, C. J. (1965). *Arch. Exptl. Path. Pharmakol.* **251**, 413–432.
Evans, A. E. and Getz, G. S. (1968). *Blood* **31**, 110–718.
Evans, W. H. (1962). *Fed. Proc.* **21**, 45.
Evtodienko, Yu. V., Skulachev, V. P. and Chistiyakov, V. V. (1965). *Chem Abstr.* **65** (13) 20391.
Farnsworth, M. W. (1964). *J. Exptl. Zool.* **157**, 345–352.
Farnsworth, M. W. (1965). *J. Exptl. Zool.* **160**, 355–361.
Fay, F. J. (1970). *Amer. J. Physiol.* **208**, 518–523.
Fedorov, I. I. and Pekus, E. N. (1970). *Lab. Delo*, No. 3, 139–141.
Fedorov, I. I., Malyuk, V. I., Maksimova, A. V. and Belanova, A. J. (1968). *Chem. Abstr.* **68** (3) 20208.
Fedorov, N. J. and Staichenko, N. I. (1969). *Chem. Abstr.* **70** (10) 85105.
Fell, B. F., Mills, C. F. and Boyne, R. (1965). *Res. Vet. Sci.* **6**, 170–177.
Fleischer, S., Fleischer, B., Azzi, A. and Chance, B. (1971). *Biochim. et Biophys. Acta* **225**, 194–200.
Flynn, E., Lynch, M. and Zannoni, V. C. (1970). *Fed. Proc.* **29**, Abstr. 3142.
Forte, J. G. and Davies, R. E. (1964). *Amer. J. Physiol.* **206**, 218–222.
Forte, J. G., Torte, G. M., Gee, R. and Saltman, P. (1967). *Biochem. Biophys. Res. Commun.* **28**, 215–221.
Fox, H. M. (1955). *Proc. Roy. Soc. B* **143**, 203–214.
Freed, J. (1965). *Comp. Biochem. Physiol.* **14**, 651–659.
Friede, R. L. (1961). *Proc. Intern. Neurochem. Sympos.* 4th, Vienna, 1960, 151–159.

Frolkis, R. A. (1968a). *Chem. Abstr.* **68** (11) 94472.
Frolkis, R. A. (1968b). *Chem. Abstr.* **69** (6) 43440.
Galeotti, T., Azzi, A. and Chance, B. (1970). *Biochim. et Biophys. Acta* **197**, 11–24.
Gallagher, C. H. and Reeve, V. E. (1971). *Austral. J. Exp. Biol. Med. Sci.* **19**, 21–31.
Gallant, S. and Brownie, A. (1969). *Arch. Biochem. Biophys.* **137**, 441–448.
Gamble, W. J., Conn, P. A., Kumar, H. E., Plenge, R. and Monroe, R. G. (1970). *Amer. J. Physiol.* **219**, 604–612.
Gaudemer, Y., Gautheron, D. and Cyrot, M. O. (1963). *Compt. rend.* **256**, 1863–1865.
Gemba, M. (1966). *Chem. Abstr.* **66** (2) 9244.
Gemba, M., Yamamoto, K. and Ueda, J. (1966). *Chem. Abstr.* **66** (9) 74028.
Ghiretti, F. and Ghiretti-Magaldi, A. (1959). *Atti accad. nazl. Linzei, Rend. Classe sci. fis. mat. e nat.* **25**, 595–602.
Ghiretti, F. and Ghiretti-Magaldi, A. and Tosi, L. (1959a). *Boll. soc. ital. biol. sper.* **35**, 2216–2218.
Ghiretti, F., Ghiretti-Magaldi, A. and Tosi, L. (1959b). *J. Gen. Physiol.* **42**, 1185–1205.
Ghiretti-Magaldi, A., Giuditta, A. and Ghiretti, F. (1957). *Biochem. J.* **66**, 306–307.
Ghiretti-Magaldi, A., Giuditta, A. and Ghiretti, F. (1958a). *J. Cell. Comp. Physiol.* **52**, 389–429.
Ghiretti-Magaldi, A., Rothschild, H. A. and Tosi, L. (1958b). *Acta Physiol. Latino-Amer.* **8**, 239–247.
Giuditta, A. and Aloj, E. (1965). *J. Neurochem.* **12**, 567–579.
Gleason, F. H. (1968). *Plant Physiol.* **43**, 597–605.
Gleason, F. H. and Unestam, T. (1968). *J. Bact.* **95**, 1599–1603.
Glick, J. L. and Bronk, J. R. (1965). *Biochim. et Biophys. Acta* **97**, 16–22.
Goldberg, E. (1957). *Exptl. Parasitol.* **6**, 367–382.
Goldin, H. H. and Farnsworth, M. W. (1966). *J. Biol. Chem.* **241**, 3590–3594.
Golubev, A. M. (1968). *Chem. Abstr.* **69** (2) 9330.
Gonze, P. H. (1959). *Fed. Proc.* **18**, 236.
Gonze, P. H. (1962). In (D. W. Bishop, ed.) "Spermatozoan Motility", p. 99, Publ. No. 72. Amer. Assoc. Adv. Sci., Washington, D.C.
Gonze, J. and Tyler, D. D. (1965). *Biochem. Biophys. Res. Commun.* **19**, 67–72.
Gordon, E. E., Ernster, L. and Dallner, G. (1967). *Cancer Res.* **27**, 1372–1377.
Gordon, M. S. (1968). *Science* **159**, 87–89.
Graham, T. P., Covell, J. M., Sonnenblick, E. H., Ross, J. and Braunwald, E. (1968). *J. Clin. Invest.* **47**, 375–385.
Gram, T. E., Guarino, A., Greene, F. E., Gigon, Ph. L. and Gillette, J. R. (1968). *Biochem. Pharmacol.* **17**, 1769–1778.
Greenawalt, J. W., Hall, D. O. and Wallis, O. C. (1967) "Methods in Enzymol" **10**, 142–147.
Greene, A. A. and Haven, F. L. (1957). *Cancer Res.* **17**, 613–617.
Greene, F. E., Stripp, B. and Gillette, J. R. (1969). *Biochem. Pharmacol.* **18**, 1531–1533.
Guarino, A. M., Gram, T. E., Schroeder, D. H., Call, J. and Guillette, J. R. (1969). *J. Pharmacol. Exptl. Therap.* **168**, 224–228.
Guillory, R. J. and Racker, E. (1968). *Biochim. et Biophys. Acta* **153**, 490–493.
Gulidova, G. P. (1967). *Chem. Abstr.* **67** (13) 114931.
Hagen, J. H. and Ball, E. G. (1960). *J. Biol. Chem.* **235**, 1545–1549.

Hanninen, O., Kivisaari, E. and Autila, K. (1969). *Biochem. Pharmacol.* **18**, 2203–2210.
Harary, I. and Farley, B. (1963). *Exptl. Cell Res.* **29**, 451–465.
Harding, B. W. and Samuels, L. T. (1961). *Biochim et Biophys. Acta* **54**, 42–51.
Harding, B. W., Nelson, D. H. and Bell, J. B. (1966). *J. Biol. Chem.* **241**, 2212–2219.
Harley, J. L. and Rees, T. (1959). *New Phytologist* **58**, 364–386.
Harvey, G. T. and Beck, S. D. (1953). *J. Biol. Chem.* **201**, 765–773.
Harvey, W. R. (1961). *Biol. Bull.* **120**, 11–28.
Harvey, W. R. and Williams, C. M. (1958). *Biol. Bull.* **114**, 23–53.
Hatt, P-Y., Moravec, J., Seroussi, S. and Swijngedauw, B. (1969). *Le Presse Méd.* **77**, 1557–1560.
Haugaard, N. (1968). *Physiol. Revs.* **48**, 311–373.
Haven, E. L., Bacon, J. B., Peck, B. B. and Bloor, W. R. (1967). *Proc. Soc. Expt. Biol. Med.* **120**, 26–28.
Hearn, G. H. (1959). *Amer. J. Physiol.* **196**, 465–466.
Heath, H. and Fiddick, R. (1965). *Biochem. J.* **94**, 114–118.
Heim, L. M. (1969). *Fed. Proc.* **28**, Abstr. 1029.
Heinz, E. (1967). *Gastroenterol.* **107**, 7–14.
Helper, E. W., Carver, N. J., Jacobi, H. P. and Smith, J. A. (1958) *Arch. Biochem. Biophys.* **76**, 354–361.
Herold, R. C. and Borei, H. (1963). *Develop. Biol.* **8**, 67–79.
Hersey, S. J. and Jöbsis, F. F. (1969). *Biochem. Biophys. Res. Commun.* **36**, 243–250.
Hess, H. H. and Pope, A. (1953). *J. Biol. Chem.* **204**, 295–306.
Hess, H. H. and Pope, A. (1954). *Fed. Proc.* **13**, 228.
Hess, H. H. and Pope, A. (1960). *J. Neurochem.* **8**, 207–217.
Hewick, D. S. and Fouts, J. R. (1970). *Biochem. Pharmacol.* **19**, 457–472.
Higashi, D. S. and Kawai, K. (1970). *Biochim. et Biophys. Acta* **216**, 274–281.
Higgins, E. S. and Friend, W. H. (1970). *Proc. Soc. Exptl. Biol. Med.* **133**, 435–438.
Hildebrandt, A. and Estabrook, R. W. (1971). *Arch. Biochem. Biophys.* **143**, 66–79.
Hill, G. and White, D. C. (1968a). *J. Bact.* **95**, 2151–2157.
Hill, G. and White, D. C. (1968b). *Chem. Abstr.* **77** (2) 10652.
Hoch, F. L. (1965). *Chem. Abstr.* **65** (3) 4206.
Hoch, F. L. (1968). *Arch. Biochem. Biophys.* **124**, 248–257.
Hoch, F. L. and Motta, M. V. (1968). *Arch. Biochem. Biophys.* **124**, 238–247.
Holloszy, J. O. (1967). *J. Biol. Chem.* **242**, 2278–2282.
Holtzman, J. L. and Gillette, J. R. (1966). *Biochem. Biophys. Res. Commun.* **24**, 639.
Holtzman, J. L. and Gillette, J. R. (1969). *Biochem. Pharmacol.* **18**, 1927–1933.
Holtzman, J. L., Gram, T. E., Gigon, P. L. and Gillette, J. R. (1968). *Biochem. J.* **110**, 407–412.
Honig, C. R., Frierson, J. L. and Nelson, C. N. (1971). *Amer. J. Physiol.* **220**, 357–363.
Honjo, I., Ozawa, K., Kitamura, O., Sakai, A. and Ohsawa, T. (1968a). *J. Biochem.* **63**, 311–320.
Honjo, I., Takasan, H. and Ozawa, K. (1968b). *J. Biochem.* **63**, 332–340.
Hook, J. E. and Barron, E. S. G. (1941). *Amer. J. Physiol.* **133**, 56–63.
Hopkinson, L. and Jackson, F. L. (1967). *Nature* **192**, 264.
Horgan, P. H. and Griffin, D. H. (1969). *Plant Physiol.* **44**, 1590–1593.
Hoshi, T. (1958). *Sci. Repts. Tohoku Univ., 4th Series* **24**, 131–136.
Howell, J. McC. and Davison, A. N. (1959). *Biochem. J.* **72**, 365–368.
Hülsmann, W. C. (1970). *Biochem. J.* **116**, 32P.

Humiczewska, M. (1966). *Chem. Abstr.* **66** (13) 113369.
Humphrey, G. F. (1947). *J. Exptl. Biol.* **24**, 352–360.
Humphrey, G. F. and Mann, T. (1948). *Nature* **161**, 352–353.
Hunter, N. W. (1965). *Expt. Mol. Pathol.* **4**, 449–455.
Hutterer, F., Bachin, P. G., Ransfield, I. H., Schenkman, J. B., Schaffner, F. and Popper, H. (1970). *Proc. Soc. Exptl. Biol. Med.* **133**, 702–706.
Hydén, H., Løvtrup, S. and Pigon, A. (1958). *J. Neurochem.* **2**, 304–311.
Hyperbaric Oxygenation (1965). Symposium (H. E. Whipple, ed.) in *Ann. N.Y. Acad. Sci.* **117**, 647–890.
Ichikawa, Y. and Yamano, T. (1970). *Biochem. Biophys. Res. Commun.* **40**, 297–305.
Ichikawa, Y., Kuroda, M. and Yamano, T. (1970). *J. Cell Biol.* **45**, 640–643.
Ikeda, K., Hozumi, M. and Sugimura, T. (1965). *J. Biochem.* **58**, 595–598.
Ikuma, H. and Bonner, W. D. (1967). *Plant Physiol.* **42**, 1535–1544.
Inouye, A. and Shinagawa, Y. (1965). *J. Neurochem.* **12**, 803–813.
Inouye, A., Shinagawa, Y. and Shinagawa, Tasuko (1966). *J. Neurochem.* **13**, 385–390.
Iosinov, K. (1966). *Chem. Abstr.* **66** (12) 103113.
Jakovcic, S., Haddock, J., Getz, G. S., Rabinovitz, M. and Swift, H. (1971). *Biochem. J.* **127**, 341–347.
Jamieson, D. and Chance, B. (1966). *Biochem. J.* **100**, 254–262.
Jansky, L. (1961). *Nature* **189**, 921–922.
Jansky, L. (1963). *Canad. J. Biochem. Physiol.* **41**, 1847–1854.
Jansky, L., Votapkova, Z. and Feiglova, E. (1969). *Physiol. Bohemoslav.* **18**, 443–457.
Jodrey, L. H. and Wilbur, K. M. (1955). *Biol. Bull.* **108**, 346–358.
Jöbsis, F. F. (1959). *Ann. N.Y. Acad. Sci.* **81**, 505–509.
Jöbsis, F. F. (1963). *J. Gen. Physiol.* **46**, 905–969.
Joel, C. D. and Ball, E. G. (1960). *Fed. Proc.* **19**, 32.
Joel, C. D. and Ball, E. G. (1962). *Biochemistry* **1**, 281–287.
Johnson, G. R., Ruhmann-Wennhold, A., Asali, N. and Nelson, D. H. (1970). *Fed. Proc.* **29**, Abstr. 2610.
Jones, E. A. and Gutfreund, A. (1959). *Biochem. J.* **72**, 31P.
Jones, G. R. N. (1965). *Brit. J. Cancer* **19**, 360–369.
Jones, J. D. (1963). In (G. A. Kerkut, ed.) "Problems in Biology" **1** (9) 11–89.
Justesen, N. P. B. (1966). *Chem. Abstr.* **66** (3) 17308.
Kadenbach, B. (1964). *Bull. soc. chim. biol.* **46**, 189–190.
Kadenbach, B. (1966b). *Biochem. Z.* **344**, 49–75.
Kadenbach, B. (1966a). In (J. M. Tager, S. Papa, E. Quagliariello and E. C. Slater, eds.) "Regulation of Metabolic Processes in Mitochondria", *Biochim. et Biophys. Acta Library* **7**, Elsevier, Amsterdam, pp. 508–517.
Kamino, K. and Inouye, A. (1970). *Biochim. et Biophys. Acta* **205**, 246–283.
Kartashova, O. Ya., Golubov, I. S. and Popova, N. A. (1966), *Chem. Abstr.* **67** (3) 30760.
Kaschnitz, R. (1970). *Z. Physiol. Chem.* **351**, 771–774.
Kasbekar, D. K. and Durbin, R. D. (1965). *Biochim. et Biophys. Acta* **105**, 472–482.
Kato, R. (1967). *Chem. Abstr.* **67** (10) 89052.
Kato, R., Onaka, K. and Takayanagi, M. (1970a). *Chem. Abstr.* **72** (13) 130879.
Kato, R., Takahashi, A. and Omori, Y. (1970b). *Biochim. et Biophys. Acta* **208**, 116–124.
Kato, R., Takanaka, A. and Takahashi, A. (1968). *Chem. Abstr.* **68** (12) 102278.

Kato, R., Takanaka, A. and Onada, K. (1969). *J. Biochem.* **66**, 739–741.
Kato, R., Takanaka, A. and Takahashi, A. (1970). *Chem. Abstr.* **73** (11) 107814.
Katsura, H. (1954). *Chem. Abstr.* **49**, 11735.
Kawai, K. (1956). *Seigaku* **28**, 379–383.
Kawai, K. (1958). *Nature* **181**, 1468.
Kawai, K. (1959). *Biol. Bull.* **117**, 125–132.
Kawai, K. (1961). *J. Biochem.* **49**, 427–435.
Kawakita, M. (1970). *J. Biochem.* **68**, 625–631.
Kawakita, M. (1971). *J. Biochem.* **69**, 35–42.
Kazakova, T. B., Golubkov, V. I., Derkachev, E. F., Leontev, V. G. and Mashanski, V. T. (1968). *Chem. Abstr.* **71** (10) 88780.
Kean, E. L., Adams, P. H., Davies, H. C., Winters, R. W. and Davies, R. E. (1962). *Biochim. et Biophys. Acta* **64**, 503–507.
Kety, S. (1955). In (K. A. C. Elliott, I. H. Page and J. H. Quastrel, eds.), "Neurochemistry" p. 294, Thomas, Springfield.
Keul, J., Doll, E., Steim, E., Hamburger, H., Kern, H. and Reindell, H. (1965a). *Pflügers Arch. ges. Physiol.* **282**, 1–27.
Keul, J., Doll, E., Steim, H., Fleer, U. and Reindell, H. (1965). *Pflügers Arch. ges. Physiol.* **282**, 43–53.
Keul, J. (1964). *Z. Naturfschg.* **193**, 409.
Khavinzon, A. G. (1968). *Chem. Abstr.* **69** (3) 17293.
Kidder, G. W. III (1970). *Amer. J. Physiol.* **219**, 641–648.
Kidder, G. W. III, Curran, P. F. and Rehm, W. S. (1966). *Amer. J. Physiol.* **211**, 513–519.
Kikuchi, G. and Ban, S. (1961). *Biochim. et Biophys. Acta* **51**, 387–389.
Kikuchi, G. and Barron, E. S. G. (1959). *Arch. Biochem. Biophys.* **84**, 96–105.
Kikuchi, G., Ramirez, J. and Barron, E. S. G. (1959). *Biochim. et Biophys. Acta* **36**, 335–342.
Kimmich, G. and Rasmussen, H. (1966). *Biochim. et Biophys. Acta* **113**, 457–466.
King, E. N. (1966). *Chem. Abstr.* **64** (9) 12990.
King, M. E. and King, D. W. (1968a). *Arch. Biochem. Biophys.* **125**, 527–531.
King, M. E. and King, D. W. (1968b). *Arch. Biochem. Biophys.* **127**, 302–309.
Kirsten, R., Brinkhoff, B. and Kirsten, E. (1970). *Pflügers Archiv. Ges. Physiol.* **314**, 231–239.
Kitsutani, Sh., Sawada, K., Osumi, M. and Nagahisa, M. (1970). *Plant Cell Physiol.* **11**, 107–118. *Chem. Abstr.* **72** (10) 97557.
Kline, E. S. and Waravdekar, V. S. (1960). *J. Biol. Chem.* **235**, 1803–1808.
Klitgaard, H. M. (1966). *Endocrinol.* **78**, 642–644.
Klouwen, H. M., Betel, I., Appelman, A. W. U. and Arts, C. (1965). *Biochim. et Biophys. Acta* **97**, 152–154.
Kmetec, E., Beaver, P. C. and Bueding, E. (1963). *Comp. Biochem. Physiol.* **9**, 115–120.
Kobayashi, H. (1965). *Chem. Abstr.* **63** (13) 18744.
Korecky, B. and Rakusan, K. (1967). *Chem. Abstr.* **66**, (9) 73732.
Korkonen, E. and Korkonen, L. K. (1966). *Acta Ophthalmol.* **44**, 577–580. *Chem. Abstr.* **65** (12) 19067.
Kornacher, M. S. and Ball, E. G. (1968). *J. Biol. Chem.* **243**, 1638–1644.
Korner, P. L. and Uther, J. B. (1970). *Austral J. Exp. Biol. Med. Sci.* **48**, 663–685.
Kuehnel, W. (1968). *Z. Zellfschg. Mikrosk. Anat.* **87**, 37–45. *Chem. Abstr.* **69** (3) 25444.
Kunkel, H. O. and Campbell, J. E. (1952). *J. Biol. Chem.* **198**, 229–236.
Kuriyama, Y. and Omura, T. (1971). *J. Biochem.* **69**, 659–669.
Kurland, G. G. and Schneiderman, H. A. (1959). *Biol. Bull.* **116**, 136–161.

Lachinova, R. I. (1967). *Chem. Abstr.* **66** (10) 83520.
Lang, C. A., Swigart, R. H. and Jefferson, D. J. (1963). *Proc. Soc. Exp. Biol. Med.* **112**, 153–155.
Lardy, A. A. and Feldott, G. (1951). *Ann. N. Y. Acad. Sci.* **54**, 636–648.
Lardy, H. A. and Phillips, P. H. (1945). *Arch. Biochem.* **6**, 53–61.
Lawrie, R. A. (1953). *Biochem. J.* **55**, 298–309.
Lee, C. H., Kim, J. S., Lee, D. Y. and Suh, S. K. (1966). *Chem. Abstr.* **67** (5) 41699.
Lee, I-Y. and Chance, B. (1968). *Biochem. Biophys. Res. Commun.* **32**, 547–553.
Lehninger, A. L. (1962). *Physiol. Revs.* **42**, 467–517.
Lemberg, R. (1969). *Physiol. Revs.* **49**, 48–121.
Lemberg, R., Morell, D. B., Lockwood, W. H., Stewart, M. and Bloomfield, B. (1956). *Chem. Ber.* **89**, 309–314.
Lemberg, R., Newton, N. and Clarke, L. (1962). *Austr. J. Exptl. Biol. Med. Sci.* **40**, 367–372.
Lennie, R. W. and Birt, L. M. (1967). *Biochem. J.* **102**, 338–350.
Lewis, S. E. and Slater, E. C. (1953). *Biochem. J.* **55**, XXVII.
Lewis, S. E. and Slater, E. C. (1954). *Biochem. J.* **58**, 207–217.
Lindall, A. and Frantz, I. D. III (1967). *J. Neurochem.* **14**, 771–774.
Lloyd, D. and Chance, B. (1968). *Biochem. J.* **107**, 829–837.
Lloyd, D., Howells, L. and Cartledge, T. G. (1970). *Biochem. J.* **116**, 24P, 25P.
Loewenstein, J., Schulte, H. R. and Wit-Peeters, E. U. (1970). *Biochim. et Biophys. Acta* **223**, 432–436.
Loewenthal, L. J. A. and Hins, S. C. (1963). *Brit. J. Dermatol.* **75**, 82–85.
Lucas, M., Raoul, B. and Michel, R. (1970). *Bull. soc. chim. biol.* **52**, 547–561.
MacDonald, M. and Spector, R. B. (1963). *Brit. J. Exp. Path.* **44**, 11–15.
McIlwain, H. and Gore, M. B. R. (1951). *Biochem. J.* **50**, 24–28.
Machino, A., Inano, H. and Tamaoki, B. (1969). *J. Steroid Biochem.* **1**, 9–16; *Chem. Abstr.* **72** (5) 40405.
Maggio, R. (1959). *Exptl. Cell Res.* **16**, 272–278.
Maggio, R. and Ghiretti-Magaldi, A. (1958). *Exptl. Cell Res.* **15**, 95–102.
Maggio, R. and Monroy, A. (1959). *Nature* **184**, 68–69.
Maggio, R., Ajello, F. and Monroy, A. (1960). *Nature* **188**, 1195–1196.
Maier, N. and Haimovici, H. (1965). *Circulation Res.* **16**, 65–73.
Makinen, M. W. and Lee, C. P. (1968). *Arch. Biochem. Biophys.* **126**, 75–82.
Maksimova, L. A., Kartashova, O. Ya., and Zarinish, L. A. (1966). *Chem. Abstr.* **67** (4) 30759.
Malessa, P. (1969). *Marine Biol.* **3**, 143–158. *Chem. Abstr.* **71** (10) 88942.
Maley, G. F. (1955). *Fed. Proc.* **14**, 250.
Maley, G. F. (1957). *Amer. J. Physiol.* **188**, 35–39.
Maley, G. F. and Lardy, H. A. (1955). *J. Biol. Chem.* **215**, 377–387.
Malyuk, V. I. (1965). *Chem. Abstr.* **63** (13) 18828.
Malyuk, V. I. (1966). *Chem. Abstr.* **65** (4) 6055.
Mangum, J. H., Klinger, M. D. and North, J. A. (1970). *Biochem. Biophys. Res. Commun.* **40**, 1520–1525.
Mann, T. (1945). *Biochem. J.* **39**, 451–458; 458–465.
Manocha, S. and Bourne, G. H. (1966). *Exptl. Brain Res.* **2**, 230–246. *Chem. Abstr.* **67** (1) 1274.
Manocha, S. and Bourne, G. H. (1967). *Histochemie* **9**, 300–319. *Chem. Abstr.* **67** (9) 80249.
Marshall, W. J. and McLean, A. E. M. (1967). *Biochem. J.* **114**, 77–78P.
Marshall, W. J. and McLean, A. E. M. (1969). *Biochem. J.* **115**, 27–28P.

Marshall, W. J. and McLean, A. E. M. (1969). *Brit. J. Exp. Path.* **50**, 578–583.
Marshall, W. J. and McLean, A. E. M. (1971). *Biochem. J.* **122**, 563–573.
Martius, C. and Hess, B. (1957). *Arch. Biochem.* **33**, 486–487.
Matova, E. E. and Sukasova, M. I. (1967). *Chem. Abstr.* **69** (5) 34183.
Matthews, H. B. and Casida, J. E. (1970). *Life Sci.* **9**, 981–1001.
Mattison, G. M. (1962). *Arkiv. Zool.* **15**, 65–70.
Mattoon, J. R. and Balcavaga, W. X. (1967) "Methods in Enzymol." **10**, 135–142.
Mattoon, J. R. and Sherman, F. (1966). *J. Biol. Chem.* **241**, 4330–4338.
Mazarean, H. H., Domjan, G. and Takass, O. (1967). *Enzymol. Biol. Clin.* **8**, 235–240. *Chem. Abstr.* **67** (12) 106663.
Mengebier, W. L. and Wood, L. (1967). *Comp. Biochem. Physiol.* **21**, 611–617.
Mercado, T. I. (1969). *J. Parasitol.* **55**, 853–858.
Mia, A. J. and Pathak, S. M. (1963). *J. Exptl. Bot.* **16**, 177–181.
Michel, R. and Leblanc, A. (1969). *Bull. soc. chim. biol.* **51**, 355–371.
Michel, R. and Pairault, J. (1969). *Compt. rend.* **268**, 1549–1551.
Michel, R., Michel, O. and Raoul, B. (1968). *Compt. rend.* **267**, 969–971.
Milanov, S. and Dashev, G. (1965). *Chem. Abstr.* **65** (4) 6030.
Millington, P. F. and Orr, M. M. (1966). *J. Physiol.* **185**, 61P.
Mills, E. and Frans, F. (1970). *Nature* **225**, 1147–1149.
Minakami, S. and Mokhova, E. N. (1968). *J. Biochem.* **64**, 561–562.
Minnaert, K. (1960). *Biochim. et Biophys. Acta* **44**, 595–597.
Miyanishi, M., Kasahara, M., Furonaka, H., Ito, M., Sekiguchi, Y. and Uriuhara. T. (1969). *Chem. Abstr.* **72** (13) 130631.
Mockel, J. and Dumont, J. E. (1969). *Arch. Internat. Physiol. Biochem.* **77**, 168.
Moguilevsky, J. A., Schiaffini, O. and Foglia, V. (1966). *Life Sci.* **5**, 447–452.
Moguilevsky, J. A., Libertun, C., Szarcfarb, B. (1969). *Experientia* **25**, 378–379.
Mohri, H. (1956). *J. Exptl. Biol.* **33**, 330–337.
Monier, R., Zajdela, F., Chaix, P. and Petit, J. F. (1959). *Cancer Res.* **19**, 927–934.
Montagna, W. and Soon Yun, J. (1961). *J. Histochem. Cytochem.* **9**, 694–698.
Moon, K. E. and Schofield, P. J. (1968). *Comp. Biochem. Physiol.* **26**, 445–448.
Moore, C. L. and Strasberg, P. M. (1969). *Handb. d. Neurochemie*, p. 53–85. *Chem. Abstr.* **73** (5) 42782.
Moraczewski, A. and Anderson, R. C. (1966). *J. Histochem. Cytochem.* **14**, 64–76.
Morgan, L. R. and Singh, R. (1969). *Comp. Biochem. Physiol.* **28**, 83–94.
Morgan, L. R., Singh, R. and Fisette, R. J. (1967). *Comp. Biochem. Physiol.* **20**, 343–349.
Mori, M. and Mizushima, T. (1965). *J. Dental Res.* **44**, 825. *Chem. Abstr.* **63** (8) 10398.
Moury, D. N., Crane, F. L. and McNeely, C. (1964). *Biochemistry* **3**, 1068–1072.
Mukasa, H., Itoh, M. and Nosch, Y. (1966). *Plant Cell Physiol.* **7**, 683–687. *Chem. Abstr.* **66** (8) 64597.
Murphy, P. J., Frank, R. M. and Williams, T. I. (1969). *Biochem. Biophys. Res. Commun.* **37**, 697–704.
Mustafa, M. G., Cowger, M. L. and King, T. E. (1969). *J. Biol. Chem.* **244**, 6403–6414.
Myers, D. K. and Slater, E. C. (1957). *Biochem. J.* **67**, 558–572; 572–579.
Nachlas, M. M., Crawford, D. T., Goldstein, T. P. and Seligman, A. M. (1958). *J. Histochem.* **6**, 445–456.
Nagata, T. (1957). *Med. J. Shinghu Univ.* **2**, 335–346.
Nakane, H. S., Weber, C. W. and Reid, B. L. (1967). *Proc. Soc. Exp. Biol. Med.* **125**, 663–664.
Nakatsugawa, Ts. (1960). *Nature* **185**, 85.

Nandy, K. and Bourne, G. H. (1964). *J. Histochem. Cytochem.* **12**, 188–193.
Nartsissov, R. P. (1968). *Chem. Abstr.* **68** (11) 93100.
Neely, J. R., Liebermeister, H., Battersby, E. J. and Morgan, H. E. (1967). *Amer. J. Physiol.* **212**, 804–874.
Nelson, L. (1955). *Biochim. et Biophys. Acta* **16**, 494–501.
Neufeld, H. A., Levay, A. N., Lucas, F. V., Martin, A. P. and Stotz, E. (1958). *J. Biol. Chem.* **233**, 209–211.
Niemeyer, M., Crane, R. K., Kennedy, P. and Lipmann, F. (1951). *Fed. Proc.* **10**, 229.
Nigam, V. N. (1966). *Biochem. J.* **99**, 413–418.
Nigro, G., Comi, L. I., Tota, B., Limongelli, F. (1969). *Biochem. Appl.* **16**, 1–21. *Chem. Abstr.* **73** (9) 85288.
Nord, F. F. and Mull, R. P. (1945). *Adv. Enzymol.* **5**, 165–205.
Obuchowicz, L. (1962). *Bull. Soc. Amis. Sci. Lettres Poznan*, Ser. D., 3–27.
Obuchowicz, L. and Klepke, A. K. (1963). *Bull. Soc. Amis Sci. Lettres Poznan*, Ser. 4D, 103–110.
Obuchowicz, L. and Zerbe, T. (1962). *Bull. Soc. Amis Sci. Lettres Poznan*, Ser. D., No. 3, 39–46.
Oda, T. and Seki, S. (1966). *6th Internat. Congress Electron Microscopy, Kyoto*, p. 387–388.
Oda, T., Seki, S. and Watanabe, S. (1969). *Acta Med. Okayama* **23**, 357–376.
Ohlin, P. (1965a). *Quart. J. Expt. Physiol.* **50**, 446–455.
Ohlin, P. (1965b). *Experientia* **21**, 447–448.
Ohnishi, T. and Hagihara, B. (1964). *J. Biochem.* **55**, 584–585; (1965) *J. Biochem.* **56**, 484–486.
Okla, J. (1965). *Chem. Abstr.* **64** (7) 10164.
Oldham, S. B., Bell, J. J. and Harding, B. W. (1968). *Arch. Biochem. Biophys.* **123**, 496–506.
Orrenius, S. and Thor, H. (1969). *Europ. J. Biochem.* **9**, 415–418.
Osanai, M. (1967). *Z. Physiol. Chem.* **348**, 469–470.
Osanai, M. and Rembold, H. (1968). *Biochim. et Biophys. Acta* **162**, 22–31.
Oshino, N. and Sato, R. (1971). *J. Biochem.* **69**, 169–180.
Oshino, N., Imai, Y. and Sato, R. (1971). *J. Biochem.* **69**, 155–167.
Pablo, I. S. and Tappel, A. L. (1961). *J. Cell Comp. Physiol.* **58**, 185–194.
Paléus, S. and Johannson, B. W. (1968). *Acta Chem. Scand.* **22**, 342–344.
Palmer, J. U. and Kalina, M. (1966). *Biochem. J.* **101**, 14P.
Panchenko, L. F., Archakov, A. I., Bokhonko, A. I. and Karmanov, P. A. (1969). *Tsitologiya* **11**, 616–621. *Chem. Abstr.* **71** (7) 56628.
Parson, P. and Basford, R. E. (1967). *J. Neurochem.* **14**, 823–840.
Penniall, R., Currie, W. D., McConnell, N. P. and Bigg, W. R. (1964). *Biochem. Biophys. Res. Commun.* **17**, 752–757.
Perry, R. T., Thorell, B., Åkerman, L. and Chance, B. (1959). *Nature* **184**, 929–935.
Person, P. and Fine, A. (1959). *Arch. Biochem. Biophys.* **84**, 123–133.
Person, P., Smarsh, A., Lipson, S. J. and Carriker, M. R. (1967). *Biol. Bull.* **133**, 4–10.
Peschel, E. and Georgiade, R. (1958). *J. Lab. Clin. Med.* **52**, 410–416.
Petrasek, R. (1961). *Experientia* **17**, 414–415.
Petrov, V. N. and Kalinin, V. I. (1969). *Chem. Abstr.* **71** (2) 10806.
Petrucci, D. (1955). *Chem. Abstr.* **50**, 17220.
Pigareva, Z. D. (1963). *Chem. Abstr.* **64** (2) 2307.
Pigareva, Z. D. and Kamysheva, A. S. (1963). *Chem. Abstr.* **64** (2) 2724.

Pigareva, Z. D., Bogolepov, N. N., Gulidova, G. P. and Dovedova, E. L. (1965). *Chem. Abstr.* **66** (3) 16672.
Pinto, R. E. and Bartley, W. (1969). *Biochem. J.* **114**, 5–9.
Pojda, S. M. (1966). *J. Atheroscler Res.* **6**, 548–554. *Chem. Abstr.* **66** (4) 26970.
Pokrovskii, A. A., Panchenko, L. F., Ivkov, N. N. and Stipakov, A. A. (1969). *Chem. Abstr.* **71** (11) 99538.
Polacov, I. and Cilento, G. (1970). *Arch. Biochem. Biophys.* **139**, 401–405.
Poluhowich, J. J. (1970). *Comp. Biochem. Physiol.* **36**, 817–821.
Popov, I. P. (1966). *Chem. Abstr.* **66** (6) 44994.
Potapov, N. G., Salamatova, T. S. and Orobysheva, N. I. (1964). *Chem. Abstr.* **65** (3) 42642.
Pressman, B. and Lardy, A. A. (1956). *Biochim. et Biophys. Acta* **21**, 458–466.
Prosser, C. L. (1955). *Biol. Rev.* **30**, 229–262.
Prusiner, S. B. (1970). *J. Biol. Chem.* **245**, 382–389.
Prusiner, S. B., Williamson, J. R., Chance, B. and Paddle, B. M. (1968). *Arch. Biochem. Biophys.* **123**, 368–377.
Purves, M. J. (1970a). *J. Physiol.* **209**, 395–416.
Purves, M. J. (1970b). *J. Physiol.* **209**, 417–431.
Quaglino, D., Emelia, G., Artusi, T. and Ferrari, G. (1965). *Bull. soc. ital. biol. sper.* **41**, 636–638.
Quigley, J. P. and Gotterer, G. S. (1969). *Biochim. et Biophys. Acta* **173**, 456–468.
Rafael, J., Hübsch, M., Stratmann, D. and Hohorst, H-J. (1970). *Z. Physiol. Chem.* **351**, 1513–1523.
Ramirez, J. (1959). *J. Physiol.* **147**, 14–23.
Ramirez, J. (1964). *Biochim. et Biophys. Acta* **88**, 648–650.
Raoul, B., Lucas, M. and Michel, R. (1970). *Bull. soc. chim. biol.* **52**, 563–576.
Rapoport, S., Hofmann, E. C. G. and Ghiretti-Magaldi, A. (1958). *Experientia* **14**, 71–72.
Rapoport, S., Ababei, L., Hinterberger, U., Kahrig, C., Künzel, W., Gerischer-Mothes, W., Raderecht, H. J., Schenck, D. and Viereck, G. (1961). *Acta biol. et Med. germ.* **7**, 589–601.
Raw, I. and da Silva, A. A. (1965). *Exptl. Cell Res.* **40**, 677–678.
Reddy, R. S. and Swami, K. S. (1964). *Current Sci.* **33**, 495–496.
Reed, N. and Fain, J. N. (1968). *J. Biol. Chem.* **243**, 2843–2848.
Reverberi, G. (1957). *Pubbl. staz. zool. Napoli* **29**, 187–212.
Ridge, J. W. (1964). *Biochem. J.* **93**, 13–14P.
Ridge, J. W. (1967). *Biochem. J.* **102**, 612–617.
Riggs, B. C. (1945). *Amer. J. Physiol.* **145**, 211–217.
Riley, V. and Pack, G. T. (1965). *Chem. Abstr.* **67** (6) 51984.
Riley, V. (1967). *Chem. Abstr.* **68** (9) 76424.
Ringler, R. L. and Singer, T. P. (1958). *Arch. Biochem. Biophys.* **77**, 229–232.
Rish, M. A. and Shcherbakova, L. I. (1965). *Doklady Vses. Selskokhoz Nauk*, 33–36.
Ritter, C. and Elkin, J. (1967). *Biochim. et Biophys. Acta* **143**, 269–272.
Ritz, E. and Kirk, J. E. (1967). *Experientia* **23**, 16.
Robertson, R. N. (1960). *Biol. Revs. Cambridge Philos. Soc.* **35**, 231–264.
Rockman, H., Clark, P. B., Lathe, G. H. and Parsons, F. M. (1967a). *Biochem. J.* **102**, 44–47.
Rockman, H., Lathe, G. H. and Level, M. J. (1967b). *Biochem. J.* **102**, 48–52.
Roels, F. (1970). *Compt. rend.* **270**, 2322–2324.
Rogers, W. P. (1949). *Austral. J. Sci. Res.* **B2**, 166–174.

Rogers, W. P. (1962). "The Nature of Parasitism", Academic Press, London and New York.
Ronin, V. S. (1966). *Chem. Abstr.* **65** (11) 17268. *Lab. Delo* 1966, 458–459.
Ronin, V. S. (1967). *Tsitologiya* **9**, 823–827. *Chem. Abstr.* **67** (13) 114860.
Roodyn, D. B., Freeman, K. B. and Tata, J. R. (1965). *Biochem. J.* **94**, 628–647.
Rothschild, Lord and Tyler, A. (1958). *Biol. Bull.* **115**, 136–146.
Ruiz, A. N., Yepes, J. L. and Vinas, J. L. R. (1965). *Chem. Abstr.* **65** (7) 11164.
Rusev, G. and Stefanov, St. (1970). *Chem. Abstr.* **73** (11) 117972.
Ruzicka, F. J. and Crane, F. L. (1971). *Biochim. et Biophys. Acta* **226**, 221–233.
Ryan, C. A. and King, T. E. (1962). *Arch. Biochem. Biophys.* **85**, 450–456.
Rybicka, K. (1967). *Exp. Parasitol.* **20**, 255–262. *Chem. Abstr.* **68** (3) 19996.
Sacktor, B. (1951). *Biol. Bull.* **100**, 229–243.
Sacktor, B. (1953a). *J. Gen. Physiol.* **36**, 371–387.
Sacktor, B. (1953b). *J. Gen. Physiol.* **37**, 343–359.
Sacktor, B. (1955). *J. Biophys. Biochem. Cytol.* **1**, 29–46.
Sacktor, B. (1964). *Biochim. et Biophys. Acta* **90**, 163–166.
Sacktor, B. and Bodenstein, D. (1952). *J. Cell Comp. Physiol.* **40**, 157–161.
Sacktor, B. and Packer, L. (1962). *J. Neurochem.* **9**, 361–382.
Sacktor, B., Packer, L. and Estabrook, R. W. (1959). *Arch. Biochem. Biophys.* **80**, 68–71.
Sakano, K., Asaki, T. and Uritani, I. (1968). *Plant Cell Physiol.* (Tokyo) **9**, 49–60. *Chem. Abstr.* **68** (13) 111354.
Sasame, H. A., Castro, J. A. and Gillette, J. R. (1968). *Biochem. Pharmacol.* **17**, 1759–1768. *Chem. Abstr.* **69** (11) 85087.
Sato, N. and Hagihara, B. (1970). *Cancer Res.* **30**, 2061–2068.
Scaife, J. F. (1966). *Canad. J. Biochem.* **44**, 433–448.
Scheitel, L. W. and Miller, J. (1969). *J. Parasitol.* **55**, 825–829.
Schmidt, C. G. (1952). *Biochem. Z.* **323**, 266–274.
Schmidt, C. G. and Schlief, H. (1955). *Naturwiss.* **42**, 105–106.
Schmidt, C. G. and Schlief, H. (1956). *Z. ges. exptl. Med.* **127**, 53–61.
Schole, J. (1959). *Z. physiol. Chem.* **317**, 281–284.
Schubert, J. and Brill, W. A. (1968). *J. Pharmacol. Exptl. Therap.* **162**, 352–359.
Seligman, A. M., Plapinger, R. E., Wasserkrug, H. L., Deb, Ch. and Hanker, J. S. (1967). *J. Cell Biol.* **34**, 787–800.
Semakov, V. G. (1961). *Biokhimiya* **26**, 556–559.
Semenova, I. G. (1967). *Chem. Abstr.* **67** (13) 114578.
Seraydarian, M. W., Harary, I. and Sato, E. (1968). *Biochim. et Biophys. Acta* **162**, 414–423.
Shakhnazarov, A. A. (1967). *Chem. Abstr.* **66** (8) 63512.
Shapiro, A. Z. (1967). *Chem. Abstr.* **66** (12) 102824.
Shappirio, D. G. and Harvey, W. R. (1965). *J. Insect Physiol.* **11**, 305–327.
Shappirio, D. G. and Williams, C. M. (1957a). *Proc. Roy. Soc.* (London) B **147**, 215–232.
Shappirio, D. G. and Williams, C. M. (1957b). *Proc. Roy. Soc.* (London) B **147**, 233–246.
Sharma, V. N. (1967). *Acta Anat.* **68**, 416–431. *Chem. Abstr.* **69** (5) 33953.
Sharma, V. N. (1968). *Acta Anat.* **69**, 349–357. *Chem. Abstr.* **69** (12) 94169.
Shichi, H. (1969). *Exptl. Eye Res.* **8**, 60–68. *Chem. Abstr.* **71** (1) 68.
Shimizu, N., Morikawa, N. and Ishi, Y. (1957). *J. Comp. Neurol.* **108**, 1–22.
Shkolovov, V. V. (1966). *Chem. Abstr.* **66** (6) 44966.
Shoheir, M. H. K. and Schreffler, D. C. (1969). *Proc. Nat. Acad. Sci.* **62**, 867–872.

Shug, A. L., Ferguson, Sh., Shrago, E., Burlington, R. F. (1971). *Biochim. et Biophys. Acta* **226**, 309–312.
Shukolyukov, S. B. (1967). *Chem. Abstr.* **67** (10) 88784.
Sies, H. and Brauser, B. (1970). *Europ. J. Biochem.* **15**, 531–540.
Simon, E. W. (1957). *J. Exptl. Bot.* **8**, 20–35.
Simpson, E. R., Cooper, D. Y. and Estabrook, R. W. (1969). *Recent Progr. Hormone Res.* **25**, 523–562.
Skelton, C. L., Pool, P. E., Seagren, S. S. and Braunwald, E. (1971). *J. Clin. Invest.* **50**, 463–473.
Smialek, M. and Hamberger, A. (1970). *Brain Res.* **17**, 269–271.
Smith, C. H. (1970). *Pediat. Res.* **4**, 328–336.
Smith, M. H. (1969). *Nature* **223**, 1129–1132.
Smith, R. E. and Horwitz, B. A. A. (1969). *Physiol. Revs.* **49**, 330–425.
Smolyar, V. I. (1968). *Chem. Abstr.* **71** (7) 58322.
Sokolov, M. V. (1966). *Chem. Abstr.* **67** (3) 30651.
Sorokina, I. N., Kuznetsova, N. I. and Pigareva, Z. D. (1966). *Chem. Abstr.* **66** (3) 17711.
Sotonyi, P., Huttner, I., Jellinek, H., Toth, A. and Makoi, Z. (1965). *Acta Histochem.* **27**, 213–218. *Chem. Abstr.* **64** (1) 1089.
Soyka, L. T. (1970). *Biochem. Pharmacol.* **19**, 945–951.
Spector, R. G. (1963). *Brit. J. Exp. Path.* **44**, 251–254.
Squame, G., Piccinino, F., Manzillo, G., Ballestrieri, G. G. and Amodio, F. (1968). *Biochem. Appl.* **15**, Suppl. 438–443. *Chem. Abstr.* **73** (13) 129096.
Stephens, N. L. and Wrogemann, K. (1970). *Amer. J. Physiol.* **219**, 1796–1801.
Stone, W. D. (1968). *J. Clin. Path.* **21**, 616–619.
Stoppani, A. O. M., Claisse, L. M. and de Pahn, E. M. (1969). *Comp. rend. soc. biol.* **163**, 65–66.
Straub, S. P. (1967). *Nature* **215**, 1196.
Strelina, A. V. (1970). *Chem. Abstr.* **72** (12) 118244.
Strittmatter, Ph. and Burch, H. B. (1963). *Biochim. et Biophys. Acta* **78**, 562–563.
Sugimura, T., Ikeda, K., Hirota, K., Hozumi, M. and Morris, H. P. (1966). *Cancer Res.* **26**, 1711–1716.
Sulkowski, E. and Wojtczak, L. (1958). *Acta Biol. Exptl.* **18**, 238–248.
Suzuki, M. (1954). *Endocrinol. Jap.* **1**, 159–166. *Chem. Abstr.* **49**, 7612.
Suzuki, M. (1966). *Gann* **57**, 155–167. *Chem. Abstr.* **65** (2) 2679.
Suzuki, M., Imai, K., Ito, A., Omura, Ts. and Sato, R. (1967). *J. Biochem.* **62**, 447–455.
Svihla, G., Dainko, J. L. and Schlenk, F. (1969). *J. Bact.* **100**, 498–564.
Sylk, S. R. (1969). *Expt. Parasitol.* **24**, 32–36. *Chem. Abstr.* **70** (10) 85311.
Symposium: (H. E. Whipple, ed.) "Hyperbaric Oxygenation". *Ann. New York Acad. Sci.* **117**, 647–690.
Takahashi, A. and Kato, R. (1969). *J. Biochem.* **65**, 325–327.
Tanaka, A. and Fukui, S. (1970). *Chem. Abstr.* **72** (13) 131076.
Tapley, D. F. and Cooper, C. (1956). *Nature* **178**, 1119.
Tappel, A. L. (1960). *J. Cell Comp. Physiol.* **55**, 111–126.
Taptykova, S. D., Kalakutskii, L. V. and Agre, N. S. (1969). *J. Gen. Appl. Microbiol.* **15**, 383–386. *Chem. Abstr.* **72** (5) 39940.
Tarshis, M. A., Uteshev, A. B. and Kusin, A. M. (1968). *Doklady* **181**, 234–236.
Tata, J. R. (1963). *Nature* **197**, 1167–1168.
Tata, J. R. (1966). In (J. M. Tager, S. Papa, E. Quagliariello and E. C. Slater, eds.) "Regulations of Metabolic Processes in Mitochondria", pp. 489–505. B.B.A. Library **7**, Elsevier, Amsterdam.

Tata, J. R., Ernster, L., Lindberg, O., Arrhenius, E., Pedersen, S. and Hedman, R. (1963). *Biochem. J.* **86**, 408–428.
Tata, J. R., Ernster, L. and Lindberg, O. (1962). *Nature* **193**, 1058–1062.
Terranova, T., Galeotti, T., Baldy, S. and Neri, G. (1967). *Biochem. Z.* **346**, 439–445.
Terui, G. and Sugimoto, M. (1969). *Chem. Abstr.* **71** (6) 46890.
Theakston, R. D. G., Howells, R. E., Fletcher, K. A., Peters, W., Fullard, J. and Moore, G. A. (1969). *Life Sci.* **8**, 521–529.
Thorell, B., Chance, B. and Legallais, V. (1965). *J. Cell Biol.* **26**, 741–746.
Titova, L. K., Bronshtein, A. A., and Lukashevich, T. P. (1962). Struktura i Funktiya Nervn. Sistemy 98–105, 346–347.
Tobin, M. B. and McIlwain, H. (1965). *Biochim. et Biophys. Acta* **105**, 191–192.
Tolani, A. J. and Talwar, G. P. (1963). *Biochem. J.* **88**, 357–362.
Tosi, L. and Ghiretti, F. (1960). *Atti Accad. nazl. Lincei. Rend. Classe Sci. fis. mat. e nat.* **29**, 207–214.
Tremblay, G. (1962). *Lab. Invest.* **11**, 514–517.
Tucker, A. (1966). *Science* **154**, 150–151.
Turner, G., Lloyd, D. and Chance, B. (1969). *Biochem. J.* **114**, 91P.
Ueda, K. (1961). *Nippon Sanshigaku Zasshi* **30**, 313–324.
Ueda, K., Matsuura, T., Date, N. and Kawai, K. (1969). *Biochem. Biophys. Res. Commun.* **34**, 322–327.
Ungar, G., Aschheim, E., Psychoyos, S. and Romano, D. V. (1956). *J. Gen. Physiol.* **40**, 635–651.
Urich, K. (1965). *Z. vergleich. Physiol.* **50**, 542–550.
Ushatinskaya, R. S. (1959). *Doklady Akad. Nauk. S.S.S.R.* **129**, 687–690.
Vallejos, R. H. and Stoppani, A. O. M. (1967a). *Compt. rend. soc. biol.* **167**, 2068–2070.
Vallejos, R. H. and Stoppani, H. O. M. (1967b). *Compt. rend. soc. biol.* **161**, 2081–2082.
Van den Bergh, S. G. and Slater, E. C. (1960). *Biochim. et Biophys. Acta* **40**, 176–177.
Van den Bergh, S. G. and Slater, E. C. (1962). *Biochem. J.* **82**, 362–377.
Van Dyke, R. A. (1969). *Chem. Abstr.* **70** (11) 95099.
Van Heyningen, W. E. (1955). *Brit. J. Exp. Path.* **36**, 381–390.
Van Rossum, D. G. V. (1965a). *Biochim. et Biophys. Acta* **110**, 221–236.
Van Rossum, D. G. V. (1965b). *Biochim. et Biophys. Acta* **110**, 237–251.
Vassiletz, I. U., Derkatchev, E. F. and Neifakh, S. A. (1967). *Exptl. Cell Res.* **46**, 419–427.
Vernberg, W. B. and Vernberg, F. J. (1968a). *Comp. Biochem. Physiol.* **26**, 499–508.
Vernberg, W. B. and Vernberg, F. J. (1968b). *Exp. Parasitol.* **23**, 347–354. *Chem. Abstr.* **70** (5) 35496.
Vitale, J. H., Jankelson, O. M., Connors, P., Hegsted, D. M. and Zamchek, N. (1956). *Amer. J. Physiol.* **187**, 427–437.
Viti, I. (1968). *Pathologica* **60**, 213–221.
Voigt, W., Fernandez, E. C. and Hsia, S. L. (1970). *Proc. Soc. Exptl. Biol. Med.* **133**, 1158–1161.
Volland, C. and Chaix, P. (1970). *Bull. soc. chim. biol.* **52**, 581–584.
Walker, J. G. (1970). *Biol. Bull.* **138**, 235–244.
Warburg, O. (1930). "Metabolism of Tumours" (translated by F. Dickens), Constable, 1930, London.
Warburg, O. (1956). *Science* **123**, 309–314; **124**, 269–270.

Ward, C. L. and Bird, M. B. (1969). *Growth* **33**, 255–258.
Wattiaux, R. and Wattiaux, de Conmink, S. (1968). *Europ. J. Cancer* **4**, 193–200. *Chem. Abstr.* **69** (6) 42534.
Weimar, V. L. and Haraguchi, K. M. (1965). *J. Histochem. Cytochem.* **13**, 239–240.
Weinhouse, J. (1956). *Science* **124**, 267–269.
Werner, S. C. and Nauman, J. A. (1968). *Ann. Rev. Physiol.* **30**, 213–242.
Williams, J. N., Jr., Jacobs, R. M. and Hurlebaus, A. J. (1966). *J. Nutrit.* **90**, 81–85, 400–404.
Williams, J. N., Jr. and Thorp, S. L. (1969). *Biochim. et Biophys. Acta* **189**, 25–28.
Willis, R. and Kratzing, C. C. (1970). *Proc. Austral. Biochem. Soc.* **3**, 29.
Wilson, D. F. and Epel, D. (1968). *Arch. Biochem. Biophys.* **126**, 83–90.
Wilson, L. D., Oldham, S. B. and Harding, B. W. (1968). *J. Clin. Endocrinol. Metab.* **28**, 1143–1152. *Chem. Abstr.* **69** (13) 104382.
Wilson, J. B. (1970). *Biochim. et Biophys. Acta* **223**, 383–387.
Wilson, M. and Greenwood, C. (1970). *Biochem. J.* **116**, 17–18P.
Wittenberg, B. A., Wittenberg, J. B., Stolzberg, S. and Valenstein, E. (1965b). *Biochim. et Biophys. Acta* **109**, 530–535.
Wittenberg, B. A., Briehl, B. W. and Wittenberg, J. B. (1965c). *Biochem. J.* **96**, 363–371.
Wittenberg, J. B. (1970). *Physiol. Revs.* **50**, 559–636.
Wittenberg, J. B., Brown, P. R. and Wittenberg, B. A. (1965a). *Biochim. et Biophys. Acta* **109**, 518–529.
Woernley, D. L., Carruthers, C., Lilga, K. R. and Baumler, A. (1959). *Arch. Biochem. Biophys.* **84**, 157–161.
Wohlrab, H. (1970). *Biochemistry* **9**, 474–479.
Wohlrab, H. and Jacobs, E. (1967). *Abstr. 7th Internat. Congress Biochem. Tokyo* **V**, H-104.
Wojtczak, L. (1952). *Acta Biol. Exp.* **16**, 199–246. *Chem. Abstr.* **48**, 1590.
Wojtczak, A. B., Chmurzynska, W. and Wojtczak, L. (1958). *Acta Biol. Exptl.* (Lodz) **18**, 249–264.
Wolff, E. C. and Ball, E. B. (1957). *J. Biol. Chem.* **224**, 1083–1098.
Wolken, J. J. (1969). *J. Cell Biol.* **43**, 354–360.
Woods, R. I. (1967). *Nature* **213**, 1240.
Yago, N., Omata, S., Kobayashi, S. and Ishii, Sh. (1967). *J. Biochem.* **62**, 339–344.
Yamada, K., Kuzuya, F. and Yamada, M. (1966). *Chem. Abstr.* **66** (6) 45310.
Yamada, Y. (1961). *Med. J. Osaka Univ.* **11**, 383–400.
Yamagata, S., Ueda, K. and Sato, R. (1966). *J. Biochem.* **60**, 160–171.
Yohro, I. and Horie, S. (1967). *J. Biochem.* **61**, 515–517.
Yunusova, Kh. A. and Bolobonkin, V. G. (1967). *Chem. Abstr.* **68** (6) 47839.
Zaguskin, S. L., Zaguskina, L. D. and Zelik, T. G. (1967). *Chem. Abstr.* **71** (13) 120011.
Zamfirescu-Gheorghi, M., Velican, D., Popescu, I., Dobreanu-Enescu, V. and Apostelescu, I. (1966). *Chem. Abstr.* **66** (4) 27214.
Ziegler, F. D. (1967). *Amer. J. Physiol.* **212**, 197–202.
Ziliotto, G. R. and Curri, S. B. (1967). *Chem. Abstr.* **68** (7) 58499.
Zittle, C. A. and Zitin, B. (1942). *J. Biol. Chem.* **144**, 99–104.
Zubovskaya, A. M. (1968). *Chem. Abstr.* **69** (1) 1698.
Zviagilskaia, P. H. and Karapetyan, N. V. (1965). *Doklady Akad. Nauk. S.S.S.R.* **163**, 497–499.

Chapter IX

Biosynthesis of cytochromes

A. Haem synthesis.
   1. The biosynthesis of porphyrins and of haem.
   2. Effects of metals other than iron.
   3. Control of haem biosynthesis.
B. Biosynthesis of cytochrome c.
C. Turnover of cytochromes.
D. Site of synthesis of cytochrome oxidase.
E. Biogenesis of mitochondria and chloroplasts.
F. Bacterial membranes.
G. Mutants and genetic studies in yeast.
H. Bacterial mutants.
I. Inductive and repressive actions of oxygen, glucose and light.
   1. Oxygen effect.
   2. Glucose repression.
   3. Effects of light.
   References.

A. HAEM SYNTHESIS

1. The biosynthesis of porphyrins and of haem

The studies of Shemin and of Neuberger established that glycine and succinate contributed the nitrogen and carbon atoms of protohaem of haemoglobin and the same is assumed to hold for the haem groups of the cytochromes. Glycine labelled with ^{14}C has been shown to be incorporated into cytochrome c haem (Yças and Drabkin, 1957) and Barrett (1969) has shown that glycine and δ-aminolaevulinic acid (ALA) is incorporated into haem a and into the protohaem of cytochrome b_2.

The overall scheme (Fig. 11) and the main intermediates of biosynthesis of protohaem have been established (for reviews see Rimington, 1957; Lascelles, 1964; Tait, 1968; Shemin, 1970). Some details have yet to be clarified, in particular the mechanism of formation of α-amino-β-oxoadipic acid, postulated to be the intermediate precursor of ALA (cf. Neuberger, 1961) and 5-amino-4-hydroxy valeric acid, a recently discovered intermediate

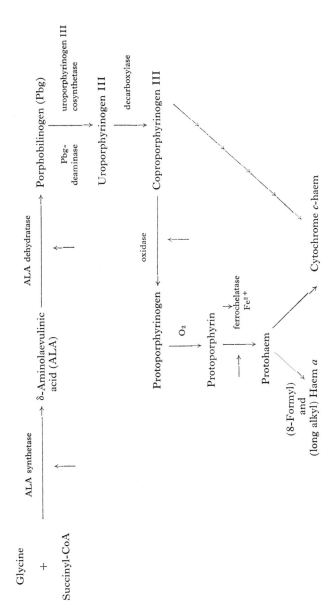

FIG. 11. Biosynthesis of haems. Arrows on reactions indicate regulation sites.

of ALA metabolism in *R. rubrum* grown under illumination (Shigesada *et al.*, 1970).

The formation of ALA is catalyzed by ALA-synthetase which requires pyridoxin-phosphate as a co-factor. Two molecules of ALA condense to form porphobilinogen (Pbg), a monopyrrole, the reaction being catalyzed by ALA-dehydratase. In the presence of the two enzymes Pbg-deaminase and uroporphyrinogen III co-synthetase, four molecules of Pbg condense to form a hexahydroporphyrinogen, uroporphyrinogen III. Disturbance of the synthesis may give the symmetrical isomer I. Pluscec and Bogorad (1970) have investigated the role of dipyrrylmethane intermediates in the formation of uroporphyrins and the formation of the macrocyclic ring in tetrapyrrole biosynthesis from di- and tri-pyrroles appears possible from theoretical consideration of molecular models (Dalton and Dougherty, 1969).

Uroporphyrinogen III is subsequently decarboxylated to coproporphyrinogen III which undergoes oxidative decarboxylation to protoporphyrin IX. This step requires molecular oxygen in all aerobic organisms and is catalyzed by coproporphyrinogen oxidase (Batlle *et al.*, 1965). As protohaem and haem *a* have been obtained from bacteria grown under vigorous anaerobiosis an alternative electron acceptor must be available in them to allow the conversion of coproporphyrinogen to protoporphyrin. Anaerobic formation of protoporphyrin IX from coproporphyrinogen was obtained with extracts of aerobically grown *Pseudomonas* (Ehteshamuddin, 1968). However the formation of protoporphyrin from coproporphyrin by extracts of anaerobically grown *E. coli* and *P. fluorescens* and *P. denitrificans* grown under denitrifying conditions required O_2 and several other electron acceptors could not replace oxygen (Jacobs *et al.*, 1970), while nitrite inhibited the conversion in the presence of oxygen (Jacobs *et al.*, 1971). Extracts from *Rps. spheroides* grown semi-anaerobically converted coproporphyrinogen to protoporphyrin when incubated anaerobically in the dark with Mg^{2+}, ATP and L-methionine (Tait, 1969). The coproporphyrinogenase obtained from the same bacteria grown under O_2 was not effective under the conditions.

That protoporphyrin or protohaem are precursors of haem *a* has been indicated by incorporation of ^{14}C labelled protohaem into haem *a* (Sinclair *et al.*, 1967) in a haem-requiring mutant. The conversion of protoporphyrin to porphyrin *a* required the oxidation of the methyl at the 8 position of the tetrapyrrole to a formyl. This type of conversion has been shown to occur in the formation of chlorophyll *b*, where the formyl is at the 3 position, from ^{14}C labelled chlorophyll *a* in soy bean homogenates in the dark (Ellsworth *et al.*, 1970). On the basis of the similarity of the structure of the long hydroxyalkyl side chain to farnesyl Seyffert *et al.* (1966) proposed that protohaem or a formyl haem, is alkylated by farnesyl pyrophosphate followed by secondary modifications of the side chain. However, attempts to demonstrate the incorporation of the farnesyl chain to form the haem a of yeast were not successful (Seyffert, personal commun.). Although formyl-containing cryptohaem a is found in cells actively synthesizing cytochrome aa_3 and the labelling of cryptoporphyrin a with ^{14}C-glycine or the ^{14}C-ALA was similar to that of haem a in yeast adapting to oxygen (Barrett, 1961a, 1969) it is not clear whether this formyl haem is an intermediate in the biosynthesis of haem a.

Insertion of iron into protoporphyrin is mediated by ferrochelatase—a membrane-bound enzyme (Porra and Jones, 1963a,b; Labbe and Hubbard, 1960). Ferrochelatase from pig liver and yeast cannot incorporate iron into porphyrin a and other formyl porphyrins (Porra and Jones, 1963b). Reduction of the formyl of porphyrin a to an alcohol did not facilitate incorporation of iron therefore steric hindrance is attributed to the long alkyl side chain. Porra and Jones (1963b) reported that ferrochelatase does not incorporate iron into the bi-cysteine adduct of protohaem (haem c). An alternative view that this final step in the synthesis of haem is not necessarily enzymatic had been advanced by Tokunaga and Sano (1966) who demonstrated rapid incorporation of iron into protophyrin in the presence of unidentified ferrous sulphur compounds. Iron is incorporated into the porphyrin as ferrous ion; hence the redox state of the immediate environment can influence the rate of formation of haem. However, insertion of iron into the chlorin-type tetrapyrroles would be considerably more difficult (Barrett, 1961b), so that in order to retain a rate of haem d synthesis comparable to that of the haems of the associated cytochromes participation of a ferrochelatase seems likely.

Lipid, probably a phospholipid, is involved in Fe-incorporation (Mazanowska *et al.*, 1966; Yoshikawa and Yoneyama, 1964; Sawada *et al.*, 1969). Phosphatidyl choline, phosphatidic acid and laureate stimulate the activity of rat liver ferrochelatase (Mazanowska and Dancewicz, 1970).

Compared to the ferrocheletase activity of haemoglobin synthesizing tissues ferrochelatase activity of yeast and bacteria and plants could be expected to be much lower. Nonetheless ferrochelatase activity has been detected not only in the mitochondria of aerobic *S. cerevisiae* and *Candida utilis* but also in the subcellular particles of anaerobic yeast (Riethmüller and Tuppy, 1964; Porra and Barrett, 1964; Porra and Ross, 1965; Labbe *et al.*, 1968) in chloroplasts of spinach, beans and oats and in potato tubers (Jones, 1968; Porra and Lascelles, 1968). *M. denitrificans* is rich in ferrochelatase (Porra and Lascelles, 1965) and it is present in the anaerobic photosynthetic organism *Chromatium* (Porra and Jones, 1963b), but the strict anaerobes *Clostridium welchii* and *Clostridium tetanomorphum* (Porra and Ross, 1965) contain little of the enzyme.

Jones and Jones (1969) by using an osmotic-shrinking method were able to separate the membranes of rat liver mitochondria and have shown that ferrochelatase is located in the inner membrane, and this is confirmed by McKay et al. (1969). No ferrochelatase activity could be detected in the microsomes, although Goudy et al. (1967) found 20% of rat liver ferrochelatase to be in the cytoplasmic fraction.

Iron deficiency has been well established to cause lowering of cytochrome content in bacteria (Tissières, 1952; Lenhoff and Kaplan, 1956). Reduction of the iron content of growth media inhibited the synthesis of cytochromes in *P. fluorescens*, but not growth. Concentrations of Fe^{2+} of $0\cdot1$–10 $\mu g/l$ stimulated cytochrome formation and growth, while higher levels suppressed growth and cytochrome formation (Kiprianova and Kornyushenko, 1969), and similar observations have been made with *Achromobacter metalcaligenes* (Doss and Philipp–Dormston, 1971). Beutler and Blaisdell (1958) obtained evidence that cytochrome oxidase, cytochrome c and catalase may be decreased in tissues of iron deficient animals even when the blood haemoglobin is not greatly lowered. Cytochrome c, (and myoglobin) of skeletal muscle and intestinal mucosa of the iron-depleted rat are decreased by about 50% (Salmon, 1962; Dallman and Schwartz, 1965). Brain and heart muscle liver and kidney are affected to a lesser degree or not at all. The restoration of full cytochrome oxidase activity and cytochrome content of cells was related to the production of new cells. The cytochrome oxidase content of buccal mucosa, which has a rapid rate of cell renewal, is lowered in latent iron deficiency, but may increase after only one day of iron therapy (Jacobs, 1961; Dagg et al., 1966). See also Chapter VIII, B4.

Iron deficiency in *Neurospora crassa* results in a very low level of ALA dehydratase and an accumulation of δ-aminolaevulinic acid (Muthukrishnan, et al., 1969). A tenfold increase in the specific activity of ferrochelatase was found in mitochondria of iron-limited *Candida utilis* (Jones and Jones, 1970). A common consequence of iron deficiency is the accumulation of coproporphyrin or protoporphyrin in bacteria, *Euglena* and plant tissues (Carell and Price, 1965; Marsh et al., 1963; Hsu and Miller, 1965). However, addition of iron to cultures of *R. spheroides* and *Tetrahymena verox* (Lascelles, 1956a,b; 1957) promoted the formation of protoporphyrin. Iron appears to promote the conversion of coproporphyrin to protoporphyrin except in cells in which the end product of iron-deficiency is protoporphyrin (Doss and Phillip–Dormston, 1971).

The absence of enterochelin, a cyclic trimer of 2,3-dihydroxy-N-benzyl serine in some strains of *E. coli* resulted in an impaired ability to take up iron with a consequent lowering of the concentration of cytochrome b_1 (Cox et al., 1970).

2. Effects of other metals

Trace amounts of copper, lead, arsenic and vanadium were reported by Fischer and Fink (1925) to stimulate porphyrin formation in yeast. Cobalt in trace amounts (0·27 ppm) stimulated porphyrin production in *M. tuberculosis avium* (Patterson, 1960) but larger amounts inhibit equally glycine incorporation into haem and globin in reticulocytes (Schulman and Jobe, 1968). The rate of incorporation of ALA into microsomal haem fell markedly after injection of rats with cobaltous chloride (60 mg/Kg) (Tephly and Trussler, 1971). Cytochrome P-450 disappeared more rapidly than cytochrome b_5 from the microsomes (Tephly and Hibbcln, 1971). At a level of 40 mg/kg $CoCl_2$ prevented the phenobarbital induction of P-450, but not of NADPH-cytochrome *c* reductase. Cytochrome oxidase in rat liver mitochondria is inhibited slightly by low concentrations of $CoCl_2$ (0·01 mM) (Tarkowski, 1967); a possible explanation of this is the polymerising effect of cobalt on proteins, as observed with globin (Schulman and Jobe, 1968).

Zinc promotes the formation of cytochromes in yeast and fungi (Ward and Nickerson, 1958). Cytochrome formation in *Ustilago sphaerogena* is optimal at 1 ppm of zinc but completely inhibited at 100 ppm (Grimm and Allen, 1954). Zinc added to cultures of *U. sphaerogena* enters the cell within one minute and the incorporation of ^{14}C-leucine into cytochrome *c* increased within 6 minutes (Mendiola and Price, 1969); there is a concomitant increase in the rates of synthesis of DNA, RNA and protein.

Komai and Neilands (1968) noted that 3×10^{-5} M Zn increased the activity of ALA-dehydratase in cultures of *U. sphaerogena*. The maximum formation of ALA-synthetase in *U. sphaerogena* requires Zn (Hirocheka and Neilands, 1968). The synthesis of iso-cytochromes *c* in respiratory adaptation of yeast from anaerobic cultures containing zinc and no zinc has been studied by Ohaniance and Chaix (1970). At the level of zinc that was optimal for growth, Minagawa (1958) observed that cytochrome *c* formation in *Saccharomyces ellipsoides* was more strongly inhibited than that of cytochrome *b*. Excess zinc causes decrease of cytochrome oxidase and catalase activity and also of haemoglobin formation, and large increases of excess coproporphyrin or protoporphyrin are observed in yeast cells (Vallee, 1959; Pretlow and Sherman, 1967). The cytochrome deficiency observed may be due to diversion of large amounts of porphyrin intermediates from the haem biosynthesis; alternatively zinc may, by combining with iron binding sites, suppress cytochrome formation by inhibiting ferrochelatase activity or secondarily blocking electron transfer in the respiratory chain (Christyakov and Gendel, 1968).

The formation of cytochrome oxidase and the associated respiratory enzymes in yeast adapting to oxygen was decreased by zinc at levels optimal for growth, but this decrease does not appear to result from suppression of porphyrin synthesis (Ohaniance and Chaix, 1966; Ohaniance and Chaix, 1968; Ohaniance, 1967). The inhibitory effect of zinc ($2·5 \times 10^{-6}$ M) on aa_3 synthesis could be overcome by the addition of 1×10^{-5} M copper to the culture. Competitive effects of Zn and Cu on the cytochrome oxidase levels in rat tissues have been reported by Van Reen (1953).

Elvehjem (1931) and Cohen and Elvehjem (1934) established the critical dependence of the occurence of cytochrome oxidase of yeast and mammalian tissues on the Cu content of the growth medium or diet, later confirmed in studies on mitochondria isolated from livers of rats raised on Cu-deficient diets (Gallagher et al., 1956a). Gubler et al. (1957) found cytochrome oxidase activity to be decreased by 80% in Cu-deficient pig hearts, in contrast to the presence of normal catalase activity in Cu-deficient pigs (Lahey et al., 1952).

Keilin and Hartree (1938) drew attention to the possibility that copper was an essential component of the oxidase and Graubard (1941) on the basis of the inhibition of cytochrome oxidase by copper chelators also proposed Cu as a component of cytochrome oxidase. Wainio (cf. Wainio, 1961) early stressed the essential role of Cu in cytochrome oxidase, and its presence in the apoprotein is now well established (see Chapter III).

Copper deficiency has an indirect effect on the biosynthesis of haem a expressed in the absence or diminution of the cytochrome aa_3 absorption bands, observed in Cu-deficient yeast and rat liver mitochondria (Cohen and Elvehjem, 1934; Wohlrab and Jacobs, 1967), or by direct analysis of haem a in Cu-deficient rats and pigs (Gallagher et al., 1956a; Lemberg et al., 1962). Protohaem synthesis in red cells of Cu-deficient pigs is diminished (Gallagher et al., 1956b) but the effect of Cu-deficiency on mitochondrial synthesis of haem has not been reported. Gubler et al. (1957) observed a slight increase of cytochrome c in cytochrome oxidase deficient tissues and cytochrome aa_3 was suppressed in yeast adapting to oxygen but not cytochrome c in the presence of Cu-chelators (Barrett, 1962). Wohlrab and Jacobs (1967) however, found that in Cu-deficient rat liver mitochondria and yeast the level of cytochrome c in addition to that of cytochrome aa_3 was decreased but not the cytochromes b and c_1.

Cytochrome oxidase in copper-deficient rats varied from 18% in skeletal tissues to 34% in intestinal mucosa (heart 27%) of the normal (Dallman and Loskutoff, 1967). On supplementation of the diet with copper, cytochrome oxidase activity was restored to intestinal mucosa in 2-3 days and in liver and skeletal muscle in 10-15 days but more slowly in heart muscle. The restoration of cytochrome oxidase activity appeared to be related to the rate of synthesis of new mitochondrial material rather than new cell formation.

3. Control of haem synthesis

End-product inhibition of the synthesis of ALA has been invoked as a principal regulatory step in the formation of haem (Burnham and Lascelles,

1963). Granick proposed that ALA-synthetase is controlled by a protein apo-repressor and that haem acts as a co-repressor; haem can be replaced by various steroids and other agents permitting an increased rate of synthesis (Kappas and Granick, 1968; Kappas et al., 1968). Haem has been shown to decrease the ALA-synthetase activity of rat liver and kidney mitochondria (Kurashima et al., 1970; Barnes et al., 1971). ALA-synthetase is localized in the matrix space or loosely bound to the inner membrane (Borst and Huijing, 1969), but evidence of Hayashi et al. (1970) indicates that it is formed on cytoplasmic ribosomes, and the cytoplasmic enzyme is also inhibited by haem (Barnes et al., 1971). Haem and haem protein complexes are potent repressors of drug-mediated induction of hepatic ALA synthetase and of microsomal cytochromes but not of amino-acetone synthetase in rats (Marver et al., 1968); a number of proteins that can bind free haem block the inhibition by haem and bilirubin of purified ALA synthetase in vitro (Scholnick et al., 1969). ALA synthetase of *Spirillum itersonii* is also inhibited by haem (Clark-Walker et al., 1967); a key role for the enzyme in this organism is shown by its fluctuation which corresponds to the variation in the levels of cytochrome under different growth conditions (Ho and Lascelles, 1971).

Several studies have been made of ALA-synthetase induction and its suppression by oxygen in photosynthetic bacteria where its role in the synthesis of bacteriochlorophyll overshadows that of the cytochrome but these studies do demonstrate that the enzyme has a rapid turnover and that as the synthesis of bacteriochlorophyll (BCHL) ceases the level of ALA-synthetase in the cells decreases very rapidly (Higuchi et al., 1965; Gorchein et al., 1967; Lascelles, 1968). Tuboi et al. (1970) have presented evidence that there are two ALA-synthetases in *Rhodopseudomonas spheroides*. Fraction I can be induced by the reduction of the O_2 tension alone, while fraction II requires in addition illumination of the organism. From a study of the biosynthesis of BCHL in *R. spheroides* Marriott (1968) has concluded that ALA is regulated by an equilibrium between ALA synthetase and an activator, this being a low M.W. compound. The ALA-synthetase of *R. spheroides* is inhibited by haem (Burnham and Lascelles, 1963); a specific and reversible inhibitor of ALA-synthetase of small M.W. which differs from the tetrapyrroles present has been found in extracts of *R. spheroides* and this may be an additional regulator (Tuboi et al., 1969). Using ^{59}Fe, Lascelles and Hatch (1969) find that in *R. spheroides* BCHL is formed fifty times faster than protohaem, and that in mutants which can form Mg-porphyrin there is an increased rate of formation of haem from glycine and succinate. Added ALA stimulated haem synthesis ten-fold without increasing BCHL production. Haem synthesis was not so sensitive to inhibition by puromycin as that of BCHL.

An earlier suggestion (Lascelles, 1968) that there is a specific form of

ALA synthetase for the iron and magnesium-porphyrin branches of the biosynthetic pathway is not sustained by later experiments (Lascelles and Altschuler, 1969). Neuwirt et al. (1969) have however, concluded from studies on the incorporation of labelled glycine into haem or rabbit reticulocytes that haem does not directly inhibit its own synthesis nor exert feedback control at the level of ALA-synthetase, and suggested that haem may influence transfer of iron into the intact cell.

Suggestions have come from the study of haemoglobin synthesis that globin, rather than haem is the regulator of ALA-synthetase (Hrinda and Goldwasser, 1968) but Ponka and Neuwirt (1969, 1971) concluded from the effect of cycloheximide on the rate of uptake of ^{14}C-glycine and ^{59}Fe into reticulocytes that haem synthesis is not directly dependent on the availability of globin. However in a mutant of *Spirillum itersonii* which could not synthetize ALA the formation of the apoprotein of the cytochromes strictly required the presence of haem (Lascelles et al., 1969b).

In several photosynthetic bacteria the intracellular level of ATP has been observed to be inversely related to BCHL synthesis both during growth and light-induced changes of BCHL content of chromatophores (Fanica–Gaignier and Clement–Metral, 1971; Schmidt and Kamen, 1971; Oelze and Kamen, 1971; Fanica–Gaignier, 1971). Fanica–Gaignier and Clement–Metral (1971) found that purified ALA-synthetase is inhibited by AMP (10^{-3} M) and Nandi and Waygood (1967) reported that ALA-dehydratase of wheat leaves is inhibited by a similar level of ATP. In contrast, the synthesis of ALA by liver mitochondria was increased significantly by addition of ATP to the assay system (Yoshida and Ishihara, 1969). Porphobilinogen desaminase-uroporphyrinogen III co-synthetase (Llambias and Batlle, 1971) and rabbit erythrocyte ferrochelatase (Palma-Carlos, 1968) have also been reported to be inhibited by ATP. Decreased levels of ATP produce abnormally high levels of porphyrins (Gajdos et al., 1968 and Gajdos–Torok, 1969) in *R. spheroides* and white rats, and the enzymic step suppressed was considered to be before ALA synthetase. Thus the effect of ATP appears not to be specific to any particular step in the haem synthesis pathway.

Vitamin E deficiency in the rat leads to decreased activity of ALA-synthetase and ALA-dehydratase in the bone marrow and lowered incorporation of ^{14}C-ALA and ^{14}C-glycine in cytochromes b_5 and P-450 of liver microsomes (Murty et al., 1970). The effect of vitamin E deficiency is specific to the cytochromes and haemoproteins and Murty et al. suggest that vitamin E deficiency functions as a regulator of haem synthesis as one of the rate-limiting steps in the pathway to haem, probably ALA-synthetase.

The increase in ALA-synthetase in liver cells induced by porphyria-inducing drugs such as allylisopropylacetamide (Granick and Urata, 1963; Granick, 1966; Sassa and Granick, 1970) does not appear to be accompanied by an increase of cytochromes (Marver, 1969). The finding of Beattie and Stuchell (1970) that the administration of allylisopropylacetamide to rats leads rapidly to a 30–50% increase of cytochrome aa_3, b and c has not been confirmed by similar experiments of Barnes et al. (1971). Other factors than the level of porphyrin synthesis therefore must control the amount of cytochrome formed in a cell. Haemin enhances synthesis of the apoprotein of cytochrome c in liver slices of very young rats; as this effect is inhibited by actinomysin, Gonzalez–Cadavid et al. (1970) conclude that the control of synthesis is at the transcription level.

B. CYTOCHROME C

Drabkin (1947a,b) in his studies of liver regeneration showed that cytochrome c formation was independent of liver regeneration synthesis and that cytochromes c and DNA were preferentially produced in regenerating tissue. He also observed no increase in cytochrome synthesis at low oxygen concentration even though haemoglobin content and erythrocytes numbers increased. In labelling experiments with ^{14}C-glycine (Marsh and Drabkin, 1957) appreciable differences in the degree of labelling were found for the cytochrome c of different rat tissues. A turnover rate of 12·7 days was calculated for the cytochrome c of regenerating liver and this was faster than that for haemoglobin. The extent of labelling of cytochrome c and tissue protein was similar in both liver and kidney but in skeletal muscle the cytochrome c was more highly labelled. Utilizing the system of anaerobic yeast in their resting phase adapting to oxygen (Ephrussi et al., 1950) it was shown by Yças and Drabkin (1957) that the cytochrome c haem and the protein were substantially synthesized de novo, though the labelling data left open the possibility that some of the apoprotein was synthesized before the haem or that there was some transfer of haem from pre-existing haemoproteins. However, Barrett (1969) found that the labelling of the haem and protein moiety in adapting yeast proceeded at substantially the same rate and comparison of the labelling of the free porphyrin formed with the haem did not indicate that there was any transfer of haem from other haemoproteins.

Since this early work of Drabkin interest has been directed to establishing the locus of synthesis of the haem and protein components of cytochrome c. Radioactive iron was rapidly taken into cytochrome c as early as two days after partial hepatectomy and the ^{59}Fe was found to be equally distributed between the mitochondrial and fluffy layer fractions (Gear, 1965) and this pattern continued over a 30 day period of regeneration. From experiments in which rats were injected with the ^{14}C-lysine and killed from 7·5–60 minutes later and the subcellular fractions (see Chapter V, B7a) analyzed for distribution of labelled cytochrome c, González–Cadavid and Campbell

(1967) concluded that cytochrome c is synthisized *in toto* by the microsomes and then transferred to the mitochondria. The microsomal fraction was separable by differential extraction with water and 0·15 M sodium chloride solution into a standard fraction and a pool of newly synthesized cytochrome c. Kadenbach (1969a) found 4% of the total cytochrome c was bound to the microsomal fraction. An extramitochondrial site for the synthesis of cytochrome c was indicated in studies of the synthesis of cytochrome c by Krebs mouse ascites tumour cells (Haldar et al., 1966) and the failure of chloramphenicol to inhibit cytochrome c synthesis (Clark–Walker and Linnane, 1966, 1967). The synthesis of cytochrome c_1 similarly is not inhibited by chloramphenicol (Mahler and Perlman, 1971). Kadenbach (1967a) demonstrated the transfer of labelled proteins from microsomes into mitochondria in vitro and later showed that labelled cytochrome appeared in mitochondria which had been incubated with ^{14}C-labelled microsomes (Kadenbach, 1967b). Kinetic studies of the incorporation of ^{59}Fe and ^{14}C-ALA into cytochrome c (Kadenbach, 1968; Davidian et al., 1969; Penniall and Davidian, 1968) led to the conclusion that cytochrome c apoprotein and the haem were synthesized in the microsome; however, the kinetics of incorporation of lysine into cytochrome c showed that only a small part of the cytochrome c extracted from isolated microsomes is involved in the biosynthetic pathway. The newly synthesized cytochrome is attached to the endoplasmic reticulum for about 75 minutes (Kadenbach, 1969c).

Mitochondria have been reported (Wojtczak and Zaluska, 1969) not to be permeable to active cytochrome c, Form I of Flatmark and Sletten, (1968) and it has been postulated that the cytochrome c is transferred into the mitochondrion as a complex with phospholipid or that haem combines with apocytochrome c in the mitochondrion (Kadenbach, 1970). Davidian and Penniall (1971), using ^{3}H-ALA as label, find that the newly synthesized cytochrome c isolated from rat liver mitochondria is the minor, inactive Form IV and propose that the endoplasmic reticulum supplies Form IV to the mitochondria. The question in which form the precursor of functional cytochrome c enters the mitochondria thus requires further clarification.

Keilin (1925) observed a high concentration of cytochrome in the flight muscle of honey bee and recently some investigators have turned to developing insects as an alternative system for the study of the biosynthesis of cytochrome c. The concentration of cytochrome c is very low (0·06–0·02 μg) in the honey bee larva and early pupal stages, but this is followed by a very rapid increase (about 40 fold) in the developing adult (Osanai and Rembold, 1970). Radioactive iron and lysine were found to be incorporated into cytochrome c, but most was bound to substances of high molecular weight. Similarly the concentration of cytochrome c is very low in the diapausing pupae of the silk worm, *Platysamia cecropia*; during the final period of development the concentration of the cytochrome in the flight muscles increases rapidly, reaching maximum levels shortly after emergence of the moth (Shappirio, 1960; Skinner, 1963). In the related Saturnid moth *Samia cynthia* a 20 fold increase of cytochrome c content occurs in the 2–3 days before the emergence of the adult and Chan and Margoliash (1966), using ^{14}C arginine showed that this increase is due to *de novo* synthesis of its protein. Injected δ-aminolaevulinic acid increased cytochrome c synthesis in the developing *Polyphemus* moth, and the results of the action of various antibiotic inhibitors of protein synthesis on this increased rate of synthesis of cytochrome c indicated that apocytochrome c was synthesized *de novo* on cytoplasmic ribosomes (Soslau et al., 1971). The messenger RNA for apocytochrome c was found to be stable over a 24 hour period.

Normal *S. cerevisiae* yeast contains two molecular species of cytochrome c, iso-1-c and iso-2-c, the latter being the minor component. Slonimski et al. (1963) postulated that a haem free precursor of iso-2-c exists in the stationary phase of the cell before induction by O_2 of cytochrome c synthesis and that this protein is a specific repressor for the synthesis of the iso-1-c polypeptide chain.

On introduction of O_2 to resting yeast cells the synthesis of iso-2-c proceeds rapidly whereas the maximum rate of synthesis of iso-1-c is initially delayed. However, the formation of iso-2-c is inhibited by only 40% by high concentrations of cycloheximide during the first hour of adaptation. The cycloheximide-resistant formation of iso-2-c was considered to be a conversion of the non-haem precursor (Fukuhara, 1966). A considerably lower rate of incorporation of radioactive amino acids into the iso-2-c than iso-1-c, a greater sensitivity of the synthesis of iso-1-c to inhibition by 5-methyltryptophan, and the non-suppression of the formation of the iso-2-c in a phenylalanine and tyrosine requiring mutant in the absence of these amino acids, while that of iso-1-c was severely affected (Fukuhara and Sels, 1966) all provide further evidence in support of Slonimski's hypothesis.

Zinc reduces the rate of synthesis of both iso-cytochromes c in the exponential phase and both rates then become parallel and the repressive effect of iso-2-c on iso-1-c is no longer apparent (Ohaniance and Chaix, 1970).

C. TURNOVER OF CYTOCHROMES

Flatmark and Sletten (1968) using ^{59}Fe as the label calculated a half life of 15·6 days for kidney cytochrome c comparable with a half-life of 9·7 days obtained by Fletcher and Sanadi (1961) using ^{35}S-methionine.

Calculating the half-life from the rate of synthesis of cytochrome c at the microsomes Kadenbach (1969a) using ^{14}C-lysine obtained a half-life of 13·2 days for liver mitochondria cytochrome c, but direct measurement of the radioactivity of the iso-1-cytochrome c fraction of liver mitochondrial cytochrome c gave a half-life of 10·5 days (Kadenbach, 1969b) while for heart and skeletal muscle the half lives of cytochrome c were 43 and 32 days respectively.

These values are much greater than the half life of 6–8 days for cytochromes and other mitochondrial components of rat liver (Lusena and Depocas, 1966) and 5–6 days for rat heart cytochrome c obtained by Aschenbrenner et al. (1970) who used ^{3}H-ALA to label the haems and (guanidino-^{14}C)-arginine to label the mitochondrial proteins. The discrepancies were thought to arise from reutilization of the labelled amino-acids following catabolism of the protein. Aschenbrenner et al. found that labelling of hepatic mitochondrial proteins with ^{3}H$_2$-leucine gave a prolonged apparent half-life. The half-lifes of haem a and cytochrome c of rat heart and liver are all similar (5–6 days) compared to cytochrome b_5, situated in the outer mitochondrial membrane which had a half-life of 4·4 days (Druyan et al., 1969). Rat heart mitochondrial DNA had a renewal half-life of 6·7 days (Aschenbrenner et al., 1970).

The microsomal cytochromes have even more rapid turnover. Druyan et al. (1969) found a half-life of 2–3 days with ^{3}H$_2$-ALA and 2·5 with ^{14}C-arginine as markers for the b_5 of rat liver microsomes, at variance with the half-life of 4–4·5 days found by Kuriyama (1968) using ^{14}C-leucine and ^{14}C-arginine as tracers, but in accord with the half-life of 1·9 days obtained by Greim et al. (1970) for the haem of b_5. Employing double labelling with ^{3}H$_2$-ALA and ^{14}C-arginine half-lifes of 1·7 and 3·5 days for the haem and protein respectively of cytochrome b_5 were obtained by Bock and Siekevitz (1970). Garner and Maclean (1969) in studies with livers of cycloheximide rats and ^{14}C-ALA as a marker, calculated that haem had a rapid turnover and that its synthesis and incorporation into the microsome was independent of protein synthesis, while Bock and Siekevitz (1970) found evidence of separate pools of haem and protein and that a slow haem exchange occurs with the cytochrome b_5 bound to the microsomal membrane. Negishi and Omura (1970) and Hara and Minakami (1970) have reported on the occurrence of b_5 apoprotein in liver microsomes.

Hara et al. (1970) using ^{59}Fe to follow haem incorporation into the microsome found that the turnover of P-450 was considerably more rapid than that of b_5. Studies with ^{14}C-labelled haem show that haem exchange occurs between P-450 and b_5 and any free haem pool (Garner and Maclean, 1971). Phenobarbital-induced increase of microsomal P-450 and b_5 is accompanied by a decrease in the breakdown of labelled haem. Half-life of P-450 haem radioactivity was 22 hours and that of cytochrome b 45 hours.

D. SITE OF SYNTHESIS OF CYTOCHROME OXIDASE

Ephrussi et al. (1950) reported that in the petite mutant of yeast the absence or diminution of enzymatically and spectroscopically detectable cytochrome oxidase is caused by irreversible alteration of the structure and function of the mitochondria and that the genetic determinant resides in the mitochondria. Since yeast grown in the presence of chloramphenicol, an inhibitor of mitochondrial biogenesis, exhibited spectral and enzymic properties similar to those of petite yeast, Clark-Walker and Linnane (1966, 1967) concluded that synthesis occurred in the mitochondria. However, immunological techniques demonstrated that petite yeast had antigenic sites common to the cytochrome oxidase fraction of the respiratory pathway of wild yeast. Mahler et al. (1964) and Kraml and Mahler (1967) detected a protein in the petite mutant which cross-reacted with antisera prepared against cytochrome oxidase from wild yeast. Tuppy and Birkmayer (1969) obtained cytochrome oxidase activity on addition of haem a to submitochondrial fractions from petite yeast devoid of cytochrome aa_3. The cytochrome oxidase activity was 30% of that in wild yeast, and this was attributed to a lesser amount of apoprotein in petite yeast mitochondria than in wild yeast. The apoprotein was similar in behaviour on disc electrophoresis and its copper content to that of wild yeast, but reconstitution of cytochrome oxidase activity of the purified apoprotein has not been reported.

That the cytoplasmic ribosomes have a major role in the synthesis of cytochrome aa_3 is well established. Isolated mitochondria of *Neurospora crassa* though able generally to incorporate amino acids into mitochondrial proteins incorporates these very poorly into cytochrome oxidase and the incorporation of amino acids is sensitive to low levels of both chloramphenicol and cycloheximide (Edwards and Woodward, 1970; Birkmayer, 1971b). Edwards and Woodward consider that the biosynthesis of cytochrome oxidase in *Neurospora* requires two copper components, one specified by a cytoplasmic gene and synthesized on mitochondrial ribosomes, the other specified by a gene of indeterminate origin, synthesized on cytoplasmic ribosomes. The induced biosynthesis of cytochrome oxidase in non-dividing anaerobically grown *S. cerevisiae* requires the participation of both the mitochondrial and the cytoplasmic (cycloheximide-sensitive) system (Chen and Charalampous, 1969) and a similar finding was obtained with dividing yeast adapting to respiration and during the lifting of glucose repression from yeast growing aerobically (Yu et al., 1968). Both cycloheximide and chloramphenicol inhibit formation of cytochrome oxidase in cultured heart cells (Kroon and De Vries, 1971).

Kinetic data on the incorporation of labelled lysine and choline into the components of subcellular membranes of rat liver suggest that both the protein and lipid components of cytochrome oxidase are synthesized in the endoplasmic reticulum (Schiefer, 1969a). Cytochrome oxidase purified from rat liver following incubation with ^{14}C-leucine did not show labelling except for a minor fraction considered to be an impurity (Beattie et al., 1970). However, Weiss et al. (1971) find that cytochrome oxidase is build up of several types of polypeptides and that one is highly labelled with ^{14}C-leucine in the presence of cycloheximide so that part of cytochrome oxidase may be synthesized by the mitochondrial system. Kadenbach (1971) has compared the pathway of synthesis in rat liver of the structurally bound cytochromes with that of cytochrome c by intraperitoneal injection of ^{14}C-lysine and ^{14}C-leucine into rats followed by injections of chloramphenicol and cycloheximide. From the differences in specific activities of the cytochrome oxidase, cytochrome c and the other soluble and insoluble proteins of the mitochondria he concluded that cytochrome oxidase had a much lower turnover than the other mitochondrial proteins or else a much longer half-life of the mitochondrial apoprotein compared to that of cytochrome c apoprotein (1·5 hours). Further inhibitor studies strengthened the view that apoprotein of cytochrome oxidase is synthesized on the cytoplasmic ribosomes, while the transport across the inner membrane and attachment of the haem a groups within the mitochondrion involves the intermediate binding of the haem group to a protein different from apoprotein and synthesized in the mitochondrion. Schiefer (1969b) has concluded from kinetic data on the incorporation of ALA into rat liver haem a and protohaem that haem a is synthesized outside the mitochondrion and enters it separately from the apoprotein.

Evidence for the dual role of the cytoplasmic ribosomes and mitochondria in the synthesis of cytochrome oxidase apoprotein has also come from the effect of the selective inhibitors on the incorporation of amino acids into mitochondrial proteins in wild and petite yeast (Vary et al., 1969, 1970; Henson et al., 1968). Succinate and malate dehydrogenase and fumarase are synthesized solely by the cytoplasmic ribosomes but the extent of their formation appears to be regulated by a mitochondrial factor (Vary et al., 1970).

Evidence from nuclear mutants with single and multiple cytochrome deficiencies suggest that nuclear genes code cytochrome apoprotein synthesis but its insertion into the mitochondrial membrane is controlled by the mitochondrion (Subik et al., 1970; Kuzela et al., 1969). Birkmayer (1971a) has presented evidence that the integrating component may be coded by mitochondrial DNA.

E. BIOGENESIS OF MITOCHONDRIA AND CHLOROPLASTS

The biogenesis of mitochondria and chloroplasts as structural entities is outside the scope of this book except in general terms and specific points which bear on the biosynthesis of the components of the respiratory or photosynthetic chain and their insertion and organisation within the organelle. The literature on biogenesis of mitochondria has been comprehensively and critically reviewed by Ashwell and Work (1970) and the "Symposium on Autonomy and Biogenesis of Mitochondria and Chloroplasts" (1971) contains authoritative contributions from leading investigators. More specialized reviews are by Schatz (1970) and Borst and Kroon (1969).

The average rate of turnover of mitochondria as distinct from the respiratory components has been estimated from the half-life of mitochondrial DNA, which is synthesized only in the mitochondrion (cf. Borst and Kroon, 1969). Thus the half-life of mitochondrial DNA of rat liver is 9·4 days, heart muscle 6·7 days and of brain 31 days (Gross and Rabinowitz, 1969). The average half-life of the outer membrane was 7 days and that of the inner membrane 8·4 days, using ^{14}C-leucine as marker (Beattie, 1969) but with ^{35}S-methionine a half-life of 4·2 and 12·6 respectively was obtained (Brunner and Neupert, 1968). In comparison microsomal protein has a half-life of 4·2 days (Beattie, 1969).

Only a twentieth of the mitochondrial protein, mainly structural and some insoluble enzymes, is synthesized by the mitochondrial ribosomes. The elucidation of the contribution of the mitochondrial ribosomes and the cytoplasmic ribosomes to the synthesis of mitochondrial proteins has been explored extensively with chloramphenicol, an effective inhibitor of mitochondrial protein synthesis (Wintersberger, 1965; Linnane *et al.*, 1967; Lamb *et al.*, 1968) and cycloheximide an inhibitor of the ribosomal protein synthesis (Siegel and Sisler, 1965; Clark–Walker and Linnane, 1967; Beattie, 1968; Hartwell *et al.*, 1970). Recently it has become apparent however that the pattern of inhibition by chloramphenicol and cycloheximide is rather complex and that neither system of protein synthesis is independent of the other (Work *et al.*, 1968; Ashwell and Work, 1970). Ethidium bromide which alters the mitochondrial genome (Slonimski *et al.*, 1968) and thus DNA and RNA synthesis (Zylber *et al.*, 1969; Perlman and Mahler, 1971) and protein synthesis (De Vries and Kroon, 1970; Perlman and Penman, 1970; Kellerman *et al.*, 1969) has been used to analyze the mitochondrial contribution to the biosynthesis of the respiratory chain in yeast (Mahler and Perlman, 1971).

The number of DNA molecules per mitochondrion is not firmly known. In vertebrate liver there are 4–5 per mitochondrion (Borst *et al.*, 1967) but more in some yeast strains (Mounolu, 1967). The available evidence indicates that the DNA molecules within the mitochondria of each species are identical. Several lines of evidence support the view that mitochondria possess their own independent mechanism for DNA synthesis (Ashwell and Work, 1970). The rate of synthesis of mitochondrial

DNA is influenced by changes in the cell environment, e.g. exposure to O_2 in yeast; it is lowered in glucose repression and elevated in rapidly dividing cells, in hormone treatment and partial hepatectomy (Tewari et al., 1966; Rabinowitz et al., 1969; Nass 1967; De Leo et al., 1969).

The mitochondrion has the capacity to synthesize some lipids but some are transferred in from outside (Kaiser and Bygrave, 1968; Wirtz and Zilversmit, 1968; Bygrave, 1970). The complex polyglycerophosphates, including cardiolipin, can be synthesized in the mitochondrion (Davidson and Stanacev, 1971; Petzold and Agranoff, 1967; Stanacev et al., 1969; Vorbeck and Martin, 1970; Daae and Bremer, 1970). Each component phospholipid has a different half-life. The acyl groups of all the phospholipids of yeast turn over at the same rate but the glycerophosphate moiety of cardiolipin turns over slowly compared with that of lecithin and phosphatidylethanolamine (Taylor et al., 1967; Bailey et al., 1967; Cuzner et al., 1966.) The biosynthesis of the lipid component of a membraneous cytochrome oxidase preparation for rat liver has been studied in vivo by Weiss and Venner (1969). Defective lipid synthesis as in pantothenate deficiency results in lack of cytochrome oxidase in yeast cells (Okuda and Takahashi, 1967).

Cytochrome oxidase formation is relatively more sensitive to disturbance of cell functions than that of the other cytochromes. Cytochrome aa_3 is absent from yeast grown in the presence of antimycin A, although synthesis of cytochrome b and c is not suppressed (Yças, 1956). Synthesis of cytochrome aa_3 is depressed to a much greater extent than that of cytochrome b and c in the parasitic protozoan *Crithidia fasciculata* grown in the presence of acriflavin (Hill and White, 1969); incorporation of thymidine into the DNA of the kinetoplast is markedly inhibited, but not into that of the nucleus. An increase of cytochrome aa_3 was found in rat liver mitochondria in response to administration of aldosterone *in vivo* (Kirsten et al., 1970). Adrenalectomy caused a decrease of 46% in the level of cytochrome oxidase, but no changes were seen in cytochromes b, c or c_1. A pronounced decrease in cytochrome aa_3 and a change in the relative amounts of b and c occurs in liver mitochondria of the ground squirrel *Citellus laterclis* during hibernation (Shug et al., 1971). Benzimidazole caused a 95% inhibition of the formation of cytochrome aa_3 in *S. cerevisiae* adapting to oxygen, whereas synthesis of enzymes of the NADH-cytochrome c oxidoreductase segment including cytochromes b and c_1, was only 40–50% inhibited (Sels and Verhulst, 1971). Biosynthesis of cytochrome c peroxidase was not affected and that of iso-1-cytochrome c, the predominant cytochrome species, only slightly.

Anaerobic yeast adapting to oxygen have been used extensively to investigate the formation of the respiratory chain (Slonimski, 1953; Lindenmayer and Smith, 1964; Chaix, 1961; Barrett, 1969; Chen and

Charalampous, 1969). Mitochondria-like particles which contained cytochromes a_1 and b-558 were obtained from anaerobic yeast (Zajdela et al., 1961), but later studies suggested that the mitochondria of the adapted yeast were derived from endoplasmic reticular membrane and not from promitochondria (Linnane et al., 1962; Wallace and Linnane, 1964; Polakis et al., 1964; see also Damsky et al., 1969 and Watson et al., 1971). It is now clearly established from the studies of Schatz (Criddle and Schatz, 1969; Paltauf and Schatz, 1969; Plattner et al., 1971) that functional mitochondria are formed in yeast adapting to oxygen by the differentiation of incomplete promitochondria. These promitochondria have been shown to have an energy transfer system which functions in the absence of virtually all respiratory carriers and this system may have distinct physiological functions (Groot et al., 1971).

Green and Perdue (1966a,b) proposed that the membrane bound electron transport system of mitochondria is formed by the stoichiometric agglomeration of large subunits of constant composition. This simplistic view has been replaced by a more sophisticated analysis of the complexity and diversity of mitochondria from various sources and of the factors which bear on the assembly of the membrane components (Green et al., 1971; Kirk, 1971).

Reconstitution of the lamellar membrane of chloroplasts and restoration of the activity of photosystem I activity by reaggregation of the components has been effected (cf. Benson et al., 1971) and the same workers have studied the assembly of the lamellar lipoprotein matrix. An excess of membrane components in the chloroplast is not likely to occur because of tight metabolic control of the biosynthesis of the chlorophylls, membrane-protein, carotenoids and the various complex lipids (Goldberg and Ohad, 1970). The concentration of cytochromes b-559 and c-553 varied in synchronously grown *Chlamydomonas reinhardi* in parallel with the development of Photosystem II (Schor et al., 1970). The development of the photoxidizability of cytochrome f followed a complex pattern and was indicative of a stepwise process in the development of the chloroplast (Schuldiner and Ohad, 1969).

F. BACTERIAL MEMBRANES

In bacteria extremely large quantitative changes in the cytochromes and other essential components of the respiratory and photosynthetic chain can occur as well as variation in the terminal oxidases and branching of the electron transport chain (cf. White and Sinclair, 1970). Frerman and White (1967) have postulated that the membrane of *Staph. aureus* is a mosaic of different respiratory components and lipids and that the electron transport system is formed either by the addition of subunits of differing composition or by modification of a common membrane. On formation of the electron transport system in *Staph. aureus* adapting from anaerobiosis to aerobiosis there is a 1·6 fold increase in the vitamin K_2 content of the membrane with a concomitant alteration in the ratios of the 35 and 45 carbon side chains of the iso-prenologues—a two-fold increase in phosphatidyl glycerol and a 1 to 6-fold increase in cardiolipin also occurs (Frerman and White, 1967; White and Frerman, 1968). The molar proportions of the 64 fatty acids mostly associated with complex lipids changed markedly. In *Haemophilus parainfluenzae* 2-demethyl vitamin K_2 is synthesized coordinately with cytochrome b_1 at a molar ratio of 15 : 1 over an 8-fold range of this cytochrome (White, 1965). During light-induced formation of the photosynthetic apparatus in *Rps. spheroides* the total phospholipids increased without much change in the proportions of the individual phospholipids (Lascelles and Szilágyi, 1965). White (1970) has concluded from pulse chase experiments using ^{32}P that phospholipid metabolism is essential in the initial stages of formation of the electron transport membrane of *H. parainfluenzae*. When the composition of the electron transport membrane is changed post-synthesis of the phospholipids and glucolipids may occur. These involve most actively the polar phosphate ester and acyl moieties of these molecules (White and Tucker, 1969a,b; Tucker and White, 1970).

The cytochrome content of *Micrococcus lysodeikticus* and *Sarcina lutea* grown in the presence of diphenylamine was normal although the stability of the plasma membrane was reduced following suppression of carotenoid synthesis (Salton and Ehtisham-ud-din, 1965). Diphenylamine caused an increase in cytochrome c-554 and b-562 in *Anaebena variabilis*, but cytochromes c-550 and b-557 were not increased (Ogawa and Vernon, 1971).

G. MUTANTS AND GENETIC STUDIES IN YEAST

Mutants, especially yeast and bacteria, have been used extensively to investigate not only the biosynthetic pathways of haem but also the differential regulation of the cytochromes.

Apart from these which have specific biochemical lesions in the porphyrin segment of the overall biosynthetic sequence, other mutants have been obtained which lack either the protein components or have a disturbance of the site of attachment of the membranes; the mitochondrion or plasma may be defective in organization.

Typical respiratory-deficient mutants (RD) of yeast lack the insoluble cytochrome aa_3 and b, the soluble cytochrome c and even c_1 being still formed by the cell irrespective of whether the mutations are cytoplasmic or nuclear. Other RD mutants of yeast have been reported including those having a deficiency in cytochrome a, but not cytochrome b and c, or a decrease in cytochrome c with a normal amount of cytochrome a and b

(Sherman and Slonimski, 1964; Reilly and Sherman, 1965; Mackler et al., 1965). An RD mutant deficient in all cytochromes except b_2 has been described by Morita and Mifuchi (1970). Not all RD mutants have a deficiency of the cytochrome system; of six independent ultraviolet-induced mutants of *Saccharomyces lactis*, all segregational petites, two possessed a normal cytochrome spectrum and another had an increased level of all the cytochromes. An oxidative phosphorylation mutant had a normal cytochrome spectrum (Kovać and Hrusovska, 1968) when grown in a complex medium with glucose as carbon source, but in synthetic medium the cytochrome aa_3 was relatively lowered, but not b or c. Cytochrome deficiency in RD mutants may be the consequence of other deficiencies in the cell apart from lesions in the porphyrin segment of the overall biosynthetic sequence. A mutant had aberrant mitochondria lacking both an organized inner membrane and the other bound enzymes of the respiratory chain (Yotsuyanagi, 1962; Roodyn and Wilkie, 1967).

Genetic and metabolic aspects of cytochrome oxidase deficiency in *S. cerevisiae* has been investigated by Slonimski and his school (cf. Surdin et al., 1969; de Robichon–Szulmajster et al., 1969) and the reader is referred to reviews by Kovać (1969) and by Linnane and co-workers (Saunders et al., 1971, and Beck et al., 1971) for assessments of the application of genetics to the elucidation of function and biogenesis of mitochondria in yeast. The yeast mutants are of two genetic classes, cytoplasmic or ρ-mutants (vegetative, petite mutants) and chromosomal or segregational mutants. Cytochrome aa_3 is absent from the cytoplasmic (ρ^-) mutants which are also largely deficient in the insoluble cytochromes b and c_1 (cf. Chapter IV, B7b2). Cytochrome c is not diminished in these. The chromosomal mutants may be further subdivided into Cy and P types; the former have an altered cytochrome spectrum, but still retain the ability to use non-fermentable carbon metabolite for growth. The P mutants may not have altered spectra but cannot grow on non-fermentable substrates such as lactate and acetate. Genetic studies of mutants deficient in cytochrome c but having normal amounts of cytochromes have indicated that six unlinked chromosomal genes, Cy_1-Cy_6, affect the production of cytochrome c (Sherman et al., 1965). Three genes Cy_1, Cy_2, Cy_3 have been mapped and are of similar size; Cy_1 encodes the primary structure of iso-1-cytochrome c (Sherman et al., 1966, 1968). Mutation sites induced by x-ray were within a very small region of the Cy_1 gene (Parker and Sherman, 1969).

Threonine-requiring mutants have been obtained which lack cytochrome oxidase. These mutants are not petites and cytochrome deficiency segregates with the auxotrophic character for threonine and methionine. Partial cytochrome oxidase deficiency is found in mutants blocked at other steps in the threonine biosynthetic pathway, indicating a metabolic relationship between the two phenotypes (Surdin et al., 1969). These strains are able to synthesize cytochrome oxidase without added threonine when the carbon source is ethanol or glycerol, but in the presence of glucose threonine or its immediate precursors are required to restore cytochrome synthesis (de Robichon-Szulmajster et al., 1969).

Respiratory deficient mutants have been obtained by several means. They may arise spontaneously in haploid yeasts with a frequency of 50–70%, but only 1% in diploid yeasts (Ephrussi and Hottinguer, 1951; Nagai, 1962), during temperature shock or by disturbance of the metabolic processes of the cell (Sherman, 1958). Cultivation of a tetraploid strain of *Saccharomyces* in the presence of allylglycine, a cysteine antagonist, induced RD mutants (Sarachek and Fowler, 1959) as did also the absence of pantothenate from aerobic cultures of the same strain (Sarachek and Fowler, 1961), the mutant frequency being about 90%. Anaerobiosis induced respiratory mutants in small numbers in a haploid strain of *Saccharomyces* (Hino and Lindegren, 1958) as did adaptation to caffeine (Lindegren, 1959). A mutant frequency of 100% was obtained with originally stable strains of *S. cerevisiae* var. *ellipsoideus* when grown in a simple medium lacking vitamins (Nagai, 1969). Incubation of the same variety of yeasts in a minimal medium with various salts added, especially strontium chloride and calcium chloride also yielded high frequencies of RD mutants (Yanagishima, 1967). A mutant grown under conditions of anoxia or glucose repression produced cells of the petite type, but otherwise produced normal respiratory cultures (Negrotti and Wilkie, 1968). Genetic analysis showed that these characteristics of the *gi* mutant were under control of a nuclear gene. A high percentage of survivors of heat-inactivation of *S. cerevisiae* at 54°C developed into RD mutants (Sherman, 1956). A temperature-sensitive RD mutant resulting from a single chromosomal gene mutation formed no cytochromes above 30°C (Gunge et al., 1967). Large amounts of coproporphyrin accumulate despite the presence in the mutant of coproporphyrinogenase activity 10 times greater than in the parent strain. Miyake and Sugimura (1968) investigated this mutant and suggest that the localization of the coproprophyrinogenase may be disorganized so that the enzyme cannot react with the substrate. Ultraviolet has been used to induce RD mutants in anaerobic and aerobic yeast (Sarachek, 1958; Kovacova et al., 1969). Yças and Starr (1953) obtained mutants in which the biochemical lesion was a defective glycine synthesis resulting in a lack of precursors of the haem.

Greater use however has been made of mutagenic chemicals to produce large numbers of petite mutants. Acriflavine yields high frequencies of cytoplasmic mutants but not proflavine (Marcovich, 1953; Ephrussi et al., 1949). Mutants of this type have been extensively studied in connection with the biogenesis of mitochondria. Other dyes have been used as mutagens and this field has been extensively reviewed by Nagai et al. (1961). Sherman (1967) has described the technique for the production of chromosomal mutants and has used a microspectroscopic method for the rapid examination at liquid nitrogen temperatures of very large numbers of colonies (Sherman, 1964). A method of estimating cytochrome c in thick pastes of yeast by absorption spectrophotometry, and at liquid nitrogen temperature is described by Claisse et al. (1970); a linear relationship existed between quantities of cytochrome c extracted and absorption measured in the cell pastes. The theoretical and technical aspects of reflectance spectrophotometry, which has also been used widely to estimate the cytochrome content of yeast, have been reviewed by Moss et al. (1971). Rickard et al. (1967) have refined the quantitative aspects by computer analysis of the spectral data.

H. BACTERIAL MUTANTS

A prototrophic respiratory-mutant of *Neurospora* derived from alteration of the cytoplasmic genetic material has an excess of cytochrome c but is deficient in cytochromes a and b (Haskins et al., 1953; Tissières et al., 1953). This mutant, known as poky, has cytochrome oxidase activity which appears to be due to a modification of cytochrome aa_3 as haem a could be extracted from the purified oxidase (Edwards and Woodward, 1969) although the spectral properties of the preparation resembled mitochrome (see Chapter III, 8e).

Haem-dependence of cytochrome formation occurs naturally in some micro-organisms (cf. Lascelles, 1961). Davis (1917) found that haemin could support growth of *Haemophilus influenzae* and Lwoff and Lwoff (1937) showed that respiratory activity was increased by haemin and that the growth of this organism was proportional to the haemin in the medium. White (1963) established that *H. influenzae*, *H. aegyptius* and *H. canis* in the presence of excess haemin formed cytochromes a_2, a, o, b and c-553. The biochemical deficiency occurs in the biosynthetic steps between ALA synthesis and formation of protoporphyrin (White and Granick, 1963). *H. aegytius* further lacks ferrochelatase, but the capacity to form haem can be induced by transformation with DNA from *H. influenzae*. On the other hand DNA from the haem-dependent *H. parainfluenzae* could not transform any of the haem-dependent species to haem-independence. An interesting variation is found in the lactic acid bacteria which grow well with oxygen as terminal electron acceptor, but normally lack haematin enzymes (Dolin, 1953, 1961). In the presence of protohaem many strains e.g. *Streptococcus faecalis* form catalase and some cytochrome type a and b (Whittenbury, 1964; Bryan-Jones and Whittenbury, 1969).

A haem-requiring streptomycin-resistant mutant of *Staph. aureus* was isolated by Jensen and Thofern 1953a,b,c, 1954; Paul and Thofern, 1960). As catalase was found to be formed within 15 minutes of adding haematin to resting cell suspensions Jensen (1957) suggested that the apoprotein was present. That apoprotein was not synthesized in that time interval is however, not excluded; disruption of the cells prevented catalase formation. Similar evidence indicated that catalase and cytochrome *b* apoproteins were present in a *Staphylococcal* mutant lacking catalase and nitrate reductase (Chang and Lascelles, 1963). Kiese *et al.* (1958) analyzed the haem content of the *Staphylococcal* mutant grown in the presence of haemin and found it to have the same haem *a* content as the wild strain. Stable haem-requiring mutants have been obtained from *Staph. aureus* using kanamycin as the mutagen (Yegian *et al.*, 1959). The mutant could form cytochrome aa_3 and *b* if coproporphyrin or coprohaem was substituted for protohaem, but not with uroporphyrin or haem *a* alone (Barrett, 1962). The haem-dependent mutants induced by kanamycin can be separated into five classes which represent five enzymatic lesions in the protohaem biosynthetic pathway; the genes are tightly linked and are arranged in the same sequence as the reactions in the biosynthetic pathway (Tien and White, 1968).

Haem-dependent mutants were isolated from *B. subtilis* after exposure to ultra-violet irradiation, or to exposure to copper ions, or chemical mutagens. In these mutants protoporphyrin could not replace haem, but other mutants were obtained which could utilize ALA (Anderson and Ivanovics, 1967). Beljanski and Beljanski (1957) isolated haem-requiring mutants from *E. coli* using streptomycin for study of catalase formation. Employing N-methyl-N'-nitro-N-nitrosoguanidine ALA-requiring mutants were obtained from *E. coli* (Wulff, 1967); anaerobic cells could on aeration develop substantial respiration in the presence of amounts of ALA which did not sustain growth. The ALA-dependent mutants of *E. coli K-12* induced by neomycin, lacked the capacity to take up haemin (Šašarman *et al.*, 1968a,b). It appears that the haem genes of *E. coli K-12* are located in at least two regions of the cytochrome, the *pur-B-trp* region (*hem A* locus) and the *lac-pur-E* region (*hem loci*).

Haem-deficient, neomycin-resistant mutants of the closely related *Salmonella typhimurium* are similarly stimulated by ALA and thus are *hem A* mutants (Šašarman *et al.*, 1970). They are not able to incorporate haemins. Two *hem* regions on their chromosomes correspond in their location to the comparable regions of *E. coli K-12* chromosomes.

Chippaux and Pichinoty (1968) have used mutants of *Hafnia sp.* to study the regulation of the biosynthesis of cytochromes by various terminal acceptors. The mutant differs from the wild strain and from other mutants in the relative suppressive effects of NO_3^- and NO_2^- on the formation of cytochrome a_2 and c-550 in particular. Two genetically distinct types of chlorate-resistant mutants of *E. coli K-12* are deficient in cytochrome electron acceptors (Azoulay et al., 1969; Azoulay and Puig, 1968; Piéchaud et al., 1969). The mutants were mutually complementary as cell extracts from each when combined restored nitrate reductase activity.

Ruiz-Herrera et al. (1969) isolated thirty eight mutants of *E. coli K-12* unable to reduce nitrate; all the mutants had variously lowered levels of cytochrome b_1 formate and nitrate reductases. Gene mapping indicated that the components of the nitrate reductase complex and their organisation are controlled by a cluster of genes linked to the tryptophan gene.

Nitrate-reductase-deficient mutants have also been obtained from *E. coli K-12, A. aerogenes* and other bacteria (Venables et al., 1968; Stouthamer, 1967; Hackenthal et al., 1964).

I. INDUCTIVE AND REPRESSIVE ACTIONS OF OXYGEN, GLUCOSE AND LIGHT

1. Oxygen effects

The effect of oxygen on the induction of cytochromes in bacteria and yeast has been extensively investigated in *E. coli* and *A. aerogenes* (Moss 1952, 1956); in *Bacillus subtilis* (Chaix and Petit, 1956); the nitrate reductase system in *A. aerogenes* and fungi (Pichinoty and D'Ornano, 1961) and in *Azotobacter* in the presence of nitrogen (Drozd and Postgate, 1970). The cytochrome oxidase activity of cultured diploid fibroblasts and Hela cells is 3–6 times greater when grown in the presence of 20% O_2 compared to 0·35% O_2 (Adebonojo et al., 1961; Hakami and Pious, 1967). Cytochromes b and $c + c_1$ were similarly increased in amount (Pious, 1970).

Experimental work in the last decade has in particular emphasized the repressive role of O_2 in cytochrome synthesis and its selectivity. Formation of cytochrome *d* in *A. aerogenes* occurs maximally at low O_2 concentrations (Moss, 1956) and similarly in *H. parainfluenzae* (White, 1962) whereas cytochrome *o* in this organism is formed maximally at higher O_2 levels. The c-550 of *Spirillum itersonii* was maximally formed at low O_2 concentrations (Clark-Walker et al., 1967). The cytochrome *c* of *E. coli K-12* and several related bacteria was almost completely repressed in aerobic cultures (Wimpenny et al., 1963). The cytochrome P-450 content of wild-type yeast grown anaerobically and an RD mutant grown semianaerobically decreased sharply on exposure of the cells to oxygen (Ishidate et al., 1969); this repressive effect was countered by glucose (see below).

Longmuir (1954) found that the Michaelis constant (Km) of different organisms for O_2 was related to their size and that the diffusion gradient across the wall of the bacteria was very small (cf. also Wimpenny, 1969). The oxygen gradient may occur over only part of the cell wall in *Bacillus subtilis* (Longmuir *et al.*, 1960). Johnson (1967) concluded from studies with *Candida utilis* that differences across the cell wall may be limiting at low O_2 concentrations but that at higher concentrations the electron transport chain may impose a restriction on the uptake of oxygen.

Wimpenny (1967, 1969) and Wimpenny and Warmsley (1968) have suggested that the redox level of the cell may determine the extent and definition of cytochrome synthesis. In chemostat cultures of *E. coli* changing of the terminal electron acceptors, NO_2^-, NO_3^-, O_2 did not alter the amount of cytochrome formed if the redox state of the cells remained constant. Nevertheless O_2 has a specific inductive effect as cytochrome *d* is not formed in *E. coli* in the absence of O_2.

A secondary role for O_2 is probably on the rate of haem synthesis, through its availibility as a substrate for the conversion of coproporphyrin to protoporphyrin. Anerobically grown *Staphylococcus epidermidis*, which excretes large amounts of coproporphyrin under anaerobiosis, has a low catalase content and respiration rate, but the addition of protohaem to resting cells under anaerobic conditions considerably increases their catalase content and respiration rate (Jacobs *et al.*, 1967).

Several workers have reported competitive inductive effects between oxygen and nitrate in bacteria which have branched electron transport pathways. Competition was found to occur in cultures of *H. parainfluenzae* (White, 1963; Sinclair and White, 1970), in *M. denitrificans* (Lam and Nicholas, 1969), in *Achromobacter* (Arima and Oka, 1965), in *E. coli* (Barrett and Sinclair, 1967) and in *Bacillus stearothermophilus* (Downey *et al.*, 1969). Pichinoty (1965a,b, 1966) has reported extensive studies on the effect of O_2 on bacterial nitrate-reductase systems.

2. Glucose repression

Formation of the electron transport system in yeast is repressed in the presence of glucose (Ephrussi *et al.*, 1956) and this suppressive effect of hexoses or hexose analogues, termed reversed or negative Pasteur effect or Crabtree effect has been widely investigated as a regulatory mechanism in the aerobic respiratory system, particularly in yeast, under steady state conditions in continuous cultures. The Crabtree effect in animal cells, including tumour-cells, has been reviewed by Ibsen (1961).

Extremely minute amounts of glucose can cause inhibition of respiration in ascites cells (Chance and Hess, 1956; Ibsen *et al.*, 1960). A concentration of 0·1 mM represses the synthesis of the cytochromes in yeast (Lester, 1967). From the alteration to the mitochondrial cytochromes and other enzymes, electron microscopy studies and the pulse labelling of mitochondria with radioactive amino acids Jayaraman *et al.* (1966) were led to postulate that during glucose repression of respiring *S. cerevisiae* disassembly of the functional mitochondria occur.

At very low concentrations of O_2 (0·1 μM–0·01 μM) glucose almost completely supressed the cytochromes of yeast whereas galactose caused only 70% repression (Tustanoff and Bartley, 1964a,b; see also Strittmatter, 1957). Glucose also was more effective than galactose in repression of the citric acid pathway (Polakis *et al.*, 1965) and the yeast grown on galactose can respire on a wider range of substrates (Polakis *et al.*, 1964). If O_2 is rigorously excluded from the medium then no cytochrome is found even in the galactose-repressed cells (Somlo and Fukuhara, 1965); hence some O_2 is required to induce cytochromes in yeast in contrast to the conclusion of Tustanoff and Bartley (1964b). In continuous cultures of *C. utilis* there is an inverse relationship between dissolved O_2 and cytochrome and between glucose concentration and cytochrome (Moss *et al.*, 1969). Above the low level of O_2 required for the induction of cytochromes, cytochrome formation in response to changes in O_2 concentration varies with the species and the glucose concentration.

Rickard *et al.* (1971a, 1971b) have concluded from the relative effects of O_2 and glucose on cytochrome synthesis during steady-state growth that aerobic cytochrome formation is more sensitive to glucose level than to oxygen, and that the most significant effect of O_2 is to regulate the rate of respiratory activity rather than the concentration of respiratory components. Cytochrome aa_3 is more responsive to glucose-oxygen effects than the cytochromes b, c and c_1 (Moss *et al.*, 1971). The relationship between the pyruvate oxidase system and pyruvic decarboxylase in the regulation of glucose repression has been examined by Polakis and Bartley (1965). Studies of the intracellular adenosine phosphates and nicotinamide nucleotides led Polakis and Bartley (1966) to conclude that adenosine diphosphate is the probable inducer of the respiratoy system in *S. cerevisiae*. De Deken (1966) has investigated the Crabtree effect in several yeasts and has discussed the directional control of aerobic and anaerobic metabolism.

Jayaraman *et al.* (1971) have shown that ALA-dehydratase, a cytoplasmic enzyme is subject to glucose repression and it may therefore have a regulatory role in the conversion of promitochondria to functional mitochondria. In contrast ALA-synthetase shows oscillatory behaviour during the glucose repression-derepression cycle. This may arise from a greater sensitivity to feedback inhibition following the spurt of haem synthesis at the beginning of the de-repression phase. Rapid oscillatory changes in cytochromes have been observed during steady growth in *C. utilis* after the end of a lag period that followed a switch from high to low dissolved O_2 (Moss *et al.*, 1969). Oscillations in the concentrations of the cytochrome induced in *C. utilis* during continuous culture by addition and withdrawal of chloramphenicol provide evidence of a feedback loop controlling cytochrome synthesis (Gray and Rogers, 1971a). Mathematical simulation of model systems indicates that a metabolic product which causes repression of RNA synthesis is necessary to explain the oscillation observed (Gray and Rogers, 1971b).

Repression by glucose of cytochrome c occurs in *Staph. aureus* (Strasters and Winkler, 1963) and in *Salmonella typhimurium* (Richmond and Maaloe, 1962) and generally in bacteria in which higher levels of O_2 promotes synthesis of the cytochromes. In contrast in *H. parainfluenzae* the formation of the cytochromes and other compounds of the respiratory chain, other than the primary dehydrogenases, is not affected by the catabolism of glucose (White, 1967). Growth of *Azotobacter vinelandii* on glucose repressed components of the respiratory chain especially cytochrome d (Daniel and Erickson, 1969), but not markedly in *Acetobacter suboxydans*, which has only cytochrome o as terminal oxidase and lacks enzymes of the tricarbocylic acid cycle. Synthesis of enzymes of the tricarboxylic acid cycle was markedly repressed in *E. coli* grown in complex medium, but less so in a synthetic medium in which *E. coli* requires the tricarboxylic acid cycle for synthesis (Gray *et al.*, 1966). In *Bacillus tricheniformis* glucose repressed nitrate reductase and cytochrome c_1 formation (Schulp and Stouthamer, 1970).

3. Effect of light on cytochromes

The role of light on the induction and repression of cytochromes in plants and photosynthetic micro-organisms has been mentioned in Chapters V and VI. In respiring micro-organisms a number of instances have been reported of the inhibitory effect of light on their respiration. Goucher and Kocholaty (1957) reported that *Azotobacter agilis* and *Azotobacter chroocuum*, compared to other *Azotobacter* species, had a high sensitivity to inactivation of respiration by light of 2537Å, and this sensitivity could be correlated with the presence of cytochrome a_1 in the cells, and a pigment absorbing at 505 nm (see Labbe *et al.*, 1967). Inhibition by white light of the oxidation of glycerol by O_2 or NO_3^- in *M. denitrificans* was associated with the photobleaching of cytochromes of a, b and c type (Harms and Engel, 1965). Light-bleaching of the cytochromes has also been observed in *Nitrosomonas europaea* and *Nitrobacter winogradski* (Bock, 1965). Irradiation of cells and isolated mitochondria of *Protheca zopfii*, a colourless eukaryotic alga, with blue light caused a destruction of cytochromes a_3, b-559 and c-551 (Epel and Butler, 1969, 1970). Cyanide protected the cytochrome a_3 from photodestruction. Cytochromes b-564, b-555 and c-549 were unaffected by irradiation treatment.

Irradiation of purified cytochrome oxidase or beef heart mitochondria in the presence of O_2 inhibited electron transport from substrate (Ninnemann, *et al.*, 1970b). Light of 405 nm or 436 nm was more effective in destroying cytochrome aa_3 than 365 nm light; the latter was more effective however in inhibiting respiration. Cytochrome oxidase, and some cytochrome b was similarly destroyed in respiring *S. cerevisiae*; cyanide protected cytochrome a_3 from photodestruction while cytochrome a was protected by azide (Ninneman et al., 1970a).

Inhibition or increased lag of respiratory adaptation in resting *S. cerevisiae* was obtained with white light, of moderate intensity but not if light had been filtered through a solution of cytochrome c (Sulkowski et al., 1964; Guerin and Sulkowski, 1966; Guerin and Jacques, 1968). Photoaction spectra of the inhibition of cytochrome formation in the adapting yeast corresponded to the spectrum of protoporphyrin and synthesis of protoporphyrin by extracts of the yeast was inhibited by light. Rapid photooxidation of protoporphyrin occurs in a lipophilic environment but slowly in a hydrophilic one, the product also being different (Barrett, 1967).

REFERENCES

Adebonojo, F. A., Bensch, K. G. and King, D. W. (1961). *Cancer Res.* **21**, 252–256.
Anderson, T. J. and Ivanovics, G. (1967). *J. Gen. Microbiol.* **49**, 31–40.
Arima, K. and Oka, T. (1965). *J. Bact.* **90**, 734–743.
Aschenbrenner, V., Druyan, R., Albin, R. and Rabinowitz, M. (1970). *Biochem. J.* **119**, 157–160.
Ashwell, M. A. and Work, T. S. (1970). *Ann. Rev. Biochem.* **39**, 251–283.
Azoulay, E. and Puig, J. (1968). *Biochem. Biophys. Res. Commun.* **33**, 1019–1024.
Azoulay, E., Puig, J. and Couchoud–Beaumont, P. (1969). *Biochim. et Biophys. Acta* **171**, 238–252.
Bailey, E., Taylor, C. B. and Bartley, W. (1967). *Biochem. J.* **104**, 1026–1032.
Barnes, R., Jones, M. S., Jones, O. T. G. and Porra, R. J. (1971). *Biochem. J.* **124**, 633–637.
Barrett, J. (1961a). *Aust. J. Sci.* **24**, 139.
Barrett, J. (1961b). "Haematin Enzymes", p. 216.
Barrett, J. (1962). M. Sc. Thesis, University of N.S.W.
Barrett, J. (1967). *Nature* **215**, 733–735.
Barrett, J. (1969). *Biochim. et Biophys. Acta* **177**, 442–455.
Barrett, J. and Sinclair, P. (1967). *Biochim. et Biophys. Acta* **143**, 279–281.
Batlle, A. M. del C., Benson, A. and Rimington, C. (1965). *Biochem. J.* **97**, 731–740.
Beattie, D. S. (1968). *Biochem. Biophys. Res. Commun.* **31**, 901–907.
Beattie, D. S. (1969). *Biochem. Biophys. Res. Commun.* **35**, 721–727.
Beattie, D. S. and Stuchell, R. N. (1970). *Arch. Biochem. Biophys.* **139**, 291–297.
Beattie, D. S., Patton, G. M. and Stuchell, R. N. (1970). *J. Biol. Chem.* **245**, 2177–2184.
Beck, J. C., Parker, J. H., Balcavage, W. X. and Mattoon, J. R. (1971). In "Autonomy and Biogenesis of Mitochondria and Chloroplasts" (eds. N. K. Boardman, A. W. Linnane and R. M. Smillie) pp. 194–204, North Holland, London.
Beljanski, M. and Beljanski, M. (1957). *Ann. Inst. Pasteur* **92**, 396–412.
Benson, A. A., Gee, R. W., Ji, T.-H. and Bowes, G. W. (1971). In "Autonomy and Biogenesis of Mitochondria and Chloroplasts" (N. K. Boardman, A. W. Linnane and R. M. Smillie, eds.) pp. 18–26. North Holland, London.
Beutler, E. and Blaisdell, R. K. (1958). *J. Lab. Clin. Med.* **52**, 694–699.
Birkmayer, G. D. (1971a). *Biochemistry* **21**, 258–263.
Birkmayer, G. D. (1971b). *Z. physiol. Chem.* **352**, 761–763.

Bock, E. (1965). *Arch. Mikrobiol.* **51**, 18–41.
Bock, K. W. and Siekevitz, P. (1970). *Biochem. Biophys. Res. Commun.* **41**, 374–380.
Borst, P. and Huijing, F. (1969). *Biochim. et Biophys. Acta* **178**, 408–411.
Borst, P. and Kroon, A. M. (1969). In "International Review of Cytology"(G. H. Bourne and J. F. Danielli, eds.) Academic Press, pp. 107–190.
Borst, P., Ruttenberg, G. J. C. M. and Kroon, A. M. (1967). *Biochim. et Biophys. Acta* **149**, 140–155.
Brunner, G. and Neupert, W. (1968). *FEBS Lett.* **1**, 153–155.
Bryan-Jones, D. G. and Whittenbury, R. (1969). *J. Gen. Microbiol.* **58**, 247–260.
Burnham, B. F. and Lascelles, J. (1963). *Biochem. J.* **87**, 462–472.
Bygrave, F. L. (1969). *J. Biol. Chem.* **244**, 3768–3772.
Carell, E. F. and Price, C. A. (1965). *Plant Physiol.* **40**, 1–7.
Chaix, P. (1961). "Haematin Enyzmes", pp. 225–233.
Chaix, P. and Petit, J. F. (1956). *Biochim. et Biophys. Acta* **22**, 66–71.
Chan, S. K. and Margoliash, E. (1966). *J. Biol. Chem.* **241**, 2252–2255.
Chance, B. and Hess, B. (1959). *J. Biol. Chem.* **234**, 2404–2412.
Chang, J. P. and Lascelles, J. (1963). *Biochem. J.* **89**, 503–510.
Chen, W. L. and Charalampous, F. C. (1969). *J. Biol. Chem.* **244**, 2767–2776.
Chippaux, M. and Pichinoty, F. (1968). *Eur. J. Biochem.* **6**, 131–134.
Chistyakov, V. V. and Gendel, L. Y. (1968). *Biokhimiya* **33**, 1200–1209.
Claisse, M. L., Péré-Aubert, G. A., Clavilier, L. P. and Slonimski, P. P. (1970). *Eur. J. Biochem.* **16**, 430–438.
Clark-Walker, G. D. and Linnane, A. W. (1966). *Biochem. Biophys. Res. Commun.* **25**, 8–13.
Clark-Walker, G. D. and Linnane, A. W. (1967). *J. Cell Biol.* **34**, 1–14.
Clark-Walker, G. D., Rittenberg, B. and Lascelles, J. (1967). *J. Bact.* **94**, 1648–1665.
Cohen, E. and Elvehjem, C. A. (1934). *J. Biol. Chem.* **107**, 97–105.
Cox, G. B., Gibson, F., Luke, R. K. J., Newton, N. A., O'Brien, I. G. and Rosenberg, H. (1970). *J. Bact.* **104**, 219–226.
Criddle, R. S. and Schatz, G. (1969). *Biochemistry* **8**, 322–334.
Cuzner, M. L., Davison, A. N. and Gregson, N. A. (1966). *Biochem. J.* **101**, 618–626.
Daae, L. N. and Bremer, J. (1970). *Biochim. et Biophys. Acta* **210**, 92–104.
Dagg, J. H., Jackson, J. M., Curry, B. and Goldberg, A. (1966). *Brit. J. Haemat.* **12**, 331–333.
Dallman, P. R. and Loskutoff, D. (1967). *J. Clin. Invest.* **46**, 1819–1827.
Dallman, P. R. and Schwartz, H. C. (1965). *J. Clin. Invest.* **44**, 1631–1638.
Dalton, J. and Dougherty, R. G. (1969). *Nature* **223**, 1151–1153.
Damsky, C. H., Nelson, W. H. and Claude, A. (1969). *J. Cell Biol.* **43**, 174–179.
Daniel, R. M. and Erickson, S. K. (1969). *Biochim. et Biophys. Acta* **180**, 63–97.
Davidian, N. and Penniall, R. (1971). *Biochem. Biophys. Res. Commun.* **44**, 15–28.
Davidian, N., Penniall, R. and Elliott, W. B. (1969). *Arch. Biochem. Biophys.* **133**, 345–348.
Davidson, J. B. and Stanacev, N. Z. (1971). *Biochem. Biophys. Res. Commun.* **42**, 1191–1199.
Davis, J. D. (1917). *J. Infect. Diseases* **21**, 392–402.
De Deken, R. H. (1966). *J. Gen. Microbiol.* **44**, 149–156.
De Leo, T., Barletta, A. and Di Meo, S. (1969). *Life Science* **8**, 747–755.
De Robichon-Szulmajster, H., Surdin, Y. and Slonimski, P. P. (1969) *Eur. J. Biochem.* **7**, 531–536.

De Vries, H. and Kroon, A. M. (1970). *Biochim. et Biophys. Acta* **204**, 531–541.
Dolin, M. I. (1953). *Arch. Biochem. Biophys.* **46**, 483–485.
Dolin, M. I. (1961). In (I. C. Gunsalus, R. Y. Stanier, eds.) "The Bacteria," Vol. II, pp. 425–460. Academic Press.
Doss, M. and Philipp-Dormston, W. K. (1971). *Z. physiol. Chem.* **352**, 43–51.
Downey, R. J., Kiszkiss, D. F. and Nuner, J. H. (1969). *J. Bacteriol.* **98**, 1056–1062.
Drabkin, D. L. (1947a,b) *J. Biol. Chem.* **171**, 395–408; 409–417.
Drozd, J. and Postgate, J. R. (1970). *J. Gen Microbiol.* **63**, 63–73.
Druyan, R., De Bernard, B. and Rabinowitz, M. (1969). *J. Biol. Chem.* **244**, 5874–5878.
Edwards, D. L. and Woodward, D. O. (1969). *FEBS Lett.* **4**, 193–196.
Edwards, D. L. and Woodward, D. O. (1970). *Fed. Proc.* **29**, Abstr. 1642.
Ehtisham-ud-din, A. F. M. (1968). *Biochem. J.* **107**, 446–447.
Ellsworth, R. K., Perkins, J. H., Detwiller, J. P. and Kris Liu (1970). *Biochim. et Biophys. Acta* **223**, 275–280.
Elvehjem, C. A. (1931). *J. Biol. Chem.* **90**, 111–132.
Epel, B. L. and Butler, W. L. (1969). *Science* **166**, 621–622.
Epel, B. L. and Butler, W. L. (1970). *Plant Physiol.* **45**, 728–734.
Ephrussi, B. and Hottinguer, H. (1951). *Cold Spring Harbour Sympos.* **16**, 75–84.
Ephrussi, B., L'Heritier, P. and Hottinguer, H. (1949). *Ann. Inst. Pasteur* **77**, 64–83.
Ephrussi, B., Slonimski, P. P. and Perrodin, G. (1950). *Biochim. et Biophys. Acta* **6**, 256–267.
Ephrussi, B., Slonimski, P. P., Yotsuyanagi, Y. and Tavlitsky, J. (1956). *Compt. Rend. Rav. Lab. Carlsberg* **26**, 87–102.
Fanica-Gaignier, M. (1971). *2nd Int. Congress Photosynthesis*, Stresa.
Fanica-Gaignier, M. and Clement-Metral, J. D. (1971). *Biochem. Biophys. Res. Commun.* **44**, 192–198.
Fischer, H. and Fink, H. (1925). *Z. physiol. Chem.* **144**, 101–122.
Flatmark, T. and Sletten, K. (1968). *J. Biol. Chem.* **243**, 1623–1629.
Fletcher, M. J. and Sanadi, D. R. (1961). *Biochim. et Biophys. Acta* **51**, 356–360.
Frerman, F. E. and White, D. C. (1967). *J. Bact.* **94**, 1868–1874.
Fukuhara, H. (1966). *J. Mol. Biol.* **17**, 334–342.
Fukuhara, H. and Sels, A. (1966). *J. Mol. Biol.* **17**, 319–333.
Gajdos, A. and Gajdos-Torok, M. (1969). *Biochem. Med.* **2**, 372–388.
Gajdos, A., Gajdos-Torok, M., Gorchein, A., Neuberger, A. and Tait, G. H. (1968). *Biochem. J.* **106**, 185–192.
Gallagher, C. H., Judah, J. D. and Rees, K. R. (1956a,b). *Proc. Roy. Soc. London* B **145**, 134–150; 195–205.
Garner, R. C. and McLean, A. E. M. (1969). *Biochem. Biophys. Res. Commun.* **37**, 883–887.
Garner, R. C. and McLean, A. E. M. (1971). *Fed. Proc.* **30**, Abstr. 496.
Gear, A. R. L. (1965). *Biochem. J.* **97**, 532–539.
Goldberg, I. and Ohad, I. (1970). *J. Cell Biol.* **44**, 563–571.
González-Cadavid, N. F. and Campbell, P. N. (1967). *Biochem. J.* **105**, 443–450.
González-Cadavid, N. F., Wecksler, M. and Bravo, M. (1970). *FEBS Lett.* **7**, 248–250.
Gorchein, A., Neuberger, A. and Tait, G. H. (1967). In (Goodwin, T. W. ed.) Biochemistry of Chloroplasts, Vol. 2, p. 411, Academic Press, London.
Goucher, C. A. and Kocholaty, W. (1957). *Arch. Biochem. Biophys.* **68**, 30–38.

Goudy, B., Dawes, E., Wolkinson, A. E. and Wills, E. D. (1967). *Eur. J. Biochem.* **3**, 208–212.
Granick, S. (1966). *J. Biol. Chem.* **241**, 1359–1375.
Granick, S. and Urata, G. (1963). *J. Biol. Chem.* **238**, 821–827.
Graubard, M. (1941). *Amer. J. Physiol.* **131**, 584–588.
Gray, C. T., Wimpenny, J. W. T. and Mossman, M. R. (1966). *Biochim. et Biophys. Acta* **117**, 33–41.
Gray, P. P. and Rogers, P. L. (1971a,b). *Biochim. et Biophys. Acta* **230**, 393–400; 401–410.
Green, D. E. and Perdue, J. F. (1966a). *Proc. Nat. Acad. Sci. U.S.* **55**, 1295–1302.
Green, D. E. and Perdue, J. F. (1966b). *Ann. N.Y. Acad. Sci.* **137**, 667–684.
Green, D. E., Korman, E. F., Vanderkooi, G., Wakabayashi, T. and Valdivia, E. (1971). In (N. K. Boardman, A. W. Linnane and R. M. Smillie, eds.) Autonomy and Biogenesis of Mitochondria and Chloroplasts, pp. 1–17. North Holland, London.
Greim, H., Schenkman, J. B., Klotzbücher, M. and Remmer, H. (1970). *Biochim. et Biophys. Acta* **201**, 20–25.
Grimm, P. W. and Allen, P. J. (1954). *Plant Physiol.* **29**, 369–377.
Groot, G. S. P., Kovàč, L. and Schatz, G. (1971). *Proc. Nat. Acad. Sci. U.S.* **68**, 308–311.
Gross, N. J. and Rabinowitz, M. (1969). *J. Biol. Chem.* **244**, 1563–1566.
Gubler, C. J., Cartwright, G. E. and Wintrobe, M. M. (1957). *J. Biol. Chem.* **224**, 533–546.
Guerin, B. and Jacques, R. (1968). *Biochim. et Biophys. Acta* **153**, 138–142.
Guerin, B. and Sulkowski, E. (1966). *Biochim. et Biophys. Acta* **129**, 193–200.
Gunge, N., Sugimura, R. and Iwasaki, M. (1967). *Genetics* **57**, 213–226.
Hackenthal, E., Mannheim, W., Hackenthal, R. and Becher, R. (1964). *Biochem. Pharmacol.* **13**, 195–206.
Hakami, N. and Pious, D. A. (1967). *Nature* **216**, 1087–1090.
Haldar, D., Freeman, K. and Work, T. S. (1966). *Nature* **211**, 9–12.
Hara, T. and Minakami, S. (1970). *J. Biochem.* **67**, 741–743.
Hara, T., Tanaka, S. and Minakami, S. (1970). *J. Biochem.* **68**, 805–1018.
Harms, H. and Engel, H. (1965). *Arch. Mikrobiol.* **52**, 224–230.
Hartwell, L. H., Hutchison, H. T., Holland, T. M. and McLaughlin, C. S. (1970). *Mol. Gen. Genet.* **166**, 347–361.
Haskins, F. A., Tissières, A., Mitchell, H. K. and Mitchell, M. B. (1953). *J. Biol. Chem.* **200**, 819–826.
Hayashi, N., Yoda, B. and Kikuchi, G. (1970). *J. Biochemistry* **67**, 839–861.
Henson, C. P., Weber, C. N. and Mahler, H. R. (1968). *Biochemistry* **7**, 4431–4444; 4445–4454.
Higuchi, M., Goto, K., Fujimoto, M., Namiki, O. and Kikuchi, G. (1965). *Biochim. et Biophys. Acta* **95**, 94–110.
Kikuchi, G. (1965). *Biochim. et Biophys. Acta* **95**, 94–110.
Hill, G. C. and White, D. C. (1969). *U.S. Govt. Res. Develop. Rep.* **69** (6), 55.
Hino, S. and Lindegren, C. C. (1958). *Exp. Cell Res.* **15**, 628–630.
Hirocheka, K. and Neilands, J. B. (1968). *Arch. Biochem. Biophys.* **124**, 450–455.
Ho, Y. K. and Lascelles, J. (1971). *Arch. Biochem. Biophys.* **144**, 734–740.
Hrinda, M. E. and Goldwasser, E. (1968). *Ann. N.Y. Acad. Sci.* **149**, 412–415.
Hsu, W.-P. and Miller, G. W. (1965). *Biochim. et Biophys. Acta* **3**, 393–402.
Ibsen, K. H. (1961). *Cancer Research* **21**, 829–841.
Ibsen, K. H., Coe, E. L. and McKee, R. W. (1960). *Cancer Research* **20**, 1399–1407.

Ishidate, K., Kawaguchi, K. and Tagawa, K. (1969). *J. Biochem.* **65**, 385–392.
Jacobs, A. (1961). *Lancet* **2**, 1331–1333.
Jacobs, N. J., Jacobs, J. M. and Brent, P. (1970). *J. Bact.* **102**, 398–403.
Jacobs, N. J., Maclosky, E. R. and Jacobs, J. M. (1967). *Biochim. et Biophys. Acta* **148**, 645–654.
Jacobs, N. J., Jacobs, J. M. and Brent, P. (1971). *J. Bact.* **107**, 203–209.
Jayaraman, J., Cotman, C., Mahler, H. R. and Sharp, C. W. (1966). *Arch. Biochem. Biophys.* **116**, 224–251.
Jayaraman, J., Padmanaban, G., Malathi, K. and Sarma, P. S. (1971). *Biochem. J.* **121**, 531–535.
Jensen, J. (1957). *J. Bact.* **73**, 324–333.
Jensen, J. and Thofern, E. (1953a,b,c). *Z. Naturf.* **8b**, 559–603; 604–607; 697.
Jensen, J. and Thofern, E. (1954). *Z. Naturf.* **9b**, 596–599.
Johnson, M. J. (1967). *J. Bact.* **94**, 101–108.
Jones, M. S. and Jones, O. T. G. (1969). *Biochem. J.* **113**, 507–514.
Jones, M. S. and Jones, O. T. G. (1970). *Biochem. J.* **116**, 19P.
Jones, O. T. G. (1968). *Biochem. J.* **107**, 113–119.
Kadenbach, B. (1967a,b). *Biochim. et Biophys. Acta* **134**, 430–442; **138**, 651–654.
Kadenbach, B. (1968). *Z. Anal. Chem.* **243**, 542–554.
Kadenbach, B. (1969a). *Eur. J. Biochem.* **10**, 312–317.
Kadenbach, B. (1969b). *Biochim. et Biophys. Acta* **186**, 399–401.
Kadenbach, B. (1969c). In (L. Ernster and Z. Drahota, eds.) Mitochondria: Structure and function, pp. 179–188. Academic Press, London.
Kadenbach, B. (1970). *Eur. J. Biochem.* **12**, 392–398.
Kadenbach, B. (1971). *Biochem. Biophys. Res. Commun.* **44**, 724–730.
Kaiser, W. and Bygrave, F. L. (1968). *Eur. J. Biochem.* **4**, 582–585.
Kappas, A. and Granick, S. (1968). *J. Biol. Chem.* **243**, 346–351.
Kappas, A., Song, C., Levere, R., Sachson, R. A. and Granick, S. (1968). *Proc. Nat. Acad. Sci.* (U.S.) **61**, 509–513.
Keilin, D. (1925). *Nature* **171**, 922–925.
Keilin, D. and Hartree, E. F. (1938). *Nature* **141**, 870–871.
Kellerman, G. M., Biggs, D. R. and Linnane, A. W. (1969). *J. Cell. Biol.* **42**, 378–391.
Kiese, M., Kurz, H. and Thofern, A. (1958). *Biochem. Z.* **230**, 541–544.
Kiprianova, E. A. and Kornyushenko, O. N. (1969). *Mikrobiol. Zh.* (Kiev) **31** (5), 530–532.
Kirk, J. T. O. (1971). In (N. K. Boardman, A. W. Linnane and R. M. Smillie, eds.) Autonomy and Biogenesis of Mitochondria and Chloroplasts, pp. 267–276, North-Holland, London.
Kirsten, R., Brinkhoff, B. and Kirsten, E. (1970). *Pfluegers Arch.* **314**, 231–239.
Komai, H. and Neilands, J. B. (1968). *Arch. Biochem. Biophys.* **124**, 456–451.
Kováč, L. (1969). In (L. Ernster and Z. Drahota, eds.) Mitochondria: Structure and Function, pp. 199–204. Academic Press, London.
Kováč, L. and Hrušovska, E. (1968). *Biochim. et Biophys. Acta* **153**, 43–54.
Kovacova, V., Vlcek, D. and Miodokova, E. (1969). *Folia Microbiologica* **14**, 554–556.
Kraml, J. and Mahler, H. R. (1967). *Immunochem.* **4**, 213–226.
Kroon, M. and De Vries, H. (1971). In (N. K. Boardman, A. W. Linnane and R. M. Smillie, eds.) Autonomy and Biogenesis of Mitochondria and Chloroplasts, pp. 318–338. North-Holland.
Kurashima, Y., Hayashi, N. and Kikuchi, G. (1970). *J. Biochem.* **67**, 863–865.

Kuriyama, Y., Omura, T., Siekevitz, P. and Palade, G. F. (1969). *J. Biol. Chem.* **244**, 2017–2026.
Kuzela, S., Šmigan, P. and Kováč, L. (1969). *Experientia* **25**, 1042–1043.
Labbe, R. F. and Hubbard, N. (1960). *Biochim. et Biophys. Acta* **41**, 185–191.
Labbe, R. F., Volland, C. and Chaix, P. (1967). *Biochim. et Biophys. Acta* **143**, 70–78.
Labbe, R. F., Volland, C. and Chaix, P. (1968). *Biochim. et Biophys. Acta* **159**, 527–539.
Lahey, M. E., Gubler, C. J., Chase, M. S., Cartwright, G. E. and Wintrobe, M. M. (1952). *Blood* **7**, 1053–1074.
Lam, Y. and Nicholas, D. J. D. (1969). *Biochim. et Biophys. Acta* **172**, 450–461.
Lamb, A. J., Clark-Walker, G. D. and Linnane, A. W. (1968). *Biochim. et Biophys. Acta* **161**, 415–427.
Lascelles, J. (1956a). *Biochem. J.* **62**, 78–93.
Lascelles, J. (1956b). *J. Gen. Microbiol.* **15**, 404–416.
Lascelles, J. (1957). *Biochem. J.* **66**, 65–72.
Lascelles, J. (1961). *Physiol. Rev.* **41**, 417–441.
Lascelles, J. (1964). "Tetrapyrrole Biosynthesis and its Regulation". (W. J. Benjamin, ed.) New York, Amsterdam.
Lascelles, J. (1968). In (A. H. Rose and J. F. Wilkinson, eds.) Advances in Microbiological Physiology, Vol. 2, pp. 1–42. Academic Press, London.
Lascelles, J. and Altschuler, T. (1969). *J. Bact.* **98**, 721–727.
Lascelles, J. and Hatch, T. P. (1969). *J. Bact.* **98**, 712–720.
Lascelles, J. and Szilagyi, J. F. (1965). *J. Gen. Microbiol.* **38**, 55–64.
Lascelles, J., Ho, Y. K. and Rittenberg, B. (1969a). *Ann. N.Y. Acad. Sci.* **165**, 305–310.
Lascelles, J., Rittenberg, B. and Clark-Walker, G. D. (1969b). *J. Bact.* **97**, 155–156.
Lemberg, R., Newton, N. and Clarke, L. (1962). *Aust. J. Exp. Biol. Med. Sci.* **40**, 367–372.
Lenhoff, H. M. and Kaplan, N. O. (1956). *J. Biol. Chem.* **220**, 967–982.
Lester, R. L. (1967). see White, D. C. (1967). *J. Bact.* **93**, 567–573.
Lindegren, C. C. (1959). *Nature* **184**, 397–400.
Lindenmayer, A. and Smith, L. (1964). *Biochim. et Biophys. Acta* **93**, 445–461.
Linnane, A. W., Biggs, D. R., Huang, M. and Clark-Walker, G. D. (1967). In (R. K. Mills, ed.) Aspects of Yeast Metabolism, p. 217. Oxford, Blackwell.
Linnane, A. W., Vitols, E. and Nowland, P. G. (1962). *J. Cell. Biol.* **13**, 345–350.
Llambias, E. B. C. and Batlle, A. M. del C. (1971). *Biochim. et Biophys. Acta* **227**, 180–191.
Longmuir, I. S. (1954). *Biochem. J.* **57**, 81–87.
Longmuir, I. S., Milesi, J. and Bourke, A. (1960). *Biochim. et Biophys. Acta* **37**, 521–522.
Lusena, C. V. and Depocas, F. (1966). *Can. J. Biochem.* **44**, 497–508.
Lwoff, A. and Lwoff, M. (1937). *Ann. Inst. Pasteur* **59**, 129–136.
McKay, R., Druyan, R., Getz, G. S. and Rabinowitz, M. (1969). *Biochem. J.* **114**, 455–461.
Mackler, B., Douglas, H. C., Will, S., Hawthorne, D. C. and Mahler, H. R. (1965). *Biochemistry* **4**, 2016–2020.
Mahler, H. R. and Perlman, P. S. (1971). *Biochemistry* **10**, 2979–2990.
Mahler, H. R., Mackler, B., Slonimski, P. P. and Grandchamp, S. (1964). *Biochem.* **3**, 677–682.
Marcovich, H. (1953). *Ann. Inst. Pasteur.* **85**, 199–216.

Marriott, J. (1968). In (T. W. Goodwin, ed.) Porphyrins and Related Compounds, pp. 61-74, Academic Press, London.
Marsh, H. Jr., Evans, H. J. and Matrone, G. (1963). *Plant Physiol.* **38**, 632-638.
Marsh, J. B. and Drabkin, D. L. (1957). *J. Biol. Chem.* **224**, 909-920.
Marver, H. S. (1969). In (J. R. Gillette, A. H. Conney, G. J. Cosmides, R. W. Estabrook, J. R. Fouts and G. J. Mannering, eds.) Microsomes and Drug Oxidations, p. 513. Academic Press.
Marver, H. S., Schmid, R., Schuetzel, H. (1968). *Biochem. Biophys. Res. Commun.* **33**, 969-974.
Mazanowska, A. and Dancewicz, A. M. (1970). *Acta Biochim. Pol.* **17**, 1-10.
Mazanowska, A. M., Neuberger, A. and Tait, G. H. (1966). *Biochem. J.* **98**, 117-127.
Mendiola, L. R. and Price, C. A. (1969). *Amer. J. Clin. Nutrit.* **22**, 1264-1267.
Minagawa, T. (1958). *Expl. Cell Res.* **14**, 333-340.
Miyake, S. and Sugimura, T. (1968). *J. Bact.* **96**, 1997-2003.
Morita, T. and Mifuchi, I. (1970). *Biochem. Biophys. Res. Commun.* **38**, 191-196.
Moss, F. J. (1952). *Aust. J. Exp. Biol. Med. Sci.* **30**, 531-540.
Moss, F. J. (1956). *Aust. J. Exp. Biol. Med. Sci.* **34**, 395-405.
Moss, F. J., Rickard, P. A. D., Beech, G. A. and Bush, F. E. (1969). *Biotech. and Bioeng.* **11**, 561-580.
Moss, F. J., Rickard, P. A. D. and Roper, G. H. (1971). In (J. R. Norris and D. W. Ribbons, eds.) Methods in Microbiology,Vol. **5B**, pp. 615-629, Academic Press, London.
Mounolou, J. C. (1967). Thesis, Paris.
Murty, H. S., Caasi, P. I., Brooks, S. K. and Nair, P. P. (1970). *J. Biol. Chem.* **245**, 5498-5504.
Muthukrishnan, S., Padmanaban, G. and Sarma, P. S. (1969). *J. Biol. Chem.* **244**, 4241-4246.
Nagai, S. (1962). *Exptl. Cell Res.* **26**, 253-259.
Nagai, S. (1969). *Mutat. Res.* **8**, 557-564.
Nagai, S., Yanagishima, N. and Nagai, H. (1961). *Bact. Revs.* **25**, 404-426.
Nandi, D. L. and Waygood, E. R. (1967). *Can. J. Biochem.* **45**, 327-336.
Nass, S. (1967). *Biochim. et Biophys. Acta* **145**, 60-67.
Negishi, M. and Omura, T. (1970). *J. Biochem.* **67**, 745-747.
Negrotti, T. and Wilkie, D. (1968). *Biochim. et Biophys. Acta* **153**, 341-349.
Neuberger, A. (1961). *Biochem. J.* **78**, 1.
Neuwirt, J., Ponka, P. and Borova, J. (1969). *Europ. J. Biochem.* **9**, 36-41.
Ninnemann, H., Butler, W. L. and Epel, B. L. (1970a,b). *Biochim. et Biophys. Acta* **205**, 499-506; 507-512.
Oelze, T. and Kamen, M. D. (1971). *Arch. Mikrobiol.* **76**, 51-64.
Ogawa, T. and Vernon, L. P. (1971). *Biochim. et Biophys. Acta* **226**, 88-97.
Ohaniance, L. (1967). Thesis Science Paris (C.R.N.S.A.O. 1454).
Ohaniance, L. and Chaix, P. (1966). *Biochim. et Biophys. Acta* **128**, 228-238.
Ohaniance, L. and Chaix, P. (1968). *Biochim. et Biophys. Acta* **170**, 435-437.
Ohaniance, L. and Chaix, P. (1970). *Bull soc. chim. biol.* **52**, 1105-1109.
Okuda, S. and Takahashi, H. (1967). *J. Biochem.* **61**, 263-264.
Osanai, M. and Rembold, H. (1970). *Z. physiol. Chem.* **351**, 643-648.
Palma-Carlos, A. G. (1968). *Nouv. Rev. Fr. Haematol.* **8**, 339-346.
Paltauf, F. and Schtaz, G. (1969). *Biochemistry* **8**, 335-339.
Parker, J. H. and Sherman, F. (1969). *Genetics* **62**, 9-22.
Patterson, D. S. P. (1960). *Nature* **185**, 57.

Paul, K. G. and Thofern, E. (1960). *Ann. Chem. Soc.* **14**, 1770–1776.
Penniall, R. and Davidian, N. M. (1968). *Fed. Eur. Biochem. Soc. Letts.* **1**, 38–41.
Perlman, P. S. and Mahler, H. R. (1971). *Nature* **231**, 12–16.
Perlman, S. and Penman, S. (1970). *Biochem. Biophys. Res. Commun.* **40**, 941–948.
Petzold, G. L. and Agranoff, B. W. (1967). *J. Biol. Chem.* **242**, 1187–1191.
Pichinoty, F. (1965a). *Colloq. Int. C.N.R.S. Marseilles* 1963, No. 24.
Pichinoty, F. (1965b). *Ann. Inst. Pasteur* **109**, 248–255.
Pichinoty, F. (1966). *Bull. Soc. Frc. Physiol. Veg.* **12**, 97.
Pichinoty, F. and D'Ornano, L. (1961). *Nature* **191**, 879–881.
Piéchaud, M., Pichinoty, F., Azoulay, E., Couchoud-Beaumont, P. and Gendre, J. (1969). *Ann. Inst. Pasteur* **116**, 276–287.
Pious, D. A. (1970). *Proc. Nat. Acad. Sci. U.S.* **65**, 1001–1008.
Plattner, H., Salpeter, M., Saltzgaber, J., Rouslin, W. and Schatz, G. (1971). In (N. K. Boardman, A. W. Linnane and R. M. Smillie, eds.) Autonomy and Biogenesis of Mitochondria and Chloroplasts, pp. 175–184. North-Holland, London.
Pluscec, J. and Bogorad, L. (1970). *Biochemistry* **9**, 4736–4743.
Polakis, E. S. and Bartley, W. (1965). *Biochem. J.* **97**, 284–297.
Polakis, E. S. and Bartley, W. (1966). *Biochem. J.* **99**, 521–533.
Polakis, E. S., Bartley, W. and Meek, G. A. (1964). *Biochem. J.* **90**, 369–374.
Polakis, E. S., Bartley, W. and Meek, G. A. (1965). *Biochem. J.* **97**, 298–302.
Ponka, P. and Neuwirt, J. (1969). *Blood* **33**, 690–707.
Ponka, P. and Neuwirt, J. (1971). *Biochim. et Biophys. Acta* **230**, 381–392.
Porra, R. J. and Barrett, J. (1964). Unpublished results.
Porra, R. J. and Jones, O. T. G. (1963a,b). *Biochem. J.* **87**, 181–185; 186–192.
Porra, R. J. and Lascelles, J. (1965). *Biochem. J.* **94**, 120–126.
Porra, R. J. and Lascelles, J. (1968). *Biochem. J.* **108**, 343–348.
Porra, R. J. and Ross, B. D. (1965). *Biochem. J.* **94**, 557–562.
Pretlow, T. and Sherman, F. (1967). *Biochim. et Biophys. Acta* **148**, 629–644.
Rabinowitz, M., Getz, G. S., Casey, J. and Swift, H. (1969). *J. Mol. Biol.* **41**, 381–400.
Reilly, C. and Sherman, F. (1965). *Biochim. et Biophys. Acta* **95**, 640–651.
Richmond, M. H. and Maaloe, O. (1962). *J. Gen. Microbiol.* **27**, 285–297.
Rickard, P. A. D., Moss, F. J. and Roper, G. H. (1967). *Biotech. and Bioeng.* **9**, 223–233.
Rickard, P. A. D., Moss, F. J., Phillips, D., Not, T. C. K. and Ganez, M. (1971a). *Biotech. and Bioeng.* **13**, 1–16.
Rickard, P. A. D., Moss, F. J. and Ganez, M. (1971b). *Biotech. and Bioeng.* **13**, 169–184.
Riethmüller, G. and Tuppy, H. (1964). *Biochem. Z.* **340**, 413–420.
Rimington, C. (1957). *Ann. Rev. Biochem.* **26**, 561–586.
Roodyn, D. B. and Wilkie, D. (1967). *Biochem. J.* **103**, 3c.
Ruiz-Herrera, J., Showe, M. K. and de Moss, J. A. (1969). *J. Bact.* **97**, 1291–1297.
Salmon, H. A. (1962). *J. Physiol.* **164**, 17–30.
Salton, M. R. J. and Ehtisham-ud-din, A. F. M. (1965). *Aust. J. Exp. Biol. Med. Sci.* **43**, 255–264.
Sarachek, A. (1958). *Cytologia* **23**, 143–158.
Sarachek, A. and Fowler, G. L. (1959). *Can. J. Microbiol.* **5**, 584–588.
Sarachek, A. and Fowler, G. L. (1961). *Nature* **190**, 792–794.
Šašarman, A. M., Surdeanu, M. and Horodniceanu, T. (1968b). *J. Bact.* **96**, 1882–1884.

Šašarman, A. M., Sanderson, M., Surdeanu, M. and Sonea, S. (1970). *J. Bact.* **102**, 531–536.
Šašarman, A. M., Surdeanu, M., Szegli, G., Horodniceanu, T., Greceanu, V. and Dumitrescu, A., (1968a). *J. Bact.* **96**, 570–572.
Sassa, S. and Granick, S. (1970). *Proc. Nat. Acad. Sci. U.S.* **67**, 517–522.
Saunders, G. W., Gingold, E. B., Trembath, M. K., Lukins, H. B. and Linnane, A. W. (1971). In (N. K. Boardman, A. W. Linnane and R. M. Smillie, eds.) Autonomy and Biogenesis of Mitochondria and Chloroplasts, pp. 185–193. North-Holland, Amsterdam, London.
Sawada, H., Takeshita, M., Sugita, Y. and Yoneyama, Y. (1969). *Biochim. et Biophys. Acta* **178**, 145–155.
Schatz, G. (1970). In (E. Racker, ed.) Structure and Functions of Membranes in Mitochondria and Chloroplasts, pp. 251–314, Van Nostrand Reinhold Co., New York.
Schiefer, H. G. (1969a,b). *Z. physiol. Chem.* **350**, 235–244; 921–928.
Schmidt, G. L. and Kamen, M. D. (1971). *Arch. Mikrobiol.* **76**, 51–64.
Scholnick, P. L., Hamaker, L. E. and Marver, H. S. (1969). *Proc. Nat. Acad. Sci. U.S.* **63**, 65–70.
Schor, S., Siekevitz, P. and Palade, G. E. (1970). *Proc. Nat. Acad. Sci. U.S.* **66**, 174–180.
Schuldiner, S. and Ohad, I. (1969). *Biochim. et Biophys. Acta* **180**, 165–177.
Schulman, H. and Jobe, A. (1968). *Biochim. et Biophys. Acta* **169**, 241–247.
Schulp, J. A. and Stouthamer, A. H. (1970). *J. Gen. Microbiol.* **64**, 195–203.
Sels, A. A. and Verhulst, A. M. (1971). *Eur. J. Biochem.* **19**, 115–123.
Seyffert, R. (1966). Personal communication.
Seyffert, R., Grassl, M. and Lynen, F. (1966) "Hemes and Hemoproteins" pp. 45–51.
Shappirio, D. G. (1960). *Ann. N.Y. Acad. Sci.* **89**, 537–548.
Shemin, D. (1970). *Naturwiss.* **57**. 185–190.
Sherman, F. (1956). *Expt. Cell Res.* **11**, 659–660.
Sherman, F. (1958). Thesis, Univ. of California, Berkeley.
Sherman, F. (1964). *Genetics* **49**, 39–48.
Sherman, F. (1967). "Methods in Enzymology" **10**, 610–616.
Sherman, F. and Slonimski, P. P. (1964). *Biochim. et Biophys. Acta* **90**, 1–15.
Sherman, F., Taber, H. and Campbell, W. (1965). *J. Mol. Biol.* **13**, 21–39.
Sherman, F., Stewart, J. W., Margoliash, E., Parker, J. and Campbell, W. (1966). *Proc. Nat. Acad. Sci. U.S.* **55**, 1498–1504.
Sherman, F., Stewart, J. W., Parker, J. H., Inhaber, E., Shipman, N. A., Putterman, G. J., Gardisky, R. L. and Margoliash, E. (1968). *J. Biol. Chem.* **243**, 5446–5456.
Shigesada, K., Ebisuno, T. and Katsuki, H. (1970). *Biochem. Biophys. Res. Commun.* **39**, 135–141.
Shug, A. L., Ferguson, S., Shrago, E. and Burlington, R. F. (1971). *Biochim. et Biophys. Acta* **226**, 309–312.
Siegel, M. R. and Sisler, H. D. (1965). *Biochim. et Biophys. Acta* **103**, 558–567.
Sinclair, P. and White, D. C. (1970). *J. Bact.* **101**, 365–372.
Sinclair, P., White, D. C. and Barrett, J. (1967). *Biochim. et Biophys. Acta* **143**, 427–428.
Skinner, D. M. (1963). *Biol. Bull.* **125**, 165–176.
Slonimski, P. P. (1953). *Actualités Biochim.* **17**, 1–203.

Slonimski, P. P., Ascher, R., Péré, G., Sels, A. A. and Somlo, M. (1963). *Colloq. Int. Centre Nat. Res. Sci. Paris* **124**, 435–461.
Slonimski, P. P., Perrodin, G. and Croft, J. H. (1968). *Biochem. Biophys. Res. Commun.* **30**, 232–239.
Somlo, M. and Fukuhara, H. (1965). *Biochem. Biophys. Res. Commun.* **19**, 587–591.
Soslau, G., Stotz, E. H. and Lockshin, R. A. (1971). *Biochemistry* **10**, 3296–3299.
Stanacev, N. Z., Stuhne-Sekalec, L., Brookes, K. B. and Davidson, J. B. (1969). *Biochim. et Biophys. Acta* **176**, 650–653.
Stouthamer, A. H. (1967). *Antonie Leeuwenhoek* **33**, 227–228.
Strasters, K. C. and Winkler, K. C. (1963). *J. Gen. Microbiol.* **33**, 213–219.
Strittmatter, C. F. (1957). *J. Gen. Microbiol.* **16**, 169–183.
Subik, J., Kuzela, S., Kolarov, J., Kovàč, L. and Lachowicz, T. M. (1970). *Biochim. et Biophys. Acta* **205**, 513–519.
Sulkowski, E., Guerin, B., Defaye, J. and Slonimski, P. P. (1964). *Nature* **202**, 36–39.
Surdin, Y., De Robichon-Szulmajster, H., Lachowicz, T. M. and Slonimski, P. P. (1969). *Eur. J. Biochem.* **7**, 526–530.
Tait, G. H. (1968) In (T. W. Goodwin, ed.) Porphyrins and Related Compounds, pp. 19–34, Academic Press, London.
Tait, G. H. (1969). *Biochem. Biophys. Res. Commun.* **37**, 116–122.
Tarkowski, S. (1967). *Med. Pracy.* **18**, 233–237.
Taylor, C. B., Bailey, E. and Bartley, W. (1967). *Biochem. J.* **105**, 605–609.
Tephly, T. R. and Hibbeln, P. (1971). *Biochem. Biophys. Res. Commun.* **42**, 589–595.
Tephly, T. R. and Trussler, P. (1971). *Fed. Proc.* **30**, Abstr. 497.
Tewari, K. K., Votsch, W., Mahler, H. R. and Mackler, B. (1966). *J. Mol. Biol.* **20**, 453–481.
Tien, W. and White, D. C. (1968). *Proc. Nat. Acad. Sci. U.S.* **61**, 1392–1398.
Tissières, A. (1952). *Biochem. J.* **50**, 279–288.
Tissières, A., Mitchell, H. K. and Haskins, F. A. (1953). *J. Biol. Chem.* **205**, 423–433.
Tokunaga, R. and Sano, S. (1966). *Biochem. Biophys. Res. Commun.* **25**, 489–494.
Tuboi, S., Kim, H. J. and Kikuchi, G. (1969). *Arch. Biochem. Biophys.* **130**, 92–100.
Tuboi, S., Kim, H. J. and Kikuchi, G. (1970). *Arch. Biochem. Biophys.* **138**, 147–154.
Tucker, A. N. and White, D. C. (1970). *J. Bact.* **101**, 508–513.
Tuppy, H. and Birkmayer, G. D. (1969). *Eur. J. Biochem.* **8**, 237–243.
Tustanoff, E. R. and Bartley, W. (1964a). *Can. J. Biochem.* **42**, 651–665.
Tustanoff, E. R. and Bartley, W. (1964b). *Biochem. J.* **91**, 595–600.
Vallee, B. L. (1959). *Physiol. Rev.* **39**, 443–490.
Van Reen, R. (1953). *Arch. Biochem. Biophys.* **46**, 337–344.
Vary, M. J., Edwards, C. L. and Stewart, P. R. (1969). *Arch. Biochem. Biophys.* **130**, 235–243.
Vary, M. J., Stewart, P. R. and Linnane, A. W. (1970). *Arch. Biochem. Biophys.* **141**, 430–439.
Venables, W. A., Wimpenny, J. W. T. and Cole, J. A. (1968). *Arch. Microbiol.* **63**, 117–121.
Vorbeck, M. L. and Martin, A. P. (1970). *Biochem. Biophys. Res. Commun.* **40**, 901–908.
Wainio, W. W. (1961). "Haematin Enzymes", pp. 281–300.

Wallace, P. G. and Linnane, A. W. (1964). *Nature* **201**, 1191–1194.
Ward, J. M. and Nickerson, W. J. (1958). *J. Gen. Physiol.* **41**, 703–704.
Watson, K., Haslam, J. M., Veitch, B. and Linnane, A. W. (1971). In (N. K. Boardman, A. W. Linnane and R. M. Smillie, eds.) Autonomy and Biogenesis of Mitochondria and Chloroplasts, pp. 162–174, North-Holland, Amsterdam and London.
Weiss, R. and Venner, H. (1969). *Z. physiol. Chem.* **350**, 230–234.
Weiss, H., Sebald, W. and Bücher, T. (1971). *Eur. J. Biochem.* **22**, 19–26.
White, D. C. (1962). *J. Bact.* **83**, 851–859.
White, D. C. (1963). *J. Biol. Chem.* **238**, 3757–3761.
White, D. C. (1965). *J. Bact.* **89**, 299–305.
White, D. C. (1967). *J. Bact.* **93**, 567–573.
White, D. C. (1970). *Proc. 10th Internat. Congress Microbiol., Mexico City.*
White, D. C. and Frerman, E. F. (1968). *J. Bact.* **95**, 2198–2209.
White, D. C. and Granick, S. (1963). *J. Bact.* **85**, 842–850.
White, D. C. and Sinclair, P. R. (1970). *Adv. in Microbial Physiol.* **5**, 173–211.
White, D. C. and Tucker, A. N. (1969a). *J. Lipid Res.* **10**, 220–233.
White, D. C. and Tucker, A. N. (1969b). *J. Bact.* **97**, 199–209.
Whittenbury, R. (1964). *J. Gen. Microbiol.* **35**, 13–26.
Wimpenny, J. W. T. (1967). *Biochem. J.* **106**, 34P.
Wimpenny, J. W. T. (1969). In "Microbial Growth", 19th Symposium of the Society for General Microbiology, pp. 161–197.
Wimpenny, J. W. T. and Warmsley, A. M. (1968). *Biochim. et Biophys. Acta* **156**, 297–303.
Wimpenny, J. W. T., Ranlett, M. R. and Gray, C. T. (1963). *Biochim. et Biophys. Acta* **73**, 170–172.
Wintersberger, E. (1965). *Biochem. Z.* **341**, 409–419.
Wirtz, K. W. and Zilversmit, D. B. (1968). *J. Biol. Chem.* **243**, 3596–3602.
Wohlrab, H. and Jacobs, E. E. (1967). *Biochem. Biophys. Res. Commun.* **28**, 991–997; 998–1002.
Wojtczak, L. and Zaluska, H. (1969). *Biochim. et Biophys. Acta* **193**, 64–72.
Work, T. S., Coote, J. L. and Ashwell, M. (1968). *Fed. Proc.* **27**, 1174–1179.
Wulff, D. L. (1967). *J. Bact.* **93**, 1473–1474.
Yanagishima, N. (1967). *Plant Cell Physiol.* **8**, 211–215.
Yças, M. (1956). *Exper. Cell Res.* **11**, 1–6.
Yças, M. and Drabkin, D. (1957). *J. Biol. Chem.* **224**, 921–933.
Yças, M. and Starr, T. J. (1953). *J. Bact.* **65**, 83–88.
Yegian, D., Gallo, G. and Toll, M. W. (1959). *J. Bact.* **78**, 10–12.
Yoshida, T. and Ishihara, N. (1969). *Tohoku J. Exp. Med.* **99**, 1–8.
Yoshikawa, H. and Yoneyama, Y. (1964). In (F. Gross, ed.) Iron Metabolism, p. 24. Berlin: Springer-Verlag.
Yotsuyanagi, Y. (1962). *J. Ultrastructure Res.* **7**, 141–158.
Young, H. L. (1968). *Nature* **219**, 1068–1069.
Yu, R., Lukins, H. B. and Linnane, A. W. (1968). In (E. C. Slater, J. M. Tager, S. Papa and E. Quagliariello, eds.) Biochemical Aspects of the Biogenesis of Mitochondria, pp. 359–366. Adriatica Editrice, Bari.
Zajdela, F., Heyman-Blanchet, T. and Chaix, P. (1961). *Compt. Rend. Nat. Acad. Sci., Paris* **253**, 1268–1270.
Zylber, E., Vesco, C. and Penman, S. (1969). *J. Mol. Biol.* **44**, 195–204.

Chapter X

The cytochromes in evolution

A. Introduction.
B. The large scale timetable of evolution.
C. Evolution of cytochrome c.
   References

A. INTRODUCTION

The purpose of this chapter is not to serve as an introduction to the theory of evolution, but to describe the contribution that the increase of knowledge of the comparative biochemistry and molecular biology of the cytochromes, in particular of cytochromes c, has made to this theory. The data themselves have been provided in preceding chapters, particularly in Chapter V, B2 and in Tables X and XI. That molecular biology of proteins in different species can make a contribution to our knowledge of evolution was at first received with some scepticism by "organismic" biologists. It is now accepted, however, that criteria from molecular biology of proteins must complement earlier organismic, morphological and biological data. This is whether one agrees with Florkin (1966), Wald (1964), Zuckerkandl (1965) and Zuckerkandl and Pauling (1965b) that opposition between the organismic and the molecular approach is without meaning and that the informational macromolecules are the causative elements in the organism, or whether one takes a more cautious approach to what the molecular approach can contribute with Simpson (1963, 1964, 1967), Richmond (1970) and Dickerson (1971). No one type of evidence, genetic, molecular or phenotypic is sufficient in itself. "In certain cases at least discrepancy with molecular data has been taken into account by competent biologists in revision of their phyletic sequences and the tree of phylogeny" (Simpson, 1967). Although some divergencies have been found (see e.g. Mayr, 1965) both approaches have led to a similar phylogenetic tree. The molecular approach has also raised some interesting questions as to the role and mode of natural selection with regard to "neutral mutations" and "non-Darwinian evolution" (see below and Kimura, 1968).

Cytochromes are, of course, not the only proteins the linear amino acid sequences of which (determined genetically by trinucleotide, usually nuclear, coding) are of evolutionary significance. They, and particularly cytochrome c, however, offer some distinct advantages at least for the study of long-term evolution. Cytochrome c has been studied more completely than any other protein with the possible exception of haemoglobin. It is a rather stable, purifiable and crystallizable protein. Its evolutionary range, from some bacteria which have some claim to be early life forms, to man is exceptionally wide.

Haemoglobin is either absent or distributed sporadically and haphazardly in invertebrates, plants (leguminous nodules) and yeasts. It is composed of several polypeptide chains, each with its own evolutionary history (Itano, 1957; Ingram, 1962, 1963; Edmundson and Hirs, 1961).

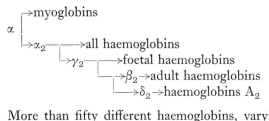

More than fifty different haemoglobins, varying greatly in their pathogenicity have been found in the human species alone, and several myoglobins (Kendrew, 1962, 1963; Kendrew et al., 1954) and haemoglobins (Perutz et al., 1965) are frequently found in other species. In contrast the primary structure of cytochrome c appears to be unique for one species, and some closely related species; man and apes, cattle, sheep and goats, chicken and turkey have the same cytochrome c. At present at least it appears improbable that the study of cytochrome c from a greater number of individuals would alter this picture significantly, although the possibility cannot yet be finally excluded (Margoliash et al., 1971). The number of "invariant" amino acid residues in cytochrome c is far greater than in haemoglobin (Fitch and Margoliash, 1967b). Cytochrome c is more essential for life right from early ontogenesis in the embryo. Disadvantageous ("malefic") mutations which are obviously not always lethal in haemoglobin, are generally lethal in cytochrome c. Compared with the comparative simple biological function of haemoglobin as a dissolved oxygen carrier, the function of cytochrome c requires a complex interaction with cytochrome oxidases and cytochromes reductases in an organized membrane system. It is thus not surprising that the rate of evolutionary change of cytochrome c, one preserved codon mutation in 26 million years, should be so much slower than that in haemoglobin, one mutation in 3·5 million years, and even faster rates in other proteins (Nolan and Margoliash, 1968; Margoliash and Fitch, 1968; Ingram, 1963; Perutz et al., 1965). This makes cytochrome

c more suitable for the study of long range evolution, haemoglobin and other proteins, fibrinopeptides, immunoglobulins, insulins and neurohypophysical hormones for the study of evolution in shorter periods and in restricted classes (Florkin, 1966; Perutz, 1965; Acher, 1966; Nolan and Margoliash, 1968; Margoliash and Fitch, 1970).

Granick (1965) sees the special evolutionary significance and fitness of the porphyrin derivatives in their ability to perform two different tasks, photochemical reactions, and after metal chelation the formation of compounds with different redox potentials.

B. THE LARGE SCALE TIME-TABLE OF EVOLUTION

This section which may be unnecessary for biologists is intended as a guide for chemists and physicists who are less familiar with the data of biological evolution. It is not intended to discuss the origin of life on earth (Calvin, 1969; Wald, 1964). In the "prebiotic" period the rate of replication may have been more important than the construction of efficient enzymes and natural selection for the survival (Black, 1970) and the enzymes may have been so slow in their action as to be undiscoverable. It has been long accepted with Oparin (Oparin and Morgulis, 1938; Oparin, 1959) that the early atmosphere of the earth lacked oxygen and that most of the oxygen was formed only by the process of photosynthesis in algae and green plants (see also Urey, 1959; Berkner and Marshall, 1965). Most of the geologists (but not all, see Davidson, 1965; Wald, 1964) agree with the biologists that the rise of oxygen in the atmosphere took place in the later part of the Precambrian period, somewhere between 2,000 and 600 million years ago. In the early Cambrian period there was a sudden evolutionary "explosion" with the rapid appearance of many forms of metazoan marine fossils, whereas only a few fossils, e.g. primitive Thallophyta have been found in Precambrian sediments which perhaps began the production of oxygen. The oldest known metazoan fossils of the Nonesuch shale and the Australian Ediacara formation (Meinschein *et al.*, 1964; Glaessner, 1966; Cloud, 1965; Symposium on the Evolution of the Earth's Atmosphere, 1965) are probably 1 billion years old. In the subsequent evolutionary explosion the development of the ozone layer is assumed to have played an essential role (cf. Fischer, 1965); it allowed marine organisms to forego the protection from the UV radiation of the sun by deep layers of water and to ascend to more shallow water where visible light could penetrate and made photosynthesis possible. The first photosynthesizing organisms were probably Cyanophyceae and the study of their cytochrome c may be of great interest.

The same holds for the phycobiliproteins (phycoerythrins and phycocyanins) of blue-green and red algae which like cytochromes c are comparatively small though aggregated soluble proteins only loosely attached to membranes. They absorb light at shorter wavelengths than chlorophyll. Although they now act only as adjuvants to chlorophyll in blue-green and red algae, they may have played an independent role in the Precambrian period before chlorophyll was evolved. Life may then have become possible in a depth of the ocean into which blue, green and yellow, but not red light could penetrate. The phycobilins may have evolved from the porphyrin ring of magnesium porphyrins (Barrett, 1967) or cytochromes (Nichols and Bogorad, 1960) in a critical interzone between the upper layers of the ocean still forbidding life owing to the penetration of ultraviolet radiation, and lower layers protected from this and allowing no red light but light of shorter wavelengths to penetrate. Raff and Raff (1970) have calculated that even at a low oxygen pressure of 10 nm Hg, perhaps available 1 billion years ago, 1 mm thick tissues spread on a metabolically inactive mesoglial layer in primitive Coelenterates and Annelids, may have had sufficient supply of oxygen for the first aerobic marine animals, but larger animals could not have obtained sufficient oxygen.

Aerobic and anaerobic photosynthetic organisms can therefore be assumed to represent early life forms and among the latter are some, such as *Rhodospirillum rubrum* and *Chromatium* which contain cytochromes c.

An instructive example of an early photosynthetic organism (Cyanophyceae) has recently been studied by Stewart and Pearson (1970). These Cyanophyceae depend on a photosynthetic process which develops oxygen from water. Yet they live best in semianaerobic conditions in which the oxygen concentration is kept low, e.g. by H_2S which cannot, however, replace water as electron donor, in contrast to the photosynthetic sulphur bacteria. The Cyanophyceae also contain oxygen-sensitive enzymes which fix nitrogen and can reduce acetylene to ethylene. These organisms can therefore be considered as living today in special niches under conditions similar to those under which they evolved at the end of Precambrian times. While they may also have undergone mutations in the subsequent millions of years, it is reasonable to assume that they are genetically conservative (Barghoorn, 1971) and underwent fewer changes in their molecular biology as well as in their physiology than other forms, which underwent profound physiological changes.

The cytochrome chain is probably of greater antiquity than cytochrome c oxidase (George, 1965; Margoliash and Smith, 1965). The evolution of the cytochrome system was possibly preceded by peroxidative systems such as contained in the peroxisomes (De Duve and Baudhuin, 1966). Yeast contains an active cytochrome c peroxidase in addition to a strong cytochrome c oxidase. Copper proteins possibly played a role, and copper is still present as separate Cu-protein in *Pseudomonas* oxidase, but as part of the cytochrome aa_3 oxidase. The ubiquity of cytochromes o in bacteria and protozoa (see Chapter VI) may indicate a protohaem cytochrome oxidase as a precursor of the cytochrome aa_3 haem a-type oxidase. According to Ablaev et al. (1968) the concentration of the cytochromes in liver, kidney and spleen has increased during the evolution from fishes to mammals.

There is much evidence to show that the mitochondrial membranes of eukaryotic organisms developed from the cytoplasmic bacterial and algal membranes (see Chapter VII and Gelman, 1964). One of the essential problems of biochemistry which remains to be solved is the point at which the information-bearing structures were linked up with the catalytic and polymeric-molecular structures, e.g. of membranes to make their effect firmly determinant (Calvin, 1969; cf. also Commoner, 1965).

Pyrophosphate may have preceded ATP in energy conservation of oxidative phosphorylation (Baltscheffsky, 1971). There are still bacterial and even yeast mutants (Kovàč et al., 1968) which have their full complement of cytochromes but lack oxidative phosphorylation.

C. EVOLUTION OF CYTOCHROME c

The major work in this field has been carried out by Margoliash and his co-workers (Margoliash, 1963, 1964, 1969; Fitch, 1966, 1970; Margoliash and Fitch, 1968, 1970; Smith, 1966, 1967). There is now strong evidence that there has been a common ancestor for all cytochromes c, certainly for all eukaryotic cytochromes c, thus evidence for a divergent homologous, not a covergent analogous evolution. According to Zuckerkandl and Pauling (1962, 1965a,b) covergent analogous evolution was also less likely for haemoglobin. The whole field of evolutionary aspects of the primary structure of proteins has been well reviewed by Nolan and Margoliash (1968). At first there appeared to be no evidence for a homology of the bacterial cytochrome c-551 from *Pseudomonas aeruginosa* (Smith and Margoliash, 1964) though Cantor and Jukes (1966b) found evidence for homology between it and the cytochrome c from *Neurospora* which, however, also differs considerably from that of other eukaryotic cytochromes in its amino acid sequence (Heller and Smith, 1966). The data of Dus et al. (1962) for cytochrome cc' of *Chromatium* and particularly those of Dus et al. (1968) on cytochrome c_2 of *Rhodospirillum rubrum* showed however, strong evidence for homology. Recently Dickerson (1971) has also found evidence for homology of the *Pseudomonas* cytochrome c-551 with horse heart cytochrome c if the deletion of 16 residues in a hairpin loop is assumed (see also Needleman and Blair, 1969). All cytochrome c contain at least once in the molecule the characteristic sequence Cys-X-Y-Cys-His-Thr bound by two thioether linkages to the haem and a coordinate linkage of the haem iron between histidine-imidazole-nitrogen and methionine-sulphur. However, the validity of the evidence for repeating shorter polypeptide sequences in the amino acid sequence of a single cytochrome *c* is doubted (see also Nolan and Margoliash, 1968). This had been assumed for bacterial cytochromes *c* by Dus *et al.* (1968), for *Pseudomonas* cytochrome *c*-551 by Needleman and Blair (1969), for some cytochromes *c* and globins by Cantor and Jukes (1966a,b), for globins by Perutz (1965) and with good evidence, for ferredoxins by Eck and Dayhoff (1966), Jukes (1966) and McLaughlin and Dayhoff (1970). The need for assuming deletions in the chains without stringent criteria where these deletions should be placed, makes it doubtful whether there is sufficient statistical evidence for repeating sequences. Although it has proved its value in comparing sequences of known homology*, Dickerson (1971) has pointed out that the method of evolutionary homology, based on the use of "Minimum Mutation Distance" (Jukes, 1963; Fitch and Margoliash, 1967a,b) has an inbuilt error when used for detecting otherwise invisible homologies and is not convincing. This is so because the DNA triplet code itself is far from random. While altering the third purine or pyrimidine base of the coding trinucleotide seldom changes the amino acid coded for, and alteration of the first base usually converts one amino acid to another of similar chemical or physical properties, the alteration of the second base is most influential in changing the chemical character of the amino acid residue.

Assuming divergent evolution from a common primordial cytochrome *c* attempts have been made by Margoliash and co-workers to reconstruct a phylogenetic tree, going back from the amino acid residues to the coding DNA bases, using the principles of the minimal mutation distance to work out from the factual amino acid sequences the hypothetical ancestral structures from the primordial cytochrome *c* to the ancestral cytochromes *c* of fungi, metazoa, insects, vertebrates, birds, mammals and primates (Fitch and Margoliash, 1967a,b; Fitch *et al.*, 1968; Margoliash *et al.*, 1968). The principles applied were to choose (1) the smallest number of codon mutations required between the actual cytochromes and the hypothetical ancestor; or the fewest segments requiring multiple mutations; (2) the fewest sequential mutations, i.e. one mutation in each of two lines is preferred to one before and one after the branch point; (3) the fewest back mutations; (4) the fewest amino acid changes. These pairs of smallest mutational distance are then joined to a set and these sets are then used for finding the most probable phylogenetic tree. A rough correlation is assumed to exist between the total number of variant codons and the time elapsed since the divergence of lines of phylogenetic descent. Only the two isocytochromes-1 and -2 of yeast are under different genetic control, the one under nuclear, the other under cytoplasmic (see Chapter V, 7g). The phylogenetic trees constructed on the assumptions is shown in Fig. 12.

---

* But cf. Epstein (1964) for myoglobin and haemoglobin.

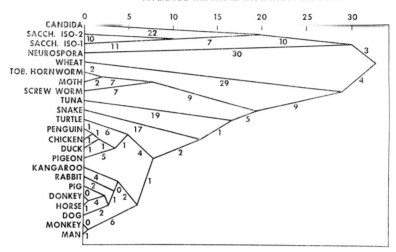

FIG. 12. Phylogenetic tree based on cytochrome c. (According to Fig. 9 of Margoliash, Fitch and Dickerson, 1968.)

On the whole, this phylogenetic tree constructed on the basis of the molecular structure of the cytochromes c, is not very different from one constructed by biologists on the basis of "organismic" assumptions, but there are a few unexpected features. Shark cytochrome c appears to be closer to that of the lamprey than that of the tuna; the turtle cytochrome c closer to that of birds than of the rattle snake; penguin c closely related to that of the domestic fowl; and kangaroo c unexpectedly close to non-primate placental mammals for a marsupial. Accordingly primates would appear to have branched off ancestral mammals before marsupials. A study of a greater number of species may decrease some of these divergencies. The structure assumed for the reconstituted primordial cytochrome c appears to have e.g. been modified between 1968 and 1970 (Margoliash and Fitch), but 80 positions of the primordial c seem to have been ascertained, a much greater number than the "invariant" residues (less than 35).

In addition to the 30 or so "invariable" positions in a total of 100–110 residues, and the 70 or so normally mutable positions, the number of mutated codons in which follow the Poisson probability distribution (Margoliash and Fitch, 1968), there are 5 "hypervariable" positions in which preserved mutations appear more frequently (9–11 per position). That the number of invariable codons in cytochrome c is greater than in haemoglobin is not surprising in view of the greater demands for specificity of its structure enumerated above. It should be noted, however, that cytochromes of other species can successfully replace the cytochrome c of the same species in its intramitochondrial reaction with its cytochrome oxidase (Jacobs and Sanadi, 1960; see also Estabrook and Margoliash, quoted by Margoliash and Smith, 1965; Margoliash et al., 1971). This interchangeability is not unlimited. Yamanaka and others, Yamanaka (1966, 1967), Yamanaka and Kamen (1965), Yoshioda (1966), Yamanaka and Okunuki (1968) and Yamanaka et al. (1969) have attempted to make evolutionary deductions from the relative rates of interaction of the cytochromes with mammalian cytochrome aa_3 on the one hand and that of *Pseudomonas* (probably a cd_1 oxidase on the other. Since the latter also acts as a nitrite reductase it may be considered a primitive precursor of the cytochrome aa_3 oxidase, which the rise in atmospheric oxygen content and its greater sensitivity to H_2O_2 may have made obsolete when the oxygen content of the atmosphere increased. Bacterial cytochromes c of many denitrifying bacteria with the exception of *Micrococcus denitrificans* do not react with mammalian oxidase, but react with *Pseudomonas* oxidase and so does cytochrome c of the protozoon *Tetrahymena*. *B. subtilis* reacts with both, but yeast cytochrome c (though not *Aspergillus c*) reacts far more rapidly with the mammalian oxidase. There is thus some justification for the Yamanaka scheme only at the lowest end of the evolutionary scale and it appears hardly justified to use it for organisms above this. It appears particularly questionable to make deductions from the rates of reaction with the two cytochrome oxidases for cytochromes c which function in photosynthesis and do not react *in vivo* with oxidases. It appears e.g. unlikely that bony fishes are earlier in the evolutionary scale than invertebrates and protozoa than yeasts, fungi and some bacteria. Kiesow (1964) and others (see Chapter VI) have found cytochromes a_1 and c in the autotrophic *Nitrobacter winogradskyi* which produces energy by nitrite oxidation coupled with NAD reduction and phosphorylation; this may be another type of evolutionary primitive organism.

What then is the selective advantage of a specific structure of cytochrome c in the organism higher in the evolutionary scale? There is no evidence for a better fittedness in the respiratory process itself which is its main function. The "malefic" effect particularly in "hypervariable" positions appears negligible yet they are also fixed in the individual species. Such considerations have led some scientists (Jukes, 1963, 1966; King and Jukes, 1969; Arnheim and Taylor, 1969 for haemoglobin) to postulate "neutral mutations" which can be fixed by "biological drift". This suggestion has understandably met with strong opposition from convinced Neodarwinian biologists (Simpson, 1964; see also Margoliash and Smith, 1965, Richmond, 1970 and the discussion after the paper of Margoliash et al., 1968). They point out that the concept of an advantageous mutation in natural selection should not be narrowed down to a direct advantage in the main function of a specific molecule, e.g. of a cytochrome c; Margoliash and Smith (1965) speculated that elapsed time can become a major factor if other factors of natural selection cancel each other out so that the number of displacements become a "molecular evolutionary clock" (Zuckerkandl and Pauling, 1965b). One can certainly postulate unknown advantages in the general complex metabolism of an organism, but the danger of any orthodoxy, scientific or religious, is that the conviction of being in possession of an indisputable truth here as there may lead to an all to easy abandonment of further search for truth. It appears more fertile to search for definite reasons for specific advantages of these mutations. Recently Margoliash et al. (1971) have brought forward some evidence which indicates that cytochrome c may be specifically selected for its role in ion-binding properties facilitating ion transport through the mitochondrial membrane rather than differences in its electron transport properties. The selective pressures on cytochrome c structure appear to have been expressed at different sites of the molecule in different taxonomic groups. Barlow and Margoliash (1966) have found differences of cytochrome c in binding ions, e.g. chloride which was assumed before as an essential function of cytochromes c by Robertson (1960, cf. Chapter VII). They point out the importance of the isoelectric point so that mutations leading to a change of charge would be expected to favour the subsequent retention of mutations that would reestablish the most favourable balance of charges. In contrast to anions like chloride, which are preferentially bound to c^{3+}, the transport cations, notably Ca^{2+}, are preferentially bound to c^{2+}. Other transported cations are bound to both valency states with the help of translocating exchange diffusion carriers, to which ADP and ATP belong. It is suggested that cytochrome c acts as a carrier for some of the ions which require specific translocation mechanisms to move in and out of the mitochondrial matrix (Margoliash et al., 1971). Cytochrome c may e.g. be reduced near the outer surface of the inner mitochondrial membrane, with discharge of Cl^- on the outer surface and bindings of cations, while near the inner surface of the membrane oxidation of c^{2+} occurs by reaction with cytochrome oxidase with discharge of cations and binding of anions. Hypotheses like this have the great advantage that they are testable in contrast to the general postulate that any fixation of a character must necessarily provide such an advantage because the Darwinian theory demands it. Adherents of "biological drift" on the other hand should provide more evidence for the way in which a "neutral" mutation could be fixed so stringently as it obviously occurs in the fixation of amino acid residues in the cytochromes c of different species (see Kimura, 1968). Boulter et al. (1970) and Thompson et al. (1970) have concluded from the primary structure of five cytochromes c of plants, using minimal mutational distance that the first flowering plants were probably primitive dicotyledons, from which monocotyledons (wheat) branched off before other dicotyledons (Leguminosae, Compositae,

Pedaliaceae and Euphorbiaceae). However, the investigation of a greater number of species including several from one and the same family would appear desirable.

So far, this subchapter has dealt with the evolution of the species-specific protein of cytochrome c. There is in addition the problem of the evolution of the covalent linkage between the prosthetic groups and the apoprotein.

TABLE XVII

Approximate time course of evolution

		Geological periods	Millions of years	Fossils	Divergence points*	
PALAEOZOIC		Precambrian	1,000–3,500	few fossils	yeast/bacteria	2,000?
					animals/yeast	1,100?
		Cambrian	500	marine animals	vertebrates/invertebrates	600
					reptiles/fish etc.	400
		Ordovician	400	early fishes	birds/reptiles	400
		Silurian	350	first land plants and animals		
		Devonian	320	first amphibians		
		Lower carboniferous (Mississippian)	285	abundant sharks	mammals/birds	280
		Upper carboniferous (Pennsylvanian)	260	reptiles, coal forests		
		Permian	235	reptiles, conifers		
MESOZOIC		Triassic	200	dinosaurs appear		
		Jurassic	160	zenith of dinosaur, early birds, mammals		
		Cretaceous	130	end of dinosaurs		
CENOZOIC	TERTIARY	Palaeocene	70	flowering plants		
		Oligocene		archiodactyls (hoofed animals) appear		
		Miocene	abt. 50	whales, monkeys		
		Pliocene		caprinae, bovidae, suidae, large carnivores		
	QUATERNARY	Pleistocene	1	man		
		Recent	0·011	Homo sapiens		

* Margoliash et al., 1968; Nolan and Margoliash, 1968.

Protoporphyrinogen has been shown readily to combine with cysteine to form thioether linkages (see Chapter IIB). It is, however, doubtful whether this reaction can have played a role in the evolution of the thioether linkages of cytochrome c. As the presence of cytochrome c in primitive anaerobic bacteria, e.g. in *Chromatium* shows, this occurred at a period before molecular oxygen was present in the atmosphere. So far at least, no formation of protoporphyrin or protoporphyrinogen has been demonstrated for which the presence of molecular oxygen is not required. It was for this reason that Lemberg (1961) suggested that two of the four propionic groups of uroporphyrinogen, or two of coproporphyrinogen may be decarboxylated and simultaneously combined with SS-groups of a primitive apocytochrome c. It is, of course, also possible that this decarboxylation occurs on combination with two SH-groups of the apoproteins accompanied by dehydrogenation by an unknown oxidizing agent different from oxygen. Such a dehydrogenation is also required for the final conversion of the porphyrinogen to the porphyrin. The iron insertion must have been an anaerobic process, as indeed it is even today in the ferrochelatase process. In this connection, it is perhaps also significant that lack of iron insertion in bacteria commonly leads to the excretion of coproporphyrin rather than of protoporphyrin.

While there is at present insufficient evidence for the evolution of other cytochromes, this can perhaps be expected in the near future. At present there appears to be more evidence for the evolution of cytochrome b_5 from cytochrome c by divergent than by covergent evolution, the similarity of the binding of its haem iron to that in haemoglobin and myoglobin support the importance of some convergent evolution in cytochrome b_5 (Nolan and Margoliash, 1968). The story of the evolution of cytochromes is thus at its beginning rather than at its end.

REFERENCES

Ablaev, N. R., Ratmanova, E. Ya and Fingerit, D. S. (1968). *Chem. Abstr.* **69** (13) 103173.
Acher, R. (1966). *Angew. Chem.* (Internat. Ed., Engl.) **5**, 798–806.
Arnheim, N. and Taylor, Ch. E. (1969). *Nature* **223**, 900–903.
Baltscheffsky, H. (1971). *Abstr. 8th Internat. Congress I.U.B. Sympos.* **2**, 142–143.
Barghoorn, E. S. (1971). *Scientific American* **224**, 30–42.
Barlow, G. H. and Margoliash, E. (1966). *J. Biol. Chem.* **241**, 1473–1477.
Barrett, J. (1967). *Nature* **215**, 733–735.
Berkner, L. V. and Marshall, L. C. (1965). *Proc. Nat. Acad. Sci. U.S.* **53**, 1215–1226.
Black, S. (1970). *Nature* **226**, 754–755.
Boulter, D., Thompson, E. W., Ramshaw, J. A. M. and Richardson, M. (1970). *Nature* **228**, 552–554.
Calvin, M. (1969). "Chemical Evolution", Clarendon Press, Oxford.

Cantor, C. R. and Jukes, T. H. (1966a). *Biochem. Biophys. Res. Commun.* **23**, 319–323.
Cantor, C. R. and Jukes, T. H. (1966b). *Proc. Nat. Acad. Sci. U.S.* **56**, 177–184.
Cloud, P. E. Jr. (1965). *Science* **148**, 27–35.
Commoner, B. (1965). *Proc. Nat. Acad. Sci. U.S.* **53**, 1183–1193.
Davidson, C. F. (1965). *Proc. Nat. Acad. Sci. U.S.* **53**, 1184–1205.
De Duve, C. and Baudhuin, P. (1966). *Physiol. Revs.* **46**, 323–357.
Dickerson, R. E. (1971). *J. Mol. Biol.* **57**, 1–15.
Dus, K., Bartsch, R. G. and Kamen, M. D. (1962). *J. Biol. Chem.* **237**, 3083–3093.
Dus, K., Sletten, K. and Kamen, M. D. (1968). *J. Biol. Chem.* **243**, 5507–5518.
Eck, P. V. and Dayhoff, M. O. (1966). *Science* **152**, 363–366.
Edmundson, A. B. and Hirs, C. H. W. (1961). *Nature* **190**, 663.
Epstein, C. J. (1964). *Nature* **203**, 1350–1352.
Fischer, A. G. (1965). *Proc. Nat. Acad. Sci. U.S.* **53**, 1205–1215.
Fitch, W. M. (1966). *J. Mol. Biol.* **16**, 9–27.
Fitch, W. M. (1970). *System Zool.* **19**, 99–113.
Fitch, W. M. and Margoliash, E. (1967a). *Science* **155**, 279–284.
Fitch, W. M. and Margoliash, E. (1967b). *Biochem. Gen.* **1**, 65–71.
Fitch, W. M., Margoliash, E. and Gould, K. S. (1968). *Brookhaven Sympos. Biol.* **21**, 217–232.
Florkin, M. (1966). "A molecular approach to phylogeny", Elsevier, Amsterdam.
Florkin, M. (1971). *Belgian-German Joint Biochem. Meeting. Z. physiol. Chem.* **352** (1), 8.
Florkin, M. Comparative biochemistry, molecular evolution. Vol. 29 of "Comprehensive Biochemistry" (M. Florkin and E. Stotz, eds.) Elsevier.
Gelman, N. S. (1964). *Probl. Evolyutsionnoi i Takhn. Biochim. Akad. Nauk S.S.S.R. Inst. Biokhim.*, 159–164.
George, P. (1965). "Amherst", pp. 3–33.
Glaessner, M. F. (1968). *Canad. J. Earth Sciences* **5**, 585–590.
Glaessner, M. F. (1971). *Geol. Rundschan* (Enke, Stuttgart) **60**, 1323–1339.
Granick, S. (1965). In (V. Bryson and H. J. Vogel, eds.) "Evolving Genes and Proteins". Symposium, Inst. Microbiol. Rutger's State Univ., Academic Press, New York and London, pp. 67–88.
Heller, J. and Smith, E. L. (1966). *J. Biol. Chem.* **241**, 3165–3180.
Ingram, V. N. (1962). *Fed. Proc.* **21** (6) 1053–1057.
Ingram, V. N. (1963). "The Haemoglobins in Genetics and Evolution", Columbia Univ. Press, New York.
Itano, H. A. (1957). *Adv. Protein Chem.* **12**, 215–268.
Jacobs, E. E. and Sanadi, D. R. (1960). *J. Biol. Chem.* **235**, 531–534.
Jukes, T. H. (1963). *Adv. Biol. Med. Phys.* **9**, 1–41.
Jukes, T. H. (1966). "Molecules and Evolution". Columbia Univ. Press, New York, pp. 191–229.
Kendrew, J. C. (1962). *Brookhaven Sympos. in Biol.* **15**, 216–228.
Kendrew, J. C. (1963). *Science* **139**, 1259–1266.
Kendrew, J. C., Parrish, R. G., Marrack, J. R. and Orlans, E. S. (1954). *Nature* **174**, 946–955.
Kiesow, L. (1964). *Proc. Nat. Acad. Sci. U.S.* **52**, 980–988.
Kimura, M. (1968). *Nature* **217**, 624–626.
King, J. L. and Jukes, T. H. (1969). *Science* **164**, 788.
Kovač, L., Lachowicz, T. M. and Slonimski, P. P. (1968). *Science* **158**, 1546–1567.

Lees, H. and Simpson, J. R. (1957). *Biochem. J.* **65**, 297–305.
Lemberg, R. (1961). "Haematin Enzymes", p. 417.
Lemberg, R. (1971). John Falk Memorial Lecture. *Search* **2**, 160–165.
McLaughlin, P. J. and Dayhoff, M. O. (1970). *Science* **168**, 1469–1470.
Margoliash, E. (1963). *Proc. Nat. Acad. Sci. U.S.* **50**, 672–679.
Margoliash, E. (1964). *Canad. J. Biochem.* **42**, 745–753.
Margoliash, E. (1969). *Science* **163**, 127.
Margoliash, E. and Fitch, W. M. (1968). *Annals N.Y. Acad. Sci.* **157**, 359–381.
Margoliash, E. and Fitch, W. M. (1970). *Miami Winter Sympos.* **1**, 33–51. North Holland Publ. Co.
Margoliash, E. and Smith, E. L. (1965). Evolving Genes and Proteins. In (V. Bryson and H. J. Vogel, eds.) Sympos. Rutgers State Univ. pp. 227–242.
Margoliash, E., Barlow, G. H. and Byers, V. (1971). *Nature* **228**, 723–726.
Margoliash, E., Fitch, W. M. and Dickerson, R. E. (1968). *Brookhaven Sympos. Biol.* **21**, 259–302 (discussion p. 302–305).
Mayr, E. (1965). In the discussion of Margoliash and Smith (1965). p. 292.
Meinscheim, W. G., Barghoorn, F. S. and Schopf, J. N. (1964). *Science* **145**, 262–263.
Needleman, S. B. and Blair, T. T. (1969). *Proc. Nat. Acad. Sci. U.S.* **63**, 1227–1233.
Nichols, K. E. and Bogorad, L. (1960). *Nature* **188**, 870.
Nolan, C. and Margoliash, E. (1968). *Ann. Rev. Biochem.* **37**, 727–790.
Oparin, A. I. (ed.) (1959). "The Origin of Life on Earth". Pergamon Press, Oxford.
Oparin, A. I. and Morgulis, S. (1938). "The Origin of Life". Macmillan, New York.
Perutz, M. (1965). *J. Mol. Biol.* **13**, 646–668.
Perutz, M., Kendrew, J. C. and Wilson, H. C. (1965). *J. Mol. Biol.* **13**, 669–678.
Raff, R. A. and Raff, E. C. (1970). *Nature* **228**, 1003–1005.
Richmond, R. C. (1970). *Nature* **225**, 1025–1028.
Simpson, G. G. (1963). *Science* **139**, 81–88.
Simpson, G. G. (1964). *Science* **146**, 1535–1538.
Simpson, G. G. (1967). "The Meaning of Evolution; a Study of the History of Life and of its Significance for Man". Revised ed. Yale Univ., New Haven.
Smith, E. L. (1966/67). Harvey Lectures **62**, 231–256.
Smith, E. L. and Margoliash, E. (1964). *Fed. Proc.* **23** (6), 1243–1247.
Stewart, W. D. and Pearson, H. W. (1970). *Proc. Roy. Soc. London* **B175**, 293–311.
Symposium on the Evolution of the Earth's Atmosphere (1965). (P. E. Cloud, Jr. ed.). *Proc. Nat. Acad. Sci. U.S.* **53**, 1169–1226.
Thompson, E. W., Richardson, M. and Boulter, D. (1970). *Biochem. J.* **121**, 439–446.
Urey, H. C. (1959). In (A. I. Oparin, ed.) "The Origin of Life on Earth". *Sympos. Internat. Union of Biochem.* Macmillan, New York, Vol. I, 16–22.
Wald, G. (1964). *Proc. Nat. Acad. Sci. U.S.* **52**, 595–611.
Yamanaka, T. (1966). *Ann. Report Biol. Works Fac. Sci. Osaka Univ.* **14**, 47–71.
Yamanaka, T. (1967). *Nature* **213**, 1183–1186.
Yamanaka, T. and Kamen, M. D. (1965). *Biochim. et Biophys. Acta* **96**, 328–330.
Yamanaka, T. and Okunuki, K. (1968). "Osaka", pp. 390–403.
Yamanaka, T., Takenami, S. and Okunuki, K. (1969). *Biochim. et Biophys. Acta* **180**, 193–195.

Yoshida, T. (1966). *Ann. Reports Biol. Works, Fac. Sci. Osaka Univ.*, **14**, 1–37.
Zuckerkandl, E. (1965). In (H. Peeters, ed.) "Protides of Biological Fluids", Elsevier, Amsterdam, pp. 102–109.
Zuckerkandl, E. and Pauling, L. (1962). In (M. Kasha and B. Pullman, eds.). "Horizons in Biochemistry", Academic Press, New York, pp. 189–225.
Zuckerkandl, E. and Pauling, L. (1965a). *J. Theor. Biol.* **8**, 357–366.
Zuckerkandl, E. and Pauling, L. (1965b). "Evolving Genes and Proteins", Sympos. Rutgers State Univ. 1964, pp. 97–166. Academic Press, New York and London.

Appendix

Some of the more important papers published after the completion of the book in 1971 and 1972 have been collected in this Appendix. Only the reference, the titles, the chapter and subchapter of the book to which they are relevant are given. This will be a guide to the more recent literature for the reader.

Chapter II

B   Hoard, J. L. (1971), *Science* **174**, 1295–1302. Stereo-chemistry of haems and other metalloporphyrins.
B   Yamagi, Y., Sekuzu, I., Orii, Y. and Okunuki, K. (1972). *J. Biochem.* **41**, 47–56. Reaction of monomeric and dimeric species of haem a with imidazole and cyanide.

Chapter III

B   Antonini, E., Brunori, M., Greenwood, C., Malmström, B. G. and Rotilio, G. C. (1971). *European J. Biochem.* **23**, 396–400. The interaction of cyanide with cytochrome oxidase.
B6 Blockziehl-Homan, M. F. J. and Van Gelder, B. F. (1971). *Biochem. et Biophys. Acta* **234**, 493–498. EPR spectrum of NO-cytochrome a_3.
B5 Chance, B. and Erecińska, M. (1971). *Arch. Biochem. Biophys.* **143**, 675–687. Flow-flash kinetics of the cytochrome a_3–O_2 reaction in coupled and uncoupled mitochondria using the liquid laser.
B4 Elliott, W. B., Holbrook, J. P. and Penniall, R. (1971). *Biochim. et Biophys. Acta* **251**, 277–280. Studies on cytochrome oxidase antibody. 3. Cross reactivity.
B4c Erecińska, M., Chance, B. and Wilson, D. F. (1971). *FEBS Letters* **16**, 284–286. The oxidation-reduction potential of the copper signal in pigeon heart mitochondria.
B5 King, T. E., Bailey, P. M. and Yong, F. C. (1971). *European J. Biochem.* **20**, 103–110. ORD and CD of cytochrome oxidase.
B5 Muijsers, A. O., Tiesjema, R. H. and Van Gelder, B. F. (1971). *Biochim. et Biophys. Acta* **234**, 481–482. Biochemical and biophysical studies on cytochrome aa_3. 2. Conformations of oxidized cytochrome aa_3.
B2 Shakespeare, P. and Mahler, H. R. (1971). *J. Biol. Chem.* **246**, 7649–7655. Purification and some properties of cytochrome c oxidase from the yeast *Saccharomyces*.
B5 Tiesjema, R. H., Muijsers, A. O. and Van Gelder, B. F. (1972). *Biochim. et Biophys. Acta* **256**, 32–42. Some properties of oxygenated cytochrome aa_3.
B4 Tsudzuki, T. and Okunuki, K. (1971). *J. Biochem.* **69**, 909–922. 19. Subunit structure and spin states of cytochrome oxidase from beef heart muscle.
B5 Tsudzuki, T. and Wilson, D. F. (1971). *Arch. Biochem. Biophys.* **145**, 149–154. Redox potentials of the haem and Cu of cytochrome oxidase from beef heart.

B5 Van Buuren, K. J. H., Zuurendonk, P. F., Van Gelder, B. F. and Muijsers, A. O. (1972). *Biochim. et Biophys. Acta* **256**, 243–257. 5. Binding of cyanide to cytochrome aa_3.
B5 Van Buuren, K. J. H., Nicholls, P. and Van Gelder, B. F. (1972). ibid. 258–276. 6. Reactions of cyanide with oxidized and reduced enzyme.
B5 Wilson, D. F., Lindsay, G. and Brocklehurst, E. S. (1972). *Biochim. et. Biophys. Acta* **256**, 277–286. Haem-haem interaction in cytochrome oxidase,

Chapter IV

F Arnon, D. I., Knaff, D. B., McSwain, B. D., Chain, R. K. and Tsujimoto, H. Y. (1971). *Photochem. and Photobiol.* **14**, 397–425. Three light reactions and two photosystems of plant photosynthesis.
G5 Baudras, A., Krupa, M. and Labeyrie, F. (1971). *European J. Biochem.* **20**, 58–64. Molecular complexes between cytochrome b_2 and cytochrome c in the crystalline state and in solution.
F Bendall, D. S. and Sofrova, D. (1971). *Biochim. et Biophys. Acta* **234**, 371–380. Reactions at 77°K in photosystem II of plants.
F Ben-Hayyim, G. and Avron, M. (1971). *Photochemistry and Photobiol.* **14**, 389–396. Light distribution and electron donation in the Z scheme.
F Bishop, N. I. (1971). *Ann. Rev. Biochem.* **40**, 197–226. Photosynthesis: the electron transport system of green plants.
B Boveris, A., Erecińska, M. and Wagner, M. (1972). *Biochim. et Biophys. Acta* **256**, 223–242. Reduction kinetics of cytochromes b.
D8 De Matteis, F. (1971). *Biochem. J.* **124**, 767–777. Loss of haem in rat liver caused by the porphyrinogenic agent 2-alkyl-2-isopropylacetamide.
B Dutton, L. P., Wilson, D. F. and Lee, C. P. (1971). *Biochem. Biophys. Res. Commun.* **43**, 1186–1191. Energy-dependence of oxidation-reduction potentials of the b and c cytochromes in beef heart submitochondrial particles.
B Erecińska, M., Chance, B., Wilson, D. F. and Dutton, P. L. (1972). *Proc. Nat. Acad. Sci. U.S.* **69**, 50–54. Aerobic reduction of cytochrome b-566 in pigeon-heart mitochondria.
F Erixon, K. and Butler, W. L. (1971). *Biochem. et Biophys. Acta* **234**, 381–389. Relationship between Q, C-550 and cytochrome b-559 in photoreactions at —196° in chloroplasts. See also *Photochem. and Photobiol.* **14**, 427–433, (1971).
B Flatmark, T. and Terland, O. (1971). *Biochim. et Biophys. Acta* **253**, 487–491. Cytochrome b-561 of the bovine chromaffine granules. A high potential b-type cytochrome.
D Gallo, M., Bertrand, J. C. and Azoulay, E. (1971). FEBS Letters **19**, 45–49. Participation of cytochrome P-450 in the oxidation of alkanes by *Candida tropicalis*.
F Garewal, H. S., Singh, J. and Wasserman, A. R. (1971). *Biochem. Biophys. Res. Commun.* **44**, 1300–1305. Purification of chloroplast cytochrome b-559.
G Grondinski, O. (1971). *Europ. J. Biochem.* **18**, 480–484. Study of haemprotein linkage in cytochrome b_2. Destruction of a crucial histidine residue by photooxidation of apo-b_2 in the presence of Rose Bengal.
C Hagihara, B. and Iizuka, T. (1971). *J. Biochem.* **69**, 355–362. Low temperature spectra of respiratory pigments. 1. Absorption spectra between liquid helium and room temperature of cytochrome b_5.

F   Hiller, R. G., Anderson, J. M. and Boardman, N. K. (1971). *Biochim. et Biophys. Acta* **245**, 349–352. Photo-oxidation of cytochrome b-559 in leaves and chloroplasts at room temperature.
F   Hiller, R. G. and Boardman, N. K. (1971). *Biochem. et Biophys. Acta* **253**, 449–458. Light driven redox changes of cytochrome f and the development of photo-systems I and II during greening of bean leaves.
D   Imai, Y. and Siekevitz, Ph. (1971). *Arch. Biochem. Biophys.* **144**, 143–159. Comparison of some properties of microsomal P-450 from normal, methylcholanthrene- and phenobarbital-treated rats.
D   Isaka, S. and Hall, P. F. (1971). *Biochem. Biophys. Res. Commun.* **43**, 747–753. Soluble cytochrome P-450 from bovine adrenocortical mitochondria.
C   Ito, A. (1971). *J. Biochem.* **70**, 1061–1064. Hepatic sulphite oxidase identified as cytochrome b_5-like pigment extractable from mitochondria by hypotonic treatment.
D8 Kaufman, L., Swanson, A. L. and Marver, H. S. (1970). *Science* **170**, 320–322. Chemically induced porphyria: prevention by prior treatment with phenobarbital.
F   Ke, B., Vernon, L. P. and Chaney, T. H. (1972). *Biochim. et Biophys. Acta* **256**, 345–357. Photoreduction of cytochrome b-559 in a photosystem II subchloroplast particle.
G   Lederer, F. and Simon, A. M. (1971). *European J. Biochem.* **20**, 469–474. Subunits of bakers' yeast cytochrome b_2. (L-lactate-cytochrome c oxidoreductase). 1. Separation, molecular weight and amino acid analysis.
G   Lederer, F. and Jacq, Cl. (1971). *ibid.* 475–481. 2. End groups and origin of microheterogeneity.
B   McLachlan, A. D. (1971). *J. Mol. Biol.* **61**, 409–424. Tests for comparing related amino acid sequences. Cytochrome c and cytochrome c-551.
C   Matthews, F. S., Levine, M. and Argos, P. (1971). *Nature: New Biology* **233**, 15–16. Structure of calf liver cytochrome b_5 at 2·8Å resolution.
D   Miyake, Y., Mori, K. and Yamano, T. (1971). *Biochem. Biophys. Res. Commun.* **44**, 564–570. Oxidation-reaction mechanisms of cytochrome P-450.
C   Nobrega, F. G. and Ozols, J. (1971). *J. Biol. Chem.* **246**, 1706–1717. Amino acid sequence of tryptic peptides of cytochrome b_5 from microsomes of human, monkey, porcine and chicken liver.
D8 Nomura, T. and Tsuchiya, Y. (1971). *Biochim. et Biophys. Acta* **244**, 117–120. The formation of verdohaemochrome from pyridine ferrihaemochrome by boiled extracts of fish liver.
D8 Pimstone, N. R., Engel, P., Tenhunen, R., Seitz, P. T., Marver, H. S. and Schmid, R. (1971). *J. Clin. Invest.* **50**, 2042–2050. Inducible haem oxygenase in the kidney: a model for the homeostatic control of haemoglobin concentration.
B   Rieske, J. S. (1971). *Arch. Biochem. Biophys.* **145**, 179–193. Changes in redox potential of cytochrome b observed in the presence of antimycin A.
G   Riesler, J. L. (1971). *Biochemistry* **10**, 2664–2669. Fluorescence and phosphorescence of yeast L-lactate dehydrogenase (cytochrome b_2). Relative orientation of the prosthetic haem and flavin.
B   Sato, N., Wilson, D. F. and Chance, B. (1971). *Biochim. et Biophys. Acta* **253**, 88–97. Spectral properties of the b-cytochromes in intact mitochondria.
C   Schnellbacher, E. and Lumper, L. (1971). *Z. physiol. Chem.* **352**, 615–628. CD and ORD spectra of cytochrome b_5 from the microsomes of pig liver.

D Simpson, E. R., Jefcoate, C. R. and Boyd, G. S. (1971). FEBS Letters **15**, 53–58. Spin state changes in cytochrome P-450 associated with cholesterol side chain cleavage in bovine adrenal cortex mitochondria.

C Spatz, L. and Strittmatter, Ph. (1971). *Proc. Nat. Acad. Sci. U.S.* **68**, 1042–1046. A form of cytochrome b_5 that contains an additional sequence of 40 amino acid residues.

D Strobel, H. W. and Coon, M. J. (1971). *J. Biol. Chem.* **246**, 7826–7829. Effect of superoxide generation and dismutation on hydroxylation reactions catalysed by liver microsomal cytochrome P-450.

D Uehlecke, H., Breyer, U., Budzies, E., Tabarelli, S. and Hellmer, K. H. (1971). *Z. physiol. Chem.* **352**, 403–411. The influence of metyrapon on different types of microsomal N-oxidations.

F Vernon, L. P., Shaw, E. R., Ogawa, T. and Raveed, D. (1971). *Photochem. and Photobiol.* **14**, 343–347. Structure of photosystems I and II of plant chloroplasts.

F Wada, K and Arnon, D. I. (1971). *Proc. Nat. Acad. Sci. U.S.* **68**, 3064–3068. Three forms of cytochrome b-559 and their relation to the photosynthetic acitivity of chloroplasts.

B Wikstrom, M. K. F. (1971). *Biochim. et Biophys. Acta* **253**, 332–345. Properties of three cytochrome b-like species in mitochondria and submitochondrial particles.

C Wohlrab, H. and Degn, H. (1972). *Biochim. et Biophys. Acta* **256**, 216–222. Retarded reduction of cytochrome b_5 following the aerobic-anaerobic transition of intact rat liver mitochondria.

Chapter V

C Amesz, J., Visser, J. W. M., Van den Engh, G. J. and Dirks, M. P. (1972). *Biochim. et Biophys. Acta* **256**, 370–380. Reaction kinetics of intermediates of the photosynthetic chain between the two photo-systems.

C2 Cusanovich, M. A., Meyer, T., Tedro, S. M. and Kamen, M. D. (1971). *Proc. Nat. Acad. Sci. U.S.* **68**, 629–631. Question of histidine content in c-type cytochromes.

B Czerlinski, G. and Bracokova, V. (1971). *Arch. Biochem. Biophys.* **147**, 707-716. Kinetics of the interconversion among the electron-transfer linked forms of ferricytochrome c.

C1,3 Fork, D. C. and Murata, N. (1971). *Photochem. and Photobiol.* **13**, 33–44. Oxidation reduction reactions of P-700 and cytochrome f in fraction I particles prepared from spinach chloroplasts by French press treatment.

B4,6 Greenwood, C. and Wilson, M. T. (1971). *Europ. J. Biochem.* **22**, 5–10. Studies on ferricytochrome c. 1. Effects of pH, ionic strength and protein denaturants on the spectra of ferricytochrome c.

B2 Gürtler, L. and Horstmann, H. J. (1971). FEBS Letters **18**, 106–108. Primary structure of cytochrome c of *Equus quagga*.

B2 Hill, G. C., Perkowski, C. A. and Mathewson, N. W. (1971). *Biochim. et Biophys. Acta* **236**, 242–245. Purification and properties of cytochrome c-550 from *Ascaris lumbricoides* var. suum.

C Larkum, A. D. and Bonner, W. D. (1972). *Biochim. et Biophys. Acta* **256**, 385–395. Light-induced oxidation of cytochrome f in isolated chloroplasts of *Pisum sativum*, see also p. 396–408.

B2 Nakayama, T., Titani, K. and Narita, K. (1971). *J. Biochem.* **70**, 311–326. The amino acid sequence of cytochrome *c* from the bonito (*Katsuwomus pelamis*, L.).

B6 Nicholls, P. and Mochan, E. (1971). *Nature* **230**, 276. A stable but functional complex of cytochrome *c* with yeast cytochrome *c* peroxidase.

B2 Ramshaw, J. A. M., Richardson, M. and Boulter, D. (1971). *Europ. J. Biochem.* **23**, 475–483. The amino acid sequence of the cytochrome *c* of *Gingko biloba*.

B7 Richardson, M., Richardson, D., Ramshaw, J. A. M. and Boulter, D. (1971). *J. Biochem.* **69**, 911–913. Improved method for purification of cytochrome *c* from higher plants.

C1 Singh, J. and Wasserman, A. R. (1971). *J. Biol. Chem.* **246**, 2532–2541. Use of disk gel electrophoresis with nonionic detergent in the purification of cytochrome *f* from spinach grana membranes.

B4 Skov, K. and Williams, G. R. (1971). *Canad. J. Biochem.* **49**, 441–447. Correlations between ORD and other spectroscopic properties of ferricytochrome *c* (horse heart).

B2 Sokolovsky, M. and Moldovan, M. (1972). *Biochemistry* **11**, 145–149. Primary structure of cytochrome *c* from the camel, *Camelus dromedarius*.

B4 Strickland, H., Horwitz, J., Kay, E., Shannon, L. M., Wilchek, M. and Billups, C. (1971). *Biochemistry* **10**, 2613–2638. Near UV absorption bands of tryptophan. Studies using horseradish peroxidase isoenzymes bovine and horse heart cytochrome *c* and N-stearyl-L-tryptophan n-hexylester.

B2,7 Sugeno, K., Narita, K. and Titani, K. (1971). *J. Biochem.* **70**, 659–682. Amino acid sequence of cytochrome *c* from *Debaromyces kloeckeri*.

B2 Thompson, E. W., Richardson, M. and Boulter, D. (1971). *Biochem. J.* **124**, 779–781. The amino acid sequence of cytochrome *c* from *Cucurbita maxima* L.; ibid. 783–785 from *Fagopyrum esculentum* Moench, (Buckwheat) and *Brassica oleracea* L. (Cauliflower).

B2 Thompson, E. W., Notton, B. A., Richardson, M. and Boulter, D. (1971). Ibid. p. 787–791 from *Abutilon theophrasti* Medic and *Gossypium barbadense* L. (Cotton).

B4,6 Wilson, M. T. and Greenwood, C. (1971). *Europ. J. Biochem.* **22**, 11–18. Studies of ferricytochrome *c*. 2. A correlation between reducibility and possession of the 695 nm absorption band.

D Yu, C. A., Yong, F. C. and King, T. E. (1971). *Biochim. et Biophys. Res. Commun.* **45**, 508–513. CD of mammalian cytochrome c_1.

Chapter VI

D4 Ambler, R. P., Bruschi, M. and Le Gall, J. (1971). *FEBS Letters* **18**, 347–350. The amino acid sequence of cytochrome c_3 from *Desulfovibrio desulfuricans* (strain EC Agheila Z, NC1B 8380).

D6 Ambler, R. P. (1971). *FEBS Letters* **18**, 351–353. The amino acid sequence of cytochrome *c*-551 from the green photosynthetic bacterium *Chloropseudomonas ethylica*.

E1 Baillie, R. D., Hon, D. and Bragg, P. D. (1971). *Biochim. et Biophys. Acta* **234**, 45–56. The preparation and properties of a soluble respiratory complex from *Escherichia coli*.

D5,6 Bartsch, R. G., Kakuno, T., Horio, T. and Kamen, M. D. (1971). *J. Biol. Chem.* **246**, 4489–4496. Preparation and properties of *Rhodospirillum rubrum* cytochromes c_a, cc' and h-557·5 and of flavin mononucleotide protein.

D6 Case, G. D. and Parson, W. W. (1971). *Biochim. et Biophys. Acta* **253**, 187–202. Thermodynamics of primary and secondary photochemical reactions in *Chromatium*.

D2 Cox, C. D. Jr., Payne, W. L. and Dervartanian, D. V. (1971). *Biochim. et Biophys. Acta* **253**, 290–294. EPR studies on the nature of haemoproteins in nitrite and NO reduction.

D5 Cusanovich, M. A. (1971). *Biochim. et Biophys. Acta* **236**, 238–241. Molecular weight of some cytochromes cc'.

D4 Dervartanian, D. V. and Le Gall, J. (1971). *Biochim. et Biophys. Acta* **243**, 53–65. EPR studies on the reaction of exogenous ligands with cytochrome c_3 from *Desulphovibrio vulgaris*.

Dus, K., Flatmark, T., de Klerk, H. and Kamen, U. D. (1970). *Biochemistry* **9**, 1984–1990. Isolation and chemical properties of two c-type cytochromes of *Rhodosperillum moliscluanum*.

D6 Eley, J. H., Knoblach, K. and Aleem, M. I. H. (1971). *Arch. Biochem. Biophys.* **147**, 419–429. Cytochrome c-linked reactions in *Rhodopseudomonas palustris* grown photosynthetically on thiosulphate.

B1 Erickson, S. K. (1971). *Biochim. et Biophys. Acta* **245**, 63–69. Respiratory system of the aerobic N-fixing, gram-positive bacterium *Mycobacterium flavum* 301.

E1 Ferrandes, B. and Chaix, P. (1972). *Biochim. et Biophys. Acta* **256**, 548–564. The cytochromes of cytoplasmic and mesosomal membranes of *B. subtilis*.

D6 Fowler, C. F., Nugent, N. A. and Fuller, R. C. (1971). *Proc. Nat. Acad. Sci. U.S.* **68**, 2278–2282. Isolation and characterization of a photochemically active complex from *Chloropseudomonas ethylica*.

E1 Hendler, R. W. (1971). *J. Cell Biol.* **51**, 664–673. On the reduction of nonhaem iron and the cytochromes by nicotinamide adenine dinucleotide and succinate (in *Escherichia coli*).

D5 Herber, R. H. (1971). *Scientific American* **225**, 86–95. Mössbauer spectroscopy.

E1 Kalra, V. K., Krishna Murti, C. R. and Brodie, A. F. (1971). *Arch. Biochem. Biophys.* **147**, 734–743. Resolution and reconstitution of the succinoxidase pathway of *Mycobacterium phlei*.

D6 Kakuno, T., Bartsch, R. G. and Kamen, M. D. (1971). *J. Biochem.* **70**, 79–94. Redox components associated with chromatophores of *Rhodospirillum rubrum*.

D6 Kennel, S. J. and Kamen, M. D. (1971). *Biochim. et Biophys. Acta* **234**, 458–467. Iron-containing proteins in *Chromatium D*. 1. Solubilization of membrane-bound cytochromes.

D6 Kennel, S. J. and Kamen, M. D. (1971). *Biochim. et Biophys. Acta* **253**, 153–166. 2. Purification and properties of cholate-solubilized cytochrome complex.

D5 Kennel, J. J., Meyer, T. E., Kamen, M. D. and Bartsch, R. G. (1972). *J. Biol. Chem.* **247**, 3432–3435. On the monohaem character of cytochrome c'.

E1 Kröger, A., Dadek, V., Klingenberg, M. and Diemer, F. (1971). *European J. Biochem.* **21**, 322–333. On the role of quinones in bacterial electron transport. Differential roles of ubiquinone and menaquinone in *Proteus rettgeri*.

B1 Oshino, R., Asakura, T., Tamura, M., Oshino, N. and Chance, B. (1972). Yeast haemoglobin reductase complex.
E1 Schnaitman, C. A. (1971). *J. Bact.* **108**, 545–552. Solubilization of the cytoplasmic membrane of *Escherichia coli* by Triton X-100.
D6 Shioi, Y., Takamiya, K.-i. and Nishimura, M. (1972). 1. Isolation and characterization of cytochromes from *Chloropseudomonas ethylica* strain 2K.
B2 Shrivastava, H. K. (1971). FEBS Letters **16**, 189–191. Carbon monoxide-reactive haemoproteins in parasitic flagellate *Crithidia oncopelti*.
E   Thornber, J. Ph. and Olson, J. M. (1971). *Photochem. and Photobiol.* **14**, 329–341. Chlorophyll-proteins and reaction center preparations from photosynthetic bacteria, algae and higher plants.
E1 Wildermuth, H. (1971). *J. Gen. Microbiol.* **68**, 53–63. The fine structure of mesosomes and plasma membranes in *Streptomyces*.
D4 Yagi, T. and Maruyama, K. (1971). *Biochim. et Biophys. Acta* **243**, 214–224. Purification and properites of cytochrome c_3 of *Desulphovibrio vulgaris*, Mijazami.
D5 Yamanaka, T. and Imai, S. (1972). *Biochem. Biophys. Res. Commun.* **46**, 150–154. A cytochrome cc'-like haemoprotein isolated from *Azotobacter vinelandi*.

Chapter VII

F   Albracht, S. P., Van Herrikhuizen, H. and Slater, E. C. (1972). *Biochim. et Biophys. Acta* **256**, 1–13. Iron-sulphur proteins in the succinate oxidase system.
H   Azzi, A. and Santalo, M. (1971). *Biochem. Biophys. Res. Commun.* **44**, 211–217. Interaction of ethidium bromide with the mitochondrial membrane. Cooperative binding and energy-linked change.
H   Azzi, A., Gherardini, P. and Santalo, M. (1971). *J. Biol. Chem.* **246**, 2035–2042. Fluorochrome interaction with the mitochondrial membrane.
E4 Berden, J. A. and Slater, E. C. (1972). *Biochim. et Biophys. Acta* **256**, 199–215. The allosteric binding of antimycin A to cytochrome b in the mitochondrial membrane.
C3 Chuang, F. T. and Crane, F. L. (1971). *Biochem. Biophys. Res. Commun.* **42**, 1076–1081. Separation of protein components of the mitochondrial complex IV and their molecular weights.
    Discussion: *Acta Histochem.* **10**, Mitochondrial membrane (Schaenstein, Hajos, Pearse *et al.*).
F   Dixon, M. (1971). *Biochim. et Biophys. Acta* **226**, 241–258. The acceptor specificity of flavins and flavoproteins.
D   Grimmelikhuizen, C. J. P. and Slater E. C. (1972). *Biochim. et Biophys. Acta* **256**, 23–31. The redox states of respiratory chain components in rat liver mitochondria. 3. Crossover points in site III.
F   Gutman, M., Coles, C. J., Singer, T. P. and Casida, J. E. (1971). *Biochemistry* **10**, 2036–2043. On the functional organization of the respiratory chain at the dehydrogenase-coenzyme Q junction.
F   Gutman, M., Kearney, E. B. and Singer, T. P. (1971). *Biochemistry* **10**, 2726–2733. Regulation of succinic dehydrogenase activity by reduced coenzyme Q_{10}. See also (1972) *Biochemistry* **11**, 556–562.
C3 Kaplan, D. M. and Criddle, R. S. (1971). *Physiol. Revs.* **51**, 249–272. Membrane structural proteins.

A   Knowles, A. F., Guillory, R. J. and Racker, E. (1971). *J. Biol. Chem.* **246**, 2672–2679. Partial resolution of enzymes catalysing oxidative phosphorylation. Factor for binding of mitochondrial ATPase to inner mitochondrial membrane.

H   Koltover, V. K., Reichman, L. M., Yasaitis, A. A. and Blumonfeld, L. A. (1971). *Biochim. et Biophys. Acta* **234**, 306–310. Specific probe solubility in mitochondrial membranes correlated with ATP-dependent conformational changes.

H   McCord, J. M., Keele, B. B. Jr. and Fridovich, I. (1971). *Proc. Nat. Acad. Sci.* **68**, 1024–1027. Enzyme-based theory of obligate anaerobiosis. The physiological function of superoxide dismutase.

F   Masters, B. S. S. and Ziegler, D. M. (1971). *Arch. Biochem. Biophys.* **145**, 358–364. The distinct nature and function of NADPH-cytochrome c reductase and the NADPH-dependent mixed function amino oxidase of porcine liver microsomes.

F   Nelson, D. D., Norling, B., Persson, B. and Ernster, L. (1971). *Biochem. Biophys. Res. Commun.* **44**, 1321–1329. Effect of certain iron chelators and antibodies on the interaction of succinate dehydrogenase and cytochrome b in ubiquinone-depleted submitochondrial particles.

D   Nishibayashi-Yamashita, H., Cunningham, C. and Racker, E. *J. Biol. Chem.* **247**, 698–704 (1972). Resolution and reconstitution of the mitochondrial electron transport system. 3. Order of reconstitution and requirement for a new factor for respiration.

F   Okamoto, H. (1971). *Biochem. Biophys. Res. Commun.* **43**, 827–833. Influence of L-thyroxine on kynurenine-3-hydroxylase, monoamine oxidase and rotenone-insensitive NADH-cytochrome c reductase in the mitochondrial outer membrane.

E   Orme-Johnson, N. R., Hansen, R. E. and Beinert, H. (1971). *Biochem. Biophys. Res. Commun.* **45**, 871–878. EPR studies of the cytochrome b-c_1 segment of the mitochondrial electron transfer system.

F   Panfili, E., Sottocasa, G. L. and De Barnard, B. (1971). *Biochim. et Biophys. Acta* **253**, 323–331. A high molecular weight form of NADH-cytochrome b_5 reductase from ox liver microsomes.

G   Sies, H., Brauser, B. and Bücher, T. (1969). *FEBS Letters* **5**, 319–323. On the state of mitochondrial in perfused liver; action of sodium azide on respiratory carriers.

A   Slater, E. C. (1971). *Quart. Rev. Biophys.* **4**, 35–71. The coupling between energy-yielding and energy-utilizing reactions in mitochondria.

F   Strittmatter, Ph. (1971). *J. Biol. Chem.* **246**, 1017–1024. Characterization and interconversion of two conformational states of cytochrome b-reductase.

A   Van Dam, K. and Meyer, A. J. (1971). *Ann. Rev. Biochem.* **40**, 115–151. Oxidation and energy conservation by mitochondria.

G   Wagner, M., Erecińska, M. and Pring, M. (1971). *Arch. Biochem Biophys.* **147**, 666–682. Theoretical and experimental studies of the electron transport chain. 1. Effect of respiratory chain inhibitors on the CO-blocked respiratory chain. 2. Fitting of the experimental data.

A-H Wainio, W. W. (1970). The mammalian mitochondrial respiratory chain. Academic Press, N.Y. Mol. Biol. Series.

C   Wikström, M. K. F. (1971). *Biochim. et Biophys. Acta* **245**, 512–516. Effect of ^2H_2O on energy-dependent oxidoreduction of cytochr ome b.

H   Wilson, D. F., Erecińska, M. and Nicholls, P. (1972). FEBS Letters **20**, 61–65. An energy-dependent transformation of a ferricytochrome of the mitochondrial respiratory chain.
A   Young, J. H., Blondin, C. A. and Green, D. E. (1971). *Proc. Nat. Acad. Sci. U.S.* **68**, 1364–1368. Conformational model of active transport: role of protons.

Chapter VIII

B3   Aithal, H. N. and Ramasarina, T. (1971). *Biochem. J.* **123**, 677–682. Changes in the liver mitochondrial oxidation of succinate during cold exposure.
C5   Basch, R. S. and Finegold, M. J. (1971). *Biochem. J.* **125**, 983–989. 3-β-hydroxysteroid dehydrogenase activity in the mitochondria of rat adrenal homogenates.
C3   Cummins, J. T. and Bull, R. (1971). *Biochim. et Biophys. Acta* **253**, 29–45. Spectrophotometric measurement of metabolic responses in isolated rat brain cortex.
C3   Flatmark, T., Lagercrantz, H., Terland, O., Helle, K. B. and Stjärne, L. (1971). *Biochim. et Biophys. Acta* **245**, 249–252. Electron carriers of the noradrenaline storage vesicles from bovine splenic nerves.
C6   Galeotti, T., Cittadini, A., Dionisi, O., Russo, M. and Terranova, T. (1971). *Biochim. et Biophys. Acta* **253**, 303–314. Pathways of intracellular hydrogen transport in the Walker carcinosarcoma 256. 1. The intramitochondrial electron transport and the translocation of reducing equivalents across the mitochondrial membrane. 2. Cittadini, A., Galeotti, T., Russo, M. and Terranova, T., ibid. 314–322. Observations on oxidoreductive changes of electron carriers in slices.
C2   Hallman, M. (1971). *Biochim. et Biophys. Acta* **253**, 360–372. Changes in mitochondrial respiratory chain proteins during perinatal development. Evidence of the importance of environmental oxygen tension.
D   Hill, G. C., Chan, S. K. and Smith, L. (1971). *Biochim. et Biophys. Acta* **253**, 78–87. Purification and properties on cytochrome c-555 from a protozoon *Crithidia fasciculata*.
D   Hill, G. C., Gutteridge, W. E. and Mathewson, N. W. (1971). *Biochim. et Biophys. Acta* **243**, 225–229. Purification and properties of cytochromes c from Trypanosomatids.
C5   Kapley, S. S. and Sanadi, D. R. (1971). *Arch. Biochem. Biophys.* **144**, 440–442. Thyroxine-induced mitochondrial protein and its effect on respiration.
C2   Kuriyama, Y. and Omura, T. (1971). *J. Biochem.* **69**, 659–669. Different turnover behaviour of phenobarbital-induced and normal NADPH-cytochrome c reductases in rat liver microsomes.
C5   McIntosh, E. N., Uzgiris, V. I., Alonson, C. and Salhanick, A. (1971). *Biochemistry* **10**, 2909–2923. Spectral properties, respiratory activity and enzyme systems of bovine corpus luteum mitochondria. (See also Uzgiris, McIntosh, Alonso and Salhanick, ibid. 2916–2923.
C3   McMilland, V. and Siejo, B. K. (1971). *Acta Physiol. Scand.* **82**, 412–414. Critical oxygen tension in the brain.
C2   Masters, B. S. S., Baron, J., Taylor, W. E., Isaacson, E. L. and Lo Spallato, J. (1971). *J. Biol. Chem.* **246**, 4143–4150. Immunochemical studies on electron transport chains involving cytochrome P-450. 1. Effect of antibodies to pig liver microsomal NADPH-cytochrome c reductase and the non-haem iron protein from adrenocortical mitochondria.

B   Raison, J. K. and Lyons, J. M. (1971). *Proc. Nat. Acad. Sci. U.S.* **68**, 2092–2093. Alteration of mitochondrial membranes as a requisite for metabolism at low temperature.

C2   Reisser, O. and Uehlecke, H. (1971). *Zeitschr. physiol. Chem.* **352**, 1048–1052. Carbon tetrachloride interactions with reduced microsomal cytochrome P-450 and haem.

C5   Robinson, J. and Stevenson, P. (1971). *Europ. J. Biochem.* **24**, 18–20. Electron flow and cholesterol side-chain cleavage in ovarian mitochondria.

B5, C4   Rupec, M. and Bruhl, H. (1970). *Histochem.* **24**, 127–129. Ultrastructural localization of cytochrome oxidase in the mitochondria of normal guinea pig epidermis.

B3, C5c   Simon, R. G., Eybel, G. E., Galster, W. and Morrison, P. (1971). *Comp. Biochem. Physiol* **40B**, 601–614. Mitochondrial involvement in cold acclimatization.

C1   Smoly, J. M., Wakabayashi, T., Addink, D. F. and Green, D. E. (1971). *Arch. Biochem. Biophys.* **143**, 6–21. Partial purification of the outer membrane fraction from sonicated heavy beef heart mitochondria.

D   Turner, G., Lloyd, D. and Chance, B. (1971). *J. Gen. Microbiol.* **65**, 359–374. Electron transport in phosphorylating mitochondria from *Tetrahymena pyriformis*, strain S.

C3   Watanabe, H. (1971). *J. Biochem.* **69**. 275–281. Removal of outer membrane from brain mitochondria.

Chapter IX

I   O'Brien, R. W. and Morris, J. G. (1971). *J. Gen. Microbiol.* **68**, 307–318. O_2 and the growth and metabolism of *Clostridium acetobutyricum*.

E   Coote, J. L. and Work, T. S. (1971). *Europ. J. Biochem.* **23**, 564–574. Proteins coded by mitochondrial DNA of mammalian cells.

E   Davidson, J. B. and Stanacev, N. Z. (1971). *Canad. J. Biochem.* **49**, 1117–1124. Biosynthesis of cardiolipin in mitochondria.

A   De Matteis, F. and Gibbs, A. (1972). *Biochem. J.* **126**, 1149–1160. Stimulation of liver δ-ALA synthetase by drugs and its relevance to drug-induced accumulation of cytochrome P-450.

A   Doss, M. and Philipp-Dormston, W. K. (1971). *Z. physiol. Chem.* **352**, 725–733. Porphyrin and haem biosynthesis from endogenous and exogenous δ-aminolaevulinic acid, in *Escherichia coli*, *Pseudomonas aeruginosa* and *Achromobacter metalcaligenes*.

E   Ellis, J. R. and Hartley, M. R. (1971). *Nature New Biology* **233**, 193–196. Site of synthesis of chloroplast proteins.

I   Erixon, K. and Butler, W. L. (1971). *Biochim. et Biophys. Acta* **253**, 483–486 Destruction of c-550 and b-559 by UV radiation.

A   Gaughan, P. L. and Krassner, S. M. (1971). *Comp. Biochem. Physiol.* **B39**, 5–18. Haemin deprivation in culture stages of the haemoflagellate *Leishmania tarentola*.

A   Gilardi, A., Djawadi-Ohaniance, L., Labbe, P. and Chaix, P. (1971). *Biochim. et Biophys. Acta* **234**, 446–467. Effect of accumulation of Zn-porphyrin by the yeast cell and the function of the respiratory system.

B   González-Cadavid, N. F., Ortega, J. P. and González, M. (1971). *Biochem. J.* **124**, 685–694. The cell-free synthesis of cytochrome c by a microsomal fraction from rat liver.

E Hallman, M. and Kankara, P. (1971). *Biochem. Biophys. Res. Commun.* **45**, 1004–1010. Cardiolipin and cytochrome aa_3 in intact liver mitochondria of rats. Evidence of successive formation of inner membrane components.
F Hamilton, J. A. and Cox, B. G. (1971). *Biochem. J.* **123**, 435–443. Ubiquinone synthesis in *E. coli*-K12. Accumulation of an octaprenol.
A2 Keyhani, E. and Chance, B. (1971). FEBS Letters **17**, 127–132. Cytochrome biosynthesis under copper-limited conditions in *Candida utilis*.
H Kiss, I., Berck, I. and Ivanovics (1971). *J. Gen. Microbiol.* **66**, 153–159. Mapping the δ-ALA synthetase locus in *Bacillus subtilis*.
I Luzikov, V. N., Zubatov, A. S., Rainina, E. I. and Bakeyeva, L. E. (1971). *Biochim. et Biophys. Acta* **245**, 321–334. Degradation and restoration of mitochondria upon deaeration and subsequent aeration of anaerobically grown *S. cerevisiae* cells.
G Michaelis, G., Douglass, St., Tsai, M.-J. and Criddle, R. S. (1971). *Biochem. Genetics* **5**, 487–495. Mitochondrial DNA and suppressiveness of Petite mutants in *Saccharomyces cerevisiae*.
C Nebert, D. W., and Grelen, J. L. (1971). *J. Biol. Chem.* **246**, 5199–5206. Aryl hydrocarbon hydroxylase induction in mammalian liver cell culture. 2. Effect of actinomycin D and cycloheximide on induction processes by phenobarbital and polycyclic hydrocarbons.
E Omura, T. and Kuriyama, Y. (1971). *J. Biochem.* **69**, 651–658. Role of rough and smooth microsomes in the biosynthesis of microsomal membranes.
A Porra, R. J., Irving E. A. and Tennick, A. M. (1972). *Arch. Biochem. Biophys.* **148**, 37–43. The nature of the inhibition of δ-aminolaevulinic acid synthetase by haemin.
D Schatz, G., Groot, G. S. P., Mason, T., Rouslin, W., Wharton, D. C. and Saltzgaber, J. (1972). *Fed. Proc.* **31**, (1), 21–29. Biogenesis of mitochondrial membranes in bakers' yeast.
E Soslau, G. and Nass, M. K. (1971). *J. Cell Biol.* **51**, 514–524. Effect of ethidium bromide on the cytochrome content and ultrastructure of L cell mitochondria.
G Stewart, J. M., Sherman, F., Shipman, N. A. and Jackson, M. (1971). *J. Biol. Chem.* **246**, 7429–7445. Identification and mutational relocation of the AUG. codon initiating translocation of isocytochrome *c* in yeast.
G Weislogel, P. O. and Butow, R. A. (1971). *J. Biol. Chem.* **246**, 5113–5119 Control of the mitochondrial genome in *S. cerevisiae*. The fate of the mitochondrial membrane proteins and mitochondrial DNA during Petite induction.

Chapter X

C Dickerson, R. E. (1971). *J. Mol. Evolution* **1**, 26–45. The structure of cytochrome *c* and the rates of molecular evolution.
C Fitch, W. M. (1971). *Biochem. Genetics* **5**, 231–241. The non-identity of invariable positions in the cytochromes *c* of different species.
C Goodman, M., Barnabas, J., Matsuda, G. and Moore, G. W. (1971). *Nature* **233**, 604–613. Molecular evolution in the descent of man (globin).
B Hall, D. O., Cammack, R. and Rao, K. K. (1971). *Nature* **233**, 136–138. Role of ferredoxins in the origin of life and biological evolution.

B Margolis, L. (1971). *Scientific American* **225**, 48–00. Symbiosis and evolution. Cell organelles such as mitochondria may have once been free-living organisms.
B Oehler, J. and Schopf, D. W. (1971). *Science* **174**, 1229–1231. Artificial microfossils: experimental studies on permineralization of the blue-green algae in silica.

Author Index

Italic numbers refer to reference lists.

A

Aasmi, K., 47, *49*
Ababei, L., 407, *441*
Ablaev, N. R., 448, *494*
Acher, R., 154, *157*, 486, *494*
Ackerman, E., 382, 399, *429*
Ackrell, B. A. C., 308, 309, 310, *312*
Ada, G. L., 127, *131*
Adams, P. H., 401, *437*
Adebonoso, F. A., 469, *473*
Adebonjo, F. O., 413, *429*
Adolfsen, R., 311, *312*
Agner, K., 134, *158*
Agranoff, B. W., 462, *480*
Agre, N. S., 429, *443*
Ahmad, K., 349, *366*
Ajello, F., 416, *438*
Aisen, P., 1, *7*
Akagi, J. M., 281, *312*, *324*
Akbari-Ford, 394, *432*
Akerman, L., 387, *440*
Åkeson, Å., 12, *16*, 127, *132*, 158, *170*, 170, 172, 177, *182*, 195, *206*
Akiyama, N., 274, *326*
Albaum, H. G., 44, *49*
Albers, R. W., 395, *429*
Alberty, R. A., 352, *376*
Albin, R., 458, *473*
Albracht, S. P. J., 352, 357, 359, *366*
Alden, R. A., 163, *169*, 297, 298, *320*
Aldridge, W. N., 361, *366*, 390, *429*
Aleem, M. I. H., 22, *54*, 224, 231, 257, 273, 274, 310, *312*, 314, 318, *320*, 322, *323*
Aleksandrova, T. A., 401, *429*
Alexander, A. G., 415, *429*
Alfimova, E. Ya., 277, *317*
Allen, P. J., 451, *476*
Allmann, D. W., 337, 347, 361, *366*, 374
Alm, B., 346, 349, 362, *376*
Aloj, E., 405, 417, 421, *434*
Altenbrunn, H. J., 407, *429*
Altschul, A. M., 342, *382*
Altschuler, T., 454, *478*
Alvares, A., 82, *92*
Alvares, A. O., 399, *429*
Alvares, A. P., 78, 79, 83, *89*, *92*

Amaral, D. F., 68, *73*
Ambe, K. S., 340, *366*
Ambike, S. H., 82, *89*
Ambler, R. P., 145, 146, 147, 148, 149, 150, 151, *155*, 267, 268, 269, 278, 279, 281, *312*
Amesz, J., 203, 204, *206*, *207*, 305, *312*
Amodio, F., 398, *443*
Anderson, J. M., 100, 104, 105, 106, 107, 108, *109*, *112*, 204, 205, *206*
Anderson, M. M., 205, *206*
Anderson, N. G., 403, *432*
Anderson, R. C., 404, *439*
Anderson, T. J., 468, *473*
Anderson, W. A., 408, 423, *429*
Ando, K., 152, *155*, 159, 162, *168*, *169*, 185, *191*
Ando, M., 184, 185, *195*, 344, *382*
Ando, O., 129, *132*, 197, *211*
Andrews, E. C., 22, 25, *51*, 330, *373*
Andrews, J. E., 76, *91*
Andrews, W. E., 68, *72*
Anfinsen, C. B., 350, *367*
Angelovic, J. W., 391, 417, *433*
Antonini, E., 40, 48, *49*, 57, 269, *312*, 343, *366*
Apostelescu, I., 398, *445*
Appelman, A. W. U., 386, 387, *437*
Appleby, C. A., 21, 22, *49*, 74, 76, 83, *89*, 112, 113, 114, 115, 116, 117, *118*, *120*, 227, 228, 231, 233, 256, 257, 259, 260, 262, 263, *312*, *315*
Aranjo, P. S., 68, 71, *72*
Aratei, H., 409, *429*
Arcasoy, M., 401, *429*, *430*
Archakov, A. I., 69, *71*, 399, 401, *429*, *440*
Archer, R., 198, *211*
Arcos, J. C., 394, *429*
Argus, M. F., 394, *429*
Arima, K., 22, *49*, 227, 236, 237, 238, 239, 240, 265, *312*, *321*, *326*, 470, *473*
Arion, W. J., 353, *366*
Arion, W. V., 341, *366*
Armstrong, J. McD., 112, 113, 114, 115, 116, 117, *118*, *120*, 198, *206*
Arnheim, N., 491, *494*
Arnon, D. I., 105, 106, *108*, *109*, 110, *111*, 204, *209*

511

Aronoff, S., 22, *56*
Arrhenius, E., 410, 412, *444*
Artman, M., 238, *316*
Arts, C., 386, 387, *437*
Artusi, T., 407, *441*
Arvanitaki, A., 405, 422, *431*
Arvy, L., 409, *429*
Asai, J. A., 347, 366, *372*, *374*, *378*
Asaki, T., 427, *442*
Asali, N., 409, *436*
Asami, K., 351, *366*
Asano, A., 309, 310, *318*
Aschenbrenner, V., 458, *473*
Ascher, R., 457, *481*
Aschheim, E., *182*, 402, *444*
Ashida, T., 129, *132*, 163, *170*, 197, *211*
Ashwell, M. A., 461, *473*, *483*
Ashworth, J. M., 426, *433*
Aston, P. R., 222, 239, 307, 309, *319*
Atanasov, B., 185, *191*
Atkinson, D. E., 282, *320*
Atkinson, M. R., 113, 117, *120*
Atsmon, A., 336, *366*
Aubert, J. P., 231, 273, *321*
Aubert, X., 405, *429*
Auer, H. E., 29, 43, *53*
Augenfeld, J. M., 417, *429*
Augsburg, G., 13, *15*
Augustin, H. W., 407, *429*
Autila, K., 400, *435*
Avers, C., 427, *429*
Avery, J., 362, *366*
Aviado, D. M., 394, *432*
Avi-Dor, Y., 341, *382*
Avigliano, L., 269, *312*
Aviram, I., 160, 161, *168*, *169*, 172, 178, *179*, 185, *191*, *194*
Avron, M., 105, 106, 108, *109*, *111*, 204, *206*, *208*, *209*, 211
Awashti, Y. C., 31, *49*, *50*
Ayers, J., 336, 363, *374*
Azoulay, E., 223, *312*, 469, *473*, *480*
Azzi, A., 64, *65*, 191, *191*, 360, 366, *369*, *384*, 398, 399, 413, *433*, *434*
Azzone, G. F., 360, 361, *367*, *375*

B

Baba, Y., 133, *155*, 341, *367*
Baccarini-Melandri, A., 311, *313*
Bacchin, P. G., 78, *91*
Bach, S. J., 112, *118*
Bachin, P. G., 401, *436*
Bacon, J. B., 413, *435*
Badano, B. N., 337, 352, *367*
Baginsky, M. L., 63, *65*, 354, 359, *373*, *682*

Bahl, O. P., 138, 139, 140, 141, 142, 143, 144, *155*
Bahr, J. T., 20, *55*, 104, *111*
Bailey, E., 462, *473*, *482*
Bailey, J. L., 105, *109*
Bailie, M., 68, *71*
Bailie, N., 408, *429*
Baker, J. E., *109*
Baker, R. M., 82, *89*
Baksky, S., 351, *367*
Balcavage, W. X., 427, *439*, 465, *473*
Baldwin, R. L., 301, *325*
Baldy, S., 413, *444*
Ball, A. G., 68, 70, *73*
Ball, E. G., 38, *49*, 62, *66*, 177, *181*, 196, *211*, 213, *215*, 336, 350, 353, *367*, *370*, 390, 411, *429*, *432*, *434*, *436*, *437*, *445*
Ballestrieri, G. G., 398, *443*
Ballou, D., 48, *49*
Balogh, U. Jr., 394, *429*
Baltscheffsky, H., 187, *191*, 353, *367*, 488, *494*
Baltscheffsky, J., 299, *324*
Baltscheffsky, M., 312, *313*, 333, 353, *367*, *368*
Ban, S., 396, 424, *437*
Bank, P., 403, *429*
Baranov, V. F., 403, *429*
Barany, M., 337, *369*
Barbashova, Z. I., 387, 404, *429*
Barcroft, J., 87, 88, *92*
Barber, R. E., 388, *429*
Barghoorn, E. S., 487, *494*
Barghoorn, F. S., 486, *496*
Barker, E. A., 401, *430*
Barker, S. B., 411, *430*
Barletta, A., 462, *474*
Barlow, G. H., 138, 139, 140, 141, 142, 143, 144, *155*, 170, *179*, 278, 280, *315*, 366, *376*, 485, 491, 492, *494*, *496*
Barlow, R. M., 391, *430*
Barnes, R., 453, 455, *473*
Baron, J., 81, 82, 86, *89*, *90*, 262, *316*
Barrett, J., 4, *6*, 11, 12, 13, 14, *14*, *15*, 25, 26, 27, 35, 37, *52*, *53*, 221, 223, 234, 236, 237, 239, 241, 243, 264, 283, 284, 291, *313*, *314*, 446, 448, 449, 452, 455, 462, 468, 470, 473, *473*, *480*, *481*, 487, *494*
Barrnett, R. J., 336, 353, *367*
Barron, E. S. G., 389, 390, 424, 428, *430*, *435*, 437
Bartholomaus, R. C., 261, *326*
Bartley, W., 189, *192*, 398, *441*, 462, 463, 471, *473*, *480*, *482*
Barton, L. L., 309, *313*

Bartsch, R. G., 5, *6*, 21, *49*, 177, *181*, 246, 249, 257, 266, 284, 285, 286, 287, 289, 290, 291, 292, 293, 296, 297, 299, 300, 301, 302, 303, 304, *313*, *315*, *318*, *319*, *321*, 488, *495*
Basa, D., 263, *322*
Baseman, J. B., 227, 231, *313*
Basford, R. E., 214, *215*, 332, *370*, 402, *440*
Batelli, F., 359, *367*
Battersby, E. J., 393, *440*
Battfay, O., 163, *168*, 362, *370*
Battle, A. M. del C., 454, *473*, *478*
Bauchop, T., 310, *313*
Baudhuin, P., 487, *495*
Baudras, A., 113, 114, 115, 116, *118*, *119*
Baum, H., 62, 63, *67*, 214, *215*, 337, 347, 349, 350, 357, *367*, *379*, *380*
Baumier, A., 413, *431*
Baumler, A., 413, *445*
Baxter, M. I., 417, *430*
Bay, Z., 362, *366*
Bayer, E., 84, *94*, 260, *313*
Bayley, M. P., 27, 30, *57*
Bayley, P. M., 14, *16*, 37, *52*
Bayne, R. A., 43, *49*
Bearden, A. J., 177, *181*, 289, 296, 297, *313*, *321*
Beattie, D. S., 455, 460, 461, *473*
Beaudreau, C. A., 297, *313*
Beaver, P. C., 396, 424, *437*
Becher, R., 469, *476*
Beck, A. B., 391, *430*
Beck, J. C., 465, *473*
Beck, S. D., 421, *435*
Becker, G. E., 224, *322*
Bedrak, E., *430*
Beech, G. A., 471, *479*
Beechey, R. B., 20, *49*, 417, *430*, *431*
Beetlestone, J., 36, *50*
Beevers, H., 104, *110*
Bégin-Heick, N., 389, *430*
Beinert, H., 31, 32, 33, 37, 38, 41, 43, 45, 47, 48, *49*, *54*, *56*, 77, 84, *95*, 174, 175, *182*, 260, 261, 275, 276, *325*, 333, 342, 343, 358, *367*, *372*, *380*
Belanova, A. J., 402, *433*
Beljanski, M., 468, *473*
Bell, J. B., 73, *91*, 410, *435*
Bell, J. J., 410, *440*
Bellet, S., 394, *432*
Belov, M. Yu., 277, *318*
Bendall, D. S., 20, 38, 46, *49*, 62, *66*, 102, 104, 105, 106, *109*, 202, *206*
Bendall, F., *110*
Benditt, E. P., 401, *429*
Ben-Gershom, E., 182, *191*
Ben Hayyim, G., 105, 106, *109*

Bennets, H. W., 391, *430*
Bensch, K. G., 413, *429*
Benson, A., 138, 139, 140, 141, 142, 143, 144, *157*, *158*, 448, *473*
Benson, A. A., 463, *473*
Bensch, K. G., 469, *473*
Beraud, T., 196, *206*
Berden, J. A., 63, *66*, *67*, *68*, 347, 348, 349, 350, *367*, *381*, *383*
Berezney, R., 386, *430*
Berg, A., 355, *378*
Berg, P., 113, *119*
Berg, W. E., 416, *430*
Bergersen, F. J., 256, *312*
Bergstrand, A., 332, 355, *381*
Bergström, L., 265, *325*
Berke, G., 214, *215*
Berkner, L. V., 486, *494*
Bernath, P., 352, 353, 358, *380*
Bernheim, F., 112, *118*
Bernstein, E. H., 213, *215*
Bernstein, I. A., 361, *377*
Bertina, R. M., 63, *67*, 348, 350, *381*
Bertolè, M. L., 108, *110*, 201, 202, *207*
Bessman, S. P., 361, *371*
Betel, I., 20, *49*, 386, 387, *430*, *437*
Betel, T., 196, *206*
Beutler, E., 391, *430*, 450, *473*
Bhagvat, K., 100, *109*
Biale, J. B., 360, *376*, 427, *431*
Bickis, I. J., 412, *430*
Bidleman, K., 79, *89*
Biedermann, M., 309, *322*
Bieri, J. G., 337, *378*
Biggins, J., 105, *109*, 203, *206*, 227, *313*
Bigg, W. R., 386, *440*
Biggs, D. R., 335, *367*, 461, *477*, *478*
Biörck, G., 195, 196, *206*
Bird, M. B., 420, *445*
Birdsell, D. C., 309, *313*
Birjuzova, V. I., 307, *323*, *324*
Birk, Y., 254, 255, *313*, *317*
Birkmayer, G. D., 13, *16*, 28, *55*, 459, 460, *473*, *482*
Birt, L. M., 397, *438*
Biserte, J., 182, *192*
Bishop, N. I., 104, *111*, 203, *210*
Black, C. C., 427, *430*
Black, S., 486, *494*
Blair, P. V., 22, *49*, 332, *371*
Blair, T. T., 488, 489, *496*
Blaisdell, R. K., 450, *473*
Blanquet, P., 187, *193*
Blauer, G., 159, *168*, 288, *313*
Blaylock, B. A., 222, *313*
Blockhuis, G. G. D., 403, *430*

Blondin, G. A., 362, 366, *384*
Bloomfield, B., 26, *52*, 392, *438*
Bloor, W. R., 413, *435*
Blow, D. M., 129, *130*
Blum, J. J., 389, *430*
Blumberg, W. E., 1, 7
Blumberg, W. H., 76, 77, 79, 84, *94*, *95*, 260, 261, *313*, *322*, *325*
Boardman, N. K., 100, 104, 105, 106, 107, 108, *109*, *112*, 128, *130*, 204, 205, *206*
Bock, E., 472, *474*
Bock, K. W., 458, *474*
Bock, R. M., 27, *50*, 337, *370*, *372*
Bodansky, O., 44, *49*
Bodenstein, D., 421, *442*
Bodo, G., 2, *6*, 123, 129, *130*, 133, 135, *158*, 159, *170*
Boeri, E., 112, 113, 114, 115, 116, 117, 118, *118*, *119*, *120*, 158, *168*, 170, 171, 177, *179*, 186, 187, *191*
Bogin, E., 311, *313*, 361, *367*
Bogolepov, N. N., 403, *430*, *441*
Bogorad, L., 448, *480*, 487, *496*
Bogucka, K., 344, 360, *367*
Bois, R., 358, 359, *367*
Bois-Poltoratsky, R., 69, 70, *72*
Bokhonko, A. I., 399, *440*
Boll, M., 309, *313*
Bolobonkin, V. G., 407, *445*
Bomstein, R., 62, *66*, 213, *215*
Bomstein, R. A., 213, *215*
Bond, E. J., 419, *430*
Bone, D. H., 271, *313*
Bongers, L., 257, *313*
Bonner, W. D., 20, 38, 46, *49*, *50*, 100, 101, 102, 103, 104, *109*, *110*, 200, 202, *206*, *207*, *208*, *209*, 329, 362, *369*, *384*, 427, 436
Bonnett, R., 87, 88, *89*
Bonnichsen, R., 187, *191*
Borchett, P., *109*
Borders, C. L., 129, 130, *130*, 162, *168*, 362, *370*
Borei, H., 47, *49*, 397, *435*
Borgo, M., 405, *432*
Borova, J., 454, *479*
Borst, P., 421, *430*, 453, 461, *474*
Bortigon, C., 361, *367*
Borukaeva, M. R., 414, *430*
Bostelmann, W., 406, *430*
Boulter, D., 145, 146, 147, 148, 149, 150, 151, *155*, *157*, *158*, 427, *430*, 492, *494*, *496*
Bouman, H., 358, *381*
Bourke, A., 470, *478*
Bourne, G. H., 403, 404, *438*, *440*
Boveris, A., 337, 352, *367*

Bowes, G. W., 463, *473*
Boxer, G. E., 412, *430*
Boyd, G. S., 79, 80, 81, *89*, *92*, *95*
Boyer, P. D., 334, 363, *367*
Boyne, R., *433*
Bragg, P. I., 238, *313*
Brame, R. G., 409, *430*
Brand, L., 399, 429
Brandt, K. G., 177, *179*
Braunwald, E., 393, 394, 411, *432*, *434*, *443*
Brauser, B., 77, *95*, 400, 402, *430*, *443*
Bravo, M., 196, *208*, 455, *475*
Bray, R. C., 48, *54*, 359, *371*
Breese, K., 183, *191*
Bremer, J., 462, *474*
Brenti P., 448, *477*
Brentani, R., 401, *430*
Briantis, J. M., 106, *109*
Briehl, B. W., 423, *445*
Brierley, G., 361, *367*, 377
Brierley, G. P., 189, *192*, 332, 337, 354, 357, *367*, *371*, *372*
Bright, J., 187, 190, *193*
Brignone, C. M., 351, *382*
Brill, A. S., 34, *49*
Brikker, V. N., 388, *430*
Brill, W. A., 404, *442*
Brinigar, W. S., 346, 363, *367*, *369*
Brinkhoff, B., 410, *437*, 462, *477*
Broberg, P. L., 21, *49*, 227, 230, 232, 254, 267, *313*, *314*
Brocklehurst, J. R., 366, *367*
Brodie, A. F., 22, *54*, 227, 230, 232, 254, 311, *313*, *322*, *323*, 361, *367*
Brody, M., 206, *207*
Brody, S. S., 206, *207*
Broeger, P., 204, *207*
Bronk, J. R., 366, *367*, 410, 411, *430*, *434*
Bronk, M. S., 411, *430*
Bronshtein, A. A., 402, 404, *444*
Brookes, K. B., 462, *482*
Brookes, S. K., 454, *479*
Brosteaux, J., 112, *119*
Brown, B. W., 82, *94*
Brown, C. B., 350, 358, *367*
Brown, F. C., 29, *49*
Brown, P. K., 20, *57*
Brown, P. R., 397, 422, 423, *445*
Brownie, A., 410, *434*
Brownie, A. C., 77, 81, *89*, *90*, 409, *431*
Broyde, S. B., 206, *207*
Brumer, P., 183, *191*
Bruni, A., 63, *66*, 337, 354, 361, *367*
Brunner, G., 332, 355, *367*, 461, *474*
Brunori, M., 40, 48, *49*, *57*, 343, *366*
Bruschi, M., 145, 146, 147, 148, 149, 150, 151, *155*

Bruschi-Heriaud, M., 278, 279, 281, *312*, *314*, *320*
Bryan-Jones, D. G., 467, *474*
Brykhanov, O. A., 388, *431*
Bryła, J., 63, *66*, 348, 349, 350, 354, *367*, *368*, *373*
Bryson, M. J., 74, 77, 79, 80, *95*, *96*
Bryson, M. L., 80, *89*
Brzhevskaya, O. N., 346, *368*
Bücher, T., 329, 331, 336, 352, 359, *368*, *374*, *380*, *383*, 460, *483*
Bücher, Th., 402, *430*
Buchner, A. J., 79, *94*
Bueding, E., 396, 424, *437*
Buikis, I., 403, *431*
Bull, H. B., 183, *191*
Bullock, E., 13, *15*
Bullock, T. H., 391, *431*
Bulmer, G. S., 307, *322*
Bulos, B., 337, *368*
Bunyan, P. J., 81, *89*
Burch, G. E. 394, *429*
Burch, H. B., 422, *443*
Burde, R. M., 307, *323*
Burgos, J., 359, *379*
Burgoyne, L. A., 113, 115, 117, *119*, *120*
Burlington, R. F., 390, *443*, 462, *481*
Burnham, B. F., 452, 453, *474*
Burrin, D. H., 20, *49*, 417, *430*, *431*
Burris, R. H., 200, *211*, 275, 276, 277, *322*, *324*, *325*
Burstein, C., 331, *368*
Burstein, S., 80, *89*, 410, *431*
Burstone, M. S., 392, *431*
Burton, W. G., 20, *53*
Bush, F. E., 471, *479*
Busnyuk, M. M., 403, 404, *431*
Butler, W. L., 25, *50*, *54*, 104, 105, 106, 108, *109*, *111*, 204, *207*, *209*, 472, 473, *475*, *479*
Butlin, K. R., 277, *314*
Butow, R. A., 346, 358, 362, *368*, *373*, *382*
Butt, D., 224, *314*
Butt, W. D., 170, 171, 172, 178, 179, *179*, 182, 183, *191*
Bydalek, T. L., 349, *371*
Byers, V., 366, *376*, 485, 491, 492, *496*
Bygrave, F. L., 332, 355, *367*, 462, *474*, 477
Byington, K. H., 331, *383*
Bykov, E. G., 394, *431*

C

Caiger, P., 26, *52*
Cairns, J. F., 160, *168*, 173, 176, *180*

Calabrese, R. L., 74, 77, *92*,
Caldwell, R. S., 391, *431*
Call, J., 409, *434*
Calvin, M., 486, 488, *494*
Camerino, P. W., 24, 44, *49*, *55*, 340, 353, *368*, *381*
Cameron, B. F., 128, *130*
Cammer, W., 73, 74, 77, 80, *89*, *90*, 409, 410, *431*
Campbell, J. E., 395, *437*
Campbell, L., 281, *317*
Campbell, L. L., 277, 278, 279, 280, *315*, *323*
Campbell, P. N., 190, *192*, 195, 196, *208*, 455, *475*
Campbell, W., 154, *157*, 198, *210*, *211*, 465, *481*
Cantor, C. R., 488, 489, *495*
Capelliere, Ch., 113, 114, 115, 116, *119*
Capèllière-Blandin, C., 113, *119*
Capilna, S., 81, *90*
Caplan, A. I., 341, *368*
Carafoli, E., 341, *377*
Cardot, J., 422, *431*
Cardini, C., 257, 263, *314*
Cardini, G., 81, *89*
Carell, E. F., 450, *474*
Carlson, K., 341, *375*
Carmeli, C., 427, *431*
Caroll, J. W., 304, *322*
Carr, N. G., 203, *209*
Carriker, M. R., 423, *440*
Carruthers, C., 413, *431*, *445*
Cartledge, T. G., 428, *438*
Cartwright, G. E., 452, *476*, *478*
Carver, N. J., 404, *435*
Casai, P. I., 454, *479*
Cascarono, J., 360, *384*
Case, G. D., 305, *314*, *322*
Casey, J., 462, *480*
Casida, J. E., 79, *93*, 359, *373*, 420, *439*
Castellino, N., 196, *207*
Castor, L. N., 11, *14*, 21, 34, *49*, 220, 221, 222, 225, 226, 227, 232, 237, 238, *314*
Castro. C. E., 71, *72*, 362, *368*
Castro, J. A., 401, *442*
Caswell, A. H., 39, *49*, 358, *368*
Catalano, G., 413, *432*
Caughey, W. S., 13, 14, *14*, *16*, 26, 43, *49*, *50*, *57*, 297, *313*
Cederstrand, C., 204, *207*
Chaffee, R. R. J., 389, 390, *431*
Chain, R. K., 106, *108*
Chaix, P., 68, *72*, 118, *119*, 198, *207*, *208*, *210*, 254, 265, 307, 308, *314*, *316*, 413, 428, *439*, *444*, 451, 457, 462, 463, 469, 472, *474*, *478*, *479*, *483*

Chalazonitis, N., 405, 422, *431*
Challoner, D. R., 393, 411, *431*
Chalmers, R., 204, *207*
Chan, S. H. P., 27, 29, 43, *50, 53*
Chan, S. K., 129, *130*, 134, 138, 139, 140, 141, 142, 143, 144, 145, 146, 147, 148, 149, 150, 151, *155*, 197, *207*, 426, *431*, 449, 457, *474*
Chance, B., 5, *6*, 11, *14*, 20, 21, 25, 34, 39, 41, 42, 43, 46, *49, 50, 51, 55, 56*, 62, 63, 64, 65, *65, 66, 67*, 68, *72*, 100, 101, 103, 104, 106, 108, *109, 110*, 115, 116, *119*, 172, *179*, 187, 191, *191*, 197, 203, 204, *206, 207, 211*, 212, *215*, 220, 221, 222, 225, 226, 228, 229, 231, 232, 237, 238, 250, 290, 292, 298, 299, 304, *314, 315, 319, 322, 324*, 329, 331, 332, 333, 334, 335, 336, 338, 339, 340, 341, 343, 344, 345, 346, 348, 349, 350, 352, 354, 355, 257, 358, 359, 360, 361, 362, 363, 364, 365, 366, *367, 368, 369*, 371, *372, 373, 375, 378, 379, 380, 382, 384*. 387, 389, 390, 396, 397, 398, 399, 402, 405, 408, 410, 412, 413, 425, 426, *429, 431, 433, 434, 436, 438*, 440, *441, 444*, 470, *474*
Chaney, T. H., 306, *319*
Chang, J. P., 468, *474*
Chang, S. B., 191, *191*, 205, *207*
Changeux, J. P., 348, 365, *377*
Chaplin, M. D., 79, *95*
Chapman, D., 191, *194*, 336, *369*
Chapman, G. B., 307, *319*,
Chappell, J. B., 361, *369*
Charalampous, F. C., 459, 462, *474*
Charles, A. M., 274, *314*
Charles, R., 361, *369*
Charney, E., 172, 176, *179*
Chase, M. S., 452, *478*
Chavez, E., 359, *371*
Cheah, K. S., 22, *50*, 65, *66*, 222, 225, 226, 227, 229, 232, 233, 255, *314*, 396, 425, *431*
Chen, W. L., 459, 462, *474*
Cheng, S. Y., 341, *369*
Cherkasova, L. S., 404, *431*
Cherniak, N. S., 404, *431*
Chidlow, J. W., 188, *191*
Chiesara, E., 81, *92*
Chin, C. H., *431*
Chin, H. C., 198, *207*
Chino, H., 419, *431*
Chippaux, M., 239, *314*, 468, *474*
Chistyakov, V. V., 336, 349, 354, *369, 380*, 398, *433*, 451, *474*
Chmurzynska, W., 418, *445*
Cho, Y. W., 393, 394, *432*

Chowdhury, D. M., 187, *193*
Christiane, M. I., 409, 411, *432*
Christman, D. R., 187, *194*
Chu, T, L. C., 137, 138, 139, 140, 141, 142, 143, 144, *157*
Chu, Y. W., *431*
Chuang, T. F., 31, *49, 50*
Cilento, G., 411, *441*
Cinti, D. L., 400, *432*
Claisse, L. M., 427, *443*
Claisse, M. L., 467, *474*
Clark, H. W., 213, *215, 216*, 340, 351, *369, 383*
Clark, P. B., 401, *441*
Clarke, E. C., 411, *432*
Clarke, L., 23, 25, 31, 32, 34, 36, 38, 43, 44, *53*, 391, *438*, 452, *478*
Clark-Walker, G. D., 240, 272, *314, 317*, 453, 454, 456, 458, 461, 469, *474, 478*
Claude, A., 463, *474*
Clauser, S. C., 163, *168*
Clavilier, L. P., 198, *207*, 467, *474*
Clayton, M. L., 304, 306, *322, 324*
Clayton, R. K., 305, *314, 322*
Clegg, R. A., 359, *371*
Cleland, K. W., 331, *369*
Clement-Metral, J. D., 454, *475*
Clemmons, J. J., *432*
Cleveland, P. S., 401, *432*
Clezy, P. S., 11, 13, 14, *14, 15*, 25, 26, 27, 35, 37, *52, 53*, 87, *89*, 264, *314*
Cloud, P. E., Jr., 486, *495*
Coates, J. H., 113, 114, *118*, 198, *206*
Coburn, R. F., 86, *89*
Cochran, D. G., 421, *432*
Cockrell, R. S., 366, *369*
Coddington, A., 356, *369*
Coe, E. L., 470, *476*
Cohen, B., 75, 76, 84, *95*
Cohen, B. S., 399, *432*
Cohen, E., 452, *474*
Colby, H. D., 77, *90*, 409, *431*
Cole, J. A., 239, 240, 283, *314, 325*, 469, *482*
Cole, J. S., 310, *314*
Coleman, H. N., 394, *432*
Coles, C., 359, *369*
Coles, H., 389, *431*
Colleran, E., 88, *90, 93*
Colli, W., 68, *72*
Collins, S., 241, *313*
Colmano, G., 103, *109*, 210, 203, *207*
Colpa-Boonstra, J. P., 62, 63, *67*, 351, 363, *381*
Comi, L. I., 395, *440*

Commission on Enzymes of the International Union of Biochemistry, 1, 3, 6, 8, *15*
Commoner, B., 48, *50*, 358, *373*, 488, *495*
Conn, P. A., 393, *434*
Connelly, C. M., 339, 345, *369*
Connelly, J. L., 360, *375*
Conney, A. H., 78, 79, 81, 82, *90*, *92*
Connolly, T. N., 108, *110*
Connors, P., 406, *444*
Conover, T. E., 20, *50*, 196, *207*, 332, 337, *369*, *379*, 386, 387, *432*
Conrad, H., 25, *55*, 340, 342, *381*
Conrad, H. E., 81, *91*
Conrad-Davies, H. C., 184, *191*
Contessa, A. R., 361, *367*
Conti, A., 405, *432*
Conti, F., 413, *432*
Conti, L. I., 390, *432*
Conti, S. F., 227, 231, 300, *319*, *321*
Cook, T. M., 274, *314*, 351, *373*
Cooke, R., 177, *179*
Coolidge, T. B., 177, *179*
Coon, J., 263, *320*
Coon, M. J., 77, 81, 84, 85, *93*, *95*, 263, *322*
Cooper, A., 138, 139, 140, 141, 142, 143, 144, 145, 146, 147, 148, 149, 150, 151, *155*
Cooper, B., 209, *209*
Cooper, C., 332, *373*, 410, *443*
Cooper, D. Y., 68, *72*, 73, 74, 75, 76, 77, 78, 80, 81, 84, 85, *90*, *91*, *93*, *94*, *95*, 410, *443*
Cooper, J. M., 80, *90*, 409, *432*
Cooper, O., 213, *215*, 336, 350, *367*, 411, *429*
Cooper, T., 394, *432*
Cooper, T. A., 346, *369*
Cooperstein, S. J., 22, 28, *50*
Coote, J. L., 461, *483*
Cope, F. W., 362, *369*
Coratelli, P., 355, *375*
Cordes, E. H., *517*, 329, 333, *376*
Corradin, G., 186, *191*
Corwin, L. M., *432*
Costa, L. E., 387, *432*
Costello, L. C., 424, *432*
Cota-Robles, E. H., 307, 309, *313*, *314*
Cotman, C., 403, *432*, 470, *477*
Couchoud-Beaumont, P., 223, *312*, 469, *473*, *480*
Coval, M. L., 268, 280, 289, *313*, *314*
Cove, D. J., 356, *369*
Covell, J. W., 393, 394, *432*, *434*
Cowger, M. L., 340, 377, 405, *439*
Cox, C. D., 227, 231, *313*

Cox, G. B., 238, 309, *314*, *315*, 357, 358, 359, *369*, *381*, 450, *474*
Cox, G. F., 102, *109*
Coy, U., 13, *15*
Cramer, W. A., 104, 105, *109*, *110*, 204, *207*, 306, *315*
Crane, F. L., 22, 23, 25, 31, *49*, *50*, *51*, *55*, 189, 190, *191*, *192*, *194*, *195*, 213, *215*, 332, 333, 340, 353, 357, 358, 359, *366*, *369*, *370*, *372*, *373*, *376*, *378*, *381*, 386, 412, *430*, *439*, *442*
Crane, R. K., 410, *440*
Cranefield, P. F., 393, *432*
Craven, P. A., 332, *370*
Crawford, D. T., 392, *439*
Cremona, T., 351, 352, 358, 361, *367*, *370*, *376*
Creshoff, E., 187, *193*
Cress, H. R., 407, *432*
Criddle, R. S., 27, *50*, 337, 340, *370*, *372*, 463, *474*
Croft, J. H., 461, *482*
Crofts, A. R., 105, *111*, 361, *369*
Cronin, J. R., 4, *6*, 12, *15*, 135, *156*, 159, 160, 162, *168*, 173, 174, *179*, 184, *191*, 301, *317*
Cross, R. J., 363, *370*
Cunningham, W., 22, 25, *51*, 330, *373*
Cunningham, W. P., 353, *370*
Curran, P. F., 196, *209*, 406, *437*
Curri, S. B., 387, *445*
Currie, W. D., 360, *370*, 386, 387, 388, *432*, *440*
Curry, B., 392, *432*, 450, *474*
Cusanovich, M. A., 33, 39, *56*, 177, *181*, 266, 287, 289, 291, 292, 293, 294, 296, 297, 300, 301, 302, 303, 304, *315*, *321*
Cutler, M. E., 4, *6*, 23, 28, 30, 32, 34, 36, 37, 38, 39, 40, 41, 42, 43, *52*, *56*, 171, *180*, 342, 343, *375*
Cutolo, E., 112, 116, 117, *118*, *119*
Cuzner, M. L., 462, *474*
Cyrot, M. O., 396, *434*
Czerlinski, G., 170, *179*, 183, *191*
Czerlinski, G. H., 177, *179*

D

Daae, L. N., 462, *474*
Dadak, V., 254, *320*
Dagg, J. H., 392, *432*, 450, *474*
Dainko, J. L., 428, *443*
d'Alberton, A., 356, *372*
D'Alessandro, L., 390, 413, *432*
Dallman, P. R., 394, *432*, 450, 452, *474*
Dallner, G., 79, 81, *91*, *94*, 412, *434*

Dalton, J., 448, *474*
Daly, J., 79, *90*
Daly, M de B., 388, *432*
Damoky, C. H., 463, *474*
Dancewicz, A. M., 398, *432*, 449, *479*
Daniel, R. M., 226, 227, 232, 250, 257, 259, 260, 262, 263, *312*, 315, 472, *474*
Danielli, J. F., 332, *370*
Danielson, L., 363, *370*
Danielsson, H., 80, *90*
Dannenberg, A., 361, *371*
Darr, K., 170, *179*
Das, M. L., 81, 82, *94*, 189, *191 192*, 340, 356, *370*, *376*
Dashev, G., 394, *439*
da Silva, A. A., 412, *441*
Date, N., 387, *444*
Datta, A., 332, *379*
Davenport, H. E., 5, *6*, 12, *15*, 100, 104, *109*, 203, *207*, *208*
Davidian, N. M., 386, 387, 388, *432*, 456, *474*, *480*
Davidson, C. F., 486, *495*
Davidson, J. B., 462, *474*, *482*
Davidson, J. T., 280, *325*
Davies, D. S., 77, 81, 85, 86, *90*, *95*
Davies, E. J., 360, *370*
Davies, H. C., 342, *370*, 401, *437*
Davies, H. D., 189, *194*
Davies, R. E., 401, 405, *433*, *437*
Davis, A. F., 189, *192*
Davis, A. M., 412, *432*
Davis, B., 413, *431*
Davis, H. F., 71, *72*, 362, *368*
Davis, J. D., 467, *474*
Davis, K. A., 68, *72*
Davis, R. P., 336, *366*
Davis, S. D., 352, *376*
Davison, A. J., 33, 39, 40, *50*, 183, 186, *192*
Davison, A. N., 391, *435*, 462, *474*
Davison, T. F., 391, *432*
Davson, H., 332, *370*
Dawes, E., 450, *476*
Dawson, A. P., 102, *109*, 361, *381*
Dawson, R. M. C., 190, *194*
Day, P., 176, *179*
Dayhoff, M. O., 489, *495*, *496*
Deb, Ch., 392, *442*
De Bernard, B., 458, *475*
De Berardinis, E., 405, *432*
Debrunner, P., 177, *179*
De Duve, C., 487, *495*
De Keken, 474
Deeb, S. S., 247, 248, 253, *315*, *317*
Defaye, J., 473, *482*
Degknitz, E., 81, *90*, 401, *432*
Degn, H., 186, *193*

De Groot, L. J., 412, *432*, *433*
De Jairala, S. W., 402, *433*
De Keken, R. H., 471, *474*
De Klerk, A., 291, 292, 293, *315*
De Klerk, H., 200, 203, *212*, 235, 242, 285, 292, 296, 297, 299, 300, *315*, *316*, *318*, *319*, *324*
De Kouchkovsky, Y., *109*
De Lange, R. J., 145, 146, 147, 148, 149, 150, 151, 153, *155*, 197, 200, *207*
De Leo, T., 462, *474*
Del Favero, A., 82, *96*
De Matteis, F., 82, *90*
De Moss, J. A., 250, *323*
Denis, M., 20, *50*
Denk, H., 78, *91*
De Otamendi, M. E., 409, *433*
Depocas, F., 390, *433*, 458, *478*
Derbyshire, E., 427, *430*
Derhachev, E. V., 394, *433*
Derkachev, E. F., 398, 428, *437*, *444*
De Robichon-Szulmajster, H., 469, 466, *474*, *482*
Der Vartanian, D. V., 133, *155*
Detwiler, T. C., 22, *50*
Detwiler, T. D., 187, *192*
Detwiller, J. P., 448, *475*
Deul, D. H., 63, *66*, 351, *370*
Devault, D., 63, *66*, 106, *109*, 204, *207*, 305, 306, *323*, *338*, 343, 348, 361, 364, 365, *369*, *370*
Devichensky, V. U., 69, *71*
Devlin, J. D., 412, *430*
De Vries, H., 459, 461, *475*, *477*
Dewey, R. S., 349, *382*
Diano, M., 20, *50*
Di Augustine, R. P., 77, *90*
Dickens, F., 112, *119*
Dickerson, E. E., 362, *370*
Dickerson, R. E., 4, *6*, 12, *15*, 129, 130, *130*, 138, 139, 140, 141, 142, 143, 144, 145, 146, 147, 148, 149, 150, 151, 153, *155*, *156*, 162, 163, 164, 165, 166, 167, *168*, *169*, 269, 279, *315*, *320*, 362, *370*, 484, 488, 489, 490, 491, 493, *495*, *496*
Dickie, J. P., 349, *382*
Dickmann, S. R., 187, *195*
Diehl, H., 69, *73*, 81, *90*
Dietrich, W. E., 227, *313*
di Franco, A., 113, 115, 116, *119*
Dikstein, S., 114, 117, *119*
Di Meo, S., 462, *474*
Dimsdale, M. J., 87, *89*
Dinescu, G., 187, *192*
Dirks, M. P., 203, *206*
Dixon, H. B. F., 128, *130*
Dixon, M., 2, *6*, 112, *118*, *119*, 389, *433*

Author Index

Do Ameral, D. F., 355, *376*, *379*
Dobreanu-Enescu, V., 398, *445*
Doebbler, G. F., 171, *180*, 339, *370*
Doeg, K., 352, *384*
Doeg, K. A., 62, *66*, *68*, 358, *370*
Doering, G., 105, *109*
Dolin, M. I., 467, *475*
Dolivo, M., 402, *430*
Doll, E., 393, *433*, *437*
Domjan, G., 401, *439*
Domnas, A., 254, *315*
Donowa, A., 310, *318*
Dontsova, G. V., 416, *433*
D'Ornano, L., 469, *480*
Döring, G., 106, *112*
Doss, M., 450, *475*
Doty, P., 173, 174, *180*, *182*
Dougall, D. K., 13, *15*
Dougherty, R. G., 448, *474*
Douglas, H. C., 198, *209*, 331, *380*, 465, *478*
Dovedova, E. L., 403, *430*, *441*
Downey, R. J., 470, *475*
Downton, W. J. S., 106, *112*
Drabkin, D. L., 87, 88, *96*, 134, *155*, 195, 196, *207*, *210*, 385, 397, 413, *433*, 446, 455, *475*, *479*, *483*
Dragoni, N., 277, *320*
Drahota, Z., 331, *370*
Drell, E. C., 311, *322*
Dreosti, I. E., 391, *433*
Drews, G., 307, 309, *315*, *316*, *322*
Drott, H. R., *168*, 177, *179*
Drozd, J., 469, *475*
Drucker, H., 277, 278, 279, 280, *315*
Druyan, R., 450, 458, *473*, *475*, *478*
Drynov, I. D., 398, *433*
Duce, E. D., 361, *383*
Ducet, G., 20, *50*
Duckworth, M. W., 196, *207*
Duffield, J., 173, 176, *181*
Dugge, E. J., 394, *432*
Dumitrescu, A., 468, *481*
Dumont, J. E., 395, *439*
Duncan, H. M., 21, 24, 27, *50*, *54*
Dunn, A. D., 412, *432*, *433*
Durbin, R. D., 405, *436*
Dus, K., 145, 146, 147, 148, 149, 150, 151, *155*, 197, *211*, 260, 281, 291, 292, 293, 296, 297, 298, 299, *314*, *315*, *316*, *320*, *324*, *325*, 488, 489, *495*
Dutcher, J. S., 77, *95*
Dutkiewicz, J. S., 196, *207*
Dutton, C., 63, *66*
Dutton, L., 261, *322*

Dutton, P. L., 39, 50, *56*, 63, 64, *66*, *68*, 304, 305, 306, *315*, *319*, *323*, 339, 345, 348, 364, 365, *369*, *370*, *384*
Duysens, L. N. M., 203, 204, *206*, *207*, *211*, 304, *315*, *325*
Dworkin, M., 227, *315*
Dyer, P. Y., 115, *119*

E

Eady, R. R., 263, *319*
Eaton, W. A., 165, *168*, 172, 176, 177, *179*, *180*, 187, *192*
Ebisuno, T., 448, *481*
Eck, P. V., 489, *495*
Edelman, N. H., 404, *431*
Edmundson, A. B., 485, *495*
Edward, D. L., 24, *50*
Edwards, C. L., 460, *482*
Edwards, D. L., 340, *370*, 428, *433*, 459, 467, *475*
Edwards, M., 304, *321*
Edwards, S. W., 336, *370*
Egami, F., 5, *6*, *7*, 264, *315*
Egidio, G., 390, *432*
Ehrenberg, A., 32, 36, *50*, 69, 70, *72*, 133, 134, *155*, 158, 159, 163, *168*, 170, 171, 177, 178, *179*, *180*, *181*, 288, 289, 290, 302, *315*
Ehtisham-Ud-Din, A. F. M., 307, *323*, 448, 464, *475*, *480*
Eichel, H. J., 227, *322*, 353, *370*, 390, *433*
Eichhorn, G. L., 160, *168*, 173, 176, *180*
Eilermann, L. J. M., 308, 310, *315*, *316*
Einarsson, K., 80, *90*
Eisenberg, D., 4, *6*, 129, *130*, 138, 139, 140, 141, 142, 143, 144, 145, 146, 147, 148, 149, 150, 151, *155*, 162, 163, 166, 167, *168*
Eisenberg, J., 12, *15*
Eisenberg, M. A., 393, *433*
Eisenberg, R. E., 362, *370*
Eisenstein, K. K., 206, *207*
Elkin, J., 413, *441*
Ellar, D. J., 307, 308, *316*, *321*, *323*
Ellery, R. W., 113, *120*
Elliott, K. A. C., 402, *433*
Elliott, W. B., 30, 41, *50*, *51*, *53*, 63, 67, 81, *89*, 171, *180*, 332, 339, *370*, *376*, 386, 387, 388, 409, *431*, *432*, 456, *474*
Ellsworth, R. K., 448, *475*
Elowe, D., 352, *376*
Elsden, S. R., 284, 296, 310, *313*, *316*
Elstner, E., 204, *207*
Elvehjem, C. A., 452, *474*, *475*
Ely, S., 249, *317*

Emelia, G., 407, *441*
Emerson, R., 204, *207*
Endo, A., 129, *132*, 187, *193*, 199, *211*, 308, *323*
Engel, D. W., 391, 417, *433*
Engel, H., 472, *476*
Engel, L. L., 80 *94*
Engel, P. S., 48, *53*
Engelberg, H., 238, *316*
Engelmann, B., 406, *430*
Enomoto, H., 187, *193*
Epel, B. L., 25, *50*, *54*, 105, *109*, *207*, 472, 473, *475*, *479*
Epel, D., 38, *56*, 409, 416, *445*
Ephrussi, B., 118, *119*, 455, 459, 466, 470, *475*
Epstein, C. J., 489, *495*
Erbes, D. L., 260, 299, *315*
Erdelt, H., 361, *383*
Erecinska, M., 63, 64, *66*, 104, *109*, 332, 333, 352, 364, 365, *369*, *370*
Erickson, R. J., 352, *376*
Erickson, S. K., 21, *50*, 309, *316*, 426, *433*, 472, *474*
Ericsson, J. L. E., 82, *94*
Ernst, B., 406, *430*
Ernster, L., 63, 64, *66*, *67*, 74, 77, 81, 82, 85, *90*, *94*, 329, 331, 332, 334, 336, 341, 346, 350, 352, 354, 355, 356, 358, 359, 360, 363, 366, *369*, 370, *371*, *373*, *375*, *377*, *378*, *379*, *381*, 410, 411, 412, *433*, *434*, *444*
Estabrook, R. W., 7, 64, 65, *66*, *67*, 68, 69, 72, 73, 74, 75, 76, 77, 78, 80, 81, 82, 84, 85, 86, *89*, *90*, *91*, *92*, *93*, *94*, *95*, 101, 103, *109*, 115, 118, *119*, 171, 172, 173, 176, *180*, *182*, 197, 198, *207*, *210*, 213, *215*, 262, 296, *316*, 332, 335, 336, 337, 338, 339, 340, 349, 351, 353, 354, 358, 359, *367*, *370*, *371*, *375*, *377*, *379*, *380*, *382*, 397, 399, 403, 409, 410, *431*, *432*, *433*, *435*, *442*, *443*
Estler, C. J., 404, *433*
Euler, H. v., 221, *316*
Evans, A. E., 407, *433*
Evans, H. J., 450, *479*
Evans, M. C. W., 358, *372*
Evans, M. G., 362, *371*
Evans, W. H., 407, *433*
Everling, F. B., 68, *72*
Evtodienko, Yu. V., 335, 349, *371*, *380*, 398, *433*
Eyring, H., 162, *169*

F

Fain, J. N., 390, *441*
Falk, J. E., 160, *168*

Fan, H. N., 105, *110*
Fanger, J. R., 135, *156*
Fanger, M. W., 4, *6*, 12, *15*, 159, 160, 162, *168*, *268*, *316*, *317*
Fanica-Gaignier, M., 454, *475*
Farley, B., 393, *435*
Farley, T. M., 349, *371*
Farnsworth, J., 79, *90*
Farnsworth, M. W., 20, *51*, 197, *208*, 420, *433*, *434*
Faure, M., 190, *192*
Fay, F. J., 388, *433*
Fechner, W., 358, *370*
Fedorov, I. I., 402, 407, *433*
Fedorov, N. J., 429, *433*
Feigelson, P., 40, *53*
Feiglova, E., 390, *436*
Feldman, D., 62, *66*
Feldott, G., 410, *438*
Fell, B. F., *433*
Fellman, J. H., 250, *316*
Felton, J. H., 23, *54*
Felton, S. P., 331, *380*
Ferguson, S., 350, *380*, 390, *443*, 462, *481*
Ferguson, S. M., 334, *375*
Fernandez, E. C., 400, *444*
Fernandez, E. P., 80, *95*
Fernandez-Moran, H., 332, *371*
Ferrandes, B., 307, 308, *316*
Ferrari, G., 407, *441*
Ferro, M., 188, *193*
Fessenden, J. M., 361, *371*
Fiddick, R., 405, *435*
Fielden, M., 48, *54*
Filimonov, M. M., 404, *431*
Finazzi-Agro, A., 269, *312*
Fine, A., 416, *440*
Fingerit, D. S., 488, *494*
Fink, H., 221, *316*, 450, *475*
Finkelstein, A., 517
Fischer, A. G., 486, *495*
Fischer, H., 12, *15*, 450, *475*
Fisette, R. J., 391, *439*
Fitch, W. M., 138, 139, 140, 141, 142, 143, 144, 153, 154, *155*, *156*, 162, 167, *169*, 279, *320*, 485, 486, 488, 489, 490, 491, 493, *495*, *496*
Fitz-James, P. C., 307, *316*
Flatmark, T., 127, 128, 129, *130*, 154, *156*, 170, 171, 173, 174, 175, 176, *180*, 184, 187, *192*, 288, 296, 297, 299, *316*, 456, 457, *475*
Fleer, U., 393, *437*
Fleischer, B., 398, 399, *433*
Fleischer, S., 189, 190, 191, *191*, *192*, 337, *367*, *371*, *376*, *379*, *384*, 398, 399, *433*
Fletcher, K. A., 425, *444*

Fletcher, M. J., 457, *475*
Florkin, M., 484, 486, *495*
Floyd, R. A., 204, *207*
Flynn, E., 77, *90*, 399, *433*
Fogel, S., 188, *194*
Foglia, V., 404, *439*
Folkers, K., 359, *372*, *373*
Fork, D. C., 205, *207*
Forrest, W. W., 311, *316*
Forte, J. G., 405, *433*
Forti, G., 108, *110*, 201, 202, *207*, 356, *371*
Foulkes, E. C., 46, *50*, *52*
Fouts, J. R., 77, *90*, 400, *435*
Fowler, C. E., 304, *324*
Fowler, C. F., 305, *316*
Fowler, G. L., 466, *480*
Fowler, L. R., 25, 30, *50*, 341, 353, 357, *371*, *372*, *379*
Fox, H. M., 415, 417, *433*
Frank, R. M., 399, *439*
Frankel, M., 135, *158*
Franklin, M. R., 75, *90*
Frans, F., 388, *439*
Frantz, I. D. III., 403, *438*
Freed, J., 391, *433*
Freeman, K., 456, *476*
Freeman, K. B., 410, 412, *442*
Freedman, R. B., 366, *367*
Freer, J. H., 307, *321*, *323*
Freer, S. T., 163, *169*
Frenkel, A. W., 309, *317*
Frerman, F. E., 463, 464, *475*, *483*
Frey, I., 77, *94*
Fridman, C., 129, *130*, 153, *156*
Fridovich, I., 48, *50*, 183, 185, 187, *192*, *193*
Friede, R. L., 404, *433*
Friend, W. H., 427, *435*
Frierson, J. L., 396, *435*
Froede, H. C., 187, *192*, 363, *371*
Fröhlich, H., 363, *371*
Frohwirt, N., 159, *169*
Frolkis, R. A., 393, *434*
Frommer, U., 81, *90*
Frowirth, N., 133, *156*, 160, 162, *169*, 171, *181*
Fugmann, U., 362, *368*
Fujishima, H., 74, 83, *91*
Fujita, A., 4, *6*, 233, *316*
Fujita, T., 247, 251, 253, 282, 283, 284, *316*
Fujita, Y., 203, 205, *208*
Fujimoto, M., 453, *476*
Fukuhara, H., 198, *210*, 457, 471, *475*, *482*
Fukui, S., 428, *443*
Fullard, J., 425, *444*
Funahashi, S., 337, *380*
Funatsu, G., 103, *111*

Fung, D., 171, *180*
Funk, L. K., 386, *430*
Furonaka, H., 394, *439*
Furuno, K., 184, *194*
Fürth, O., 112, *119*
Fynn, G. H., 337, 350, 357, *371*, *379*

G

Gajdos, A., 454, *475*
Gajdos-Torok, M., 454, *475*
Galeotti, T., 413, *434*, *444*
Gallagher, C. H., 391, 434, 452, *475*
Gallant, S., 409, 410, *431*, *434*
Gallenkamp, H., 355, *380*
Galliazzo, G., 161, *168*
Gallo, G., 468, *483*
Galzigna, L., *371*
Gamble, W. J., 393, *434*
Ganapathy, K., 31, 48, *49*, *53*
Ganelina, I. E., 388, *430*
Ganez, M., 471, *480*
Gangal, S. W., 361, *371*
Ganguli, B. N., 74, 77, 81, *92*, 257, 260, 261, *319*, *326*
Garcia, A. P., 402, *433*
Gardas, A., 22, *51*
Gardisky, R. L., 154, *157*, 198, *211*, 465, *481*
Gardner, G. M., 304, *322*
Garette, R. H., 102, *110*
Garfinkel, D., 69, *72*, 73, *90*, 339, *371*
Garfinkel, L., 339, *371*
Garland, P. B., 329, 346, 352, 353, 354, 358, 359, *369*, *371*, *379*
Garner, R. C., 81, 82, *90*, 458, *475*
Garnier, J., 106, *110*
Garrard, W. T., Jr., 272, *317*
Garrett, R. H., 22, *50*
Garver, J. C., 200, *211*
Gaudemer, Y., 396, *434*
Gauthcron, D., 396, *434*
Gautheron, D. C., 363, *371*
Gawron, A., 357, *371*
Gay, A., 336, *378*
Gaylor, J. L., 74, 75, 76, 77, 78, 81, 82, 83, 84, *90*, *92*, *93*
Gear, A. R. L., 455, *475*
Gee, R., 405, *433*
Gee, R. W., 463, *473*
Gelboin, H. V., 78, 82, *93*
Geller, D. M., 292, 298, *316*
Gel'man, N. S., 307, 308, 309, 310, *316*, *323*, *324*, 488, *495*
Gemba, M., 402, *434*
Gendel, L. Y., 451, *474*
Gendre, J., 469, *480*

Gennari, G., 161, *168*
George, P., 36, *50*, 159, 160, 162, 163, *168*, *169*, 171, 172, 173, 176, 177, 178, *180*, *181*, 182, 185, 187, *192*, *193*, *194*, 199, 209, 487, *495*
Georgiade, R., 393, *440*
Gergely, J., 362, *371*
Gerischer-Mothes, W., 407, *441*
Gerischer, W., 233, *321*
Gerrits, J. P., 310, *317*
Gershbein, L. L., 351, *367*
Gest, H., 311, *313*
Getz, G. S., 189, *192*, 398, 407, *433*, *436*, 450, 462, *478*, *480*
Gewitz, H. S., 13, *16*
Geyer, D., 187, *192*
Ghiretti, F., 20, *50*, 65, *66*, 97, 98, 99, *99*, *100*, 197, *208*, 421, 422, 423, *434*, *444*
Ghiretti-Magaldi, A., 20, *50*, 65, *66*, 97, 99, *99*, 117, *119*, 197, *208*, 416, 421, 422, 423, *434*, *438*, *441*
Gibson, F., 238, 309, *314*, *315*, 357, 358, *369*, 450, *474*
Gibson, J., 302, 303, 304, *316*, *321*
Gibson, K. D., 309, *316*
Gibson, Q. H., 23, 24, 25, 32, 33, 34, 36, 39, 41, 42, 43, 44, 51, *56*, 85, *92*, 188, *193*, 289, 290, *315*, *316*, 342, 343, 355, 361, 365, *371*, *372*, *376*
Giesebrecht, P., 307, *315*, *316*
Gigon, P. L., 77, 78, 85, 86, *90*, 399, 400, *434*, *435*
Gillette, J. R., 77, 78, 81, 85, 86, *90*, 94, 356, *373*, *380*, 399, 400, 401, *434*, *435*, *442*
Gillies, N. E., 127, *131*, 133, 137, *156*
Gilman, A. G., 81, 82, *90*
Gilmour, M. V., 11, *15*, 23, 28, 33, 34, 35, 36, 38, 39, 40, 41, 46, 47, *51*, *52*, *56*, 344, 345, 362, *371*, *384*
Gingold, E. B., 465, *481*
Giuditta, A., 105, 417, 420, *434*
Givan, A. L., 105, 106, *110*
Glaessner, M. F., 486, *495*
Glaid, A. J., III, 357, *371*
Glauert, A. M., 307, *317*
Glaumann, H., 79, *91*
Glauser, S. C., 171, 172, 178, *180*, *181*, 182, *194*
Glaze, R. P., 213, *215*
Glazer, A. N., 145, 146, 147, 148, 149, 150, 151, 153, *155*, *157*, 197, 200, *207*, *211*
Gleason, F. H., 236, 251, *323*, 428, *434*
Glenn, J. L., 22, *50*, 213, 214, *215*, 357, *370*
Glick, J. L., 410, *434*
Gnosspelius, Y., 81, 82, *94*
Goddard, D. R., 126, *130*

Goedheer, J. C., 204, *208*
Gogleva, T. V., 277, *318*
Gola, N., 422, *431*
Goldberg, A., 392, *432*, 450, *474*
Goldberg, E., 424, *434*
Goldberg, I., 463, *475*
Goldberger, R., 59, 62, *66*, 213, 214, *215*
Goldblatt, P. J., 332, *370*
Goldin, H. H., 20, *51*, 197, *208*, 420, *434*
Goldman, D. S., 254, *319*
Goldstein, T. P., 392, *439*
Goldstone, A., 138, 139, 140, 141, 142, 143, 144, *156*
Goldwasser, E., 454, *476*
Golubev, A. M., 393, *434*
Golubov, I. S., 398, *436*
Golubkov, V. I., 428, *437*
Gomez-Pyou, A., 359, *371*
González-Cadavid, N. F., 190, *192*, 195, 196, *208*, 455, *475*
Gonze, J., 336, 358, *371*, *380*, *382*, 411, *434*
Gonze, P. H., 408, 409, *434*
Good, N. E., 108, *110*
Gorchein, A., 190, *195*, 309, *317*, 453, 454, *475*
Gordon, A., 332, *383*
Gordon, E. E., 412, *434*
Gordon, M. S., 395, *434*
Gore, M. B. R., 402, *438*
Gorman, D. S., 106, 108, *111*, 203, 204, *208*, *209*
Goto, K., 453, *476*
Goto, M., 81, *96*
Gotterer, G. S., 406, *441*
Goucher, C. A., 472, *475*
Goudy, B., 450, *476*
Gould, K. S., 489, *495*
Gouthier, D. K., 272, *317*
Govindjee, R., 305, *317*
Graae, J., 356, *371*
Graham, T. P., 393, *434*
Gram, T. E., 77, 78, *90*, 399, 400, 409, *434*, *435*
Grandchamp, S., 116, 117, 118, *120*, 356, *376*, 459, *478*
Granick, S., 12, *16*, 154, *157*, 453, 454, 467, *476*, 477, *481*, *483*, 486, *495*
Grant, N. G., 254, *315*
Grasse, F. R., 81, *91*
Grassl, M., 13, *15*, *16*, 448, *481*
Graubard, M., 452, *476*
Gravante, G., 195, *210*
Gray, C. H., 82, 88, 89, *91*, *93*, *94*, *96*
Gray, C. T., 249, 282, 283, *317*, *325*, 469, 472, *476*, *483*
Gray, P. P., 471, 472, *476*
Grebner, D., 30, 33, 40, *56*

Greceanu, V., 468, *481*
Green, D. E., 22, *50*, *53*, 59, *66*, 112, *119*, 189, 190, *192*, 213, 214, *215*, 329, 332, 337, 338, 340, 347, 352, 353, 354, 357, 361, 362, 364, 365, 366, *366*, *370*, *371*, *372*, *374*, *376*, *378*, *379*, *383*, *384*, 463, *476*
Green, D. H., 307, *317*
Green, S. B., 339, *371*
Greenawalt, J. W., 341, *368*, 427, *434*
Greene, A. A., 413, *434*
Greene, F. E., 400, 401, *434*
Greengard, P., 81, *91*
Greenlees, J., 43, 44, *56*
Greenspan, K., 393, *432*
Greenwood, C., 23, 24, 25, 27, 29, 32, 33, 34, 41, 42, 43, 48, *49*, *51*, *57*, 171, 172, 173, *180*, 342, 343, 361, *371*, *372*, *445*
Greenwood, M., 343, *366*
Gregg, C. T., 360, *370*
Gregolin, C., 112, 117, 118, *119*, *120*, 356, *372*
Gregson, N. A., 462, *474*
Grehel, C., 308, *316*
Greim, H., 7, 70, *72*, 74, 78, 79, 81, 82, *91*, *94*, *95*, 458, *476*
Griffin, B., 258, 261, *317*
Griffin, D. H., 428, *435*
Griffith, J. S., *372*
Griffiths, D. E., 23, 24, 27, 30, 31, 32, *49*, *51*, 353, 354, 357, 359, *369*, *372*
Griffiths, J. S., 36, *50*
Griffiths, S. K., 104, *110*
Grigoreva, G. I., 404, *429*
Grimm, P. W., 451, *476*
Groot, G. S., 361, *381*
Groot, G. S. P., 360, *372*, 463, *476*
Gross, J. A., 201, 203, *208*, *211*
Gross, N. J., 461, *476*
Grondinski, O., 113, 115, 116, *119*, *120*
Guarino, A. M., 400, 409, *434*
Gubler, C. J., 452, *476*, *478*
Guerin, B., 473, *476*, *482*
Guerrieri, P., 269, *312*
Guiditta, A., 20, *50*, 65, *66*
Guillory, R. J., 390, *434*
Gulidova, G. P., 403, 404, *434*, *441*
Gulik-Krzywicki, T., 190, *192*
Gunge, N., 221, *324*, 466, *476*
Gunsalus, I. C., 74, 75, 77, 81, 83, 84, *92*, *93*, *95*, *96*, 257, 258, 259, 260, 261, 262, 299, *315*, *317*, *319*, *325*, *326*
Gunsalus, J. C., 261, *321*
Gunsalus, T. C., 81, *91*
Gupta, R. K., 178, *180*
Guroff, G., 79, *90*

Gürtler, L., 128, *130*, 137, 138, 139, 140, 141, 142, 143, 144, *156*
Gutfreund, A., 408, *436*
Gutman, M., 359, 360, *372*, *374*
Gvozdev, R. I., 277, *317*
Gwodzz, B., 196, *207*

H

Haak, D., 189, *191*
Haas, D. W., 104, *110*
Haas, E., 220, *320*, *325*, 355, *372*
Haavik, A. G., 350, 353, 354, 357, *372*
Hackenbrock, C. R., 366, *372*
Hackenthal, E., 469, *476*
Hackenthal, R., 469, *476*
Hackett, D. P., 20, 21, *50*, *51*, *53*, 65, *67*, 101, 102, 103, 104, *109*, *110*, *111*, *112*, *211*, 227, 229, 254, *322*, *325*
Haddock, B. A., 359, *371*
Haddock, J., 398, *436*
Hadjipetrou, L. P., 310, *317*
Hagen, J. H., 390, *434*
Hager, L. H., 252, 253, *325*
Hager, L. P., 65, *66*, 160, *170*, 247, 248, 251, 252, 253, *315*, *317*, *318*, *325*
Hagihara, A., 4, *6*
Hagihara, B., 4, *6*, 39, *54*, 65, *67*, 68, 69, 70, 71, *72*, *73*, 76, *91*, 118, *119*, 127, 128, 129, *130*, *131*, 221, 222, *318*, 331, *378*, 413, 414, 427, *440*, *442*
Hahn, C., 407, *429*
Haimovici, H., 406, *438*
Hakami, N., 469, *476*
Haldar, D., 456, *476*
Hale, J. H., 21, *54*
Hall, C., 359, *372*, *373*
Hall, D. O., 358, *372*, 427, *434*
Hall, P. F., 77, 79, 80, *96*
Hamaker, L. E., 453, *481*
Hamaguchi, K., 145, 146, 147, 148, 149, 150, 151, *157*, 174, 175, *180*, 184, *193*, 288, *318*
Hamberger, A., 404, *443*
Hamburger, H., 393, *437*
Hamilton, J. A., 238, *314*, 357, 358, *369*
Hamilton, W. K., 388, *429*
Hammaker, L., 82, *95*
Hammond, R. K., 263, *317*
Hanahan, D. V., 337, *373*
Hanania, G. I. H., *180*
Hancock, D. J., 366, *367*
Handler, P., 48, *50*
Haneishi, R., 129, *132*
Haneishi, T., 199, *208*, *211*

Hanker, J. S., 392, *442*
Hanninen, O., 400, *435*
Hansen, M., 365, *383*
Hansen, R. E., 37, *49*, *56*, 357, 358, *379*
Hara, T., 71, *72*, 355, *372*, 458, *476*
Haraguchi, K. M., 405, *445*
Harary, I., 360, 363, *372*, 393, *435*, *442*
Harbury, H. A., 4, *6*, *12*, *15*, 133, 134, 135, *156*, 159, 160, 162, *168*, *170*, 171, 173, 174, 175, *179*, *180*, *181*, *182*, 184, 186, *191*, 268, 269, *316*, *317*, *325*
Hardesty, B. A., 191, *192*
Harding, B. W., 73, 75, 76, 77, 78, 79, 80, 81, 84, *91*, *93*, *96*, 408, 410, *435*, *440*, *445*
Harley, J. L., 427, *435*
Harley, R., 391, *430*
Harmon, E. A., 108, *110*, 204, 205, *209*
Harms, H., 472, *476*
Harper, L., 239, *319*
Harpley, C. H., 236, 246, 282, *319*
Harrap, B. S., 14, *16*
Harrer, C. J., 355, *372*
Harris, E. J., 366, *369*
Harris, J., 33, *51*
Harris, R. A., 366, *372*, *378*
Harris, R. H., 361, *366*
Harrison, D. E. F., 311, *317*
Hart, C. J., 189, *192*
Hartree, E. F., 3, *6*, 20, 34, 46, 47, *51*, 68, *72*, 100, *110*, 126, 127, *131*, 170, 171, 178, *180*, 199, *208*, 212, *215*, 265, *319*, 340, 349, *374*, 452, *477*
Hartwell, L. H., 461, *476*
Hartzell, C. R., 31, 32, 41, 48, *49*
Harvey, G. T., 421, *435*
Harvey, W. R., 418, 419, *435*, *442*
Hasegawa, H., 116, 117, *119*, *120*
Hashimoto, Y., 76, 81, 83, *91*, *93*
Haske, R., 281, *317*
Haskins, F. A., 467, *476*, *482*
Haslam, J. M., 463, *483*
Hassan, M., 341, 344, *375*
Hassinen, I., 352, *372*
Hatch, T. P., 453, *478*
Hatefi, Y., 25, 30, *50*, 63, *65*, 350, 353, 354, 357, 358, 359, *367*, *370*, *372*, *382*
Hatt, P-Y., 393, *435*
Haugaard, N., 388, 389, *435*
Hauge, J. G., 248, 253, *317*
Hauska, G. A., 107, 108, *110*
Haustein, W. G., 359, *372*
Haven, E. L., 413, *435*
Haven, F. L., *434*
Havez, K., 182, *192*
Hawthorne, D. C., 198, *209*, 465, *478*
Hayaishi, H., 23, *54*

Hayaishi, O., 2, *6*, 40, *51*, 74, *91*
Hayano, M., 74, 80, *91*
Hayashi, N., 453, *476*, *477*
Hayem-Levy, A., 182, *192*
Hearn, G. H., 396, *435*
Heath, H., 405, *435*
Heatherington, D. C., 407, *432*
Heber, U. W., 205, *207*
Hechter, O., 332, 353, *371*
Hedegaard, J., 260, *317*
Hedman, R., 410, 412, *444*
Hegsted, D. M., 406, *444*
Heidema, J., 84, *95*
Heim, A. H., 254, 255, *313*, *317*
Heim, L. M., 404, *435*
Heinen, W., 358, *367*
Heiney, R. E., 336, 363, *374*
Heinz, E., 406, *435*
Held, H., 82, *91*
Heldt, H. W., 361, *372*
Heller, J., 145, 146, 147, 148, 149, 150, 151, *156*, 488, *495*
Hellström, H., 221, *316*
Helper, E. W., 404, *435*
Hemmerich, P., 31, *51*
Hempfling, W. P., 231, 311, *317*, *318*
Henderson, R. W., 127, *131*, 171, 177, *180*, 293, 300, *317*
Hendler, R. W., 331, *373*
Hendlin, D., 351, *373*
Hendricks, R., 301, *317*
Hennig, A., 137, *156*
Henson, C. P., 460, *476*
Heppel, L. A., 307, *317*
Herbert, D., 177, *180*
Hernandez, P. H., 356, *373*
Herold, R. C., 397, *435*
Hersey, S. J., 406, *435*
Hess, B., 63, 65, *67*, 338, 339, 349, *368*, *375*, 410, 412, *431*, *439*, 470, *474*
Hess, G. G., 403, 404, *435*
Hess, G. P., 170, 177, *179*
Hess, H. W. I., 392, 403, 404, *435*
Hettinger, K., 4, *6*
Hettinger, M. W., 135, *156*
Hettinger, T. P., 12, *15*, 159, 160, 162, *168*, 268, *316*, *317*
Hewick, D. S., 400, *435*
Heyman-Blanchet, T., 198, *208*, 463, *483*
Hibbeln, P., 451, *482*
Hickman, D. D., 309, *317*
Higa, H., 138, 139, 140, 141, 142, 143, 144, *157*, *158*
Higashi, D. S., 408, 409, *435*
Higashi, T., 12, *15*, 76, *96*, *121*, 128, *131*, 235, 241, 242, 244, 268, 269, 311, *313*, *318*, 361, *367*

Higgins, E. S., 427, *435*
Higgins, J., 339, 345, *369*
High, D. F., 163, *169*
Higuchi, M., 453, *476*
Hildebrand, D. C., 236, 251, *323*
Hildebrandt, A. G., 75, 77, 78, 81, 86, *90, 91, 92*, 262, *316*
Hildreth, W. W., 106, *109*, 203, 204, *207, 208*
Hill, G., 426, *435*
Hill, G. C., 426, *431*, 462, *476*
Hill, H. A. O., 84, *91*, 260, *313*
Hill, J. E., 224, *317*
Hill, J. M., 183, *192*
Hill, R., 5, *6*, 11, 12, *15*, 62, *66*, 100, 102, 103, 104, 105, 106, *109, 110*, 134, *156*, 199, 201, 202, 203, *207, 208*
Hill, R. L., 138, 139, 140, 141, 142, 143, 144, *156*
Hiller, R. G., 105, 107, 108, *109*. 204, 205, *206*
Hillman, K., 160, *170*
Hind, G., 104, 105, 106, *110*, 204, 205, *208*, 304, *317*
Hinkle, P. C., 346, 362, *373*
Hinkson, J. W., 117, *119*
Hino, S., 466, *476*
Hins, S. C., 407, *438*
Hinterberger, U., 407, *441*
Hirai, K., 264, *321*
Hiramatsu, A.,197, *208*
Hirata, K., 76, 79, 80, 81, *96*
Hiratsuka, H., 189, *191*
Hirocheka, K., 451, *476*
Hiromi, K., 117, *119*, 301, *317*
Hirota, K., 414, *443*
Hirota, S., 21, 24, *55*
Hirota, T., 342, *380*
Hirs, C. H. W., 485, *495*
Hiyama, T., 104, 105, 106, *110*, 203, 208
Hlavica, P., 84, *91*
Ho, Y. K., 453, 454, *476, 478*
Hoch, F. L., 411, *435*
Hoch, G. E., 105, *111*, 205, *209, 210*
Hochster, R. M., 251, *317*
Hochstrasser, R. M., 165, *168*, 172, 176, *180*
Hofmann, E. C. G., 416, *441*
Hofmann, T., 173, *182*, 185, *194*
Hogeboom, G. H., 331, *373, 380*
Hogness, T. R., 342, 355, *372, 382*
Hohorst, H-J., 389, *441*
Holden, H. F., 83, *91*
Holl, J. Y. C., 336, 363, *374*
Holland, T. M., 461, *476*
Hollis, D. P., 13, *14*
Hollocher, T., 128, *131*, 171, *181*
Hollocher, T. C., 48, *50*, 342, 358, *373*

Holloszy, J. O., 395, *435*
Holloway, P. W., 355, *373*
Hollunger, G., 334, 349, 359, 362, *368, 369*
Holmes, W., 339, 345, *369*
Holowinski, A., 332, *371*, 397, *433*
Holroyd, J. D., 77, 79, 80, *96*
Holt, S. C., 309, *317, 319*
Holton, R. W., 200, 203, *208*
Holtzman, J. L., 399, 401, *435*
Holzwarth, G., *180*
Hommersand, M. H., 21, *51*
Hommes, F. H., 64, *66*
Honig, C. R., 396, *435*
Honjo, I., 337, *373*, 406, *435*
Honya, M., 280, *320*
Hook, J. E., 390, *435*
Hopkinson, L., 408, *435*
Hoppel, C., 332, *373*
Hoppe-Seyler, F., 3, *6*
Horecker, B. L., 44, 45, 46, *55*, 172, 178, *180*, 355, *372, 373*
Horgan, D. J., 359, *373*
Horgan, P. H., 428, *435*
Hori, H., 130, *131*
Hori, K., 250, 264, 273, *317*
Horie, S., 24, 26, 27, 31, 35, 43, 44, 46, *51, 53*, 73, 74, 75, 78, 80, *91, 92, 93, 96*, 409, *445*
Horinishi, H., 158, *168*
Horio, R., 12, *15*, 268, *318*
Horio, T., 4, *6*, 34, 38, 39, *51, 54*, 65, *68*, 102, *112*, 112, 114, 117, *119, 120, 121*, 127, 128, 129, *130, 131, 132*, 161, *169*, 171, 172, *182*, 183, 185, *195*, 197, *211*, 228, 229, 232, 235, 240, 241, 242, 244, 245, 246, 249, 257, 268, 269, 278, 280, 285, 288, 290, 291, 292, 293, 296, 297, 300, 303, 304, 306, 312, *313, 314, 315*, *317, 318, 319, 322, 326*, 342, 343, 345, *373*
Horodniceanu, T., 468, *480, 481*
Horowitz, M. G., 336, 363, *374*
Horstman, L. L., 361, *379*
Horstmann, H. J., 128, *130*, 137, 138, 139, 140, 141, 142, 143, 144, *156*
Horton, A. A., 128, *131*
Horton, A. H., 200, 203, *208*
Horwitz, B. A. A., 389, *443*
Hoshi, T., 414, 421, *435*
Hosokawa, M., 184, 185, *195*, 344, *382*
Hottinguer, H., 466, 475
Hovmoller, S., 355, *375*
Howard, R. L., 352, 353, *374*
Howell, J. McC., 391, *435*
Howell, L. G., 48, *53*, 183, *192*
Howells, L., 428, *438*
Howells, R. E., 425, *444*

Howland, C. J. L., 360, *382*
Howland, J. L., 21, *54*, 254, 311, *320*, 344, 349, 350, 351, 358, 362, *367*, *373*, *375*, *382*
Hoyt, R. F., Jr., 394, *429*
Hozumi, M., 414, *436*, *443*
Hrinda, M. E., 454, *476*
Hrušovska, E., 63, 65, *67*, 349, 360, *375*, 465, *477*
Hsia, S. L., 80, *95*, 400, *444*
Hsiao, Y. Y., 133, *158*
Hsu, W-P., 450, *476*
Huang, K., 346, *373*
Huang, M., 461, *478*
Hubbard, N., 449, *478*
Hübsch, M., 389, *441*
Huennekens, F. M., 117, *119*
Huet, J., 361, *383*
Hughes, D. E., 282, *317*
Huijing, F., 360, *373*, 453, *474*
Hülsmann, W. C., 411, *435*
Hultin, E., 354, *373*
Hultin, H. O., 82, *95*, 337, *379*
Hultquist, D. E., 11, *15*, 68, 69, *72*, 76, *91*
Hume, R., 79, 81, *92*
Humiczewska, M., 425, *436*
Humphrey, G. F., 397, 408, 414, *436*
Hunter, F. E. Jr., 187, *192*, *194*, 363, *371*, *378*
Hunter, N. W., 398, *436*
Huntley, T. E., 69, 71, *73*
Hurlebaus, A. J., 392, *445*
Hutchinson, D. W., 359, *369*
Hutchinson, H. T., 461, *476*
Hutterer, F., 78, *91*, 401, *436*
Huttner, I., 406, *443*
Hydén, H., 404, *436*

I

Ibsen, K. H., 470, *476*
Ichii, S., 74, 80, *91*, *96*
Ichikawa, Y., 68, *72*, 74, 76, 83, *91*. 356, *373*, 399, 410, *436*
Ida, S., 129, *131*
Idelman, C. S., 74, *94*
Igo, R. P., 337, *373*
Ikeda, K., 13, *16*, 174, 175, *180*, 288, *318*, 414, *436*, *443*
Ikeda, M., 81, *90*
Ikegami, I., 104, 105, *110*, 201, 202, 203, *208*
Ikuma, H., *110*, 231, *318*, 427, *436*
Imai, K., 285, 288, 309, 310, *318*, 411, *443*
Imai, Y., 74, 75, 76, 77, 79, 83, *91*, 285, 288, *318*, 399, *440*

Inano, H., 409, *438*
Ingram, V. N., 485, *495*
Inhaber, E., 154, *157*, 198, *211*, 465, *481*
Inoue, Y., 218, 255, *318*
Inouye, A., 68, *72*, 405, *436*
Iosinov, K., 408, *436*
Isaev, P. I., 311, *318*
Isenberg, I., 362, *382*
Ishaque, H., 310, *318*
Ishaque, M., 231, 257, 310, *318*, *320*
Ishi, Y., 403, 404, *442*
Ishida, M., 308, 321
Ishidate, K., 74, 82, *91*, 118, *119*, 221, 222, *318*, 469, *477*
Ishihara, N., 454, *483*
Ishihura, H., 129, *131*
Ishii, Sh., 410, *445*
Ishikawa, Sh., 359, *374*
Ishikawa, W., 311, *318*
Ishikura, H., 133, 152, *157*, *158*, *181*, 184, *192*, *193*, *194*, 198, *209*
Ishimoto, M., 5, *6*, 277, 280, *318*
Ishimura, Y., 40, *51*, 86, *91*, 262, *318*
Israel, H. W., 306, *323*
Israels, L. G., 87, *92*
Issakyan, L. A., 388, *430*
Itagaki, E., 65, *66*, 247, 248, 251, 252, 253, *316*, *317*, *318*, *324*
Itahishi, M., 264, *315*
Itano, H. A., 47, *51*, 485, *495*
Ito, A., 69, *72*, 341, *367*, 411, *443*
Ito, M., 394, *439*
Itoh, M., 427, *439*
Ivanov, I. D., 277, *318*
Ivanovics, G., 468, *473*
Ivkov, N. N., 392, *441*
Iwai, Y., 74, *96*
Iwasaki, H., 227, 229, 230, 235, 240, 241, 245, 249, 270, 271, 293, 294, 295, 296, *318*, *320*, *324*
Iwasaki, M., 466, *476*
Iwatsobu, M., 113, 114, 115, 116, 118, *119*
Iyanagi, F., 355, 356, *373*
Izawa, S., 105, 108, *110*

J

Jackson, A. H., 87, *91*, *92*
Jackson, F. L., 253, *319*, 350, *376*, 408, *435*
Jackson, J. F., 113, *119*
Jackson, J. M., 392, *432*, 450, *474*
Jacob, F., 365, *377*
Jacobi, H. P., 404, *435*
Jacobs, A., 450, *477*
Jacobs, E. E., 22, 25, *51*, *55*, 190, *192*, 330, 340, 341, 345, 361, *373*, *376*, *380*, *384*, 391, *445*, 452, *483*, 491, *495*

Jacobs, H., 361, *372*
Jacobs, J. M., 448, 470, *477*
Jacobs, N. J., 227, 231, *319*, 448, 470, *477*
Jacobs, R. M., 185, *195*, 392, *445*
Jacobs, S. S., 23, *55*,
Jacobson, M., 78, 82, *92*
Jacobson, S., 79, *92*
Jacq, C., 114, *119*
Jacques, R., 473, *476*
Jacquet-Armand, Y., 113, 115, 116, *119*
Jagendorf, A. T., 108, *110*, 208
Jagow, G., 331, *383*
Jakovcic, S., 398, *436*
Jakubiak, M., 22, *51*
Jalling, O., 359, *373*
James, S. P., 81, *91*
James, W. O., 100, 102, 104, *110*, 199, *208*
Jamieson, D., 389, *431*, *436*
Jankelson, O. M., 406, *444*
Jansky, L., 390, *436*
Jarina, F., 79, *90*
Jarman, T. R., 263, *319*
Järnefelt, J., 213, 214, *215*
Jasaitis, A., 336, 353, 354, *369*, *373*
Jasper, D. K., 366, *367*
Jayaraman, J., 470, 471, *477*
Jefcoate, C. R. E., 74, 76, 77, 78, 79, 81, 82, 83, 84, *92*
Jefferson, D. J., 393, *438*
Jellinek, H., 406, *443*
Jeng, M., 359, *373*
Jensen, J., 468, *477*
Ji, T-H., 463, *473*
Jobe, A., 451, *481*
Jöbsis, F. F., 395, 402, 406, 410, *431*, *435*, *436*
Jodrey, L. H., 414, *436*
Joel, C. D., 390, *436*
Johansson, B., 341, 353, *375*
Johansson, B. W., 196, *210*, 390, *440*
John, P., 310, *319*
Johnson, D., 360, *375*
Johnson, G. R., 409, *436*
Johnson, L. V. R., 81, *94*
Johnson, M. J., 470, *477*
Jolchine, G., *110*, 305, *319*
Jones, C. W., 222, 233, 235, 237, 238, 239, 240, 276, 277, 307, 308, 309, 310, 311, *312*, *319*
Jones, E. A., 408, *436*
Jones, G. R. N., 413, *436*
Jones, H. E., 239, 264, *319*
Jones, J. D., 397, *436*
Jones, M. S., 449, 450, 453, 455, *473*, *477*
Jones, O. T. G., 257, 299, *325*, 449, 450, 453, 455, *473*, *477*, *480*
Jones, P. D., 355, *373*

Jonsson, J., 188, *192*
Jonxis, J. H. P., 161, *168*
Jori, G., 161, *168*
Judah, J. D., 345, 360, *373*, 452, *475*
Jukes, T. H., 488, 489, 491, *495*
Jung, F., 177, *181*
Junk, K. W., 81, *93*
Jurtschuk, P., 81, *89*, 222, 239, 257, 263, 307, 309, *314*, *319*, 350, 353, 354, *372*
Justesen, N. P. B., 423, *436*

K

Kabel, B. S., 30, *51*, 342, *377*
Kadenbach, B., 410, 411, *436*, 456, 458, 460, *477*
Kagawa, Y., 332, 361, *373*
Kaegi, J. H. R., 161, *170*, 183, *192*
Kagihara, T., 68, 71, *73*
Kahrig, C., 407, *441*
Kaiser, W., 462, *477*
Kajihara, T., 68, 69, 70, *72*, *73*
Kakinuma, K., 352, 361, *376*
Kakuda, M., 163, *170*
Kakudo, M., 197, *211*
Kakudo, N., 129, *132*
Kakuno, T., 249, 257, 292, 303, 306, *313*, *319*, *322*
Kalakutskii, L. V., 429, *443*
Kalina, M., 427, *440*
Kalinin, V. I., 392, *440*
Kallai, D. B., 138, 139, 140, 141, 142, 143, 144, 145, 146, 147, 148, 149, 150, 151, *155*
Kallai, O. B., 162, 163, 164, 165, *168*
Kamdydind, T., 280, *318*
Kamen, M. D., 5, *6*, *7*, 10, 12, *14*, *16*, *53*, *54*, 102, 104, *110*, *111*, 188, *194*, 197, 200, 201, 203, 206, *208*, *210*, *212*, 227, 228, 229, 233, 235, 242, 246, 249, 257, 267, 268, 271, 272, 278, 280, 284, 285, 287, 288, 289, 290, 291, 292, 293, 294, 296, 297, 298, 299, 300, 301, 302, 303, 304, 305, 311, 312, *313*, *314*, *315*, *316*, *318*, *319*, *322*, *324*, *325*, *326*, 356, *384*, 454, *479*, *481*, 488, 489, 491, *495*, *496*
Kamin, H., 85, *92*, 355, *376*, *383*
Kamino, K., 405, *436*
Kaminsky, L. S., 183, *192*
Kamp, B. M., 203, 204, *207*
Kamysheva, A. S., 403, *440*
Kaniuga, Z., 22, *51*, 63, *66*, 348, 349, 350, 352, 354, *367*, *368*, *373*
Kaplan, N. O., 267, *320*, 450, *478*
Kappas, A., 453, *477*
Karapetyan, N. V., 212, 231, 256, *321*, 428, *445*

Karmanov, P. A., 399, *440*
Kartashova, O. Ya., 398, *436*, *438*
Karuzina, I. I., 401, *429*
Kasahara, M., 394, *439*
Kasbekar, D. K., 405, *436*
Kaschnitz, R., 401, *436*
Kasinsky, H. E., 20, *51*, 103, *110*, *111*
Kassner, R., 206, *208*
Katagiri, M., 74, 77, 81, *92*, 257, 260, 299, *315*, *319*
Katano, H., 198, 199, *209*, *212*
Kataoka, N., 179, *180*
Katchalski, E., 129, *130*, 153, *156*
Kato, R., 79, 81, *92*, 399, 400, 414, *436*, *437*, *443*
Katoh, S., 102, 104, 105, *110*, 128, 129, *131*, 201, 202, 203, 205, *208*
Katsuki, H., 448, *481*
Katsumata, Y., 366, *374*
Katsura, H., 411, *437*
Kattermann, R., 118, *119*
Kawaguchi, K., 74, 82, *91*, 118, *119*, 221, 222, *318*, 331, *378*, 469, *477*
Kawai, K., 65, *66*, 97, 98, 99, *99*, 387, 408, 409, 414, 421, 422, *435*, *437*, *444*
Kawai, Y., 70, *72*
Kawakita, A., 358, *374*
Kawakita, M., 428, *437*
Kay, C. M., 30, *56*
Kaye, J. J., 307, *319*
Kayra, A., 402, *433*
Kazakova, T. B., 428, *437*
Kazini, G. M., 78, 80, *96*
Ke, B., 205, *208*, 306, *319*
Kean, E. A., 359, *374*
Kean, E. L., 401, *437*
Kearney, E. B., 112, 118, *119*, *120*, 254, *319*, 352, 353, 358, *370*, *380*, *383*
Keilin, D., 3, 4, *6*, 11, 12, *15*, 20, 34, 45, 46, 47, *51*, 59, *67*, 68, *72*, 100, *110*, 123, 126, 127, *131*, 170, 171, 172, 178, 179, *179*, *180*, 182, 183, *191*, 212, *215*, 219, 220, 221, 233, 236, 246, 265, 277, 282, 289, *319*, 331, 340, 342, 349, 353, 359, *374*, 452, 456, *477*
Keilin, J., 11, *15*, 97, 98, 99, *99*, *100*
Keister, D. L., 311, *319*
Kellerman, G. M., 461, *477*
Kelley, A. B., 205, *209*
Kelly, D. P., 274, *325*
Kelly, J., 204, 205, *209*
Kemp, A., 361, *374*
Kemp, A. Jr., 361, *381*
Kendrew, J. C., 252, 253, *319*, 485, *495*, *496*
Kennedy, P., 410, *440*
Kennel, S., 298, *325*

Kenner, G. W., 87, *91*, *92*
Kerekjarto, B., 74, 78, 81, 83, *95*
Kern, H., 393, *437*
Kersten, H., 355, *374*
Kersten, W., 355, *374*
Kertesz, D., 187, *192*
Ketchem, P. A., 309, *319*
Kety, S., 402, *437*
Keul, J., 393, *433*, *437*
Keylock, M. J., *209*
Keynes, R. D., 405, *429*
Khavinzon, A. G., 408, *437*
Khorana, H. G., 113, *119*
Kidder, G. W., 43, *57*, 196, *209*, 406, *437*
Kiese, M., 25, 26, *51*, *52*, 342, *374*, 468, *477*
Kiesow, L., 223, 224, 310, *319*, 491, *495*
Kihara, T., 304, 306, *315*, *319*
Kijimoto, S., 241, 242, 244, *320*, *326*
Kikuchi, G., 22, *52*, 227, 228, 232, 233, 292, 298, *320*, *323*, 396, 424, 428, *437*, 453, *476*, *477*, *482*
Kim, H. J., 453, *482*
Kim, J. S., 387, *438*
Kimelberg, H. K., 25, 42, 46, 47, 48, *52*, *53*, 189, 190, *192*, 333, 341, 342, 343, *375*, *377*
Kimmel, J. R., 138, 139, 140, 141, 142, 143, 144, *156*
Kimmich, G., 412, *437*
Kimura, M., 145, 146, 147, 148, 149, 150, 151, 154, *157*, *158*, 484, 492, *495*
Kimura, T., 85, *92*
King, D. W., 413, *429*, *437*, 469, *473*
King, E. N., 417, *437*
King, H. K., 21, *54*
King, J. L., 491, *495*
King, M. E., 413, *437*
King, T. E., 14, *16*, 22, 25, 27, 29, 30, 33, 35, 37, 43, 44, *49*, *52*, *53*, *55*, *57*, 62, *67*, 84, *96*, 213, 214, 301, 302, *326*, 329, 338, 340, 341, 342, 347, 352, 353, 357, 358, 359, 365, *368*, *374*, *375*, *376*, *377*, *379*, *382*, *383*, *384*, 405, 421, *439*, *442*
Kinoshita, T., 73, 74, 75, *91*, *92*
Kiprianova, E. A., 450, *477*
Kirk, J. E., 406, *441*
Kirk, J. T. O., 463, *477*
Kirkpatrick, F. A., 25, *51*
Kirschbaum, J., 28, *52*, 62, 63, *67*, 351, *374*
Kirsten, E., 410, *437*, 462, *477*
Kirsten, R., 410, *437*, 462, *477*
Kirt, J. T. P., *209*
Kiszkiss, D. F., 470, *475*
Kitagawa, M., 188, *195*

Kitahara, N., 187, *193*
Kitamura, O., 406, *435*
Kitsutani, Sh., 427, *437*
Kivisaari, E., 400, *435*
Kjeldgaard, N. O., 284, *323*
Kleenan, T. W., 31, *49*
Klemme, J. H., 232, *320*
Klepke, A. K., 422, *440*
Kline, E. S., 425, *437*
Klingenberg, M., 46, *54*, 62, *67*, 73, *92*, 115, *119*, 195, *210*, 329, 331, 333, 335, 339, 344, 357, 360, 361, 362, *372, 374, 375, 378, 383*
Klinger, M. D., 402, *438*
Klingler, M. D., 68, *72*
Klitgaard, H. M., 409, 411, *432, 437*
Klotz, I. M., 336, 363, *374*
Klotzbücher, M., 82, *91*, 458, *476*
Klouwen, H. M., 20, *49*, 189, *192*, 386, 387, *430, 437*
Kmetec, E., 396, 424, *437*
Knaff, D. B., 105, 106, *108, 110*, 204, *209*, 346, 363, *367*
Knobloch, K., 310, *320*
Knowles, C., 310, *320*
Knowles, C. J., 239, 277, *320*
Kobayashi, H., 398, 426, *437*
Kobayashi, M., 68, 71, *73*
Kobayashi, S., 74, 80, *91*, *96*, 410, *445*
Kocholaty, W., 472, *475*
Kodama, T., 4, *6*, 233, 250, 269, 270, 287, 295, *316, 320*
Kok, B., 104, 108, *110*, 203, 204, 205, *209*, 305, *324*
Kolarov, J., 460, *482*
Kolesnik, L. V., 199, *209*
Kolk, A. H. J., 310, *316*
Koller, K., 335, *374*
Komai, II., 451, *477*
Kompa, L., 40, *57*
Kon, H., 179, *180*
Kondo, J., 280, *318*
Kondo, M., 187, *193*
Kopaczyk, K. C., 332, 335, 338, 347, *374*
Kopka, M. L., 4, *6*, 12, *15*, 129, 130, *130*, 162, 166, 167, *168*, 362, *370*
Korecky, B., 391, *437*
Korkonen, E., 405, *437*
Korkonen, L. K., 405, *437*
Korman, E. F., 463, *476*
Kornacher, M. S., 390, *437*
Kornberg, A., 113, *119*, 172, 178, *180*
Kornberg, R. D., 113, *119*, 332, *374*
Korner, P. L., 388, *437*
Kornyushenko, O. N., 450, *477*
Korshunova, V. S., 108, *110*
Kosaka, T., 359, *374*

Koshland, D. E., *6*, 365, *374*
Kossel, A., 187, *193*
Kotani, M., 288, 289, *324*
Kouchkovsky, 101
Kovàč, L., 63, 65, *67*, 349, 360, *375*, 460, 463, 465, *476, 477, 478, 482*, 488, *495*
Kovacova, V., 466, *477*
Koval, G. V., 395, *429*
Kowalsky, A., 135, *156*, 161, 168, *169*, 177, *180*, 183, 187, *193*
Koyaka, J., 277, *318*
Kraayenhof, R., 308, *320*
Kraml, J., 459, *477*
Krasnovskii, A. A., 205, 206, *209*
Kratz, F., 81, *92*
Kratzing, C. C., 389, *445*
Kraut, J., 163, *168*, 297, 298, *320*
Kreil, G., 68, *72*, 128, *131*, 134, 138, 139, 140, 141, 142, 143, 144, 152, *156, 157*
Krejcarek, G. E., 298, *320*
Krendeleva, T. E., 108, *110*
Kretovich, V. L., 231, 256, *321, 323*
Kretsinger, R. H., 70, *72*
Kreutz, W., 105, *109*
Krioskaya, V., 388, *430*
Krisch, K., 68, 69, *72*
Kris Liu, 448, *475*
Kröger, A., 254, *320*, 339, 357, *374, 375*
Krogmann, D. W., 203, 204, 205, *209, 211*
Krogstad, D. J., 358, *375*
Kroon, A. M., 461, *474, 475*
Kroon, M., 459, *477*
Krüger, S., 62, *66*
Krukenberg, C. F. W., 97, *100*
Krunsz, L. M., 160, *170*
Ksiezak, H., 348, 349, *368*
Kubo, H., 248, 255, *318*
Kubowitz, F., 220, *320*
Kuboyama, M., 35, *52*, 342, 353, *374, 375*
Kuby, S. A., 129, *131*
Kuehnel, W., 408, *437*
Kufe, D. W., 254, 311, *320*
Kumaoka, H., 118, *121*
Kumar, H. E., 393, *434*
Kunkel, H. O., 395, *437*
Kuntzman, R., 78, 79, 81, 82, 83, *89, 90, 92*, 399, *429*
Künzel, W., 407, *437*
Kupfer, D., 78, *92*, 94
Kurashima, Y., 453, *477*
Kurihara, K., 155, 158, *168*
Kurihara, M., 159, *169*, 187, *193*
Kuriyama, Y., 400, *437*, 458, *478*
Kurland, G. G., 418, *437*
Kuroda, M., 410, *436*
Kuroda, Y., 62, 65, *67*
Kurokawa, H., 74, *96*

Kurup, C. K. R., 334, *375*
Kurz, H., 25, 26, *51*, *52*, 468, *477*
Kusai, K., 268, 269, *318*
Kusai, M., 128, *131*
Kusaka, T., 251, *320*
Kusel, J. P., 197, *209*
Kusin, A. M., 387, *443*
Kutscher, F., 187, *193*
Kuylenstierna, B., 79, *91*
Kuzela, S., 460, *478*, *482*
Kuznetsova, N. I., 404, *443*
Kuzuya, F., 407, *445*
Kuylenstierna, B., *91*, 332, 336, 355, *370*, *375*, *381*
Kyo, S., 68, 71, *73*

L

Labbe, R. F., 118, *119*, 340, *377*, 449, 472, *477*
Labeyrie, F., 112, 113, 114, 115, 116, 117, 118, *118*, *119*, *120*
Lachinova, R. I., 420, *438*
Lachowicz, T. M., 460, 465, 466, *482*, 488, *495*
Ladkany, D., 135, *158*
Lahey, M. E., 452, *478*
Lahiri, S., 404, *431*
Lam, K. W., 361, *375*
Lam, Y., 240, 241, 245, *320*, 470, *478*
Lamb, A. J., 461, *478*
Lambertson, C. I., 388, *432*
Lance, C., 101, 103, 104, *110*, 200, *209*
Landgraf, W. L., 73, 76, 80, *93*
Landriscina, C., 355, *375*
Lang, C. A., 355, *375*, 393, *438*
Lang, G., 177, *180*, 199, *209*
Lang, R. W., 30, *51*, *53*
Langdon, R. G., 355, *378*
Lange, P. J., 263, *319*
Lanyi, J. K., 22, *52*, 222, *320*
Lanyi, J. L., 222, 255, *320*
Laparra, J., 187, *193*
Lardy, A. A., 390, 410, *438*, *441*
Lardy, H. A., 334, 345, 360, *375*, *376*, 408, 410, *438*
Lascelles, J., 231, 272, 304, 309, *314*, *317*, *319*, *323*, 446, 449, 450, 452, 453, 454, 464, 467, 468, 469, *474*, *476*, *478*, *480*
Laser, H., 336, *375*
Lathe, G. H., 401, *441*
Laudelut, H., 224, *325*
Lauwers, A., 337, 361, *366*
Lawrence, G. A., 189, *192*
Lawrie, R. A., 395, *438*

Lawton, V. D., 253, *319*
Laycock, M. V., 145, 146, 147, 148, 149, 150, 151, *158*
Lazzarini, R. A., 282, *320*
Leach, K. C., 203, *209*
Leaf, G., 127, *131*, 133, 137, *156*
Leblanc, A., 411, *439*
Lederer, F., 114, *119*
Lee, A. Y. H., 84, *95*
Lee, C., 331, 332, 338, 361, 362, *369*
Lee, C. H., 387, *438*
Lee, C. P., 44, *50*, *52*, 63, 64, *66*, 67, *168*, 172, *179*, 189, 190, *192*, 329, 332, 333, 334, 338, 339, 340, 341, 345, 346, 348, 350, 352, 353, 354, 358, 359, 360, 361, 366, *367*, *369*, *370*, *371*, *375*, *381*, 395, 408, 411, *431*, *433*, *438*
Lee, D. Y., 387, *438*
Lee, I-Y., 63, *66*, 350, *370*, 396, *438*
Lee, J., 388, *429*
Lee, S. S., 203, 205, *209*
Lee, W., 358, *367*
Leech, P. M., 101, *110*
Leech, R. M., 100, 102, 104, *110*, 199, *208*
Lees, H., 224, *314*, *320*, *495*
Le Gall, J., 145, 146, 147, 148, 149, 150, 151, *155*, 277, 278, 279, 281, 309, *312*, *313*, *314*, *320*
Legallais, V., 104, *109*, *207*, 387, *444*
Legge, J. W., 11, 12, *15*, 20, *52*, 83, 86, 87, 88, *92*, 97, *100*, 123, 126, *131*, 170, *180*, 195, *209*
Le Guen, M., 281, *314*
Lehninger, A., 361, *384*
Lehninger, A. L., 5, *6*, 331, 334, 341, 344, *375*, *377*, 410, *438*
Leibman, K. C., 75, 77, 78, 81, 90, *91*, *92*, 262, *316*
Lemberg, M. R., 3, 4, *6*, 10, 11, 12, 13, 14, *14*, *15*, *16*, 20, 23, 24, 25, 26, 27, 28, 29, 30, 31, 32, 33, 34, 35, 36, 37, 38, 39, 40, 41, 42, 43, 44, 45, 46, 47, 48, *50*, *51*, *52*, *53*, *56*, 83, 86, 87, 88, *91*, *92*, 97, *100*, 123, 126, *131*, 165, *169*, 170, 171, *180*, 187, 189, *192*, *193*, 195, 196, 205, *209*, 264, *314*, *320*, 335, 336, 338, 339, 340, 342, 343, 344, 362, 363, 364, *371*, *375*, 391, 392, *438*, 452, 458, *478*, 494, *496*
Lenanz, G., 341, *376*
Lenaz, G., 128, *131*, 190, *193*, 337, 361, *366*
Lenhoff, H. M., 267, *320*, 450, *478*
Lennie, R. W., 397, *438*
Leontev, V. G., 428, *437*
Leporda, G., 409, *429*
Leslie, R. B., 189, 191, *192*, *194*, 336, *369*

Lester, R. L., 337, 357, 358, *370*, *372*, *376*, 470, *478*
Levay, A. N., 413, *440*
Levchuk, T. P., 199, *209*
Level, M. J., 401, *441*
Levere, R., 453, *477*
Levin, E. Y., 88, *92*
Levin, K., 297, *320*
Levin, S. S., 73, 74, 75, 84, 85, *90*
Levin, W., 78, 79, 81, 82, 83, *89*, *90*, *92*, 399, *429*
Levine, R. P., 104, 105, 106, 108, *110*, *111*, 200, 203, 204, 205, *208*, *209*, 211
Levine, W. G., 183, *191*
Levitt, M., 87, *92*
Levy, L., 3, *6*
Lewis, S. E., 397, *438*
L'Heritier, P., 466, *475*
Li, W-C., 127, 128, *131*, *132*
Lieben, F., 112, *119*
Liberman, E. A., 311, *318*
Libertun, C., 404, *439*
Liebermeister, H., 393, *440*
Liem, H. H., 83, *93*, 261, *321*
Light, P. A., 329, 346, 352, 358, 359, *369*, *371*
Lightbody, J. J., 203, 204, *209*
Lightbrown, J. W., 350, *376*
Lilga, K. R., 413, *445*
Liljeqvist, G., 129, *132*, 197, *210*, 354, *373*
Limbach, D. A., 205, *211*
Limetti, M., 357, *371*
Limongelli, F., 395, *440*
Lin, F., 427, *429*
Lindall, A., 403, *438*
Lindberg, O., 359, *373*, 410, 412, *444*
Lindegren, C. C., 466, *476*, *478*
Lindenmayer, A., 74, *92*, 115, 118, *119*, 221, 225, *320*, 462, *478*
Linkins, A. E., 304, *322*
Links, J., 358, *381*
Linnane, A. W., 118, *120*, 246, 325, 332, 335, *367*, *376*, 456, 458, 459, 460, 461, 463, 465, *474*, *477*, *478*, *481*, *482*, *483*
Lipmann, F., 360, *376*, 410, *440*
Lips, S. H., 360, *376*
Lipscomb, J. D., 258, *317*
Lipson, S. J., 423, *440*
Lipton, H. J., 62, 63, *67*
Lipton, S. H., 214, *215*, 337, 347, 349, 350, 357, *367*, *379*
Lipton, S. J. H., 337, *380*
Lis, H., 129, *130*, 153, *156*
Little, C., 154, *156*
Llambias, E. B. C., 454, *478*
Lloyd, D., 197, *211*, 425, 426, 428, *438*, *444*

Loach, P. A., 133, 134, *156*, 306, *320*
Lockshin, R. A., 457, *482*
Lockwood, W. H., 26, *52*, 87, 88, *92*, 335, *375*, 392, *438*
Loewenstein, J., 397, *438*
Loewenthal, L. J. A., 407, *438*
Lofrumento, N. E., 334, *378*
London, J., 274, *320*
Long, R. T., 82, *92*
Longmuir, I. S., 469, 470, *478*
Loomans, M. E., 349, *382*
Loomis, W. I., 360, *376*
Looney, F. D., 87, *89*
Lorenc, R., 118, *119*
Lorin, M., 360, *378*
Losinov, A. B., 199, *209*
Loskutoff, D., 452, *474*
Love, B., 27, 29, 43, *50*, *53*
Loverde, A., 68, 71, *72*, 355, *376*
Løvtrup, S., 404, *436*
Löw, H., 340, 346, 349, 357, 358, 359, 362, *370*, *373*, *376*, *383*
Lu, A. Y. H., 77, 81, 85, *93*
Lucas, F. V., 413, *440*
Lucas, M., 412, *438*, *441*
Luciani, S., 361, *367*
Luft, D., 401, *432*
Lukashevich, T. P., 402, 404, *444*
Luke, R. K. J., 450, *474*
Lukins, H. B., 459, 465, *481*, *483*
Lukoyanova, M. A., 251, 307, 308, 309, 310, *316*, *320*, *323*, *324*
Lumper, L., 69, *72*
Lumry, R., 2, *6*, 161, 162, *169*, 177, *180*
Lundblad, G., 354, *373*
Lundegårdh, H., 100, 101, 102, 103, 104, *110*, *111*, 199, 204, *209*, 333, *376*
Lundin, K., 191, *191*, 205, *207*
Lurie, D., 189, *192*
Lusena, C. V., 458, *478*
Lustgarten, J., 127, *131*, 182, 188, *193*
Luthy, L., 352, *376*
Luzzati, V., 190, *192*
Luzzatti, M., 112, 117, *118*, *119*
Lwoff, A., 467, *478*
Lwoff, M., 467, *478*
Lynch, M., 77, *90*, 399, *433*
Lynen, F., 13, *15*, *16*, 448, *481*
Lynn, W. D., 362, *382*
Lyrie, R. M., 274, *320*
Lyster, R. L. J., 163, *168*

M

Maaloe, O., 472, *480*
McCarty, R. E., 107, 108, *110*, 205, *206*

McConnell, D. G., *53*, 74, 83, *93*, 308, *320*, 332, 353, 354, *376*, *383*
McConnell, H. M., 332, *374*
McConnell, N. P., 386, *440*
McCord, J. M., 187, *193*
McCoy, S., 13, *14*, *16*, 26, 43, *49*, 57
McCray, J. A., 306, *315*
McDermott, P., 160, 162, *169*, 177, *180*
McDonagh, A. F., 87, 88, *89*
McDonald, C. C., 4, *6*, 12, *15*, 161, *169*
MacDonald, M., 404, *438*
MacDonald, S. F., 13, *15*
McDowall, M. A., 138, 139, 140, 141, 142, 143, 144, *156*
McGillivray, G., 87, *91*, *92*
McGowan, E. B., 162, *170*, 184, *193*
McGuiness, E. T., 342, *376*
Machinist, H. M., 189, *191*
Machinist, J. M., 189, *191*, 340, *376*
Machino, A., 409, *438*
McIlwain, H., 361, *382*, 402, *438*, *444*
McIntosh, E. N., 77, *93*
McKay, R., 450, *478*
McKee, R. W., 470, *476*
McKenna, E. J., 263, *320*
Mackler, B., 21, 24, 27, *50*, *54*, 116, 117, 118, *120*, 198, *209*, 331, 337, 340, 352, 355, 356, *371*, *372*, *373*, *376*, *378*, *380*
McLain, G., 271, *323*
McLaughlin, C. S., 461, *476*
McLaughlin, J., 362, *832*
McLaughlin, P. J., 489, *496*
McLean, A. E. M., 77, 81, 82, *90*, *93*, 392, 400, 401, *438*, *439*, 458, *475*
MacLennan, D. H., 22, 30, 32, 33, *53*, *55*, 74, 83, *93*, 190, *193*, 308, *320*, 331, 332, 337, 341, 353, 354, 361, *376*, *380*, *383*
MacLennan, D. W., 128, *131*
MacLennan, J. H., 343, *383*
Mackler, B., 459, 462, 465, *478*, *482*
Maclosky, E. R., 470, *477*
MacMunn, C. A., 3, 7, 20, *53*, 100
McMurray, W. C., 360, *375*
McNeely, C., 412, *439*
McSwain, B. D., 106, *108*, *109*, *111*
Maeno, H., 40, *53*
Maes, A. A., 303. *322*
Maggio, R., 20, *53*, 416, *438*
Mahagan, K. P., 357, *371*
Mahler, H. R., 5, 7, 68, *72*, 73, 113, 116, 117, 118, *119*, *120*, 198, *209*, 329, 333, 340, 352, 355, 356, *372*, *376*, *379*, 403, *432*, 456, 459, 460, 461, 462, 465, 470, *476*, *477*, *478*, 480, *482*
Maier, N., 406, *438*
Maitra, P. K. C., 311, *317*
Maiwald, C., 393, *433*

Makinen, M. W., 395, *438*
Makoi, Z., 406, *443*
Maksimova, A. V., 402, *433*
Maksimova, L. A., 398, *438*
Malathi, K., 471, *477*
Malessa, P., 391, *438*
Maley, G. F., 345, *376*, 410, 411, *438*
Malmström, B. G., 48, *49*, 343, *366*
Malviya, A. N., 63, *67*, 332, 336, *376*, 377
Malyuk, V. I., 387, 402, *433*, *438*
Mandel, J. E., 80, *93*
Mangnall, D., 403, *429*
Mangum, J. H., 265, *325*, 402, *438*
Mangum, V. M., 68, *72*
Mann, T., 408, *436*, *438*
Mannering, G. J., 78, 79, 81, 82, *89*, *94* 95
Mannheim, W., 469, *476*
Manocha, S., 403, 404, *438*
Manoilov, S. E., 388, 394, *431*, *433*
Mansley, G. E., 23, 24, 29, 34, 39, 40, 41, 43, 45, *52*, *53*
Manzillo, G., 398, *443*
Mapson, L. W., 20, *53*
Marcovich, H., 466, *478*
Margalit, R., 161, *169*, *180*, *181*, 185, *193*, *194*, 199, *209*
Margoliash, E., 4, *6*, 12, *15*, 35, *53*, 126, 127, 128, 129, 130, *130*, *131*, 133, 134, 135, 138, 139, 140, 141, 142, 143, 144, 145, 146, 147, 148, 149, 150, 151, 152, 153, 154, 155, *155*, *156*, *157*, 158, 159, 160, 162, 164, 166, 167, *168*, *169*, 170, 171, 172, *179*, *180*, *181*, 182, 184, 188, 189, *193*, *194*, 197, 198, 199, *207*, *209*, *210*, 211, 269, 278, 279, 280, *315*, *320*, 362, 366, *370*, *376*, 457, 465, *474*, *481*, 485, 486, 487, 488, 489, 490, 491, 492, 493, 494, *494*, *495*, *496*
Margulis, S. I., 340, *377*
Marinetti, G., 128, *131*, 171, 173, *181*
Mark, L. C., 399, *429*
Marks, G. S., 13, *15*
Maroc, J., 106, *110*
Marque, C., 198, *210*
Marquez, E. D., 230, 232, 254, *323*
Marr, A. G., 309, *317*
Marrack, J. R., 485, *495*
Marriott, J., 453, *479*
Marsh, H., Jr., 450, *479*
Marsh, J. B., 455, *479*
Marshall, L. C., 486, *494*
Marshall, W. J., 82, *93*, 392, 400, 401, *438*, *439*
Marsho, T. V., 205, *209*
Marsubara, H., 185, *194*
Martin, A. P., 413, *440*, 462, *482*
Martin, E. M., 102, 103, 104, *111*

Martius, C., 410, *439*
Maruo, B., 307, *321*
Marver, H. S., 82, 86, 87, 88, 89, *93*, *95*, 453, 455, *479*, *481*
Mashanski, V. T., 428, *437*
Mason, H. S., 24, 31, 48, *49*, *53*, 74, 75, 76, 81, 83, 84, 85, *90*, *91*, *93*, *96*, 365, *380*
Massey, B., 48, 53
Massey, V., 127, *131*, 171, *181*
Massey, Y., 48, *49*
Masters, B. S. S., 85, *92*, 355, *376*
Masuda, T., 187, *193*
Mathewson, H. J., 266, 301, 302, 303, *321*
Matkhanov, G. I., 277, *318*
Matova, E. E., 406, *439*
Matrone, G., 450, *479*
Matsubara, A., 138, 139, 140, 141, 142, 143, 144, *158*
Matsubara, H., 27, 28, *53*, *121*, 128, 129, *131*, 134, 137, 138, 139, 140, 141, 142, 143, 144, 152, *155*, *157*, *158*, 159, 162, 166, *168*, *169*, 171, *181*, 214, *215*, 235, 241, 242, 244, 268, *318*, 344, *378*
Matsubara, J. K., 308, *320*
Matsubara, T., 235, 240, 241, 245, 296, *318*, *320*
Matsuda, K., 187, *193*
Matsumura, X., 21, 24, *55*
Matsumura, Y., 21, 27, 33, *54*, 342, *380*
Matsuura, T., 387, *444*
Matthews, H. B., 79, *93*, 420, *439*
Matthews, R. G., 48, *53*
Matthijsen, C., 80, *93*
Mattison, G. M., 416, 417, *439*
Mattoon, J. R., 341, *376*, 427, 428, *439*, 465, *473*
Mauleon, P., 409, *429*
May, A. K., 222, 239, 307, 309, *319*
May, L., 160, 162, *169*, 177, *180*
Mayer, D., 403, *429*
Mayer, G., 69, 72, *73*
Mayhew, S. G., 48, *53*
Maynard, J. M., 389, *433*
Mayr, E., 484, *496*
Mayr, M., 360, *372*
Mazanoswka, A., 398, *432*, 449, *479*
Mazanowska, A. M., 449, *479*
Mazarean, H. H., 401, *439*
Mazel, P., 356, *373*
Mazza, G., 277, *320*
Mazzarella, L., 355, *375*
Meek, G. A., 463, 471, *480*
Mehl, T. D., 352, *376*
Meigs, R. A., 80, 81, *93*
Meinscheim, W. G., 486, *496*
Mela, L., 172, *179*, 331, 332, 338, 361, 362, *369*

Melik-Sarkisyan, S. S., 231, 256, *321*
Mendiola, L. R., 451, *479*
Mengebier, W. L., 414, 421, *439*
Mercado, T. I., 399, *439*
Merker, H. J., 81, *94*
Merola, A. J., 337, *367*
Mersmann, H., 352, *376*
Merz, R., 360, *384*
Meyer, H., 11, 12, *16*
Meyer, T. E., 266, 300, 301, 302, 303, *313*, *321*
Mia, A. J., 415, *439*
Michel, O., 411, 412, *439*
Michel, R., 411, 412, *438*, *439*, *441*
Mifuchi, I., 118, *120*, 465, *479*
Miginiac-Maslow, M., *376*
Mikelsaar, H. M., 308, *316*
Mikelsaar, K. L. N., 308, *324*
Mikhailova, E. S., 205, *209*
Miki, K., 21, *53*, 197, *212*, 254, 265, 266, 267, *321*
Miki, T., 271, *326*
Milanov, S., 394, *439*
Milesi, J., 470, *478*
Milhaud, G., 231, 273, *321*
Miller, G. W., 450, *476*
Miller, J., 425, *442*
Miller, P. A., 21, *54*
Miller, R. W., 187, *193*, 359, *376*
Millet, J., 231, 273, *321*
Millington, P. F., 396, *439*
Mills, C. F., *433*
Mills, E., 388, *439*
Mills, R. C., 250, *316*
Milnes, L., 42, *51*
Minagawa, T., 451, *479*
Minakami, S., 71, *72*, 129, *131*, 133, 137, 152, *157*, *158*, 177, *181*, 184, *192*, *193*, *194*, 198, *209*, 337, 352, 355, 361, *372*, *375*, *377*, *379*, 407, *439*, 458, *476*
Minnaert, K., 38, 41, 46, *53*, 342, 343, *377*, *381*, 398, *439*
Minton, N. Y., 311, *319*
Miodokova, E., 466, *477*
Mirsky, R., 159, 160, 162, *169*, 173, 175, 176, *181*, 185, *193*, 199, *209*, 307, *321*
Misaka, E., 198, *209*
Mitani, F., 78, 80, *93*
Mitchell, H. K., 191, *192*, 197, *210*, 467, *476*, *482*
Mitchell, M. B., 467, *476*
Mitchell, P., 5, 7, 333, 363, *377*
Mitra, R. S., 361, *377*
Mitral, T., 308, *321*
Mitsui, A., 129. *131*, 203, *209*
Miura, T., 307, 308, *321*
Miyajima, T., 183, *195*

S

Miyake, S., 466, *479*
Miyake, Y., 74, 75, 76, *93*
Miyanishi, M., 394, *439*
Miyata, M., 270, 271, *321*
Mizoguchi, T., 145, 146, 147, 148, 149, 150, 151, *157*, 184, *193*
Mizou, J., 182, *192*
Mizushima, A., 187, *193*
Mizushima, H., 21, 24, *55*, *57*, 133, *155*, 161, *169*, *182*, 197, 199, *212*, 341, *367*
Mizushima, M., *121*
Mizushima, S., 238, 240, 307, 308, *321*, 342, *380*
Mizushima, T., 127, 128, *132*, 408, *439*
Mochan, B. S., 30, *53*
Mochan, E., 178, *181*, 186, *193*, 333, 342, 343, *377*
Mockel, J., 395, *439*
Moguilevsky, J. A., 404, *439*
Mohri, H., 408, *439*
Mok, T. C. K., 221, 231, *321*
Mokhova, E., 335, *371*
Mokhova, E. N., 338, *377*, 398, 407, *433*, *439*
Molinari, R., 68, *73*, 355, *376*, *379*
Moller, K. M., 197. *209*
Momose, K., 350, *378*
Monder, E., 187, *193*
Mondovi, G., 269, *312*
Monier, R., 413, *439*
Monod, J., 348, 365, *377*
Monroe, R. G., 393, *434*
Monroy, A., 416, *438*
Montagna, W., 407, *439*
Montague, M. D., 113, *120*
Monteilhet, C., 115, *120*
Montonaga, K., 198, 199, *209*
Moon, K. E., 424, *439*
Moore, C. L., 402, *439*
Moore, C. W. D., 100, 103, 104, *111*
Moore, G. A., 425, *444*
Mora, P. T., 187, *193*
Moraczewski, A., 404, *439*
Moravec, J., 393, *435*
Morell, D. B., 13, 14, *15*, 25, 26, 27, 35, 37, *52*, *53*, 87, 88, *93*, 335, *375*, 392, *438*
Moret, V., 360, *378*
Moreth, C. M. C., 304, *322*
Morey, A. V., 280, *323*
Morgan, H. E., 393, *440*
Morgan, L. R., 391, *439*
Morgulis, S., 486, *496*
Mori, K., 75, *93*
Mori, M., 408, *439*
Mori, T., 264, 270, 271, 287, 295, *315*, *318*, *320*, *321*
Moriarty, D. J. W., 256, 274, *321*

Morikawa, I., 127, 128, 129, *130*, *131*, *132*, 203, *212*, 266, *326*
Morikawa, J., 4, *6*
Morikawa, K., 128, *131*
Morikawa, N., 403, 404, *442*
Morimoto, H., 130, *131*
Morita, R., 118, *120*
Morita, S., 292, 300, 304, 305, *321*, *324*
Morita, T., 465, *479*
Morita, Y., 129, *131*
Morris, H. P., 414, *443*
Morrison, M., 11, *15*, 21, 24, 26, 31, 32, 35, 43, 44, 46, *51*, *53*, *55*, 128, *131*, 171, 173, 176, *181*, 187, 190, *193*, 213, *215*, 226, 227, 228, *324*
Morrow, P. F. W., 389, *433*
Morton, R. A., 357, *377*
Morton, R. K., 68, *71*, 102, 103, 104, *111*, *112*, 112, 113, 114, 115, 116, 117, *118*, *120*, 126, *131*, 198, 199, *206*, *211*, 264, *321*, 408, *429*
Morton, T. C., *15*
Moss, F., 236, 237, *321*
Moss, F. J., 221, 231, *321*, 467, 469, 471, *479*, *480*
Moss, J. A., de, 469, *480*
Moss, T. H., 177, *181*, 289, 296, 297, *313*, *321*
Mossman, M. R., 472, *476*
Motokawa, Y., 22, *52*, 227, 228, 232, 233, 292, 298, *320*, *323*
Motta, M. V., 411, *435*
Moudrianakis, E. N., 311, *312*
Mounolou, J. C., 461, *479*
Moury, D. N., 412, *439*
Moyle, J., 5, 7, 363, *377*
Msu, M-C., 262, *317*
Muijsers, A. O., 23, 31, 32, 46, 47, *53*, 56
Mujica, A., 345, *379*
Mukass, H., 427, *439*
Mull, R. P., 428, *440*
Müller, A., 104, 106, *111*, *112*, 204, 210, *211*
Muller-Eberhard, Y., 83, *93*, 261, *321*
Mulrow, P. J., 81, *94*
Munoz, E., 307, *316*, *132*
Murakami, H., 145, 146, 147, 148, 149, 150, 151, 154, *157*, *158*
Murano, F., 205, *208*
Muraoka, S., 187, *193*, 336, 339, 344, 345, 360, *377*
Murata, E., 129, *132*
Murata, R., 200, 202, 203, *211*
Murata, T., 104, *111*
Murayama, K., 187, *193*
Murphy, A. J., 4, *6*, 12, *15*, 134, 135, *156*, 159, 160, 162, *168*, 173, 175, *180*, *181*

Murphy, P. J., 399, *439*
Murphy, R. F., 88, *93*
Murphy, T. P., 135, *156*
Murray, R., 128, *131*, 171, *181*
Murray, R. G. E., 307, *321*
Murty, H. S. 454, *479*
Muscatello, U., 341, *377*
Mustafa, M. G., 340, 350, *377*, *378*, 405, *439*
Muthukrishnan, S., 450, *479*
Myer, A. J., 135, *156*
Myer, Y. P., 4, *6*, 12, *15*, 29, 30, 33, 37, *43*, *53*, 134, 135, *156*, *157*, 159, 160, 162, *168*, 173, 174, 175, *180*, *181*, 183, *193*
Myers, D. E., 189, *192*
Myers, D. K., 397, *439*
Myers, J., 200, 203, 205, *208*
Myers, L., 427, *430*

N

Nachbar, M. S., 307, 308, *321*, *323*
Nachlas, M. M., 340, *377*, 392, *439*
Nadler, K. D., 304, *322*
Nagahisa, M., 427, *437*
Nagai, H., 467, *479*
Nagai, J., 277, *318*
Nagai, S., 462, 467, *479*
Nagata, J., 241, 242, 244, *321*
Nagata, T., 413, *439*
Nagata, Y., 197, *212*, 307, *321*
Nagel, K. H., 406, *430*
Nagishi, M., 71, 72
Nagoa, M., 221, *324*
Nair, P. M., 24, 31, *53*
Nair, P. P., 454, *479*
Nakai, M., 128, *131*, 268, 269, *318*
Nakajima, H., 88, *93*
Nakajima, E., 198, 199, *209*
Nakajima, O., 88, 89, *91*, *93*
Nakamura, H., 205, *210*
Nakamura, S., 187, *193*
Nakamura, T., 85. *93*, 116, 117, *120*
Nakamura, Y., 85, *93*
Nakane, H. S., 398, *439*
Nakanishi, K., 198, 199, *209*
Nakao, K., 76, 80, 81, 82, *96*
Nakashima, T., 138, 139, 140, 141, 142, 143, 144, *157*, *158*
Nakatani, H. Y., 105, *110*
Nakatani, M., 159, *169*
Nakatsugawa, Ts., 397, *439*
Nakayama, N., 129, *132*, 199, *211*
Nakayama, T., 138, 139, 140, 141, 142, 143, 144, *157*

Namiki, O., 453, *476*
Nandi, D. L., 454, *479*
Nandy, K., 404, *440*
Nankiville, D. D., 293, 300, *317*
Nanni, G., 188, *193*
Nanzyo, N., 13, *16*, 161, *169*
Narasimhulu, S., 68, *72*, 73, 74, 75, 77, 78, 80, 81, *90*, *93*, 94
Narita, K., 133, *158*, 184, *193*, 197, *209*, *212*
Narita, K., 134, 138, 139, 140, 141, 142, 143, 144, 145, 146, 147, 148, 149, 150, 151, 154, *157*, *158*, 174, 175, *180*
Nason, A., 22, *50*, *55*
Nartsissov, R. P., 407, *440*
Naslin, L., 112, 113, 115, 116, 118, *119*
Nason, A., 74, *94*, 102, *110*, 187, *192*, 222, 223, 224, 225, 257, *312*, *313*, *322*, *323*, *324*, 355, *375*
Nass, S., 462, *479*
Nauman, J. A., 410, 411, *445*
Nebert, D. W., 78, 79, 81, 82, *93*
Nedelina, O. S., 346, *368*
Needleman, S. B., 138, 139, 140, 141, 142, 143, 144, 152, 154, *155*, *157*, 488, 489, *496*
Neely, J. R., 393, *440*
Neet, K. E., *6*
Negelein, E., 3, 4, 7, 11, 13, *15*, *16*, 43, *56*, 220, 225, 233, *321*, *325*
Negishi, M., 458, *479*
Negrotti, T., 466, *479*
Neifakh, A. A., 416, *433*
Neifakh, S. A., 398, *444*
Neilands, J. B., 12, *15*, 127, 128, *132*, 171, *181*, 199, *209*, 451, *476*, *477*
Nelson, C. N., 396, *435*
Nelson, D. H., 73, *91*, 409, 410, *435*, *436*
Nelson, D. M., 81, *94*
Nelson, L., 408, *440*
Nelson, W. H., 463, *474*
Neri, G., 413, *444*
Netter, K. J., 75, 86, *90*, 262, *316*
Neuberger, A., 190, *195*, 309, *317*, 446, 449, 453, 454, *475*, *479*
Neufeld, H. A., 213, *215*, *216*, 340, 351, *369*, *383*, 413, *440*
Neumann, N. P., 275, *322*
Neupert, W., 461, *474*
Neuwirt, J., 454, *479*, *480*
Newton, J. D., 5, *7*
Newton, J. W., *53*, 280, 301, *322*
Newton, N. A., 12, 13, 14, *15*, *16*, 21, 22, 23, 25, 26, 27, 31, 32, 34, 36, 38, 43, 44, *52*, *53*, *55*, 234, 238, 240, 241, 242, 243, 245, 272, *313*, *314*, *322*, *324*, 344, 357, 358, *381*, 391, *438*, 450, 452, *474*, *478*

Nichol, A. W., 14, *16*, 86, 87, 88, *89*, *93*
Nicholas, D. J. D., 224, 231, 240, 241, 244, 245, 256, 267, 274, *320*, *321*, *323*, *325*, 170, *478*
Nicholaus, R. A., 88, *91*
Nicholas, D. G., 355, *375*
Nicholls, P., 1, 7, 25, 40, 41, 42, 43, 45, 46, 47, 48, *51*, *52*, *53*, 63, *67*, 178, *181*, 186, *193*, 329, 333, 336, 340, 341, 342, 343, 364, *377*, 409, *431*
Nicholls, R. G., 113, 117, *120*
Nichols, K. E., 487, *496*
Nicholson, D. C., 88, *91*, *94*
Nickerson, W. J.,451, *483*
Niederpruem, D. J., 21, *53*, 104, *110*, 227, 254, *315*, *322*
Nielsen, S. O., 344, *377*
Niemann, R. H., 205, *210*
Niemeyer, M., 410, *440*
Nieradt-Hiebsch, Ch., 355, *383*
Nigam, V. N., 412, *440*
Nigro, G., 195, *210*, 395, *440*
Nijs, P., 47, *53*, 350, *377*
Nilsson, R., 48, *54*, 356, *377*
Ninnemann, H., 25, *54*, 472, 473, *479*
Nishibayashi, H., 73, 74, 75, 76, 79, 81, *93*, *94*, 355, 356, *377*
Nishikawa, K., 249, 303, *319*
Nishimura, I., 199, *212*
Nishimura, M., 104, 106, 107, *109*, *110*, *111*, 201, 203, 204, *207*, *208*, *210*, 304, *314*, *322*
Nishimura, O., 159, *169*
Nisinoff, F., 188, 189, *193*, *194*
Nisman, B., 277, *323*
Niyazi, A., 366, *374*
Noble, R. W., 40, *57*, 188, *193*
Nobrega, F. G., 68, 71, 72
Nogueira, O. C., 68, 69, *72*, *73*
Nolan, C. 138, 139, 140, 141, 142, 143, 144, 154, *157*, 485, 486, 488, 489, 493, 494, *496*
Nord, F. F., 428, *440*
Nordenbrand, K., 350, 366, *375*, *377*
Norling, B., 63, *66*, 350, 352, 358, *370*, *379*
North, J. A., 68, *72*, 402, *438*
North, J. C., 74, 81, 83, *93*
North, J. E., 76, 81, 83, *93*
Northcote, D. W., 203, *208*
Norton, J. E., 307, *322*
Nosch, Y., 427, *439*
Notton, B. M., 189, *192*
Novack, B., 74, 75, *90*
Novikov, Yu, G., 394, *431*
Nowland, P. G., 463, *478*
Nozaki, M., 4, 6, *91*, 114, *119*, *121*, 127, 128, *130*, *132*, 161, *169*, *182*, 182, *193*

Nozaki, N., 40, *51*
Nozzolio, C. G., 251, *317*
Nuner, J. H., 470, *475*
Nygaard, A. P., 112, 113, 118, *120*, 356, *371*, *377*

O

O'Brien, I. G., 450, *474*
O'Brien, P. J. O., 154, *156*, 160, 161, *169*, 183, 186, *193*, *194*
O'Brien, R. L. O., 361, *367*, *377*
Obuchowitz, L., 20, *54*, 417, 422, *440*
O'Carra, P. O., 88, *90*, *93*
O'Connell, D. J., 23, *54*
Oda, T., 23, 45, *54*, *56*, 62, *67*, *215*, 332, 347, 354, *371*, *374*, *377*, 406, *440*
O'Donnell, V. J., 77, 78, 81, *96*
Oelze, J., 309, *322*
Oelze, T., 454, *479*
O'Farrell, A., 33, 40, *56*
Ogamo, A., 357, *377*
Ogawa, T., 464, *479*
Ogunmole, G. B., 364, *384*
Ogura, Y., 116, 117, *119*, *120*, 358, *374*
Ohad, I., 463, *475*, *481*
O'Hagan, J. E., 13, *15*, 25, 26, 27, *52*, *53*
Ohaniance, L., 198, *208*, *210*, 451, 457, *479*
O'hEocha, C., 88, *93*
Ohkawa, J., 34, 38, *51*, 342, 343, 345, *373*
Ohlin, P., 408, *440*
Ohnishi, K., 62, 65, *67*, 347, *378*
Ohnishi, T., 329, 331, 346, 352, 358, 359, *369*, 427, *440*
Ohsawa, T., 406, *435*
Oka, T., 22, *49*, 227, 236, 237, 238, 239, 240, *312*, *321*, 470, *473*
Okabe, K., 221, *324*
Okad, I., 203, *210*
Okada, Y., 65, *67*, 188, *194*, *195*
Okayama, S., 306, *322*
Okazaki, H., 129, *132*, 187, *193*, 199, *211*
O'Kelley, J. C., 223, 224, *322*
Okla, J., 409, *440*
Okuda, S., 462, *479*
Okui, S., 350, 357, *377*, *378*
Okunuki, K., 2, 4, *6*, 7, 12, *15*, *16*, 21, 22, 23, 24, 27, 28, 29, 30, 31, 32, *33*, 36, 37, 39, 40, 41, 42, 43, 45, 47, 48, *53*, *54*, *55*, *57*, 62, 65, *67*, *68*, 102, *111*, *112*, 112, 114, 117, *119*, *120*, *121*, 127, 128, *129*, *130*, *131*, *132*, 133, 134, 152, *155*, *158*, 159, 161, 162, 166, *168*, *169*, 171, 172, *182*, 183, 184, 185, *191*, *194*, *195*, 197, 199, 200, 202, 203, *211*, *212*, 212, 213, 214, *215*, *216*, 235, 241, 242,

243, 244, 245, 254, 265, 266, 267, 268, 269, 271, 274, 303, *318*, *321*, *326*, 339, 340, 342, 343, 344, *378*, *380*, *382*, *384*, 491, *496*
Old, F., 239, *319*
von Oldershausen, H. F., 82, *91*
Oldham, S., 84, *96*, 365, *384*
Oldham, S. B., 73, 76, 77, 79, 80, 81, *93*, *96*, 410, *440*, *445*
Olson, J. M., 104, 105, 106, *110*. 111, 203, 205, *208*, *210*, 292, 302, 303, 304, 305, 306, *317*, *322*, *324*
Oltzik, R., 360, *372*
Omata, S., 80, *91*, 410, *445*
Omori, Y., 399, *436*
Omura, T., 11, *16*, 69, 71, *72*, 73, 74, 75, 76, 79, 80, 81, 85, *93*, *94*, 263, *322*, 355, 356, *377*, *378*, *382*, 400, 411, *437*, *443*, 458, *478*, *479*
Onada, K., 400, *437*
Onaka, K., 399, 400, *436*
Oparin, A. I., 486, *496*
Oprisor, M., 409, *429*
Orgel, L. E., 289, *322*
Orii, Y., 21, 22, 23, 24, 27, 28, 29, 33, 39, 40, 41, 42, 45, 47, 48, *53*, *54*, *55*, *57*, 159, *168*, 185, *191*, 213, 214, *215*, 340, 343, *378*
Orlando, J., 160, 162, *169*, 177, *180*
Orlando, J. A., 249, 257, 303, *322*
Orlans, E., 11, *15*
Orlans, E. S., 485, *495*
Orlans, P., 97, 98, *100*
Orme, T. W., 230, 311, *322*
Orme-Johnson, W. H., 32, 33, 37, 43, 45, 48, *49*, *54*, *56*, 77, 84, *95*, 260, 261, *325*
Orme-Johnson, W. R., 31, 32, 41, 48, *49*
Orobysheva, N. I., 427, *441*
Orr, M. M., 396, *439*
Orrenius, S., 74, 77, 78, 79, 81, 82, 85, *90*, *92*, *94*, *95*, 355, 356, *370*, *378*, 401, *440*
Orth, H., 12, *15*
Osanai, M., 419, 420, *440*, 456, *479*
Oshino, N., 390, 399, *440*
Oshino, R., 408, *431*
Osmond, C. B., 106, *112*
Ostrovskii, D. N., 307, 308, 309, 310, *316*
Osumi, M., 427, *437*
Ota, A., 244, *326*
Otsuka, H., 85, *93*
Otsuka, J., 288, 289, *324*
Overby, J. R., 409, *430*
Oya, H., 424, *432*
Ozawa, H., 337, *384*
Ozawa, K., 337, 373, 406, *435*

Ozawa, T., 366, *374*
Ozols, J., 68, 69, 70, 71, *72*, *73*, 252, *322*

P

Pablo, I. S., 20, *54*, 417, 421, *440*
Pachecka, J., 114, *120*
Pack, G. T., 414, *441*
Packer, L., 350, *378*, 397, 402, 403, *442*
Paddle, B. M., 389, 390, *441*
Padmanaban, G., 450, 471, *477*, *479*
Pahn, E. M., de, 427, *443*
Painter, A. A., 187, *194*, 363. *378*
Pairault, J., 411, *439*
Pajot, P., 114, 115, *120*
Palade, G. E., 331, 366, *378*, 458, 463, *478*, *481*
Palamarczyk, G., 114, 118, *119*, *120*
Paléus, S., 4, *7*, 126, 127, 128, 129, *131*, *132*, 133, 134, 135, *157*, *158*, 158, 159, *169*, 171, 177, *181*, 188, *192*, 196, 197, *206*, *210*, 297, *322*, 354, *373*, 390, *440*
Palma-Carlos, A. G., 454, *479*
Palmer, G., 21, 23, 24, 25, 31. 32, 33, 43, 44, 48, *49*, *51*, *54*, 172, *180*, 331, 342. 343, 352, 358, *367*, *371*, *372*, *376*, *380*
Palmer, J. U., 427, *440*
Palmieri, F., 46, *54*, 344, 360, *378*
Paltauf, F., 463, *479*
Panchenko, L. F., 392, 399, 401, *429*, *440*, *441*
Pandit-Hovenkamp, H. G., 310, *316*
Panfil, B., 398, *432*
Papa, S., 334, 355, *375*, *378*
Pappenheimer, A. M., 21, *54*, 68, *72*, 212, *215*
Paradies, G., 334, *378*
Parisi, B., 108, *110*, 201, *207*
Park, R. B., 205, *210*
Parke, D. V., 81, *94*
Parker, G. L., 21, *50*, 309, *316*
Parker, J. H., 154, *157*, 198, 210, *211*, 465, *473*, *479*, *481*
Parker, M. J., 264, *314*
Parkes, J. G., 332, *378*
Parks, P. C., 177, *179*
Parli, C. J., 79, *94*
Parrish, R. G., 485, *495*
Parsa, B., 63, *67*, 332, *376*
Parson, P., 402, *440*
Parson, W. W., 305, 306, *314*, *322*
Parsons, D. F., 65, *66*, 331, 332, 341, 355, *369*, *378*, *379*
Parsons, F. M., 401, *441*
Parsons, W. W., 106, *109*, 204, *207*
Partridge, S. M., 128, *130*

Pasetto, M., 68, 71, *72*
Passon, P. G., 68, 69, *72*, 76, *91*
Paterson, J. A., 258, 261, *317*
Pathak, S. M., 415, *439*
Patterson, D. S. P., 451, *479*
Patton, G. M., 460, *473*
Paul, K. G., 12, *16*, 126, 127, 129, *131*, *132*, 158, *168*, *169*, 170, 171, 177, *179*, *181*, 187, *191*, 269, *322*, 468, *480*
Pauling, L., 484, 488, 492, *497*
Pawar, S., 75, *90*
Pawar, S. S., *94*
Peak, H. D., 280, *325*
Pearson, A. M., 187, *194*
Pearson, H. W., 487, *496*
Peck, B. B., 413, *435*
Peck, H. D., 224, 274, 309, *313*, *322*
Pedersen, S., 410, 412, *444*
Peeters, T., 231, 257, *322*
Peisach, J., 1, 7, 76, 77, 79, 84, *94*, *95*, 183, *191*, 260, 261, *313*, *322*, *325*
Pekus, E. N., 407, *433*
Pell, K. S., 114, *120*
Peña, G., 359, *371*
Penefsky, H. S., 332, 362, *378*, *379*
Penman, S., 461, *480*, *483*
Penn, N., 340, 355, *376*, *378*
Penniall, R., 334, *378*, 386, 387, 388, *432*, *440*, 456, *474*, *480*
Penniston, J. T., 332, 354, 366, *372*, *378*
Perdue, J. F., 332, 338, *371*, *374*, 463, *476*
Péré, G. A., 198, *207*, *210*, *211*, 457, *481*
Péré-Aubert, G. A., 467, *474*
Pereira, H. da S., 113, *119*
Perini, F., 20, *54*, 102, 104, *111*, 201, 203, *210*
Perkins, J. H., 448, *475*
Perlish, J. S., 227, *322*
Perlman, P. S., 456, 461, *478*, *480*
Perrodin, G., 455, 459, 461, *475*, *482*
Perry, A. S., 79, *94*
Perry, R. T., 387, *440*
Person, P., 23, 29, *54*, 187, *193*, 416, 423, *440*
Personne, P., 423, *429*
Persson, B., 63, *66*, 350, 352, 358, *370*, *379*
Perutz, M. T., 365, *378*, 485, 486, 489, *496*
Peschel, E., 393, *440*
Peters, W., 425, *444*
Peterson, J. A., 86, *91*, 260, 261, 262, 263, *318*, *322*
Petit, J. F., 254, 265, *314*, 413, *439*, 469, *474*
Petragnani, N., 68, 69, *72*, 73
Petrasek, R., 395, *440*
Petrov, V. N., 392, *440*

Petrucci, D., 423, *440*
Petryka, Z., 88, *94*
Pettegrew, J. W., 135, *158*
Petukhov, E. V., 398, *433*
Petzold, G. L., 462, *480*
Pfeiffer, U., 401, *432*
Pfennig, N., 304, *322*
Pharo, R. L., 358, *380*
Philipp-Dormston, W. K., 450, *475*
Phillips, A. H., 355, *378*
Phillips, P. G., 311, *322*
Phillips, P. H., 408, *438*
Phillips, W. D., 4, *6*, 12, *15*, 161, *169*
Philpott, D. E., 23, *54*
Piatelli, M., 13, *16*
Piccinino, F., 398, *443*
Pichinoty, F., 239, *314*, 468, 469, 470, *474*, *480*
Pick, F. M., 48, *54*
Piechaud, M., 469, *480*
Pigareva, Z. D., 403, 404, *440*, *441*, *443*
Pigon, A., 404, *436*
Pilger, T. B. G., 23, 25, 29, 31, *32*, 34, 36, 38, 43, 44, *52*, *53*
Pinchot, G. B., 311, *323*
Pinna, L. A., 360, *378*
Pinto, R. E., 398, *441*
Pious, D. A., 469, *476*, *480*
Pirrie, R., 127, *131*, 137, *156*
Pistorius, E., 204, *207*
Plapinger, R. E., 392, *442*
Plattner, H., 463, *480*
Plenge, R., 393, *434*
Plesnicar, M., 20, 38, *49*, 102, 104, *109*, 200, *206*
Pluscec, J., 448, *480*
Pojda, S. M., 395, *441*
Pokrovskii, A. A., 392, *441*
Polacov, I., 411, *441*
Polakis, E. S., 463, 471, *480*
Polglase, W. J., 238, *313*
Pollard, C. J., 337, *378*
Polonovsky, J., 336, *378*
Poltoratsky-Bois, R., 69, *72*
Poluhowich, J. J., 425, *441*
Ponka, P., 454, *479*, *480*
Pool, P. E., 411, *443*
Pope, A., 392, 403, 404, *435*
Pope, L. M., 222, 239, 307, 309, *319*
Popescu, I., 398, *445*
Popov, I. P., 387, *441*
Popova, N. A., 398, *436*
Popper, H., 78, *91*, 401, *436*
Popper, T. L., 13, *16*
Porath, J., 127, *132*
Porra, R. J., 231, *323*, 449, 453, 455, *473*, *480*

Porter, G., 205, *209*
Porter, H. K., *207*
Postgate, J. A., 307, 308, *316*
Postgate, J. R., 5, 7, 277, 278, 279, 280, *314, 323*, 469, *475*
Postow, E., 362, *379*
Potapov, N. G., 427, *441*
Potter, V. R., 178, *181*, 349, 350, 351, *378, 379*
Powls, R., 104, *111*, 203, *210*
Pozbindowski, K. S., 23, *55*
Prabhakararao, K., 231, *323*
Pragay, D. A., 394, *429*
Pressman, B. C., 358, 366, *368, 369, 378*, 390, *441*
Pretlow, T. P., 221, *323*, 451, *480*
Prezbindowski, K. S., 333, *378*
Price, C. A., 450, 461, *474, 479*
Pring, M., 42, *54*, 339, *371*
Prokopenko, S. M., 403, *429*
Prosser, C. L., 391, *441*
Prough, R. A., 81, *96*
Prusiner, S. B., 389, 390, *441*
Psychoyos, S., 81, *91*, 402, *444*
Puig, J., 469, *473*
Pullman, M. E., 32, 362, 363, *378, 379*
Pumphrey, A. M., 337, 351, 357, *379*
Pumphrey, A. N., 62, *66*
Pumphrey, H. M., 63, *67*
Purves, M. J., 388, *441*
Putterman, C. G., 198, *211*
Putterman, G. J., 154, *157*, 211, 465, *481*
Putterman, G. P., 154, *157*

Q

Quagliariello, E., 334, 355, 360, *375, 378*
Quaglino, D., 407, *441*
Quastel, J. H., 412, *430*
Quigley, J. P., 406, *441*
Quinn, J. R., 187, *194*
Quinn, P. J., 190, *194*

R

Rabinovitz, M., 398, *436*, 450, 458, 461, 462, *473, 475, 476, 478, 480*
Racker, E., 62, 63, *66, 68*, 107, 108, *110*, 329, 331, 332, 337, 338, 346, 347, 349, 353, 354, 358, 359, 361, 362, *366, 367, 368, 371, 373, 379, 380, 384*, 390, *434*
Radda, G. K., 366, *367*
Raderecht, H. J., 407, *441*
Rafael, J., 389, *441*
Raff, E. C., 487, *496*
Raff, R. A., 487, *496*
Ragan, C. I., 329, 346, 352, 353, 358, 359, *369, 371, 379*
Rahman, H., 81, *94*
Raich, W., 80, *90*
Raikhman, L. M., 414, *430*
Raison, J. K., 21, *54*, 102, *111*, 201, 203, *210*
Rajehandary, U. L., 113, *119*
Rakano, T., 162, 163, 165, *168*
Rakusan, K., 391, *437*
Ramirez, J., 299, 304, 310, *323, 324*, 345, *379*, 394, 424, *437, 441*
Ramseyer, J., 76, 77, 78, 80, 84, *96*
Ramshaw, J. A. M., 145, 146, 147, 148, 149, 150, 151, *155, 157, 158*, 492, *494*
Rancourt, U., 427, *429*
Randles, J., 105, *111*, 205, *210*
Ranlett, M. R., 282, 283, *317, 325*, 469, *483*
Ransfield, I. H., 401, *436*
Raoul, B., 411, 412, *438, 439, 441*
Rapoport, S., 355, *379, 383*, 407, 416, *429, 441*
Rareed, D., 306, *323*
Rashid, M. A., 68, 71, *73*
Rasia, M. L., 402, *433*
Rasmussen, H., 412, *437*
Ratmanova, E., Ya, 488, *494*
Raw, I., 68, 69, 71, *72, 73*, 355, *376, 379*, 401, 412, *430, 441*
Rawlinson, W. A., 21, *54*, 127, *131*, 171, 177, *180*
Ray, G. S., 44, 46, *57*, 340, 342, 344, *384*
Reaveley, D. A., 308, *323*
Reddy, R. S., 97, *100*, 422, *441*
Redfearn, E. R., 222, 233, 235, 237, 238, 239, 240, 276, 277, 307, *319, 320*, 337, 351, 357, 359, *368, 379*
Redfield, A. G., 178, *180*
Reed, D. W., 11, *15*, 68, *72*, 76, *91*, 305, 306, *323*
Reed, N., 390, *441*
Rees, K. R., 452, *475*
Rees, M., 74, *94*, 257, *323*
Rees, T., 427, *435*
Reeve, V. E., 391, *434*
Rehm, W. S., 196, *209*, 406, *437*
Reich, M., 189, *194*
Reichlin, M., 188, 189, *193, 194*
Reid, B. L., 398, *439*
Reif, A. E., 349, 350, 351, *378, 379*
Reilly, C., 465, *480*
Rein, H., 177, *181*
Reindell, H., 393, *433, 437*
Reinosa, J. A., 133, *158*
Reinwald, E., 105, *109*
Reinwein, D., 342, *374*

Reis, W., 127, *132*, 170, 171, *182*
Reiss-Husson, F., *110*, 305, *319*
Rembold, H., 420, *440*, 456, *479*
Remmer, H., 7, 74, 77, 78, 79, 81, 82, *90*, *91*, *94*, *95*, 458, *476*
Repaske, R., 340, *372*
Report of the Commission on Enzymes of the I.U.B., 1, 3, *6*, 8, *15*, 201, *210*
Renger, G., 105, *109*
Rere, G., 154, *157*
Reuter, F., 12, *16*, 127, *132*, 161, *170*
Reverberi, G., 416, *441*
Revsin, B., 22. *54*, 227, 230, 232, 254, 311, *322*, *323*
Reyerson, L. H., 161, *169*, 177, *180*
Richardson, M., 145, 146, 147, 148, 149, 150, 151, *155*, *158*, 492, *494*, *496*
Richardson, S. H., 337, 341, 353, *371*, *379*
Richardson, S. W., 25, 30, *50*
Richmond, M., 284, *323*
Richmond, M. H., 472, *480*
Richmond, R. C., *210*, 484, 491, *496*
Rickard, P. A. D., 221, *321*, 467, 471, *479*, *480*
Ridge, J. W., 404, *441*
Riemersma, J. C., 360, *379*
Rieske, J. S., 27, *55*, 62, 63, *67*, 213, 214, *215*, 332, 337, 347, 349, 350, 357, 358, *367*, *372*, *379*, *380*
Riethmüller, G., 449, *480*
Riggs, B. C., 389, *441*
Riley, V., 414, *441*
Rimington, C., 13, *16*, 154, 446, 448, *473*, *480*
Ringler, R. L., 352, 361, *376*, *379*, 403, *441*
Rippa, M., 112, 113, 114, 116, 117, 118, *118*, *120*
Rish, M. A., *441*
Risler, J. L., 115, 118, *119*, *120*
Ristau, O., 177, *181*
Ritchie, K., 33, *51*
Rittenberg, B., 272, *314*, 453, 454, 469, *474*, *478*
Ritter, C., 413, *441*
Ritz, E., 406, *441*
Rivas, E., 190, *192*
Roberts, J. C., 389, 390, *431*
Robertson, R. N., 5, 7, 102, 103, *112*, 199, *211*, 333, 366, *379*, 405, *441*
Robinson, A. B., 12, *16*, 173, 174, 175, *180*, 188, *194*, 288, 296, 297, 300, 301, 302, *313*, *316*
Robinson, S. H., 82, *94*
Rockman, H., 401, *441*
Rode, W., 348, 349, *368*
Röder, A., 84, *91*, 260, *313*
Rodkey, F. L., 177, *181*

Rodman, N. F., 386, 387, 388, *432*
Roeder, A., 84, *94*
Roels, F., 417, *441*
Rogers, H. J., 308, *323*
Rogers, L. J., *209*
Rogers, P. L., 471, 472, *476*
Rogers, W. P., 414, *441*, *442*
Romano, D. V., *182*, 402, *444*
Romano, V., 195, *210*
Romanoff, V. L., 256, *323*
Ronin, V. S., 407, *442*
Roodyn, D. B., 410, 412, *442*, 465, *480*
Roper, G. H., 467, 471, *479*, *480*
Rosenberg, B., 362, *379*
Rosenberg, H., 450, *474*
Rosenthal, O., 68, *72*, 73, 74, 75, 76, 77, 78, 80, 81, 84, 85, *90*, *91*, *93*, *94*, *95*, 196, *210*, 413, *433*
Ross, A. J., 22, *54*
Ross, B. D., 449, *480*
Ross, J., 393, 394, *432*, *434*
Ross, M. E., 87, 88, 89, *94*
Rossi, E., 358, *379*
Rossmann, M. G., 129, *130*
Rothfield, C., 308, *323*
Rothfield, L., 5, 7
Rothfus, J. A., 138, 139, 140, 141, 142, 143, 144, *157*
Rothschild, H. A., 421, *434*
Rothschild, Lord, 423, *442*
Rotilio, G., 269, *312*
Roughton, F. J. W., 289, *316*
Rouser, G., 187, 190, *193*
Rouslin, W., 463, *480*
Roy, B. L., 356, *380*
Roy, S. B., 304, *322*
Rubin, A. B., 108, *110*
Rudescu, K., 128, *132*
Rudney, H., 358, *381*
Ruhmann-Wennhold, A., 81, *94*, 409, *436*
Ruiz, A. N., 387, *442*
Ruiz-Herrera, J., 250, *323*, 469, *480*
Rumberg, B., 104, 105, 106, 107, *109*, *111*, *112*, 204, *210*, *211*
Rummel, W., 81, *90*
Rupley, J. A., 184, *194*
Rurainski, H. J., 105, 108, *110*, *111*, 204, 205, *209*, *210*
Rusev, G., 390, *442*
Russell, J. R., 350, 358, *367*
Ruttenberg, G. J. C. M., 461, *474*
Ruzicka, F. J., 333, *378*, *442*
Ryan, C. A., 421, *442*
Ryan, K. J., 80, 81, *93*, *94*
Rybicka, K., 425, *442*
Ryter, A., 307, *316*

S

Sach, G. S., 87, *91*
Sachson, R. A., 453, *477*
Sacktor, B., 65, *66*, 197, *207*, 338, *368*, 397, 402, 403, 407, 420, 421, *431*, *433*, *442*
Sadana, J. C., 280, *323*
Sadkov, A. P., 277, *317*
Sager, R., 104, *109*, 203, *207*
Saito, Y., 22, *52*, 232, 292, 298, *320*
Sakagishi, P., 76, 81, 83, *93*
Sakai, A., 406, *435*
Sakai, H., *180*
Sakakibara, S., 13, *16*
Sakamoto, Y., 76, 79, 80, 81, *96*
Sakano, K., 427, *442*
Salamatova, T. S., 427, *441*
Salhanick, H. A., 77, *93*
Salmon, H. A., 450, *480*
Salpeter, M., 463, *480*
Saltman, P., 405, *433*
Salton, M. R. J., 307, 308, *316*, *317*, *321*, *323*, 464, *480*
Saltzgaber, J., 463, *480*
Saludjian, P., 160, *169*, 171, 176, *182*
Samoiloff, M. R., *430*
Samson, L., 138, 139, 140, 141, 142, 143, 144, 145, 146, 147, 148, 149, 150, 151, *155*, 162, 163, 164, 165, *168*, 362, *370*
Samuels, L. T., 408, *435*
Samuilov, V. D., 311, *318*
Sanadi, D. R., 64, *67*, 190, *192*, 334, 340, 341, 345, 352, 358, 361, 365, *373*, *375*, *380*, 457, *475*, 491, *495*
Sanbon, R. C., 68, *73*
Sanders, E., 74, 80, 85, *94*
Sanderson, M., 468, *480*
Sandoval, F., 359, *371*
Sandri, G., 74, *95*
Sands, D. C., 236, 251, *323*
Sands, R. H., 333, *380*
Sands, R. W., 32, *49*
Sano, S., 12, 13, *16*, 154, 155, *157*, 159, 161, *169*, 187, *193*, 449, *482*
San Pietro, A., 102, 104, *110*, 202, 203, 205, *208*, *211*
San Pietro, L., 311, *313*
Sarachek, A., 466, *480*
Sarago, E., 350, *380*
Saris, N. E. L., 360, *383*
Sarkar, N. K., 352, *376*
Sarma, P. S., 450, 471, *477*, *479*
Sartorelli, L., *371*
Sasada, Y., 129, *132*, 197, *211*
Sasagawa, M., 128, *131*, 268, 269, *318*
Sasaki, T., 227, 228, 233, *323*
Sasame, H. A., 77, 78, *90*, *94*, 356, 380, 401 *442*
Šašarman, A. M., 468, *480*, *481*
Sassa, S., 454, *481*
Satake, K., 198, *209*
Sato, C., 145, 146, 147, 148, 149, 150, 151, *157*
Sato, E., 393, *442*
Sato, N., 65, *67*, 413, 414, *442*
Sato, R., 5, 7, 11, *16*, 69, *72*, 73, 74, 75, 76, 77, 79, 80, 81, 83, 84, 85, *91*, *93*, *94*, 196, *212*, 247, 251, 263, 264, 282, 283, 284, 285, 309, 310, *315*, *316*, *318*, *320*, *322*, 355, 356, *377*, *378*, 387, 390, 399, 411, *440*, *443*, *445*
Satre, M., 74, *94*
Sauer, K., 204, 205, *209*, *210*
Sauer, L. A., 81, *94*, 333, *380*
Saunders, E., 74, 75, *90*
Saunders, G. W., 465, *481*
Sawada, H., 449, *481*
Sawada, K., 427, *437*
Scaife, J. F., 402, *442*
Scallela, P., 356, 361, *367*, *372*
Scarisbrick, R., 5, *6*, 100, 103, *110*, 201, *208*
Schachter, B. A., 87, *92*
Schaeffer, P., 277, *323*
Schaffner, F., 401, *436*
Schaffner, J., 78, *91*
Schatz, G., 358, 359, 362, 363, *378*, *380*, 461, 463, *474*, *476*, *479*, *480*, *481*
Schejter, A., 126, 127, 129, *131*, 153, *156*, 159, 160, 161, 162, 163, *168*, *169*, 171, 172, 173, 176, 178, *179*, *180*, *181*, 182, 184, 185, *191*, *193*, *194*, 199, *209*, 214, 215, 280, *320*
Scheitel, L. W., 425, *442*
Schellman, J. A., 30, *52*
Schenck, D., 407, *441*
Schenkman, J. B., 7, 74, 77, 78, 79, 81, 82, 84, *90*, *91*, *94*, *95*, 400, 401, *432*, *436*, 458, *476*
Schiaffini, O., 404, *439*
Schiefer, H. G., 460, *481*
Schiff, J. A., 20, *54*, 102, 104, *111*, 201, 203, *210*
Schilling, G. R., 78, 79, *89*, 399, *429*
Schindler, F. J., 25, *50*, 337, 338, *368*, *377*
Schlegel, H. G., 232, *320*
Schlenk, F., 428, *443*
Schleyer, H., 76, 84, *95*, 104, *109*, 186, *195*, *207*, 304, *322*
Schleyer, M. D., 85, *90*
Schlief, H., 396, 413, *442*
Schmeling, U., 69, *73*

Schmid, R., 82, 86, 87, 88, 89, *93, 94, 95,* 453, *479*
Schmidt, C. G., 196, *210,* 396, 413, *442*
Schmidt, C. L., 151, *181*
Schmidt, W. R., 138, 139, 140, 141, 142, 143, 144, *156*
Schmidt-Mende, P., 106, *112*
Schneider, H. G., 349, *366*
Schneider, W. C., 331, *380*
Schneiderman, H. A., 418, *437*
Schoener, B., 64, *66,* 402, 410, *431*
Schoener, D., 338, 343, *369*
Schoenhoff, M. I., 22, *54*
Schofield, P. J., 424, *439*
Scholan, N. A., 80, *95*
Schole, J., 411, *442*
Scholnick, P. L., 453, *481*
Scholes, P. B., 21, 22, *54, 55,* 230, 271, 272, *323, 324,* 344, *381*
Schollmeyer, P., 195, *210,* 362, *374*
Scholz, R., 352, 359, *380*
Schonbaum, G. R., 30, *56*
Schopf, J. N., 486, *496*
Schor, M. R., 307, *321*
Schor, S., 463, *481*
Schreffler, D. C., 391, *442*
Schroeder, D. H., 409, *434*
Schubert, J., 404, *442*
Schuetzel, H., 453, *479*
Schuldiner, S., 203, *210,* 463, *481*
Schulman, H., 451, *481*
Schulman, J., 40, *56*
Schulp, J. A., 472, *481*
Schulte, H. R., 397, *438*
Schulze, H. U., 355, *380*
Schuman, M., 48, *53*
Schützel, H., 82, *93*
Schwartz, H. C., 450, *474*
Schwarz, K., *432*
Schwarzberg, M., 135, *158*
Schweitzer, A., 388, *432*
Scocca, J. J., 311, *323*
Scoffone, E., 161, *168*
Scott, W. A., 197, *210*
Seagren, S. S., 411, *443*
Sebald, W., 460, *483*
Sedar, A. W., 307, *323*
Seibert, M., 305, 306, *323*
Seifert, K., 105, *109*
Seki, S., 23, *54,* 353, *380,* 406, *440*
Sekiguchi, Y., 394, *439*
Sekiyama, S., 74, *96*
Sekura, D. L., 306, *320*
Sekuzu, I., 4, *6,* 12, *15,* 21, 24, 27, 28, 30, 31, 32, 33, 39, 40, 41, 43, 48, *53, 54, 55,* 62, *67,* 127, 128, 129, *130, 131, 132,* 184, 185, *194, 195,* 203, 212, 213, 214, 215, 254, 265, 266, *321, 326,* 342, 343, 344, *378, 380, 382*
Seligman, A. M., 340, *377,* 392, *439, 442*
Sels, A. A., 128, *132,* 154, *157,* 198, *210, 211,* 457, 462, *475, 481*
Selwyn, M. J., 102, *109,* 361, *381*
Semakov, V. G., 408, *442*
Semenova, I. G., 391, *442*
Senior, A. E., 337, *380*
Seraydarian, M. W., 393, *442*
Seroussi, S., 393, *435*
Severina, V. A., 69, *71*
Sewell, D. L., 224, *323*
Seyffert, R., 13, *15, 16,* 448, 449, *481*
Shabalkin, V. A., 414, *430*
Shakhnazarov, A. A., 394, *442*
Shapiro, A. Z., 421, *442*
Shappiro, D. G., 197, *210,* 418, 419, *442,* 457, *481*
Sharma, V. N., 403, 404, *442*
Sharon, N., 129, *130,* 153, *156*
Sharonov, Y. A., 185, *191*
Sharp, Ch., 352, *376*
Sharp, C. W., 331, *380,* 470, *477*
Shashoua, V. E., 173, 176, *181, 182,* 186, *194,* 198, *210*
Shaw, E., 205, *211*
Shaw, E. K., 302, 303, *322*
Shaw, E. R., 205, *208*
Scherbakova, L. I., *441*
Shechter, E., 160, *169,* 171, 176, *182,* 190, *192*
Shemin, D., 446, *481*
Shepley, K., 113, 114, 115, 116, 117, *120,* 264, *321*
Sherman, F., 117, *120,* 154, *157,* 198, 199, *210, 211,* 221, *323,* 341, *376,* 428, *439,* 451, 465, 466, 467, *479, 480, 481*
Shibata, K., 158, *168*
Shibata, M., 81, *96*
Shibiya, I., 307, *321*
Shichi, H., 20, *51,* 62, 65, *67,* 103, *110, 111,* 337, *380,* 405, *442*
Shidara, S., 250, 269, 270, 271, 293, 294, 295, *318, 320*
Shigesada, K., 448, *481*
Shimakawa, H., 355, *383*
Shimazono, N., 73, 74, 75, *91, 92*
Shimizu, N., 403, 404, *442*
Shimoda, K., 270, *320*
Shin, M., 4, *6,* 102, *111,* 128, 129, *130, 131*
Shinagawa, Y., 65, *67,* 68, *72,* 73, 405, *436*
Shinagawa, Ya., 68, *72,* 73
Shinagana, Tasuko, 405, *436*
Shipley, G. G., 191, *194*
Shipman, N. A., 154, *157,* 198, *211,* 465, *481*

Shirasaka, M., 129, *132*, 199, *208*, 211
Shivaki, M., 280, *318*
Shkolovov, V. V., 399, *442*
Shneyour, A., 204, *206*
Shoeman, D. W., 79, *95*
Shoheir, M. H. K., 391, *442*
Shoji, K., 251, *320*
Shore, J. D., 62, *67*
Shore, J. P., 63, *67*
Showe, M. K., 469, *480*
Shrago, E., 390, *443*, 462, *481*
Shug, A. L., 350, *380*, 390, *443*, 462, *481*
Shukolyukov, S. B., 405, *443*
Siebert, G., 196, *207*
Siegel, G. V., 395, *429*
Siegel, M. R., 461, *481*
Sieker, L. C., 163, *169*
Siekevitz, P., 353, *380*, 458, 463, *474*, *478*, *481*
Sies, H., 77, *95*, 400, *443*
Siggel, U., 105, 106, *109*, *112*
Sih, C. J., 80, 85, 89, *95*
Siliprandi, N., 360, *378*
Silman, H. I., 214, *215*, 329, 337, 347, 349, 350, 353, 357, *367*, *372*, *379*, *380*
Silver, A. W. S., 254, 255, *313*
Silver, W. S., 254, *317*
Simmonds, D. H., 114, *118*
Simakova, I. M., 307, 308, *323*, *324*
Simokova, J. M., 308, *316*
Simon, E. W., 104, *111*, 415, *443*
Simon, R., 195, *206*
Simpson, E. R., 74, 80, 81, *89*, *95*, 410, *443*
Simpson, G. G., 484, 491, *496*
Simpson, J. R., 224, *320*, *495*
Sinclair, P. R., 219, 220, 223, 231, 237, 238, 239, 283, 284, 308, *313*, *324*, *325*, 448, 463, 467, 470, *473*, *481*, *483*
Singer, T. P., 112, 118, *119*, *120*, 351, 352, 353, 355, 358, 359, 360, 361, *367*, *370*, *372*, *373*, *374*, *376*, *379*, *380*, *383*, 403, *441*
Singh, R., 391, *439*
Singh, S., 163, *169*, 297, 298, *320*
Sironval, C., *211*
Sisler, H. D., 461, *481*
Sistrom, W. R., 305, *314*
Sjöbstrand, T., 87, *95*
Sjöstrand, F. S., 331, *380*
Skelton, C. L., 411, *443*
Skelton, F. R., 81, *89*, 409, *431*
Skerra, B., *112*
Skiladov, V. P., 336. 354, *369*
Skimp, N. F., 31, *56*
Skinner, D. M., 457, *481*
Skov, K., 128, *132*, 172, 173, *182*, 185, *194*

Skulachev, V. P., 311, *318*, 349, *380*, 398, *433*
Skyrme, J. E., 359, *371*
Slade, A., 74, 75, 80, *90*
Sladek, N. E., 78, 79, 81, 82, *95*
Slater, E. C., 5, 7, 31, 32, 46, 47, *53*, *56*, 62, 63, 66, *67*, *68*, 102, *109*, 127, *132*, 171, *182*, 182, 189, *194*, 212, *215*, 277, *319*, 329, 331, 333, 334, 336, 339, 340, 342, 343, 344, 345, 347, 348, 349, 350, 351, 352, 357, 358, 359, 360, 361, 362, 363, 364, *366*, *367*, *368*, *369*, *372*, *373*, *374*, *375*, *377*, *380*, *381*, *382*, *383*, 397, *438*, *439*, *444*
Slautterback, D. B., 189, *192*
Sletten, K., 145, 146, 147, 148, 149, 150, 151, 154, *155*, *156*, 291, 296, 297, 298, *315*, *324*, 456, 457, *475*, 488, 489, *495*
Slonimski, P. P., 112, 113, 116, 117, 118, *119*, *120*, 154, *157*, 198, *207*, *210*, *211*, 356, *376*, 455, 457, 459, 461, 462, 465, 466, 467, 470, 473, *474*, *475*, *478*, *481*, *482*, 488, *495*
Slooten, L., *211*
Smarsh, A., 423, *440*
Smialek, M., 404, *443*
Šmigan, P., 63, 65, *67*, 349, 360, *375*, 460, *478*
Smillie, R. M., 21, *54*, 102, 104, *111*, 200, 201, 203, *210*, *211*
Smirnova, E. G., 336, 349, 354, *369*, *380*
Smith, A. L., 62, *66*, 365, *383*
Smith, C. H., 398, *443*
Smith, D. W., 176, *179*
Smith, E. L., 134, 135, 138, 139, 140, 141, 142, 143, 144, 145, 146, 147, 148, 149, 150, 151, 152, 153, 154, 155, *155*, *156*, *157*, 197, 200, *207*, *211*, 487, 488, 491, *495*, *496*
Smith, H., 188, *191*
Smith, J. A., 404, *435*
Smith, K-M., 87, *91*, *92*
Smith, L., 5, 7, 20, 21, 22, 23, 24, 25, *49*, *54*, *55*, 74, *92*, *95*, 184, 189, *191*, *194*, *211*, 221, 222, 223, 225, 226, 227, 230, 231, 232, 236, 238, 240, 250, 254, 267, 271, 272, 298, 299, 304, 308, 309, 310, *313*, *314*, *320*, *323*, *324*, *325*, 340, 342, 344, *368*, *370*, *381*, 462, *478*
Smith, L. L., 80, *95*
Smith, M. H., 414, *443*
Smith, R. E., 389, *443*
Smith, W., 424, *432*
Smolyar, V. I., 392, *443*
Smuckler, E. A., 82, *95*, 401, *429*, *430*, *432*

Smythe, G. A., 13, 14, *16*, 87, *89*
Snoswell, A. M., 238, 309, *314*, *315*, 357, 358, 359, 360, *369*, *381*, *382*
Snyder, A. L., 88, *95*
Sokal, R. S., 394, *429*
Sokatch, J. R., 307, *322*
Sokolov, W. V., 394, *443*
Sokolovsky, M., 161, *169*, 185, *194*
Solbakken, A., 161, *169*, 177, *180*
Somlo, M., 154, *157*, 198, *211*, 356, *381*, 457, 471, *481*, *482*
Sonea, S., 468, *480*
Song, C., 453, *477*
Sonnenblick, E. H., 393, 394, *432*, *434*
Soon, Yun, J., 407, *439*
Sorby, H. C., 97, *100*
Sordahl, L. H., 358, *380*
Sorger, G. J., 356, *381*
Sorokina, I. N., 404, *443*
Soru, E., 128, *312*
Soslau, G., 457, *482*
Sotonyi, P., 406, *443*
Sottocasa, G. L., 74, *95*, 332, 346, 353, 355, *370*, *375*, *381*
Souchleris, I., 134, *158*
Souverijn, J. M., 361, *381*
Soyka, L. T., 399, *443*
Spector, R. B., 404, *438*
Spector, R. G., 404, *443*
Spencer, E. L., Jr., 336, 339, *368*
Spett, K., 196, *207*
Spiock, F., 196, *207*
Spiro, M. J., 196, *211*
Spots, A., 407, *432*
Squame, G., 398, *443*
Stahl, G., 354, *373*
Staichenko, N. I., 429, *433*
Stanacev, N. Z., 462, *474*, *482*
Stanbury, J. T., 23, 24, 29, 35, 39, 40, 41, 43, *52*, *53*
Stangherlin, P., 413, *432*
Stannard, J. N., 44, 45, 46, *55*, 172, 178, *180*
Starr, T. J., 466, *483*
Stasny, J. T., 332, *381*
Staudinger, Hj., 68, 69, 70, *72*, *73*, 74, 78, 81, 83, *90*, *92*, *95*, 355, 374, *380*, 401, *432*
Steele, W-F., 187, *193*
Stefanov, St., 390, *442*
Steim, E., 393, *437*
Steim, H., 393, *433*, *437*
Stellwagen, E., 135, *157*, 159, 160, 161, 162, 163, *169*, *170*, 172, 178, *182*, 183, 184, 185, 186, *193*, *194*
Stempel, K. E., 357, 359, *372*
Stephen, J., 188, *191*
Stephens, N. L., 396, *443*

Stephenson, G. F., 87, *89*
Stern, L., 359, *367*
Stevens, F. C., 145, 146, 147, 148, 149, 150, 151, *157*, *211*
Stewart, J. M., 138, 139, 140, 141, 142, 143, 144, 154, *157*
Stewart, J. W., 129, *131*, 138, 139, 140, 141, 142, 143, 144, 152, 154, *155*, *157*, 189, *194*, 198, 199, *210*, *211*, 465, *481*
Stewart, M., 13, 14, *15*, 25, *52*, 335, *375*, 392, *438*
Stewart, P. R., 460, *482*
Stewart, W. D., 487, *496*
Stiel, H. A., 105, 106, *112*
Stiehl, H. H., *109*
Stipakov, A. A., 392, *441*
Stockdale, M., 361, *381*
Stoeckenius, W., 332, 338, *382*
Stolzberg, S., 20, *57*, 397, 422, 423, *445*
Stone, W. D., 406, *443*
Stoner, C. D., 62, 63, *67*, 347, 350, *379*
Stoppani, A. O. M., 337, 351, 352, *367*, *382*, 409, 427, *433*, *443*, *444*
Storck, J., 188, *194*
Storey, B. T., 20, 39, 45, *55*, 100, 101, 103, 104, *109*, *111*, *211*, 214, *215*, 329, 345, 352, 357, 365, *369*, *370*, *382*
Stotz, E., 22, 25, 26, 27, 29, 43, *50*, *53*, *55*, *56*, 128, *131*, 171, *181*, 213, *215*, *216*, 340, 342, 351, *369*, *382*, *383*, 413, *440*
Stotz, E. H., 457, *482*
Stouthamer, A. H., 310, *317*, 469, 472, *481*, *482*
Straat, P. A., 22, *55*, 224, 225, *324*
Strasberg, P. M., 402, *439*
Strasters, K. C., 472, *482*
Stratmann, D., 389, *441*
Straub, F. B., 22, *55*
Straub, J. P., 62, *67*
Straub, K. D., 362, *369*, *382*
Straub, S. P., 389, *443*
Strecker, H., 183, *191*
Street, B. W., 361, 366, 390, *429*
Streichman, S., 341, *382*
Strelina, A. V., 406, *443*
Strickland, E. H., *382*
Strickland, S., 48, *53*
Stripp, B., 401, *434*
Strittmatter, C. F., 68, 70, *73*, 82, *95*, 471, *482*
Strittmatter, P., 68, 69, 70, 71, *72*, *73*, 252, *322*, 355, *376*, *382*, 422, *443*
Strobel, H. W., 77, 81, 84, 85, *93*, *95*
Strong, F. M., 349, *366*, *371*, *382*
Stryer, L., 366, *383*
Stuart, A., 113, *119*
Stuchell, R. N., 455, 460, *473*

Stuhne-Sekalec, L., 462, *482*
Stumm-Tegethoff, B., 30, *55*
Sturani, E., 356, *371*
Sturtevant, J. M., 114, 115, 116, 117, 118, *119*, *120*, 170, *182*, 183, *195*, 301, *317*
Subik, J., 460, *482*
Sudduth, H. C., 341, 344, *375*
Sugeno, K., 145, 146, 147, 148, 149, 150, 151, *157*, *180*, 184, *193*
Sugihara, A., 163, *170*, 197, *211*
Sugijara, A., 129, *312*
Sugimoto, M., 428, *444*
Sugimura, R., 466, *476*
Sugimura, T., 221, *324*, 414, *436*, *443*, 466, *479*
Sugimura, Y., 104, *111*, 129, *132*, 200, 202, 203, *211*, *212*, 266, *324*, *326*, 337, *380*
Sugita, Y., 449, *481*
Sugiyama, M., 187, *193*
Suh, B., 281, *324*
Suh, S. K., 387, *438*
Sukasova, M. I., 406, *439*
Sulkowski, E., 418, *443*, 473, *476*, *482*
Sullivan, J., 161, *169*, 177, *180*
Sullivan, P. A., 48, *53*
Sun, F. F., 22, 23, *55*, 190, *194*, 333, *378*
Sun, S. C., 394, *429*
Surdeanu, M., 468, *480*, *481*
Surdin, Y., 465, 466, *474*, *482*
Suriano, J. R., 197, *209*
Susor, W. A., 203, *211*
Sutherland, I. W., 128, *132*, 272, 273, *324*
Sutin, N., 187, *194*
Suzuki, H., 117, *120*, 293, 295, *318*, *324*
Suzuki, I., 74, *96*, 274, *314*, *320*
Suzuki, K., 85, *92*
Suzuki, M., 411, 413, *443*
Suzuki, Y., 305, *324*, 350, 357, *377*, *378*
Svihla, G., 428, *443*
Swami, K. S., 97, *100*, 422, *441*
Swank, R. T., 275, 276, 277, *324*
Swann, J. C., 359, *371*
Sweat, M. L., 74, 77, 79, 80, *89*, *95*, *96*
Sweetman, A. J., 359, *369*
Swift, H., 398, *436*, 462, *480*
Swigart, R. H., 393, *438*
Swijngedauw, B., 393, *435*
Sybesma, C., 304, 305. *316*, *317*, *324*
Sylk, S. R., 424, *443*
Symbas, P. N., 393, *431*
Symons, R. H., 113, 115, 117, 119, *120*
Symposium, 388, *443*, 461, *482*, 486, *496*
Szarcfarb, B., 404, *439*
Szarkowska, L., 190, *193*, 357, *382*
Szcepkowski, T. W., 178, *180*, 290, *315*
Szegli, G., 468, *481*

Szent-Gyorgyi, A. E., 362, *366*, *382*
Szilagyi, J. F., 464, *478*

T

Taber, H., 154, *157*, 198, *211*, 226, 227, 228, *324*, 465, *481*
Taber, H. W., 21, *55*
Taborsky, G., 173, *182*
Tada, K., 76, *96*
Tagawa, B., 118, *119*
Tagawa, K., 4, *6*, 74, 82, *91*, 102, *111*, 128, 129, *130*, *131*, 469, *477*
Tagawa, M., 129, *130*
Tager, J. M., 334, 360, 362, *382*
Tait, G. H., 309, *317*, 446, 448, 449, 453, 454, *475*, *479*, *482*
Taits, M. Yu., 404, *431*
Takahashi, A., 399, 414, *436*, *437*, *443*
Takahashi, H., 359, *372*, 462, *479*
Takahashi, K., 129, *131*, 137, *158*, 184, *192*, *193*, *194*
Takahashi, N., 359, *373*
Takamiya, A., 104, 105, *110*, 201, 202, 203, *208*, *210*, 305, *324*
Takanaka, A., 79, *92*, 400, 414, 428, *436*, *437*
Takano, T., 129, *132*, 138, 139, 140, 141, 142, 143, 144, 145, 146, 147, 148, 149, 150, 151, *155*, 162, 163, 164, *168*, *170*, 197, *211*, 362, *370*
Takasan, H., 406, *435*
Takass, O., 401, *439*
Takayanagi, M., 399, 400, *436*
Takeda, Y., 267, *319*
Takemori, S., 25, 27, 28, 29, 30, 31, 32, 35, 40, 41, 43, 48, *52*, *54*, *55*, 159, 166, *168*, *169*, 184, 185, *191*, *194*, *195*, 200, 203, *212*, 213, 214, 342, 344, 353, 358, *375*, *378*, *382*
Takemura, T., 88, *93*
Takenami, S., 274, *326*, 491, *496*
Takeshita, M., 449, *481*
Takesue, S., 69, *72*, 355, 356, *378*, *382*
Tallan, H. H., 81, *91*
Talwar, G. P., 404, *444*
Tamaoki, B., 409, *438*
Tamaoki, B-I., 85, *93*
Tamaoki, V-I., 85, *93*
Tamiguchi, S., 5, *6*
Tamiya, H., 4, 7, 219, 233, 265, *324*, *326*
Tamiya, N., 280, *326*
Tamura, G., 265, *326*
Tamura, M., 40, *51*, *57*
Tamura, S., 359, *372*, *373*
Tanaka, A., 399, 400, 414, 428, *443*

Tanaka, C., 65, *67*, 58, *73*, 155
Tanaka, K., 155, *157*
Tanaka, S., 458, *476*
Tani, S., 68, 71, *73*
Taniguchi, S., 227, 228, 229, 232, 247, 257, 289, 290, 292, 298, *314*, *324*
Tapley, D. F., 410, *443*
Tappel, A. L., 20, *54*, 359, *382*, 417, 421, *440*, *443*
Taptykova, S. D., 251, 256, 308, *316*, *320*, *323*, *324*, 429, *443*
Taquini, A. C., 387, *432*
Tarkowski, S., 451, *482*
Tarshis, M. A., 387, *443*
Tasaki, A., 288, 289, *324*
Tata, J. R., 410, 412, *442*, *443*, *444*
Taube, H., 187, *195*
Tavlitsky, J., 470, *475*
Tawaga, K., 221, 222, *318*
Taylor, A., 81, *89*
Taylor, C. B., 462, *473*, *482*
Taylor, Ch. E., 491, *496*
Taylor, C. P. S., 129, *130*, 224, *317*, 362, *382*
Taylor, D. N., 13, *16*, 26, *57*
Taylor, G. P. S., 228, 232, 291, *318*
Tchan, Y. T., 236, *321*
Tedro, S., 287, 293, 294, *315*
Teeter, M. E., 359, *382*
Tenhunen, R., 86, 87, 88, 89, *95*
Tephly, T. R., 81, 82, *89*, 451, *482*
Tepperman, J., 44, *49*
Teraoka, A., 65, *67*, 68, *73*
Terranova, T., 413, *444*
Terui, G., 428, *444*
Terwelle, H. F., 360, *381*
Teulings, F. A. G., 310, *317*
Tewari, K. K., 462, *482*
Theakston, R. D. G., 425, *444*
Theodoropoulos, D., 134, *158*
Theogaraj, T., 394, *432*
Theorell, H., 4, 7, 11, 12, *16*, 126, 127, 128, 129, *131*, *132*, 133, 134, *155*, *158*, 158, 163, *168*, *170*, 170, 171, 172, 177, *179*, *181*, *182*
Thimann, K. V., 21, *51*
Thofern, A., 468, 477
Thofern, E., 468, *477*, *480*
Thomas, J., *111*
Thomas, P., 80, *90*, 409, *432*
Thomas, T. D., 307, 308, *316*
Thompson, C. M., 128, *130*
Thompson, E. W., 145, 146, 147, 148, 149, 150, 151, *155*, *157*, *158*, 492, *494*, *496*
Thompson, W., 331, 332, 355, *378*
Thor, S., 79, *92*, 401, *440*
Thorell, B., 387, *440*, *444*
Thorgeirsson, S. S., 77, *95*

Thorn, M. B., 63, *66*, 351, *370*, *382*
Thornber, J. P., 306, *315*, *324*
Thornber, P. J., 305, *314*
Thorne, S. W., 106, *112*
Thornley, M. J., *307*, *317*
Thorp, S. L., 341, *383*, 392, 398, *445*
Thurman, R. G., 352, 359, *380*
Tien, W., 468, *482*
Tikhonova, G. V., 308, *316*, *324*
Tint, H., 127, *132*, 170, 171, *182*
Tisdale, H. D., 62, *66*, *67*, 213, 214, *215*, 331, 332, 337, *370*, *372*, *382*
Tissières, A., 99, *100*, 221, 222, 236, 237, 238, 246, 275, 276, *319*, *324*, *325*, 450, 467, *476*, *482*
Titani, K., 129, *131*, 133, 134, 137, 138, 139, 140, 141, 142, 143, 144, 145, 146, 147, 148, 149, 150, 151, 152, 154, *157*, *158*, *181*, 184, *192*, *193*, *194*, 197, *209*, *212*
Titchener, E. B., 332, *376*
Titova, L. K., 402, 404, *444*
Tixier, R., 188, *194*
Tobin, M. B., 402, *444*
Tobin, R. B., 361, *382*
Toda, F., 104, *111*, 129, *132*, 200, 202, 203, *211*
Tokugama, S., 197, *212*
Tokuma, Y., 163, *170*
Tokunaga, R., 449, *482*
Tolani, A. J., 404, *444*
Toll, M. W., 468, *483*
Toroff, O., 74, 75, 80, *90*
Torte, G. M., 405, *433*
Tosi, L., 20, *50*, 65, *66*, 97, 98, 99, *100*, 112, 114, 116, 117, *118*, *119*, 177, *182*, 186, *191*, 197, *208*, 421, 422, 423, *434*, *444*
Tota, B., 129, *132*, 197, *210*, 354, *373*, 395, *440*
Toth, A., 406, *443*
Towsend, M. G., 81, *89*
Trachtman, M., 177, *180*, 187, *192*
Traniello, M., 114, *120*
Trebst, A., 204, *207*
Trembath, M. K., 465, *481*
Tremblay, G., 412, *444*
Trousil, E. B., 278, 279, 280, *315*
Trudgill, P. W., 81, *91*
Trudinger, P. A., 256, 273, 274, *325*
Trussler, P., 451, *482*
Tsai, H., 159, *170*
Tsai, H. J., 159, 160, *170*
Tsai, R. L., 77, 84, *95*, 260, 261, 262, *317*, *325*
Tsofina, L. M., 311, *318*
Tsong, M., 82, *94*

Tsong, T. Y., 114, 115, 118, *120*
Tsou, C. L., 127, 128, *132*, 133, *158*, 163, *168*, 178, *180*, *182*, 189, *195*, 196, *211*, 340, *382*
Tsou, S-L., 127, *131*
Tsudzuki, T., 21, 24, 27, 28, 29, 33, 36, 37, 39, 40, 45, 47. *54*, *55*
Tsugita, A., 68, 70, 71, *72*, *73*,
Tsujimoto, H. Y., 106, *108*, *109*
Tsukihara, T., 163, *170*
Tsushima, K., 129, *131*, 183, *195*, 203, *209*
Tu, A. T. T., 133, *158*
Tu, H-T., 341, *369*
Tu, Shu-I., 205, *211*, 346, *382*
Tuboi, S., 453, *482*
Tucker, A., 395, *444*
Tucker, A. N., 464, *482*, *483*
Tuena, M., 359, *371*
Tullross, I., 138, 139, 140, 141, 142, 143, 144, *155*
Tuppy, H., 12, 13, *15*, *16*, 28, *55*, 133, 134, 135, 138, 139, 140, 141, 142, 143, 144, 152, 153, *156*, *157*, *158*, 159, *170*, 177, *181*, 197, *211*, 297, *322*, 449, 459, *480*, *482*
Turner, G., 197, *211*, *212*, *444*
Turner, L., 298, *320*, *325*
Tustanoff, E. R., 471, *482*
Tuzimura, K., 256, *325*
Tyler, A., 423, *442*
Tyler, D. D., 63, 64, *67*, 332, 333, 336, 358, *371*, *379*, *380*, *382*, 411, *434*
Tyrtyshina, G. F., 404, *431*
Tysarowski, W., 112, 118, *120*
Tyson, C. A., 258, 262, *317*
Tzagoloff, A., 22, 27, 29, 30, 32, 33, 43, *53*, *55*, *56*, 74, 83, *93*, 308, *320*, 331, 332, 338, 343, 353, 354, 361, *372*, *376*, *383*

U

Ueda, J., 402, *434*
Ueda, K., 196, 197, *211*, *212*, 387, 419, *444*, *445*
Ueda, S., 198, *209*
Ueki, T., 163, *170*
Uemura, T., 76, *91*
Uglova, N. N., 387, *429*
Ullrich, V., 69, *72*, *73*, 74, 75, 76, 78, 81, 83, 84, 86, *90*, *91*, *95*, 262, *318*
Ulmer, D. D., 135, *158*, 161, *170*, 173, 174, 175, 176, *182*, 183, 184, 185, 190, *192*, *195*
Umberger, F. T., 82, *95*
Umbreit, W. W., 274, *314*
Unestam, T., 428, *434*
Ungar, G., 402, *444*

Urata, G., 82, *96*, 454, *476*
Urban, P. F., 62, *67*
Urey, H. C., 486, *496*
Uribe, E., 108, *110*
Urich, K., 423, *444*
Uritani, I., 427, *442*
Uriuhara, T., 394, *439*
Urry, D. W., 30, *55*, *56*, 134, 135, *158*, 173, 174, 175, 176, *182*, 275, 276, 288, 301, *325*
Ushatinskaya, R. S., 420, *444*
Uteshev, A. B., 387, *443*
Uther, J. B., 388, *437*
Utsumi, K., 45, *56*
Uzan, A., 188, *194*

V

Valdivia, E., 463, *476*
Valenstein, E., 20, *57*, 397, 422, 423, *445*
Vallee, B. L., 190, *195*, 451, *482*
Vallejos, R. H., 409, *444*
Vallin, I., 334, 340, 346, 349, 358, 362, *376*, *383*
Van Buuren, K. J. H., *53*
Van den Bergh, S. G., 361, *369*
Vanderkooi, G., 25, 26, *56*, 362, 366, *383*, *384*, 463, *476*
Van der Wende, C., 31, *56*
Van den Bergh, S. G., 397, *444*
Van Dyke, R. A., 405, *444*
Vane, F., 79, *95*
Vanecek, J., 145, 146, 147, 148, 149, 150, 151, 154, *157*, *158*
Van Gelder, B. F., 23, 30, 31, 32, 33, 37, 38, 41, 45, 46, 47, 48, *49*, *53*, *55*, *56*, 127, *132*, 171, 174, 175, *182*, 275, 276, *325*, 342, 343, *381*
Van Gool, A., 224, *325*
Van Gool, H., 224, *325*
Van Groningen, H. E. M., 361, *383*
Van Heyningen, W. E., *444*
Van Holde, K. E., 301, *325*
Van Lier, J. E., 80, *95*
Vanneste, M. T., 34, *56*, 74, 81, 83, *93*
Vanneste, W. H., 24, 32, 34, 35, 36, 37, 38, 43, *56*, *335*, *383*
Vannotti, A., 196, *206*
Van Reen, R., 452, *482*
Van Rooyan, S., 161, *170*, 186, *194*
Van Rossum, G. D. V., 408, *431*, *444*
Van Tamelen, E. E., 349, *382*
Varnum, J. C., 4, *6*, 12, *15*, 129, 130, *130*, 162, 166, 167, *168*, 362, *370*
Vary, M. J., 460, *482*
Vasilev, P. V., 387, 429

Vassanelli, P., 81, *92*
Vassiletz, I. U., 398, *444*
Vater, J., 105, *109*, *112*
Vazquez-Colon, L., 338, *383*
Veeger, C., 352, *373*
Veitch, B., 463, *483*
Veldsema-Currie, R. D., 334, *382*
Veldstra, H., 403, *430*
Velican, D., 398, *445*
Velick, S. F., 69, 70, *73*, 355, *382*
Velins, A., 14, *15*, 26, *52*, 205, *209*
Venables, W. A., 469, *482*
Ven den Bos, P., 203, *206*
Venner, H., 462, *483*
Vennesland, B., 205, *210*
Verhulst, A. M., 462, *481*
Vernberg, F. J., 391, *444*
Vernberg, W. B., 391, *444*
Vernon, L. P., 10, *16*, 105, *111*, 205, *208*, *211*, *212*, 250, 265, 267, 271, 272, 284, 285, 296, 298, 300, 304, 311, *316*, *319*, *325*, 354, *376*, 464, *479*
Vesco, C., 461, *483*
Vidal, J. C., 337, 352, *367*
Viereck, G., 407, *441*
Vigers, G. A., 360, *383*
Vignais, P. M., 361, *383*
Vignais, P. V., 74, *94*, 361, *383*
Villavicencio, M., 352, *370*
Vinas, J. L. R., 387, *442*
Vinogradov, S. N., 4, *6*, 12, *15*, 134, 135, *156*, 159, 160, 161, 162, *168*, *169*, *170*, 171, 173, 175, *180*, *182*, 268, 269, *317*, *325*
Vitale, J. H., 406, *444*
Viti, I., 394, *444*
Vitols, E., 118, *120*, 463, *478*
Vlcek, D., 466, *477*
Voigt, W., 80, *95*, 400, *444*
Volkenstein, M. V., 185, *191*
Volland, C., 428, *444*, 449, 472, *478*
Volpe, J. A., 28, *51*
Volpert, E. E., 388, *430*
Vorbeck, M. L., 462, *482*
Voss, D. O., 200, *207*
Votapkova, Z., 390, *436*
Votsch, W., 462, *482*
Vredenberg, W. J., 203, 211, 304, 305, *312*, *325*

W

Wada, F., 76, 79, 80, 81, *96*, 355, *383*
Wada, H., 344, *378*
Wada, K., 134, *158*, 166, *169*, 184, 185, *194*, *195*, 214, *215*, 344, *382*, *383*
Wada, O., 82, *96*
Wadkins, C. L., 5, *6*, 334, *375*
Wagenknecht, C., 355, *379*, *383*
Waggoner, A. S., 366, *383*
Wainio, W. W., 22, 24, 28, 30, 31, *33*, 38, 39, 40, 43, 44, *50*, *52*, *56*, 62, 63, *66*, *67*, 160, *170*, 189, *194*, 213, *215*, 342, 351, *374*, *376*, 452, *482*
Wakabayashi, T., 463, *476*
Wakil, S. J., 355, *373*
Walasek, O. F., 35, *53*, 127, 128, *131*, 138, 139, 140, 141, 142, 143, 144, *155*
Wald, G., 484, 486, *496*
Walker, D. A., 105, *111*
Walker, G. C., 244, *325*
Walker, J. G., 389, *444*
Wallace, P. G., 463, *482*
Wallace, W., 224, *325*
Wallach, D. F. H., 332, *383*
Wallis, O. C., 427, *434*
Wang, J. H., 205, *211*, 346, 363, *367*, *369*, *370*, *373*, *382*, *383*
Wang, J. K., 206, *207*
Waravdekar, V. S., 425, *437*
Warburg, O., 3, 4, 7, 13, *16*, 43, *56*, 220, 225, *325*, 359, *383*, 412, *444*
Ward, C. L., 420, *445*
Ward, J. M., 451, *483*
Waring, R. H., 81, *91*
Warme, P. K., 160, *170*, 252, 253, *325*
Warmsley, A. M., 470, *483*
Warshaw, J. B., 361, *375*
Wasserkrug, H. L., 392, *442*
Wasserman, A. R., 184, *191*, 200, *211*, 342, *370*
Watanabe, H., 133, *155*, 341, *367*
Watanabe, I., 256, *325*
Watanabe, S., 188, *194*, *195*, 406, *440*
Watari, H., 70, 115, *120*, 352, *383*
Waterman, M. R., 84, *96*, 365, *384*
Watersfield, M. D., 82, *96*
Watson, K., 463, *483*
Watson, M. L., 353, *380*
Watson, S. W., 307, *321*
Watt, G. D., 170, *182*, 183, *195*
Wattiaux, R., 414, *445*
Wattiaux de Conmink, S., 414, *445*
Waygood, E. R., 454, *479*
Weaver, E. C., 204, *211*
Webb, E. C., 2, *6*
Webb, J. L., 359, *384*
Weber, A., 396, *431*
Weber, C. N., 460, *476*
Weber, C. W., 398, *439*
Weber, H., 70, *73*
Weber, M. W., 197, *209*
Webster, D. A. 102, *112*, *211*, 227, 229, *325*

Author Index

Webster, G., 364, 365, *383*
Wecksler, M., 455, *475*
Wedgwood, R. J., 352, *376*
Wegdam, H. J., 63, *67*, *68*, 348, 349, *381*, *383*
Weibull, C., 265, *325*
Weidemann, M. J., 361, *383*
Weigers, P. J., 361, *381*
Weikard, J., 105, 106, *109*, 112
Weimar, V. L., 405, *445*
Weinhouse, J., 412, *445*
Weinzierl, J. E., 4, *6*, 12, *15*, 129, 130, *130*, 162, 166, 167, *168*, 362, *370*
Weiser, M., 308, *323*
Weiss, H., 331, *383*, 460, *483*
Weiss, J., 2, 7
Weiss, R., 462, *483*
Weiss, W., 68, 70, *72*, *73*
Welsch, M., 21, *49*, 254, *314*
Werner, S. C., 410, 411, *445*
Westcott, W. L., 187, *195*
Whale, F. R., 257, 299, *325*
Wharton, D. C., 22, 23, 24, 25, 27, 29, 30, 31, 32, 33, 36, 39, 41, 43, 44, 48, *49*, *51*, *55*, *56*, 332, 342, 343, 353, *371*, *372*
Whatley, F. R., 5, *6*, *207*, 310, *319*
Wheeldon, L. W., 351, *383*
Whipple, H. E., 388, *436*
White, D. C., 219, 220, 223, 227, 231, 236, 238, 239, 240, 263, 308, *317*, *324*, *325*, 426, *435*, 448, 462, 463, 464, 467, 468, 469, 470, 472, *475*, *476*, *478*, *481*, *482*, *483*
White, J. B., Jr., 29, *49*
Whittaker, P. A., 357, 359, *379*
Whittenbury, R., 467, *474*, *483*
Whysner, J. A., 73, 77, 78, 80, 84, *96*
Widmer, C., 189, *195*, 213, *215*, *216*, 340, 351, 357, 358, *369*, *370*, *372*, *383*
Wiener, E., 133, *156*, 159, *169*
Wikstrom, M. K. F., 360, *383*
Wilbur, K. M., 414, *436*
Wilkie, D., 465, 466, *479*, *480*
Wilkinson, J. F., 273, *324*
Will, S., 198, *209*, 465. *478*
Williams, C. H., 85, *92*, 355, *376*, *383*
Williams, C. M., 68, *73*, 197, *210*, 418, *435*, *442*
Williams, D. M., 306, *324*
Williams, F. R., 247, *325*
Williams, G. P., 160, *170*
Williams, G. R., 28, 30, 36, 39, 40, 41, 42, 43, *50*, *65*, 62, *66*, *68*, 72, 128, *132*, 159, *170*, 172, 173, *179*, 182, 185, 187, *194*, *195*, 331, 332, 333, 334, 335, 339, 341, 345, 355, 361, 363, *368*, *369*, 378

Williams, J. N., Jr., 26, *56*, 185, *195*, 196, *211*, 214, *216*, 335, 341, *383*, 392, 398, *445*
Williams, J. P., 280, *325*
Williams, R. G., 5, *6*,
Williams, R. J. P., 2, 7, 34, 36, *49*, *56*, 83, 84, *91*, *96*, 172, 175, 176, 177, *179*, *182*, 260, *313*, 364, *384*
Williams, T. I., 399, *439*
Williamson, D. G., 77, 78, 81, *96*
Williamson, J. R., 352, 359, *380*, 389, 390, *431*, *441*
Willick, G. E., 30, *56*
Willis, R., 389, *445*
Wills, E. D., 450, *476*
Wilson, D. F., 34, 35, 38, 39, *46*, 47, *50*, *56*, 63, 64, *66*, *68*, 172, *179*, 331, 339, 345, 348, 353, 355, 360, 364, 365, *369*, *370*, *378*, *384*, 409, 416, *445*
Wilson, D. L., 76, *80*, 93
Wilson, H. C., 485, *496*
Wilson, J. B., 415, *445*
Wilson, L. D., 73, 75, 76, 77, 79, 80, 81, *96*, 410, *445*
Wilson, M., 27, 29, 33, *57*, 360, *384*, *445*
Wilson, S. B., 362, *384*
Wimpenny, J. W. T., 239, 240, 282, 283, *314*, *317*, *325*, 469, 470, 472, *476*, *482*, *483*
Winget, G. D., 108, *110*
Winkler, H. H., 361, *384*
Winkler, K. C., 472, *482*
Winters, R. W., 401, *437*
Wintersberger, E., 461, *483*
Wintrobe, M. M., 452, *476*, *478*
Wirtz, K. W., 462, *483*
Wise, C. D., 87, 88, *96*
Wisenberg, D., 163, 164, *168*
Wiskich, J. T., 101, 102, 103, *112*, 199, *211*
Wit-Peeters, E. U., 397, *438*
Witt, H. T., 104, 105, 106, *109*, *111*, *112*, 204, *210*, 211
Wittenberg, B. A., 20, 40, *57*, 397, 422, 423, *445*
Wittenberg, J. B., 20, 40, *57*, 394, 397, 422, 423, *445*
Wittkop, J. A., 81, *96*
Woernley, D. L., 413, *431*, *445*
Wohlrab, H., 42, *57*, 330, 338, 350, 361, 364, *373*, *384*, 391, *445*, 452, *483*
Wojtczak, A. B., 418, *445*
Wojtczak, L., 341, 344, 360, *367*. *384*, 418, *443*, *445*, 456, *483*
Wolff, E. C., 411, *445*
Wolken, J. J., 103, *109*, 201, 203, *207*, *208*, 211, 429, *445*

Wolkinson, A. E., 450, *476*
Wolman, Y., 135, *158*
Wong, D., 329, 346, 352, 354, 358, 359, *369*, *371*
Wong, J., 104, *111*, 203, *210*
Wong, S. H., 73, *91*
Woo, K. C., 106, *112*
Wood, L., 414, 421, *439*
Woods, R. I., 388, *445*
Woodward, D. O., 24, *50*, 459, 467, *475*
Woodward, D. P., 428, *433*
Wordy, R. W., 280, *315*
Work, T. S., 456, 461, *473*, *476*, *483*
Wosilait, W. D., 187, *195*
Wright, B. V., 341, *366*
Wright, R. L., 183, *192*
Wrigley, C. W., 246, *325*
Wrogemann, K., 396, *443*
Wu, M., 359, *372*
Wulff, D. L., 468, *483*
Wüthrich, K., 160, 161, 162, *170*
Wyman, J., 40, *57*, 348, 365, *377*
Wyndham, R. A., 87, *92*

Y

Yagi, T., 280, *318*, *326*
Yago, N., 74, *96*, 410, *445*
Yajima, H., 159, *169*
Yakushiji, E., 22, 41, 48, *54*, *57*, 104, *111*, 129, *132*, 200, 202, 203, *211*, *212*, 212, 213, *216*, 266, *324*, *326*, 339, 344, *384*
Yamada, K., 407, *445*
Yamada, M., 407, *445*
Yamada, Y., 403, *445*
Yamagata, S., 196, *212*, 387, *445*
Yamaguchi, S., 233, 265, *324*
Yamagutchi, S., 282, *326*
Yamagutchi, T., 265, *326*
Yamamoto, H., 205, *211*
Yamamoto, K., 402, *434*
Yamamoto, T., 37, *57*
Yamamura, Y., 188, *194*
Yamanaka, T., 12, *16*, 65, *68*, 102, *112*, 112, 114, *119*, *120*, *121*, 127, 128, *132*, 161, *169*, 171, 172, *182*, 183, 185, 187, *193*, *195*, 197, 199, 200, 202, 203, *211*, 235, 241, 242, 243, 244, 245, 266, 271, 274, 303, *318*, *321*, *326*, 344, 356, *384*, 491, *496*
Yamano, T., 68, *72*, 74, *75*, 76, 81, 83, *91*, *93*, 356, *373*, 399, 410, *436*
Yamaoka, K., 88, *93*
Yamasaki, H., 187, *193*

Yamashita, H., 356, *377*
Yamashita, J., 4, *6*, 112, 114, 117, *119*, *120*, *121*, 127, 129, *130*, 311, 312, *326*
Yamashita, S., 62, *68*, 317, *349*, 358, *384*
Yamazaki, I., 40, *51*, *57*, 187, *193*, 355, 356, *373*
Yanagishima, N., 466, 467, *479*, *483*
Yanasugondia, D., 359, *384*
Yang, P. C., 27, *55*
Yang, W. C., 359, *384*
Yano, Y., 82, *96*, 332, *378*
Yaoi, H., 4, 7, 219, 233, 265, *326*
Yaoi, Y., 4, 145, 146, 147, 148, 149, 150, 151, 154, *157*, 197, 198, *212*
Yasaitis, A. A., 349, *380*
Yasunobu, K. T., 128, 129, *131*, 137, 138, 139, 140, 141, 142, 143, 144, *157*, *158*, 171, *181*
Yates, M. G., 277, 310, *326*
Yças, M., 221, *326*, 446, 455, 462, 466, *483*
Yegian, D., 468, *483*
Yeh, C. C., 206, *207*
Yepes, J. L., 387, *442*
Yocum, C. S., 104, *112*
Yoda, B., 453, *476*
Yodaiken, R. E., 63, *67*
Yohno, T., 73, *92*
Yohro, I., 409, *445*
Yohro, T., 74, *96*
Yokaiden, R. E., 332, *376*
Yokolev, V. A., 277, *317*
Yokota, K., 40, *57*
Yoneda, M., 129, *130*
Yonetani, T., 20, 22, 23, 24, 27, 30, 32, 33, 34, 36, 40, 41, 43, 44, 46, *50*, *54*, *55*, *57*, *168*, 177, *180*, 185, 186, *195*, 199, *209*, 213, *215*, 340, 342, 344, *384*
Yoneyama, Y., 70, *72*, 449, *481*, *483*
Yong, F. C., 14, *16*, 27, 30, 33, 37, 43, *52*, *57*, 84, *96*, 301, 302, *326*, 365, *384*
York, J. L., 13, *14*, *16*, 26, *50*, *57*
Yoshida, T., 454. *483*, 491, *496*
Yoshida, Y., 68, 71, *73*, 118, *121*
Yoshikawa, H., 68, 70, *72*, *73*, 352, 361, *376*, 449, *483*
Yoshikawa, S., 47, *57*
Yotsuyanagi, Y., 465, 470, *475*, *483*
Young, A. M., 203, 205, *209*
Young, D. G., 79, 80, *96*
Young, J. H., 362, 366, *384*
Young, R. B., 74, 77, *95*
Yu, C.-A., 73, 77, 84, *95*, *96*, 260, 261, 262, 299, *315*, *317*, *321*, *325*, *326*
Yu, R., 459, *483*
Yubisui, T., 21, 24, 28, *55*, *57*, 342, *380*
Yunusova, Kh, A., 407, *445*

Z

Zaguskin, S. L., 417, *445*
Zaguskina, L. D., 417, *445*
Zahler, W. L., 337, *384*
Zajdela, F., 463, *483*
Zalind, N. D., 82, *89*
Zajdela, F., 413, *483*
Zalkin, H., 337, *384*
Zaluska., H., 341, *384*, 456, *483*
Zamchek, N., 406, *444*
Zamfirescu-Gheorghi, M., 398, *445*
Zand, R., 173, 175, *182*
Zanen, J., 128, *132*, 198, *210*
Zanetti, G., 202, *207*
Zange, M., 78, 79, *95*
Zannoni, V. C., 77, *90*, 399, *433*
Zarinish, L. A., 398, *438*

Zaugg, W. S., 349, 350, 357, 358, *379*
Zeile, K., 11, 12, *16*, 127, *132*, 161, *170*
Zelik, T. G., 417, *445*
Zerbe, T., 20, *54*, 422, *440*
Zerfas, L. G., 112, *118*, *119*
Ziegler, D. M., 62, *66*, *68*, 352, *384*
Ziegler, F. D., 360, *383*, 387, *445*
Ziliotto, G. R., 387, *445*
Zilversmit, D. B., 462, *483*
Zipper, H., 23, 29, *54*
Zipurski, A., 87, *92*
Zitin, B., 408, *445*
Zittle, C. A., 408, *445*
Zubovskaya, A. M., 394, *445*
Zubrzycki, Z., 69, *72*, 74, 78, 81, 83, *95*
Zuckerkandl, E., 484, 488, 492, *497*
Zviagilskaya, P. H., 199, *212*, 428, *445*
Zylber, E., 461, *483*

SUBJECT INDEX

A

Absorption spectra
  charge transfer bands 34, 35, 36, 177
  low temperature 10, 99, 101, 115, 171,
      172, 184, 197, 198, 200, 202,
      242, 251, 255, 256, 265, 266,
      303, 339, 396, 414, 416–418,
      420, 421, 425
  of haemochromes 8
  reflectance spectroscopy 287, 467
Acclimatization VIII B3 (389–391)
Acetylsalicylic acid 398, 401
Acriflavin as mutagen 466
  effect on cytochrome B-574 264
  and trypanosomes 426
Actinomycetes cytochrome in 429
Actinomycin 455
Action spectra
  cytochrome a_1 220
  cytochrome aa_3 (oxidase) 3, 21
  cytochrome d 238, 240
  cytochrome o 226, 228, 229
  inhibition of cytochrome formation in
      adapting yeast 473
  oxidation of cytochrome 555 (556) of
      Chromatium D 304
Adenosinediphosphate (ADP) (see
      Respiratory Control)
  inducer of respiration 471
  passage through mitochondrial
      membrane 361, 390
Adenosine monophosphate (AMP) cyclic
      390, 454
Adenosine triphosphate (ATP) (see also
      *Energized state, Phosphorylation,
      Reversed electron flow,
      Transhydrogenase*)
  effects on bacteriochlorophyll
      synthesis 454
    on ferrochelatase 454
    on porphobilinogen deaminase
      447, 454
    on porphyrin biosynthesis 454
    on uroporphyrinogen III
      synthetase 454
  formation as last step of energy
      conservation 363, 364

Adenosine triphosphate—continued
  in brain 402
  in evolution 488
  in gastric mucosa 405
  in heart 363, 393
  in muscular contraction 395
Adipose tissue and cold acclimatization
      389, 390
  mitochondria in white and brown – 389
  P/O ratio 389
Adrenalectomy 404, 410, 462
Adrenals cortex mitochondria 73, 409, 410
  cytochrome oxidase 409, 410, 462
  effect on P-450 in microsomes
      355, 400, 410
Adrenocorticotrophic hormone 410
δ-*Aminolaevulinic acid (ALA)* 446, 447,
      450, 468
  incorporation of labelled – in
      cytochromes 456, 458, 460
ALA-dehydratase 446, 447, 450, 451,
      454, 471
ALA-synthetase 446, 447, 450, 451,
      453, 454, 471
Aldimines of haem a and cytochromes a
      26, 264, 428
Aldosterone 410, 462
Algae blue-green 105, 106, 200, 203, 204,
      468, 467
  brown 129, 200, 202, 203
  colourless 105
  diatoms 200, 202
  green 105, 106, 129, 200–204
  red 105, 129, 200, 202, 203, 468
  precambrian 486
Alkane see *Cytochrome P-450*
2-*n-Alkyl-4-hydroxyquinoline-N-oxide*
      331, 347, VII E2b (350 f.)
Allostereoisomerism 348, 365 (see also
      Conformational Change)
  in cytochromes a and a_3 364, 365
  in cytochrome b 365
  in cytochrome bc_1 and other cytochrome
      complexes 348, 350, 365
  in haemoglobin 365
Alloxan 400
Allylglycine 466
Allylisopropylacetamide 454

Amino acid residues (see also individual cytochromes)
  deletions 153
  symbols 137
5-Amino-4-hydroxyvaleric acid 446
α-Amino-β-oxoadipic acid 446
Amnion 409
Amoebae 426
Amytal see *Barbiturates*
Anaemia of iron deficiency 391
Anaerobic photosynthesis 487
Anaerobiosis
  in boring organs of neogastropode 423
  in early life on earth 486
  in protohaem formation 494
  in stages of *Ascaris* development 424
  producing yeast mutants 466
Androgens 400, 409
  and P-450 409
Aniline hydroxylation 400 (see also *Cytochrome P-450*)
Annelids 423, 487
Anostraca 417
Anoxia VIII B2 (387–389), 466
Antigenic properties 30, V B6d (188–189), 230, 341
Antimycin A insensitive fungal oxidase 428
  insensitive NADH-dehydrogenase 352
  insensitive NADH-cytochrome *c* reductase VIII Fl (354–356), 419, 421
  insensitive NADPH-cytochrome *c* reductase 419, 421
  insensitive succinic-cytochrome *c* reductase 422
  sensitive NADH-cytochrome *c* reductase 353, 412, 424
  sensitive succinate-cytochrome *c* reductase 424
  sensitive TMPD-oxidase 396
  and cytochrome *b* 63, 102, 250, 255, 256, 267, 309, 347–350, 409, 426, 462
  and cytochrome bc_1 VII E2a, 347, 349, 350, 354, 358
  as growth stimulant in *Tetrahymena* 350
  inhibitor of cytochrome oxidase synthesis in yeast 462
  inhibitor of respiratory chain, 330, 331, 347, 349, 352, 361, 415, 425
Antisera against brain mitochondria 404
Aplysia 422
Arginine in myoglobin 253
  labelled—incorporation in cytochrome b_5 and *c* 458
  in moth pupae 457

Arteries VIII C4b (406)
Arteriosclerosis 406
Arthropods 415, 417–421
Ascaris muscle cytochromes 414, 415
  (see also *Worms, Nematodes*)
Ascidiaceae 416
Ascites tumour cytochromes in Ehrlich – 413
  in mouse and rat – 413, 414, 456
  in hyperbaric oxygen 389
Ascorbate-DCIP 226, 277
  in liver 401
  in retina 405
  -oxidase in fungi 429
  -TMPD 230, 239, 255, 428
Atmungsferment 3
ATP see *Adenosine triphosphate*
ATP-ase Na^+K^+ stimulated in intestinal mucosa 406
  in nerve endings 403
  in respiratory epithelium 407
  oligomycin-sensitive in yeast 331, 332, 334, 361
  in bacteria 307, 308
  in myocarditis 394
  in reactivation by phospholids 337
Atractyloside 360, 361
Azide as inhibitor of cytochrome oxidase 45–47, 228, 344, 427
  as inhibitor of phosphorylation-coupled respiration 331, 360
  in amphibian muscle 396
  in Bivalva 421
  in insect metamorphosis 415
  protecting cytochrome *a* from light 473
Azoreductase 356
Azurin 242, 244, 273

B

Bacteria continuous culture 470, 471
  coupled phosphorylation in VI E2 (309–312)
  mutants 220
  spheroblast 307
Bacterial cytochromes VI (217 ff.)
  type A III B1 (21–22), 18, VI B1 (217–225)
  type B IV A (59–61), VI C (225 ff.)
  type C V A (123–126), VI D (265 ff, 487)
  oxidases VI 220 ff.
  evolution of cytochrome *c* X C (488–494)
  repeated subunits in cytochrome *c* 488, 489

Bacterial cytochromes—continued
  localization and organization of
    cytochrome complex VI E1
    (306–309)
  reaction of – with oxidases 491
Bacteriochlorophyll 292, 300, 303–306,
    453, 454
BAL see *Dimercaptopropanol*
Barbiturates and NADH-cytochrome *c*
    reductase 352, 353
  in avian salt glands 408
  inhibition, direct and indirect 359
  of respiratory chain 330, 351, 359
Barnacles 417
Bees 420
  muscles 395, 397, 414, 456
Beetles see *Coleoptera* 419, 420
Benzimidazole 462
Bile duct obstruction 398, 401
Bile salts and open system 340, 342
  in preparation of cytochrome aa_3 22, 23
Bilirubin effect on ALA synthetase 453
  brain respiration 405
Biliverdin formation IV D (86–89)
Biosynthesis (see also *Cytochrome aa_3*,
    Cytochromes (General))
  feedback 471
  induction and repression by O_2,
    glucose and light IX I
    (469–473)
  of chlorophyll 463
  of cytochromes IX (466 ff.)
  of cytochrome a_1 223
  of cytochrome *c* IX B (455–457)
  of cytochrome d_1c 223
  of cytochrome oxidase IX D (459, 460)
  of cytoplasmic membranes IX F
    (463–464)
  of haem 447, IX A (446 ff.)
  – regulation 447, IX A3 (452–455)
  – effects of metals IX A2 (450–452)
  of haem *a*, effect of Cu 452
  of mitochondria 460–463
Bivalva 421, 422
Blattodea 421
Blood, cytochromes and cytochrome
    oxidase in – VIII C4c (407)
B.M.R. see *Metabolism, basic*
Brachyurea 417
Brain cytochromes in – VIII C3 (402–405)
  distribution in different parts of –
    404–405
  iron deficiency, effect on – 450
  respiration and oligomycin 360
  thyroxine effect on – 411
Branching of respiratory chain in cellular
    respiration 329, 361, 463

Branching—continued
  in bacteria 470
  in nematodes 424
Branchiopods 417
Butterflies see *Lepidoptera* 418–419
By-pass see *Respiratory Chain*

C

Caffeine 466
Cambrian biological explosion 486
Camphor as P-450$_{Cam}$ substrate 261
Carbon monoxide effect on cellular
    respiration 196, 361
  inhibition in invertebrates 417–419,
    421–424, 426
  reaction with cytochromes and
    cytochromoids 2, 10, 59
    (see also individual cytochromes)
  with haem *a* compounds 27
Carbon tetrachloride effects in liver
    398, 401
Carcinogen detoxication 414
Carcinoma adeno- 413
  of mice breast 413
  squamous 413
  Walker-256 413, 414
Cardiolipin 25, 31, 190, 336, 462, 464
Carotenoids 463, 464
Carotid body chemoreceptors 388
  cytochrome a_3 function in – 388
  ligation of – 404
Castration 401, 404
Catalse 450, 451, 468, 470
Catechols 359
Catecholamines and heat production 390
Cell nuclei cytochromes 20, 386, 387
  respiratory chain and oxidative
    phosphorylation 387
Cellular respiration 3, VII (327 ff.) see
    also *Electron transport* and
    Respiratory chain, *Heart* and
    other organs.
  branching see *Branching*
  complexity of 327–329
  cyanide-insensitive 339
  endogenous 408, 414
  inhibitors 330, 331, 334, VII D 5 (344),
    VII E5 (349–351), VII F4
    (359–360), 408
    (see also *Barbiturates, Rotenone,
    Piericidin, Thenoyl-trifluoro-
    acetone, Antimycin A, 2-n-
    Alkyl-4-hydroxyquinoline
    N-oxide, Naphthoquinones,
    Dimercaptopropanol, Cyanide,
    Azide* and *Oligomycin*)
  mobile carriers 329

Subject Index

Cellular respiration—continued
  non-phosphorylating – in heart
      muscle 393
  states III and IV 334
  stoichiometry of cytochromes and of
      cytochrome complexes 329,
      VII B (335)
  theories of – 329–331, VII H (361–366)
Cephalopods 421
Charge transfer bands see *Absorption
      Spectra*
Chemiosmotic (electrochemical theory)
      (see also *Phosphorylation*) 332,
      333, 346, 363, 405, 406
Chemiautotrophs and oxidative
      phosphorylation 303, 310, 312
      (see also *Nitrate and nitrite
      reduction, Sulphate reduction,
      Sulphite oxidoreduction*)
Chemostat 470, 471
Chloramphenicol 456, 459, 461, 471
Chlorate-resistant fungal mutants 469
Chlorin-iron in cytochrome *d* and (?) d_1
      243
Chlorocruorohaem 13
p-Chloromercuriphenylsulphonate 400
3-(p-Chlorophenyl)-1,1-dimethylurea,
      inhibitor of fungal respiration
      427
Chlorophyll a (see *Chloroplasts, Electron
      Transport, Photosynthesis*)
  in evolution 487
  in green plants IV F3 (104–108) and
      V C (201–206)
Chlorophyll b 448
Chloroplasts 5, 101, 206, 308, 332, 463
Chlorpromazine 404, 427
Cholate in cytochrome aa_3 preparation 22
Choline 459
Chromatophores 307, 309
Cilia of respiratory tract 407
Circular dichroism (see *Cytochrome c*)
Cirripedia 417
Cobalt effect on cytochrome oxidase 451
  and formation of porphyrins 450, 451
  and P-450 formation and disappearance
      451
Coccidia (see *Parasites*, Malaria) 425
Cochlea cytochrome oxidase in 405
Cockroaches (see *Blattodea*) 421
Coding trinucleotide – for amino acids
      485, 489
Coelenterata (see also *Nematocysts*)
      425, 487
Coenzyme Q (Ubiquinone) Q_8 238, 274
  Q_{10} 330, 337, 349–352, VII F3
      (356–359)

Coenzyme Q—continued
  Q of short chains 350
  as central point of respiratory chain
      354, 357
  in complex III 347, 352, 357
  in reduction of cytochrome *b* 358
  in submitochondrial particles 357
Coeruloplasmin 29
Cold acclimatization (see also *Temperature*)
      VIII B3 (389–391)
Coleoptera metamorphosis 420
Complexes I–IV see *Respiratory Chain*
Conchostraca 417
Conformation changes 363–366
Co-operativity see *Allostereoisomerism*
Copper (see also *Cytochrome aa_3*)
  absorption band 830 nm in cytochrome
      oxidase 32
  as mutagen 468
  -deficiency 391, 450, 452
  effect of biosynthesis of cytochromes
      452
  EPR 31, 32
  in cytochromes 2, 3, 391
  in nitrite reductase of *Pseudomonas
      aeruginosa* 244
  in phospholipid formation 391
  -protein azurin 242, 244, 487
Coprohaem as substitute for protohaem
      468
Coproporphyrin accumulation in Fe-
      deficient bacteria and algae 450
  as substitute for protohaem 468
  conversion to protoporphyrin and
      protohaem 448, 450
  in yeast mutant 466
  in Zn excess 451
Coproporphyrinogen III 447, 448, 494
Coproporphyrinogenase (anaerobic) 448,
      466, 470
Core protein (see *Structural Protein*)
Corpora lutea 409
Corti organ 405
Corticosteroids 409, 410
Coupled oxidation of haemoglobin 86–89
Coupled respiration see *Respiratory
      Control* 333, 334
Coupling factor see *Phosphorylation* 337
  CF_0 347
  F_1 and ATPase 361
Crabtree effect see *Glucose repression*
Crevice structure 2, 71, 163, 166, 172,
      177, 178
Cristae see *Mitochondria, inner membrane*.
Crossover points 339, 345, 394, 408
Crustacea 417
Cryptocytochrome c VI D5i (293–295)

Subject Index

Cryptohaem compounds 11, 264, 449
Cyanide (see also individual Cytochromes
    and Cytochrome oxidase) 44, 45
  as inhibitor of cellular respiration
    330, 331, 428
  as inhibitor of plant oxidases 104, 415
  as protector of cytochrome a_3 against
    light 473
  in insect metamorphosis 415, 418–420
  inhibitor in *Coleoptera*, *Diptera*,
    Gastropoda and Worms 419–424
  inhibitor in *Ascaris* muscle 396
  -poisoning 404
  -tolerance in anoxia 387
Cyanophyceae (see also *Algae, blue-green*
    and *Phycobiliproteins*) 486
Cybernetic control 328, 452–454, 471
Cycloheximide 454, 457–461
Cysteine (see *Cytochrome c*, *Haem c*)
  -antagonist in yeast 466
Cytochromes (general) absorption
    spectrum 3
  and allylisopropylacetamide 454
  and oxidases 2
  as enzymes 2
  autoxidizability 1, 2, 276
  bacterial (see Bacterial Cytochromes)
    4, 18, 21, 32, VI (217 ff.)
  bioenergetics 5
  biosynthesis IX (446 ff.), 397
    of apoenzymes and incorporation
      of haems 460
  classification II A (8–12)
  cytochromoids 10, VI D5 (284–296)
    (see *Cytochromes c'* and *cc'*)
  distribution in the cell V B7a (195–201),
    VIII B1 (386–387)
  definition 1
  evolution X (484 ff.)
  history of discovery 3–5
  hydrophobic drop structure 163
  in adipose tissue 390
  in cell nuclei 20, 386, 387
  in fungi 427–429
  in inner mitochondrial membrane 331
  in insects VIII D 414, 415, 418–426
  in insect metamorphosis 397
  in malignant tumours VIII C6
    (412–414)
  in plants VIII E (415, 427–429)
  in temperature regulation and
    acclimatization VIII B3
    (389–391)
  in yeast 415, 427–429 (see also *Yeast*)
  magnetochemistry 2, 133
  nutrition and – VIII B4 (391–392)
  physiological aspects VIII (385)

Cytochromes—*continued*
  primary structure 4
  prosthetic groups IIB (12–14)
  purification and crystallization 4
  ratios aa_3/c in mitochondria 335
    $aa_3/c + c_1$ in inner mitochondrial
      membrane 335
    c_1/c; $c + c_1/b$ in respiratory chain
      VII B (335)
    in rat kidney 410
    in sea urchin spermatozoa 408
  redox potentials 227
  reductases VII E3 (351–354), VII F
    (354–360) (see also *Cytochrome
    b-* and *c-reductases*) 4, 331
  steady state reduction VII C4 (338–339)
  stoichiometry in *Amoeba* 426
  tertiary structure in leukocytes 4, 163
  turnover number 24, 25, 340, 397
Cytochromes of type a 8, 9, 11, III A (17 ff.),
    397, 407, 428, 464
Cytochrome a 2, 18, III 5a (33–35),
    38, 467
  in brain 402
  interaction with cytochrome c VIII
    D3 (343, 344), 427
  stoichiometry 343
  position in respiratory chain 339, 344
  protection against light by azide 473
  redox potential 38, 39, 339, 345
  steady state reduction in 343, 402
Cytochrome a-601 of *Bacillus subtilis* 266
Cytochrome a_1 10, 18, 22, VI B1a
    (220–225), 238
  a_1^{2+} 221
  a_1CO 221, VI BIe (225)
  a_1CN 221, VI BIe (225)
  a_1b of *E. coli* 223
  absorption bands 223
  in *Acetobacter pasteurianum* 220
  in bacterial particulate preparations
    VI BIc (222, 223), 255
  in formate oxidase of *Nitrobacter* 223
  in yeast VI BIb (221, 222), 463
  inhibitors 223
  interaction with ferricytochrome
    c (horse) 223
  light repression of respiration 472
  nitrate and nitrite reductase 220,
    223, 224
  terminal oxidase 220, 222, 224
Cytochrome a_2 see *Cytochrome d*
  VI B3 (233–240)
Cytochrome $a_3(a')$ III B5a (33–35),
    10, 18
  destruction by borohydride 35
  in amphibian and lobster hearts 394

Subject Index

Cytochromes of type a—continued
  in bacteria 222, 224
  in inner mitochondrial membrane 333
  redox potential 39
  in mitochondrial membrane 345
Cytochrome aa_3 (Cytochrome oxidase, O_2: Cytochrome c oxidoreductase, EC 1.9.3.1.) 2, 4, 11, III B (20 ff.)
  absorption spectrum 18, 35–38
  activation energy 362
  activity assay III B3 (24–25)
  amino acid composition 28
  antibody formation 30, 459
  apoenzyme 28, 459, 460
  autoxidation in absence of cytochrome c 362
  biosynthesis 447–452, IX D (459–460)
    effect of cobalt 451
    effect of Cu 452
    effect of Zn 451, 452
    immunology of 451, 452
  chromatography 30
  circular dichroism (CD) and optical rotatory dispersion (ORD) spectra 30, 37, 38
  compounds with ligands III B6 (42–47)
  complexes with cytochromes VII D2 (342, 343), 353
  complex IV of respiratory chain 330
  conformational change 37, 48, VII H (361–366)
  copper 19, 29, 31–33, 37, 48, 345, 452, 459, 487
  effect of temperature acclimatization 390
  electrophoresis 30
  ferric oxidase, different forms of 35–37
    biphasic reduction 41
  in adrenal mitochondria, effect of androgen 409
  in algae 20, 21
  in anoxia 387, 404
  in arteries 406
  in *Ascidians* 416
  in bacteria 21, 22, 230, 231, 237, 254, 272
  in brain 387, 402–405
  in carotid body 388
  in *Echinodermata* 416
  in erythrocytes 407
  in eye 405
  in fungi and yeast 21, 24, 28, 427–429, 459, 462, 464
  in gastric mucosa 405
  in heart VIII C1 (392–395)
  in inner mitochondrial membrane 331, 333, 340–342

Cytochrome aa_3—continued
  in insects 20, 395, 397, 418–421
  in intestine 406
  in invertebrates 20, VIII D (414–415), 416–426
  in kidney 401
  in leukocytes 407
  in liver 397
  in muscles red and white 395–397
  in flight muscles of birds 395
    of insects 397
  in myocarditis 394
  in nerves 402, 403
  in nuclear membrane 386
  in ovaries 409
  in plants 20, 406, 415, 427
  in poikilothermic animals 390, 391
  in rat tissues 392
  in skin and glands 407
  in the thyroid 411, 412
  in tumours, malignant 413
  in the uterus 409
  in yeast (see in *Fungi*)
  inhibition of III B6 (42–47), VII D5 (344)
  isoelectric point of yeast 28
  kinetics 40–42, 342
  lipids in – 20, 25, 27, 31, 462
  magnetochemistry 36, 37, 47
  mechanism of reaction III B7 (47, 48)
  modifications 23, 24, 29, 30
    mitochrome 29
    aldimine formation 29, 343
  molecular weight 24, 27
  occurrence III B1 (20–22)
  oxidized (see *ferric-*)
  oxygenated III B5f (39–41), 4, 18, 20, 28, 343
  preparations III B2 (22–24)
  reaction with cytochrome c VII D1 (340–342), 485
  redox potential III B5e (38–39)
  relation to body weight and surface 389
  stability to light 25, 472, 473
  sulphydryl groups 28
  theories (aa_3 and "unitarian") III B5a (33–35)
  turnover 460
Cytochromes of type B 2, 8, 9, 11, 59–61
  effect of light 472
  in bacteria VII C (245–264)
    differentiation of b_1 and b 245, 246
    (see also individual cytochromes b)
  in photosynthesis 100, 101, 246 (see also Cytochromes b_3, b_6, b-559)
  molar ratio to chlorophyll a in chloroplasts 108

Cytochromes of type B—continued
  in plant microsomes 103, 104 (see
    also Cytochromes b_5, b-559,
    b-561)
  in plant respiration 100, 104, 346
    (see also Cytochromes b, b-555,
    b-559, b-564)
Cytochrome b
  b_T and $b_1(b')$ 64, 346, 348, 349
  absorption spectra 59, 60, 63, 65,
    102, 347, 348
  apoenzyme 468
  autoxidizability 59
  biosynthesis 369
  energized forms 63–65, 348, 349
  in algae 102
  in bacteria 253–257
    aerobic 253–256
    chemoautotrophs 256, 257
    photosynthetic 257
    Rhizobium 256
  in brain 402
  in complex II 353
  in fungi 65, 428, 464
  in invertebrates 65, 416–426
  in mitochondrial inner membrane
    331, 333
  in muscle contraction 395, 396
  in plants, 65, 100, 102, 346
  in uterus 409
  inhibition 349–351
  modification 62
  preparation 62
    separation from cytochrome c_1 347
  reactions with CO and CN 59, 60
  redox potential 60, 62, 64, 348
  reduction by succinate and NADH
    347
  role in electron transport 62–63,
    330, 346
  spectrophotometric determination 338
Cytochrome b reductases 352, 354
  fatty acid-cytochrome b reductase 354
  microsomal cytochrome b_5-reductase
    355, 400
Cytochrome b_1 VI C2 (246–251),
    238, 469
  absorption spectra 248
  low temperature spectra 251
  autoxidizability 247
  biosynthesis 239, 469
  history 246, 247
  in *Azotobacter*, red and green
    particles 237
  in *Escherichia coli* VI C2a (246–250),
    450

Cytochromes of type B—continued
  in other organisms VI C2b (250–251),
    237, 464
  induction 250, 283
  molecular weight 247
  multiple forms 247
  purification and crystallization 247
  redox potential 247
Cytochrome b_2 (see *Lactate dehydro-
    genases*) 11, 60, IV G (112–118)
  absorption spectra 115
  acceptor specificity 117
  amino acid composition 114
  "b_2" of Okunuki 112
  b_2-core 113, 115
  biological function 117, 118
  biosynthesis 446
  CD and ORD spectra 115
  DNA in – 113
  EPR-spectra 115
  flavin in – 301
  in *Hansenula* 118
  in yeast petite 116, 117
  in other respiratory-deficient
    mutants 465
  kinetics 116, 117
  modifications: see b_2-core
  molecular weight 113, 114
  purification and crystallization 113
  reaction mechanism 116, 117
  redox potential 116
  structure 114
  substrate specificity 117
    reduction with bacterial cyto-
      chrome c-552 270
Cytochrome b_3 60, 100, 103
  purification 100
  in fungi? 428
Cytochrome "b_4" VI C6 (296–303),
    61, 68
  absorption spectrum 264
  haem of – 264
  conversion to c-type compound 264
Cytochrome b_5 IV C (68–71), 61, 65
  absorption spectra 61, 103
  amino acid sequence 70
  and ACTH 410
  and thyroxine 412
  apoenzyme 70, 71, 458
  autoxidizability 69
  electron transfer through 69
  evolution 494
  haem-iron-linkage 71, 494
  half-life 458
  in adrenals 68, 410
  in insect metamorphosis 397
  in kidney 402

Subject Index

Cytochromes of type B—continued
  in liver of rat 398, 399, 401
  in mammary gland 408
  in microsomes 68, 103
  in outer mitochondrial membrane
    68, 458
  in protein synthesis 397
  in silkworm 68, 419
  in tumours 414
  induction by phenobarbitone 458
  nomenclature 68
  physical properties 69, 70
  purification and crystallization
    69, 70, 399
  redox potential 70
  reduction by microsomal reductase
    355, 399
    in fatty acid desaturation 399, 400
    by NADPH-cytochrome c
      reductase 356
  role in cytochrome P-450 metabolism
    399
  species-specificity 69, 71
Cytochrome b_6 5, 61, 100, 104, 107
  in algae 104, 106, 108
Cytochrome b_7 61, 100, 102, 104
Cytochrome b-555* of colourless alga
  Zopfia 472
  of plant microsomes and mitochondria
    103
  of tapeworm 426
  of trypanosomes 427
Cytochrome b-556 of Micrococcus
  lysodeicticus 308
Cytochrome b-557 of Bacillus megaterium
  254
  of Thiobacillus neopolitanus 256
Cytochrome b-558 of Streptomyces
  griseus 255
    absorption spectrum table 248
    purification and crystallization 255
  of Acetobacter suboxydans 249
  of Pseudomonas stutzeri 270
  of yeast (anaerobic) 463
Cytochrome b-559 of chloroplasts
  61, 100, 104, 107
  in algae 104, 105, 463, 472
  redox potential 105
  of photosynthetic bacteria 249, 257
  of plant microsomes 103
  of plant mitochondria 102, 103
  of Halobacteria 255, 256
  of Rhizobium japonicum free-living
    and bacteroids 256

Cytochrome b-560
  of Micrococcus lysodeictictus 308
Cytochrome b-561
  of Halobacteria 255
  of plant microsomes 103
Cytochrome b-562
  of Anabaena (see Cytochrome b)
  of Bacillus anitratum 248
  of Enterobacteriaceae VI C3 (251–253)
    60, 65, 245, 246
  absorption spectra (E. coli) 248
  amino acid sequence 252, 253
  haem-linkage 252
  homology with myoglobin 253
  ligands bound to haem-iron 252
  molecular weight 252
  preparation and crystallization
    251, 252
  role in electron transport 253
  of Rhodopseudomonas spheroides 305
Cytochrome b-563 of bacteria 255–257
Cytochrome b-564
  of Acetobacter suboxydans 250
  of Halobacteria 255
Cytochrome b-567
  of denitrifying Pseudomonas 249
Cytochrome "b(?)-574" see Cytochrome
  "b_4" 264
Cytochrome bc_1 complex 59, 63, 214,
  VII E1 (341–349), 333, 462
  and coenzyme Q-cytochrome
    c-reductase 347
  inhibition of respiratory chain
    between
    cytochromes b and c_1 VII E2
    (349–351)
    by hydroxyalkylnaphthoquinones
    350
  preparation IV B (59), 346, 347
  polymer states R and T 348, 349
  reconstitution 347
  role of coenzyme Q and of non-haem
    iron proteins 357, 358
Cytochromes of type C 8, V (122 ff.)
  124, 125
  formation from cytochrome b_2 115
  in bacteria VI D (265 ff.)
Cytochrome c V B (126 ff.) (see also
  Amino acid residues and
  individual amino acids,
  Evolution, Haem c,
  Isocytochromes c, Cytochromes c
  (bacterial)

---

* Note that the position of the α-band cannot always be considered as definitely established and that the identity of these cytochromes b and c remains somewhat doubtful.

Cytochromes of type C—continued
  absorption spectra of c^{2+} and c^{3+}
    127, 128, 160, V B4b (171–176)
    as criterion of purity 171
    band 695 nm 160, 172, 173, 183, 269
    low temperature spectrum 171, 172, 200
    molecular extinction 124, 171
  acetylation 184, 344
  amino acid composition V B2b (133–137)
  amino acid residues
    chemical modifications of – V B6b (183–186)
    clusters of hydrophobic and hydrophilic 152
    positions, variant, hypervariant and invariant 137, 152, 485, 490, 491
    replacements, conservative and radical 137, 490
  amino acid sequence V B2c (137–154), 126, 133, 134, 491, 492
    in amphibians 136–144
    in birds 129, 136, 144
    in fishes 129, 130, 136–144, 197
    in fungi 129, 145–151, 153, 175, 197
    in invertebrates 129, 136, 145–151, 153
    in mammals 136–144, 152, 162
    in marsupials 138–144, 490
    in plants 5, 126, 129, 136, 145–151, 153, 492, 493
    in primordial – 489–491
    in reptiles 136–144
    in vertebrates 136, 138–144, 152
    in whales 129, 138–144, 152
    in yeast 5, 129, 136, 145–151, 154, 172, 197, 198
    N-terminal 133, 166, 167
    C-terminal (see *Haemopeptides*)
    and thyroxine 411
    as model in photosynthesis V C3 (205, 206)
      photooxidase 205
      -reducing factor (CRF) 205
  binding of – by c-deficient yeast 428
  biological role 123
  biosynthesis of – IX B (455–457), 167, 196, 387, 452, 455
    of apoprotein 455, 456
    Cu-deficiency effect on – 452
    effect of oxygen on – – 469
    haem as organizing principle 164, 167
    in insect metamorphosis 456
    in regenerating liver 455

Cytochromes of type C—continued
  locus of – 456
  transfer into mitochondria 456
  carboxymethylation 159, 162, 172, 176, 185
  complex with oxidase VII D2 (342–343)
  c^{2+} and c^{3+} conformation change with valency V B3b (161–162), 175, 183, 189, 191, 364, 365
  presence of Cl in c^{3+} 171
  c^{3+} at various pH V B4a (170, 171) 198
  CD, ORD and MORD spectra 161, 162, 168, 173–177, 198
  covalent linkages to cysteine (see also *Haem c*) V B2a (133–135), 489, 493, 494
  crystallization V B1 (126–130), 162, 163
  deficiency in yeast 428
  denaturation 174, 175, 182, 183
  evolution of – X C (488–494)
    repeated small amino acid sequences in – 488, 489
    primordial – 340
  exogenous and endogenous 189, 340, 341, 356, 428
  extractability from mitochondria 190, 341
  forms I – IV V B6b (183–186), 456 (see also modification of amino acid residues)
  guanidinylation 185, 344
  haem linkage V B3a (158–161), (162–168), 126, 133, 134, 162, 174, 488 (see also covalent linkage to cysteine)
    in crevice 163, 165, 172
  haemopeptides V B2a (133–135), 297
    absorption spectra 134
    CD and ORD spectra 135, 174, 175
    synthesis 135
  histidine in – 488
  history 123
  homogeneity 127
  inhomogeneity (partial hydrolysis) 127–129, 176, 184
  immunological reactions V B6d (188, 189), 341
    in adipose tissue 196, 389
    in adrenals 196
    in arteriosclerosis 407
    in *Candida* 428
    in cell nuclei 196
    in eukaryotic organisms 126, 136, 138–151, 428, 429

Subject Index 561

Cytochromes of type C—continued
  in fungi 145–151, 197, 199
    Neurospora 136, 145–151, 488
  in gastric mucosa 196, 406
  in heart 195, 196, 390, 394
  in insect metamorphosis 20, 197, 397, 418–420 (Lepidoptera)
  in invertebrates 197
  in kidney 401
  in leukaemia 413
  in leukocytes 407
  in liver 196
  in microsomes 456
  in mitochondria 333, 340, 492
  in muscles, skeletal, of mammals 195, 197, 390, 395, 396
    flight – of birds 195
    of insects 421, 397
  in oxidation of cysteine 187
  in plants V B7c (199–206), 492
  in protochordata 197
  in protozoa 197
  in tumours, malignant 413
  in various parts of cell V B7a (196), 456
  in yeast 197–199, 467
  induced fit theory 365
  iron content 126, 127
  isoelectric point 170, 171, 492
  ligands V 5b (178, 179)
    azide 178
    cyanide 159, 160, 162, 178
    fluoride 179
    nitric oxide (or NO) 162, 179
  "lipid" – V B6e (189–191), 387
    in lipid monolayers 190
  magnetochemistry V B4c (177)
  modifications V B6a (182–183)
    reactions of modified – with CO and O_2 178, 182, 185, 186
  Mössbauer spectra 161, 177
  nuclear paramagnetic resonance 161, 177, 178
  peroxidase of yeast 221, 462, 487
  pH-forms of mammalian – V B4a (170, 171)
  polymerization 127, 128, 172, 175, 182
  preparation and purification V B1 (126–130)
  reactions with ascorbate 172, 183, 397
    with cysteine 187
    with cytochrome a VII D3 (343, 344), 428
    with cytochrome b_2 116, 117

Cytochromes of type C—continued
  with cytochrome oxidases 34, 38, 39, 152, 166, 199, 200, 245, VII D1 (340–342), VII D2 (342, 343), 415, 427, 485, 491
  cytochrome c peroxidase 186, 187
  ferricyanide 187
  hydrogen peroxide 186
  lipoperoxides 187
  phospholipids 189, 191
  reductases 166, 331, 337, 351
    (see Cytochrome c reductases)
  redox potential 177, 198, 200, 201
  removal of iron by HF 188
  role in ion transport 366
  species-specificity of enzymic activity 490, 492
  specificity of structure VII D4 (344)
  stability 127, 170, 172
  steady state reduction 196, 201, 411
  structure and collision theory of action 361, 362
    secondary structure (α-helix) 168, 174
    tertiary structure V B3a (158–165), 12, 488
  succinylation 184, 344
  synthesis (chemical) V B2d (154–155), 126
  therapy with – 387, 407
  thioether bonds 488 (see Haem c and covalent linkages to cysteine)
  trichloroacetylation 344
  trifluoroacetylation 159, 185
  trinitrophenylation 159, 185, 344
  turnover 24, 340, 341, IX B (455–457), IX C (457–458)
  x-ray diffraction of – V B3c (162–168)
  xanthydrylation 187
Cytochrome c-reductases VII Ec (351–354), 331, 337, 393, 400, 410, 485 (see also NADH-cytochrome c and NADPH-cytochrome c reductases)
  in crustacea 417
  in muscle denervation 396
  in temperature acclimatization 391
  lactate-cytochrome c-reductase 356
  succinate-cytochrome c reductase 352, 391
    decrease in hyperbaric oxygen 387
    reconstitution 353
Cytochrome c_1 V D (212–214), 124, 197, 330, 347, 409 (see also Cytochrome bc_1 complex)
  absorption spectrum 214

Cytochromes of type C—continued
  biological role in mammalian tissues
    V D4 (214)
  in mung bean mitochondria 214
  biosynthesis 456, 472
  crevice structure 214
  determination 214, 335
  discovery 212 (cytochrome "e")
  c_1 and f distinction 213
  haemopeptide 214
  in *Acetobacter* ? 213
  in bacteria ? 267, 272, 277
  in fly larvae 420
  in *Lepidoptera* 419
  in yeast and fungi 213, 428
  in methionine deficiency 392
  isoelectric point 213
  ligands CN 214
    CO with modified cytochrome
      c_1 214
  molecular weight 213
  modified – 214
  position in inner mitochondrial
    membrane 333
    in respiratory chain 213, 214, 330,
      339
  preparation 213
  redox potential 213
  role in reaction of BAL with O_2 351
Cytochrome c (bacterial) VI (265 ff.)
  Cytochrome c in bacilli VI D1
    (265–267)
    absorption spectrum 265, 266
    isoelectric point 266
    location in membrane 265
    molecular weight 266
    purification 265
    reaction with oxidases aa_3 and d_1c 267
    redox potential 265, 266
    reducibility with NADH 267
  in *Bordatella* 272
    (absorption spectrum, raction with
      azurin, redox potential)
  in *Chlorobium* 302
  in denitrifyers VI D2 (267–273), 491
  in *Halobacteria* 255, VI D2g (273)
  in halotolerant *Micrococcus* 273
  in *Micrococcus denitrificans* 271, 344
    amino acids, crystallization,
      tertiary structure, reaction with
      oxidases
  in primitive bacteria 494
  in *Rhizobium* bacteroids 256
  in *Spirillum itersonii* 272
    absorption spectrum, mol. weight,
      redox potential, histidine
  in *Thiobacilli* 273, 274

Cytochromes of type C—continued
  Cytochrome c_2 of *Rhodospirillum rubrum*
    VI D6 (296–299), 124, 145–154,
    163
    absorption spectrum 296–297
    amino acid sequence 297, 298
    biological function 298, 299
    CD spectrum 175, 176, 297
    crystallization 163
    haempeptide 297
    homology 488
    isoelectric point 297
    Mössbauer spectrum 297
    molecular weight 296, 297
    preparation 296
    reaction with cytochrome d_1c 245
      with cytochrome o 228
    redox potential 296, 297, 299
    role in photosynthetic electron
      transport 304
    x-ray diffraction 298
  Cytochrome c_2
    of *Chromatium c*-553, 500 300
    of *Rhodomicrobium vanielli* 300
    of *Rhodospirillum molischianum* 299
    of *Rhodopseudomonas palustris* 300
    of *Rhodopseudomonas spheroides* 299
  Cytochrome c_3 of *Desulphovibrio* VI D4
    (277–281), 124, 145–151
    absorption spectrum 278
    amino acid sequence 279, 145–151
    biological function 280, 281
    CD spectra 280
    crystallization 278
    immunological reactions 280
    iron content 278
    isoelectric point 279
    ligands 280
    molecular weight 278
    purification 277, 278
    redox potential 277, 278
    structure, primary and tertiary
      279, 280
  Cytochrome cc_3 of *Desulphovibrio* 281
  Cytochrome c_4 and c_5 VI D3 (275–277),
    125, 222
    absorption spectra 177, 275, 276
    acetylene – reduction 277
    amino acid composition 276
    biosynthesis 276, 277
    CD spectrum 175, 276
    crystallization 275
    isoelectric point 276
    molecular weight 275, 276
    preparation 275
    reaction with cytochrome d 237, 239,
      276, 277

Cytochromes of type C—continued
  redox potential 276
  reduction by ascorbate 222, 277
  role in nitrogen fixation 277
Cytochrome c_6 see Cytochrome f
Cytochromes c-551*
  of Chlorobium thiosulphatophilum 302
  of Chloropseudomonas ethylicum 302
    role in photosynthetic electron transport 304
  of Pseudomonas VI D2a (267, 268), 144–151, 154, 160
    absorption spectrum 268, 269
    amino acid sequence 144–151, 268
    homology with horse cytochrome c 488
    iron content and molecular weight 268
    purification and crystallization 268
    reaction with azurin 269
      with dithionite and ferricyanide 269
    redox potential 268
    removal of prosthetic group 269
    structure, tertiary 269
Cytochrome c-552 of Enterobacteriaceae VI D4g (282–284)
  absorption spectrum 282
  autoxidizability 282, 283
  biological function 283, 284
  biosynthesis 239, 282, 283, 469
  isoelectric point 282, 283
  molecular weight 282
  occurrence 282, 283
  purification 282, 283
  redox potential 282
Cytochrome c-552 of Pseudomonas species 267–271
  of Rhizobium bacteroids 256
  of Rhodopseudomonas spheroides 304–306
Cytochrome c-552 (flavocytochrome)
  of Chromatium VI D6c (300–302)
    absorption, CD and ORD spectrum 301, 302
    crystallization 300
    flavin binding 301
    isoelectric point 302
    magnetochemistry 302
    molecular weight 301
    redox potential 301
    role in photosynthetic electron transport 304–306, 344
Cytochrome c-553 (flavocytochrome)
  of Chlorobium VI D6e (301–303)

* See footnote p. 21.

Cytochromes of type C—continued
  Cytochrome c-553 (see also Cryptocytochrome c)
    of Desulphovibrio 281
    of Haemophilus 467
    of Pseudomonas denitrificans 270, 271
      crystallization, physical properties and role in denitrification 271
    of Rhodopseudomonas 304
    of Thiobacilli c-553–5 273, 304, 344
  Cytochrome c-554 (78°K) of Bacillus subtilis 265–267
    absorption and CD spectrum 265, 266
    isoelectric point 266
    molecular weight 266
    purification 265
    reactions with cytochrome oxidases 267
    redox potential 266
    reducibility by NADH 266
    relation to cytochrome c_1 267
  Cytochromes c-554 of halotolerant Micrococcus (Cytochromes 1 and 2)
    reactions with cytochrome b_1 250, 255, 273, 344
    of Pseudomonas 269
    of Chlorobium (non-flavin) 302, 303
    physical properties and reaction with cytochrome d_1c 302, 303
    of Chloropseudomonas ethylicum 303
  Cytochrome c-555 of fly larvae 420
    of Chromatium 305, 306
  Cytochrome c-556 of Rhodopseudomonas palustris 303–306
  Cytochromes c-557
    of clam sperm 409
    of Thiobacilli 273
  Cytochrome c-575,551 of Alkaligenes faecalis 296
  Cytochrome c-558,552 of Pseudomonas stutzeri VI D5j (295–296), 287
  Cytochromes c' and cc' (see also Cryptocytochrome c and Cytochrome c-558,552) 2, 10, 12, 125, 228, VI D5 (284–296)
    absorption and CD spectra 285–288, 292
    amino acid composition and sequence 291, 293
    biological function as terminal oxidase 291
      in bacterial photosynthesis 292
    crystallization 285
    homology with other cytochromes c 488

Cytochromes of type C—continued
  immunological detection 292
  isoelectric point 293
  ligands 286, 287, 289, 290, 292
  magnetochemistry 288, 289
  models 185
  modification 294
  molecular weight 292, 293
  nomenclature 284
  occurrence 284, 292 (*c'* in *Rhodopseudomonas palustris*, *cc'* in *Chromatium*, *Pseudomonas denitrificans* and *Rhodospirillum*)
  preparation 285
  redox potentials 292
Cytochromes d VI B3 (233–240), 8–11, 222, 277, 467, 469 (see also *Cytochrome* "a_2")
  absorption spectra 234–236
  biosynthesis 239, 270
    glucose repression 472
    induction by oxygen 469, 470
  cyanide compound 240
  cyanide haemochrome 234, 235
  in red and green particles of *Azobacter vinelandii* 237
  in *Tetrahymena* 426
  lability 236, 237
  nitrate and sulphate reductase VI B3e (239, 240)
  occurrence 233, 236, 237
  preparations 236
  prosthetic group 236 (see also *Haem d*)
  reaction with mammalian cytochrome *c* 236
    with cytochromes c_4 and c_5 237, 239
    with D(−) lactate 239
  terminal oxidase VI B3d (237–239)
Cytochrome d_1c (*Pseudomonas* oxidase) VI B4 (240–245), 236
  absorption spectra 234, 235, 242
  as complex of two haemoproteins 244
  haem *c*-protein moiety 242
  H_2O_2-sensitivity and evolution 491
  isoelectric point 241, 242
  ligands (CN, CO, NO) 235, 243
  molecular weight 241
  nitrite reductase 244, 245
  occurrence (*Alkaligenes faecalis*, *Nitrosomonas europaea*, *Pseudomonas*, *Micrococcus*) 240–245
  preparation and crystallization 241
  prosthetic groups 242, 243

Cytochrome d_1c—continued
  reaction with mammalian cytochrome *c* 245, 491
    with bacterial and protozoan cytochromes *c* 245, 271, 491
  reconstitution 244
  redox potentials 242
  terminal oxidase 245
Cytochrome "*dh*" 101
Cytochrome e (see *Cytochrome c_1*)
Cytochrome f V C1 (201, 202), 5, 12, 107, 125
  absorption spectrum 202
  algal cytochromes of type *f* V C2 (202–205), 463, 464
    absorption spectrum 203
    chlorophyll-deficient mutants 204
    crystallization 129
    in *Euglena* 201
    in *Navicula c*-554 245
    molecular weight 203
    purification 128
  isoelectric point 202
  molar ratio in chloroplasts
    to chlorophyll *a* and P-700 108
    to cytochromes *b* 108
  molecular weight 202
  preparation 201
  reaction with cytochrome aa_3 202
  redox potential 202
  relation to cytochrome c_2 296
  role in photosynthesis V F3 (205, 206), 107, 204
  stability 202
Cytochrome h and h' IV E (97–99), 11, 59, 61, 421, 422
  absorption spectrum 98, 99
  amino acid composition 98
  molecular weight 98
  purification 97, 98
Cytochrome oxidases (see also *Cytochrome aa_3, d_1c and o*)
  of *Acetobacter oxydans*, reactions with cytochrome *h* 98
  of bacteria, plurality of 219, 463
  in invertebrates 416–426
Cytochromes o VI B2 (225–233), 4, 10, 22, 61, 222, 223, 246, 250, 256, 291, 487, 488
  absorption spectra
    $\Delta Fe^{2+}-Fe^{3+}$ 226, 227
    $\Delta Fe^{2+}CO-Fe^{2+}$ 226, 227, 229, 231
  biosynthesis, effect of O_2 on 469
  CN and CO compounds VI B2c (232, 233)
  definition 225, 226, 228

Subject Index

Cytochrome o—continued
  estimation 226
  in *Acetobacter suboxydans* (*b*-558) 227, 229, 230, 233, 472
  in anaerobic bacteria 231
  in bacteria 227
  in *Haemophilus* 227, 231, 467
  in invertebrates (worms, protozoa) 227, 415, 424, 425
  in *Pseudomonas oleovorans* 263
  in *Rhizobium* 228, 231
  in *Rhodopseudomonas spheroides* 227, 228, 232
  in *Rhodospirillum* 227, 228, 232
  in *Spirillum itersonii* 272
  in worms and yeast ? 231, 232
  inhibitors 228
  particulate and soluble preparations VI B2b (226–231)
  terminal oxidase in *Cyanophyta* 227, 229, 231
  in *Staphylococcus aureus* 227, 228
Cytochrome P-420 74, 76, 79, 103, 263, 401
  in bacteria 263
Cytochrome P-450 IV D (73–89), 2, 4, 10, 11, 59, 61
  absorption 75
  and haem synthesis 82
  and haemoglobin catabolism IV D8 (86–89)
  biosynthesis, effect of oxygen on 469
  competition with cytochrome aa_3 333 (see also *Adrenals* and *Ovaries*)
  in adrenal mitochondria 73, 410
  in bacteria VI C5 (405–412), 74 (see *Cytochrome P-450, bacterial*)
  in brain microsomes 405
  in corpora lutea 409
  in fungi 74, 82, 222
  in insect thoracic muscles 420
  in liver 399–401
  in microsomes, smooth and rough 73, 104, 397,
    Fe_x 76, 77
  in mitochondria of adrenals and ovaries 73, 74, 409, 410
  in ovaries 74, 409
  in testes 409
  in thyroid 412
  in tumours, malignant 414
  in uterus 409
  induction by phenobarbitone and hexobarbitone 74, 78, 79, 82, 83, 392, 399, 400, 451, 452

Cytochrome P-450—continued
  by methylcholanthrene 78, 79, 82, 83, 399, 400
  iron incorporation into liver – microsomal 458
  ligands CN 75
    CO 75, 76, 400
    isocyanide 75, 76
    NO 75
    O_2 75, 85, 86, 262
  magnetochemistry 76–78, 84
  multiple cytochromes P-450 76, 78, 79, 263
  purification 74, 75
  reaction mechanism 84–86, 262
  redox potential 84
  role in drug detoxication 81, 82, 399, 400
    in sterol metabolism 80, 81, 409
    of cytochrome b_5 in P-450 metabolism 399, 400
  structure 73, 83, 84
  substrates (see also role in drug and sterol metabolism)
    combination with – 77, 78, 399, 414
    types I and II of spectral alteration by substrates 77, 78, 399, 400, 409
  turnover 458
Cytochrome P-450, bacterial VI C5 (257–263)
  in *Corynebacterium* 257, 263
  in *Nitrosomonas europaea* 257
  in *Rhizobia* 73, 257, 259, 262, 263
  absorption spectra 75, 259, 263
  in free-living bacteria and bacteroids 263
  ligands (CO, isocyanide) 259, 263
  preparation 262
Cytochrome P-450$_{Cam}$ of *Pseudomonas putida* VI C5 (259–262)
  absorption spectra 258–260
  amino acid composition 260
  crystalline complex with camphor 259, 260
  glucosamine content 260
  induction by camphor 257
  isoelectric point 260
  ligands CO, NO 258, 262
    O_2 258, 262
    isocyanide 261, 263
  magnetochemistry 260, 261
  methionine-haem iron linkage 261
  reaction mechanism 262
  reaction with apomyoglobin (haem exchange) 261

Cytochrome P-450—continued
 with substrates 258, 259, 261, 262
 redox potential 261
 structure and molecular weight 260
Cytochrome P-450 reductase (see
  *NADPH-cytochrome P-450
  reductase*) IV D7 (84–86),
  262, 356, 412
 inhibition by cytochrome *c* 356
Cytodeuteroporphyrin 13
Cytoplasmic endothelium 403, 405, 418,
  419, 463
 interaction with mitochondria in
  tumour cells 443
 membranes 425, 428, 488
 synthesis in – of ALA-synthetase 453
 of cytochrome *a* IX D (459, 460)
 of cytochrome *c* 387, 400, IX B
  (455–459)
Cytoplasmic membrane of bacteria VI E1
  (306–309), IX F (463–464),
  21, 220, 331
 biosynthesis of – IX F (463, 464)
 mesosomes 307, 308

D

D_2O see *Heavy water*
DDT resistance and cytochromes 420
Decapoda cytochromes in – 417
Dehydrogenases (see individual substrates)
  328, 336, 472
 and thyroxine 411, 412
 as cytochrome *b* reductase 351, 352
 inhibitors 359, 360
 pacemaker – 412
Deletions in amino acid sequences 488, 489
Denaturation of cytochrome *c* 174, 175,
  182, 183
Deoxycorticosterone hydroxylase 409
Deoxyribonuclease 262
Detergents and respiratory chain in
  bacteria 308
 and surface structure of cytoplasmic
  membrane 307
 in preparation of cytochrome aa_3
  III B2 (22–24)
  (see also other individual
  cytochromes)
Diapause of insect eggs 415, 418–420
 of insect pupae 415, 418–420
Dicotyledon plants 492
Dicumarol 360, 408
Digitonin in preparation of
  submitochondrial particles
  311, 332
 in separation of photosystems I and II
  106, 108, 205

Dihydrostreptomycin 350
Dimercaptopropanol inhibitor of
 respiratory chain with O_2
  VII E2b (351), 331, 350, 398,
  401, 427
Dinitrophenols as uncouplers 331, 334
 in the kidney 401
Diphenylamine 464
Diphtheria toxin 394
Diptera eggs and pupae 420
Dipyrrylmethane as intermediate in
  tetrapyrrole synthesis 448
Dithiocarbamate 414
DNA 413, 451, 458, 461, 462, 467
Drift biological 492

E

Echinoidea eggs, sperms 415, 416
Ediacara formation 486
Eggs of invertebrates 416, 419, 420, 422,
  424
Electric organs of fish, cytochromes in 405
Electrocardiogram 387
Electron paramagnetic resonance (*EPR*)
  31, 32, 77, 130
Electron transport (see *Respiratory chain,
  Photosynthesis, Reversed electron
  flow, Transhydrogenase*) IV F3
  (104, 108), V C (205–206), 1,
  5, 62, 63, 204, 205
 alternative pathways in bacteria 219,
  220, 308
 branching 329, 361, 424, 463
 in photosynthesis 104–108, 204, 205
 cytochrome *c* as model V C3
  (205, 206)
 cytochromes as primary oxidants
  104–108, 304
 "E" (C-550, P-550) 106, 107, 306
 P-700 107, 205
 photosystems 105–108
 theories of – VII H (361–366)
 complexes of cytochromes 362
 conduction band 362, 363
 conformation changes 363–366
 linear chain, collisions 361, 362
 thermodynamic relations to energy
  conservation 365
 thyroxine and – 411
"*Elementary particles*" (see *ATPase*) 332
Embryos of invertebrates 425 (see also
  Ontogenesis)
Endoplasmic reticulum (see *Cytoplasmic
  endothelium*)
Energized state (see *Respiratory states*)
 and oxidative phosphorylation 365, 366

Subject Index

Energised state—continued
  in bacteria 308
  in nerve tissue 402
  kinetics of energization 365, 366
  of cytochrome aa_3 364, 366 (see also *Cytochrome a_3*)
  of cytochrome b 365 (see also *Cytochrome b*)
  of mitochondrial membranes 364, 366
  fluorescent probes of – 366
Energy conservation 328 (see also *Oxidative phosphorylation*)
Energy conversion in thyroid 411
  theories of VII H (361–366)
"Energy pressure" 364
Energy rich intermediates (see *Cytochrome b* and *Phosphorylation*) 334, 336, 346, 360, 363
Enterochelin and cytochrome b_1 450
Enterochrome-556 59, 61, IV E (97–99)
  absorption spectrum 98
Enterochrome-566 61, IV E (97–99)
Enzootic ataxia in lambs 391
Enzymes in early evolution 486
Epithelium (see *Skin and Glands, Mucosa*)
Erythrocytes 407, 452, 454, 455
Ethidium bromide 366, 461
Ethyl ether antimycin A extraction 350
Evolution (see also *Anaerobiosis, Light, Precambrian, Oxygen*) 5, 123, 126, 154, 279, 296, 298
  cytochromes in – X (484 ff.)
  cytochrome c reaction with oxidases 491
  divergence points of phyla from primary structure of cytochrome c 490, 493
  divergent homologous and convergent analogous 488
  "explosion" in Cambrian 486, 487
  fitness of porphyrin derivatives for 486
  homology of eukaryotic and bacterial cytochromes 488
  molecular – clock 485, 492
  nitrite oxidation in – 491
  non-Darwinian – 484, 491
  of covalent linkages in cytochromes c 448, 493, 494
  of cytochromes c 485, 486, X C (488 ff.), 492
  of phycobilins 486, 487
  prebiotic – 486
    earth atmosphere 486, 487
  precambrian 486, 487
  Pseudomonas oxidase and – 491
  time course of – X B (486–488), 124, 125, 493

F

Farnesyl pyrophosphate 448
Fatty acids deficiency and cytochrome P-450 401
  desaturation in adipose tissue 390
  in adipose tissue and heat regulation 390
  in cytochrome c biosynthesis 428
  in cytoplasmic succinoxidase 464
  oxidation and laurate 401
Feedback see *Cybernetic control*
Ferredoxins 107, 206, 285, 489
  photoreduction by H-donors 206
Ferricyanide reaction with cytochromes c, c_1 and b 333
  with NADH-dehydrogenase 352
Ferrochelatase 447, 449, 450, 494
  effect of ATP on – 454
  in anaerobic yeast 449
  in bacteria 449, 467
  in chloroplasts 449
  in cytochrome and haemoglobin synthesis 449
  in formation of haems a, c and d 449
  in fungi 450
  in inner membrane of mitochondria 449
  in microsomes 449
  role of phospholipid 449
Fertilization of sea urchin eggs 416
Fibrillation ventricular 387
Fibrinopeptides 486
Flagellae 486
Flavocytochromes 2, 11, 60, IV G (112–118) (see also cytochrome b_2)
  cytochrome c-552 of *Chromatium* 300–302
  cytochrome c-553 of *Chlorobium* 300–302
Flavins and flavoproteins FAD 355
  FMN 355
  Zn-FAD 356
  FpD_1 and FpD_2 359
  in cytochrome reductases 352, 354–356, 358, 409
  in hyperbaric oxygen 389
  in thyroxine action 411
  terminal flavoprotein oxidase in nematode larvae 424
Flies (see *Diptera*) 420
Fluorescence probes for energization of mitochondrial membranes 366, 410
Formic hydrogenase 280, 283, 469
Formyl group formation from methyl in haem a and chlorophyll b 448, 449

Fossils metazoa 486
Fumarase 460
Fungi (see also Cytochrome *c*) VIII E
    (415, 427–429)
  cytochrome *c*, amino acid composition
      136
    amino acid sequence 145–151
    iron deficiency 450
    kinetics of respiratory chain of
      Candida 338
    mitochondrial cytochromes of 335,
      415, 427
    poky mutation of Neurospora 467
    respiration and uncoupling 472

G

Galactolipids 191, 205
Galactose repression 471
Gastric HCl-secretion 405, 406
Gastropods 422, 423
Genetic control (see *Mutants*)
  mutants and genetic studies in yeast
      IX G (464–467)
  nuclear, – – of apoprotein biosynthesis
      of cytochromes 460
  of cellular respiration 328, 413, 471
  of cytochrome oxidase biosynthesis
      459, 460
Geological periods 493
Gills of arthropods and molluscs 416
Glands (see also *Prostate*) VIII C4d
    (407, 408)
  mammary 408, 413
  thyroid and parathyroid VIII C5c
      (410–412) (see also *Thyroxine*)
    adenoma 412
    Askanazi-Hürthle cells 412
Globin see *Haemoglobin*
Glucosamine in P-450$_{Cam}$ 260
Glucose repression in ascites cells 470
  of citric acid pathway 470
  of cytochrome synthesis IX 12
      (470–472), 284, 466
  of cytochrome oxidase synthesis
      459, 471
  of DNA formation 462
Glutathione in hyperbaric oxygen 389
  in liver oxygen uptake 398
  in oxidative phosphorylation 363
Glycerol effect on cellular respiration 336
α-*Glycerophosphate* as substrate in
      brain mitochondria 403
  in insect muscle sarcosomes 397
    dehydrogenase in brain 403
    moiety of phospholipids 462

Glycine (see *Amino acid residues*)
  in cytochrome *c* biosynthesis 455, 466
  in haem biosynthesis 446, 447, 451
  in haem *a* biosynthesis 446, 447
Glycolysis aerobic, in tumour cells
      412, 413
  in heart muscle 392
Golgi membranes 399
Gonadotrophin human chorionic 409
Gramicidin A 311, 405
Guanidine (see also *Modification*)
  and bacterial cytochromes 293
  and cytochrome *c* 184, 185
  and NADH- and succinic oxidases 340
  derivatives as inhibitors of cytochrome
      reductases 359

H

Haem, Haemin see *Protohaem*
Haem a (*Cytohaem a*) 8, 9, 11, II A
    (8–12)
  as basis for counting number of
      respiratory chains 397
  biosynthesis formation from
      protohaem 448
  half-life 458
  iron incorporation 449, 458
  role of Cu 391
  estimation 26
  formyl group aldimine 26, 27
    role in oxidative phosphorylation 346
  haemochromes, imidazole 26, 35
    proteins 26, 27
    aldimines 26, 27
    pyridine 25, 26
  in *Staphylococcus* mutant 467
  in yeast 221
Haem c 8–12
  biosynthesis IX B (455–457)
  iron incorporation 449
  removal from protein 269
Haem (*Cytohaem*) *d* 8–12, 14, 233, 234,
    236
  absorption spectra 234
  compounds 234
Haem d_1 VI B4e (243), 235, 236
  compounds with ligands 235, 236, 242,
    243
Haematoporphyrin and ferredoxin 206
Haemochromes 3, 158, 159, 201, 206, 264
Haemoglobin α, β, γ, δ chains 485
  autoxidizability 2
  biological role 19
  biosynthesis 446, 451
  catabolism and stereospecificity IV D7
      (86–89)

Subject Index

Haemoglobin—continued
　conformation and allosteric changes 365
　evolution 154, 485, 486, 488–490
　foetal and adult 485
　in anoxia 387
　in carotid body 388
　in invertebrates 415
　in yeast 221
　iron incorporation into 449
　magnetochemistry, compared with that of cytochrome cc' 288, 289
　pathogenic haemoglobins 485
　regulation of ALA synthetase 454
　valency states 2
Haemoproteins differentiation from cytochromes 2
　iron valency 1, 2
　magnetochemistry 2
"Headpieces" in mitochondrial membrane 332 (see *ATPase*)
Heart arterio-venous oxygen difference 393
　cells, beating of 360, 363
　　and cytochrome oxidase 387, 391, VIII C1 (392–395), 459
　contraction 388, 389, 393
　cytochrome c reductases in 394
　digitalis effect on 394
　electric tissue 393
　invertebrate 418, 421, 422
　iron deficiency 450
　mitochondria, cytochromes of 335, 393, 394
　　structural protein 337
　ouabain, effect on 394, 395
　serotonin, effect on 394
　thyroid hormones, effect on 411
　turnover of cytochrome c in 458
　　of haem a in 458
Heavy water influence on cellular respiration 336
　on oxidative phosphorylation 336
Helical content of proteins 167, 168, 174, 228, 298
Helicorubin IV E (97–99), 11, 61
　absorption spectrum 98
Hepatectomy ⎫
Hepatitis ⎬ see *Liver*
Hepatoma cytochrome c in Ascites 413
　cytochrome oxidase 413
　microsomal cytochromes 414
Hepatopancreas 97, 99, 417, 421, 422
Hexobarbitone 400
Hexoestrol 409
Hibernation 196, 389, 390, 462
　of insect eggs 419

High energy intermediates see *Energized State*
Hill reaction 205
Histamine 394, 406
Histidine in haem linkage 158, 159, 268, 272, 276, 280, 293, 297, 299
Histochemistry VIII B5 (392), 396, 407, 423, 425, 427
Holothuroidea 416
HOQNO 2-heptyl-4-hydroxyquinoline-N-oxide VII E2b (530, 531)
Hormones adrenal VIII C5 (410) see also *Adrenals*
　sexual VIII C5a (409, 410)
　thyroid 410–412 (see *Thyroxin*)
　effects on DNA synthesis 461
　on respiratory enzymes VIII C5 (409–412)
Hydrazoa 420
Hydrazoic acid see *Azide*
Hydrocortisone 404
Hydrogen ion transport in photosynthetic bacteria 309 (see also *Oxidative phosphorylation*)
Hydrogen sulphide metabolism in evolution 487
Hydroperoxidases 1
Hydrophobic drop structure of respiratory enzymes 163, 336
β-*Hydroxybutyrate* oxidation coupled with phosphorylation 310
7-α-*Hydroxycholesterol* oxidation 355
Hydroxylamine 360
　reductase 428
Hydroxylases 2, 74–76, 80, 263, 409
Hymenoptera 420
Hyperbaric oxygen VIII B2 (388, 389) (see also *Flavins*, *Sulphydryl groups*, *NADH* and *NADPH* oxidation)
Hypercholesteraemia and cytochrome c 407
Hyperosmolarity effect on cellular respiration 336
Hyperoxia (see also *Hyperbaric oxygen*, *Oxygen* (*molecular*) and *Oxygen* (*toxicity*)) VIII B2 (387–389)
Hypervariability of amino acid residues 137, 152, 490
Hypophysectomy 408, 410
Hypothalamus 404
Hypoxia (see *Anoxia*) VIII B2 (387, 388)

I

Imidazole (see also *Histidine*) linkage to haem a 26, 269
Immunoglobins 486
Indoanilines in histochemical methods 392

Indophenoloxidase 3
Induced fit theory for cytochrome *c* 365
Infarction cardiac 393, 394
Informational macromolecules 484
   and membranes 488
Inhibitors of cellular respiration (see also *Cellular Respiration*)
   of electron transport in photosynthesis (see *Electron transport*)
   of oxidative phosphorylation (see *Oxidative Phosphorylation*)
   of cytochrome oxidase VII D5 (344)
   of cytochrome reductase VII F4 (359, 360), 427
   of protein biosynthesis 461
   of respiratory chain between cytochromes *b* and *c* VII E2 (349–351)
   of respiratory chain-oxidative phosphorylation interaction VII G (360, 361)
Insects 418–421
   metamorphosis 20, 397, 415, 418–420 (see also cytochromes aa_3 and b_5)
Intestines rat, cytochromes 335
Invertebrates 391, VIII D (414, 415), 416–426
Ion transport 5, 328, 404
   and oligomycin 361
   and trialkyltin 361
   in bacterial membranes 307
   in parathyroid 412
   ionophoretic antibiotics 331, 336
   role of cytochrome *c* in – 366, 492
Iron and haemoprotein formation 450
   and porphyrin formation 450
   deficiency anaemia 391
   in cytochrome biosynthesis IX A1 (499, 450)
   in fungi, protozoa and bacteria 450
   incorporation 494 (see also *Ferrochelatase*)
      role of phospholipids in – – 449
   into catalase 450
   into cytochromes 450, 455, 456
   into cytochrome *c* 455, 456, 457
   into cytochrome oxidase 391, 392, 450
   into haem *a* 458
   into P-450 458
   transfer and haem 454
   uptake in reticulocytes 454
Iron-sulphur proteins (see *Non-haem iron proteins*, *Ferredoxins*)
   role in iron incorporation 449

Isocytochromes 154, 160, 161, 198, 428, 451, 457, 462, 465
   crevice structure 178
   in yeast mutants 198, 199, 460
   role of -2 in synthesis of -1 457
Isoleucine as antigenic determinant in cytochrome *c* 189
Isooctane lipid extractant 337, 357

J

Japanese landsnail 99

K

Kanamycin 468
Keilin–Hartree muscle preparation (see also *Mitochondria*, *Submitochondrial particles*) 22, 329
Kernicterus 405
Kidney and thyroid hormones 411
   cytochromes VIII C2c (401–402)
   cytochrome *c*, renal clearance 401
      in iron deficiency 450
   cytochrome oxidase 387, 391, 401
   haem and ALA synthetase in – 453
   perfusion 401
   structural protein in – mitochondria 337
   turnover of cytochrome *c* in – 455, 457

L

L-cells 413
Lactate dehydrogenases and cytochrome c reductases of yeast IV G (112–118), 356
   EC 1.1.2.5. L(+)lactate: cytochrome *c* oxidoreductase (see *Cytochrome b_2*) 112 f.
   EC 1.1.2.4 D(−)lactate: cytochrome *c* oxidoreductase 112 f., 356
Larvae (see also *Nauplia*)
   of invertebrates 416–420, 424
Laurate 79, 356, 401
Lecithin 462
Lepidoptera 418, 419
Leucine [3] or ^{14}C-labelled in synthesis 458, 460, 461
Leukaemia 407, 413
Leukocytes 407
Light bleaching of cytochromes IX I3 (472, 473)
   inhibition of ALA synthetase in *Rhodopseudomonas* 453

Light—continued
  inhibition of protoporphyrin synthesis
    in yeast extracts 472
    of respiration 472
  penetration into ocean 486, 487
  photooxidation of protoporphyrin 473
Limulus gills 416
Lipids (see also *Cardiolipin, Phospholipids*)
    400, 459
  biosynthesis 462, 463
  in bacterial membranes 307, 463
Liver cytochromes of mitochondria
    335, 386, 397–399
  cytochrome c turnover 455, 458
  cytochrome oxidase 2, 31, 391, 397–399,
    409, 452, 459, 462
  fatty degeneration 395
  foetal liver mitochondria 398
  haem and ALA synthetase 452, 453
  hepatectomy (partial) 455, 462
  hepatitis 398, 407
  iron deficiency 450
  microsomes VIII C2b (399–401), 458
    species difference 399
  mitochondria VIII C2a (397–399), 454
    guinea pig 335
    half life 461
    ox, structural protein 337
    rat, haem a and protohaem 460
  plasma membranes, cytochromes in
    389, 459
  regeneration and turnover 462
  succinoxidase 389, 398, 411
  thyroxine effect on 411
Low temperature spectra see *Absorption spectra*
Lymph nodes 696
Lysine in cytochrome aa_3 27
  in cytochrome c 159, 162, 166, 175,
    280, 343, 344
  acetylation 184
  guanidination 159, 175, 183, 268, 269
  succinylation 184
  trifluoroacetylation 159, 175, 268
  trinitrophenylation 183, 185
  in cytochrome c synthesis 456
  in cytochrome b_5 and reaction with
    NADH-b_5 reductase 355
  in cytochrome b-562 252
  incorporation into liver membranes
    and cytochrome a biosynthesis
    459
Lysolecithin effect on mitochondrial
    respiration 337
  in pancreas 406
Lysosomes 397
Lysozyme 307

M

Magnetic optical rotatory dispersion
    (*MORD*) see *Cytochrome c,*
    CD, ORD, and MORD
Malate dehydrogenase biosynthesis 460
  of bacteria 309
Malignancy see *Tumours* 412
Mantle of Bivalva 421
Matrix of mitochondria 331, 337, 408,
    453
Meissner corpuscles 408
Membranes (see also *Mitochondria,
    Cytoplasmic membranes,
    Chloroplasts* and *Biosynthesis*)
    328, 486
  permeability 341
Membrane potentials see *Chemosmotic
    theory*
Menadione 358
Menaquinone 311
Mercurials 346
Mesoglia 487
Mesohaem sulphuric acid anhydrides
    as cytochrome c models 160
  reaction with apocytochrome
    b-562 252
Mesosomes 307 (see Cytoplasmic
    Membranes)
Metabolism, basal (BMR) 410
Methaemoglobin 404
Methionine deficiency and cytochrome
    c_1 392
  in bacteriochlorophyll formation 448
  in cytochrome b-562 252
  in cytochrome c synthesis 457, 458
  in cytochrome c-554 of *Chlorobium* 303
  in cytochrome P-450 iron linkage 261
  oxidation by N-bromosuccinimide 186
  -sulphur in haem linkage 159–162,
    164, 172, 268, 276, 279, 293,
    297–299, 388
  ^{35}S-labelled and mitochondrial turnover
    461
3-Methyl-4-dimethylaminoazobenzene 401
*Methylene hydroxylase system of
    Pseudomonas putida* 260, 261
"α-*Methylene oxygenase*" 86
5-Methyltryptophan 457
Microscopic intracellular structure 329,
    331
Microsomes (see also *Cytochromes b_5,
    P-450*)
  cytochrome c reductases VII F2
    (355, 356)
  cytochrome c synthesis in – 456
  turnover in – 458

Subject Index

Microsomes—continued
  depression of – enzymes by haem 453
    in malignant tumours 414
  Echinodermata 416
  Fe_x see *Cytochrome P-450*
  Lepidoptera eggs, 419
  NADH-cytochrome b_5 reductase
      VII F2a (355)
    reaction with cytochrome c 355
    in 7-hydroxycholesterol oxidation 355
  NADPH-cytochrome c reductase
      VII F2b (355, 356)
  NADP-reducing systems and
    mitochondria in tumours 413
  phospholipids 337
Mitochondria 5, 111, 385
  amino acid incorporation into –
      459, 461
  biogenesis IX E (460)
  compartmentation 413
  cytochromes, position in – membranes
      VII A3 (331–333)
    stoichiometry 335
  DNA 461
  energized – 364, 366
  EPR 333
  haem a-binding protein 460
  half-life 458, 460
  history 331
  in *Amoeba* and other protozoa 425, 426
  in *Ascaris* and other worms 424, 425
  in brain and nerves 402, 403
  in *Echinoderms* 416
  in fungi 427, 428, 459
  in gastropod spermatozoa 423
  in pancreas 406
  in plants 352, 427
  in thyroid 411
  in tumours 413
  in yeasts 427
  interaction with cytoplasmic factors
      413, 460
  membranes, changes 366
    permeability to cytochrome c 465
    separation of inner and outer 449
    inner, cytochromes see
        Cytochromes (general)
      evolution 488
      ferrochelatase in – 444
      half-life 461
      matrix 339
      snail spermatogenesis 403
      subunits 332 (see *ATPase*)
      in yeast 331, 335, 340
    outer 331, 332, 341, 458
      half-life 461
      cytochrome c-reductases 354, 355

Mitochondria—continued
  phospholipids 336, 337
  preparation 331
  protein synthesis in – 460, 461
    glucose repression 471, 472
    role in cytochrome biosynthesis
        IX E (460–463), 459
    haem incorporation 460
    in cytochrome oxidase biosynthesis
        459, 460
  separation from microsomes in plants
      100, 101
  submitochondrial particles (see also
      *Keilin–Hartree preparation,
      Oxidative phosphorylation,
      Sonication, Digitonin*) 63, 332,
      334, 341, 366
  subunit theory 332, 463
  swelling by thyroxine 410
Mitochrome (see *Cytochrome aa_3*) 29, 467
Mixed function oxidases (see *Hydroxylases*)
Mobile carriers in electron transport
    329, 330, 333, 354
Modification of cytochrome aa_3 24, 27,
    29, 30
  of cytochrome c V B6a (182, 183)
Mössbauer spectra 261, 288, 289, 296, 297
Molecular biology 484, 487, 490
Monoamine oxidase 404
Monocotyledon-plants 492
Morphine 400
Mucosa buccal in iron deficiency 392, 450
  gastric VIII C4a (405, 406)
    energy supply for HCl-secretion 406
  intestinal VIII C4a (406)
    in copper deficiency 452
    in iron deficiency 450
Multienzyme systems 328
Multiheaded enzymes cytochrome aa_3
    19, 35
  cytochrome b_2 IV G (112–118)
  cytochrome cc' VI D5 (284–292)
Muscles (see also *Heart*)
  amphibial 396
  avian flight 395
  contraction and cytochrome b 395
  cytochrome contents in – of mammals
      395
    ratios 335
  oxidase 387, 395–397, 452 (see also
      Invertebrates)
    in slow moving *Ascaris* – 414, 415
    in invertebrates 396, 424
  denervation 396
  exercise, effect on reduction of
      cytochromes by NADH 395
  insect thoracic flight – 395, 397, 414

Muscles—continued
  invertebrate – cytochromes 397, 416, 417, 420–424
    cytochrome oxidase 397, 417, 420, 421, 423–426
    in *Aplysia* 422
    in slow moving *Ascaris* 396, 414
    in Prosobranch snails 423
    in snail radula 422
  mitochondria 395, 396, 412
  oxygen consumption 395, 396
  red and white – cytochromes 391, 395, 397
  sarcosomes in slow and fast twitch – 396
  succinoxidase 389
  thyroid hormones, effect on – 411, 412
  turnover of cytochrome a 458
    of cytochrome c in – 455, 458
Mussels see *Bivalva, gills, mantle* 421
Mutagen (N-methyl-N-nitro-N-nitrosoguanidine) 468 see also *Streptomycin*
Mutants haem-dependent bacterial – 464, 468
  lacking nitrate reductase 469
  lacking oxidative phosphorylation 465, 488
  nuclear, with cytochrome deficiencies 460–469
  of bacteria IX H (467–469)
    of *Rhodopseudomonas spheroides* 453
  of fungi 467–469 (see also *Fungi*)
  of yeast IX G (464–467) (see also *Yeast*)
Mutations advantageous – in cytochrome c 491, 492
  average – distance 489, 490
  and variability of amino acid residues 490, 491
  conservative 487
  "malefic" 485, 491
  "neutral" 484, 487, 491
  of codons 485, 489
Myeloperoxidase 407
Myocarditis 394
Myoglobin autoxidizability 2
  CO-compound as model for cytochrome a_1CO 225
  effect of BAL + O_2 on – 351
  EPR spectrum compared with that of cytochrome P-450 262
  evolution 154, 485, 489
  exchange of haem with cytochrome P-450 261
  in anoxia 387
  in red and white muscles 395, 397
  iron incorporation 450

Myoglobin—continued
  resemblance to cytochromes c' 284, 288, 290, 294
  -like compounds in invertebrates 417, 422, 423
  variability of amino acid sequence 153

N

N-phenylimidazole 261
NADH-coenzyme Q reductase 358
  -cytochrome c reductase EC 1.6.2.1. 354, 355
  and thyroxine 411
  biosynthesis 462
  decrease in hyperbaric oxygen 389
  effect on pyrogens 394
  in adrenals 410
  in brain 402, 403
  in invertebrates 416–419, 422, 424–426
  in liver 399
  in muscles 395
  in plants 427
  in thyroid gland 412
  in uterus 409
  in yeasts 427
  mitochondrial VII F1 (354, 355)
NADH-cytochrome b_5 reductase in microsomes VII F2 (355), 399, 400, 412, 414
NADH-dehydrogenase (see also *Microsomes, Mitochondria*)
  and Complex I 353
  and non-haem-iron protein 358, 359
  -flavin 358
  in bacteria 308, 309
  in cytochrome c reductase 352, 353
  in cytoplasm 355
  in muscle 395
NADH-oxidase and thyroxine 411
  BAL-inactivation 351
  in fungi 428
  in *Lepidoptera* eggs 419
  in nerve endings 403
  in protozoa 353
  in retina 405
  in thyroid 412
  influence of D_2O 336
  "open" and "closed" 340
  reconstitution with cytochrome c 352, 353
NADPH in biosynthetic reactions 362, 410
  in hyperbaric oxygen 389

NADPH-cytochrome c reductase (EC 1.6.2.3) VII F2b (353–356), 399, 400, 410, 414, 451
 in invertebrates 419, 421
 in thyroid gland 412
 inhibition by NADPH-pyrophosphates 356
 isolation 356
 occurrence 356
NADPH-cytochrome c-554 reductase in algae 356
NADPH-cytochrome f reductase 356
NADPH-cytochrome P-450 reductase 84–86, 356, 412
Naevus malignant blue, cytochrome oxidase in 414
Naphthoquinones 2-dimethyl vitamin K_2 in cytoplasmic membranes 464
 2-hydroxy-3-alkyl-1,4-naphthoquinones, inhibition of respiratory chain VII E2b (350), 331, 344
 vitamin K_3 in cytochrome b_5 reduction 356, 358
Natural selection 484, 486
 and primary structure of cytochrome c 491
NMR spectra 366 (see also *Cytochrome c_1 magnetochemistry*)
Narcotics see *Urethane*
Nauplia 417
Nebenkern cytochromes and NADH in – of grasshopper spermatids 387
Nematocysts of hydrozoa 425
Neomycin 468
Nerves action potentials 402
 cytochromes in – VIII C3 (402–405), 408
 denervation (see *Muscles, Heart, Glands* (submaxillary))
 endings 403
 in invertebrates 417, 421–423
 mitochondria in – 402, 403
 neuroglia 403
 neurons 403, 422
 osmotic regulation 402
Nitrate and nitrite reductases (see *Cytochrome d_1c*) 5, 11, 102, 239, 240, 244, 270, 273, 283, 284, 356, 468–470.
 competitive inductive effects of nitrate and oxygen
 in bacteria 470
 evolution 491
 glucose repression 472
 in fungi 428, 468
 nitrate effect on cytochromes b-560 and "b-574" 264

Nitrite as agent against cyanide poisoning 404
 as suppressor of cytochrome formation in fungi 468
Nitroaniline hydroxylation 400
NO-compounds (see individual cytochromes)
 NO-haemoprotein-carcinogen interaction 414
Nomenclature 1, 3, 101, 123
Nonactin 366, 405
Nonesuch shale 486
Non-haem iron protein (*NHI*) and antimycin A 349
 EPR 358, 359
 in coenzyme Q-cytochrome c reductase 358
 in energy conversion 359
 in *Escherichia coli* 358, 359
 in flavin dehydrogenases 358, 359
 in NADH-cytochrome b reductase 357
 in NADH-cytochrome b_5 reductase 355
Non-haem iron proteins
 in respiratory chain VII F3 (356–359), 330, 349, 354
 in succinic-coenzyme Q reductase 358
Nutrition atherogenic diet 395
 cytochromes VIII B4 (391, 392)

O

Oestradiol effect on cytochrome oxidase of hypothalamus 404
 in uterus 409
Oestrus 409
Oligomycin as indirect inhibitor of the respiratory chain 330, 331, 334
 as inhibitor of oxidative phosphorylation 334, 360, 361, 393
 as inhibitor of brain respiration 361, 402
 -sensitive ATPase 331
 in bacterial respiration 309
 required for energy coupling in submitochondrial particles 366
Ontogenesis 416–426, 485
Optical rotatory dispersion see *Cytochrome c, CD, ORD, and MORD*
Origin of life 486
Orotic acid 401
Orthodoxy 492
Oscillations of ALA synthetase 471
 of cytochromes 471, 472
Ouabain 394, 402
Ovaries and thyroxine 410
 dual cytochrome system 471, 472
Oxalacetate 411

Oxidation-reduction potentials see
   Redox potentials
Oxidative decarboxylation of porphyrin
   propionic side chains 494
Oxidative phosphorylation (see *Energy conservation*; *Phosphorylation*)
   coupled to oxidation in bacteria
      I C (5), 299, 328, 333–335
   in the aorta 406
   in *Arum spadix* 415
   in muscle 396
   in solution without membranes 363
   in trypanosomes 426
   inhibitors (see also *Azide, Oligomycin, Uncouplers*) 331, 334, 359, 360, 415, 427
   1-phosphoimidazole in – 346
   P/O ratios 345, 389, 412, 428
   role of protein SH and SS groups 363
   sites of 330
      site I 360, 390
      site II 346, 347
      site III VII D6 (344–346), 394
   theories (see also *Chemiosmotic Theory, Energy-rich intermediates*) 329, 363, 364
Oxidosomes 237
Oxygen consumption in basal metabolism 410
   in bee metamorphosis 420
   in birds flight muscles 395
   in blood cells 407
   in brain 402
   in heart 393
   in insect muscles 395
   in kidney 393, 401
   in liver 393
   in spleen 393
   in yeast and cytochrome *c* 428
   of tissue, effect of thyroid hormones 411
   molecular absence in early atmosphere 486, 491, 494
      depression of ALA synthetase in photosynthetic bacteria 453
      diffusion gradient through bacterial cell wall 469
      effect on biosynthesis of cytochromes IX I1 (469, 470)
      of cytochrome *c* 455
      formation by photosynthesis, role in evolution 486
      oxidative decarboxylation of copro- and uroporphyrins 447, 470
   pressure in plant oxidative phosphorylation 415, 427

Oxygen pressure—continued
   of aqueous environment of invertebrates 415, 417
   precambrian 487
   toxicity (see *Hyperbaric oxygen*) 388, 389
      in *Tubifex* worms 389
Oysters 414, 421
Ozone layer 486

P

Pancreas 406
Pantothenate 462, 466
Paraquat 427
Parasites and cytochromes of *Bivalva* 421
   malaria – 425, 462
Parathyroid see *Glands*
Pasteur effect, negative see *Glucose repression*
Pentane extractant of coenzyme Q 357
Peptides (see also *Cytochrome c haemopeptides*)
   evolutionary building stones of cytochromes 488, 489
   of ferredoxin 489
Peroxidases (see *Cytochrome c peroxidase*) 274, 413, 487
   in iodination process in thyroid 412
   in liver fluke 425
Peroxisomes 487
Phenazine methosulphate 339, 346
Phenylalanine-lacking yeast mutants 457
Phenyldiimide 351
Phenylhydrazine as inhibitor of cellular respiration 351
Phosphatidyl-ethanolamine 462
   -glycerol 464
Phospholipases 263, 340, 352, 406
Phospholipids biosynthesis 462
   effect on respiratory chain VII C2 (336, 337)
   in cytoplasmic membrane 464
   in membranes 332
   in microsomes 337
   in mitochondrial membranes 336
   in NADH-cytochrome b_5 reductase 355
   in NADPH-cytochrome *c* reductase 355
   in photosynthetic bacteria 464
   in protein deficiency 392
   reactivation of cytochrome *c* reductases 337, 352
   of ATPase 337
   role of Cu in formation of – 391
Phosphorylation (see also *Oxidative phosphorylation, Respiratory control* and *Bacterial cytochromes*)
   energized state VII A4 (333–335)

Phosphorylation—continued
  coupled to oxidation in bacteria VI 2E (309–312)
  coupling factors 311
  efficiency 309–311
  phosphorylation sites 310, 311
  P/O ratios 309
  phosphorylation enzymes in inner mitochondrial membranes 332
    coupled to nitrite oxidation in *Nitrobacter* 491
    coupled to nitrate reduction 309–311
    coupled to bacterial photosynthesis 311
Phosphovitin complex with cytochrome c 173
Photosynthesis algal 5, 486, 487
  bacterial 219, 220, 292, 299, VI D6e (303–306) (see also *Bacteriochlorophyll, Electron Transport*)
  plant IV F3 (104–108), V C (201–206), 5, 107, 463, 491 (see also *Chlorophyll, Chloroplasts*)
  cytochrome oxidation, light-induced 304
  evolution of – 486
  photosystems I and II 107, 204, 205, 463
  P-700 107
  plastocyanin 107
  reaction centres, bacterial 305, 306
Phycobiliproteins phycoerythrins, phycocyanins 486, 487
Phylogeny and *Phylogenetic tree* 484, 489, 490
Piericidin in respiratory chain 359, 425
Pilocarpin 408
Pituitary see *Hypophysis*
Plasmocid 394
Plastocyanin IV F3 (104–108), 204, 205
Plastoquinone 205
Pollen 102
Porphobilinogen 446, 447
  -deaminase 446, 447, 454
Porphyrins biosynthesis and ATP 454
  and allylisopropylacetamide 454
  excretion in Fe-deficient organisms 450
  stimulation by metals 450
  fitness for evolution 486
Porphyrin a 9, 13, 14
Precambrian importance in early evolution 486, 487
Pregnancy 409
Promitochondria of yeast 462, 463, 471
Prostate adenomatous 408
  and thyroxine 411

Prostate—continued
  citric acid excretion 408
  cytochrome c 408
  endogenous respiration 408
Proteins deficiency and cytochromes 392
Proteolytic enzymes 126
Prothoracic gland hormone 418 (see also *Insects*, metamorphosis)
Protohaem 8, 9, 11, 447 (see also *Iron* incorporation)
  biosynthesis and turnover 446–448, 452, 454, 458, 460
  co-repressor of porphyrin biosynthesis 453
  effect on respiration of anaerobically grown *Staphylococcus* 470
  exchange between – of cytochromes b_5 and P-450 458
  genes in *E. coli* 468
  prosthetic group of cytochromes b II A (8, 9, 11), IV A (59)
  -proteolipid 337
  requirement in bacterial mutants 467, 468
Proton migration 5 (see *Oxidative phosphorylation*)
Protoporphyrin 447–451, 467, 468, 473, 494
  accumulation in iron deficiency 449
  in excess of Zn 451
  increase by iron in *Rhodopseudomonas spheroides* and *Tetrahymena* 459
  inhibition of cytochrome formation in yeast by – 473
  photooxidation 473
  role of alternative electron acceptors in bacterial – formation 448, 491, 494
  role of oxygen in – formation 448
Protoporphyrinogen 12, 154, 155, 447, 448, 494
Protozoa 350, 391, 425, 426
Pseudomonas oxidase (*nitrite reductase*) 11, 12 (see *Cytochrome d_1c*)
Puromycin in bacteriochlorophyll synthesis 453
  in *Echinoidea* 416
Putidaredoxin 260, 262
  -reductase 260
Pyridoxin-phosphate cofactor of ALA synthetase 448
Pyrogens 394
Pyrophosphate in evolution 488
Pyruvate in glucose repression 471
  reductant in muscle 395, 396

Q

Q_{O_2} see *Oxygen consumption*

R

Rack mechanism 2, 162
Reconstitution of the respiratory chain 341, 342, 352, 353
Redox potentials (see individual compounds)
  in evolution 486
  redox state of cell and cytochrome synthesis 469, 470
Respiration see *Cellular respiration*
  pulmonary 388
Respiratory chain (see also *Electron transport*, *Steady state*) VII A2 (329-331), 60, 62, 68
  bacterial 308
  bypasses of – 339, 340, 346
  complexes in – 329, 330, 353
    complex I 330, 353 ⎫ in cytochrome
    complex II 330, 353 ⎭ reduction
    complex III 330, 346, 347, 353 (see also *Cytochrome bc_1*)
    complex IV 330, 353 (see also *Cytochrome aa_3*)
    stoichiometry 353, 362
  cytochromes, stoichiometry VII V (335)
  influence on total respiratory chain VII C (335-338)
  effect of acetone on – 337
  of isooctane on – 337
  kinetics VII C4 (338-340)
  mitochondrial inner membrane 331
  reducing equivalents between cytochrome b, cytochrome c and oxygen 350
  role of phospholipids VII C2 (336-337)
  of water 335
  theories of mechanism VII H (361-366) 329, 330, 335
  of complex-interaction 362
  thermodynamics 365
Respiratory control (see *Respiratory States* and *Oxidative Phosphorylation*, *Salicylic acid*) VII A (333, 334), 5, 338, 339, 353, 387, 395-397, 406, 408, 421
Respiratory states 334, 338, 339, 360, 406, 409, 411, 427
Retina cytochromes 405
  microsomes 405
Reticulocytes 405, 407
Rivanol 241
Reversed electron flow 362, 402, 408
  and antimycin A 349

Reversed electron flow—continued
  and ATP 360, 363
  in *Nitrobacter winogradskii* 223
  in photosynthetic bacteria 309
  reconstitution of – 353
  site of – 346
RHP (*Rhodospirillum haemoprotein*) 10, 284 (see *Cytochromes c' and cc'*)
Rhein inhibitor of NADH-dehydrogenase 359
Ribosomes 453, 457, 459, 461
Ricketts 392
Rivanol 241
RNA 451, 457, 461, 472
Roots of lupins, role of cytochromes and succinic dehydrogenase in absorption from soil 427
Rotenone and non-haem iron protein 358-360
Rutamycin 360

S

Salicylic acid and respiratory control 334
Salicylanilide 390
Salinity effect on cytochrome oxidase in crabs and nauplia 417
Sarcoma 413, 414
Sarcosomes of insect muscles 391, 397
  of mammalian muscles 396
Schiff base see *Aldimines*
Scurvy 401
Sea urchin eggs 416
Sedormid see *Allylisopropylacetamide* 455
Serotonin 394
Serum albumin abolishing alkylhydroxynaphthoquinone inhibition of respiration 351
  activation energy for electron transport 362
  increase of P/O ratio in brown adipose tissue 390
  inhibition of phospholipase 406
Sex differences in invertebrates 421
  in liver 401
Shivering 389
Shock, anaphylactic 387, 394
Sialic acid as impurity in cytochrome c 127
Skin cytochromes in – VIII C4d (407, 408)
  in squamous carcinoma 413
SN 5949 see *Naphthoquinones* 350, 351
Snails see *Gastropoda* 422, 423 (see also *Japanese landsnail*)

Snake venoms and cytochrome *b* 62
   and NADH-cytochrome *c* reductase 352
   effect on cytochrome P-450 263
   on mitochondrial respiration 337
Sodium pump 402
Sonication 349, 366
Spadix (Arum) 102, 104, 415
Spectrophotometric methods (see also
      Cytochromes a and c_3, c_1)
      219, 338, 392, 398
Spermatozoa cytochromes and cytochrome
      oxidase VIII C4e (408, 409)
   flagellae 408
   of *Bivalva* and *Gastropoda* 422, 423
   of sea urchin 408, 416
   spermatogenesis in snails 408
Spleen cytochromes 401, 402, 407
   thyroxin effect on 411
Spores 265, 429
Starvation 392, 395, 400, 401, 418
Steady state of cytochrome oxidation
      VII C4 (338–340), 5, 398
   freezing in of – 339, 390
   in brain mitochondria 402
Stearyl coenzyme A desaturase 355, 400
Stilboestrol 409
Stomach see *Gastric mucosa*
   in snails 422
Streptomycin as mutagen 468
   -dependent strains of *E. coli* 238
   in preparation of cytochrome d_1c 241
Stria vascularis of Cochlea 405
Structural protein VII C3 (337, 338), 347
Submicroscopic intracellular structure
      328, 329
Submitochondrial particles see
      Mitochondria
Substrates (see also *NADH-oxidase*,
      Dehydrogenase,
      Glycerophosphate, *Lactate*, etc.)
   in brain mitochondria 403
   in liver mitochondria 398
   NAD-linked 427
Succinic coenzyme Q reductase 358
Succinic cytochrome c reductase VII E3
      (351–354)
Succinic dehydrogenase
   cytochrome *b* in – – 330, 354
   complex II 353
   non-haem iron protein in – – 359
   biosynthesis 460
   in adenocarcinoma 413
      bacterial membranes 308
      brain 403, 404
      heart 394
      invertebrates 416, 420
      leukocytes 407

Succinic dehydrogenase—continued
      muscle 359, 360
      plants 427
      retina 405
      spermatozoa 408
      thyroid 412
      uterus 409
Succinic oxidase bioelectrode concept 345
   effect of D_2O on – 336
   of thyroxine 411
   in bacteria 307, 309
      brain 402, 404
      fungi 428
      heart 393
      invertebrates 416–418, 421, 423–425
      liver 404
      muscle 395, 396
      plants 427
      reticulocytes 407
      temperature acclimatization 389, 390
   inhibition by antimycin A 349, 350
      by alkylhydroxyquinoline-N-oxide
         350
      by BAL 351
   in nematocysts of hydrazoa 425
   "open" and "closed" 340
   reconstitution 342, 353
Succinyl-coenzyme A 446, 447
Sulphate (and thiosulphate reduction)
      5, 273, 274
   cytochrome c_3 in – VI D4d (280–281)
Sulphydryl groups (see also individual
      cytochromes)
   in BAL–O_2 inhibition 351
   in hyperbaric oxygen 389
Sulphite oxidoreductase in *c*-552 (*E. coli*)
      284
   in *Thiobacilli* 273, 274
Superoxide anion (O_2^-) in cytochrome
      c–xanthine oxidase reaction 187
   in flavin autoxidations 359
   in haemochrome autoxidation 346
Swayback in lambs 391

T

Tapeworm 425
Taurochenodeoxycholate-6-hydroxylase 400
Temperature, biological effects (see
      Acclimatization) VIII B3,
      (389–391), 466
Testes and thyroxine 411
   cytochrome P-450 in microsomes 409
Testosterone 404, 408, 409, 414
Tetrachloroquinols 330
Tetrahymena 426
   cytochrome *c* 491

Subject Index

Tetramethyl-p-phenylene diamine (TMPD)
   by-pass of cytochrome bc_1 346, 350
   -oxidase of *Ascaris* muscles 396
   in fungi 428
Tetrapyrrole biosynthesis see *Biosynthesis*
Tetrazolium salts 427
Thallophyta see *Algae*
Thenoyltrifluoroacetone 309, 330, 359
Thiocyanate 406
Thioglycolate 351
Thiosulphate 404
Threonine 465
Thymidine 462
Thymus 402, 411
   nuclei 196, 387
Thyroxine 400, 408
   action on biogenesis 412
   and cytochromes 411
   and cytochrome oxidase 411, 412
   and NADH-cytochrome c reductase 411
   binding of ^{131}I 412
   effect of dehydrogenase 411, 412
   on microsomal enzymes 412
   on NADH oxidation 411
   hypothyroid state 411, 412
   iodination in thyroid 412
   thyrotoxicosis 411, 412
   uncoupling 410
Tocopherols 358
Toxins diphtheria 394
   endotoxins, effect of cytochrome oxidase 398
   streptococcal 394
Toxohormone 414
Tranquilizer drugs 404
Transcription 455
Transhydrogenase 354, 360, 362, 363, 407, 408, 410, 411
Trialkyl (aryl)tin 361
Tricarboxylic acid cycle 472
Triiodothyronine 411
Trimethyllysine residues in fungal cytochrome c 135, 136, 149, 150, 153, 197, 200
   in plant cytochrome c 135, 136, 149, 153, 200
Tripyrroles in tetrapyrrole synthesis 448
Trypanosomes 426
Trypsin-sensitivity of NADH-cytochrome reductases 355
Tryptophan formylation 185
   in cytochrome c 155, 159, 162, 172, 174, 185, 189, 276
   reaction with N-bromosuccinimide 185
Tumours malignant, cytochromes in, VIII C6 (412–414)
   depression of microsomal enzymes 414

Tumours—continued
   mitochondrial, microsomal and cytoplasmic interaction 413
   Warburg theory 412
Tunicata 416
Tyrosine in brain 402
   conformation in cytochrome c 162, 172, 174, 184, 276
   iodination of—in cytochrome c 184
   nitration of—in cytochrome c 185
   invariant residues in cytochrome c and c_3 279
   -lacking yeast mutants 457
   -oxidase in fungi 429

U

Ubiquinones see *Coenzyme Q*
Ultrasonic treatment see *Sonication* and *Mitochondria*
Ultraviolet radiation as mutagen 466, 468
Uncouplers (see also *Dinitrophenol* and *Azide*) 331, 334, 339, 345, 360, 408, 427
   and alkylhydroxynaphthoquinone 350, 351
   and alkylhydroxyquinoline-N-oxide 350
   and antimycin A 349
   free fatty acids 390
   in parathyroid 412
   in tumour metabolism 413
   salicylanilide in hibernation 390
   thyroxine as – 410
Urethane 359
Uroporphyrin 447, 448, 467, 494
   III decarboxylation 447
Urophorphyrinogen I 448
   III cosynthetase 447, 448, 454
Uterus cytochrome oxidase 409
   NADH-cytochrome c reductase 409
   succinic-dehydrogenase 409

V

Valinomycin 331, 405
Vitamins B_{12} and cytochrome c in rabbit heart 195
   E 454
   K see *Naphthoquinones*
   induction of RD-mutants by – deficiency 466

W

Warburg theory of malignancy 412, 413
Wilson's disease 391
Worms 415, 423–425

X

X-irradiation and cytochrome b_5 410
  effect on cytochromes in heart 394
    in liver 394, 398
  on mitochondria 402

Y

Yeast (see also *Glucose repression*)
  adaptation to aeration 356, 427, 428, 451, 462, 463
    effect of light on – 472, 473
  buds 428
  copper deficiency 391
  cytochromes in anaerobic yeast 427, 462, 463
  cytochrome *c* and viable count 428
    binding by – 428
    biosynthesis in – 455
    -deficient – 428, 465
    -peroxidase 186, 487
  cytochrome oxidase 21, 24, 28, 427, 428, 459, 460, 462, 465
  cytochrome P-450 biosynthesis in—
    and O_2 469
    glucose repression 470, 471
    induction of site I phosphorylation by aeration 356
  lactate–cytochrome *c* reductases in IV G5 (116–117), VII F2c (356)
  D(–) lactate dehydrogenase in anaerobic and aerobic yeast 356
  D(+) lactate dehydrogenase in aerobic yeast, induction 356
  mitochondria 427, 465

Yeast—continued
  biosynthesis of respiratory chain 462
  in petite 427, 459 (see *Promitochondria*)
  mutants CY, P 465–467
    lacking oxidative phosphorylation 465
    lacking oxidative phosphorylation 465
    petite 356, 427, 465, 466
    RD (respiration-deficient) 464–466 469
    threonine-deficient 465, 466
  NADH-cytochrome *c* reductase 352
  ploidy 466
  promitochondria 463
  spheroblast 428
  Zn coproporphyrin in 221

Z

Zinc effect on ALA dehydratase and synthetase 451
  on cellular respiration 336, 354
  on cytochrome formation in fungi 451
  on cytochrome *b* and *c* formation in yeast 451
  on cytochrome oxidase biosynthesis 451, 452
  on formation of free porphyrins in yeast 451
  on isocytochromes *c* biosynthesis in yeast 451, 457
  on synthesis of protein, DNA and RNA 451
  excess, effect on haemoprotein biosynthesis 451